Human Development

A LIFE-SPAN PERSPECTIVE

Human Development
A LIFE-SPAN PERSPECTIVE

Richard M. Lerner
David F. Hultsch
The Pennsylvania State University

McGraw Hill Book Company

New York   St. Louis   San Francisco
Auckland   Bogotá   Hamburg   Johannesburg
London   Madrid   Mexico   Montreal
New Delhi   Panama   Paris   São Paulo
Singapore   Sydney   Tokyo   Toronto

To
Justin Samuel Lerner
and
Cindy E. Hultsch

HUMAN DEVELOPMENT: A Life-Span Perspective
Copyright © 1983 by McGraw-Hill, Inc.
All rights reserved.
Printed in the United States of America.
Except as permitted under the United States Copyright Act of 1976,
no part of this publication may be reproduced
or distributed in any form or by any means,
or stored in a data base or retrieval system,
without the prior written permission of the publisher.

1 2 3 4 5 6 7 8 9 0   DOCDOC   8 9 8 7 6 5 4 3

ISBN 0-07-037216-0

This book was set in Caledonia by Progressive Typographers.
The editors were Patricia S. Nave, Annette Hall, and David Dunham;
the designer was Charles A. Carson;
the production supervisor was Phil Galea.
The photo editor was Linda Gutierrez.
The cover, part, and chapter illustrator was David Brooks.
New drawings were done by Fine Line Illustrations, Inc.
R. R. Donnelley & Sons Company was printer and binder.

Library of Congress Cataloging in Publication Data

Lerner, Richard M.
　　Human development, a life-span perspective.

　　Bibliography: p.
　　Includes indexes.
　　1. Developmental psychology.   I. Hultsch,
David F.   II. Title.
BF713.L48 1983　　　155　　　82-13078
ISBN 0-07-037216-0

Contents

Part Two

PRENATAL AND INFANT DEVELOPMENT

Part Three

CHILD AND ADOLESCENT DEVELOPMENT

Part Four

ADULTHOOD AND AGING

For more than a decade the behavioral sciences have been involved in exciting changes in perspective. Instead of scientists working in relative disciplinary isolation, greater efforts at multidisciplinary integration have occurred. In addition, there has been a concern with the ecology of human behavior; with the role of the changing social context in individual development; with mutual influences between people and their world; and with the potential of humans to change across their entire lives. These changes in perspective may be summarized and synthesized by a view of human development that has been labeled the *life-span perspective.*

The objective of this text is to examine human development from conception to death within the context of this perspective. This text is intended primarily for beginning undergraduates in the behavioral sciences, particularly those in psychology and human development. However, the text should be accessible to students with little or no background in behavioral or social science since the presentation assumes no prior coursework in those areas.

Point of View

Texts often present information in an encyclopedic manner; that is, they set forth the "facts." However, the facts available about any given topic are generally diverse and often conflicting. In our view, an understanding of a topic requires a framework within which information may be integrated. In this text, we will use the emerging life-span developmental perspective to integrate the theory and research related to human development. The life-span approach represents a conceptual orientation characterized by emphasis on:

1. *The Description, Explanation, and Optimization of Behavior.* The questions of what behaviors occur, why behaviors occur, and how to alter behaviors are considered equally important for understanding development.

2. *Multidirectional Development.* Processes of growth and decline occur at all points in the life cycle. Development, therefore, is not defined as just a continuous process of growth, and aging is not defined as a continuous process of decline; both constancy *and* change are emphasized.

3. *The Individual in Context.* Development is the result of interactions among multiple processes and events, including those related to age and to history. In order to understand development, we must look at the interaction between the individual and the contexts in which he or she lives (e.g., physical, familial, cultural, historical).

4. *Multiple Theories and Methods.* An adequate understanding of development cannot be obtained from any one theory or methodology, nor can it be obtained from a cataloging of empirical facts.

Organization

Individuals attempting to teach a course or write a text on human development immediately confront a difficult choice: Should the material be organized by focusing on developmental processes or by examining the various "stages" of the life span? This choice is a difficult one, since there are positive features to both strategies. For example, a focus on processes allows one to appreciate the changes over time in the individual's biological, psychological, and social functioning. On the other hand, a focus on age stages allows one to understand the processes and life events which characterize particular portions of the human life cycle: infancy, childhood, adolescence, young adulthood, middle age, and old age.

In this text, we capitalize on the strengths of both approaches. We have adopted a "process-within-stage" approach. That is, within each of several succeeding portions of the life span—prenatal development, birth, infancy, childhood, adolescence, adulthood, and the aged years—we present the key developmental processes and contextual influences that, in their combination, make each period of life a special, unique one.

In our view, this organization follows logically from the life-span developmental perspective. The life-span perspective directs our attention to behavior-change processes and to the role that interactive contextual influences play in moderating development at each portion of the life cycle. Thus we first examine basic theoretical and empirical information related to these changes over the entire age range involved. We then turn to our processes-within-stages presentation. The book is thus organized as follows:

In Part 1—"Philosophical, Historical, Social Scientific, and Biological Bases" (Chapters 1, 2, and 3)—we describe the life-span developmental perspective and review its philosophical and historical bases and the research methods used to investigate development. In addition, we discuss the biological bases of human development across life. These chapters provide a framework within which the information on human development may be integrated and evaluated.

In Part 2—"Prenatal and Infant Development" (Chapters 4, 5, and 6)—we discuss the physical, cognitive, personality, and social processes of human development, as they interrelate with the social (e.g., familial) and historical features of the context, in these initial portions of the human life course.

Part 3—"Child and Adolescent Development" (Chapters 7 through 11)—continues with these emphases and pays particular attention to bidirectional influences between the developing person and his or her social world, as well as to the bases and implications of sex differences in development during these portions of the life cycle.

Finally, in Part 4—"Adulthood and Aging" (Chapters 12 through 15)—we examine changes in learning, memory, personality, and social functioning. We consider the integration of processes, person, and context within a life-course perspective by outlining a life-event and life-transition framework. In these chapters the transitions and events unique to young adulthood, middle age, and old age, and finally to death and dying—the final event in life—are discussed.

Special Features

We have included a number of special features that will make the book more interesting and helpful to the student.

Chapter Overviews. Each chapter begins with a brief statement about its focus, a listing of the first- and second-order headings, and a list of issues to consider when reading the chapter. These overviews should orient the student to the chapters.

Chapter Summaries. Each chapter ends with a summary of the main points of the chapter. These summaries should aid the student in reviewing the chapter.

Boxed Inserts. Throughout the text, we have highlighted certain material by placing it in boxed inserts. These inserts are designed to accomplish various ends. Some are designed to provide an in-depth look at topics introduced in the text or to illustrate different techniques for measuring development. Others are "asides" designed to provide case examples or to illustrate interesting new ideas or controversies in the field.

Glossary. A glossary of specialized terms that have been used in the text is provided at the end of the book.

References. An extensive reference list at the end of the text permits the instructor and student to follow up topics of interest.

Instructor's Manual. A manual summarizing the main points of each chapter and including an extensive list of test questions, film suggestions, and other instructional aids is available to all instructors who adopt the text.

Acknowledgments

The journey from idea to published text is a long and difficult one. Many people helped us along the way and we wish to express our appreciation to them.

Several of our colleagues at Penn State and other institutions provided valuable criticism of various manuscript drafts and invaluable stimulation about our ideas. We thank this group of scholars and scientists: Paul B. Baltes, Max-Planck Institut für Bildungsforschung, Berlin; Jay Belsky, Penn State University; Nancy Busch-Rossnagel, Colorado State University; Don C. Charles, Iowa State University; Rita S. Heberer, Belleville Area College; Philip C. Kendall, University of Minnesota; Jacqueline V. Lerner, Penn State University; Lynn S. Liben, University of Pittsburgh; Susan M. McHale, Penn State University; Paul Muhs, University of Wisconsin—Green Bay; John R. Nesselroade, Penn State University; Douglas Sawin, University of Texas—Austin; and Patricia A. Self, University of Oklahoma—Norman. We are particularly grateful to Graham B. Spanier and Francine Deutsch, coauthors of our respective texts on adolescence and adulthood and aging, for permitting us to incorporate many of their ideas from these earlier books. We also wish to thank Susan Friedman for her valuable contributions to the early stages of the project.

Our secretarial staff typed numerous drafts of the manuscript and, in addition, served as invaluable aides in completing the myriad of other tasks necessary for the finishing of a textbook. Thus we extend our sincere gratitude to Kathie Hooven, Joy

Barger, Barbara Huntley, and, especially, Rebecca Gorsuch—who organized, double-checked, and coordinated all our tasks.

We are also especially grateful to our able graduate assistant, Bernadette Reidy, who coordinated the completion of the References and Glossary; to Linda Gutierrez, who researched and selected the photographs for the text; to Annette Hall and David Dunham, our editors, and indeed to all the staff at McGraw-Hill, for helping turn our material into a finished product.

A major debt of thanks is due to the staff of the Center for Advanced Study in the Behavioral Sciences. Much of Richard Lerner's work on this book was completed while he was a Fellow at the Center for Advanced Study in the Behavioral Sciences for the 1980–1981 academic year. In addition to being grateful to the Center's staff for their assistance and support, he is also grateful for financial support provided by National Institute of Mental Health grant #5-T32-MH 14581-05 and by the John D. and Catherine T. MacArthur Foundation.

Finally, we are indebted to our spouses and children—Jackie and Justin; and Cindy, David, Debby, and Amy—for their constant love, support, and devotion during all the hours we were either absent and writing or present and brooding.

Richard M. Lerner

David F. Hultsch

Human Development

A LIFE-SPAN PERSPECTIVE

Part One

Human Development:
philosophical, historical, social scientific, and biological bases

A community is a group of people who, while perhaps from different backgrounds and having different abilities, live and interact in a common setting; often the members of a community have common interests. A college or university is a community of scholars from different disciplines; these scholars have common interests: to advance appreciation of and knowledge about the world within which they live.

Just as the members of a community may interact to complete some community project, often the disciplines—the areas of knowledge—studied within the university work together to advance understanding of a particular topic. Human development is a topic studied by scholars from multiple disciplines. In other words, the discussions of human development that will be found in this book are drawn from a *multidisciplinary* knowledge base. We will draw on information from philosophy, biology, medicine, psychology, sociology, anthropology, political science, economics, law, and history, for example, because scholars from all these disciplines have made important contributions to the study of human development. All, in their own ways, have been concerned with describing and explaining the nature of human life and the changes in life—across a person's age, across cultures, and across history.

1

Given the complexity of even one human life—as it begins at conception and proceeds to unfold over the course of 9 in utero months and, usually, from 70 to 80 years after the day of birth—it is fortunate that people from so many academic areas study the life span. Without their past and continued contributions, it would not be possible to be making even the modest strides we are now making toward recognizing and understanding the complexity of human development.

In turn, of course, a student who is initiating his or her study of human development has a task perhaps even more formidable than scholars whose work they will be reviewing. A newcomer to the study of human development must become acquainted with all the knowledge bases that contribute to our understanding of the life course.

While we recognize that this is indeed a difficult task, we also believe that today's student is fortunate. Events in the biological and social sciences over the last decade or so have led to the emergence of a perspective about the study of the human life course, a perspective which is a useful intellectual tool for approaching the study of human development. This perspective is the life-span view of human development. It is the perspective maintained by your authors.

The life-span view helps one understand *why* it is necessary to study human development from a multidisciplinary perspective. In addition, the life-span view helps one see *how* such study may be accomplished. By presenting the nature and bases of the life-span view of human development, therefore, we will be able to acquaint you with some of the major philosophical, historical, social scientific, and biological bases of the human life course. It is the purpose of this first section of chapters to provide you with these multidisciplinary bases for the study of human development.

chapter 1

A Life-Span Approach to Human Development

CHAPTER OVERVIEW

This chapter presents key features of the concept of development and introduces several approaches to studying development. We emphasize the characteristics of a life-span approach to human development. We stress the ideas that the potential for development exists across life and that a multidisciplinary approach to studying human life is useful. We also present the research methods used in the study of human development.

CHAPTER OUTLINE

WHAT IS DEVELOPMENT?

ALTERNATIVE VIEWS OF HUMAN DEVELOPMENT

The Psychological View
The Developmental View
The Life-Span View
Developmental Issues
The Role of Chronological Age, Cohort, and Life Transitions
Optimization
The Multidisciplinary Emphasis

SCIENTIFIC RESEARCH AND THE STUDY OF HUMAN DEVELOPMENT

The Validity of Research
Internal Validity
External Validity

DIMENSIONS OF RESEARCH METHODS IN HUMAN DEVELOPMENT

DESIGNS OF DEVELOPMENTAL RESEARCH

The Longitudinal Design
The Cross-Sectional Design
The Time-Lag Design
Sequential Strategies of Design
A Sequential Study of Adolescent Personality Development
A Sequential Study of Intelligence in Adulthood

ETHICAL ISSUES IN RESEARCH

THE PLAN OF THIS BOOK

CHAPTER SUMMARY

ISSUES TO CONSIDER

How are developmental changes different from other types of changes?
What are the foci of a life-span view of human development, and how do they
    differ from other approaches to studying development?
Why do life-span developmentalists stress the use of a multidisciplinary orien-
    tation to studying human development?
Why are external validity and internal validity of concern to scientists?
What are the key dimensions of developmental research?
What are the assets and limitations of longitudinal, cross-sectional, and time-
    lag research designs?
Why do life-span developmentalists see sequential designs as most useful?

*I*f, as Shakespeare said, all the world is a stage and all its people are players, then the human life span is the drama that all of us enact. For many of us the drama begins when our parents share the pain and the joy of the childbirth experience. The acts that follow this opening scene are familiar to us all: the toddler making his or her first tentative steps; the child, on the first day of school, with one eye on the new teacher and the other on the all-too-quickly departing parent; the early adolescent spending hour after hour gazing in the mirror wondering about complexion, about hair style, and about love; two young adults making a commitment to share their lives together; the middle-aged adult, striving for success on the job, trying to find time for family, friends, and relaxation; the aged person, retired and perhaps alone, contemplating the end of his or her life cycle.

Thus the life span is filled with events that are common to all people (birth, puberty, death), with events that occur to those of us who exist in particular groups (e.g., regarding particular religious or social ceremonies), and with events that are specific to each of us as individuals (e.g., meeting particular people, experiencing certain illnesses). Some portions of our life span are filled with success and happiness; others seem to be times of sadness. Most of life has both positive and negative features. All of us develop in ways that are similiar to others and in ways that are unique to us. Indeed, with each new generation, humans develop in a world which shares some features with the world of their parents and grandparents and, in turn, has features special to their own time in history. Thus we all develop as changing individuals in a changing world.

In this book we will focus on such changes, as they occur across the entire life span. We will explore those aspects of development that appear common to people and those that serve to make for differences between people. Our discussion will lead to our addressing several key questions as a consequence of our concern with development across the entire life span:

What is the possible significance of each person's developing unique or general characteristics of behavior?

Will the differences between individuals have different significances for them?

What will happen to infants and children as they develop into adolescence and adulthood? Will they show particular types of behavior styles?

What will be the impact of marriage, career, parenthood, illness, and economic factors in society on people's development across life?

Why do the changes we see different individuals going through occur?

What is the basis of individuality and of development?

Answers to these and a host of other questions are sought by human developmentalists, especially those who subscribe to the life-span view of human development. But what *is* development? What is developmental psychology? And what is the life-span view of human development? In this chapter we begin to address these questions.

WHAT IS DEVELOPMENT?

It is often the case in science that there are debates about the meaning of a term such as *development* (see Harris, 1957; Lerner, 1976). However, there is some consensus that development refers to *change*. But not all changes are developmental ones; the ups and downs of the balance in a person's checkbook are not developmental changes. Random, unsystematic, or unorganized changes are not development either. Developmental changes are systematic, or organized, changes. The entity or thing that changes systematically may be a culture, a society, a group, or an individual. In the study of human development we are concerned with all such changes, but we are most often interested in those involving individuals. We are most concerned with the character of the changes a person goes through and want to know what happens within an individual (or, more technically, intraindividually) from one time to the next. Thus development refers to systematic intraindividual changes.

ALTERNATIVE VIEWS OF HUMAN DEVELOPMENT

There are many different approaches to understanding the person at various portions of the life span. Currently, there are three major perspectives. Consistent with discussions in the psychological and developmental literatures devoted to childhood (Gollin, 1965; Huston-Stein & Baltes, 1976; Reese & Lipsitt, 1970), these approaches may be labeled the psychological view, the developmental view, and the life-span view.

The Psychological View

The *psychological approach* emphasizes how people of a particular age group perform on a specific task (Gollin, 1965). For instance, in this approach a researcher might ask how a 5-year-old performs on a learning task; however, there is no attention to how learning in infancy and earlier childhood periods contributed to learning at age 5. There would be a corresponding lack of concern with the way this learning influenced learning in later childhood, adolescence, and adulthood. Instead, in the psychological approach there is more of a focus on the task and therefore little, if any, concern with the history of changes in the person that contributed to task performance over time.

The Developmental View

The *developmental approach* emphasizes the history of the person. While the psychological view is ahistorical in nature, the developmental view focuses on the relation between person and task as that relation changes across the history of the person, across the person's age levels. Thus the developmental view is one which considers how people of different age groups perform on a given task (Gollin, 1965). This approach goes beyond a primary focus on the task. It considers the changes associated with the person over time. Thus here there would be an interest in the antecedents of learning at age 5 years and, also, in some of its consequences.

There is a qualification, however. Most of the existing developmental views of behavior focus just on childhood (as in Flavell, 1970). As such, long-term outcomes of developments in early periods of life are not typically a focus of the developmental view (e.g., Hurlock, 1973). As suggested in Chapter 2, a major reason for this exclusion appears to be theoretical in nature. There are some who believe that the important changes that characterize development emerge during infancy and childhood or no later than adolescence. Consequently, anyone who follows such a perspective would not likely be very concerned with the period of life after adolescence, insofar as the important *developments* of life are concerned. Said another way, this view of development has emphasized maturation and growth during infancy, childhood, and adolescence. In turn, from this perspective adulthood is characterized by stability, and old age by degeneration and decline.

Piaget (1970), for instance, has proposed an important theory of the development of knowledge. However, in this theory he stresses the view that after some point in adolescence, no new components (or, in his terms, "structures") of thought develop (Piaget, 1972). Thus although those following Piaget's ideas in their research work would certainly look at the childhood (and even the early adolescent) antecedents of adolescent cognition (knowledge), they would not be especially concerned with how adolescent cognition influences changes in later life. This lack of concern would occur because of their theoretical view that there is no major structural change in cognition in adulthood (Flavell, 1970).

Thus the developmental view takes a historical view of the person. However, it is often limited in the *amount* of the person's history considered. In addition, there are other limitations with the developmental view. To see all these most clearly, it will be useful to turn to a discussion of the characteristics of a life-span view.

The Life-Span View

A number of writers have attempted to articulate the characteristics of the life-span developmental approach (Baltes, 1973; Baltes, Reese, & Lipsitt, 1980; Baltes & Willis, 1977; Lerner & Ryff, 1978). A hallmark of this approach is that development may occur at all points along the life span, from conception to death. This is an important assertion because, as we have seen, traditional views of human development have emphasized maturation and growth during infancy, childhood, and adolescence; stability during adulthood; and degeneration and decline during old age. The emphasis on the extent of change throughout the life span, however, does not completely define the life-span developmental approach. However, the following definition of the life-span view of human development may be offered: *The life-span view of human development attempts to describe, explain, and optimize intraindividual change in behavior and interindividual differences in such change across the life span, that is, from conception to death.* A life-span approach to human development has several attributes, one of the most important ones being that any portion of life is seen as just one part of the entire life span. Another is that the knowledge from many academic disciplines is used to study development. Much of what is presented in this book is contributed by the disciplines of psychology and sociology. But information is also derived from history, biology, anthropology, law, and medicine. This is a very realistic approach since what happens in our lives is inherently *multidisciplinary* (involving many disciplines). Now let us examine each component of our definition of the life-span view.

Description, Explanation, and Optimization. Life-span developmentalists focus on three tasks: description, explanation, and optimization (Baltes, 1973; Huston-Stein & Baltes, 1976; Baltes & Willis, 1977). First, the changes which characterize development must be *described*. We must be able to depict how people change. Second, the changes characterizing development must be *explained*. One must be able to show how antecedent or current events make behavior take the form that it does over time. Third, once development has been described and explained, it should be *optimized*. One should try to prevent unhealthy development and to foster change as helpfully as possible.

For instance, during adolescence and young adulthood, males and females choose roles—socially defined forms of behavior—that will influence much of their coming lives. As presented in a later chapter, males and females typically do not enter into roles of equal status in areas such as vocational activities. Males typically enter higher-status vocations, females lower-status ones (Tangri, 1972). The life-span developmentalist would be concerned with describing the patterns of role choice in adolescence and young adulthood, their relation to antecedent, childhood role-related behavior, and their relation to consequent, adult role-related behavior. In addition, the life-span developmentalist would want to explain why the antecedent and current events led people to behave as they did in adolescence and young adulthood, and then how such behavior shapes the rest of adult life. The life-span developmentalist might, through explanatory attempts, come up with accounts of why males and females differ in the described ways. If so, the developmentalist might next attempt to intervene into the developmental process in order to foster different role-choice behavior. The developmentalist might attempt to promote higher-status role choices in females and thus enhance, or optimize, their development.

Optimization may involve encouraging alternative vocational choices for women. (ILO photo by J. Mohr; distributed by United Nations.)

Intraindividual Change and Interindividual Differences. In focusing on the description, explanation, and optimization of behavior change across the life span, two phenomena—intraindividual change and interindividual differences—are of central interest to life-span developmentalists. *Intraindividual* change refers to change within an individual over time. *Interindividual* differences refer to differences between individuals at a given point in time. These concepts are illustrated in Exhibit 1.1. This figure represents the scores of five hypothetical individuals (A–E) on a hypothetical behavior obtained at three points in time. The scores of four of the individuals (A–D) exhibit some change over the forty-year period. This change within individuals is what is meant by intraindividual change (within-person change). One individual (E) does not exhibit such change. At each point in time, there are also differences in scores between individuals. For example, at time 1 (20 years of age), A scores higher than B, who scores lower than C, and so on. These differences between individuals are what we mean by interindividual differences (between-person differences). Finally, the rate or direction of change may be different for various individuals. For example, from time 1 to time 2, the scores of persons A–D increase by the same amount. There are no differences in rate or direction of change. From time 2 to time 3, however, such differences exist. For example, the scores of individuals A and B continue to increase, while those of individuals C and D decrease. Similarly, the score of individual A increases more than that of B. Thus there are many (or multiple) directions of change possible in life. In other words, change may be *multidirectional.* These differential changes are what we mean by interindividual differences in intraindividual change (between-person differences in within-person change).

Developmental Issues

Within the context of the above general definition, life-span developmentalists have placed particular emphasis on certain developmental issues.

Multidirectional Change. It can be argued that traditional psychological and developmental perspectives of change are based on a biological concept of development

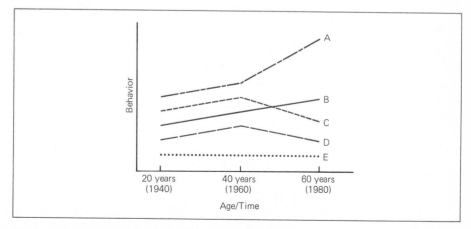

EXHIBIT 1.1   Within-person (intraindividual changes) in behavior may take different forms (as depicted by lines A–E); thus there are differences between people (interindividual differences) in within-person change.

(Harris, 1957). In the traditional developmental approach, for example, development is seen as unidirectional, irreversible, goal-directed, and universal—characteristics which apply to the biological maturation of all organisms. Development ceases once maturity is reached (Freud, 1949) and is followed by stability and eventual decline as the organism ages (Hall, 1922).

From a life-span developmental perspective, however, such a biologically based definition of development is too narrow. Rather, development occurs at all points in the life span of the individual. That is, life-span development may be characterized by multiple patterns of change differing in terms of onset, direction, duration, and termination. This view is depicted in Exhibit 1.2.

As a specific example of multidirectionality, consider Horn's (1970, 1978) work on intelligence. Horn distinguishes between two types of intelligence which exhibit different patterns of change during adulthood. On the one hand, crystallized intelligence reflects the degree to which the individual has acquired the knowledge and skills of the culture. It is indexed by tests which measure such knowledge, e.g., defining English words. On the other hand, fluid intelligence reflects the degree to which the individual has acquired qualities of thinking independent of cultural content. It is indexed by tests which minimize the role of such content, e.g., discovering a rule that makes a series of simple geometric figures go together. He states, "Intelligence may thus both decrease and increase in adulthood; crystallized intelligence, if properly measured in samples of people who remain in the stream of acculturation, increases; fluid intelligence decreases" (Horn, 1970, pp. 465–466).

The concept of multidirectional development departs most sharply from traditional views when it suggests the potential for growth during adulthood and old age. We have just mentioned that measures of crystallized intelligence increase with increasing chronological age during adulthood. Another example of the potential for growth during adulthood is provided by Butler's (1963) work on the life review process. Butler proposes that older adults engage in an inner experience or mental process of reviewing their life. This process is a response to the fact of impending death and does not reflect mental deterioration or personality dysfunction. Rather, it is a constructive process in which past experiences, particularly unresolved conflicts, are reviewed and reintegrated (recombined). Although the life review may lead to depression, guilt, and anxiety depending on a number of factors, the process often also leads to revised or expanded understanding. This understanding can give new

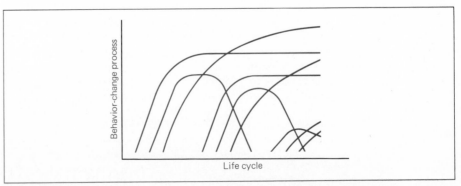

EXHIBIT 1.2   Illustration of hypothetical examples of multidirectional developmental change. Behavior-change processes differ in terms of their onset, directionality, duration, and termination over the course of the life cycle. (Based on Baltes, 1979.)

meaning to life and mitigate the fear of death. Butler (1963) comments: "It seems likely that in the majority of elderly a substantial reorganization of the personality does occur. This may help to account for the evolution of such qualities as wisdom and serenity, long noted in some of the aged" (p. 69). Butler's description of the potential for significant personality growth in the very old illustrates our point that development occurs over the entire life span.

From a life-span developmental perspective, therefore, development is seen as involving multiple patterns of change for different behaviors and different individuals at different points in time. One result of such multidirectionality is that as within-person changes become increasingly divergent (as people become more individualistic), between-person differences tend to increase over time.

The Role of Chronological Age, Cohort, and Life Transitions

Chronological age is a dimension along which behavior change functions may be charted. However, from a life-span developmental perspective, chronological age is not necessarily the most useful descriptive variable available (Baltes, 1973; Baltes & Willis, 1977). This may appear to be a contradictory statement since we typically think of development in relationship to age. However, chronological age is only useful as a descriptive variable to the extent that within-person change patterns are homogeneous enough to produce a high correlation between age and behavior change (Baltes & Willis, 1977). If large between-person differences in these patterns exist, then the use of chronological age as an organizing dimension is likely to be unproductive. Moreover, as we mentioned earlier, between-person differences in change tend to increase during adulthood as a function of the multidirectional nature of development. As a result, the life-span approach emphasizes the usefulness of examining several descriptive dimensions. In addition to chronological age, at least two other variables—cohort and life transitions—appear useful.

Cohort may be defined as a group of persons experiencing some event in common, for example, year of birth (Schaie, 1965). Someone born in 1942, for example, is a member of the 1942 *birth cohort*, or, more simply, the 1942 cohort. The specific range of time involved is arbitrary and may be variable (e.g., a month, year, decade, and so on). In other words, an individual is considered a member of a given cohort when born within a given range of time. Like age, cohort is a useful organizing variable to the extent to which interindividual differences exhibit a high correlation with variables that change with historical time. Examples of such potential cohort-related variables include historical changes in nutrition, health status, education, and family structure (Baltes, Cornelius, & Nesselroade, 1978).

The usefulness of the cohort variable is demonstrated by research on intellectual development during adulthood (Schaie, 1970; Schaie & Strother, 1968; Nesselroade, Schaie, & Baltes, 1972). For example, Nesselroade, Schaie, and Baltes (1972) analyzed measures of several dimensions of intelligence obtained from members of eight birth cohorts taken at two points in time (1956 and 1963), in which 59-year-olds measured in 1963 (1904 cohort) scored higher than 59-year-olds measured in 1956 (1897 cohort). Cohort differences were reported as large as, if not larger than, the age changes.

Life transitions may be defined by reference to normative life events such as marriage, birth of children, and retirement. While not all individuals experience these events, they are sufficiently normative and uncorrelated with chronological age to serve as potentially useful organizing variables. For example, Lowenthal, Thurnher, and Chiriboga (1975) examined four life transitions defined by graduation from high school, marriage, the last child leaving home, and retirement. They found these significant events to be very useful as dimensions within which to investigate adult development; changes in adult life were linked to these transitions.

The Changing Individual in a Changing World. Traditional views of development have emphasized the changing individual in a *static* world. That is, they have failed to recognize that individual development occurs within a changing historical context. In contrast, the life-span approach emphasizes a dynamic and interactive view of the relationship between the individual and the world (Lerner & Spanier, 1978a).

How can this dynamic interaction between person and environment be explained? Baltes, Reese, and Lipsitt (1980) specify three major influence patterns that affect this relationship: (1) normative, age-graded influences; (2) normative, history-graded influences; and (3) nonnormative, life-event influences. *Normative, age-graded influences* consist of biological and environmental determinants that are correlated with chronological age. They are normative to the extent that their timing, duration, and clustering are similar for many individuals. Examples include maturational events (changes in height, endocrine system function, and central nervous system function) and socialization events (marriage, childbirth, and retirement). *Normative, history-graded influences* consist of biological and environmental determinants that are correlated with historical time. They are normative to the extent that they are experienced by most members of a cohort. In this sense, they tend to

Significant features of children's environments change over historical time. (Courtesy of Atari, Inc.)

define the developmental context of a given cohort. Examples include historic events (wars, epidemics, and periods of economic depression or prosperity) and sociocultural evolution (changes in sex-role expectations, the educational system, and childrearing practices). Both age-graded and history-graded influences *covary* (change together) with time. *Nonnormative, life-event influences*—the third system —are not directly indexed by time since they do not occur for all people, or even for most people. In addition, when nonnormative influences do occur, they are likely to differ significantly in terms of their clustering, timing, and duration. Examples of nonnormative events include such items as illness, divorce, promotion, death of a spouse, and so on.

Thus variables from several sources, or dimensions, influence development. As such, life-span developmentalists stress that human development is *multidimensional* in character. In other words, variables from many dimensions are involved in developmental change.

In general, the relationships among these influence sources—normative, age graded; normative, history graded; and nonnormative, life-event—are *dynamic* and *reciprocal*. They are dynamic in that they are continually changing; and they are related reciprocally because each influence has an effect on the other and, in turn, is affected by it.

A simplified example of the role of these multiple influence systems on development is shown in Exhibit 1.3. This figure depicts the flow of three cohorts of individuals through time. The various life periods of each cohort occur at different

Nonnormative life influences can have a major impact on peoples' lives. The Kott family, Love Canal radiation-contamination refugees, salvaging items from a home in which they will never live again. (United Press International, Inc.)

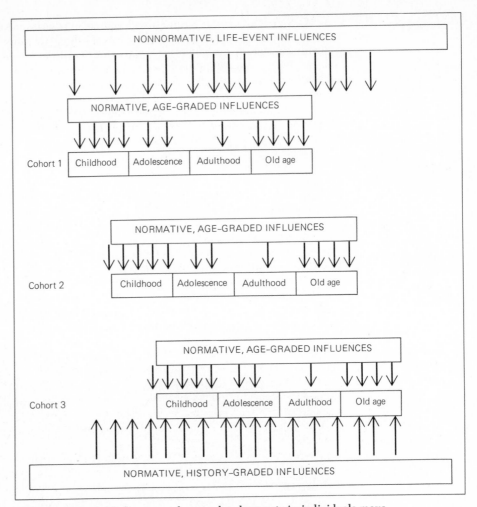

EXHIBIT 1.3   Multiple sources of influence on human development. As individuals move through the life cycle, they are exposed to multiple sources of influence. Different cohorts are exposed to different history-graded influences (e.g., wars). Within this context, members of the cohort are exposed to age-graded influences (e.g., puberty) and life-event influences (e.g., accidents). The interaction among these influences produces unique patterns of development for different cohorts and different individuals within cohorts.

points in historical time; thus each cohort is exposed to different normative, history-graded influences. Within this context, members of the cohort are exposed to normative, age-graded influences—the nature of which may vary from one cohort to another. Finally, members of the cohort are exposed to unique clusters of life events which vary in terms of their occurrence, timing, and duration for different individuals. The interaction between these multiple sources of influence produces different patterns of development for different cohorts and different individuals within cohorts.

Baltes, Reese, and Lipsitt (1980) have speculated that these three sources of influence exhibit different profiles over the life cycle. Normative, age-graded influences are postulated to be particularly significant in childhood and again in old age, while normative, history-graded influences are thought to be more important in adolescence and the years immediately following it; this is thought to reflect the importance of the sociocultural context as the individual begins adult life. Finally, non-normative, life-event influences are postulated to be particularly significant during middle adulthood and old age, reflecting increasing divergence as individuals experience unique life events. Such a perspective is consonant with the concept of multi-directional development across the life span.

Optimization

Traditionally, *interventionists* (scientists who try to change human functioning in order to optimize it) have focused on the alleviation of dysfunctions. That is, they have waited until a problem has appeared and then have tried to correct it (*remediation*) or reduce its intensity or duration (*rehabilitation*). Both these strategies are important. However, emphasizing the potential for growth, life-span developmentalists have promoted the usefulness of interventions focused on *prevention* and *enrichment* (Baltes & Danish, 1980; Lerner & Ryff, 1978). These refer to attempts to (1) reduce the likelihood that problems will occur (prevention) and (2) optimize the individual's knowledge, skills, and development (enhancement). The likelihood of the former is often increased by achieving the latter.

Prevention and enrichment strategies are often educational in nature. For example, Danish and D'Augelli (1980) have described a life development training program focused on the teaching of generic life skills such as goal assessment, decision making, risk taking, and self-development planning. These skills are assumed to be important resources for coping with various normative and nonnormative life events such as retirement and divorce. Similarly, parent-effectiveness training is designed to teach adults how to respond to the developmental needs of their children (Guerney, 1977).

The optimization strategy suggested by the life-span developmental approach presupposes the existance of an agreed-upon definition of optimum development. While this may be based on a theory of behavioral development, it is necessary to recognize that intervention involves the issues of values. When should we intervene? And for what purposes should we intervene? Whenever professionals develop and implement optimization procedures, they must be aware of their own values and "guard against the tendency to decide what others should be like" (Vondracek & Urban, 1977, p. 418). People involved with human development intervention have become aware of the value conflict associated with decisions to intervene. Baltes and Danish (1980) offer an appropriate illustration of value conflict in their discussion of optimal aging in individuals:

> In the medical sciences the criterion of lengthening life has a long tradition as a valuable general goal for intervention work aimed at optimization, both on the level of research and research application. Behavioral scientists have not yet agreed on a similiar guiding principle. Life satisfaction measures come closest; however, the assessment of life satisfaction in aging is at a preliminary stage. . . . What is desirable, on a personal, professional, societal, and philosophical level? Moreover, length of life and life satisfaction do not always correlate well. This is particularly true for advanced age and situations of death and dying. (pp. 70–71)

Moreover, value conflict can occur at a societal level. For instance, allocating resources for preventive intervention with adults and the elderly may conflict with alternative types of programs for other target groups such as infants and youth. Therefore, specialists must not only be aware of the possibility of their own well-intended but misguided efforts to impose their own standards on those they serve, but they also must understand their reasoning in making decisions to intervene at other levels (e.g., political, social, etc.).

We have mentioned that the life-span developmental approach emphasizes multidirectional change produced by multiple influences. As a result, the life-span interventionist begins with the assumption that multiple methods of intervention are required depending on the context. That is, the effects of a given intervention are likely to produce different results for different individuals. The interventionist must examine an array of methods in order to determine which will be most useful in the particular context. Lerner and Ryff (1978) conclude: "We must address the difficult issues of which behaviors, for which individuals, in which contexts, and in what time periods are most likely to lead to optimal developmental outcomes. . . . In sum, there is no single yellow brick road to developmental optimization." (p. 14)

The Multidisciplinary Emphasis

Our definition and specification of the life-span view raise several key issues which combine to promote another major aspect of this approach to development, its multidisciplinary emphasis. First, in contrasting within-person change with between-person differences in such change, researchers following the life-span view must study the group as well as the individual. Understanding between-person differences in change necessitates comparing an individual with others. Such integration requires merging ideas from within those disciplines that focus primarily on the individual (biology, molecular biology, genetics, and psychology) with those that focus primarily on the group (social psychology, sociology, and anthropology).

Second, because those taking a life-span view relate change to processes other than those associated with age, the ideas of many disciplines again are brought into use. Events associated with different birth cohorts, such as wars, politics, economic upheavals, and urbanization, may require understanding the works of historians, economists, political scientists, lawyers, and urban planners. Similarly, understanding how critical life events, such as marriage and breaking parental ties, influence personal change may necessitate integrating the ideas of family sociology with these other disciplines. In other words, because of the multidimensional nature of development, the most useful route to complete understanding of change lies along multidisciplinary paths.

A life-span view of any particular period of development, then, involves much more than *just* considering antecedent and consequent events. To illustrate, consider a psychologist interested in studying cognitive change during childhood and early adolescence. The psychologist may notice that the child moves from being concrete to being abstract in his or her thinking. But the sociologist may view childhood and adolescence as periods of change in roles and social relationships, and the economist may view them as periods of transition from having few financial resources to having the potential for many. The lawyer may consider the changing legal rights of the person, and the physiologist and the physician may be concerned with the changing hormonal balance and muscle and fat constitution of the person during these periods, and with the diseases or physical problems associated with such physiological and physical alterations. Finally, both the cultural anthropologist and the historian would also be interested in these changes. The former scholar might address issues relating to cross-cultural differences in the meaning of childhood and of adolescence. The latter might be concerned with time-related changes in the meaning of the periods.

These examples indicate that although one may be prone to think of a period of life as a developmental phenomenon of concern to scholars of only one discipline, such as psychology, such an orientation is limited. Said another way, the events and phenomena of the human life-span involve changes along several dimensions. These dimensions range from the biological to the psychological and social to the historical. Consequently, to understand development adequately one should understand *all* changes and transitions involved; one should have a multidisciplinary orientation. All portions of the life span ought to be studied this way. This approach allows one to understand how changes at all levels of analysis influence any one particular period, and how changes in this period affect subsequent periods.

A life-span approach to understanding human development means more than just looking at all dimensions of change prior to, during, and after any one period of life. It means considering the potentially interrelated influence of these dimensions of change. In other words, the many dimensions of change the person goes through across the life-span do not necessarily occur independently. All changes "meet" in some way at the level of the person. For example, one's biological functioning may affect one's psychological and social functioning, and one's psychological and social functioning may affect one's biological functioning.

Differences in likelihood of disease and quality of nutrition, for instance, are associated with both social class differences and with ability to perform intellectual tasks. People who are sick and/or hungry will function differently psychologically (e.g., in regard to mood, attentional behavior, and cognitive behavior) than will healthy and/or well-fed people. Moreover, such biological and psychological linkages are not as likely to occur with equal probability at all levels of society. Poor children are likely to be underfed and more disease-prone. They are more likely to suffer the damaging psychological effects of illness and malnutrition than are children of the affluent. Accordingly, children suffering the negative psychological impact of poor health and low social status are not as likely to profit from educational programs necessary for higher-status roles in society as are children not suffering these negative psychological impacts.

Thus different role behaviors in later life might arise from the merger of biological, social, and psychological differences in earlier life. These earlier differences may lead to adult role differences, for instance in regard to status, and hence the economic return for playing one's role. These adult status and economic differences will feed back on the person's other (e.g., biological and psychological) dimensions as well. People with different incomes can afford different degrees of health care and recreation time, for instance.

Moreover, although this illustration indicates that biological, social, and psychological dimensions merge across the life span, the illustration can be extended to include the cultural and historical influences that also always exist. Different cultures have different distributions of poor and affluent, because of differences in economic, political, and natural resources. The contrasting distributions have implications for the importance attached in society to linkages among biological, psychological, and social factors. A country with very few poor probably does not need to revise societal institutions or functions in order to attend to problems relating nutrition and psychological functioning. However, a country with a significant number of poor people might declare a "War on Poverty" and institute a "Project

Head Start" as a preventive intervention for such problems. People in the society might also institute a "Youth Job Corps" to alter chronic adolescent and young-adult underemployment among poor, undereducated, disadvantaged youth. Of course, all these latter events happened in the United States in the 1960s and 1970s.

Moreover, these changes in the United States occurred in relation to evolving economic and political events such as the civil rights movement and the 1965 Voting Rights Act. To understand even the effect of nutrition on school performance among children of a certain social class, it is useful to conceptualize all these changes as embedded in a changing historical context.

In sum, not only do many of the biological, psychological, and social characteristics of the person come together to shape the individual's behavior at one point in the life span, but also such behavior contributes to the nature of the joint influence between biological, psychological, and sociological factors at other points in the life span. Moreover, the nature of these interactions between all these dimensions of change is textured by the changing historical context. As people are changed over the course of time, these alterations accumulate and constitute the very changes known as history.

In conclusion, the life-span view, with its multidisciplinary perspective about change, leads to the notion that all dimensions of change are interrelated. Changes in one dimension are related to changes in all others; change in one aspect of development provides a basis of change in all others. Furthermore, changes in a given dimension (e.g., biology or history) feed back onto themselves; they provide a basis of their own change by changing their context, the other dimensions they are embedded within. Development thus involves a series of "feedback loops," an interrelated network of circles. In other words, the person is *both a product* of his or her biological, sociological, psychological, and historical world—*and a producer* of it! This conceptualization of the nature of human development is a complex one, involving a dynamic interaction between all dimensions of development across the entire life span.

SCIENTIFIC RESEARCH AND THE STUDY OF HUMAN DEVELOPMENT

The life-span view of human development is a complex one. It requires information from a diverse array of sources, potentially from several disciplines, to be gathered and integrated. In fact, one must observe such information not only at just one point in life, but because one portion of life may influence others, observations at multiple times of life are needed. But how are such observations made in ways that will be of use to scientists studying human development? This key question raises the issue of what the scientific methods of investigation available to human developmentalists are.

Students sometimes find research methods mystifying. For example, following a lecture several years ago, one of us recalls a student complaining, "Why bother to describe all of this research stuff? Just tell us the way things are." Scientific research, of course, is simply a particular approach to discovering "the way things are."

But why must the study of development involve scientific research? What is wrong if people just sit back in their armchairs and tell others what people are like, and never actually study if what they say corresponds to actual behavior? Why would a scientist be more likely to believe a statement about development based on the observation of people as opposed to a view which is in no way backed up by observations?

The answers to all these questions rest on the commitment to the *scientific method*, an approach used by researchers to study the phenomena of the world. The basic attribute of the scientific enterprise is *empiricism*, a view that knowledge is achieved through the systematic observation and analysis of data. This attribute is the major difference between philosophy and science. Those who are not scientists may find knowledge in ways independent of empirical research (e.g., through chiefly speculative rather than observational means). However, for a scientist to know something about the development of humans, people must be examined, questioned, interviewed, or in some way observed. One cannot rely *just* on what one believes or wants to believe about behavior. Rather, one's beliefs must be tested by determining whether such beliefs are supported when actual behavior is studied. The set of specific procedures by which a science makes observations and collects and examines data may be termed its *research methods*.

The purpose of research is to examine relationships among variables (Kerlinger, 1973). It is the specification of such relationships that allows us to achieve the goals of description, explanation, and optimization described earlier. The difficulty is that relationships may be specified either accurately or inaccurately. As a result, our descriptions and explanations of phenomena may be valid or invalid, and our efforts at optimization may be effective or ineffective. How are we to maximize the accuracy and minimize the inaccuracy in our discovery of relationships among variables and, consequently, in our efforts at description, explanation, and modification? This is accomplished by the application of research methods. The purpose of research methods is simply to reduce the degree of error in stating relations among variables. Research methods constitute a set of "rules" and procedures to help us make valid inferences about phenomena.

The Validity of Research

Many types of errors are possible as we attempt to infer relations among variables (Campbell & Stanley, 1963; Cook & Campbell, 1975). Two issues, however, are of particular concern: (1) the issue of whether an observed relationship is accurately identified and interpreted—the issue of *internal validity*—and (2) the issue of whether a relationship observed in one set of data may be observed in other sets of data—which might have been observed but were not—the issue of *external validity*.

Let us outline a hypothetical situation in order to provide a context within which to illustrate these two issues:

A researcher is contacted by the administrator of a small residential home for the aged. The administrator is concerned about the poor morale (e.g., feeling dissatisfied, unhappy, and depressed) of the residents. Following a visit to the home, the researcher hypothesizes that the physical environment of the institution is not conducive to high morale (e.g., areas, such as the lounge, are arranged in ways that discourage social interaction; the residents' rooms offer little privacy; the rooms are drab, etc). The researcher proposes, therefore, that a change in the physical environment of the home will result in an increase in morale among the residents. The administrator agrees to make the changes suggested and to assist in the study. The study is conducted as follows. Arrangements are made to send the residents to the homes of friends and relatives for one week while the institution is renovated. One week prior to departure, each resident is asked to complete a questionnaire designed to measure morale. One week following the return to the newly renovated home, each resident is again asked to complete the questionnaire. Comparison of the first and second morale scores shows a dramatic increase in morale following their return to the renovated home. The researcher and administrator conclude that renovating the home led to an increase in morale. Are they right?

This hypothetical study attempts to examine the relationship between two variables—physical environment and morale. It appears to suggest that a change in the former produces a change in the latter; however, this influence is subject to error. Let us examine the two major types of inferences involved.

Internal Validity

Internal validity refers to the adequacy with which relationships among variables are identified or interpreted (Campbell & Stanley, 1963; Cook & Campbell, 1975). Of particular concern is the establishment of accurate causal relationships where the objective is to determine whether one set of events produced or caused another. Such causal relations are required for the explanation of phenomena.

Internal validity requires that there be no viable explanation for an outcome other than the presumed causal event. In our example, then, there should be no explanation for the increase in morale other than the remodeling of the home. Internal validity is jeopardized when plausible alternative interpretations exist such that an apparent causal relationship between two variables may be due to a third variable or variables. Are there any such "third variable" explanations in our example? Yes. For instance, while the home was being renovated, the residents visited the homes of friends and relatives. This change of location and consequent opportunity to interact with others may have caused the increase in morale rather than the change in the environment of the home. Thus there was an event (visiting friends and relatives) which was occurring at the same time as the event we were interested in (renovation of the home). As a result, there is no way to determine which of these two plausible explanations (and there are others) is accurate.

External Validity

External validity refers to the adequacy with which relationships among variables may be generalized. In most cases, the observations made in a research study are a subset of some larger set of potential observations that might have been made but were not. Thus observations are made on a subset of potential persons, in a subset of potential settings, using a subset of potential treatments and measuring instruments, at a subset of potential times in history, and so forth. External validity requires that relationships among variables remain robust across these different potential data sets. External validity is jeopardized, too, when interactions exist such that the relationship is modified by the status of other variables. That is, there is a relationship between the variables, but only for certain types of people, or only in certain types of settings, or only at certain points in history, and so forth.

External validity assumes the existence of internal validity. If a relationship has not been accurately identified or interpreted, the question of whether it may be duplicated in other data sets is meaningless. Returning to our example, then, let us assume that the renovation of the home actually did cause an increase in morale among the residents. How robust is this relationship? For example, there may be an interaction between the number of residents and the characteristics of the physical environment. Changes in the environment may affect morale in a small group of residents but not in a large group of residents. Thus what was an effective manipulation in a small, proprietary nursing home may not be effective in a large, state-operated nursing home.

As in the case of internal validity, there are many potential threats to external validity. It is important to recognize that the issue of external validity involves more than generalization across a sample of persons to a population of persons. Rather, as Baltes, Reese, and Nesselroade (1977) note, it "applies to inferences made from a sample of observations to a population of potential observations" (p. 51). Each observation represents a unique combination of person, setting, measurement, treatment, and historical time variables.

DIMENSIONS OF RESEARCH METHODS IN HUMAN DEVELOPMENT

In this portion of the chapter the various research methods available in the study of development will be illustrated. It is useful to present these methods as varying along four dimensions. A *dimension* is an imaginary line continuously running between two endpoints. A person can have a characteristic at either extreme or anywhere in between the two ends on any dimension. For instance, psychologists often talk of personality dimensions like active-passive or dominant-submissive. Any given person can fall, or be located, at one point on one dimension (e.g., the middle) and at some other point on another (e.g., the extreme). A second person could have different locations along these dimensions, depending on his or her individual characteristics.

Any developmental study has characteristics which allow it to be located along some point of four dimensions of research (see Exhibit 1.4). Location at one point on one dimension does not necessarily imply a location on the other dimensions. The fact that any one study could fall along different points of each dimension means an almost limitless array of strategies of research is available to the developmental researcher.

EXHIBIT 1.4   **Four Dimensions of Studies in Human Development Research**

1. Normative Studies ——————————— (Descriptions of averages, frequencies, and norms of behavior.)	Explanatory Studies (Assessment of the causes or bases of behaviors.)
2. Naturalistic Studies ——————————— (Studies of people in their actual, "real life" [ecologically valid] settings.)	Manipulative Studies (Conditions of the setting are controlled, e.g., as in a laboratory setting. The design of such research is often experimental.)
3. Atheoretical ——————————————— (Studies designed to answer practical problems, verify casual observations, or satisfy curiosity.)	Theoretical (Studies designed to test ideas— hypotheses—derived from a theory.)
4. Ahistorical ——————————————— (Studies of relations among variables which have been measured at the same time. No assessment of the antecedents and/or the consequences of the relation that exists at one point in time.)	Historical (Studies of the antecedents and/or consequences of behavior: a focus on the history of the behavior.)

Source: Based on McCandless (1967, p. 60).

DESIGNS OF DEVELOPMENTAL RESEARCH

Many people who attempt to understand the basis of an individual's development do so by specifying age-related developmental progressions. An example is attributing storm and stress to the adolescent "stage of life." Although age-related progressions, or "stage" progressions, may be one source of a person's development, they are not the only processes which provide a basis for change. For example, if a prominent event occurs, as for instance the Watergate political crisis or the assassination of President Kennedy, behaviors of people might be affected despite what stage or age of development they are in. If one were measuring attitudes toward the government during the time of the Watergate political crisis, the events in society at this time of measurement may have influenced children, adolescents, and adults. As such, it is possible that time of measurement, as well as age-related phenomena, can influence development.

In addition, not only may age and a particular historical event influence behavior, but people may change as a consequence of being exposed to a particular *series* of historical events. Again imagine that attitudes toward government were being measured and that the subjects of the study were people born during the Great Depression in the United States (1929 through the late 1930s). During this historical era, many of the institutions designed to afford economic security to American citi-

Members of a birth cohort experience particular events in common. Children in Northern Ireland are developing in the midst of civil strife. (United Press International, Inc.)

zens (banks, for example) failed, and existing governmental policies were not able to deal with this situation. Accordingly, it may be expected that people who are members of the 1920s birth cohort, and thus experienced the effects of the Depression during childhood, might have developed differently than people born well before or well after this historical era. Indeed, research has found this to be true (Edler, 1974). Such birth cohort–related influences can affect the character of behavior people show across their lives. People who were children during the Great Depression may continue to be more wary about the stability of the economy and the ability of the government to safeguard citizens than may people who were children during eras of affluence and prosperity (the late 1950s in the United States, for example). Because of membership in a certain birth cohort, people may continue to differ from those of other cohorts, no matter at what age they are measured or what exists in the sociocultural setting at a particular time of measurement.

It may be seen, then, that there are at least three components of developmental change: *birth cohort–related events, time of measurement,* and *age-related phenomena.* Recognizing that reference is always made to phenomena that change in relation to these components, we may label these components *cohort, time,* and *age* for convenience. Thus when within-person change is seen from one point in the life span to another, one must be able to determine how processes associated with each of these three components may influence change.

Until relatively recently (Baltes, 1968; Schaie, 1965), the three commonly used designs for developmental research—longitudinal, cross-sectional, and time lag— did not allow for an adequate determination of the contributions of the age, time, and cohort components. Developmental research typically involves a *confounding* of two of the three components of change, and as a consequence, the designs' utility is severely limited. When a variable is confounded, its influence on behavior cannot be separated from that of another variable that could be simultaneously influencing behavior.

For instance, if one wanted to know whether males or females could score higher on a test of reading comprehension, one would not want all the males to be college-educated, all the females grade school–educated. If one did not equate the two sex groups on education level (if one did not "control" for the contributions of education), then one would not know if differences between the groups were influenced by their sex or by their educational disparities (or some combination of the two). Thus sex could be confounded with education. In other words, one could not separate the effects of the two variables. When the separate influence of two variables cannot be determined, these variables may be confounded, and any study that involves such a confounding has a potential methodological flaw.

In Exhibit 1.5 the particular confounding factors in each of the three commonly used designs are presented. Reference to this exhibit will be useful as discussion turns to an examination of the characteristics of each of these designs and an explanation of why they confound what they do.

EXHIBIT 1.5   **Some Characteristics of Longitudinal, Cross-Sectional, and Time-Lag Designs of Developmental Research**

Design	Study Involves:	Confounded Components of Developmental Change
Longitudinal	One birth cohort	Age with time
Cross-sectional	One time of measurement	Age with birth cohort
Time lag	One age	Time with birth cohort

The Longitudinal Design

The *longitudinal design* (also known as a *panel design*) involves observing the same group of people at more than one point in time. The main asset of this approach is that since the same people are studied over time, the similarities or changes in behavior across their development can be directly ascertained. If one did not repeatedly observe the same people, through use of longitudinal observations, one would not be able to know how a given behavior seen in a person early in life tends to be expressed by that same person later in life. Without longitudinal measurement, one would not know whether behavior stays the same or changes in a person. However, this asset of the longitudinal method leads directly to some limitations.

It obviously takes a relatively long time to do some longitudinal studies. If researchers wanted, for instance, to do a longitudinal study of personality development from birth through late adolescence, they would have to devote about twenty years of their own lives to such a research endeavor. Such a commitment would be expensive, as well as time-consuming, and thus it may be easily seen why relatively few long-term longitudinal studies have been done (e.g., see Jones, 1958; Kagan & Moss, 1962; Livson & Peskin, 1980; Thomas & Chess, 1977).

Other limitations of longitudinal studies pertain to the nature of the people studied and to problems with the measurements that may be used. Not everyone would be willing to be a subject in a study that required their continual observation over the course of many months or years of their lives. Hence samples tend to be small in such studies. Those people who are willing to take part may not be representative of most people. Thus longitudinal studies often involve unrepresentative, or "biased," samples of people. Results of such studies may not be easily applied, or "generalized," to a broader population; that is, such studies may not be externally valid. In addition, longitudinal samples typically become increasingly biased as the study continues. Some people drop out of the group, and one cannot assume that the remaining people are identical to the former group. After all, the people who stay may be different just by virtue of the fact that they continue to participate.

Another problem with longitudinal studies is that after some time people may become accustomed to the tests of their behavior. They may learn "how to respond," or they may respond differently than they would if they had never been exposed to the test. Hence the meaning of a particular test to the subjects may be altered over time through the repeated use of the instrument with the same sample. Such an occurrence would make it difficult to say that the same variable was actually being measured at different times in the subjects' lives.

Often the purpose of using this method is to ascertain the developmental time course for a particular type of behavior or psychological function. One also wants information which may be applied to understanding development about future generations of people. Yet with a longitudinal study, one is only studying people who are born in one historical era and who are measured at certain points in time. One does not know whether findings about this one cohort can be generalized to people in other cohorts.

A confounding of age and time exists during such a study. Since a longitudinal design involves assessing one particular cohort of people (for example, a group of males and females born in 1965), such people can be age 15 at only one time of measurement (1980 in this case). Thus their behavior at age 15 may be due to age-related phenomena *or* to phenomena present at the time of measurement (*or* to both). Similarly, members of one birth cohort can only be age 20 at one time of measurement. Thus, as noted in Exhibit 1.5, age and time are confounded in a longitudinal study.

One does not know if results of a longitudinal study can be applied to other 15- or 20-year-olds who are measured at other times.

Hence the findings about development that one gains from a longitudinal study reflect age-related changes, *or,* alternatively, they may reflect only characteristics of people born and studied at particular points in time. One does not know in a longitudinal study whether the findings are due to universal rules, or "laws," of development (i.e., rules which describe a person's development no matter when it occurs), *or* to particular times the subjects are measured, *or* to some combination of all these influences.

To summarize, although longitudinal studies are useful for describing development as it occurs in a group of people, such studies have expense, sampling, and measurement problems, and may present results not applicable to similarly aged people who grew up in different historical eras or who were measured at different points in time. Because of such problems, alternatives to the longitudinal method are often used.

The Cross-Sectional Design

The most widely used developmental research design is the *cross-sectional design*. Here different groups of people are studied at one point in time, and hence all observations can be completed relatively quickly. The design is less expensive than longitudinal research and requires less time. Because of these characteristics, some have argued that the method allows for a more efficiently derived description of development. However, there are important limitations of cross-sectional research.

If one wanted to study the development of aggression in individuals who range in age from 2 to 20 years, one could use the cross-sectional method. Instead of observing one group of people every year, for example, for eighteen years, groups of individuals at each age between 2 and 20 at one point in time could be observed. The first group of people would be 2-year-olds, a second group would be 3-year-olds, and so on through the last group which would be 20-year-olds.

However, it is difficult to fully and adequately control for all variables that may affect behavior differences. One may not be certain whether differences between the various age groups are reflections of real age changes or merely reflections of the groups not being really identical to begin with.

Sometimes the researcher attempts to match the individuals on a number of important variables other than age (e.g., race, father's or mother's educational background, income level, or type of housing) to ensure some degree of comparability. However, such comparability is difficult to achieve. Moreover, although it is possible to get less biased, more representative samples for cross-sectional research than for longitudinal studies (people will cooperate more readily since they are only committed to be observed or interviewed once), such better sampling may still not lead to a useful description of the components of developmental change. This failure occurs because of a flaw in the rationale for the use of a cross-sectional method instead of a longitudinal one.

The expectation in some cross-sectional studies is that they yield results comparable to those obtained from studying the same group of people over time, and do so more efficiently, as long as the only differences among cross-sectional groups are their ages. However, despite how adequately subjects are matched, it is rarely the case that the results of cross-sectional and longitudinal studies are consistent (Schaie & Strother, 1968).

For example, when studying intellectual development with a cross-sectional design, most researchers report that highest performance occurs in the early twenties or thirties and considerable decreases in performance levels occur after this period (e.g., Horn & Cattell, 1966). With longitudinal studies of these same variables, however, often no decrease in performance is seen at all. In fact, some studies (e.g., Bayley & Oden, 1955) have found some increase in performance levels into the fifties. As has been pointed out, it may be suggested that the nature of subjects typically used in the longitudinal design is considerably different from that of the subjects used in the cross-sectional study.

Longitudinal studies, as has been noted, may be composed of a select sample to begin with, and as the study proceeds, some people will drop out of the research. Such attrition may not be random. Rather, it may be due to the fact that the subjects of lower intellectual ability leave the study. Hence in the example of research on intellectual development, this bias could account for lack of decreases in level of performance. In addition, as Schaie and Strother (1968) point out, these longitudinal studies have not assessed intellectual development in the sixties and seventies, the age periods during which the greatest performance decreases have been seen in the cross-sectional studies (e.g., Jones, 1959). Thus comparisons of the age-associated changes found with the two methods are not appropriate.

On the other hand, cross-sectional samples have not escaped criticism. Schaie (1959) has argued that such samples do not give the researcher a good indication of age-associated changes because of the fact that it is difficult to control for extraneous variables in the samples used to represent people of widely different age ranges.

Although these arguments may be appropriately used to reconcile the discrepancies (or perhaps to explain them away), Schaie (1965) suggests that these arguments miss an essential point: they do not show a recognition of an essential methodological problem involved in the consideration of longitudinal and cross-sectional designs. Just as longitudinal studies are confounded (between age and time), cross-sectional studies also are confounded. As seen in Exhibit 1.5, the confounding is between age and cohort. Because the two types of studies involve different confounding, it is unlikely that they will reveal the same results.

The confounding of age and cohort which exists in cross-sectional studies occurs because at any one time of measurement (e.g., 1980) people who are of different ages can only be so because they were born in different years. To be 20 in 1980, one has to have been born in 1960, while to be 25 at this time of measurement, one has to be a member of the 1955 birth cohort. Consequently, because cross-sectional studies focus only on one time of measurement, there is no way of telling whether differences between age groups are due to age-related changes or to differences associated with being born in historically different eras.

The Time-Lag Design

Although not as frequently used in research as the cross-sectional or longitudinal designs, the *time-lag design* allows a researcher to see differences in behavior associated with particular ages at various times in history. That is, in contrast to focusing on one cohort or one time of measurement, the time-lag design considers only one age level and looks at characteristics associated with being a particular age at different times in history.

For example, when the focus of research is to discern the characteristics associated with being a particular age (e.g., 15 years old) at different times of measurement (e.g., 1950, 1960, 1970, and 1980), a time-lag design is implied.

Of course, such a design involves cross-sections of people and has all the problems of control, matching, and sampling associated with such designs. But there are also additional problems. As indicated in Exhibit 1.5, because only one age is studied at different times, the different groups are members of different birth cohorts. Thus in a time-lag design, time and birth cohort are confounded, and one does not know, for example, whether the behaviors of 15-year-olds studied at two points in time are associated with events acting on all people—no matter what their age—at a particular test time or are due to historical events associated with membership in a specific cohort.

In sum, the three types of conventional designs of developmental research do not allow for the unconfounded assessment of the contributions of the three components of developmental change. Because of these shortcomings, it is difficult to decide which method gives a more useful depiction of developmental changes. Each method may potentially introduce serious, but different, distortions into measures of developmental changes. This is perhaps the major reason why information about developmental changes derived from the three techniques often is not consistent (Schaie & Strother, 1968). Although each design has some advantages, the problem of each places limitations on the ability of developmental researchers to describe adequately how individual, sociocultural, or historical influences can influence change. This might lead some to conclude that a bleak picture exists for the study of human development, since the three conventional designs of developmental research have some methodological problems. But, of course, no research method is without its limitation. As such, our view is that *all* the conventional methods of developmental research may be used to enhance understanding *if* they are employed with a recognition of their limitations. These limitations, however, can be transcended, as is described in the following section. While these methods also have their limitations (e.g., they typically require large samples of people and are quite expensive to implement), they offer a useful alternative to traditional approaches.

Sequential Strategies of Design

Due to the influence of K. Warner Schaie (1965), Paul B. Baltes (1968), and John R. Nesselroade (Nesselroade & Baltes, 1974), the problems of confounding involved in cross-sectional, longitudinal, and time-lag designs may be resolved. Schaie (1965) demonstrated how the conventional methods were part of a more *general developmental model*. Presentation of this model allowed him to offer a new type of approach to designing developmental research: sequential methods.

Sequential methods combine features of longitudinal and cross-sectional designs, whereby the researcher may assess relative contributions of age, cohort, and time in one study, to know what differences (or portion of the differences) between groups are due to age differences, to cohort (historical) differences, or to time of testing differences. In addition, a sequential design allows these sources of differences to be ascertained in a relatively short period of time.

Research based on sequential designs is complex, due in part to the usual involvement of *multivariate* (many variable) statistical analyses and the numerous measurements that have to be taken of different groups. But a simplified example of such a design may be offered. It will suggest how use of such a design allows the developmental researcher to avoid the potential confounding involved with traditional cross-sectional and longitudinal approaches.

Basically, a sequential design involves the remeasurement of a cross-sectional

K. Warner Schaie. (Photo by S. Willis.)

sample of people after a given, fixed interval of time has passed. A researcher selects a cross-sectional sample composed of various cohorts and measures each cohort longitudinally (with the provision that each set of measurements occurs at about the same point in time for each cohort). In addition, this design calls for obtaining repeated measures from each of the different cohort groups included in a given cross-sectional sample and for obtaining data from retest control groups to assess effects of retesting. For example, if three times of testing are included as the longitudinal component of the design, then control cohort groups (in this case, assessed only at the third testing time) may be used to control for any retesting effects. The researcher is thus in a position to make statements about the relative influences of age, cohort, and times of measurement on any observed developmental functions in the results.

Cross-sectional and longitudinal sequences consist of sequences of either simple cross-sectional or longitudinal designs. The successive application of these strategies permits us to describe the extent to which behavior change is associated with age-related or history-related influences. Exhibit 1.6 provides a contrast between the simple and sequential strategies. The top portion of the figure shows the simple cross-sectional and longitudinal designs described earlier, while the bottom portion of the figure shows the two sequential strategies. Cross-sectional sequences involve successions of two or more cross-sectional studies completed at different times of measurement. Longitudinal sequences involve successions of longitudinal studies begun at different times of measurement. The strategies differ in that cross-sectional sequences involve independent measures on different individuals, while longitudinal sequences involve repeated measures of the same individuals. In practice, one can apply both strategies simultaneously. In any event, the application of sequential strategies permits the discrimination of within- and between-cohort sources of change, thus increasing the validity of our descriptive efforts.

To see how this works, it is useful to consider a sample design of such a sequential study. Such a design is presented in Exhibit 1.7 and recast in the form of a matrix in Exhibit 1.8. Different cohort levels are composed of different groups of people born at different historical periods (1954, 1955, 1956, or 1957). Thus at the time of the first testing (1970 for this design), the study has the attributes of a cross-sectional study. Indeed, there are three such cross-sectional studies in this particular design,

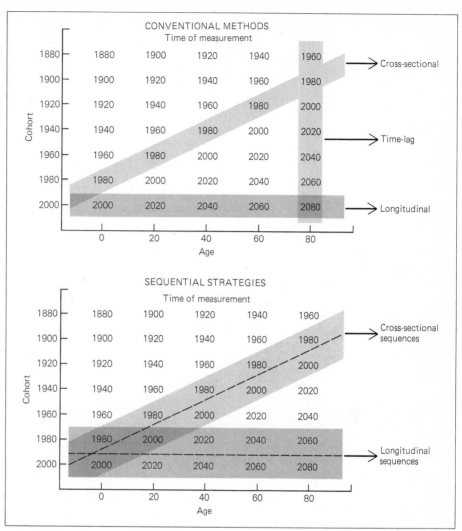

EXHIBIT 1.6   Illustration of simple cross-sectional, longitudinal, and time-lag designs (top) and cross-sectional and longitudinal sequences (bottom). (Source: Baltes, Reese, and Nesselroade, 1977.)

one for each time of measurement (see Exhibit 1.8). However, the sequential feature is introduced when these same subjects are again measured in 1971 and 1972. Thus for each cohort there is now a longitudinal study. As seen in Exhibit 1.8, each cohort in a sequential design of this sort is involved in its own short-term longitudinal study (there are four of these in the design shown in Exhibit 1.8). Additionally, it should be noted that the diagonals of the design matrix of Exhibit 1.8 represent time-lag studies; people of the same age are studied at different times. Thus a sequential study involves all combinations of observations of other designs in one integrated matrix of observations.

With such a matrix, the researcher can answer a number of questions involving the potentially interrelated influences of cohort, age, and time. Referring to Exhibits 1.7 and 1.8, if, for example, the cohort composed of people born in 1955 underwent

EXHIBIT 1.7   **The Design of a Sequential Study**

Birth Cohort	Time of Measure-ment 1	Age at Time 1	Time of Measure-ment 2	Age at Time 2	Time of Measure-ment 3	Age at Time 3	Time of Measure-ment of Retest Control Group	Age of Control Group
1957	1970	13	1971	14	1972	15	1972	15
1956	1970	14	1971	15	1972	16	1972	16
1955	1970	15	1971	16	1972	17	1972	17
1954	1970	16	1971	17	1972	18	1972	18

changes between times of measurement 1 and 2 and were found to be different at age 16 from the people in the 1954 cohort group when they were 16, then there must be some historical difference between these two cohort levels. In other words, if differences were due simply to age-related changes, then one should see the same performance for every cohort group, no matter when measured. If only age-related changes matter, then people of the same age should perform the same no matter what cohort they are from or when they are measured. A younger cohort group should perform similarly to that of an older cohort group as members of each group age *if* there were no historical differences between cohorts and if time of testing did not matter. Again from Exhibits 1.7 and 1.8, the 1957 cohort should show a level of performance on its second measurement comparable to that of the first measurement for the 1956 cohort *if* there were no historical differences between the generations.

In turn, if time of testing were a source of change, then people should respond the same despite their age or cohort. If events in 1972 were the strongest influence on behavior, then one should see that people of all cohorts represented in Exhibits 1.7 and 1.8 respond the same way.

Finally, of course, if birth cohort were of most importance, then people of a particular cohort should respond in a given way no matter what age they are and no matter at what time they are measured. As illustrated by the example of children

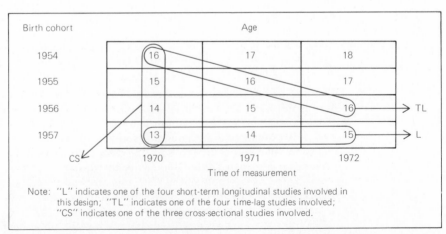

Note: "L" indicates one of the four short-term longitudinal studies involved in this design; "TL" indicates one of the four time-lag studies involved; "CS" indicates one of the three cross-sectional studies involved.

EXHIBIT 1.8   The design of a sequential study put into the form of a matrix (the same design as that shown in Exhibit 1.7 is presented).

born in the Great Depression (Elder, 1974), membership in a particular cohort would override influences due to age or time of measurement.

Additionally, it should be noted that by including groups of subjects to be tested for the first time at the end of the study (see Exhibit 1.7), sequential researchers provide a basis of assessing the issue of repeated use of the measuring instruments, noted earlier. If subjects in the core sample did not respond differently as a consequence of their having been repeatedly measured (for example, by the same tests of personality or IQ), then their behavior at the end of the study should be comparable to a group of subjects matched in every way with them except for the fact that no repeated testing was given. If there are differences, however, between the core sample and these "retest" controls, then there are statistical techniques available to researchers to measure the effects of retesting (Nesselroade & Baltes, 1974).

Despite the complexity of data analysis, and the more complex research design and reasoning process associated with it, sequential methodology has desirable attributes not associated with other techniques. It allows for the unconfounding of the components of developmental change functions in one descriptive effort. As such, it allows the contributions of variables associated with multiple levels of influence to be evaluated adequately.

In fact, although sequential research studies are relatively few in number, the design illustrated in Exhibits 1.7 and 1.8 was used because it corresponds to the one used by Nesselroade and Baltes (1974) in their sequential study of adolescent personality development. Let us look at this study in more detail.

A Sequential Study of Adolescent Personality Development

Noting that most conceptions of adolescent personality development suggest that age-related progressions are influential in this period of life, Nesselroade and Baltes (1974) argue that historical (cohort) and specific sociocultural (time) influences may also be involved. As such, they applied sequential methodology to see how these three components contributed to changes in personality in the period from 1970 to 1972.

About 1800 West Virginian male and female adolescents were measured in 1970, 1971, and 1972. These adolescents were from birth cohorts 1954 to 1957, and thus, as in Exhibits 1.7 and 1.8, ranged in age at the time of first measurement from 13 to 16. Personality questionnaires and measures of intelligence were administered to these subjects.

Contrary to what is stressed by those theorists who focus on personological components of adolescent development (e.g., Anna Freud, 1969), Nesselroade and Baltes found that change at this time of life was quite responsive to sociocultural-historical influences. In fact, age per se was not found to be a very influential contributor to change. Rather, for these groups of adolescents developmental change was influenced more by cultural changes over the two-year historical period than by age-related sequences.

For instance, adolescents as a whole, *despite their age or birth cohort*, decreased in "superego strength," "social-emotional anxiety," and achievement during the 1970–1972 period. Moreover, most adolescents, regardless of age or cohort, increased in independence during this period.

Accordingly, the Nesselroade and Baltes (1974) data show that it was the time at which all these differently aged adolescents were measured that was most influential in their changes. Perhaps due to the events in the society of that time, e.g., those associated with the Vietnamese war, all adolescents performed similarly in regard to these personality domains. Despite where they were upon "entering" the 1970–1972 historical era, members of different cohorts changed in similar directions due to events surrounding them at the times they were tested.

Without sequential methodology, the role of the specific sociocultural setting on adolescents at that time could not have been suggested. This suggestion is supported by data derived from other sequential studies which have shown the influence of birth cohort on intellectual developments in children (Baltes, Baltes, & Reinert, 1970) and adults (Schaie, Labouvie, & Buech, 1973). These data imply that to understand developmental change, one should consider the interactions between individual and sociocultural-historical processes. A similar conclusion can be drawn from another example of sequential research.

A Sequential Study of Intelligence in Adulthood

The most extensive application of sequential methodology is Schaie's study of intelligence in adulthood. This study spans a period of twenty-one years and has resulted in many analyses. However, for purposes of illustration, we will focus on the initial seven-year segment and on a particular analysis of these data (Nesselroade, Schaie, & Baltes, 1972). This analysis is based on seven-year longitudinal observations (obtained at the two points in time 1956 and 1963) of eight cohorts (birth dates averaging from 1886 to 1932) ranging in age from 21 to 77 years. The basic design is shown in Exhibit 1.9.

A total of 304 individuals were tested at both times of measurement on thirteen variables. These thirteen variables were reduced by statistical analysis to four composite variables. Exhibit 1.10 shows the outcome for visualization—the ability to manipulate the images of spatial patterns into other arrangements (e.g., mentally rotate, trim, fold, or invert images of objects or parts of objects according to explicit directions). This exhibit contains the results of two "cross sections" based on longi-

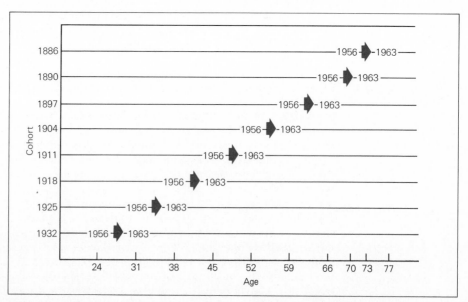

EXHIBIT 1.9   Initial seven-year sequential design used by Schaie (arrows indicate longitudinal sequences).

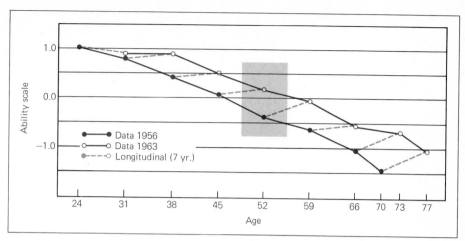

EXHIBIT 1.10   Longitudinal and cross-sectional gradients for visualization. (Source: Nesselroade, Schaie, and Baltes, 1972.)

tudinal sequences. Note that there is little longitudinal change over the seven years
for most age groups (dashed lines). The impact of cohort differences may be estimated by comparing same-aged persons measured in 1956 and 1963. For example,
the boxed area of Exhibit 1.10 shows the performance of 52-year-olds measured in
1956 and 1963 (cohorts 1904 versus 1911). Note the relatively large differences compared to the absence of seven-year age changes.

    In sum, depending on the issues which one is interested in, different types of
developmental research may be used. However, we have seen that if one is interested in intraindividual change—as are most developmentalists—then some sort
of repeated measurement (i.e., historical designs, such as longitudinal or longitudinal sequential design) is necessary. Sequential methodology represents a most useful instance of such repeated measurement designs. We will see additional applications of these designs in future chapters.

ETHICAL ISSUES IN RESEARCH

The previous section of this chapter has focused on the methods of research—techniques involved in the observation and manipulation of variables and the validity of
these techniques. However, research also involves an interaction between people—
the investigator and the participants.

    In recent years, researchers, government officials, and the public have become
increasingly concerned with the ethics of research. In particular, concern has been
directed toward protecting the rights of the individuals who serve as the "subjects"
of research. This concern has been formalized in a variety of ways. For example, the
U.S. Department of Health and Human Services, a major source of funds for research, has developed a set of regulations which all recipients of federal research
grants and contracts must follow. Similarly, many scientific societies, such as the
American Psychological Association, have developed codes of ethics for the conduct
of research. Much of this concern evolved out of past abuses. Medical research has
been particularly problematic; however, abuses occur in behavioral research as
well. Below is a list of ethically questionable practices.

1 Involving individuals in research without their knowledge or consent
2 Failing to inform participants about the true nature of the research
3 Misinforming participants about the true nature of the research
4 Coercing individuals to participate in research
5 Failing to honor promises or commitments to participants
6 Exposing participants to undue physical or mental stress
7 Causing physical or psychological harm to participants
8 Invading the participants' privacy
9 Failing to maintain confidentiality of information received
10 Withholding benefits from participants in control groups

These problems may arise not because the investigator is evil or uncaring. Rather, the nature of the problems, variables, and people themselves may cause problems. For example, it may be very difficult to study a significant event, such as the death of a spouse, without exposing the individual to mental stress. Yet knowledge of such an event is important both for its own value and for attempts to help individuals cope with such events. A critical issue, then, is the cost of research versus its benefits. All research has a cost, in other than monetary terms, to the individuals involved—time, stress, and so forth. Research may also produce benefits—knowledge. Obviously, benefits should exceed the cost, but ethical practice requires more than this. At some point, the cost may be too great no matter what the benefit. In other words, the end does not justify the means. Allowing the end to justify the means was the argument which permitted the Nazis to immerse concentration camp victims in frigid water until they died in order to determine how to better protect flyers who had to parachute into the North Sea.

Such extreme examples are relatively easy to judge as unethical. However, for most research on behavior there are no absolute answers. One justifiable approach appears to take the form of seeking the advice of others concerning the ethical acceptability of research. This may take place at two levels—internal consultation with colleagues and representatives from potential participant groups (e.g., students) and formal consultation with institutional committees set up for this purpose. The latter is required for research studies conducted on campuses where there is any research funded by the Department of Health and Human Services. However, none of these practices changes the fact that the ultimate responsibility for the ethical nature of the research rests with the investigator. One set of principles for coping with this important and difficult task is shown in Box 1.1.

BOX 1.1

ETHICAL PRINCIPLES FOR THE CONDUCT OF RESEARCH

The following principles for ethical research are taken from the *Manual of Ethical Principles* published by the American Psychological Association (1981).

Principle 1. In planning a study the investigator has the personal responsibility to make a careful evaluation of its ethical acceptability, taking into account these Principles for research with human beings. To the extent that this appraisal, weighing scientific and humane values, suggests a deviation from any Principle, the investigator incurs an increasingly serious obligation to seek ethical advice and to observe more stringent safeguards to protect the rights of the human research participants. (p. 21)

Principle 2. Responsibility for the establishment

and maintenance of acceptable ethical practice in research always remains with the individual investigator. The investigator is also responsible for the ethical treatment of research participants by collaborators, assistants, students, and employees, all of whom, however, incur parallel obligations. (p. 22)

Principle 3. Ethical practice requires the investigator to inform the participant of all features of the research that reasonably might be expected to influence willingness to participate, and to explain all other aspects of the research about which the participant inquires. Failure to make full disclosure gives added emphasis to the investigator's responsibility to protect the welfare and dignity of the research participant. (p. 29)

Principle 4. Openness and honesty are essential characteristics of the relationship between investigator and research participant. When the methodological requirements of a study necessitate concealment or deception, the investigator is required to ensure the participant's understanding of the reasons for this action and to restore the quality of the relationship with the investigator. (p. 29)

Principle 5. Ethical research practice requires the investigator to respect the individual's freedom to decline to participate in research or to discontinue participation at any time. The obligation to protect this freedom requires special vigilance when the investigator is in a position of power over the participant. The decision to limit this freedom increases the investigator's responsibility to protect the participant's dignity and welfare. (p. 42)

Principle 6. Ethically acceptable research begins with the establishment of a clear and fair agreement between the investigator and the research participant that clarifies the responsibilities of each. The investigator has the obligation to honor all promises and commitments included in that agreement. (p. 54)

Principle 7. The ethical investigator protects participants from physical and mental discomfort, harm and danger. If the risk of such consequences exists, the investigator is required to inform the participant of that fact, secure consent before proceeding, and take all possible measures to minimize distress. A research procedure may not be used if it is likely to cause serious and lasting harm to participants. (p. 61)

Principle 8. After the data are collected, ethical practice requires the investigator to provide the participant with a full clarification of the nature of the study and to remove any misconceptions that may have arisen. Where scientific or humane values justify delaying or withholding information, the investigator acquires a special responsibility to assure that there are no damaging consequences for the participant. (p. 77)

Principle 9. Where research procedures may result in undesirable consequences for the participant, the investigator has the responsibility to detect and remove or correct these consequences including, where relevant, long-term after effects. (p. 83)

Principle 10. Information obtained about the research participants during the course of an investigation is confidential. When the possibility exists that others may obtain access to such information, ethical research practice requires that this possibility, together with the plans for protecting confidentiality, be explained to the participants as a part of the procedure for obtaining informed consent. (p. 89)

THE PLAN OF THIS BOOK

In this chapter we have seen some of the major attributes of and definitions involved in our life-span view of human development. In addition we have seen some of the dimensions and pitfalls of developmental research, and how those scientists studying human development from a life-span, multidisciplinary view think about and design developmental research. In the next chapter we will review some of the philosophical and theoretical bases of a life-span view of human development. This next chapter is designed to indicate why life-span developmentalists must attend to and appreciate an array of perspectives about human development. Such appreciation and use of an array of perspectives is sometimes called *pluralism*.

We will learn that one reason life-span developmentalists are pluralistic is because there are numerous processes contributing to an individual's development. These processes involve the person's biology, his or her individual psychological functioning, and the familial and broader sociocultural contexts of life as well. These processes combine in different ways at different portions of the life span to make specific periods of life special and unique.

Accordingly, to get a full understanding of the life span, one must know something of basic processes of the individual, how these processes interrelate to promote particular periods and transitions in life, and how all this is moderated by the broader contexts (the family or the culture, for instance) of human development. As such, as we discuss each of the major portions of the life span, we will always consider major processes and products of individual development and, second, the character and influence of the contexts of human development. After we discuss the biological processes involved in human development in the next chapter, we will, in succeeding chapters, discuss each successive portion of life—from conception to death—by focusing on these two themes: *processes* and *contexts* of human development across life.

CHAPTER SUMMARY

Developmental changes are systematic, or organized, changes. They involve changes within the person (intraindividual changes). The life-span view of human development attempts to describe, explain, and optimize intraindividual change in behavior and interindividual differences in such changes across the life span, that is, from conception to death. These emphases are different from those found in traditional psychological and developmental views.

Life-span developmentalists emphasize that change may occur across life and may be multidirectional in form. Such change may be linked to several variables. In addition to chronological age, variables linked to cohort (particularly birth cohort) and life transitions are focused on by life-span developmentalists.

These emphases underscore the life-span developmentalist's concern with the changing individual in a changing world. Both people and their contexts change in interrelated ways. Normative, age-graded influences; normative, history-graded influences; and nonnormative, life-event influences may contribute to these relations. Given that variables from all these sources may influence development, life-span developmentalists stress that human life is necessarily multidimensional and must be studied in a multivariate framework.

Interest in multiple levels of influence allows one to plan strategies for optimizing human life which include preventive and enhancement, as well as remediational, interventions. Moreover, recognition of multiple influences on human life leads life-span developmentalists to suggest a multidisciplinary orientation to knowledge. Knowledge of biology, sociology, medicine, anthropology, economics, and history should be combined with psychological knowledge to understand fully all the complex and interacting sources of human development.

Scientific understanding of human development requires the use of the scientific method, an approach to knowledge based on empiricism. The research generated by the application of the methods of science has as its purpose the valid understanding of relationships among variables. Research must have both internal and external validity for this understanding to be adequate. There are several types of research methods available to human developmentalists to reach such understanding. Developmental research methods may be described as falling along four dimensions: normative-explanatory; naturalistic-manipulative; atheoretical-theoretical; and ahistorical-historical.

Developmental research should be historical in character in order to best appraise change. There are three conventional types of research designs one may use to study development. A longitudinal design repeatedly studies one cohort, but confounds age and time effects. A cross-sectional design studies differently aged people at one point in time, but confounds age and cohort. A time-lag design studies one age level at different times in history, but confounds time and cohort effects. Alternative developmental research designs, as found in the sequential methodologies developed by Schaie, Baltes, and Nesselroade, avoid the confounds in other, traditional developmental designs. In such sequential designs successive cross-sectional cohorts are repeatedly studied.

Sequential designs allow for the description of multiple influences on development, from levels of influence which include the developing person's changing sociocultural context. They allow one to describe changes in both the processes and the contexts of development. These two concerns—processes and contexts—will be our major interest as, in the rest of the book, we explore the characteristics of development across the life span.

chapter 2

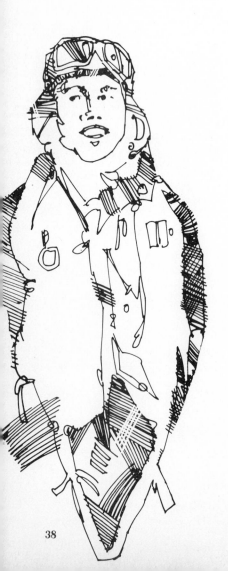

Models and Theories of Human Development

CHAPTER OVERVIEW

In this chapter we discuss some of the philosophical, historical, and theoretical bases of human development. Scientists make philosophical assumptions about the nature of the world. These models of human development influence the types of theories scientists devise. Different philosophical models and theories have been prominent over the course of the history of human development. We review this history in regard to a core issue involved in it: the nature-nurture controversy. Finally, we consider different current theories of human development.

CHAPTER OUTLINE

SCIENCE AND KNOWLEDGE

MODELS OF HUMAN DEVELOPMENT

The Mechanistic World View and the Reactive Organism
The Organismic World View and the Active Organism
The Contextual World View and the Changing Organism in a Changing World

IMPLICATIONS OF DEVELOPMENTAL MODELS FOR SCIENTIFIC ACTIVITY

HISTORICAL ROOTS OF CONTEMPORARY VIEWS OF HUMAN DEVELOPMENT

The Historical Role of the Nature-Nurture Issue
Charles Darwin (1809–1882)
G. Stanley Hall (1844–1924)
The Contributions of Terman and Gesell
Behaviorism and Learning Theory
World War II
The 1950s and 1960s
The 1970s through Today

THEORIES OF HUMAN DEVELOPMENT

NATURE THEORIES OF HUMAN DEVELOPMENT

Hall's Theory Revisited
The Instinctual Theory of Konrad Lorenz

NURTURE THEORIES OF HUMAN DEVELOPMENT

Some Dimensions of Diversity in Nurture Theories
Social Learning Theories of Human Development
The Theory of McCandless

INTERACTION THEORIES

Sigmund Freud and Psychoanalysis
A Critique of Freud's Ideas
Structures of the Personality
Erik Erikson and Psychosocial Development
Jean Piaget and Cognitive Developmental (Structural) Theory
Lawrence Kohlberg and Cognitive Developmental Moral Reasoning Theory
Klaus Riegel and Dialectical Theory

CHAPTER SUMMARY

ISSUES TO CONSIDER

Why is it important to know the philosophical assumptions involved in science?
What are the major characteristics of the mechanistic, the organismic, and the contextual world views?
What is the nature-nurture controversy? What roles has it played in the history of the study of human development?
What are the major changes in the role of theories in the study of human development in this century?
Why are the theories of Hall and of Lorenz characterized as nature theories?
What are the major features of nurture theories of development?
Why do we distinguish among weak, moderate, and strong interaction theories?
What are major examples of weak, moderate, and strong interaction theories?

*A*lthough we have indicated that human developmentalists have many research methods to use, we have not indicated *how* and *when* research methods are in fact used. Human developmentalists use research methods to answer scientific questions. Ideally, the questions that are asked are relevant to important theoretical issues. But what may be an important theoretical issue to one scientist may be irrelevant to another scientist. Differences among scientists in their theoretical preferences relate, ultimately, to philosophical issues.

SCIENCE AND KNOWLEDGE

Scientists use a specific set of rules to determine reality. Through the observations of phenomena, the fundamental task of scientists, behaviors are measured objectively under particular conditions. Then scientists attempt to identify regularities among observations. Such regular, predictable relationships among variables are called *laws*. Finally, a *theory*—a set of propositions consisting of defined and interrelated constructs integrating these laws—is developed. Besides integrating knowledge, a theory serves the function of guiding further research.

Even though we can determine reality by relying upon an empirical approach to knowledge and can delineate what steps—observation, laws, and theory—are involved, a glance at the scientific literature shows that scientists do not agree about their observations, laws, and theories. This is the case primarily because they make different philosophical assumptions about the nature of the world. Thus science is relative rather than absolute. Facts are not viewed as naturally occurring events awaiting discovery. According to Kuhn (1962), science

Throughout history there have been views of the world. Medieval philosophers had a unique view of the cosmos. (The Bettman Archive, Inc.)

> *. . . seems an attempt to force nature into [a] preformed and relatively inflexi-*
> *ble box. . . . No part of the aim of normal science is to call forth new sorts of*
> *. . . phenomena; indeed those that will not fit the box are often not seen at all.*
> *Nor do scientists normally aim to invent new theories, and they are often intol-*
> *erant of those invented by others. (p. 24)*

Life-span developmentalists contend that an adequate understanding of human development cannot be obtained from any one theory or methodology, nor can it be obtained from a cataloging of empirical "facts." To us, then, theory and research only have meaning as they are developed and interpreted within the context of a given perspective. Thus we need to describe the different philosophical assumptions on which the study of development can be based. We need to examine the models, or world views, that are used today in the study of human development.

MODELS OF HUMAN DEVELOPMENT

We can classify three different models of reality (Pepper, 1942). Each is associated with particular sets or families of theories, and each contributes to our understanding of how people develop.

The Mechanistic World View and the Reactive Organism

Some scientists adhere to the world view that humans are not qualitatively different from other phenomena that exist in the natural world. They believe that humans are controlled by the same forces that control all natural phenomena. All such phenomena are, at their most basic level, composed of the same empirical units—for example, atoms and molecules. Thus the laws which govern these basic units—the laws of chemistry and physics—should provide information about all phenomena in the natural world, including humans. Humans can be reduced to atoms and molecules, and all that is seen as necessary for the attainment of knowledge is to learn the mechanical workings of chemistry and physics. In sum, scientists who hold this view of the nature of humans—of the place of human existence in the world—are said to follow a *mechanistic* world view (Bertalanffy, 1962; Harris, 1957; Lerner, 1976; Overton & Reese, 1973; Reese and Overton, 1970). Thus the basic metaphor of the mechanistic model is a machine. In this model, reality is represented as being composed of discrete parts located in time and space. Within this world view, complex phenomena are ultimately reducible to these elementary parts and their relationships. In other words, the whole is equal to the sum of the parts. For the mechanist, movement of the parts depends on the application of forces outside the person which results in chainlike sequences of events. Forces are conceptualized not only as always being external to individuals but also as preceding an event.

Applied to the study of human development, the mechanistic world view yields a *reactive model* of development. Simply, humans are seen as essentially passive organisms who only react to events. From this perspective, the individual is seen as inherently at rest. Activity is the result of external forces, and therefore, change is explained by such forces. Further, change is quantitative rather than qualitative. Complex activities such as emotions and problem solving are ultimately reducible to simple elements (e.g., stimulus-response connections).

The Organismic World View and the Active Organism

Other world views exist. For example, there is an *organismic* world view. This view proposes that knowledge of humans cannot be gained by reducing events to the atoms and molecules studied by physicists and chemists. Such reduction is inappropriate because, although atoms and molecules certainly exist within humans, when they combine to form the whole organism, something special emerges. When the constituent parts combine, they produce *in their combination* characteristics that did not exist in the parts in isolation. Hence because of the belief that something new emerges through combination and interaction, reduction to atoms and molecules is seen as inappropriate. Knowledge of the human would be lost if the parts were studied in isolation. Thus people who follow an organismic world view see humans as appropriate to study as a whole. This is the case because of the unique characteristics that define the human level, the "whole" (Bertalanffy, 1933; Reese & Overton, 1970). For organismic researchers, individuals are considered as being more like biological organisms than machines. Reality is depicted as complex, interrelated parts located in time and space in an organized system. The whole is always equal to more than the sum of the parts because complex phenomena are never reducible to elementary parts and their relationships. The parts of the organism only have meaning when they are considered in the context of the system of which they are a part. Individuals are considered as dynamic—in a constant state of activity. Thus the movement of the parts depends on the organism's activity and its interaction with the environment.

The organismic world view applied to the study of human development, therefore, produces an active model of development. Change is explained by the organism's action on the environment. Moreover, change is considered as qualitative rather than quantitative.

The Contextual World View and the Changing Organism in a Changing World

Additionally, although several other world views exist, a third philosophical model, termed the *contextual* world view, has gained currency in social science (Pepper, 1942). This view considers more than the interaction of the constituent parts comprising the human organism. Human development is seen as arising from the continuous interaction between all the different *levels of organization* existing in the world. That is, from this view one should not only look at inner-physical or inner-biological phenomena (e.g., atoms and molecules, or tissues and organs), or individual phenomena characterizing the whole person (e.g., his or her psychological functioning). One also should consider outer-physical phenomena (weather, environmental pollution) and cultural and historical events (wars, political movements) in order to understand human development fully (Pepper, 1942; Riegel, 1976a).

Reduction of one of these dimensions to a molecular level is not appropriate because each level is seen as different. Alternatively, the type of interaction involved in organismic conceptions is not seen as appropriate. The interaction of constituent parts to produce the whole organism, emphasized in organismic conceptions, is quite different from the type of interaction depicted in a contextual perspective (Lerner & Spanier, 1978b).

First, in the contextual view, all levels of analysis are constantly changing. This is not necessarily the case in the organismic conception, where once the parts combine to produce the whole, the parts are not necessarily involved in further changes. Second, in the contextual view, all levels from the atom to the organism are related reciprocally. In other words, each level of analysis is actually providing a basis for and is actually part of the other levels with which it interacts.

For instance, inner-biological phenomena (e.g., the liver) could not exist without the total functioning of the higher level (the individual). Similarly, the individual could not function without a liver. Neither is ever independent of the other. Each is a part of the other, and changes in one will always produce changes in the other. A buildup of bile in the liver will affect the person, and drinking too much alcohol will damage the liver.

Of course, similar examples could be drawn between the individual and the physical environment, the individual and history, the physical environment and inner biology, or any combination of the four levels noted above. The key point is that the constituent levels are never independent, and the parts interact reciprocally. Each provides a basis of its "opposite," of the very thing with which it constantly and continually interacts. Thus the world is seen as a place of constant change, and humans are viewed as products of the reciprocal interactions. Such continuous and reciprocal interactions might appropriately be described as *dynamic interactions*.

Having just seen that there are at least three different philosophical views of human functioning and development, one may ask how they relate to theories and research about human development.

IMPLICATIONS OF DEVELOPMENTAL MODELS FOR SCIENTIFIC ACTIVITY

Although world views are not capable of being evaluated in terms of whether or not they are correct, they shape the theories that scientists use to interpret the facts they derive from their studies of the "real" world. Moreover, world views, in shaping theories, shape the very questions scientists ask in their study of the real world. The questions which follow from different theories are likely to be quite different, and in turn, the data generated to answer these contrasting questions are not likely to provide comparable answers.

For instance, a mechanistic theorist who derives a theory of human development may try to reduce behavior to learning principles, common to people of all ages. Thus he or she might seek to discover those environment-behavior relations which remain identical from infancy through adolescence and adulthood. Alternatively, an organismically oriented theorist would attempt to find those phenomena which are unique to and representative of particular age periods. In turn, a contextualist would look at the relation of an event to others at earlier times in the life cycle, as well as to current cultural, environmental, and long-term historical influences. Moreover, the reciprocal nature of these interactions would be considered.

Our point is that scientific activity which is derived from alternative world views asks different questions about development. Consequently, scientists committed to alternative world views may collect data on different topics. One scientist is not necessarily functioning correctly, and another, incorrectly. The issue is *not* one of deciding which theory is best, or which leads to truth and which does not. Theories from different world views ask different questions because the very nature of reality is conceived of differently. Thus what is a true depiction of reality for one world view may be irrelevant for another.

One major implication of the nature of the philosophy-science relation discussed above is that a criterion other than truth must be used to evaluate interpretations of development. Moreover, since one theory is not seen to be intrinsically better than another, all theories may be said to have an equal opportunity to advance our understanding of human development. Accordingly, it may be suggested that theories should be evaluated on the basis of criteria of usefulness for description, explanation, and optimization.

To address issues such as these we need to consider the history of a scientific view of development. This will allow us to see its basis and rationale. Such historical analysis involves consideration of initial philosophical statements pertinent to development, statements more than 2000 years old.

HISTORICAL ROOTS OF CONTEMPORARY VIEWS OF HUMAN DEVELOPMENT

Have people always believed that humans develop? Have people always said that infants are different from children and that both are different from adolescents? Indeed, have "special" portions of the life span, perhaps like adolescence or the aged years, always been held to exist? In fact, have people always believed that there is such a phenomenon as human development, and if not, when and why did such a belief arise? Why was a life-span, multidisciplinary perspective about development promoted as useful?

Many of the central questions and controversies about human development are quite old, with roots in ancient Greece and traditions of western philosophy. Indeed, in both the 2000 years of philosophy and about 100 years of pertinent science, ideas advanced to explain development revolve around the same few issues. These issues represent the core concepts in any discussion of development, and differences among philosophers and scientists can be understood by seeing the stances they take in regard to such basic conceptual issues. These issues pertain most directly to one issue: the *nature-nurture controversy*. Although definable in several ways, this controversy relates to the relative roles of inborn characteristics versus experiential influences in human development. Thus in order to organize and understand the evolution involved in the history of ideas of development, it is necessary first to introduce briefly some definition of the nature-nurture issue.

The Historical Role of the Nature-Nurture Issue

The very first idea ever elaborated about human development involved what is still the most basic issue in development today: the *nature-nurture issue*. Basically, this issue pertains to the source of human behavior and development. Simply, a question is raised about where behavior and development come from. As soon as the very first idea was formulated about what human behavior was, and where it came from, a stance was taken in regard to this issue.

Philosophical Roots of the Nature-Nurture Issue. More than 2000 years ago, Plato (427?–347 B.C.) philosophized that there exists a "realm of ideas," a spiritual place where souls reside. Upon human birth, however, the body "traps" a particular soul. The soul remains in the body for the life of the person and returns to the realm of ideas when the person dies. However, it is important to see that since the soul resides perpetually in the realm of ideas, it enters the body with these ideas at birth. That is, the person is born with *innate ideas*, with preexisting, preformed knowledge.

Similarly, as exemplified by John Calvin (1509–1564) and the American Puritans (e.g., the Pilgrims of the ship *Mayflower*), the medieval Christians had a religious philosophy which stressed the innate characteristics of humans. Based on portions of the Book of Genesis, this philosophy stressed the idea of original sin. Humans were said to be born with sin in them. They are born basically evil. A second belief was that humans are basically depraved. The sin in humans will be, they believed, compounded by the inborn tendency to continue to commit sinful acts. In short, the medieval Christian view of human development was a nature one. It stressed that a human is basically innately bad and will continue to commit sinful acts.

Although these ideas suffice to indicate that this view is appropriately classified as a nature one, the complete *nativistic* (instinctual or inborn) orientation of this position is best illustrated by the reason given for the presence of innate sin and innate tendencies toward continued badness. The medieval Christian believed in the *homunculus* idea of creation. The reason for innate sin was that there was believed to be a homunculus—a full-grown but miniature adult—present from birth in the newborn's head. Instantly created with the child, this homunculus contains the sin and the basic depravity.

Moreover, in a manner similiar to Plato, René Descartes (1596–1650) said that the soul, existing independent of the body, imbues it with knowledge upon interacting at the pineal gland. Thus, like Plato, Descartes believed in innate ideas. As such, although considered a modern philosopher, he returned to the nativistic conception of human functioning first put forth by Plato.

In turn, there were several British philosophers who held generally similar views regarding the need to stress ideas based on the observation of the world, i.e., *empirical ideas* (e.g., Thomas Hobbes, James Mill, John Stuart Mill, David Hume, David Hartley, Alexander Bain, and John Locke). One may focus on the ideas of John Locke (1632–1704) as an example of the British school's position and, because of the prominence of his ideas, as an influence on later scientific thinking.

Locke rejected the idea that the mind was composed of innate ideas. Instead, he said that at birth, the mind is like a blank state, or, to use his Latin term, a *tabula rasa*. Any knowledge that the mind attains is derived from experience. And experience makes its impression on the mind—it writes on the blank state—by entering the body through the avenues of the senses. Thus because we experience, or sense, certain observable events—for example, visual, auditory, and tactual stimulation—our minds change from having no ideas to having knowledge.

Accordingly, here we finally have a philosophical statement which stresses a concept of *ontogenetic* development, i.e., development across an individual's life span. Moreover, it does so through emphasizing, for the first time, nurture. Experiences from the environment provide the basis of development. The newborn is different from the adult because the former has no knowledge and the latter does. Thus there is development—change in knowledge in this case—and the basis of development is seen to reside in nurture.

In so stressing the role of nurture variables such as sensory stimulation in shaping behavior (or knowledge), Locke is providing a philosophical view quite consistent with a major theory in psychology: the behavioristic, learning approach to development. People like Skinner (1938) and Bijou and Baer (1961) stress that behavior changes can be understood in terms of environmentally based stimulus-response relations. Thus modern learning theorists are in this regard quite like Locke. In fact, if Locke were alive today, he would be likely to have views comparable to those of modern learning theory psychologists. These more modern views will be discussed later in the chapter.

The Nature-Nurture Issue: An Initial Formulation. In its most extreme form the issue pertains to whether behavior and development derive from *nature*, or, in modern terms, *heredity, maturation,* and *genes,* or, at the other extreme, whether behavior and development derive from *nurture,* or, in more modern terms, *environment, experience,* and *learning.* However, whatever terms are applied, the issue raises questions about how inborn, intrinsic, native, or, in short, nature characteristics (e.g., genes) may contribute to development, and/or, in turn, how acquired, socialized, environmental, experienced, or, in short, nurture characteristics (e.g., stimulus-response connections) may play a role in development. Exhibit 2.1 lists some of the terms to be met in this chapter that pertain to nature and to nurture contributions, respectively. This table will be a useful reference point in much of the historical presentation to follow.

Either explicitly or implicitly, any formulation about behavior and development says something about *if* and *how* heredity, genes, or maturation (nature), on the one hand, and experience or environment (nurture), on the other, contribute to development. Accordingly, as our historical review proceeds, it may be seen that scientists advance ideas about development which pertain to nature, to nurture, *or* to some combination of the two. A detailed discussion of the nature-nurture issue will be reserved for Chapter 3. However, as we now proceed to review the history of the study of development, we will see that the nature-nurture issue represents a recurring theme in the ideas of many scientists.

Charles Darwin (1809–1882)

There are several key ideas in Darwin's theory of evolution. The environment in which a type of animal (a species) exists places demands on that animal. If the only food for that animal in that environment exists on the leaves of tall trees, then the animal must be able to reach the leaves in order to survive. The environment "demands" that the animal possess some characteristics that will allow it to reach the high leaves. If the animal has that characteristic, it will fit in with its environment, get food, and survive. If not, it will die.

Imagine, for example, that two species of giraffe existed, one with a long neck (as *is* the case) and the other with a short one. Only that species with the long neck would have a characteristic allowing it to meet the demands of the environment. Because the long-neck giraffe would have a characteristic fitting in with the demands of the particular environment, it would survive. Of course, if the setting changed— if, for example, only food very low on the ground were available and the short-neck giraffe were to have the characteristics best fitting the environment—the outcome

EXHIBIT 2.1   **Terms Associated with Nature or Nuture Conceptions of Development**

Nature Terms	Nurture Terms
Innate	Experience
Preformed	Environment
Nativism	Empiricism
Instinct	Acquired
Inborn	Learning
Genetic	Socialization
Heredity	Education
Maturation	Acculturation

would be revised. The point Darwin stressed is that the characteristics of the natural setting determine what organism characteristics will lead to survival and which ones will not! Thus it is the natural environment which selects organisms for survival. This is termed *natural selection*.

Hence Darwin advances the idea of *survival of the fittest*. Those organisms possessing characteristics fitting the survival requirements for a particular environmental setting will survive. In other words, certain characteristics in certain settings have *fundamental biological significance;* they allow the organism to survive. Characteristics that meet the demands of the environment (and hence allow survival) are adaptive characteristics.

The giraffe example stresses that various physical characteristics of an organism may be *functional.* In an evolutionary sense, something is functional if it is adaptive, if it aids survival. Thus the *structure* of an organism (its physical makeup, its constitution, its morphological or bodily characteristics) may be functional. The function of structure is adaptation to the environment; it is survival. However, while Darwin in 1859 emphasized the function of physical structures of species, he later (Darwin, 1872) pointed out that behavior, too, has survival value. Showing the emotion of fear when a dangerous bear approaches us or being able to learn to avoid certain stimuli (snakes) and to approach others (food) are examples of behaviors which, if shown, would be adaptive; they would aid our survival.

Thus behavior also has a function. The adaptive significance of behavior became the focus of much social scientific concern. This concern was reflected not only in the ideas of those interested in the *phylogeny* (evolution) of behavior. Additionally, the idea that the behavior changes characterizing *ontogeny* (the development within an individual) could be understood on the basis of its adaptive signifi-

Charles Darwin. (The Bettmann Archive, Inc.)

cance was promoted. Thus the functional significance of behavior became a concern providing a basis of *all* of American psychology (White, 1968) and plays a core role in the ideas of theorists as diverse as Hall (1904), Freud (1949), Piaget (1950), Erikson (1959), and Skinner (1938, 1950). However, before the role of ontogenetic progressions in adaptation, and hence survival, can be completely discussed, it is useful to return to Darwin's ideas about survival and see how they reflect a concern not with ontogeny but with phylogeny.

Not all species survive. There are several reasons why this might happen. Due to natural selection, some species will come to have adaptive characteristics and some will not. The natural environment may change, putting different demands on species. Species members having characteristics which meet these demands will therefore pass on those characteristics to their offspring, and the species will continue. Other species, lacking the adaptive characteristics, will not be fit to survive, and they will die out. Another reason one species might survive over another is that due to some change in the genetic material (e.g., through mutation or crossbreeding), new characteristics might arise, and these might favor survival. In either of these illustrations, however, evolution would proceed on the basis of the transmission of adaptive characteristics from parents to offspring. Species would evolve—change with history—as a consequence of the continual interplay between natural selection and survival of the fittest.

The basis of an organism's survival then does not primarily depend on adaptive characteristics acquired over the course of its ontogeny. Rather, its potential for adaptive functioning is transmitted to it by the parents. Accordingly, adaptation is a heredity, or nature, phenomenon. Thus on the basis of evolution—the history of changes in the species, its phylogenetic development—a member of that species either will or will not be born with adaptive characteristics. Darwin thus presents a nature theory of phylogenetic development.

In summary, based on his observations, Darwin presented the first major scientific theory of development. This evolutionary view of species development had profound effects on areas of concern other than science. But it is possible to remain within the scientific realm in order to gauge not just the impact of Darwin's ideas on those concerned with nature, phylogenetic issues, but also with issues pertinent to ontogeny and, finally, to human development as well. Darwin's ideas provided a major influence on the person who both founded the field of developmental psychology and devised the first scientific theory of human development. This man was G. Stanley Hall, and a consideration of his work will bring our discussion—after more than 2000 years—to a scientific concern with human development.

G. Stanley Hall (1844–1924)

The person who organized the American Psychological Association, and became its first president, was the same person who started the first American journal of psychology (aptly titled *The American Journal of Psychology*) and the first scientific journal devoted to human development (first titled *Pedagogical Seminary*, and then given its present name, *The Journal of Genetic Psychology*). This was also the same person who authored the first psychological texts on adolescence (a two-volume work, titled *Adolescence*, 1904) and on old age (*Senescence*, 1922). This man was G. Stanley Hall.

Hall was one of the most influential and prominent psychologists at the turn of this century (Misiak & Sexton, 1966). As such, he did much to shape the nature and direction of the relatively new science of psychology. (The birthday of modern psychology is dated as 1879, with the opening of the first psychological laboratory in Leipzig, Germany, by Wilheim Wundt.)

Hall had his most specific influence in shaping developmental psychology. As implied by the title of one of the journals he founded—*The Journal of Genetic Psychology*—Hall saw development from a nature point of view. Although not many people (including his students, e.g., Gesell and Terman) adopted his specific nature-based theory of development, they did adhere to his general nature, or organismic, orientation. Consequently, Hall's influence was to start scientific concern with human development but to do so from a predominantly nature perspective.

In devising his nature viewpoint, Hall was profoundly influenced by Darwin. In fact, Hall saw himself as the "Darwin of the mind." Hall attempted to translate Darwin's phylogenetic evolutionary principles into conceptions relevant to ontogeny.

Hall's Recapitulationist Theory. Hall believed that the changes characterizing the human life cycle are a repetition of the sequence of changes the species followed during its evolution. However, although arguing that during the years from birth through sexual maturity the person was repeating the history of the species, as had been done prenatally, Hall believed that the postnatal recapitulation was more limited (Gallatin, 1975). In fact, according to Gallatin (1975, pp. 26–27), Hall believed that:

> *Rather than reflecting the entire sweep of evolution, childhood was supposed to proceed in stages, each of which mirrored a primitive stage of the human species. Very early childhood might correspond, Hall speculated, to a monkey-*

G. Stanley Hall. (The Bettmann Archive, Inc.)

like ancestor of the human race that had reached sexual maturity around the age of six. The years between eight and twelve allegedly represented a reenactment of a more advanced, but still prehistoric form of mankind, possibly a species that had managed to survive by hunting and fishing.

Furthermore, Hall believed that adolescence represented a specific period in ontogeny after childhood. As such, not only was Hall the first person, within a scientific theory of development, to conceive of adolescence as a distinct portion of the life span (the term had, however, first appeared in the first half of the fifteenth century; Muuss, 1975a), but he did so in a manner consistent with a life-span view of development. That is, Hall saw the capacities and changes of childhood continuing into adolescence *but* at a more rapid and heightened pace. Additionally, he saw adolescence as a period of transition between childhood and adulthood. That is, the stages of life previous to adolescence stressed the innate characteristics of humans held "in common with animals" (Hall, 1904, I, p. 39). However, the stage of life following adolescence was said to raise a human "above them and make him most distinctively human" (Hall, 1904, I, p. 39). In short, adolescence was a period of transition from being essentially beastlike to being essentially humanlike (i.e., civilized and mature). This ontogenetic transition mirrored the evolutionary change involved when humans moved, Hall thought, from being essentially like the apes to becoming civilized.

Thus Hall saw adolescence as a period wherein the person changed to become civilized and, as such, entered into and contributed to civilization's institutions. Human evolution, Hall believed, moved the person through this ontogenetic period and, thus, put the person in the position of being able to contribute to humans' highest level of evolutionary attainment: civilization. Hence Hall (1904, II, p. 71) said that "early adolescence is thus the infancy of man's higher nature, when he receives from the great all-mother his last capital of energy and evolutionary momentum." However, because of the acceleration and heightening capacities in adolescence, and also because of the difficulty in casting off the characteristics of animallike behavior while attaining the characteristics of civilization, the adolescent period was stressful and difficult. In short, because adolescence was this ontogenetically and evolutionarily crucial "betwixt and between" (Gallatin, 1975) phase of human development, it was necessarily, to Hall, a period of storm and stress.

Criticisms of the Recapitulationist Idea. Hall's theory of ontogenetic development was not generally accepted. The recapitulationist ideas of Hall met criticism that severely diminished the usefulness of the ideas.

First, Hall's recapitulationist application of Darwinian evolutionary ideas to ontogeny was based on a totally incorrect understanding of the meaning of evolution. Darwin's theory of evolution states that humans did not always exist as they presently do. Rather, previous forms of being existed, and through natural selection, some forms were adaptive, at least for a time, and continued to evolve until, eventually, the human species as it is presently known came to exist. However, this evolutionary process occurs for all animals, not just humans. Thus all animals that exist today have an evolutionary history. They are as currently adaptive as is the human, albeit to their own environmental settings (or, in evolutionary terms, to their own *ecological milieus*).

Thus a rat, or a monkey, or an ape, cannot be an ancestor of a human because all are existing today. All are equal in evolutionary status; all are equally as evolved and as adaptive. This is not to say that there are no differences among the species, for there are, of course, such differences (e.g., in level of complexity). However, that is not the point.

The point is that no human had a rat or a monkey as an ancestor; and when one looks at a monkey embryo or at monkey behavior, one is not therefore looking at humans as they existed in a former, lower-evolutionary status (Hodos & Campbell, 1969). Although humans and monkeys may have had—millions of years ago—a common ancestor, that creature is long extinct. It no longer exists because it was not fit to survive. Although other forms evolved from it—and some *could* have led to present-day monkeys, *while others* could have led to present-day humans—whatever exists today is, by definition, fit to survive and thus adaptive at this point in time (Hodos & Campbell, 1969). Thus Hall did not appropriately understand the evolutionary process. Humans were never monkeylike in their embryos or in their behavior, since both species have their own evolutionary history and are currently alive and well. Human ontogeny cannot, therefore, repeat a phase of its evolutionary history if that phase simply never occurred.

Second, even as an analogy, a recapitulationist description of human ontogeny is inappropriate. As initially pointed out by Thorndike (1904), and reemphasized by Gallatin (1975), by 2 or 3 years of age a human child has already exceeded the capacities of monkeys, apes, or prehistoric humanlike organisms (e.g., the Neanderthals). Sensorimotor, verbal, and social behavior, for example, are all more advanced in the 3-year-old than in the "adults" of any of these other organisms. Additionally, there is no evidence whatsoever that the events in adolescence are but a mirror of the history of evolution.

In summary, then, Hall's theory that ontogeny recapitulates phylogeny simply was untenable. Thus although his work is historically quite important in that his is the first scientific theory of human ontogenetic development and of adolescence in particular, it did not, as we have said, lead many to adopt it specifically. However, his ideas were clearly nature ones (human developmental events arise out of the genetic transmission of past evolutionary occurrences). And because of his influence and position in American psychology, his general nature orientation to ontogeny *was* followed, while his specific nature theory was dropped. Thus Hall influenced his students to use a nature orientation in their work in human development. Since two of his students—Lewis Terman and Arnold Gesell—were among the most important developmentalists in the first third of this century, Hall's influence was to start the scientific study of human development on a nature theoretical basis.

The Contributions of Terman and Gesell

Hall's most prominent students were Terman and Gesell. Their contributions illustrate much of the character of the interest in ontogenetic development through the first three decades of this century. Terman was interested in mental measurement. The first intelligence test was constructed by Binet (Binet & Simon, 1905a, 1905b) in Paris. Terman was one of the first people to translate this test into English (H. H. Goddard, in 1910, was the first). Terman, a professor at Stanford University, published the test as the Stanford-Binet (1916) and adopted the intelligence quotient (IQ), suggested by the German psychologist Stern, to express people's performance on the test (IQ equals mental age divided by chronological age, multiplied by 100 to remove the fraction).

Terman's interest in measuring intellectual ability was not based just on a concern with *describing* how people differ (i.e., with individual differences), although, in part, it was based on this. His interest was also a theoretical one. He believed that intelligence had a strong (if not exclusive) nature component to it. Accordingly, not only did he develop an instrument to describe individual differences in intelligence, but he also attempted to devise research indicating the genetic component of this mental ability. One such study was his monumental *Genetic Studies of Genius*, a longitudinal study of intellectually gifted children from 1921 onward (Terman, 1925; Terman & Oden, 1959). A longitudinal study is one in which the same persons are repeatedly measured over time, and Terman's was one of the first such studies begun in this country (Sears, 1975).

Although not proving that intelligence is genetically *determined* (for reasons we shall explore in later chapters), Terman's work, involving nearly fifty years of study and reported in five published volumes over this span (see Terman & Oden, 1959), was quite important for several reasons. First, it was an impetus to several longitudinal studies of human development. These studies provide data relevant to life-span changes and, additionally, invariably include much information relevant to human development within a life-span framework. Data from these longitudinal studies will be cited throughout this book. Second, Terman's findings did much to dispel myths about the psychological and social characteristics of intellectually gifted people. Although such people may be stereotyped as weak, sickly, maladjusted, or socially inept, Terman provided data showing them to be healthy, physically fit and athletic, and personally and socially adjusted.

Third, Terman's work did much to make developmental psychology a descriptive, normative discipline. His work with the IQ test and his descriptions of the developments of the gifted involved his making *normative* statements. A *norm* is an average, a typical, or a modal characteristic for a particular group. If nature is the source of human development and environment plays no primary role, then all one need do is describe the typical developments of people in order to be dealing with information pertinent to the inevitable (because of its biological, nature basis) pattern of ontogeny. Although there are serious problems with this reasoning, Hall's other prominent student, Arnold Gesell, based his work more explicitly on this reasoning than did Terman and, accordingly, did even more to make developmental psychology a normative, descriptive field.

Gesell proposed a theory that can be understood by his term *maturational readiness*. This nature-based theory said that maturational changes are independent of learning (Gesell's conception of nurture). Sensorimotor behavior and even many cognitive abilities (e.g., vocabulary development) were under the *primary* control of maturation. This means that their pattern of development was maturationally determined. Thus an organism would develop when it was maturationally ready to, and attempts to teach the child before this time could not be helpful.

Hence in his writing and research (Gesell, 1929, 1931, 1934, 1946, 1954), Gesell stressed the need for the careful and systematic cataloging of growth norms. His work has provided science with much useful knowledge about the expected sequence for, and times of, emergence of numerous physical and mental developments of children and adolescents of particular demographic backgrounds. These descriptions would allow people to know, he believed, the nature-based sequence and timing of development and, as such, the point at which the person was maturationally ready for learning. While this belief will be evaluated in a succeeding chapter, the present point is that Gesell's theory and research did much to make developmental psychology not only a nature-based discipline but also one whose major, if not exclusive, focus was only descriptive. However, a nurture-based theory of behavior arose to counteract the predominant nature focus.

Behaviorism and Learning Theory

Just as the pendulum swung between nature and nurture in philosophy, one may argue that it moved similarly in science. In the second decade of this century and continuing through the 1950s, American psychology, as well as other areas of social science (e.g., sociology; Homans, 1961) came to be quite influenced by a particular conceptual-theoretical movement: a mechanistic, behavioristic, learning theory view of behavior. Although this view was not developed from a *primary* concern with children or human development, extensive application of this movement to human development was made. In fact, no learning theory has ever been devised on the basis of information derived primarily from children (White, 1970). Philosophically consistent with Locke's empiricist views, this movement stressed that in order for psychology to be an objective science, ideas about behavior had to be derived from empirically verifiable sources.

John B. Watson, stressing this orientation, developed his point of view under the label *behaviorism* (Watson, 1913, 1918). He stressed that stimuli and responses combined under certain lawful, empirical conditions—the laws of *classical and operant* conditioning, types of learning which will be discussed in later chapters. By focusing on how stimuli in the environment gained control over the behavior of organisms, one could know how behavior was acquired and, by implication, developed. That is, development was seen just as the cumulative acquisition of objective and empirical stimulus-response relations, and all one had to understand in order to deal with humans' development was the way behavior was controlled by the mechanistic laws of conditioning. Watson applied these ideas to children, both in his research (Watson & Raynor, 1920) and in his prescriptions for child care (Watson, *Psychological Care of Infant and Child,* 1928).

The nurture view of behaviorism gave psychologists a position which allowed them to be viewed as objective scientists, like their colleagues in the natural sciences (e.g., physics and chemistry). As such, behaviorism and variants and extensions of it (Hull, 1929; Skinner, 1938) became the predominant conceptual focus in American psychology. As was the case with Watson's work, applications of ideas and principles derived not primarily from humans, but from other organisms, usually rats (Beach, 1950; Herrnstein, 1977), were made to humans. As such, ideas pertinent to human development arose. Thus ideas about how humans acquire behavior consistent with the rules of society—that is, how they are *socialized*—were formulated. Such *social learning* theories were not only pertinent to a nurture view of development but also, at times, represented some attempt to reinterpret nature conceptions of development in nurture terms (Dollard, Doob, Miller, Mowrer, & Sears, 1939; Miller & Dollard, 1941).

However, this nurture view of development was, in its major impact, quite distinct from integration with nature concerns with development. In fact, through the early 1940s there was little integration of efforts by nature-oriented and nurture-oriented workers. The learning-oriented workers were doing *manipulative* studies —that is, varying stimuli to ascertain the effect on responses—and their work tended to concentrate on readily observable aspects of behavior development (e.g., aggressive behaviors). This work constituted a compendium, and an elaborate and fairly precise one, of how variations in particular stimulus characteristics were related to variations in responses of, however, only certain groups of children (basically white, middle-class children of highly educated parents).

Thus through the 1940s proponents within nature or nurture camps continued to work, but with little concern for integration with the endeavors of the other. A major historical event served to alter this and to move developmental science from a primarily descriptive to a primarily theoretical, explanatory-oriented field. The event was World War II.

World War II

The events surrounding World War II irrevocably altered the nature of American social science. First, even before the United States entered the war in December 1941, effects of events in Europe were felt. Nazi persecution led many Jewish intellectuals to flee Europe, and many sought refuge and a new start for their careers in the United States. To the credit of American academicians, great pains were taken to find these refugees positions in American universities and associated institutions, despite the fact that these people brought with them ideas counter to those predominating the American academic scene (i.e., behaviorism and learning theory).

For instance, although Freud himself settled in London (had died there in 1939), many psychoanalytically oriented people, some trained by Freud and/or his daughter Anna, did come to this country. Some of these people, for example, Peter Blos and, most notably, Erik Erikson, brought with them psychoanalytic ideas about human development.

In addition, once America entered the war and numerous soldiers needed to be treated for psychological as well as physical trauma, the federal government gave universities large amounts of money to train clinical psychologists. This program opened the door for many professionals with psychoanalytic orientations to become faculty in universities previously dominated by behaviorists (Misiak & Sexton, 1966), because they were the people with backgrounds appropriate for teaching clinical skills to the new and larger groups of future clinicians that were needed.

Thus one impact of World War II was to encourage psychoanalytic thinking in many psychology departments. This orientation represented not only the introduction of nature-based thinking into departments where behaviorists previously resided in total control of the intellectual domain (Gengerelli, 1976). Additionally, it represented just *one* of many different theoretical accounts of human functioning—accounts which stressed either nature or both nature and nurture as sources of behavior and development—that were now making inroads into American thinking.

As such, nativistic ideas about perception and learning, introduced by psychologists who believed in the holistic aspects of behavior, were juxtaposed with learning ideas of the behaviorists. The *Gestalt* (meaning "totality") views represented by these Europeans (people like Wertheimer, Koffka, Kohler, Goldstein, and Lewin) were shown to be pertinent also to areas of concern such as brain functioning, group dynamics, and social problems (Sears, 1975). Ideas explicitly relevant to development were also introduced. For example, Heinz Werner (1948) presented to Americans a view of development involving continual nature-nurture interactions.

The outcome of these changes in the complexion of intellectual ideas about development, fostered by events relating to World War II, was to provide a pluralism of ideas about development. Now there were present numerous interpretations of behavior and development, interpretations that were based on substantially different conceptions of the source of human behavior and development. Any given behavior, then, was seen as interpretable by quite different alternatives, and these different alternatives were advanced by respected advocates often working in the same academic contexts (Gengerelli, 1976). The simultaneous presentation of diverse interpretations promoted a move away from a mere focus on description and toward a primary concern with theoretical interpretations of development. This focus on explanation was heightened in the post-World War II era, in the late 1950s and 1960s.

The 1950s and 1960s

Because of the *pluralism* of perspectives promoted by events surrounding World War II, developmentalists became less concerned with just collecting data descriptive of development. Rather, they were focusing more on the interpretation, the meaning, of development. As such, they primarily became concerned with the comparative use and evaluation of various theories in putting the facts of development together into an understandable whole. One measure of this change of focus is in the rediscovery of the theory of Jean Piaget.

Piaget's organismic theory of the development of cognition was known in America in the 1920s (Piaget, 1923). Yet because of the "clinical" nature of his research methods, his nonstatistical style of data analysis, and the abstract constructs he was concerned with—all of which ran counter to predominant trends in the United States—his theory and research work were not given much attention in this century until the late 1950s. However, due to the European intellectual influences on American thinking occurring from events related to World War II, greater attention was turned by Americans to the intellectual resources present in Europe. Thus the Swiss scientist Piaget was rediscovered, and it can be fairly said that concern with the abstract and conceptual ideas of his theory came to predominate American developmental psychology throughout the 1960s. In fact, his influence continues to this writing as a result both of further substantiation of his ideas and of promoting discussions of alternative theoretical conceptualizations (Brainerd, 1978; Siegel & Brainerd, 1977).

Interest in adult development and aging also began to grow rapidly in the 1960s. As explained by Baltes (1979), this interest provided a recent, major impetus to the current concern with life-span development because studies of adult development and aging moved scientific interest beyond the child and adolescent years. Major research and theoretical contributions to the study of adult development and aging were provided by Bernice Neugarten and Robert Havighurst, at the University of Chicago, in their longitudinal research beginning in the 1950s.

However, as Havighurst (1973) himself pointed out, this work had an intellectual debt to some earlier work done in the 1930s and 1940s. Except for one early work—an article by Sanford (in the *American Journal of Psychology*, 1902) termed "Mental Growth and Decay"—interest in life-span changes and in researching the nature of life-span development did not really exist at all before the 1920s. In fact, except for Hall's (1922) text, *Senescence*, and that by Hollingworth (1927), it was the 1930s that saw the growth in interests related to life-span development. At this time Else Frenkel-Brunswik began a series of studies at the University of California (Berkeley) on the basis of an interest in life-span development, the work of Charlotte Bühler (1933) in Germany was published and began to become well known, and a book by Pressey, Janney, and Kuhlen (1939) was published. However, the scientists involved in these respective endeavors worked largely in isolation from one another, often unaware of or at least not making reference to the contributions of the others (see Baltes, 1979).

Thus it was not until the 1950s, when the work of Neugarten and Havighurst really began, and the climate of the intellectual times in the United States favored conceptual integration and pluralism, that these seeds of life-span interest really took hold; it was the fostering of research and theory in adult development and aging at that time which laid another portion of the foundation for the trends in human development seen in the decades following the 1950s and 1960s. Nevertheless, even before that period there was a long historical tradition behind the perspective that we today label the life-span view of human development. Box 2.1 recounts this history.

Thus, by the 1960s, concern with development involved focus on various theories of development *and* with phenomena of development (e.g., the cognitive—thought—changes studied by Piaget) that were not only overt, behavioral ones. Bronfenbrenner (1963), in a review of the history of developmental science, similarly notes that from the 1930s to the early 1960s, there was a continuing shift from studies involving the mere collection of data toward research concerned with abstract processes and constructs. Accordingly, in depicting the status of the field in 1963, Bronfenbrenner said that "first and foremost, the gathering of data for data's sake seems to have lost favor. The major concern in today's developmental research is clearly with inferred processes and constructs" (p. 257).

Similarly, in a more recent review, Looft (1972) found a continuation of the trends noted by Bronfenbrenner. Looft's review, like Bronfenbrenner's, was based on an analysis of major handbooks of developmental psychology published from the 1930s through the time of the review (1972). Each handbook represented a reflection of the current content, emphasis, and concerns of the field. Looft found that in the first handbook (Murchison, 1931), developmental psychology was largely descriptive. Consistent with our analysis and with Bronfenbrenner's conclusions, Looft saw workers devoting their time essentially to the collection of norms. However, a shift toward more general integrative concerns was seen by 1946 (after World War II), and the trend continued through 1963 (Bronfenbrenner, 1963) to 1972 (Looft, 1972). Indeed, as a case in point we may note that the editor of a 1970 handbook (Mussen, 1970) pointed out that "The major contemporary empirical and theoretical emphases in the field of developmental psychology, however, seem to be on *explanations* of the psychological changes that occur, the mechanisms and processes accounting for growth and development" (Mussen, 1970, p. vii).

In sum, it may be seen that a multiplicity of theories, concern with explanation and with processes of development, came to be predominant foci by the beginning of the 1970s. Such concerns lead to the recognition that there is not just one way (one theory) to follow in attempting to put the facts (the descriptions) of development together. Rather, a pluralistic approach to such integration is needed. When followed, it may indicate that more descriptions may be necessary. Thus although observation (empiricism) is the basic feature of the scientific *method*, theoretical concerns guide descriptive endeavors. One gathers facts because one knows they will have a meaning within a particular theory. Moreover, since such theory-based research may proceed from any theoretical base, the data generated must be evaluated in terms of their use in advancing understanding of change processes. If these conclusions sound like those reached in the presentation of the life-span view of human development in Chapter 1, this is as it should be. The trends emerging and taking hold by the early 1970s combined to promote the life-span view in that decade.

The 1970s through Today

The prominence of theory, the evaluation of theories on the basis of criteria of usefulness, and findings that developmental changes take many different forms at different points in time (and that such changes are necessary to understand from a di-

THE LIFE-SPAN DEVELOPMENTAL APPROACH: OLD WINE IN A NEW BOTTLE

The life-span developmental approach has had a major impact on the field of developmental psychology within the last ten years. However, like other significant ideas, the approach is actually quite old, as illustrated by the following statement:

We mean by "development" all those changes which occur to constitute the life history of the individual. The infant does not merely start out from birth to become a school child or to attain voting age; his aim is far more complex than that. He sets out, from the moment of conception rather than from the moment of birth, and his goal is the production of an old man or woman; indeed, the ultimate goal of development is death.

Thus growth and decay are both developmental phenomena. The prenatal influences which prepare the way for an individual, and the enduring influences which he leaves behind him after his departure from life, should also be included in a study of his development. Perhaps the most important questions in the science of psychology, as well as in our general understanding of human affairs is, "How do people come to be what they are?" This, in the main, is the problem of developmental psychology. (Hollingworth, 1927, p. 2)

As you can see, this quotation is over fifty years old and is found in a text entitled *Mental Growth and Decline: A Survey of Developmental Psychology* written by H. L. Hollingworth. The life-span perspective, then, is not new.

Baltes (1979) recently has reviewed the history of the life-span perspective. His analysis suggests that it is grounded in three major works published in the eighteenth and nineteenth centuries—by Tetens in 1777, Carus in 1808, and

Quetelet in 1835. In particular, Quetelet's volume entitled *A Treatise on Man and the Development of His Faculties* anticipates many current theoretical and methodological issues. With one exception (Sanford, 1902), Baltes (1979) notes that the life-span perspective lay dormant during the late nineteenth and early twentieth centuries, but it reappeared again in the 1920s and 1930s with the publication of three books, one by Hollingworth in 1927, one by Bühler in 1933, and one by Pressey, Janney, and Kuhlen in 1939. In addition to emphasizing the basic concept of change over the entire life span, these texts are process-oriented and focus on many of the issues discussed in this text, such as multidirectional development, contextual influences, and social-evolutionary change.

What is the significance of the roots of the life-span perspective? Baltes (1979) points out that the significance goes beyond historical curiosity. "It indicates that the involvement of current life-span researchers in such themes as cohort effects, social change, and other macro-level features might be intrinsic to a life-span orientation rather than a reflection of the personal interest of the individual researchers" (p. 261). That is, the historical consistency and longevity suggest that these issues are central and significant ones.

While the writings of early proponents such as Hollingworth, and Pressey, Janney, and Kuhlen, laid the foundation for a life-span perspective, their work received limited recognition. In our view, the principal reason for this lack of interest on the part of other researchers is that the ideas of these life-span proponents were inconsistent with the predominant world views and paradigms of the time. Since the 1940s, however, and particularly during the last ten years, the life-span view has gained increasing recognition.

verse array of explanatory stances) led to the evolution of the life-span view of human development (Baltes & Schaie, 1973; Datan & Ginsberg, 1975; Datan & Reese, 1977; Goulet & Baltes, 1970; Nesselroade & Reese, 1973; Turner & Reese, 1980). In the conferences and books associated with this orientation, the need for a pluralism of theoretical perspectives was bolstered by the emphasis on how world views of development shape lower-order theoretical and empirical endeavors (Overton & Reese, 1973; Reese & Overton, 1970).

Thus the life-span approach has fostered a broadened conceptual and methodological awareness among developmentalists (see McCall, 1977), as well as the view that human change may be understood in the context of the changing world within which it is embedded (Baltes & Schaie, 1976; Riegel, 1975). This view has led to the call for understanding human development in its real-world (McCall, 1977; Riegel, 1976a), or ecological (Bronfenbrenner, 1977), context. This context is seen both to provide a basis of human development and, simultaneously, to be shaped by the changing human (Lerner & Spanier, 1978b; Riegel, 1976a, 1976b). The exchanges between the individual and his or her world include processes relevant to biological, psychological, social, and historical changes. For instance, Elder (1974) has shown that children developing during the years of the Great Depression (1929–1930s) in the United States differed from children developing before or after this historical era.

THEORIES OF HUMAN DEVELOPMENT

Our historical review indicates that in order to study human development most usefully one must understand changes in human functioning that can span the entire life course and that can involve influences from a wide array of sources. Thus many different biological and environmental influences need to be considered (in other words, nature and nurture influences have to be understood), and this consideration has to be capable of engaging phenomena across the entire life span. Our historical review thus led us to argue for the need to appreciate the array of theories forwarded about human development in order to best understand change from a life-span, multidisciplinary perspective. Accordingly, in this section we will briefly discuss some of the general characteristics of several of these theories. Because of the salience of the nature-nurture issue, we will examine some of the significant theories of human development from the perspective of where they stand on the nature-nuture controversy. Of course, remember that this controversy is just one important common theme. Remember, too, that life-span developmentalists prefer to focus on the interaction of nature and nurture but that they consider every theory potentially useful. One of their tasks is to decide when and how a theory is useful to them for explaining, describing, and changing behavior. We will now consider some of the major contributions to the various theories widely used today in the study of human development.

NATURE THEORIES OF HUMAN DEVELOPMENT

Hall's Theory Revisited

In turning first to nature theories of development, it may be noted that although such an approach was historically the first to be put forth (in the recapitulationist ideas of Hall), such an approach does not find wide support today. Hall viewed the human life span as mirroring evolutionary transitions of the human species. The events of ontogeny were determined, he believed, by phylogenetic history. In their nature humans had the basis of their developmental course. Hall's theory, although the exemplar of a nature-based theory, quickly lost favor in social science.

There were two major reasons for discarding this recapitulationist nature theory. First, it was based on an inappropriate understanding of the character of evolution (Hodos & Campbell, 1969); and second, human ontogeny is more complex at early age levels (e.g., 3 years of age) than is even high primate behavior at adult levels (Gallatin, 1975; Thorndike, 1904). Thus because of such major problems, Hall's specific nature theory was not tenable.

The Instinctual Theory of Konrad Lorenz

Because of limitations mentioned, a completely nature-based theory has not become widely popular. However, some nature-based conceptions have been influential. For example, the nature-based theory of Wilson (1975), labeled *sociobiology*, has attracted a lot of attention (e.g., Caplan, 1978), although there has not been, to date, much human developmental research associated with this theory. In turn, however, the position of Konrad Lorenz has influenced people to believe that human behavior, even at the adolescent and adult levels, is governed by *instincts*. Lorenz believes that instincts are preformed, innate potentials for behavior and in no way influenced by environment, experience, learning, or any variable tied to nurture (Lorenz, 1965).

Lorenz is best known today for his study of instincts in *precocial* birds (birds that can walk upon hatching), such as ducks or geese. He coined the term *imprinting* to describe early social and emotional attachments formed by such birds to the first moving object they see (usually, but not necessarily, their mother). He explains such an attachment on the basis of the presence of an instinct for attachment. However, he says that proof for the innate basis of imprinting is that attachments seem to be invariably formed; but he also says that the reason the attachments are formed is because of the presence of an instinct. Why does imprinting exist? He reasons that it exists because there is an instinct for it. How does one know that the instinct is present? Because one sees imprinting.

Thus Lorenz uses tautological, or circular, reasoning. In addition to this logical problem, there exists ample evidence (Moltz & Stettner, 1961; Schneirla & Rosenblatt, 1961, 1963) that imprinting is quite capable of being influenced by experiential variation. Accordingly, one could dismiss Lorenz's ideas of instincts for these reasons, and not bother to discuss them here, if it were not for the fact that Lorenz *applies his ideas to humans.*

Lorenz has argued that aggression is instinctual in humans. In fact, his book, titled *On Aggression* (Lorenz, 1966), not only sees social conflict as an innate component of all humans but makes specific statements about adolescents' instinctual "militant enthusiasm" (Lorenz, 1966). Although in this writing Lorenz suggests that civilization must find methods for ritualizing and containing aggressive instincts (for instance, by having people of different races get acquainted), he basically is suggesting we need to make the best of an inevitably negative situation (Schneirla, 1966). He sees humans, and perhaps particularly youth, as universally and unalterably destructively aggressive.

Although Lorenz calls for social reform, for example, to get youth involved in "genuine causes that are worth serving in the modern world" (Lorenz, 1966), and although we would not dispute such a broad social value, we do contend that his call is based on a totally false premise. As argued in Chapter 3, there is no evidence for solely nature-based critical periods or instincts in nonhuman animals, and there is no evidence for it among humans. Lorenz's speculations about innate aggressive tendencies in adolescents are as faulty logically and empirically as his ideas about

Konrad Lorenz. (United Press International, Inc.)

instinctual imprinting in geese. Rather than scientifically studying *how* nature and nurture interact to provide a basis of behavior development, and attempting to discover which combination of variables may lead to aggression, Lorenz simply asserts that aggression is innately present. In sum, theories based solely on the influence of nature are not especially useful for science. Furthermore, when they take the form that Lorenz's argument took, they may have dire social consequences. We agree with a review of Lorenz's *On Aggression* written by T. C. Schneirla, a person who led the way in showing the shortcomings of positions such as that of Lorenz. Schneirla (1966) wrote:

> *It is as heavy a responsibility to inform man about aggressive tendencies assumed to be present on an inborn basis as it is to inform him about "original sin," which Lorenz admits in effect. A corollary risk is advising societies to base their programs of social training on attempts to inhibit hypothetical innate aggressions, instead of continuing positive measures for constructive behavior.*

NURTURE THEORIES OF HUMAN DEVELOPMENT

Nurture theories of development stress the influence of experiential and environmental variables on behavior change. Most generally, such theories view behavior as a response to stimulation. The linkage between stimulus and response is conceptualized in terms of learning, or conditioning, and thus may be seen as consistent with a mechanistic model of development. However, it is important to understand that such a conception of nurture theories is a broad one. Not all nurture theorists would call themselves learning theorists, nor would they have a commitment to precisely the same view of what learning is. In other words, they might differ about what is involved in the acquisition of behavior as a consequence of experience.

Some Dimensions of Diversity in Nurture Theories

Hence there is great diversity among nurture theorists. Some, for instance, tie their accounts of behavior to the "laws" of classical and operant conditioning, and thus tie their views of behavior change to functional relations between externally manipulable stimuli and objectively verifiable responses (Bijou, 1976; Bijou & Baer, 1961). Such an approach is often called a *functional analysis of behavior* (Bijou, 1976, 1977) because behavior (actually responses, R) is seen to be a function (f) of stimulation (S), or simply $R = f(S)$.

Because of the commitment to completely empirical stimulus-response (S-R) relations, nurture theorists who take a functional analysis approach to behavior do not necessarily consider themselves "learning theorists" (Bijou, 1977). Learning theorists are, instead, those who include in their systems phenomena which are not directly (extrinsically) empirically observable. For instance, such theorists might talk about internal "mental" or "cognitive" responses and stimuli being involved in linking external S-R connections (e.g., Hull, 1952; Kendler & Kendler, 1962). Thus by including phenomena in their theories which, although derived from previous experience, are not present for extrinsic empirical observation, such learning theorists differ from those nurture theorists who take a functional analysis view.

Moreover, further differentiations can be made among those committed to a nurture view. One can differentiate between those theorists who view behavior change as involving the necessary role of *reinforcement* (any stimulus which produces or maintains behavior, or , in other words, makes a behavior more probable; Skinner, 1950) and those who see nurture capable of controlling behavior independent of the presence of any reinforcing stimulus (e.g., Bandura & Walters, 1959, 1963). In other words, one can differentiate on the basis of the role given to reinforcement or to other nurture phenomena—for instance, the experience of observing others—in the control of behavior.

Bandura and Walters (1963), for example, assert that by observing others behave, a person can acquire new responses not previously in the behavior repertoire or come again to show a response that, because of previous experience, had been dropped from the response repertoire. By observing a model behave, they argue, a person can show such changes, although there is no direct reinforcement to the person. Seeing the response consequences to the person who serves as the *model* of behavior (i.e., the person who displays the behavior a subject observes), and the social status of the model, can affect the observer's behavior independently of any direct reinforcement (Bandura & Walters, 1963). Thus free of specific reinforcement for a particular response, nurture theorists of this bent say an observer may imitate a model's behavior if the status of the model is high (Bandura & Huston, 1961) and the response consequences to the model are positive. Such theorists are often termed *observational learning* ones. Alternatively, however, those who consider themselves reinforcement theorists say that an observer being influenced by a model does indeed involve reinforcement of the observer's behavior, albeit of a more general response than that involved in a specific, single act by a model (Gewirtz & Stingle, 1968).

This discussion of the diversity among those who have nurture theories neither raises nor resolves all the many issues dividing such theorists. Rather, it suggests that a range of views exists within the nurture theoretical camp and highlights the difficulty, if not the impossibility, of defining the precise attributes of "the" nurture view. In fact, differences between the functional analysis position and the learning theory position, on the one hand, and between reinforcement theory and observational learning theory, on the other, are just two of the many dimensions of disparity that might be mentioned.

Moreover, whatever theoretical statements are made about the nurture basis of human development are derived from data primarily collected from organisms other than humans. Complicating this situation is the fact that there has never been a nurture theory devised *solely* to address phenomena of either human development, in general, or any portion of the life span, in particular (White, 1970).

Social Learning Theories of Human Development

In the 1940s and 1950s, some theorists (e.g., Dollard & Miller, 1950) used nurture-based learning principles to explain how the social environment comes to control human behavior. Although the principles eventually generated in such *social learning* interpretations of human behavior were, in the main, derived from data collected on laboratory rats, they represent influential views of how nurture is involved in human functioning and do provide us with nurture conceptions of development.

As with the nurture view in general, there is no one social learning theory of human functioning. For instance, one such position is the theory of socialized anxiety (Davis, 1944). Davis (1944) argued that what people learn in their social environment is to anticipate reward for approved behavior and punishment for disapproved behavior. Anticipation of punishment represents an unpleasant feeling state—termed *socialized* (learned) *anxiety*—for the person, and people behave in ways which will diminish or avoid this anxiety. For instance, Davis argues that, in adolescence, the behaviors society will now reward and/or punish are less certain for the person, and thus there is no definite way to decrease socialized anxiety. Accordingly, storm and stress are involved in this period.

The Theory of McCandless

Although Davis's (1944) theory is a nurture one, other social learning theories have different components. McCandless (1970), a person whose ideas we shall meet throughout this book, has proposed a "drive theory" of human behavior. He views a drive as an energizer of behavior and specifies that there exist several drives in the person (e.g., a hunger drive, a drive to avoid pain, and, emerging in adolescence for the first time in life, a sex drive). The nurture component of McCandless's theory involves the direction that behavior takes as a consequence of being energized by a particular drive.

One learns in society, through the principles of classical and operant conditioning, that certain behaviors appropriately will *reduce* drive states. Certain behaviors will reduce hunger, or pain, or sexual arousal. McCandless asserts that those behaviors that reduce drives are most likely to be learned (repeated again in comparable social situations). Drive-reducing behaviors are those that are rewarded (as a consequence of having the drive diminished), while behaviors which do not diminish a drive state are, therefore, punished behaviors. Such behaviors are less likely to be repeated.

As a consequence of repeating behavior, people form habits. Thus when a particular drive state is aroused, it is probable an individual will behave by emitting the habit. Thus, like Davis, McCandless says people learn to show certain behaviors (habits) because they reduce an internal state. However, unlike Davis, these states involve more than just socialized anxiety, which is a learned association to behavior. To McCandless, the states individuals act to reduce are drive states, and these states (e.g., hunger or sex drive) need not be learned. They may be biologically based. Nevertheless, what people do in response to the arousal of *any* drive is a socially learned phenomenon.

Like Davis, McCandless sees a specific relevance of his ideas to adolescent development; again, however, this relevance differs from the one that Davis specifies. For instance, as already implied, adolescence is special for McCandless because of the emergence of a new drive: sex. New social learning, new habits, must be formed to reduce this drive in a way that will be rewarded. However, McCandless specifies that society does not reward males and females for the same drive-reducing habits; consequently, males and females are channeled into developing those habits which are socially prescribed as sex-appropriate. Thus because of the sex drive, adolescence becomes a time of defining oneself in terms of new, sex-specific habits— habits which are socially defined as leading to reward for people of one or the other sex. Until new habits are formed, and thus new self-definitions attained, McCandless (1970) would argue that adolescence could very likely be a period of undiminished drives and hence of stress and emotional turmoil.

Later chapters will present a more detailed discussion of how the social learning theory of McCandless leads to a nurture conception of personality and self-concept development, and of sex differences in these developments.

Boyd R. McCandless.

INTERACTION THEORIES

Not all theories of development take a strong position about nature to the near exclusion of nurture or vice versa. We can classify a theory according to the stress it places on nature and nurture (Lerner & Spanier, 1978b).

A *weak interaction theory* places primary stress on one source (usually nature) as the determiner of the sequence and character of development. In such conceptions, development proceeds in accordance with a "maturational ground plan" (Erikson, 1959) or with some other nature-based construct. Nurture is seen as only a facilitator or an inhibitor of these primarily intrinsic trends (Emmerich, 1968). Nurture may speed up, slow down, or even stop ("fixate") development, but it can in no way alter the ordering or characteristics of change.

A *moderate interaction theory* places equal stress on nature and nurture but sees the two sources as independent of each other. Although they interact to provide a basis for any and all behavior development, one does not change the quality of the other over the course of their interaction (Sameroff, 1975).

Finally, a *strong interaction theory* is one which not only sees both nature and nurture as inextricably involved in all behavior development but sees the two as reciprocally related. Part of that which we call nature is nurture and vice versa (Overton, 1973). This strong view of nature-nurture interaction is consistent with notions of dynamic interactionism.

The Role of Weak Interaction Theories. Perhaps no other type of theory has been more influential in the study of human development than has the weak interaction type. Here the contributions of Sigmund Freud and Erik Erikson stand out. Piaget's influential cognitive development theory also is presented in this chapter, and in greater detail in other chapters, to be the exemplar of a moderate interactional theory. Another example of such a theory we will discuss here, and in other places in this book, is the one proposed by Lawrence Kohlberg. Finally, the ideas of Klaus Riegel will be used to illustrate a strong interaction theory.

As noted, weak interaction theories have in the past been most popular among people who identified themselves as primarily interested in human development. The classic examples of such weak interaction theories are found in the psychoanalytic literature. The theory of psychosexual development of Sigmund Freud is an example of such a theory. Additionally, the theory of psychosocial development of Erik Erikson (1959, 1963, 1964, 1968) is also a prime example of this type of theory. Because Erikson's ideas build on those of Sigmund Freud, let us review first the ideas of Freud.

Sigmund Freud and Psychoanalysis

Sigmund Freud was born in Freiberg, Moravia, in 1856 and died in London, England, in 1939. He lived most of his life, however, in Vienna, where in 1881 he obtained his medical degree.

Freud was well acquainted with many areas of scientific inquiry. In the field of physics, one idea—the law of the conservation of energy—had a profound influence on him. This principle states that physical energy can be neither created nor destroyed, but only transformed. For example, steam energy from beneath the earth's surface can generate mechanical energy in turbines, which, in turn, transform the mechanical energy into electrical energy, which can create heat energy, boiling water in a pot on a stove to create steam energy again.

The Concept of Libido. Freud saw a parallel between the transformation of energy in the physical world and events that occur in mental life. Life, he hypothesized, is governed by a human mental (or psychic) energy, termed *libido.* Libido cannot be created or destroyed. Humans are born with a finite amount of libido, and instead of its being transformed into another type of energy, its area of localization within the body changes over the course of development.

Freud said that there was a universal, biologically necessary sequence of libido progression. Libido is centered in particular body zones for predetermined lengths of time. Where the libido is concentrated determines what stimulation is appropriate, i.e., gratifying, and what sort of stimulation is not appropriate.

Experiences of the person determine if appropriate stimulation is forthcoming; however, such experiences do not alter the order of bodily concentration of libido— or, in Freud's terms, the stage of psychosexual development. Thus experience can facilitate or hinder the development of certain feeling states, but it is nature which determines where and when libido moves. Freud specified that there are five stages of psychosexual development.

The Oral Stage. The first psychosexual stage is the *oral stage.* During the first year of life, gratifying stimulation occurs in the mouth region; pleasure is obtained by sucking and biting. If the baby's attempts at oral gratification are frustrated often enough, some of the libido may become *fixated* (arrested) at this stage of development. That is, not all the libido will be free to progress to the next stage; Freud believed adult emotional problems would result from such fixation and would involve attempts to obtain the missed gratification of the oral stage. As an adult, the individual might, for example, overeat. Fixation is a potential problem at any stage.

The Anal Stage. This stage lasts from about ages 1 to 3 years and occurs when libido centers in the anal area. Children get pleasure through exercising their anal muscles, expelling or holding bowel movements. The anal stage corresponds to the pe-

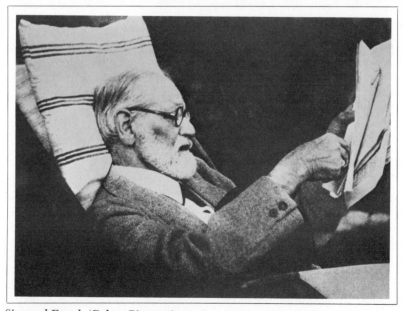

Sigmund Freud. (Culver Pictures.)

riod when children in many western societies are toilet-trained, and frustration resulting from severe toilet training can make people either too "loose"—messy and wasteful—or "uptight"—holding back everything, including their feelings.

The Phallic Stage. For both sexes, the phallic stage, which to Freud spans the period from about the third through the fifth years, involves the libido moving to the genital area. However, it is necessary to discuss the sexes separately because of the structural differences in their genitalia.

The Male Phallic Stage. The libido moves to the boy's genital area, and gratification is obtained through manipulation and stimulation of the genitals. Freud believed that the boy's mother is most likely to provide this stimulation. Because his mother is providing this stimulation, the boy comes to desire his mother sexually (incestuously), but he recognizes that his father stands in the way of his incestuous desires, and this arouses negative feelings for the father.

Freud labels this emotional reaction the *Oedipus complex.* Oedipus was a character in Greek mythology who (unknowingly) killed his father and then married his mother, and Freud saw a parallel between this myth and events in the lives of all humans.

When the boy realizes that his father is his rival, the boy comes to fear that his father will punish him by performing castration. As a result, the boy experiences *castration anxiety.* Because of the power of this anxiety, the boy gives up his desires for his mother; in turn, he identifies with his father.

Identification is a most important development. As a result of identification, the boy comes to model himself after the father. The boy forms a structure of his personality which Freud terms the *superego.* The superego has two components. The first, the *ego-ideal,* is the representation of the perfect, or ideal, man (the "father figure"); in modeling himself after father, the boy thus becomes a "man" in his society. The second part of the superego is the *conscience,* the internalization of society's standards, ethics, and morals. Thus this internalization brings about moral development (Bronfenbrenner, 1960).

If for some reason the boy does not successfully resolve his Oedipal complex, he may not give up his love for his mother and might incorporate part of the mother's superego, possibly in regard to choice of sexual partner in adulthood. Thus one possible outcome of an unresolved Oedipal complex would be male homosexuality.

The Female Phallic Stage. Freud himself was never satisfied with his own formulation of the female phallic stage (Bronfenbrenner, 1960). As the libido moves to the genital area, gratification is obtained through manipulation and stimulation of the genitals. Although presumably it is the mother who provides the major source of this stimulation for the girl, the girl (for reasons not perfectly clear even to Freud himself) falls in love with her father. Then, analogous to what occurs with boys, she desires to possess incestuously her father but realizes that her mother stands in her way. At this point, however, the similarity with male development is markedly altered.

The female is afraid that the mother will punish her for the incestuous desires she maintains toward the father. Although it is possible that the girl first fears that this punishment will take the form of castration, her awareness of her own genital structure causes her to realize that, in a sense, she already has been punished. That is, the girl perceives that she does not have a penis but only has an inferior (to Freud) organ, a clitoris. Hence the girl experiences *penis envy.*

Penis envy impels the girl to resolve the Oedipal conflict. She relinquishes her incestuous love for her father and identifies with mother. She, too, forms a superego —but Freud believed that only castration anxiety could lead to complete superego development, and, thus, because females do not experience castration anxiety, they do not attain full superego development. This lack, Freud believed, takes the form of incomplete conscience development. In short, to Freud (1950), females are never as morally developed as males. We will consider this provocative idea in later chapters. Finally, as with males, difficulties in the female's phallic stage could have effects on adult psychosexual functioning, e.g., female homosexuality could result.

The Latency Stage. After the end of the phallic stage—at about 5 years of age—and until puberty occurs—typically at about 12 years of age—Freud said that the libido is latent. It is not localized in any bodily zone; thus no erogenous zones emerge or exist.

The Genital Stage. At puberty the libido reemerges, again in the genital area, but now it takes a mature, or adult, form. If the person has not been too severely restricted in psychosexual development in the first five years of life, adult sexuality may occur. Sexuality can now be directed to heterosexual union and reproduction.

A Critique of Freud's Ideas

For Freud, the form of adolescent and adult development was determined in early life. To him, the first five years of life, involving the first three psychosexual stages, were critical stages for functioning in later life. Objections to these ideas can be raised along several dimensions. Freud was a critical period theorist, and he saw nature as having a primary role in development independent of the contribution of nurture. Why such a conception in inadequate on both logical and empirical grounds is discussed in Chapter 3. Indeed, one may object to Freud's ideas precisely because of the quality of the sources of information he used to form his ideas.

Freud had a very biased source of "data." He worked in Victorian Europe, a period of history noted for its repressive views about sexuality. As a practicing psychiatrist, he had as his main source of data the memories of *adult* neurotic patients, people who came for treatment of emotional and behavioral problems interfering with their everyday functioning. Freud used his psychoanalytic therapy methods to discover the source of his patients' emotional problems. Through work with such patients, he attempted to construct a theory of *early* development, but these patients were adults from one particular historical period—*not* children. Thus Freud constructed a theory about early development in children without actually observing them.

Freud's adult patients reconstructed their early, long-gone past through *retrospection*. With Freud's help, they tried to remember what happened to them when they were 1, 2, or 3 years of age. This is how Freud obtained the information to build his theory. But it is quite possible that adults may forget, distort, or misremember early memories; therefore, because his data were unchecked for failures of early memories, Freud likely obtained biased information. Furthermore, Freud's patients cannot necessarily be viewed as representative of other, nonneurotic Victorian European adults or, for that matter, of all other humans living during other times in history.

Finally, even if one were to ignore all the above criticisms, one might question whether Freud's ideas represented all the possible developmental phenomena that could occur in each stage of life. Is latency necessarily a period of relative quiescence, a time when few significant events occur for the child? Is adolescence just a time when events in preceding life make themselves evident? Or are there characteristics of adolescence that are special to that period? Interestingly, although accepting most of Freud's ideas as correct, other psychoanalytically oriented thinkers are the ones who led the way in showing that Freud was incomplete in his depiction of developmental phenomena within and across stages. Erikson (1959, 1963), for instance, showed that by attending to the demands placed on the individual by society as the person developed, important phenomena could be identified in latency, adolescence, and across the rest of the life span as well.

Erikson did not so much contradict Freud as he did transcend him. He reached this point not by adding anything new to Freud's basic ideas but by focusing on the implications of one aspect of Freud's theory to which Freud himself did not greatly attend. We now consider Freud's ideas about the structure of personality. This presentation is intended as preparation for the more heuristic ideas provided by Erikson as a consequence of his change in focus.

Structures of the Personality

Freud believed that several different mental structures comprise the human personality. One of these structures, the superego, arises out of the resolution of the Oedipal conflict. Other personality structures exist, however. One of these, termed the *id,* was defined as an innate structure of the personality. The id "contains" all the person's libido. Because the id is the center for the libido and since the libido creates tensions necessitating appropriate stimulation, resulting in pleasure, Freud said that the id functions in accordance with the *pleasure principle.* Freud emphasized the implications of the gratification of libidinal energy and, thus, emphasized implications of the biologically based id on human functioning.

In addition to the superego and the id, Freud specified a third structure of the personality, the *ego.* The function of the id is solely to obtain pleasure. Thus the id impels a person in the oral stage to seek appropriate stimulation—for example, the mother's nipple. But when the nipple is not present, the id has available a particular type of functioning, or process, which Freud termed the *primary process.* When the mother's nipple is not present, the child imagines that it is there. Simply, the primary process is a fantasy, or an imaginary, process; but such fantasies are not sufficient to allow the child to obtain appropriate stimulation. One must interact with reality. Accordingly, another structure of the personality is formed, a structure whose sole function is to adapt to reality, to allow the person actually to obtain needed stimulation and hence survive. This other structure of the personality is the ego.

Because the ego develops only to deal with reality, to allow the person to adjust to the demands of the real world and hence survive, Freud said the ego functions in accordance with the *reality principle.* The ego has processes that enable it to adjust to and deal with reality. This *secondary process* involves such things as cognition and perception. Through the functioning of these processes, the ego is capable of perceiving and knowing the real world and hence adapting to it.

Although Freud spoke about the implications of all three structures of the mind —the id, the ego, and the superego, the sum of which comprises a person's personality structures—he emphasized the implications of the id on human functioning. Hence in describing human psychosexual development, Freud was viewing human beings as essentially biological and psychological in nature; on the other hand, Freud did not spend a good deal of time discussing the implications of the ego. This focus is what Erikson (1963) provides.

Erik Erikson and Psychosocial Development

Erikson also proposes a stage theory of development; however, unlike Freud, Erikson's theory speaks of *psychosocial* development. That is, while Freud focused primarily on the contributions of the id to development, Erikson focuses on the role of the ego. This perspective leads Erikson toward an emphasis on the role of society in determining what the ego must do to fulfill its function of adapting to the demands of reality, a reality which is textured and shaped by one's society.

Society does not expect the same things of all people at all times of their lives. We may expect a 1-year-old to do no more than eat and sleep, but we expect a 21-year-old to prepare for a career, and we expect a 31-year-old to be engaging in his or her career. The ego must develop different capabilities to meet each of these successively different demands. The development of ego capabilities is seen as progressing through eight stages of psychosocial development. Erikson believes that the time one has to spend at each stage is fixed by a maturational "ground plan" for development. Within each stage a particular capability of the ego must be developed if the ego is to meet the adjustment demands placed on it by society. Therefore, each stage is a *critical* one in the life span; since stage development will proceed in spite of whether the capability is developed, if a person does not develop what is supposed to be developed within the time limits imposed by the person's maturational ground plan, then development will proceed on, and that part of the ego (that capability) will not have another chance for adequate development. Thus within each stage of psychosocial development there is a crisis of development, and for the person to feel that he or she is progressing through the stages in an adaptive manner, the person must sense—says Erikson—that he or she has moved closer to developing the needed ego capability than moved away from developing it. In other words, within each stage there is a crisis between developing a feeling of the appropriate capability and developing the feeling that the appropriate capability is lacking.

Within stage 1 of psychosocial development—termed by Erikson the *oral-sensory*—there exists a crisis between developing a sense of basic trust and developing a sense of mistrust toward one's world. If favorable sensory experiences are encountered, the child will develop toward the former feeling state, while if unfavorable experiences occur, a feeling closer to the latter state will develop. Thus in this stage, as well as in the remaining seven others, a person typically develops somewhere in between the two extreme ends of the crisis state. As with this first stage, where the crisis is between trust and mistrust, healthy ego development will proceed if the ratio of the former to the latter is greater than one, while unhealthy ego development will proceed if this ratio is less than one.

Stage 2 is termed *anal-musculature*, and here the crisis is between developing toward a sense of autonomy (e.g., being able to use one's muscles, for instance, to feed oneself, by oneself) and a sense of shame and doubt (and not being able, therefore, to feed oneself, for instance). Stage 3 is termed *genital-locomotor*, and the crisis involved here is between initiative (e.g., being able to walk around your world, to find your way, without having the parents completely directing you) and guilt (and feeling you are inappropriately "tied to the parents' apron strings"). In stage 4, *latency*, the crisis is between developing toward a sense of industry (e.g., learning how to read and write in our society) and a sense of inferiority (e.g., not being able to do what you feel others can do). In stage 5, *puberty and adolescence*, the crisis is between developing a sense of identity (e.g., knowing one's role in society, knowing what one believes in) and role confusion or identity diffusion (e.g., not knowing what one believes in or what one can do with oneself in one's society). Erikson divides adult life into three succeeding stages. In stage 6, *young adulthood*, there is a crisis between developing toward a sense of intimacy (e.g., establishing a strong love relationship) and a sense of isolation (e.g., feeling alone and unable to be close to anyone). In stage 7, *adulthood*, the crisis is between generativity (e.g., being able to produce those goods or services associated with one's role) versus stagnation (or, for example, being unable to continue to produce what is associated with one's role). Stage 8, the final stage of psychosocial development, is *maturity*. Here the crisis is between ego integrity (e.g., feeling that one has led a full, rich life) and despair (e.g., feeling that one's life is ending before one can have a good life).

In sum, Erik Erikson's stage theory of psychosocial development offers descriptions of personality development that are consistent with a life-span perspective. Although Erikson's explanations of psychosocial development are limited on both conceptual and empirical grounds, his descriptions have served as a major basis for research about human personality, and we will focus on his ideas in later chapters.

Jean Piaget and Cognitive Developmental (Structural) Theory

The moderate interaction theory of Swiss psychologist Jean Piaget (1896–1980) considers both heredity and environment to be independently existing sources of cognitive development. Piaget had been writing since the 1920s, but because of his unique methods of observing and describing his theories, he wasn't well regarded in this country until the 1950s. Whereas Erikson's theory covers the entire life span, Piaget's stops at adolescence. Like Freud, Piaget believed that the stages he described are universal and their sequence, unvarying, but unlike Freud, Piaget believed they unfold in response to environmental, as well as to biological, forces. Specifically, cognitive development is said to arise from the action of the organism on the environment *and* from the action of the environment on the organism. In this way, Piaget proposed that cognition develops through four stages: the sensorimotor, preoperational, concrete operational, and formal operational. One's progression through these stages involves the person's attempt to maintain a balance, an equilibration, between *assimilation* (changing external stimulation to fit already existing knowledge) and *accommodation* (changing already existing knowledge to fit external stimulation).

The *sensorimotor stage* involves developing the knowledge that external stimulation (objects) continues to exist even when not sensed by the person. When the person mentally can represent absent objects and can use symbols to represent objects, the major characteristics of the *preoperational stage* are evident. However, although thought about objects in the world is internalized mentally, it is not yet reversible.

Knowledge about the reversibility of actions occurs in the *concrete operational stage*. Operations—internalized actions which are reversible—now exist, but thought is limited in that the person can think only about objects which have a concrete, real existence. The person does not have the ability to deal adequately with counterfactual, hypothetical phenomena and does not recognize the subjective and arbitrary nature of thought.

Such recognition characterizes the *formal operational stage*, the stage most representative of adolescents and adults in modern western society. Here the person can think of all possible combinations of elements of a problem to find a solution—the real and the imaginary. Because the person centers so much on this newly emerged thinking ability, some scientists believe the adolescent is characterized by a particular type of egocentrism (Elkind, 1967). An adolescent may believe others are as preoccupied with the object of his or her own thoughts—himself or herself—as he or she is (this is termed *imaginary audience*). Because of the attention, the adolescent comes to believe he or she is a special, unique person (this is termed *personal fable*).

Thus Piaget's (1950, 1970) theory is one which does seem to accept a fuller role for experience in providing an interactive basis of stage progression and, thus, may be seen as a more moderate stance in regard to interaction. However, under close analysis, there is not as full a consideration of environmental or experiential changes as there is of organism changes.

Although Piaget continually emphasized that developmental structural change arises from an interaction between organism and environment processes, it is only the organism that is seen as going through changes. Although the organism's *conception* of the environment changes, constant flux of the physical and social world was never systematically considered by Piaget. Additionally, the impact of these physical and social environmental changes on the nature of the organism's progression is totally disregarded insofar as the structural changes comprising stage progression are concerned. Environmentally based interindividual differences can only pertain to rate of change and to final stage of development reached, but any changes in the environment are irrelevant to the sequence and characteristics of the stages themselves. Thus, as Riegel (1976b) asserts, for Piaget, the organism changes but the environment does not. A similar problem exists in another cognitive developmental theory, one based very much on Piaget's views.

Lawrence Kohlberg and Cognitive Developmental Moral Reasoning Theory

Lawrence Kohlberg's (1927–   ) theory of the development of moral reasoning is derived from a cognitive developmental viewpoint, as represented in Piaget's theory.

Kohlberg (1963, 1970) has proposed a theory of moral reasoning development that today serves as the impetus for most research and discussion about moral development. Kohlberg presents a moral dilemma—a situation where there is no right or wrong answer—and asks subjects to give their reasons for their particular responses to the situation. He has described one progression in moral reasoning which involves: *preconventional morality*, involving the stage of punishment and obedience reasoning and the stage of naive egoistic reasoning; *conventional morality*, involving the good-person orientation to morality and the stage of social order and institutional maintenance morality; and *postconventional morality*, involving the stage of contractual legalistic moral reasoning and the stage of conscience and principle orientation.

We will have reason throughout this text to return to more detailed discussions of the work of Kohlberg and of Piaget. Here, however, we may note that by minimizing the impact of continual social and physical environmental alterations, Piaget and Kohlberg do not sufficiently attend to what is claimed to be the source of developmental progression: person-environment interactions. If only the organism's changes are considered, and the environment's are not, then no variation in structural change may be attributed to variation in the environment, and nothing can be related to environmental change. The only source of variation can be the organism, and only organism changes can predict organism changes. Accordingly, Piaget's and Kohlberg's supposed interaction is reduced to an account of development which stresses primarily, if not exclusively, organism processes to the omission of environmental ones. Piaget and Kohlberg are left with only being able to account for organism changes on the basis of organism changes. A problem of circular logic is raised because of Piaget's and Kohlberg's failure to consider environmental changes as contributing to development as actively as organism changes.

Klaus Riegel and Dialectical Theory

An initial attempt to formulate a theory of development consistent with a strong interactional position has been made by Klaus Riegel (1975, 1976a, 1976b). Riegel proposes a dialectical theory of development which attempts to understand the changing individual in a changing social world. Such a theory is concerned with short-term situational changes—for instance, in the interaction between mother and child—*and* with long-term developmental changes—for instance, in the career development of husband and wife. Rather than stressing that development moves toward a balance, at which the person is "at rest," Riegel emphasizes that development involves continuous changes brought about by inner and outer conflicts between the various dimensions that comprise humans and their social world.

These conflicts may occur among any of the four levels of human existence that are represented in Exhibit 2.2. These conflicts, which create crises for the person in interaction, may be resolved positively or negatively. For instance, as seen in the exhibit, a crisis between the inner-biological level and the cultural-sociological one can result in a negative outcome, such as the widespread incidence of a disease (an epidemic) affecting humanity, or it can result in "cultivation," for instance, the discovery of a cure for the disease and, thus, the betterment of humanity. Similarly, a crisis between the individual-psychological level and the outer-physical level can result in the destruction of humanity—as, for example, when a huge flood envelops an area—or in the creation of an entity—a system of dams, for instance, that will curb the flood and help maintain human civilization.

Although research from Riegel's perspective is quite limited at this writing, one virtue of his ideas is to remind human developmentalists of the need to see all human functioning in context. That is, all human development occurs amid the continual changes that characterize the person in his or her world. And since the human

EXHIBIT 2.2 **Riegel's (1975) Model of a Dialectical Theory of Human Development: Crises with negative (upper lines) and positive (lower lines) outcomes generated by conflicts along four dimensions of human development.**

	Inner-Biological	Individual-Psychological	Cultural-Sociological	Outer-Physical
Inner-biological	Infection Fertilization	Illness Maturation	Epidemic Cultivation	Deterioration Vitalization
Individual-psychological	Disorder Control	Discordance Concordance	Dissidence Organization	Destruction Creation
Cultural-sociological	Distortion Adaptation	Exploitation Acculturation	Conflict Cooperation	Devastation Conversation
Outer-physical	Annihilation Nutrition	Catastrophy Welfare	Disaster Enrichment	Chaos Harmony

Source: Adapted from Riegel (1975, p. 58).

can never, in reality, be separated from his or her changing world, the person's changes must be understood as an outcome of *and* as an effect on these contextual changes. In other words, because of the strong (inseparable) interactions between individuals and their context, the person is both a product and a producer of his or her changing world.

CHAPTER SUMMARY

Scientific theories are based on philosophical assumptions about the nature of the world. In the study of human development the sets of assumptions, on which scientists base their theoretical and empirical work, are termed models, or world views. Several models have been influential in human development.

The mechanistic model is based on reductionism and continuity and is associated with a reactive view of humans. The organismic world view stresses emergence and discontinuity and is associated with an active view of humans. The contextual world view stresses continual, interrelated changes as characterizing human life and is associated with the view that active people interact with their active world.

Different world views lead to scientists proposing different theories of development, doing different sorts of empirical research, and, ultimately, interpreting the world differently. Since different theories are based on different assumptions, it is not appropriate to speak of theories as true or false. Rather, theories should be evaluated on the basis of their usefulness.

The current character of human development has a long history, and much of it is reviewed in this chapter. The nature-nurture issue, a controversy about the source of behavior (e.g., heredity versus environment, or maturation versus experience) has been a central concern throughout this history. Philosophers advanced ideas about human development that were on both sides of this issue, and this debate continued when the study of human development became a scientific concern.

Darwin's evolutionary theory provided a quite important basis of many of the theories of human development that followed it. Many of Darwin's ideas stressed the role of hereditary mechanisms in evolution, and when his ideas were translated into the first scientific theory of human development, proposed by G. Stanley Hall, this emphasis was enhanced. Hall's students (Terman and Gesell) continued this nature tradition. However, a nurture-based, behavioristic view arose to counter it.

World War II brought with it an influx of scholars into the United States who promoted ideas of human development relating to nature, to nurture, and to the interaction between the two. The 1950s and 1960s were characterized by an appreciation of the need to evaluate explanations of development and thus by a growing interest in theories of development.

Today, we recognize the use of a pluralism of theories, and in this chapter we briefly reviewed theories of human development stressing nature, nurture, and various types of interaction. G. Stanley Hall's recapitulationist theory and the instinctual theory of Konrad Lorenz are seen as instances of nature theories. Various nurture theories, stressing the role of stimulus-response connections and of social learning (and the theory of McCandless), were reviewed. Finally, weak, moderate, and strong interaction theories were discussed. The contributions of Sigmund Freud and Erik Erikson (weak interaction theories), of Jean Piaget and Lawrence Kohlberg (moderate interaction theories), and of Klaus Riegel (strong interaction theory) were briefly presented.

chapter 3

Biological Bases of Human Development

CHAPTER OVERVIEW

In this chapter we review the nature of genetic inheritance and of the way in which this contribution to behavior and development interacts with environmental and experiential contributions to influence the person. Our biological processes exist in a complex interaction with our behavior and our context across our lives. We illustrate this interaction with examples drawn from the infancy through aged periods of life.

CHAPTER OUTLINE

GENES AND HEREDITARY MECHANISMS

Mechanisms of Heredity
Genes

THE NATURE-NURTURE ISSUE

Nature Interactions with Nurture
Levels of the Environment
Nurture Interactions with Nature
The Critical Periods Hypothesis

BIOLOGICAL MECHANISMS ACROSS THE LIFE SPAN: GROWTH AND AGING

Growth
Aging
Biological Theories of Aging

BIOLOGY-BEHAVIOR-CONTEXT INTERACTIONS

Nutrition and Infant and Child Development
Venereal Disease in Adolescence and Young Adulthood
Personality and Cardiovascular Disease in Middle Age
The Eye and Visual Perception in Old Age

CHAPTER SUMMARY

ISSUES TO CONSIDER

What is genetic inheritance? What are chromosomes and genes?
What are the various views about the roles of nature and nurture variables in behavior and development?
What is the evidence for a strong interaction view of nature and nurture?
What is the critical periods hypothesis?
What are the major theories of growth and of aging?
What are the effects of nutrition on infant and child development?
Why is venereal disease a major problem in adolescence and young adulthood?
What is the relation between personality and cardiovascular disease in middle age?
What is the nature of the eye and of visual perception in old age?

*H*uman life involves events occurring within our bodies (inner events) and events between us and the world within which we live (outer events). Both sorts of events involve biology. Biology is the study of those processes that allow the living organism to survive and reproduce. Thus processes occurring primarily within the organism's body—for example, involving cell division or the exchange of chemicals across a cell's membrane—serve biological functions; they allow the organism to grow and to obtain nutrients, respectively. Processes involving the organism and its environment—for example, how it finds sufficient food and shelter within the particular setting within which it lives, both for itself and for its children —also are biological processes. Both types of processes—the inner and the outer ones—are biological because they relate to organisms' survival and perpetuation (reproduction). They maintain and further life.

This and other connections among inner- and outer-biological processes exist. Many will be discussed in this chapter. Indeed, in this chapter on biological processes involved in development across the life span, we will need to consider topics like genes and hereditary mechanisms, the role of the environmental context in influencing hereditary and other biological mechanisms, and the role of biological mechanisms in changing the context of human development. Thus our treatment of biological processes will be one which sees such changes as necessarily and inevitably linked with contextual changes, a treatment consistent with the views of many leading biologists (e.g., Lewontin, 1976; Mayr, 1978; Tobach, 1978).

Of course, other treatments of biological developmental processes are possible. In fact, many theories of psychological development are based very directly on another biological model of development; i.e., one that suggests that human development follows a single, normative direction, for example, of growth from conception to maturity and then inevitable decline throughout the aged years. However, in Chapters 1 and 2 we saw that such characterizations of development (be they aimed at biological or psychological changes) are based on just one interpretation of change and, in any event, are not consistent with the information which shows multidirectionality of change and the potential for plasticity throughout the life span.

As such, the life-span view draws attention to the multiple conceptions of biology that exist and to data that show that given different organism-environment interactions, alternative directions may occur in the development of life-supporting and life-perpetuating processes. Accordingly, we treat biology from this interactionist framework, and as we now turn to some of the initial biological processes involved in human development, we will have reason to reiterate this framework.

GENES AND HEREDITARY MECHANISMS

Development begins at *conception*. Conception refers to the time when the reproductive or germ cell (the sperm) of the father fertilizes the germ cell (the ovum) of the mother. With this fertilization a new cell is formed—a *zygote*—and from this new cell another organism will eventually develop. These germ cells that unite at conception are "special" in that they each contain only half the chemical units— termed *chromosomes*—necessary to form a new member of the species. In humans there are typically 46 chromosomes in every cell of the body. Each cell of the skin, the heart, the liver, and the brain, for example, contains this number of units. However, in adult humans capable of reproduction, each spermatozoon or ovum has only 23 chromosomes. Thus in the zygote these two sets of chromosomes pair up with each other to form a cell that has the normal human number of chromosomes. This

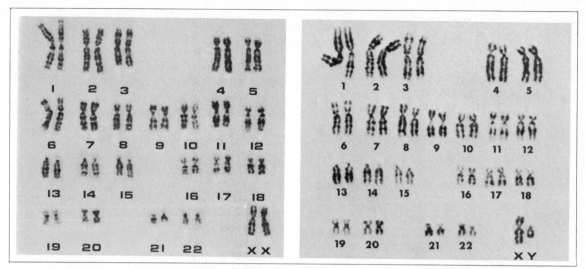

EXHIBIT 3.1   Human chromosomes, normally arranged in 23 pairs, in a female, left, and in a male, right. (Courtesy of Dr. Grumbach, Department of Pediatrics, University of California, San Francisco.)

process, wherein each parent gives the offspring half the necessary number of chromosomes, ensures that each parent will contribute *a* basis of the offspring's characteristics and future development. This process is termed *genetic inheritance,* and the 23 pairs of chromosomes we inherit from our parents will eventually be duplicated in every cell of the body.

An important illustration of the way that this genetic inheritance is expressed is in our sex determination. One of each set of transmitted chromosomes is sex-linked. That is, 1 of the 23 chromosomes in the male germ cells is either an *X* or a *Y* chromosome, while in the female germ cells there are only X chromosomes. Whether an X combines with a Y is largely a chance event; but if such a combination does occur, then a male offspring will be produced. If the chance combination involves a match between an X from the mother and an X from the father, then a female offspring will be produced.

Chromosomes influence characteristics other than sex type. In fact, as we will discuss in detail at a later point in this chapter, all human characteristics are *influenced* by our genetic inheritance. These numerous, complex influences occur through combinations involving the smaller chemical units found in each chromosome. These smaller, hereditary transmitted units are called *genes.* Although we are today sure that genes make up each chromosome, there is presently no way of exactly determining the number of genes within each chromosome. Yet, through indirect means, we can tell how genes influence behavior and can speak of how genetic differences between people contribute to behavioral differences between people. Notice, however, that at this point in our discussion we are being careful to avoid speaking of either chromosomes or genes as *determining,* or *fixing,* any of our behavioral characteristics. We are speaking of our genetic inheritance as an influence on these characteristics, and this phrasing implies that genetic inheritance alone, in and of itself, does not suffice to explain our behavior or its development. This implication will be made explicit later in the chapter.

In sum, then, development starts at conception with the genetic inheritance, and proceeds in utero (in the mother's uterus) for about nine months until birth; thereafter, an infant emerges from the womb. The understanding of the complex patterns of behaviors that precede and follow this point becomes the task of the human developmentalist.

Mechanisms of Heredity

Every living thing—from the pygmy shrew to the blue whale and from the microscopic *E. coli* to the giant redwood—begins life as a tiny, protected cell. Human development is no exception. As we have seen, it begins with a cell formed when the relatively large, immobile egg cell produced in the female's ovaries (the germ cell of the mother) unites with a smaller, tadpolelike sperm cell from the male's testes (the germ cell of the father). While the result—the fertilized ovum, the zygote —is the largest cell in the body, this new cell weighs only $\frac{1}{20}$ millionth of an ounce and is no bigger than a pinhead. Nevertheless, it embraces one of the great secrets of life, for it contributes to forming the new organism.

The "competition" among sperm cells for the egg is fierce; millions of them are available in a single ejaculation of semen from the male during sexual intercourse. The egg merges with the first sperm to penetrate it. If a sperm fails to obtain the nourishment contained in the egg, it self-destructs. A single sperm's chance of fertilization are about a hundred million to one. The united sperm and egg begin the processes of reproduction almost immediately. They provide a set of instructions, or a code, for the continually dividing zygote, until a configuration of trillions of specialized cells is formed.

Genes

The inheritance of traits is the work of a unit more fundamental than the chromosome. As noted, this is the *gene*, or unit of heredity. Each chromosome in a human cell is outfitted with thousands of genes, which are believed to line up along the threads of the chromosome. Scientists believe that each gene—or pair of genes or group of paired genes—carries a coded message that is responsible for a phase of development. This code is carried in the large molecules of protein, called *deoxyribonucleic acid* (DNA), that are found in the *nucleus* of every cell. Information from the DNA is transmitted from the nucleus to the rest of the cell, termed the *cytoplasm*. This transmission occurs through the influence of DNA on *ribonucleic acid* (RNA). RNA is a messenger; it carries the "code" contained in the DNA to the cytoplasm.

One pair of the estimated thousands of genes in each human cell is believed involved in the development of skin color, another pair is believed involved in the control of eye color; still another is believed involved in control of the individual's hair color. And although the question is quite controversial, there may be genes that are involved with IQ, athletic ability, temperament, and musical aptitude (Wilson, 1975).

In any case, the vast number of combinations of genes—and the number runs into the trillions—contributes to the great diversity of the species; indeed, the chance of any two people receiving the same combination of genes in the lottery of life and thus coming out exactly alike is virtually impossible: less than 1 in 70 trillion (Hirsch, 1970)! The only exceptions are identical, or *monozygotic*, twins— two children who come out of the same egg cell (*mono* = single; *zygote* = egg) that splits into two separate ones after conception. Each thus carries the same number

BOX 3.1

CHROMOSOMAL DEFECTS

Chromosomal difference can make the man, and on occasion it can undo the woman. In 1967 a Polish track star named Ewa Kiobusowska was ruled ineligible for international competition in female sporting events because doctors found that she had one chromosome too many and therefore was chromosomally not a woman but a man. Even though the 21-year-old objected that she felt and acted like a woman, the athletics judges accepted the evidence of the chromosomes—the first time this had ever been done—and ruled that she was a man as far as athletics were concerned.

The chromosomes can also explain why some disorders show up only in the male. If a defect occurs on one of a pair of X chromosomes, the other member of the pair is still likely to be normal and can salvage the situation. But in the male a defect in the X chromosome paired with the Y chromosome generally cannot be compensated for. Therefore the defect shows up.

One rare abnormality in men involves an extra chromosome in the package—an XYY combination. The abnormality created a good deal of excitement a few years ago when a disproportionate number of inmates in prisons and mental hospitals were found to have it—about 3.5 percent to as high as 20 percent in those studied compared to 0.01 percent to 3 percent in the general population—suggesting to some observers that it may have an important connection with aberrant behavior. The "criminal chromosome," as it came to be called, made headlines in the case, in Paris, of a stablehand who was a killer and, in the United States, in the case of Richard Speck who was put on trial for the mass murder of eight nurses in Chicago. Speck's defense was insanity, based in part on the existence of the XYY chromosome in his

blood cells, which the defense asserted was what made him a murderer. The jury, however, found Speck guilty.

Although the possession of these or any particular set of genetic characteristics cannot be said to invariantly lead to a specific behavior, because of the excitement generated by the discovery of XYY, investigators turned up the records of a number of XYY criminals and found some common features: The XYY criminal seemed to be someone who was tall, pimply, and a dullard and had a long history of criminal or bizarre behavior. Enough exceptions soon turned up to cast doubt on the influence of the abnormal chromosome on crime, and at present psychologists generally agree that the connection is very thin if existent at all. However, some psychologists have come to believe that XYY males do suffer from learning disabilities; unfortunately, a pilot project for screening XYY children in Boston and following them through the years was aborted in the spring of 1975 after protests by groups that believe those children would be stigmatized for life.

On the other hand, there is no argument that eye color, and hence color blindness, is carried by the X chromosome. Considerable experimentation shows that if an abnormal X chromosome combines with a Y, the boy will be color blind. When two X's combine to make a female and one of them contains color blindness, the other, normal X will cancel it out. However, if a female inherits color blindness in both her X chromosomes, we get the rare case of a color-blind female. Other sex-linked inherited traits associated with the X chromosome include baldness and bleeding (hemophilia). Rarely do females suffer from these afflictions.

and combination of chromosomes and genes as the other: They are the same sex, they look very much alike, and, to some extent, they behave alike.

Often two genes may be slightly different, though they form a pair and together handle an inherited characteristic. These genes are known as *alleles*. Simply, an allele is a form of a gene. Since a child gets half of his or her genes from each parent, a child has two alleles of each gene. If the alleles of both parents match, the child is homozygotic for that gene. If the alleles contributed by the parents are different, the child is heterozygotic for that gene.

In sum, all humans receive a complement of genes at their conception, at the time of fertilization. This complement is termed our *genotype*. It is this genotype which constitutes our *genetic inheritance*. Of course, all organisms have an environment within which they exist, and we may now ask, "How do these two things—heredity and environment—contribute to human development?"

THE NATURE-NURTURE ISSUE

Conventional wisdom tells us both "like father, like son" and "experience is the best teacher." Both expressions reflect a view about the basis of behavior. The former represents a nature view, the latter a nurture one. The simultaneous existence of two such contradictory statements should not be surprising, given the philosophical and scientific debate about the relative contributions of nature and nurture to behavior development reviewed in the previous chapter. Across history, views of development have been based on various stances on this issue ranging from extreme nature to extreme nurture views. However, there have also been some conceptions stressing interactions between these two sources.

This section focuses on this core conceptual issue, nature and nurture, because it is our opinion that any conception of development, pertinent to any and all portions of life, must speak to this topic. Any conception of development will be composed of statements about processes which relate to nature, to nurture, or to some relation between the two. In addition, other core issues derive from the nature-nurture one. Accordingly, another reason for focus on the nature-nurture issue is that it provides a basis for an integrative presentation of all core concepts of development. Focus on nature-nurture allows one to understand the interdependence of all issues of development.

In Chapter 2 we saw that scientists tended to take extreme, opposite positions in regard to the nature and nurture bases of development. Some stressed nature to the exclusion of nurture (e.g., Gesell), while others promoted a nurture view and excluded nature conceptions (e.g., Watson). Although others spoke of both sources of development (e.g., Riegel), the *predominant* way that the nature-nurture issue was conceptualized across history was in an "either-or" formulation (Anastasi, 1958).

The initial and historically predominant way the issue was formulated may be understood in the form of a question: "Which one (nature *or* nurture) provides the basis of behavior and development?" Anne Anastasi (1958), whose ideas did much to clarify the problems involved in appropriately conceptualizing the nature-nurture issue, was the first to cast the initial formulation of the issue in terms of this question. She argued that although one could view most philosophers and scientists as approaching the problem from this perspective of "Which one?" one had to recognize that these efforts by this majority were based on the wrong question.

The question "Which one?" is based on an *independent, isolated action* assumption of the relation between nature and nurture (Lerner, 1976). To use the terms focused on by Anastasi, nature is *heredity* and nurture is *environment*. The question "Which one?" assumes that nature and nurture are independent, separable sources of influence, and as such, one can exert an influence in isolation from the other. But Anastasi (1958) pointed out that such an assumption is illogical. This is so because there would be no one in an environment without heredity, and furthermore, there would be no place to see the effects of heredity without environment. Genes do not exist in a vacuum; they exert their influence on behavior in an environment. At the same time, however, if there were no genes (and consequently no heredity), the environment would not have an organism in it to influence. Accordingly, nature and nurture are inextricably tied together. In life they never exist independent of the other. As such, Anastasi (1958) argued that *any* formulation of development, in order to be logical and accurately reflective of life situations (i.e., to have *ecological validity*), must stress that nature and nurture are always involved in all behavior.

Moreover, she cautioned that it was necessary to understand the simultaneous contribution of nature and nurture in a particular way. Although some persons recognize that nature and nurture (heredity and environment) are both involved in behavior, many still formulate their interrelation in illogical and ecologically invalid ways. It is possible to ask another wrong question about nature and nurture. Some scientists had recognized that both nature and nurture were involved in development but had conceptualized their relation in terms of the question "How much of each is needed for behavior?" That is, they viewed the issue in terms of an *independent, additive action* assumption (Lerner, 1976). They asked, for any given behavior, how much of it was formed by hereditary and how much by environment. In so doing they were working under the belief that, as with the first question ("Which one?"), nature and nurture existed independent of each other. Here, however, instead of behavior being either a 100/0 percent or a 0/100 percent isolated action, some additive contribution of less than these extremes was thought to be involved. For example, people asked whether intelligence was 80 percent heredity and 20 percent environment or whether personality was 60 percent heredity and 40 percent environment.

This conception of "How much of each?" may be considered as illogical as the question "Which one?" since it is just a less extreme formulation of the first question. For instance, for the 80 percent of intelligence that might be thought to be nature, one may ask where that 80 percent exerts its influence if not in an environment. And for the 20 percent of intelligence thought to be derived from nurture, how does an organism get that 20 percent acted on if not first having inherited genes?

Thus, the question "How much of each?" does not seem to be a useful way to understand the simultaneous contribution of nature and nurture. Since both must be fully present and involved in all behavior, one needs *100 percent of each* for any and all behavior development. To conceptualize their simultaneous contribution, one must ask the question "How?" One must ask how (100 percent of) nature and (100 percent of) nurture interact to provide the basis of behavior development (Anastasi, 1958). Thus the quality of the relation between nature and nurture may be viewed as involving the *interdependent, interactive action* assumption. From this view heredity and environment do not add together to contribute to behavior; instead, development is seen as a *product* of nature-nurture interaction.

An analogy will help illustrate this. The area of a rectangle is determined by a formula which multiplies the length by the width (area = length × width). To know the area of a given rectangle, then, one has to look at the product of a multiplicative relation. It is simply incorrect to ask "Which one?" length or width, determines the area, because the rectangle would not exist unless it were a figure having both length and width. Similarly, it is incorrect to ask "How much of each?" is necessary to have an area because the two dimensions cannot merely be added. Both a length and a width must be fully present and must be seen to interrelate in a multiplicative manner in order to produce a rectangle.

Of course, although length and width must always be completely present in order to have a rectangle, different values of each will lead to different products (or areas). Thus in determining a particular product (a given area) of a length × width interaction, one must ask *how* a specific value of length in interaction (in multiplication) with a specific value of width produced a rectangle of a given area. More generally, then, one must recognize that the same width would lead to different areas in interaction with varying lengths and, in turn, the same length would lead to different areas in interaction with varying widths.

By moving from this analogy back to the question "How?" in regard to the nature-nurture issue, it may be seen that comparable statements may be made. There would be no product—no behavior development—if nature and nurture were not 100 percent present. Thus the assumptions underlying the questions "Which one?" and "How much of each?" are rejected, and it is recognized that any behavior development is the result of a multiplicative, interactive relation between specific hereditary and environmental influences. Moreover, it should be noted that this means that the same hereditary influence will lead to different behavioral products in interaction with varying environmental contexts; furthermore, the same environmental context will lead to varying behavioral outcomes in interaction with different hereditary influences.

This means that heredity and environment *never* function independently of each other. Nature (e.g., genes) never affects behavior directly; it always acts in the context of environment. Environment (e.g., stimulation) never directly evokes behavior either; it will have variation in its effects depending on the hereditary-related characteristics of the organism on which it acts.

These statements about the reciprocal interdependency of nature and nurture are not just casual matters. It has been argued that major philosophers and scientists have tried to conceptualize behavior and development in terms which are inconsistent with the view reflected by the interactive action conception. Indeed, succeeding chapters will consider theorists who stress that various components of human development (e.g., cognition, personality, parent-child relations) can be understood by ideas which stress *either* nature *or* nurture (i.e., the question "Which one?"). Thus it is important to point out that others do not necessarily agree that the formulation we favor is the best or most useful one.

Accordingly, because of the continuing controversy between nature and nurture interpretations of human development, it will be useful for us to provide a discussion of concepts involved in the resolution of the issues and our arguments about why we believe that our position in regard to the nature-nurture issue, which derives from our contextual, life-span orientation, has utility.

Nature Interactions with Nurture

Discussion of the implications of the question "How?" led to the assertion that there are never any direct hereditary effects on an organism. This view means that heredity never directly gives one structural characteristics, such as body build, or behavioral functions, such as intelligence or personality. Heredity always interacts with an environment, and this means that the same hereditary contribution will lead to different behavioral outcomes under different environmental conditions.

To illustrate, let us represent hereditary contributions by the letter G (for genes), environmental contributions with the letter E, and behavior outcomes with the letter B. As shown in Exhibit 3.2(*a*), it is possible to conceptualize the contribution of heredity to behavior as being direct. Here, a particular combination of genes (G_1) will invariably lead to a particular behavior outcome (B_1). However, it has been argued that this conceptualization is not appropriate. As such, an interactive idea of nature and nurture, illustrated in Exhibit 3.2(*b*), has been advanced. Here the same hereditary contribution (G_1) can be associated with an infinity of behavioral outcomes (B_1 through B_n) as a consequence of interaction in the infinity of environments (E_1 through E_n) that could exist. A basis of plasticity (i.e., of intraindividual change) in development is thus promoted. Consider as an example the case of a child born with Down's syndrome. Children born with Down's syndrome have a genetic anomaly. There is an extra chromosome in the twenty-first pair—three chromosomes instead of two.

Thus children with Down's syndrome have a specific genetic inheritance; the Down's syndrome child has a specific genotype. Yet even though the genotype remains the same for any such child, the behavioral outcomes associated with this genotype are changeable.

As recently as thirty years ago, Down's syndrome children, who typically are recognized by certain particular physical (particularly facial) characteristics, were expected to have life spans of no more than twelve years or so. They also were expected to have quite low IQ scores. They typically were classified into a group of people who, because of low intelligence, required custodial (typically, institutional) care. Today, however, Down's syndrome children often live well beyond adolescence. Additionally, they lead more self-reliant lives. Their IQs are now typically higher, often falling in the range allowing for education, training, and sometimes even employment.

How did these vast differences come about? Certainly, the genotype did not change; rather, what changed was the environment of these children. Instead of their invariably being put into institutions, different and more advanced special education techniques were applied to the people, often on an outpatient basis. These contrasts in interventions led to variations in behavioral outcomes despite hereditary constancy (i.e., *genotypic invariancy*).

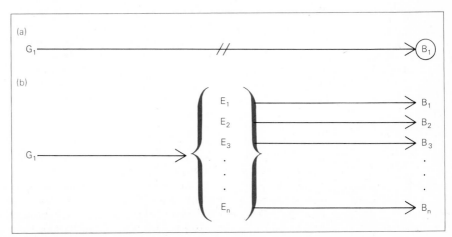

EXHIBIT 3.2 (*a*) Heredity (G) does *not* directly lead to behavior (B). (*b*) Rather, the effects of heredity on behavior will be different under different environmental (E) conditions.

That heredity always exerts its effects indirectly through environment in the development of physical as well as behavioral characteristics may also be illustrated. First, consider the disease termed *phenylketonuria* (PKU). This disorder, involving an inability to metabolize fatty substances because of the absence of a particular digestive enzyme, led to a child developing distorted physical features and severe mental retardation. It was discovered that the lack of the necessary enzyme was the result of the absence of a particular gene being present in the genotype, and as such, PKU is another instance of a disease associated with a specific genotype.

Today, however, many people—perhaps even some college students reading this book—may have the PKU genotype and, yet, have neither the physical nor the behavioral deficits formerly associated with the disease. It was discovered that if the missing enzyme was put into the diets of newborns discovered to have the disease, *all* negative effects could be avoided. Again, an environmental intervention has changed the outcome. In fact, the dietary supplement was found to be able to be terminated at about 1 year of age, since the body either no longer needed the enzyme to metabolize fat or the enzyme was produced in another way. Here again it may be seen that the same genotype will lead to alternative outcomes, both physical and behavioral, when it interacts in contrasting environmental settings.

Another example may illustrate this further and, more important, may provide a basis for specifying the parameters of environmental variation within which hereditary contributions are embedded. First, imagine that a particular experiment (one improbable for ethical and technological reasons) were done. Say a mother is pregnant with monozygotic (or identical) twins. As noted, these are twins who arise from the same fertilized egg, the same zygote, that splits after conception. Hence the two zygotes have the same genotype. But, importantly, because the zygotes implant on somewhat different parts of the wall of the mother's uterus, there exist somewhat different environments. Further imagine that it were possible immediately after the zygote splt into two to take one of them and implant it in another woman, who would carry the organism through to birth. Finally, imagine that the first woman, Mother A, had lived for the last several years on a diet of chocolate bars, potato chips, and soda pop, had smoked two packs of cigarettes a day, and drank a pint of alcohol each evening. On the other hand, say the second woman, Mother B, had maintained a well-balanced diet and neither smoked nor drank. In all other respects the women are alike.

Here is a situation wherein two genotypically identical organisms are developing in quite different uterine environments. Such differences are known to relate to prenatal, birth (perinatal), and postnatal behavior on the part of the offspring and even have implications for the mother. Thus despite the genotype identity, the offspring of Mother A would be more likely to be smaller, less alert, and more hyperactive (because of smoking and alcohol intake) than the offspring of Mother B (Hurlock, 1973; Mussen, Conger, & Kagan, 1974).

Although the above study with Mother A and Mother B is imaginary, the influence of the uterine environment on the offspring is not at all fanciful. The imaginary example was used to illustrate that variations in the environment will effect significant physical and behavioral changes in an offspring, despite the genotype. Even physical characteristics such as eye or skin color may be influenced by environmental variations (albeit extreme ones) no matter what genes are inherited. If mothers are exposed to extreme radiation or dangerous chemicals (as in the case of mothers in Britain in the 1950s who took the drug thalidomide), pigmentation of the eyes or the skin can be radically altered.

In summary, then, it is argued that in order to understand the contributions of

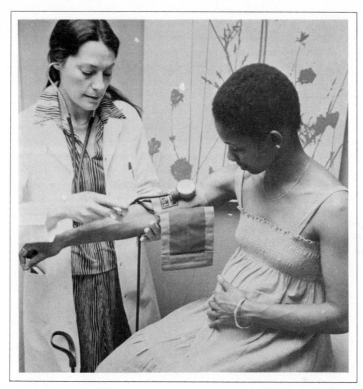

Appropriate prenatal care can reduce the stress of pregnancy for the mother and for the developing in utero child. (Paul Conklin/Monkmeyer Press Photo Service.)

heredity to development, one needs to recognize that genes influence physical and behavioral characteristics indirectly, by acting in the context of a specific environment. If the same genetic contribution were to be expressed in an environment having other specific characteristics, the same genes would be associated with an alternative behavioral outcome. Accordingly, in order to understand completely nature interactions with nurture, one should know all the ways in which the environment can vary (and, as will be argued below, the reverse of this need is also the case).

Certainly, however, there is an infinity of possible environmental variations; and today one cannot even begin to identify all the chemical, nutritional, psychological, and social variables which may vary in the environment, much less identify the ways in which they provide a significant context for development. However, one may at this point note that the environment may be thought of as existing at many levels. One can look at the environment in molecular terms—and talk of chemicals in the body of the mother. Or one can use molar terms—and talk of noise and pollution levels in particular (e.g., urban) settings. Consequently, it is useful to specify levels of the environment because it will (1) allow discussion about where the variables providing the context for nature interactions may lie and (2) allow for a consideration of nurture interactions. As such, levels of environmental variation are now considered.

Levels of the Environment

An organism does not exist independent of an environment, and it has been argued that as much as the organism is shaped by the environment, the organism shapes the environment. As a consequence of this interdependency, both organism and environment continually change, and it has been suggested that this change involves multiple levels of analysis. These levels—the inner biological, individual psychological, physical environmental, and sociocultural historical (Riegel, 1975, 1976a)—are parameters of both organism and environment. They are used to denote the types of nurture-related variables that may provide the context for nature interactions.

The Inner-Biological Level. The genotype first is expressed in utero, in the mother's body. Hence the chemical and physical makeup of the mother can affect the offspring. Chemicals in the mother's bloodstream can enter that of the offspring through the *umbilical cord*, the attachment between mother and offspring. As already noted, poor nutrition, excessive smoking or alcohol intake, and other drug intake can affect the unborn child. Additionally, the contraction of diseases (e.g., rubella) can lead to malformations of the fetus's heart or limbs and/or can affect the function of sensory organs (the eyes or ears).

The Individual-Psychological Level. Independent of her diet, her smoking or drinking habits, or her health status, the psychological functioning of the mother can affect the unborn child. Excessive maternal stress (e.g., "nervousness" about the pregnancy) can affect the offspring. Mothers who have excessive stress in about the third month of pregnancy are more likely to have children with certain birth defects (e.g., cleft palate, harelip) than are mothers not so stressed (Sutton-Smith, 1973). To illustrate the interrelation between all of the levels of the environment, it may be suggested that the way maternal stress exerts an influence on the unborn child is by altering the chemicals (e.g., adrenalin) in her blood— at the inner-biological level— at a time in the embryological period when certain organs are being formed.

In addition, previous child-rearing experiences can play a part on the individual-psychological level. Experienced parents (those who already have a child) are not the same people they were before they had a child. (Firstborns thus have different parents than later offspring, even though the parents involved may be biologically the same.) Thus an experienced mother may be less likely to be stressed by a second pregnancy. Not only might this affect the chemicals in her bloodstream, but also, in being less "nervous," she might be less likely to engage in behaviors—for example, smoking—sometimes done when nervous.

Of course, as more information about prenatal care is generated, and as cultural values change, effects on maternal stress and behaviors will change. Thus, again, one level of environment is related to another, the individual psychological to the sociocultural historical. Before turning to the latter level, however, let us consider the physical-environment level.

The Physical-Environment Level. Different physical settings have differences in such variables as quality of the air, purity of the water, levels of noise, population density, crowding, and general pollution of the environment. Such variables can affect the inner-biological functioning of the person by producing variations in the likelihood of contracting certain diseases or anomalies (Willems, 1973) and can affect the individual-psychological level, as well, by producing various levels of

The physical-environmental level of the context and the person's individual-psychological functioning may be influenced by the presence of nuclear reactors such as those at Three Mile Island. (United Press International, Inc.)

stress, for instance (Gump, 1975). In turn, the quality of the physical setting may be seen as both a product and a producer of changes in the sociocultural setting across time. If values regarding industrialization in the United States had not existed as they did through the early 1960s and if huge levels of industrial waste had not polluted the air, land, water, and wildlife, there would have been no basis for the general emergence of countervailing values in the late 1960s and 1970s regarding environmentalism, ecology, and reduction of pollution. The physical-environmental level is not independent of the sociocultural-historical one, and thus this last level may be discussed.

The Sociocultural-Historical Level. It has been argued that changes in attitudes toward smoking, knowledge about prenatal health care for the mother, and values about pollution may change across time to influence the unborn child. Thus as presented in Chapter 1, developments related to nature are embedded in history. With advances in education (remember the case of children with Down's syndrome), medicine and science (remember the case of children with PKU), and attitudes, values, and behaviors (regarding smoking, drinking, drug use, and pollution of the environment, for example), the outcome of any given hereditary contribution to development will be altered.

In sum, it may be seen that a diversity of behavioral outcomes may result from nature interactions with a multilevel environment. Development is an outcome of hereditary contributions dynamically interacting with the combined influence of biological through historical changes in an environmental context. Thus a genotype is not a blueprint for *a* final behavioral outcome. Rather, *numerous outcomes of genotype-environment interactions* may result from any one hereditary contribution. There is no one-to-one relation between genotype (our genetic inheritance) and phenotype (the observed outcome of development; the outcome of a specific genotype-environment interrelation). Rather, the genotype in actuality represents a range of potentially infinite interactions with environments, which is termed the *norm of reaction* (Hirsch, 1970).

Just as the effects of heredity on behavior can best be understood in relation to environment, simultaneously, the effects of the environment on behavior can best be understood in the context of the nature of the organism. Up to this point in the chapter we have stressed nature interactions with nurture; the reciprocal portion of this relation may also be considered.

Nurture Interactions with Nature

Just as the effects of nature on behavior may be construed to act in the reciprocal context of nurture, environmental contributions to behavior may be seen as moderated by the nature of the organism. From this view, the same environmental event (e.g., contraction of a disease, exposure to a particular stimulus) or group of events (e.g., those associated with middle-class as opposed to upper-class membership) will lead to different behavioral outcomes depending on the nature of the organism. Using the same symbols as in Exhibit 3.2, one may see this view illustrated in Exhibit 3.3. As shown in Exhibit 3.3(a), it is possible to conceptualize the contribution of environment to behavior as being direct. Thus a particular environmental event, or set of events (E_1), is seen as directly leading to a particular behavioral outcome (B_1). However, as with the former argument regarding nature contributions, this view is not tenable. We have argued for a strong interactive view of nature and nurture, and the environmental contribution component of this view is illustrated in

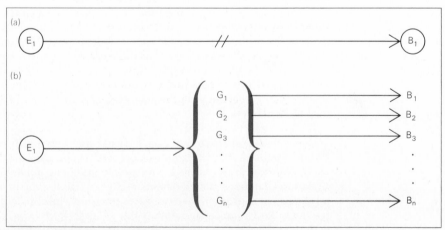

EXHIBIT 3.3   (a) Environment (E) does *not* directly lead to behavior (B). (b) Rather, the effects of environment on behavior (B) will be different in interaction with organisms having different natures (G).

Exhibit 3.3(*b*). Here the same environmental contribution (E_1) can be associated with an array of behavioral outcomes (B_1 through B_n) as a consequence of interaction with organisms having different natures (G_1 through G_n). This may be illustrated in several ways.

First, consider a very general set of experiential events associated with being a child of upper-middle-class parents. Imagine if such parents had two children, *dizygote twins*, also called *fraternal twins*. Such siblings are born of the same pregnancy but are from two separate ova that were fertilized at the same time. Although born together, these siblings have different genotypes (unlike monozygotic twins). If one of these twins were to be born with the genetic anomaly discussed earlier, Down's syndrome, while the other was born with a normal complement of genes, then a situation would result wherein children born of the same parents, at the same point in time, could potentially be exposed to the same environmental events.

However, despite whatever experiences the Down's syndrome child encountered, the effects of those experiences could not be expected to result in behaviors falling within a range identical to that of the sibling. Despite advances in special education, noted earlier, one still cannot expect the Down's syndrome child to have an IQ score within the normal range (i.e., 85 to 115), and one would not expect the child to attain a high-status vocation. However, such expectations could appropriately be maintained in regard to the sibling born with the normal genotype. Thus the hereditary nature of the organism imposes limits on the possible contributions of environment.

Other illustrations of this interaction may be drawn from the information presented above about the prenatal maternal environment. It was noted that if the mother contracted rubella during pregnancy, adverse physical and functional outcomes for the infant might follow. However, this same experience (contraction of the disease) may or may not lead to these outcomes depending on the maturational level of the organism. If the experience occurs during the embryological period, these negative effects are likely to occur; if it happens in the late fetal period, they are not likely to happen. Similarly, excessive maternal stress will or will not be more likely to lead to certain physical deformities (like cleft palate) depending on the maturational level of the organism. Here again the nature of the organism moderates the influence of experience on development.

It may be concluded, then, that even if one is talking of very narrow sorts of environmental experiences (as in encountering a specific stimulus in a specific transitory situation) or very broad types of experiences (as are associated with membership in one culture versus another), the contributions of these environmental influences would not be the same if they interacted with hereditarily (genotypically) different organisms. This lack of invariancy would occur even if it were possible to ensure that the experimental history of these different organisms were identical. As long as the nature of the organism is different, the contributions of experience will vary, and recall that we noted there are over 70 trillion potential genotypes (Hirsch, 1970).

Moreover, we should note that a genotype immediately becomes a phenotype at the moment of conception. The genotype is expressed in one and only one environment. Hence, although a norm of reaction existed for that genotype, once it was expressed in one particular context, all the other alternative phenotypes it *could* have resulted in are excluded. Even identical twins become, at least a bit, phenotypically unique from each other at the moment of implantation.

To summarize, because of genotypic uniqueness, all individuals will interact with their environments (be they the same or different) in unique, specific ways.

Even children who are identical twins may develop different phenotypic characteristics because of contrasting histories of organism-environment interaction. (United Press International, Inc.)

Thus the environment always contributes to behavior, but the precise direction and outcome of this influence can only be completely understood in the context of an appreciation of the genetic individuality of the person. This view agrees with that of Hirsch (1970, p. 70), who argued that "extreme environmentalists were wrong to hope that one law or set of laws described universal features of modifiability. Extreme hereditarians were wrong to ignore the norm of reaction."

To illustrate our view—that an interactionist view is most useful in discussions of the nature-nurture issue—we now turn our attention to a major idea in human development that, most basically, involves the nature-nurture issue. This idea— the critical periods hypothesis—has led to considerable debate.

The Critical Periods Hypothesis

As noted, other major issues of development are associated with the nature-nurture issue. An important one is the critical periods hypothesis, which rests on two ideas. First it is held that there are qualitatively distinct periods in life. Second, it is held that the presence, quality, and impact of these periods for later life are fixed and universal.

Theorists like Lorenz (1963; and discussed in Chapter 2) believe such periods exist in regard to social relations formed in some species of birds (through a process he terms imprinting). Similarly, Erikson (1959, 1964) believes that the life span is characterized by these critical periods. Such theorists believe that not all portions of life are the same. Rather, there are times in life when, it is held, certain events must occur, certain developments must take place, in order for development to be "normal."

This view is the general conception behind the critical periods hypothesis. Although there are several ways of formulating this hypothesis (Caldwell, 1962;

Lerner, 1976; Oyama, 1979; Sluckin, 1965), the thrust of it remains identical. The hypothesis suggests that the potential for development is *not* identical across the life span. Rather, it is argued that there are certain times during which certain developments must occur if those developments are ever to occur optimally or adaptively; or, it is argued that there are times in life *beyond* which certain developments will not occur optimally or adaptively if they have not already occurred. Thus it is critical for these particular developments to occur within their necessary period if development is to be optimal and adaptive. Simply, the critical periods hypothesis states that if you do not develop what you should develop when you should develop it, then you will never develop it!

However, because theorists using a critical periods hypothesis are advancing a notion saying that the same experience will lead to different outcomes depending on the time of life the experience occurs, a specific stance is taken in regard to the nature-nurture issue. The critical periods hypothesis indicates that it is not experience (or nurture) which is the primary determinant of what outcomes will arise from developmental progressions. Rather, the critical periods hypothesis indicates that there are phenomena, independent of experience, that determine what influences experience can have and what outcomes experience can result in (i.e., adaptive or maladaptive). In other words, there are phenomena imposing limits on experience, and as such, the phenomena are put into an experience-independent (a nurture-independent) category. They are put into a nature category.

Thus the critical periods theorists typically take a particular stance in regard to the nature-nurture issue and stress the primacy of nature over nurture. For example, Erikson (1959, 1964) divides the life span into eight critical periods. He says that certain phenomena must develop within each respective period in order for healthy development to proceed. However, a person does not have an infinite length of time to develop what is supposed to be developed within each critical period. Rather, the person's development is governed by a maturational ground plan, or, in Erikson's (1959) terms, the "epigenetic principle." This maturational plan is innate in each person. Such genetically based limits may differ from person to person, because of hereditary differences, but in all people this innate plan sets the time limits of each period. Each person has a particular length of time, innately set for him or her, to spend in each period. If the person does not develop as he or she is supposed to develop within the fixed time, then in accordance with the maturational ground plan, development will move on; experiences that could have led to healthy development if they had occurred before development became refocused will not now be able to have such an adaptive influence.

For instance, Erikson (1959) says that in the critical period of life termed latency, which is typically thought to occur in the childhood years of from about 6 to 11 or 12, one has to develop toward a sense of industry. If one does not, however, achieve this during one's innately fixed time limit—a time limit which, Erikson (1959) implies, cannot be known in advance (because different people do have different, but not measurable, inheritances)—then industry will *never* be adequately achieved. Additionally, the rest of development will be unfavorably altered (Erikson, 1959). Thus, to Erikson, it *is* quite critical to develop appropriately within a critical period.

As hypothesized by Erikson (1959), the basis of a period being seen as critical lies in the idea that *nature acts independently of nurture* in imposing limits on development at certain times of life. However, it has been argued that nature and nurture are dynamically interactive and that conceptions of development which contend that one source influences behavior independently of complete interaction with the other are not consistent with data showing the plasticity of behavior development across the entire life cycle. In our opinion, a conceptual problem with the critical periods hypothesis exists because it stresses fixity in development as a consequence of the independent, isolated action of nature. Indeed, it is not surprising to learn that when critical period ideas are put to empirical test, they are found to be untenable interpretations of either animal behavior (Moltz & Stettner, 1961; Sluckin, 1965) or human behavior (Schneirla & Rosenblass, 1961, 1963).

However, despite the limitations of the critical periods hypothesis, theorists like Erikson (1959) have provided ideas of some use. Erikson has discussed how antecedent developmental events (developments at earlier periods of life) influence consequent events (developments at later periods). Although events in earlier periods may not be critical determinants of development in later life—in the sense of inevitable sources of particular behavior—antecedents may promote development in one direction as opposed to another. A set of particular negative experiences with a dentist in early life may make it less likely that dentistry will be chosen as a profession in later life. Or exposure to traditional socialization pressures in childhood that relate to the sex-appropriateness of behavior (e.g., females become nurses; males become physicians) may make it differentially likely that males and females will enter the same vocation after adolescence. Yet although events in antecedent

(a)   (b)

Even though children develop in a traditional manner in one period of life (a) they may, because of human plasticity, show nontraditional behaviors in later portions of life (b). (a: Burk Uzzle/Magnam Photos; b: Vic Cox/Peter Arnold, Inc.)

periods may make one differentially *sensitive* to particular events in consequent periods, such sensitivity does not necessarily imply that intervention in these later periods cannot alter the developmental course. Positive later experiences with a dentist or alterations in the socialization pressures in society (due to revised social values, new laws, or other historical alterations) can lead to modification of development.

BIOLOGICAL MECHANISMS ACROSS THE LIFE SPAN: GROWTH AND AGING

From birth to death organisms do not remain the same biologically. The structure and function of one's biological processes change across life. Some of the changes are labeled growth, or maturation; others are labeled as aging.

Growth

Often the terms growth and maturation are used interchangeably. However, some scientists draw a distinction between the two. For example, Schneirla (1957) uses the term *maturation* in a somewhat broader sense than growth. For Schneirla (1957) maturation involves both growth and differentiation. *Growth* refers to changes by way of cell, tissue, or organ accretion, or enlargement; simply, things get bigger. *Differentiation* refers to changes in the interrelations among cells, tissues, organs, or their parts; such changes in relationships are labeled as structural changes. More and/or different cells developing from other cells are examples of differentiation. When growth and differentiation are completed, the person is said to be mature.

McElroy, Swanson, and Macey (1975) expand on these ideas. Keeping their discussion at the level of the cell, they note that cells can do one of three things: (1) they can grow and divide; (2) they can differentiate; or (3) they can die. In addition, they note that the coordination over time of growth and differentiation may be labeled as integration and that all the above aspects of cell function may be labeled as cell development.

Still other scientists see growth as a process that is broader than maturation which, to them, is linked with reproductive capability (cf. Katchadourian, 1977). Here maturity is reached when one is fertile (i.e., capable of getting pregnant or of impregnating someone), but growth (e.g., further increases in the length of bones) can continue after this point. Humans are a species that continues to grow after reproductive maturity is reached and, thus, may be said to have *open growth systems* (Orians, 1969). Accordingly, since our interest is in life-span change, we focus here on issues related to growth.

Processes of growth are often thought of as irreversible (Orians, 1969). For instance, growth is defined as an irreversible increase in size involving the production of new *protoplasm* (i.e., the living material that makes up cells; Orians, 1969). Increases in size resulting from accumulation of nonliving substances and from water absorption may be reversible, often in regular (e.g., daily or monthly) cycles and are, therefore, not considered growth.

Most humans finally reach a maximum size beyond which there is no further growth, although different people reach this maximum at different rates. But what causes the growth of a given person to slow down and then to eventually stop is not precisely known. Yet there are some general reasons that are offered for the cessation of growth (see Orians, 1969). As will become clear below, these reasons are often very similar to those offered to account for aging.

1 *Cell growth limitations.* One reason offered for the end of growth involves ideas about the mathematical relation existing between the surface area of a cell and its volume. Some biologists believe that after a certain surface area is reached, further increases in volume are not highly probable.

2 *Nutrition.* Factors related to food intake may be related to growth. A nutritional regimen lacking sufficient amounts of essential vitamins may lead to the cessation of growth.

3 *Hormones.* The body has organs called glands that release chemicals. Some glands are ductless, i.e., they secrete chemicals directly into the bloodstream. Such glands are *endocrine glands.* The chemical substances they secrete are termed hormones. One endocrine gland, the pituitary, secretes a *growth-stimulating hormone,* or somatotropin; this hormone enhances the growth of all body systems. Some scientists suggest that when this hormone is not in sufficient supply, growth will be halted. In addition, hormones released from other glands, most notably the hormone released from the thyroid gland, are known to be involved in the magnitude, rate, and termination of growth (Orians, 1969).

Aging

But what happens when growth stops? Is this the point that aging begins? Generally, biologists define aging as a set of deleterious changes which decrease the probability of the individual's survival (Comfort, 1964; Strehler, 1962). However, life-span developmentalists—including those trained in biology—stress that aging is really a process involving the entire life course. In fact, they see growth as a part of aging. Riley (1979), for instance, states, "Aging is a life-long process of growing up and growing old. It starts with birth (or conception) and ends with death" (p. 4). Moreover, consistent with our interactionist, contextual perspective, Riley (1979) notes, "Aging consists of three sets of processes—biological, psychological, and social; and these three processes are all systematically interactive with one another over the life course" (p. 4).

Thus although everyone ages, the life-span view is that the nature and rate of aging are products resulting from psychological, social, and historical interactions with biology. But what are some of the characteristics of aging, at least insofar as current cohorts are concerned? Remembering the idea that changes associated with biological aging are those associated with a lessened chance of survival, we can point out that today, at age 25, a white male in the United States can expect to live an additional forty-seven years; by age 45, this figure has been reduced to about twenty-nine years; and by age 85, it is only about six years. This reduced probability of survival is accompanied by a host of deleterious physical changes; for example, wrinkled skin, poorer vision and hearing, less muscle strength, reduced cardiac and lung capacity, and loss of brain weight. What causes these deleterious changes? Can they be slowed or reversed? How are they related to behavior? Can the human life span be lengthened?

Some idea of the scope of these changes on the population as a whole may be gained by examining the life tables which summarize mortality rates. There are two types of life tables. The *cohort life table* provides a longitudinal perspective. It follows the mortality experience of a single cohort; for example, mortality rates for all persons born in 1900 are plotted until no one from this group remains alive. The *current life table* provides a cross-sectional perspective. It considers the age-specific mortality rates for an actual population at a given time. For example, the current life table for 1976 tabulates the mortality rates for individuals of different ages in 1976. The current life table is more practical to construct than the cohort life table and is the method used by the National Center of Health Statistics. Exhibit 3.4 shows two indices from the 1976 U.S. Life Tables for men and women. These data illustrate rather dramatically the effects of aging on the population, in general, and its differential effects on men and women, in particular.

Biological Theories of Aging

What causes the decreasing survival potential revealed in the mortality data of Exhibit 3.4? Why does the human life span appear to have an upper limit as summarized in Box 3.2? There are a number of biological theories which focus on these questions, and these can be elaborated in somewhat greater detail than the ideas associated with growth, especially since we have seen that the idea of aging can subsume that of growth from a life-span perspective. Indeed, many of these theories of aging are just extensions of ideas associated with why growth slows down and stops. Each of the theories of aging has had some success in explaining a portion of the aging phenomenon, but none of them has achieved general acceptance.

All modern biological theories of aging have a genetic basis (Shock, 1977). It is assumed that the life span of a species is ultimately determined by a program built into the genes of the species. Thus, the human has a maximum life span of about 115 years, the horse 62 years, the cat 28 years, and the mouse only 3.5 years.

Support for the genetic basis of aging comes from Hayflick's (1965) work which shows that certain cells of the body, grown under culture conditions (in vitro), are only able to divide a limited number of times. Previous to this discovery, it had been

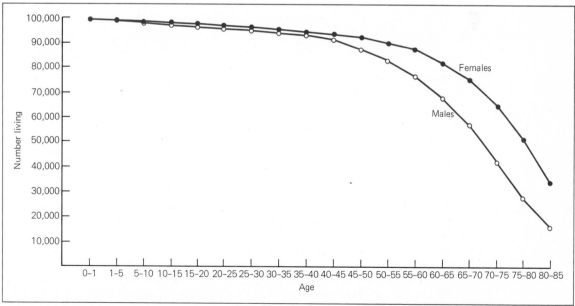

EXHIBIT 3.4(a) You can see here how many people out of 100,000 born in a given year survive until very old age. For example, of 100,000 female babies born alive, 98,631 will complete the first year of life and enter the second, 97,868 will reach age 20, and 65,138 will reach age 75.

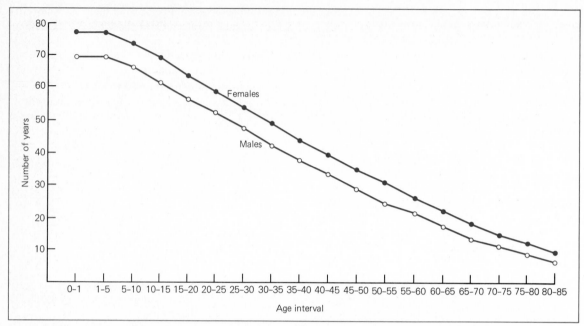

EXHIBIT 3.4(*b*)   Here you can see how long you can expect to live once you have survived to a particular age. For example, at age 20 a woman can expect to live another 58.2 years, and at age 80 she can expect to live another 8.7 years.

thought that cells could divide indefinitely under culture conditions. However, Hayflick found that fibroblast cells (connective tissue cells which normally divide) taken from human embryonic tissue undergo only about fifty doublings before cell division ceases. Further, the older the individual from whom the cell samples are obtained, the fewer doublings the cells will undergo. However, the cells of a very elderly individual will still divide, suggesting that people rarely live out their potential life span. Hayflick's work, then, suggests that the aging of the organism may be programmed into the genetic code of the cell.

While it is clear that aging has a genetic basis, the various biological theories may be divided into three major groups for the purpose of our discussion: genetic cellular theories, nongenetic cellular theories, and physiological theories (Shock, 1977). Let us briefly examine some examples from each of these groups.

Genetic Cellular Theories. These theories suggest that aging results from damage to the genetic information involved in the formation of cellular proteins. As we have mentioned, the information stored in the genetic code is determined by the structure of the DNA molecule which controls the formation of essential proteins required by the cell. The information from the DNA is transcribed and transferred to another location in the cell by the RNA molecule where the actual assembly of the proteins occurs. Several theorists have argued that it is the breakdown of these basic genetic mechanisms which causes aging.

Sinex (1974), for instance, has suggested that mutations damage the DNA at a faster rate than it can be repaired. This process continues until the cell dies. Such damage to the DNA would be particularly deleterious to nondividing cells, such as

BOX 3.2

DID METHUSELAH REALLY LIVE 969 YEARS?

The purported longevity of Methuselah is undoubtedly a myth. But how long can human beings live? In examining this question it is important to distinguish between *life expectancy* and *maximum life span*. Life expectancy refers to the average length of life. Maximum life span refers to the extreme upper limit of human life.

Life expectancy fluctuates dramatically as a function of nutrition, disease, sanitation, health care, and other environmental factors. For example, it is estimated that the average life expectancy at birth in ancient Rome was as low as 22 years. Estimates place the life expectancy figure during the Middle Ages in the 30- to 35-year range. This figure changed little during the sixteenth and seventeenth centuries, increasing only to around 40 by the middle of the nineteenth century. The greatest gains were made during this century, with life expectancy at birth in the United States increasing from 47.3 years in 1900 to 72.8 years in 1976. Note, however, that increases in life expectancy at maturity and beyond are much smaller. For example, in 1900, average life expectancy at age 20 in the United States was 42.8 years. By 1976 it had increased to only 47.3 years (U.S. Department of Health, Education and Welfare, 1978). The dramatic increase in life expectancy at birth is largely a function of health-related advances which, in particular, have led to a reduction of infant mortality. As a result, a greater proportion of all persons are reaching older ages. For example, in 1900 only 40.9 percent of the cohort born sixty-five years earlier were still alive; in 1976, 75.1 percent were still alive. The increase in life expectancy, then, has resulted largely from improved environmental conditions which have reduced the incidence of premature deaths, particularly in infancy and childhood. However, there is little evidence to suggest that the maximum life span of human beings has changed. This has been estimated at between 110 and 120 years.

A few years ago there were a number of reports in the popular press of extraordinary longevity in several parts of the world. These included the Republic of Georgia in the U.S.S.R., the Vilcabamba Valley in Ecuador, and the province of Hunza in Kashmir. These areas were pur-

While there have been reports of extraordinary longevity in the Republic of Georgia in the U.S.S.R., there is little evidence that the maximum life span of human beings is greater than about 110 years. (Photo Researchers, Inc.)

ported to contain far more persons over the age of 100 than one would expect. For example, while about 3 centenarians per 100,000 population would be normal, regions in the Caucasus mountain area of the U.S.S.R. report about 400 centenarians per 100,000 population. In these areas, reports of persons aged 100 to 120 are common, and reports of persons aged 120 to 170 are not unusual.

While these claims make "good press," a careful examination of the evidence suggests they are unfounded (Medvedev, 1974). Generally, the documentation is very poor. There was no birth registration, and few of the very old are able to produce reliable documents of other types (e.g., military, marriage, education). Usually, the reports have been substantiated by recall of significant past events and by interviews with other residents of the village. Medvedev (1974) notes that social, rather than biological, factors may account for the phenomenon. For example, in these regions older persons receive a great deal of honor and respect. Centenarians often hold special positions in the community and engage in special activities (e.g., chairperson of local celebrations). Such positive valuing of the extremely old may result in exaggerations of chronological age. There may also be more practical reasons involved, as Medvedev relates:

The famous man from Yakutia who was found to be 130 years old received especially great publicity because he lived in the place with the most terrible climate. . . . When publicity about him became all-national and a large article with a picture of this outstanding man was published in the central government newspaper, Isvestia, the puzzle was quickly solved. A letter was received from a group of Ukranian villagers who recognized this centenarian as a fellow villager who deserted from the army during the First World War and forged documents or used his father's (most usual method of falsification) to escape remobilization. It was found this man was really only 78 years old. (Medvedev, 1974, p. 387)

Thus all indications point to an upper limit on the human life span of about 110 to 120 years. However, some writers speculate that this limit may be exceeded in the not too distant future as researchers discover the basic mechanisms of the aging process (Medvedev, 1975). Such a breakthrough, however, would probably raise a host of ethical issues. What problems do you foresee?

nerve and muscle cells. Similarly, Curtis (1966) has suggested that cells undergo mutations, most of which are deleterious. His work was based on experiments with rats and mice which showed the incidence of abnormal chromosomes increased markedly after exposure to radiation and progressively with age. Nevertheless, calculations suggest that the rate at which mutations occur in the absence of radiation is too small to account for the overall age changes observed.

Other theorists suggest that aging is the result of errors involved in the transmission of information from the DNA to the final protein product (Medvedev, 1964). Unlike DNA molecules, which are highly stable, RNA molecules are being formed continuously. Therefore, Hahn (1970) has suggested that aging results from faulty transcription of the RNA from the DNA. Similarly, Orgel (1963) has suggested the faulty selection of amino acids by enzymes is a key factor in cell death. Although the initial frequency of errors may be low, it would increase over time. Eventually, this accumulation of errors would result in an "error catastrophe" and cell death. Strehler (1978) has proposed that the loss of sequential multiple copies of the same genetic information is critical. A particularly important group of genes that are present in multiple copies are those responsible for the production of the RNA required for protein synthesis. Strehler and his colleagues have found that the number of copies of these genes in the tissues of both dogs and humans decreases with age. Such a loss would result in increased error and, eventually, cell death.

Nongenetic Cellular Theories. These theories focus on changes that take place in the cellular proteins after they have been formed.

Accumulation theories suggest that aging results from the accumulation of deleterious substances in the cells of the organism. Almost 100 years ago, investigators discovered that old cells universally contained dark-colored inclusions called *lipofuscin.* These waste pigments, or cellular "garbage," have been shown to increase at a constant rate with time. The accumulation of lipofuscin is particularly prevalent in the cells which do not undergo division, such as nerve and muscle cells, especially cardiac muscle cells. It has been argued that the presence of this insoluble waste material interferes with cellular metabolism and ultimately may result in cell death (Carpenter, 1965). There is limited evidence to suggest the mechanism by which this impairment may occur. Recently, however, it has been shown that the amount of RNA in nerve cells decreases in direct proportion to the increase in the age pigment content of the cells (Mann & Yates; reported in Strehler, 1978). Additional evidence isolating the exact nature of the toxic substances involved and demonstrating how the presence of lipofuscin impairs cellular functioning is required. Nevertheless, as Strehler (1978) points out, it is probably impossible to clutter up as much as 75 percent of a cell's interior with lipofuscin without interfering with its functioning.

Other nongenetic cellular theories suggest that aging occurs as a result of damage to cell proteins. For example, according to the cross-linkage theory, *cross-linkages,* or bonds, develop with passage of time either between components of the same molecule or between molecules. Bjorksten (1968) suggested that cross-linking leads to progressive biochemical failure and represents a primary cause of aging. Most of the work on this theory has involved the study of extracellular proteins such as elastin and collagen. It has been well established that these proteins which surround the blood vessels and cells slowly cross-link with age. It has been proposed that such changes may have far-reaching effects on the functioning of cells—e.g., they may lead to oxygen deficiency. Bjorksten (1968) has suggested that similar changes may occur in intracellular proteins such as the DNA molecules. Indeed, cross-linking of DNA has been observed in both culture tissues (in vitro) and in living tissue (in vivo). To the extent that cross-linking of the DNA molecules occurs in the cells with increasing age, a mechanism for explaining cell failure may be available.

Free radical theory is related to cross-linked theory and may be viewed as one way in which cross-links are produced. *Free radicals* are unstable chemical compounds which react with other molecules in their vicinity. As a result of these reactions, cells may be altered in structure and function. Again, alteration of key proteins (e.g., DNA) could result in eventual cell death (Harman, 1968). Free radicals are generated by common substances such as various foods and tobacco smoke. The formation of free radicals is accelerated by radiation and inhibited by antioxidants. Harman has presented evidence suggesting that mice fed antioxidants show an increased average life span. As a result, various antioxidant dietary supplements, such as vitamin E, have been proposed to combat free radical damage; however, tests of these claims remain inconclusive.

Physiological Theories. These theories suggest that aging results from the failue of some physiological-coordinating system, such as the immunological or endocrine system, to integrate bodily functions properly.

For example, the immune system protects the body against mutant cells (e.g., cancer cells) and invading microorganisms. Two mechanisms are involved: (1) the immune system forms cells which engulf and digest foreign cells and (2) it generates antibodies which react with the proteins of foreign organisms. Deleterious changes occur in this system with age; for example, the production of antibodies declines with age after a peak in adolescence. The immune system may also lose its ability to recognize mutated cells; for example, the great increase in cancer in older adults is probably related to failures in the immune system. Further, Walford (1969) has noted that there is an increase with age in *autoimmunity* in which the body, in effect, attacks itself. This may result from a failure of the immune system to recognize normal cells as normal or from errors in antibody formation, so that the antibodies react to normal cells as well as foreign ones. This theory is supported by findings which indicate increase of autoimmune antibodies in the blood with increasing age (Walford, 1969).

Similarly, there is considerable evidence to show that the endocrine system declines in its function with age. For example, with increasing age, the pancreas fails to release sufficient insulin quickly when blood sugar rises. Related to this, the incidence of adult onset diabetes increases with age, becoming particularly prevalent in the elderly. Similarly, age-related changes occur in the gonads, andrenals, and other sites of hormone production. None of the site-specific changes noted above appears to account for aging. However, Finch (1976) has suggested that aging may be controlled in similar fashion as developmental events such as puberty and menopause. In particular, he has suggested cellular aging is regulated by hormonal events. The hormonal changes of aging are controlled by the brain, particularly the pituitary gland and the hypothalamus. Finch argues that aging pacemakers in these control centers initiate a cascade of neurological and hormonal events which regulate the aging process.

Integration of Theories. While each of these theories focuses on a different aspect of the aging process, they are not necessarily incompatible. Strehler (1978), for instance, suggests the following integration.

1 The primary event in aging is the loss of the ability of the cell to divide. This event is programmed in the genetic code of the cell.

2 After cell division has stopped, the cell is particularly susceptible to environmental damage. Further, the degree to which the cell can repair itself depends on other genes. A key defect that accumulates is the loss of some of the copies of a group of genes responsible for protein synthesis.

3 Decreased RNA results in an increased accumulation of damaged enzymes and lipofuscin which impair cell function.

4 Decreased normal functioning at the cellular level results in decreased integration among the physiological systems of the body. Ultimately, the systems break down and the individual dies.

BIOLOGY-BEHAVIOR-CONTEXT INTERACTIONS

We have briefly reviewed some of the molecular, cellular, and physiological changes that may be responsible for growth and aging processes. These changes affect virtually all major systems of the human body: skeletal, muscle, skin, pulmonary, cardiovascular, neural, endocrine, reproductive, gastrointestinal, and excretory. The growth and aging rates associated with some of these parts of the human body are illustrated in Exhibit 3.5(*a*) and 3.5(*b*), respectively. From the change rates displayed in this exhibit, we can draw two conclusions.

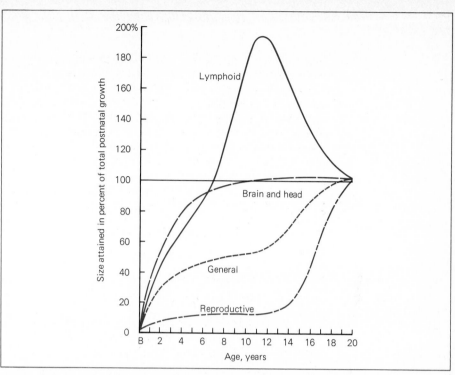

EXHIBIT 3.5(*a*)   Differential growth of some parts of the human body during development. All
curves indicate percentage of size at maturity (20 years), i.e., size at age 20 is
100 percent on the vertical scale. (From J. M. Tanner, *Growth at Adolescence*,
2d ed., Oxford: Blackwell's Scientific Publications Ltd., 1962, p. 11, Fig. 4.)

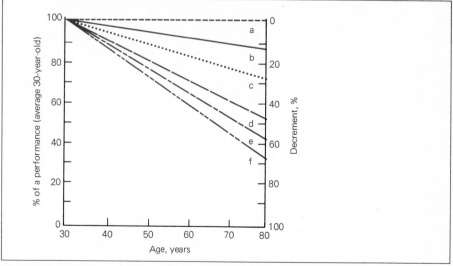

EXHIBIT 3.5(*b*)   Age decrements in physiological functions in males. Mean values for 20- to
35-year-old subjects are taken as 100 percent. Decrements shown are schematic linear projections: (*a*) fasting blood glucose; (*b*) nerve conduction velocity and cellular enzymes; (*c*) resting cardiac index; (*d*) vital capacity and
renal blood flow; (*e*) maximum breathing capacity; and (*f*) maximum work
rate and maximum oxygen uptake. (From Shock, 1972.)

First, while the biological changes associated with growth and aging are inevitable and universal, they do not affect all systems equally. Exhibit 3.5 shows the wide variation in rate of growth and in amount of decline which occurs in several indexes with age. Second, while growth and aging may be inevitable and universal, they are not unmodifiable. Interventions, such as diet and physical exercise, can modify their effects.

But what relationship do these changes have to behavior—the basic focus of our inquiry in this text? In the following sections, we will illustrate several biology-behavior interactions in order to illustrate the linkage between these domains. Note that such interactions may be of several types (e.g., biology influences behavior; behavior influences biology). We will attempt to illustrate some of these possibilities in our examples. Of course, such interactions are extensive. Our intent is to be illustrative rather than exhaustive.

Nutrition and Infant and Child Development

Physical growth and maturation are not processes that simply unfold independent of the environment. The nutritional adequacy of one's social context has a profound influence on one's growth.

When biologists refer to the *requirements for growth,* they often discuss those chemicals, ingested by the organism, which are needed for growth. Such chemicals may be labeled as *food,* and the study of growth requirements is termed *nutrition;* i.e., nutrition comprises the processes in which food is used (Harrison, Weiner, Tanner, & Barnicot, 1977). Sometimes people do not take in those foods necessary for growth. Such a condition may be labeled as *malnutrition.*

In infancy and early childhood malnutrition may lead to a disease which involves the halting of growth, tissue decay, and even death. Children in their early years may suffer from abdominal swelling, hair loss, changes in skin color, and lowered resistance to disease.

Malnutrition delays physical growth. But children have great recuperative powers, provided the adverse conditions leading to malnutrition are not carried too far or continued too long (Harrison et al., 1977). During a short period of malnutrition, the organism slows its growth and, as Harrison et al. phrase it, "waits for better times" (p. 343). When the better times come, growth takes place unusually fast, until it reaches the normative trajectory. During the "catch up" phase, weight, height, and skeletal development seem to catch up at about the same rate.

Females appear to be better buffered against the effects of malnutrition or illness. They are less easily "thrown off" their normative growth trajectories than are males.

The earlier the malnutrition, the worse its effects. Relative to body size, the caloric requirements for children are higher than those of adults. Thus infants' need for food is in this relative sense greater than that of adults; infants need to grow while adults need to maintain. A 3-week-old baby quietly sleeping a good deal of the time has a caloric requirement per unit of body weight that is more than twice that for an adult engaged in moderately heavy labor (Katchadourian, 1977). During an infant's first month of life, caloric requirements are highest—from 100 to 120 calories per kilogram of body weight. (These figures decline slowly until age 16 in males and 13 or 14 in females, and more rapidly thereafter until the adult requirement of 40 to 50 calories per kilogram is reached; Holt, 1972; Katchadourian, 1977).

Since it is known that severe malnutrition may interfere with nervous system (e.g., the brain) development—the specific form of damage depending on the age when malnutrition occurs—it may be the case that nutritional deficits interfere with intellectual development in infancy (cf. Hetherington & Parke, 1979). For example, research done in Central America on the effects of malnutrition in infancy has found that malnourished children do not develop language as early as do better-nourished children (Cravioto & DeLicardie, 1975).

Impairment of intellectual development in children associated with prebirth malnutrition seems most pronounced when the mother herself has a history of poor diet, when the malnutrition has been severe and long-lasting, and when poor nutritional factors are combined with poor postbirth social and economic factors (Hetherington & Parke, 1979, p. 87). In addition, it is known that conditions of illness which preclude proper nutritional intake can impair learning abilities such as short-term memory and attention (Klein, Forbes, & Nader, 1975).

While malnutrition affects physical growth, not all aspects of growth are affected equally. Katchadourian (1977) points out that a child's size and rate of growth seem to be more affected by malnutrition than is the shape of his or her body parts. Moreover, different nutritional backgrounds do not seem to affect body proportions. For instance, Greulich (1957) reports that children of Japanese ancestry who are reared in the United States are taller at all ages than Japanese children born and raised in Japan; but the ratio of length of leg to total height is the same in the two groups.

As with physical growth, the effects of malnutrition can be counteracted. Early malnourished Korean children, adopted and raised by American parents in American homes providing good nutrition, performed at least as well on intellectual and achievement tests, given later in life, as did their American-born peers. But those Koreans who had suffered the most extreme early malnutrition performed less well than those in the group who, while still malnourished, were somewhat less so (Winick, Karnig, & Harris, 1975). During the infant period itself, Lester (1975) found that well-nourished, Guatemalan 1-year-olds showed an orientation response to a sound, while malnourished babies showed less of one; in some cases the malnourished infants did not show this important response at all. In turn, Brody and Brody (1976) found that 1-year-olds whose mothers had received protein supplements during pregnancy showed faster habituation (a form of learning) and longer attention than did children whose mothers had either received no supplements or one consisting of calories with less protein. Lloyd-Still, Hurwitz, Wolff, and Schwachmore (1974) found that the psychological effects of severe malnutrition during the first six months of life can be reversed if the infant is given adequate nutrition, stimulation, and a stable socioeconomic environment. In sum, the effects of malnutrition on infant and child physical growth and intellectual development occur in relation to a complex set of factors derived from other biological factors and the interpersonal, physical, and sociocultural context of the infant and child, and these same effects can be remedied if appropriate contextual variation occurs.

Venereal Disease in Adolescence and Young Adulthood

The biological changes associated with adolescence—to be discussed in Chapter 9—lead to changes in physical appearance, bodily sensations, and emotions. These biological changes are often associated with increased sexual behavior on the part of adolescents. While this relation between biology and behavior probably comes as little surprise, what may not be recognized is that the behavior, which is a "product" of one's biology, can often feed back and "produce" other aspects of one's biology. Certainly, becoming pregnant changes one's biology and, of course, pregnancy results from behavior. But the sexual behaviors in which adolescents engage can affect other aspects of their biological functioning, and can do so for males as well as females. Venereal disease is a case in point.

When an infection is transmitted primarily by sexual intercourse, it is called a *venereal disease.* Such diseases are named after Venus, the goddess of love, because sexual activity is involved (Katchadourian, 1977). There are several common infections of the genital tract which are transmitted through sexual intercourse. Trichomoniasis, herpes simplex virus type 2, vaginal thrush, the "crabs," and prostatitis are common infections of the genital tract that are often transmitted by sexual intercourse with an infected person. These infections induce discharges and/or irritations which are a nuisance and sometimes difficult to treat (Goldstein, 1976). Space does not permit going into detail regarding the organisms that cause these diseases or their symptoms, cures, and effects (but see Lerner & Spanier, 1980). Our point of focus will be the risk of contracting one of the two most serious types of venereal diseases, *syphilis* and *gonorrhea.*

Syphilis and gonorrhea are both very contagious. Of reported cases of syphilis, 95 percent are contracted through intercourse, and the other 5 percent are contracted by kissing when there is a cut or scratch in the mucous membrane of the lips or mouth, by oral-genital contact, by prenatal transmission, or by transfusion. Gonorrhea is almost always contracted through sexual intercourse. Reports of contagion by means of bathtubs, swimming pools, and toilet seats are without foundation in fact. It is difficult to get accurate statistics on the incidence of syphilis and gonorrhea be-

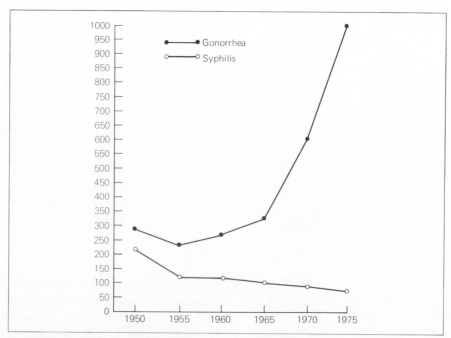

EXHIBIT 3.6   Cases of gonorrhea and syphilis reported in the United States, 1950 to 1975. (Source: U.S. Bureau of the Census, *Statistical Abstract,* 1977a, p. 114.)

cause only a proportion of cases are reported. This fact must be kept in mind in interpreting figures indicating changes in incidence. It is possible that part, although by no means all, of any apparent increase in incidence is due to better reporting. Thus the data reported in Exhibit 3.6 suggest both a possible marked increase in gonorrhea and, in addition, perhaps a greater willingness to bring these diseases to the attention of health authorities. The reported incidence of syphilis, however, has declined.

There is evidence that the incidence of venereal disease, particularly gonorrhea, is increasing dramatically among adoloscents and young adults. In 1975, there were 1 million *reported* cases of gonorrhea for the population as a whole, an increase of more than 100,000 over the previous year. There were 80,000 new cases of syphilis reported (U.S. Bureau of the Census, 1977a). There are no accurate data available on other forms of venereal disease. A major portion of the increase in venereal disease is accounted for by adolescents. There has been a particularly striking increase among males and females in the 10 to 14 age category, although the prevalence for individuals of this age group is small compared to the 15 to 19 and 20 to 24 age groups. The majority of the cases of venereal disease reported in the United States occur among persons younger than 25 years (Johns, Sutton, & Webster, 1975); 1 in 10 in this age group contract venereal disease in a given year.

There is now what may be interpreted as an epidemic. The World Health Organization ranks gonorrhea as the most prevalent communicable disease after the common cold. The situation concerning venereal disease is serious any way it is considered, but there are several facts that further complicate it. Reported cases of gonorrhea and syphilis are about twice as numerous among males as among females. As many as 80 percent of females and a small proportion of males who have gonorrhea are asymptomatic; that is, they have the disease and can transmit it to others, but they themselves show no symptoms of it (ASHA, 1975). Of course, the presence of the disease can be ascertained by laboratory tests. Asymptomatic individuals are generally not aware of the fact that they have gonorrhea. Syphilis may become asymptomatic only after the primary and secondary states of the disease have passed, and this may be several years after the original infection (ASHA, 1975). There is a not uncommon tendency to treat gonorrhea as being no more serious than, say, a cold. But complications of the disease in women account for 175,000 hospital admissions totaling 1,200,000 hospital-patient days. On any given day, an average of 3200 patients are hospitalized and an estimated 5750 girls are absent from school as a result of gonorrhea infections (ASHA, 1975).

At one time it was thought that penicillin would prove to be the ultimate cure for syphilis and gonorrhea. Although penicillin is still effective in most cases, some gonorrhea organisms are becoming partially resistant to it. Other antibiotics are being used with some success, but the final answer has not yet been found (Neumann & Baecker, 1972; Schroeter & Lucas, 1972; Rudolph, 1972).

The only contraceptive that is even partially effective in preventing venereal disease contagion is the condom. Fiumara (1972) mentions conditions (some of which would be likely to make intercourse less than satisfactory) to be met before the condom is effective against gonorrhea. There must be no preliminary sex play involving the genitals before putting the condom on; the condom must be intact before and after use; it must be put on and taken off correctly. Even if all these conditions are met, the condom gives more protection against gonorrhea than against syphilis because it does not cover all the areas of the couple's bodies that may come into contact (Hart, 1975). With the increasingly widespread use of contraceptive methods (especially the pill) designed for women, a reasonable guess would be that more men will leave the responsibility for contraception to women; consequently, fewer men will use the condom. Some investigators believe that this is now happening and that it is one of the reasons for the rising incidence of venereal infection.

Personality and Cardiovascular Disease in Middle Age

The incidence of cardiovascular disease is related to many factors including diet, smoking, obesity, and psychosocial variables. In the case of the psychosocial variables, linkages have been found between cardiovascular disease and both personality and situational stress. Of particular interest, however, has been the work of Rosenman and Friedman who have described a behavior pattern associated with high coronary risk (Friedman & Rosenman, 1974; Rosenman, 1974; Rosenman & Friedman, 1971). This behavior pattern has been labeled *Type A behavior pattern*, and it is contrasted with *Type B behavior pattern* which is associated with low coronary risk.

The Type A Behavior Pattern. As described in Box 3.3, the Type A pattern is characterized by a number of behavioral tendencies including excessive competitiveness, acceleration of the pace of ordinary activities, impatience with the rate at which most events take place, thinking about or doing several things simultaneously, hostility, and feelings of struggling against time and the environment. The converse of the Type A pattern is characterized by the relative absence of these behavioral tendencies. It is important to note that these patterns are a matter of degree rather than a true typology. Rosenman and Friedman report that only about 10 percent of their

BOX 3.3

ARE YOU TYPE A OR TYPE B?

Although only 10 percent of their subjects were wholly one type or the other, Rosenman and Friedman (1974) defined two patterns of behavior. Type A had far more tendency toward heart disease than Type B. You're Type A if:

1 You explosively accent key words when you speak and tend to speed up toward the end of your sentences. **2** You always move, talk, and eat rapidly. **3** You're often impatient with how slowly things happen. You try to hurry people through what they're saying. You're furious at cars on the road that hold you back, at having to wait in line. You fume at repetitive chores. You race through every book. **4** You do or think about two or more things at once. You worry about work on your day off and think about unrelated topics when someone else is talking. **5** You prefer to talk about what *you* want and only pretend to listen to other people's topics. **6** You feel vaguely guilty when you relax. **7** You don't see the most important, most interesting, or loveliest things you come into contact with. **8** You're preoccupied with *getting* rather then *being*. **9** You always feel a sense of urgency. **10** You feel hostile and challenged by other Type A's. **11** You have habitual gestures or tics—fist clenching, teeth grinding, etc.—that suggest an inner struggle. **12** You're afraid to stop hurrying because you think that's what makes you successful.

You're Type B if:

1 You're free of all the Type A habits and traits. **2** You feel no general sense of hostility or competitiveness. **3** You play to relax rather than to prove yourself. **4** You can relax without guilt and work calmly.

subjects had fully developed Type A or B behavior; the remainder had less fully developed forms of these patterns. It is also important to note that these behavior patterns are not context-free personality traits. Rather, they appear to come into play when a predisposed individual is challenged by a suitable environment. Finally, these behavior patterns are not correlated with intelligence, occupational category, socioeconomic status, and other indexes of "success." Thus a bank president may be a Type B or a Type A and, conversely, a janitor may be a Type A or a Type B.

Type A Behavior and Cardiovascular Disease. There have been numerous studies of the linkage between Type A behavior and coronary disease. A primary investigation, however, was a prospective study of over 3500 men aged 39 to 59 begun in 1960 (Rosenman, Friedman, Straus, Jenkins, Zyzanski, Wurm, & Kositchek, 1970; Rosenman, Friedman, Straus, Wurm, Jenkins, Messinger, Kositchek, Hahn, & Werthessen, 1966; Rosenman, Friedman, Straus, Wurm, Kositchek, Hahn, & Werthessen, 1964). These men were free from coronary disease when the initial measures of behavior pattern and physiological indexes were obtained. Follow-up evaluations were completed at the end of $2\frac{1}{2}$, $4\frac{1}{2}$, and $8\frac{1}{2}$ years.

The data indicated that coronary heart disease was much more prevalent among Type A men than among Type B men. The figures from the $4\frac{1}{2}$-year follow-up (Rosenman et al., 1970) are shown in Exhibit 3.7. It can be seen that 71.5 percent of the 39- to 49-year-old men who developed coronary heart disease were Type A, while only 28.5 percent were Type B. In cases where coronary heart disease was not present, 46.8 percent were Type A and 53.2 percent were Type B. The figures for the older age group are similar. Type A men also show chemical changes associated with coronary heart disease prior to the onset of clinical signs, including elevated serum cholesterol and triglycerides, accelerated blood coagulation, and increased daytime excretion of catecholamines.

Rosenman and Friedman's data also indicate that coronary heart disease was predicted by other risk factors, such as smoking, diabetes, high blood pressure, and high blood cholesterol levels. Indeed, many of the men showing Type A behavior also had high levels of these risk factors. However, statistical control of these other risk factors did not eliminate the impact of Type A behavior in young and middle-aged men. Even with these factors controlled, Type A men were almost three times as likely to develop coronary heart disease as Type B men. However, the older the individual, the more other risk factors begin to play a role. In other words, the link between Type A behavior and coronary heart disease is particularly significant for those in their forties.

Thus Rosenman and Friedman's research indicates that the biochemical abnor-

EXHIBIT 3.7   **Relationship of Type A and B Behavior Patterns to Coronary Heart Disease (CHD) in Two Age Groups of Men, in %**

Behavior Pattern	Age			
	39–49 Years		50–59 Years	
	CHD Present	CHD Absent	CHD Present	CHD Absent
Type A	71.5	46.8	70.0	56.3
Type B	28.5	53.2	30.0	43.7

Source: From Rosenman et al., 1970.

BOX 3.4

CARDIOVASCULAR DISEASE: THE NUMBER ONE KILLER

Diseases of the cardiovascular system represent the major causes of death in the United States. The data summarized in the table below show that in 1977, for instance, diseases of the heart were the leading cause of death, accounting for 37.8 percent of all deaths. When one adds cerebrovascular and arteriosclerotic diseases, almost 50 percent of all deaths are related to the cardiovascular system. What are some of the causes of these deaths?

Arteriosclerosis literally means "hardening of the arteries," but more accurately, it refers to a group of processes which have in common thickening and loss of elasticity of the arterial walls (Kohn, 1977). Prominent among these is *atherosclerosis*, a term often used interchangeably with arteriosclerosis. In atherosclerosis, arterial lesions develop, and fat, cholesterol, and collagen accu-

mulate at the site of these lesions. With time, the lesions become raised and begin to close off the space (lumen) inside the vessel. Such raised lesions are called plaques. The plaques become increasingly scarlike, consisting mostly of collagen. Eventually, they may ulcerate, hemorrhage, or lead to the formation of blood clots in the artery. Artery damage may be present for years without causing any noticeable symptoms. However, as the interior diameter of the vessel is decreased or the wall of the vessel loses its elasticity, a variety of problems may occur, including coronary heart disease. In this case, the degenerative changes of the artery wall related to coronary artery disease reduce the volume of blood supplied to the heart muscle and increase the possibility of the formation of blood clots.

These conditions may lead to several results.

Death Rates for Fifteen Leading Causes of Death, United States, 1977

Rank	Cause of Death	Death Rate (per 100,000 Population)	Percent of Total Deaths
	All causes	878.1	100.0
1	Disease of heart	332.3	37.8
2	Malignant neoplasms, including neoplasms of lymphatic and hematopoietic tissues (cancer)	178.7	20.4
3	Cerebrovascular diseases	84.1	9.6
4	Accidents	47.7	5.4
	Motor vehicle accidents	22.9	
	All other accidents	24.8	
5	Influenza and pneumonia	23.7	2.7
6	Diabetes mellitus	15.2	1.7
7	Cirrhosis of liver	14.3	1.6
8	Arteriosclerosis	13.3	1.5
9	Suicide	13.3	1.5
10	Certain causes of mortality in early infancy	10.8	1.2
11	Bronchitis, emphysema, and asthma	10.3	1.2
12	Homicide	9.2	1.0
13	Congenital anomalies	6.0	0.7
14	Nephritis and nephrosis	3.9	0.4
15	Septicemia	3.3	0.4
	All other causes	112.0	12.8

Source: From U.S. Department of Health, Education and Welfare, 1979.

Angina pectoris is characterized by agonizing pain felt in the region of the heart, left shoulder, and arm. It occurs when the heart muscle suffers from a lack of oxygen due to an inadequate blood supply. The pain of angina pectoris often occurs when the individual is under significant physical or emotional stress. The pain, however, rarely lasts for more than a few minutes once the stress that brought it on is reduced.

A *myocardial infarction* is an area of dead tissue in the heart muscle. It occurs because the muscle tissue has been deprived of oxygen for too long a period of time. In most cases this occurs be- cause of the formation of a blood clot, or *thrombis*, in the coronary artery which has blocked the blood supply. Depending on the size and lo- cation of the infarction, the individual may die or recover with varying degrees of impairment.

The degeneration of the cardiovascular sys- tem increases with age, and incipient signs of ath- erosclerosis are common in adolescents and young adults (Kohn, 1977). Risk factors associated with cardiovascular disease, in addition to Type A be- havior, include smoking, elevated serum lipids, excess body weight, diabetes mellitus, hyperten- sion, and prolonged hemodialysis.

malities associated with coronary heart disease and the incidence of coronary heart disease are substantially higher in individuals exhibiting Type A behavior than in individuals exhibiting Type B behavior. Of course, these data are still correlational in nature. Research with animals, however, has provided additional evidence that Type A behavior is causally related to coronary heart disease. In these studies, ag- gressive behavior was induced in rats by producing lesions in the hypothalamus. Such experimentally altered animals were more likely to develop coronary heart disease than control animals. Friedman and Rosenman (1974) conclude, "There is extraordinarily good scientific evidence on hand to indicate that Type A Behavior Pattern is the major cause of premature coronary artery disease in this country" (p. 266).

The Eye and Visual Perception in Old Age

Surveys of the incidence of blindness and the loss of visual acuity show that both of these difficulties are associated with increasing chronological age. For example, the prevalence of *legal blindness* (corrected distance vision of 20/200 or worse in the better eye, or a visual field limited to 20° in its greatest diameter) increases from less than 100 cases per 100,000 for individuals under age 21 to more than 1400 cases per 100,000 for individuals over age 69 (National Society for the Prevention of Blind- ness, 1966). Similarly, the incidence of poor *visual acuity* also increases with chron- ological age (Anderson & Palmore, 1974; U.S. National Health Survey, 1968). Ex- hibit 3.8 shows the corrected visual acuity in the better eye for 213 participants, aged 60 to 90 years, in the Duke Longitudinal Study (Anderson & Palmore, 1974). Visual acuity of 20/50 or worse indicates impairment sufficient to limit activities such as reading and driving. These data show that while 57 percent of those aged 60 to 69 had optimal visual acuity, only 27 percent of those aged 70 to 79 and only 14 percent of those aged 80 and over functioned at this level. Anderson and Palmore (1974) also reported that *cataracts* (opacities of the lens that obstruct light waves) were present in 9 percent of those aged 60 to 69, 18 percent of those aged 70 to 79, and 36 percent of those aged 80 and over. What causes the increasing visual impair- ment associated with age?

Structural Changes in the Eye. The visual problems summarized above reflect the impact of two sets of age-related changes in the eye (Fozard, Wolf, Bell, McFarland, & Podolsky, 1977).

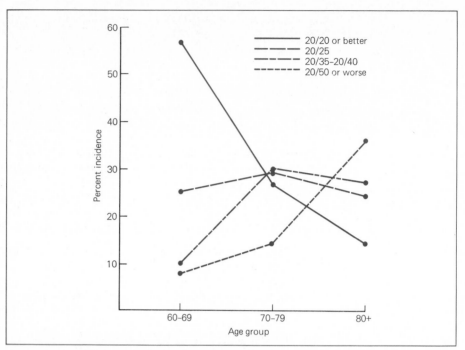

EXHIBIT 3.8   Corrected best-distance vision in the better eye. (Based on Anderson and Palmore, 1974, Table 2-7, p. 27.)

The first set of structural changes becomes evident in the thirties and forties. These changes affect the *transmissiveness* (the amount of light reaching the eye) and *accommodative power* (the ability to focus and maintain an image on the retina) of the eye. Several factors reduce the amount of light reaching the retina. These include decreased pupil size and increased clouding and yellowing of the lens (Fozard et al., 1977). As a result of these changes, middle-aged or older adults require more illumination than younger adults in order to maintain the same degree of visual discrimination. At the same time, the increasing opacity of the lens results in an increased susceptibility to glare (Wolf, 1960). In addition to reducing the amount of light, the yellowing of the lens also reduces the sensitivity of the eye to the blue-violet end of the spectrum. As a result, reds and yellows are discriminated better than blues and violets. By old age, color sensitivity over the entire spectrum declines markedly (Gilbert, 1957).

During the thirties and forties, the lens also becomes thicker and less elastic. As a result, it cannot change shape as readily; this impairs the ability of the eye to focus and maintain an image on the retina. In particular, it becomes difficult for middle-aged and older adults to view close objects clearly *(presbyopia),* which can result in the need for bifocals. The sharpest decline in accommodation occurs between 40 and 59 years of age (Brückner, 1967).

The second set of structural changes becomes evident in the fifties and sixties. These changes affect the retina itself. They appear to be primarily the result of vascular insufficiencies which result in reduced blood supply to the retina and, consequently, metabolic deficiencies and cell loss (Fozard et al., 1977). One result of these changes is a decrease in the size of the visual field and an increase in the size of the blind spot. The retina also becomes less sensitive to low levels of illumination. For example, *dark adaptation* (the ability to adjust vision when moving from high to low levels of illumination) varies with age. The time required to adapt increases, and the final level of functioning decreases, particularly for elderly adults (Domey, McFarland, & Chadwick, 1960).

Visual Perception. These structural changes in the eye have a significant impact on adult behavior in a variety of contexts (Fozard et al., 1977; Fozard & Popkin, 1978). For example, we noted that older adults are less sensitive to light at low levels of illumination. Many tasks, such as driving at twilight or night, require partial dark adaptation. This involves crossing over from rod to cone vision and vice versa. Research suggests that under such conditions the decrease in acuity is much greater for older adults than for younger adults (Richards, 1966). Similarly, abrupt changes in illumination can cause difficulty for older adults. In particular, accidental falls may be related to such changes. Archea (reported in Fozard & Popkin, 1978) found that most falls occur on the step at the top of the landing. Fozard and Popkin (1978) note:

> *In many houses, the stairway between the first and second floors is located at the end of the hall. As people walk toward the end of the hall, they typically look into windows that permit a great deal of light to enter. They then turn 90° to start down the "dark" stairs—precisely the situation in which Archea has found most falls to occur. (p. 979)*

Even at middle levels of illumination, older adults require more illumination than younger adults. For example, a study by Hughes (reported in Fozard & Popkin, 1978) examined the effect of illumination level on the work performance of younger (19 to 27 years) and middle-aged (46 to 57 years) office workers. The task involved a search for 10 target numbers printed on sheets containing a total of 420 numbers. Each worker performed several searches under three levels of illumination. The results are diagrammed in Exhibit 3.9. This figure shows that increased levels of illumination resulted in greater efficiency for both younger and middle-aged workers. However, the middle-aged workers benefited more from the increase than the younger workers.

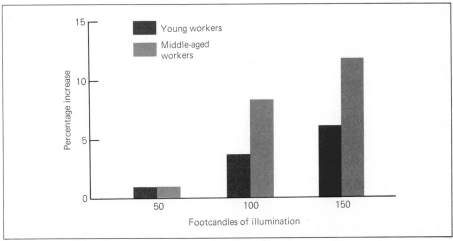

EXHIBIT 3.9   **Productivity of clerical workers as a function of illumination level. (Based on Fozard and Popkin, 1978, Fig. 2, p. 979.)**

Finally, using an intriguing approach, Pastalan, Mautz, and Merrill (1973) have attempted to simulate the effects of structural changes in the eye such as opacities and yellowing of the lens. These investigators took photographs of various scenes with normal and with coated camera lenses. The coatings on the lenses were designed to simulate the scenes as they would appear to the eyes of a person in his or her late seventies or early eighties. Exhibit 3.10 shows the same scenes photographed through the normal and the coated lenses. The coated lens simulates an acuity level of approximately 20/40. These photographs provide a dramatic illustration of the difficulties faced by many older adults.

Our brief review suggests that the structural changes which occur in the eye with increasing age have marked behavioral significance. The research also suggests that many of the visual problems of older adults may be ameliorated through appropriate environmental design. Fozard and Popkin (1978), for instance, suggest that increased local lighting of work areas, steps, and ramps and increased contrast in visual information displays would compensate for many visual problems of the elderly.

EXHIBIT 3.10   A simulation of a scene as seen by modal person in the late seventies or early eighties is shown in the photo on the right. This simulation illustrates the increased susceptibility to glare by the elderly. The prints were made from color slides prepared by Dr. Leon Pastalan. (From Fozard et al., 1977.)

CHAPTER SUMMARY

Humans typically have 46 chromosomes in every cell in the body. However, in adult humans capable of reproduction each sperm or ovum has only 23 chromosomes. When the sperm fertilizes the ovum and conception occurs, the new organism thus receives its genetic inheritance.

The contribution of heredity to development involves the gene, or unit of heredity. It is believed that each gene, or pair of genes, or group of paired genes, carries a coded message that influences development. This code is carried in deoxyribonucleic acid, or DNA, a chemical found in the nucleus of every cell.

In the view of the authors this genetic inheritance does not provide a direct basis of behavior or development. Genes interact with the environment to shape all structures and functions of the person. This assertion raises the issue of the nature-nurture controversy. Philosophers and scientists have differed in regard to the role in development they associated with nature (heredity, genes, or maturation) and nurture (environment or experience). Evidence is presented that a strong interaction exists between nature and nurture. The same genetic inheritance will lead to different developments when expressed in different environments, and the same environment will have different effects on behavior when it acts on genetically and/or maturationally different organisms. This view of the reciprocal relation between nature and nurture is useful in critically evaluating ideas which stress the role or major role of either nature or nurture variables in behavior and development. For example, the critical periods hypothesis, which stresses the primary role of nature variables, is seen to be problematic because of a failure to appreciate the strong interaction between nature and nurture.

There are various theories of growth and aging. Review of these theories suggests that while the biological changes associated with growth and aging are universal, there is wide variation in the rate of growth and in the amount of decline in aging. In addition, both growth and aging are modifiable by appropriate interventions (e.g., diet and exercise). This modifiability suggests that biology, behavior, and the context of development exist in a complex interaction. Several instances of this interaction were discussed in this chapter: the relation between nutrition and infant and child development; the occurrence of venereal disease in adolescence and young adulthood; the relation between personality and cardiovascular disease in middle age; and the nature of the eye and of visual perception in old age. Each illustration indicated that biological changes in growth and/or aging both influence and are influenced by the behavior of the developing person and by the context within which the person exists.

Part Two

Prenatal and Infant Development

Prior to the advent of the life-span view, an almost unchallenged assumption among developmentalists was in the critical, formative bases of early development for later life. This belief in "connectivity" (Kagan, 1980, in press)—the view that there is a necessary link (a direct connection) between phenomena in early life and those of later development—has been challenged by theory and research conducted from a life-span perspective; this information stresses that change and modification of the life course remain possibilities throughout life and, often, despite the character of events in early developmental periods.

However, one positive feature of the stress on early life was that a considerable amount of important information has been gathered on the nature of development from conception through the end of the years of infancy. An introduction to this information will be presented in the chapters in this section.

It is important to recognize that the life-span view is neither mute toward nor unconcerned with the infancy period. Indeed, proponents of the life-span view often are involved in the study of development in this period (e.g., Belsky, 1981; Belsky & Tolan, 1981; Lamb, 1978). However, their approach to the study of infancy is to see the child at this age in context—that is, as a product and a producer of the world around him or her. For example, the effects of an infant on his or her mother, father, and siblings are as much studied as the more traditionally studied reverse side of this situation, i.e., the effects of the family on the infant.

In addition, the life-span perspective suggests that one study infancy from the vantage point of social change and from a multidisciplinary perspective. In what ways have our understanding and treatment of infants changed over the years? How has such change affected the nature of infant development? In turn, how have changes in the ways disciplines deal with fetuses or infants—for example, in obstetrical and pediatric medical specialties—affected the treatment of the developing fetus, the parents, the nature of the way infants are born, and the sorts of medical treatments they undergo in their early weeks and months? Have changes in these practices affected developments other than those involved with the infant's health? Have the infant's behavior or social relations been influenced?

Information relevant to these issues, as well as information about the nature of basic physical, cognitive, personality, and social processes of development during the years of infancy, is presented in the chapters in this section. Our interest is in indicating that the infant is an active, interactive organism, competent to engage his or her world and to as much affect those in it as be affected by them.

chapter 4

Prenatal Development and Birth

CHAPTER OVERVIEW

Each human life span starts at conception—when a sperm fertilizes an ovum. Typically, a human infant is born nine months later. In this chapter we will describe the events surrounding conception and prebirth development. We will also discuss the characteristics of the birth experience.

CHAPTER OUTLINE

PRENATAL DEVELOPMENT

Conception
The Period of the Ovum (Zygote)
The Period of the Embryo
The Period of the Fetus
Influences on Prenatal Development

THE BIRTH EXPERIENCE: THE PERINATAL PERIOD

Effects of Obstetrical Medication
Effects of Prematurity
Postmaturity

LABOR AND DELIVERY

The Prelude to Labor
The Onset of Labor
First-Stage Labor
Second-Stage Labor
Third-Stage Labor
Types of Delivery

CHAPTER SUMMARY

ISSUES TO CONSIDER

What are the three periods of prenatal development? What are the key features of development within each of these periods?
What are the major influences on prenatal development?
What are the assets and potential problems associated with the use of obstetrical medication?
What are the effects of prematurity on infant development?
Why is postmaturity a concern for physicians?
What are the characteristics of the onset of labor?
What are the key events within each of the stages of labor?
What are the types of delivery?
What are the reasons for cesarean delivery?

A human life span begins with the uniting of the largest cell in the human body—the germ cell of the mother, the ovum—with the smallest cell in the human body—the germ cell of the father, the sperm. When such uniting occurs—that is, when *fertilization*, or *conception*, takes place—about 280 days will pass before the new human is born. Yet developmental changes occur from the moment of conception on. These changes are influenced by changes in the world surrounding the unborn child—for example, by events affecting the mother. In turn, changes associated with the unborn child affect the mother as well. Thus by the time the infant's prebirth period of development is over—that is, by the time this period of *gestation* is completed—significant interactions between the infant and his or her context will have occurred. Accordingly, in this chapter we will focus on the prebirth (i.e., *prenatal*) events that characterize the earliest portion of a new human life span.

PRENATAL DEVELOPMENT

Conception

As noted in Chapter 3, there are two general types of cells in the adult human body. Somatic cells normally contain 46 chromosomes, arranged in 23 pairs. The germ cells—ova of females and sperms of males—contain only 23 chromosomes. These germ cells are the ones that unite to form a new human organism. Thus when a sperm fertilizes an ovum, this new organism—termed a *zygote*—will have the 46 chromosomes typical of the somatic cells in the human body (*soma* means body).

At conception the zygote receives half its complement of chromosomes, half its genes, from each parent. Each human genotype is therefore a product of a 50 percent contribution of each parent. The zygote is formed as a one-celled organism, but through a process termed *mitosis*, or cell division, new cells are formed. However, created from the same initial "mold," all these new cells have the same chromosomes within them.

As soon as the zygote is formed, development begins. The changes that occur for the next 9 months (or 40 weeks, or 280 days, on the average) are often rapid and are quite significant. These changes, occurring within three periods, move us from a one-celled zygote to a sensing, perceiving, cognizing, acting individual (cf. Meredith, 1975).

The Period of the Ovum (Zygote)

As noted, the ovum is the largest cell in the human body, but it is still quite small; its diameter is only $\frac{1}{75}$ inch! However, from the moment of fertilization on, changes begin to occur. Indeed, mitosis begins within a day or so after fertilization.

Taking about ten to fourteen days, the period from fertilization until the zygote becomes attached to the uterine wall of the mother is labeled the period of the ovum, or zygote. At the end of this period, the zygote is made up of several dozen cells and has developed tendrils by which it will attach to the uterine wall. When it does so, the next period of *in utero* (within the uterus) development begins.

The Period of the Embryo

Once implantation occurs, development is quite fast. Indeed Hetherington and Parke (1979) note that from the time of fertilization to the end of the embryonic period the organism increases in size 2 million percent!

119

The rapidly increasing number of cells begins to differentiate in this period. Three distinct layers of cells are formed. The outer layer of cells is termed the *ecto-derm*. From these cells the skin, nervous system, hair, and nails will eventually develop. The middle layer of cells is termed the *mesoderm*. From these cells the muscles and bones of the body eventually develop. In addition, the circulatory and excretory systems develop from these cells. The inner layer of cells is the *endo-derm*. These cells eventually develop important components of the gastrointestinal system, the liver, lungs, several glands, and adipose tissue.

Also developing during this period is a membrane covering the developing embryo—the *amniotic sac*—and this sac is filled with amniotic fluid. The fluid acts as a cushion against shocks or shoves the embryo may experience when, for instance, the mother may fall or bump into something.

Also formed in this period is the *umbilical cord*. It is an organ connecting the embryo to the uterine wall. The site of connection is termed the *placenta*. The function of the umbilical cord is to serve as a passageway for certain chemicals from mother to infant and from infant to mother. However, because there is a membrane in the cord that does not let all substances pass between mother and infant, the connection between the mother's bloodstream and that of the infant is indirect. For example, the mother's blood cells cannot pass through, but certain vitamins, drugs, or disease-carrying organisms can. Thus, as we will explore in more detail below, the health of the mother, and certain chemicals she may ingest, can affect the healthy development of the embryo.

The period of the embryo lasts from about the end of the second week after conception through to the end of the tenth week. As suggested above, during these eight weeks many significant developments occur. By the third week after conception, the head and the tail of the embryo can be seen, and the heart is formed and begins to beat. By the fourth week after conception, the embryo reaches a length of 0.2 inch. Despite this relatively small size, the mouth, liver, and gastrointestinal tract can be seen. By the fifth week after conception, arms and legs can be identified, as can the beginning of eyes.

The embryo reaches a length of about 1 inch by the eighth week after conception, and the face, mouth, eyes, and ears are well formed and identifiable. The formation of sex organs begins at this time, and fingers and toes can be identified. During the ninth and tenth weeks after conception, the muscles and cartilage develop, and organs like the pancreas, lungs, and kidneys can be seen. The liver produces red blood cells during this time.

The Period of the Fetus

This last period of prenatal development lasts from about the end of the second month after conception until birth. Further growth and differentiation of the bodily organs and systems developed in the embryonic period continue here. In addition, these organs all begin to function. For example, the fetus shows motor (muscular) behavior: it moves its head and trunk and it can even respond to tactile (touch) stimulation. Spontaneous movements of the arms and legs occur by the end of the third month.

Mothers in their fourth or fifth month of pregnancy can often recognize this movement, labeling it as "fluttering." By the sixth month and beyond mothers experience easily recognizable "kicks." Indeed, if the infant does not show such "expected" behavior, the mother can become quite anxious about the infant's health. Thus the infant in this way can affect the mother. In turn, if the mother is made anx-

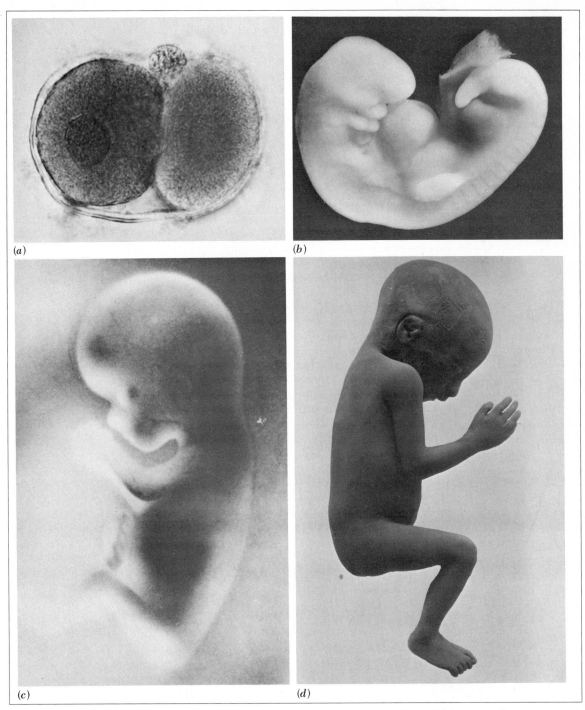

The period of the fetus involves an array of important changes. (*a*) A two-cell embryo; (*b*) 28-day embryo; (*c*) 3-month fetus; (*d*) 6-month fetus. (*a* and *b*: Carnegie Institution of Washington, Department of Embryology, Davis Division; *c*: Wide World Photos; *d*: Russ Kinne/Photo Researchers, Inc.)

ious, her adrenal glands can produce certain chemicals which can enter the fetal bloodstream through the placenta and adversely affect the infant. Thus the fetus's effect on the mother can feed back on itself—a circular process we discuss in Chapter 6 in our presentation of reciprocal socialization between infant and mother.

Throughout the fetal period numerous major developments occur. For example, by two months after conception, the reproductive system begins to develop, and during the third month after conception, eyelids and nails begin to form. During this month, the fetus reaches a length of about 3 inches and weighs about 0.75 ounce. Similarly, the fourth and fifth months after conception are times of rapid changes. During the fourth month after conception, the fetus attains a length of about 4.5 inches. But by the time the fifth month has ended, the fetus reaches a length of about 10 inches (and a weight of about 8 ounces). Moreover, during the fourth month, some expectant mothers can begin to feel some movement of the fetus—a feeling that we have noted is labeled fluttering, or "quickening." During the fifth month, the activity of the fetus is even more pronounced. Eye blinks occur, the hands can close and grip, and the sweat glands begin to develop. Also during the fifth month, hair appears on the head and body.

The eyes become completely formed and taste buds appear during the sixth month of gestation. The seventh month is an important transitional one for the fetus. During the seventh month, the fetus reaches a length of about 16 inches and a weight of about 4 pounds. These values represent about 80 percent of the final length attained by most fetuses and more than 50 percent of the final weight. If born prematurely—that is, before the complete nine-month term of a typical pregnancy—the fetus is *viable;* it can live if born.

During the eighth month of gestation, the typical fetus attains a length of about eighteen inches and a weight of about 5.5 pounds. During the ninth month, the typical fetus completes its in utero growth, reaching a length of about 20 inches and a weight of about 7.0 to 7.5 pounds. Also during the ninth month, the head of the fetus typically settles in a head-down position. The top of its head moves down toward the expectant mother's cervix, and its face is directed toward the expectant mother's back. When the head of the fetus settles into this position, it is said that the head is *engaged.*

As discussed later in this chapter, not all fetal heads become engaged in this way. Some infants are born with their faces directed toward the front of their mothers. In other cases the head does not engage for delivery. The fetus may settle into a feet-first position. As will be explained below, this is a situation requiring a breech delivery. In addition, some fetuses settle in a transverse, or crosswise, position in the uterus. Today, such fetuses are typically not delivered vaginally. Instead, a cesarean delivery is performed. We will discuss such a procedure in detail below.

We have been describing typical, or normative, events of prenatal development. However, it is clear that these generalizations do not apply equally to all cases. There is variation in the nature and outcomes of prenatal developments. This variation has been linked to several types of variables that exert an influence on the mother and her in utero child. We now turn to a discussion of these influences.

Influences on Prenatal Development

We have noted that the mother and the infant are reciprocally related. In addition, we noted that as a consequence of certain chemicals passing through the placental wall and into the umbilical cord, events and phenomena in the environment, outside the amniotic sac, surrounding the infant can affect its development. Delivered through the mother, these effects often involve chemical agents that are harmful to the infant. Such agents are termed *teratogens;* often they are associated with physical malformations and with behavioral disorders. Some of these teratogens, in the form of obstetrical medication, will be discussed in greater detail in a later section of this chapter. There are also other characteristics associated with the mother, for example, pertaining to her age, that can affect the infant's prenatal development. However, it is important to note that the time at which these potential influences act, in regard to when in the in utero developmental sequence they occur, plays an important role in their effect on the developing organism.

Effects of Maternal Nutrition. Poor nutrition in mothers is associated with several negative developments, both prenatally and postnatally. Mothers who have poor diets during their pregnancy tend to be in poorer general health during this time than are mothers on good diets (Mussen, Conger, & Kagan, 1979). But the mother's diet is associated with effects on her baby. Mothers with poor diets (e.g., little protein, a lot of fat) more often have miscarriages, premature births, stillbirths, longer labors, babies more likely to die in early infancy, and birth conplications such as anemia and toxemia (a disorder involving high blood pressure) than is the case with mothers having good diets (Hetherington & Parke, 1979; Mussen et al., 1979; Tompkins, 1948). In addition, babies born to mothers who have chronically poor diets (e.g., a long history of low protein intake) are more likely to have serious diseases and to have problems with nervous system development than is the case with babies born to mothers who are not chronically malnourished.

The newborns of poorly nourished mothers are also likely to be malnourished. Indeed, there are several maternal factors associated with malnutrition in the fetus. These include poor maternal weight gain, lack of prenatal visits to a physician or midwife, chronic major illness, and obesity (Miller & Hassanein, 1974). In one study two or more of these factors occurred in 51 percent of pregnancies involving malnourished infants (Miller & Hassanein, 1974).

Alcohol and Tobacco Consumption. Estimates are that over 80 percent of pregnant women in the United States drink alcohol and that 57 percent of pregnant women smoke (Hetherington & Parke, 1979). Each of these behaviors is associated with problems, prenatally and postnatally. The rate of premature birth, aborted pregnancies, and low-birthweight babies is higher for mothers who smoke than for those who do not (Frazier, Davis, Goldstein, & Goldberg, 1961). Similar differences are found when mothers who drink are contrasted with those who do not. Moreover, since the early 1970s it has been known that chronic drinking by a pregnant woman can produce a syndrome (or cluster) of malformations termed the *fetal alcohol syndrome* (Jones & Smith, 1973). Infants born to alcoholic mothers have been found to have a high incidence of face, heart, eye, ear, and limb defects, retarded prenatal growth, premature birth, and an abnormally small head. Other behavioral deficits associated with this syndrome include mental retardation, disturbed sleep patterns, excessive irritability, and hyperactivity (Jones, Smith, Ulleland, & Streissguth, 1973; Streissguth, 1977).

Even moderate maternal drinking during pregnancy has been associated with problem behavior in infants (Landesman-Dwyen, Keller, & Streissguth, 1977). When pregnant women both drink and smoke, they further increase the chance of their child's having physical and behavioral problems.

Smoking during pregnancy can have harmful effects on the fetus. (Mimi Forsyth/Monkmeyer Press Photo Service.)

Effects of Drug Use. In the late 1950s and 1960s many women took a drug—thalidomide—in order to increase their fertility. However, the ingestion of this drug produced gross anatomical defects in the limbs of their infants. Similarly, at about this same time many women took the drug *diethylstilbestrol* (DES) in order to prevent miscarriage. However, it has recently been found that daughters of women who took DES have a higher-than-average probability of developing cancer of the cervix when reaching adolescence. In addition, if and when these daughters become pregnant, they have a higher-than-average probability of pregnancy or birth complications. Thus drug ingestion by the pregnant mother can have several long-lasting effects on her offspring—not only during the offspring's infancy but across the life span as well.

Today, many other drugs taken by pregnant women have become suspected of producing physical or behavioral problems in their infants. Less commonly used drugs are known to cause problems too. Mothers who are addicted to heroin or to morphine have infants who are addicted also. Often these infants are of low birthweight or are premature and are much more excitable and irritable than babies born of nonaddicted mothers (Brazelton, 1970; Strauss, Lessen-Firestone, Starr, & Ostrea, 1975).

Maternal Diseases. Contraction of virus-caused diseases by the pregnant woman can produce severe defects in the infant. For example, even a mild contraction of *rubella* (or German measles) in early pregnancy can result in heart defects, deafness, cataracts, blindness, or mental retardation (Illingworth, 1975). Toxemia, a disorder involving high blood pressure, retention of fluid, and rapid and high weight gain, results in death of 13 percent of pregnant mothers who contract it; about 50 percent of the unborn infants whose mothers contract this disease die (Illingworth, 1975; Lubchenco, 1976). Children who survive have an increased likelihood of mental retardation.

When mothers and their infants have differences in a particular chemical in their bloodstreams—termed *RH factor*—complications can arise as a consequence of this incompatibility. The mother's blood can produce chemicals which interfere with the unborn infant's ability to carry oxygen in its own blood. Such a situation can result in effects ranging from anemia through mental retardation or death. This problem rarely occurs in firstborn children, because it takes several pregnancies for the mother's blood to build up sufficient chemicals to affect the infant's blood. In addition, mothers who may have an Rh factor incompatibility with their infants can now be treated with a particular chemical that appears to eliminate the problems. Injection of this chemical within seventy-two hours after a delivery (or a miscarriage) can avoid the problems of Rh incompatibility in future pregnancies.

Maternal Emotional States. Earlier we noted that the infant's in utero behavior can affect the mother's emotions; if the baby does not "kick" when the mother expects it to, she may become quite concerned about the baby's health. We noted also that as a consequence of her concern certain chemicals (acetylcholine, for example) may be produced by the mother, enter her bloodstream, and have a negative effect on the infant. Even though there is no direct neural or blood connection between the mother and the in utero infant, the mother's emotional state can influence the infant's prenatal environment and development.

Emotional disturbance during pregnancy is associated with prematurity, abortion, prolonged labor, delivery complications, and infant hyperactivity, irritability, sleep problems, and irregular eating (Despres, 1937; Sontag, 1941; Ferreira, 1969; Joffe, 1969). Sontag (1941, 1944) reported that fetal body movements increased vastly if mothers were experiencing emotional stress.

Maternal Age. The age of the mother when pregnant is associated with differences in the likelihood of prenatal and postnatal problems. Although the incidence of infant mortality in the United States has decreased in recent years—from 47 per 1000 births in 1940 to less than 17 per 1000 births in 1980—about 6 percent of first babies born to mothers under the age of 15 years die in the first year of their lives—a rate exceeding by 2.5 times that for first-time mothers in their early twenties (U.S. Bureau of the Census, 1980). Mothers under 15 years of age also face a greater risk of dying themselves; their rate of death is 60 percent higher than that for mothers in their early twenties. Walters (1975) notes that if infants are born to adolescent mothers, the chances of their having birth defects are much greater than for infants born to postadolescent mothers.

In turn, women who deliver their first baby when over 35 years of age have much greater likelihoods of having more problematic pregnancies, labors, and deliveries than is the case for younger women. However, it is important to point out that most mothers—at all ages—will have normal, uncomplicated pregnancies and deliveries. Although factors such as age and teratogens (alcohol and heroin) greatly increase the likelihood of problems, the majority of mothers do not encounter these problems.

Teratogens Acting through Males. Throughout this section we have emphasized influences on prenatal development associated with the mother. Thus chemical agents harmful to prenatal development—i.e., teratogens, as encountered when expectant mothers smoke, drink excessively, or take certain drugs—have been seen to affect the fetus and infant. On the other hand, however, it is rarely the case that the husbands or partners of the expectant mothers are cautioned to avoid these substances because of a potential negative influence on the unborn child's prenatal development. Yet there is some evidence that among males who are exposed to certain teratogens *before* conception takes place, there is a greater incidence of birth defects among their offspring (Kolata, 1978).

Nonhuman male animals (e.g., rabbits or rats) who are given drugs, such as thalidomide or methadone, before breeding have offspring having a greater-than-normal incidence of death, birth defects, or abnormal behavior (e.g., Lutwak-Mann, cited in Kolata, 1978). Similarly, there is some evidence that human males exposed to certain drugs (e.g., anesthetic gases encountered by men who work in hospital operating rooms) have wives who have significantly increased rates of spontaneous abortions, and have babies who are more likely to have congenital defects, than is the case with men not so exposed (Kolata, 1978).

There is no certainty how such effects operate. Some hypotheses are that the teratogens damage the sperm cells or that the teratogens are excreted in the male's semen and then enter the female's circulatory system by passing through her vaginal walls. Although these and other hypotheses are likely to be the object of future research, some commentators believe that, however the influence is eventually found to operate, teratogens delivered through the male can cause a clinically significant number of birth defects (Kolata, 1978).

THE BIRTH EXPERIENCE: THE PERINATAL PERIOD

At the end of about 280 days labor begins. The amniotic sac breaks, and with it the birth process is christened. The process ends with the delivery of the infant—the *neonate* (meaning newborn). Although human developmentalists have always shown concern about the events within this neonatal period, which is considered to last for about seven to ten days, a great deal of attention has recently been devoted to the relatively brief period surrounding birth, the *perinatal period.* The events involved in (at least) the latter stages of labor and in the delivery itself are those that comprise the perinatal period.

The vast majority of infants do not experience any complications or suffer any serious impairment as a consequence of their birth experience. Less than 10 percent of all infants (the rate is slightly higher in males and lower in females) have any abnormalities, and many of these disappear, in any event, over the course of development (Hetherington & Parke, 1979). Nevertheless, the degree of difficulty associated with the delivery and the time it takes for the infant to start to breathe, once born, can affect its development. Although all infants experience some degree of oxygen deprivation (anoxia) during the birth process, it is only when such loss is severe that problems (e.g., brain or intellectual defects) can result. Mild anoxia is associated with neonatal irritability (Graham, Matarezzo, & Caldwell, 1956) and some other mild motor disturbances in later infancy; but by the middle childhood such effects tend to disappear (Mussen et al., 1979). Thus Gottfried (1973) notes that if intellectual impairment resulting from anoxia is found, it is seen more often among infants and preschoolers than in older children. Accordingly, while anoxia may increase the probability of being mentally retarded, anoxic children as a group are not mentally retarded, and there are no known deficits in specific intellectual abilities (Gottfried, 1973).

The physical and behavioral conditions of the newborn in the perinatal period can be measured by use of a system developed by pediatrician Virginia Apgar. An *Apgar score* is given an infant one minute and five minutes after birth. The attending obstetrician or nurse rates the infant in the following way: *heart rate* (0 = absent; 1 = less than 100 beats per minute; 2 = 100 to 140 beats per minute); *respiratory effort* (0 = no breathing for more than one minute; 1 = slow and irregular breathing; 2 = good breathing with normal crying); *muscle tone* (0 = limp and flaccid; 1 = some flexion of the extremities; 2 = good flexion, active motion); *body color* (0 = blue or pale body and extremities; 1 = body pink with blue extremities; 2 = pink all over); and *reflex irritability* (also rated on a 0, 1, or 2 scale).

As may be obvious, the higher the score the better the physical and behavioral condition of the infant. Apgar scores of between 7 and 10 are indicative of a good condition, a score of 4 is indicative of possible difficulties, and a score of 3 or below is an indication that the infant's survival may be threatened; immediate intervention in such a situation is required. Apgar scores may be related to development even beyond the early infancy period. Serunian and Broman (1975) report that infants with 0 to 3 Apgar scores at one minute had significantly lower eight-month mental and motor scores than infants with 7 to 10 scores and significantly lower mental, but not motor, scores than infants with Apgar scores between 4 and 6.

Furthermore, although quite tentative, there are some reports that perinatal events can influence developments well into childhood. For example, Balow, Rubin, and Rosen (1976) point to a possible link between perinatal problems and reading disability in school-age children. Indeed, often babies born at risk do not survive. Box 4.1 presents details about the syndrome which is the largest cause of infant deaths. In the next section other links between specific chemicals encountered during the perinatal period and later development are discussed.

Effects of Obstetrical Medication

In the United States most infants are delivered by physicians in hospital settings. Labor and delivery are treated as medical situations. It is the norm in this country to provide medication to the mother, to ease her pain and to sedate her during delivery. As noted in Box 4.2, such obstetrical medication may act as a teratogen. That is, recent research has begun to indicate that obstetrical medication, especially when delivered in from moderate to heavy doses, may have effects on the newborn (Aleksandrowicz & Aleksandrowicz, 1974; Vander Maelan, Strauss, & Starr, 1975; Yang, Zweig, Douthitt, & Federman, 1976). Some short-term effects that have been associated with such medication are decreased attentional capacity (Stechler, 1964), disruption and/or potentially problematic changes in reflex activity (Aleksandrowicz, 1974; Brackbill, 1977; Brazelton, 1961; Conway & Brackbill, 1970; Moreau & Birch, 1974), and general neonatal depression (Schnider & Moya, 1964).

Differences in the potency of drugs used during labor and delivery lead to later differences in stimulus-elicited heart-rate response (Brackbill, 1977). Heart-rate deceleration is a useful index of the motor-orienting reflex, often regarded as a precursor of information processing. Heart-rate acceleration is associated with a "defense reflex." In the context of repeated auditory stimulation at 8 months, infants born without medication tended to show the heart-rate deceleration associated with the orienting reflex, while those born with high medication levels (general or regional

BOX 4.1

SUDDEN INFANT DEATH SYNDROME

Sudden Infant Death Syndrome (SIDS) is the largest single cause of death during infancy, and it accounts for approximately one-third of all infant deaths between the ages of 1 week and 1 year (Beckwith, 1977). It is second only to accidents as the largest killer of children under age 15 (Bergman, Ray, Pomeroy, Wahl, & Beckwith, 1972). The National SIDS foundation distributes literature which reports that *more children die each year of SIDS than of cystic fibrosis, childhood cancer, child abuse, and childhood heart disease combined.* Sudden Infant Death Syndrome is a specific disease entity, defined as "the sudden death of any infant or young child, which is unexpected by history, and in which a thorough postmortem examination fails to demonstrate an adequate cause for death" (Beckwith, 1970). Although it has been suggested that its incidence may recently be decreasing (Valdes-Dapena, 1977), it nonetheless has been estimated by the National SIDS Foundation as well as by other sources that between 8000 and 10,000 babies die of SIDS in the United States alone each year (Beckwith, 1977). The main epidemiological factors associated with SIDS are that SIDS: (1) affects babies between 1 month and 1 year with a peak incidence between 2 and 4 months of age; (2) occurs predominantly in the cold winter months; (3) has a higher incidence in prematurely born babies and in those from low socioeconomic groups; (4) is frequently associated with a cold which is often preceded by fussiness, irritability, and sleep deprivation, but the infant is otherwise presumed to be healthy; (5) results in its victims being suddenly and unexpectedly discovered dead in their cribs where they are presumed to be asleep (usually in the early morning hours); and (6) involves no sound, cry, or stridor being heard; rather, the death is silent. In the continuing search for etiologic factors in SIDS, recent evidence has shown that several babies who subsequently died of SIDS presented atypical developmental characteristics in the early neonatal period. In studies of SIDS infants and their normal siblings, temperamental characteristics in SIDS infants have been suggested to be important, as have certain additional physical and behavioral characteristics. These include differences in cardiac activity; lower responsiveness to auditory and visual stimuli; abnormal crying, breathing patterns, and vocalization; and abnormal behavior characteristics, particularly a lack of stability in state behaviors (Thoman, 1975; Steinschneider, 1975). However, it should be noted that at present these data do not establish an etiologic link between certain neonatal abnormalities and SIDS. Rather, they are indicative of the fact that there may be one or more detectable physical or behavioral anomalies which are problematic for SIDS, contrary to the earlier view that SIDS infants are entirely normal.

anesthesia) changed from initial deceleration to the acceleration associated with the defense reflex (Brackbill, 1977).

Such drug-behavior relations appear to depend on the levels of obstetrical medication administered (Horowitz, Ashton, Culp, Gaddis, Levin, & Reichmann, 1977). Light levels of medication did not have significant effects on neonatal behavior in Uruguayan or Israeli infants, but significant early behavioral differences were observed between no-drug and light-drug Israeli and Uruguayan samples as compared with a Kansas group of infants whose mothers received medication typical of that administered during most labors in the United States—i.e., moderate to heavy medication or general anesthesia. The magnitude of differences between the Kansas high-drug infants and those from both unmedicated or light-drug American, Israeli, and Uruguayan groups was the largest on comparisons of visual and auditory orientation, alertness, consolability, and self-quieting behaviors as measured by the

Brazelton (1973) Neonatal Assessment Scale (Horowitz et al., 1977). All these behaviors may reflect coping ability in response to stress. Additional evidence for the possibility that there are long-term effects after high doses of obstetrical medication is discussed in Box 4.2.

Effects of Prematurity

If an infant is born before 26 weeks of gestation have been completed and if it weighs less than 4.5 pounds, it has a lowered chance of survival. Length of prebirth gestation and of birthweight are highly related. As the two variables increase, so that they both approach the normative levels of 40 weeks and 7 or so pounds, the dangers of prematurity lessen. Indeed, infants born only slightly premature (between 34 and 38 weeks of age) resemble full-term babies to a great degree. Moreover, even for those babies born between 28 weeks and 33 weeks, recent medical interventions have enhanced their chance of survival. Most of these infants go on to develop quite normally, both physically and behaviorally. Nevertheless, it is still the case that extreme prematurity and/or low birthweight is associated with a greater-than-average likelihood of several disorders, for example, low IQ (Caputo & Mandel, 1970).

Indeed, low birthweight in and of itself is associated with both short- and long-term problems. Several studies report negative associations between low birth-

BOX 4.2

ARE THERE LONG-TERM EFFECTS OF OBSTETRICAL MEDICATION? DO OBSTETRICAL MEDICATIONS ACT AS TERATOGENS?

As noted, Brackbill and her colleagues have been the scientists most recently involved in trying to answer these questions. Brackbill estimates that women are given medication in 95 percent of all labors and deliveries in the United States. The research of Brackbill and Broman has provided evidence that such a situation may have negative effects on most of the children in this country. That is, they report evidence that obstetrical medications affect children's behaviors.

Brackbill and Broman obtained data from 3500 healthy full-term infants born to women taking part in the Longitudinal Collaborative Perinatal Project. They studied only very healthy women who, in addition, had the most uncomplicated pregnancies, labors, and deliveries and who, furthermore, had the healthiest infants in the sample. They chose this highly healthy group to minimize the possibility that if effects associated with taking obstetrical medications were seen, they could be alternately attributed to the mother's or baby's lack of health and not to the drugs.

Brackbill and Broman report that even in this highly healthy group there was an association between obstetrical medications and children's behavior at least up through 7 years of age. Moreover, the greatest associations were seen in the children of women who received the strongest drugs or the highest total doses. What were some of the associations that Brackbill and Broman identified?

When tested at 4, 8, and 12 months of age, the infants of the more heavily medicated mothers showed lags in their development of the ability to sit, to stand, and to move about (in comparison to the children of less heavily medicated mothers). The former group also showed lags in their ability to inhibit certain responses and in their tendency to stop crying when comforted and to stop responding to distracting stimuli. At older age levels the children of the heavily medicated mothers also showed a lag in language development and other cognitive skills.

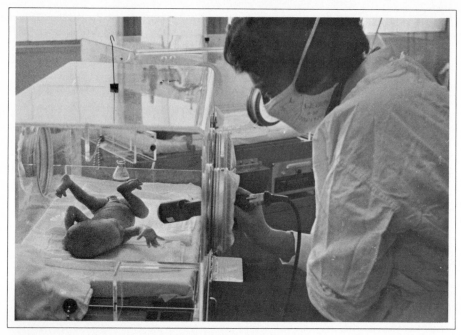

Infants born prematurely require special, intensive care.
(Lew Merrim/Monkmeyer Press Photo Service.)

weight and cognitive and learning disabilities in school and between low birth-weight and growth retardation during infancy (e.g., Francis-Williams & Davies, 1974; Rubin, Rosenblatt, & Balow, 1973). Nevertheless, as with prematurity, early interventions—here involving intensive care and extra stimulation—can improve the development of low birthweight infants and avoid the chance of serious handicap (Powell, 1974; Scarr-Salapatek & Williams, 1973; Stewart & Reynolds, 1974).

Postmaturity

Although most research attention has focused on the effects of premature birth on later psychosocial development, there is increasing concern with the implications of postmaturity. A postmature birth occurs when the infant is born at a point after its optimal gestational period has occurred. About 10 percent of all babies are postmature, i.e., born more than two weeks after the expected date of delivery (the "due date").

It is now known that the placenta may begin to lose its efficiency or even break away from the uterine wall, even though the fetus has not been born. Such a problem may occur in the postmature situation. In such a case the fetus would be in danger—due to a lack of oxygen or nutrients. Indeed, it is often the case that postmature babies weigh less than they would have if they had been born after exactly nine months. As a consequence of the potential danger of postmaturity, today many obstetricians do not allow women to go more than about two weeks beyond their expected date of delivery without intervening.

Postmaturity has been associated with several problems in delivery and thereafter. Postmature babies often experience more difficult deliveries than babies born about "on time," have a higher frequency of mortality than these other babies, and, similarly, are more likely to experience neurological or some behavioral (e.g., eating) problems (Illingworth, 1975; Lubchenco, 1976).

LABOR AND DELIVERY

We have emphasized that the time span from conception to birth is typically about nine months. However, the relatively brief time before the end of this span—the segment involving labor and delivery—is often an extremely significant one for the parents of the to-be-born infant. Indeed, today, many parents are taking childbirth education classes to learn what to expect and how to prepare for labor and for delivery. Education involves learning the implications of the variations that can occur in this period for the mother, for the infant, and for the father.

And there is great variation! No two labor and delivery experiences are exactly alike—even when they are gone through by the same mother at two different times in her life. In this section we will give an overview of some of the more general characteristics of labor and of delivery. Details about what is involved in childbirth education are found in Box 4.3.

BOX 4.3

CHILDBIRTH EDUCATION AND NATURAL CHILDBIRTH

"Natural childbirth" is often a misleading term. Having natural childbirth does *not* mean having a baby in a primitive setting or in a setting unattended by medical personnel. Indeed, natural childbirth may occur in a modern hospital and may involve the administration of medication to the mother as well as the administration of other forms of medical intervention. Natural childbirth has come to represent a means by which parents prepare for the birth of their child by experiencing education (e.g., in what is involved in labor and delivery) and special training (e.g., in breathing and relaxation techniques designed to help the woman through the pains of her contractions). In addition, natural childbirth has come to mean involving the father in these educational and training experiences so that he can assist the mother during labor and delivery. Often these preparatory experiences occur in the context of Childbirth Education Association classes offered in many communities throughout the United States.

The current meaning of natural childbirth has evolved over many years. Originally, the term referred to the method promoted by an obstetrician, Grantly Dick-Read. Dick-Read believed that

women learned that birth was a painful experience. He believed that by instructing women in the facts of labor and in techniques of relaxation, they could avoid both pain and medication.

Although this technique worked well for many women, others found that passively letting labor occur did not eliminate their pain sufficiently. While some women thus opted for medication, others turned to techniques promoted by Lamaze. The Lamaze method, often termed psychoprophylaxis, instructs the woman in various breathing techniques that are useful in allaying the pain at various stages of labor. With these breathing techniques the woman can play a more active role in getting through the pain of her labor. In addition, the husband is also encouraged to learn the breathing techniques so that he can serve as a coach for his wife.

Medication, when necessary, is accepted as an appropriate part of labor experience in this method. In addition, cooperation with the attending medical and nursing staff is encouraged, so that a healthy baby may be born and positive experiences by the entire family may be achieved.

Of course, learning breathing techniques does not mean that relaxation methods have to be

Childbirth education classes aid both parents in learning about the birth process.
(United Press International, Inc.)

eliminated. Indeed, many childbirth education classes encourage both techniques to be used simultaneously.

Finally, it must be emphasized that no matter what natural childbirth techniques are used, the woman will experience some degree of discomfort. However, if both the mother and the father work hard at their preparations, e.g., by practicing their breathing and relaxation techniques regularly, then they may feel more confident that they have some skills to use in this trying but exciting experience.

The Prelude to Labor

There are various physical and emotional reactions to labor. No mother experiences all the reactions that another mother does. Moreover, if medication. or other medical intervention, occurs, reactions may be altered.

About two weeks before the birth of first babies, an experience called *lightening* occurs. The fetus drops into its position in the uterus, typically moving its head closer to the cervix. Such lightening may not occur until the onset of actual labor in other than first deliveries.

Often throughout the last trimester of pregnancy there are contractions of the uterus. However, these contractions, termed *Braxton-Hicks contractions*, are not those that occur in true labor. Any discomfort associated with these contractions (and there may not be any) may usually be relieved by movement on the part of the woman. During the prelude to the labor period, there is an increase in the number of Braxton-Hicks contractions.

During this period, there is also an increase in discharges of vaginal mucus and there is an increased feeling of pressure in the pelvic region. The fetus may also be less active during this period than it has been in previous weeks. The woman may also have a weight *loss* of 2 to 3 pounds, typically about three to four days before labor, and she may experience a spurt of energy one to two days before labor begins. In addition, the woman may experience menstrual-like cramps several days before actual labor begins.

The Onset of Labor

Any one or a combination of several events can signal the onset of true labor. One sign is the beginning of rhythmic contractions. These contractions will become progressively more frequent and will not be terminated by changes in position (as is typically the case with Braxton-Hicks contractions). These true contractions may feel—at first—like menstrual cramps, gas, backache, pelvic pressure, or tightening by the pubic bone.

Another sign of the onset of labor is what is often termed a *bloody show.* This sign occurs when the mucus plug, blocking the opening of the cervix from the vaginal canal, is lost. The cervix, beginning to become *effaced,* or thinner, and to dilate as labor commences, is now too large to keep the mucus plug in its former place. The plug becomes dislodged, and as it does, small blood vessels surrounding it are ruptured; consequently, there may be a good deal of light-colored blood and mucus in the urine.

A final sign of the onset of labor may be the rupturing of the amniotic sac surrounding the fetus. Often, this is referred to as the breaking of one's water. Indeed, many women experience the loss of a good deal of fluid with this event. For others, however, this clear, warm fluid is lost in small amounts.

As noted, not all of these signs need occur as labor begins. Indeed, many women do not have the amniotic membranes rupture spontaneously, although they are well into their labor. Often the attending physician or nurse-midwife will intervene to rupture the membranes. Frequently, this is done to speed up the process of labor and delivery. However, the chance for infection is greater once these membranes have been ruptured, and current obstetrical practice is to deliver the infant within twenty-four hours after the membranes are broken. Often, then, this intervention is an important step in the labor and delivery experience. If the mother's labor is not progressing satisfactorily, in ways we will describe below, and/or a vaginal delivery appears to be problematic, the physician and the parents may have to decide to have a cesarean delivery in order to ensure the safe birth of the infant in the appropriate period of time.

We will return to a discussion of the types of delivery below. Here we need to continue our presentation of the process of labor. Our discussion will be based on a description of the events leading up to a normal, vaginal delivery. More than 80 percent of all births in the United States are vaginal ones.

First-Stage Labor

This first stage is the longest, usually lasting from ten to fourteen hours (and in some cases it may be even longer). There are substages that may be identified within this stage.

Early First-Stage Labor. Regularly spaced contractions become established during this period. The cervix continues to become effaced (thinned), and, typically, it dilates (opens) from its initial closed state to about 4 centimeters.

The woman may experience lower backache or abdominal cramps. Often, the bloody show or the rupturing of the amniotic sac may occur in this period. Frequently, women are quite alert during this portion of labor and may want to walk, talk, or read between contractions. However, as the labor progresses women typically become quieter, less communicative, and more centered on their labor experience.

Middle First-Stage Labor. During this portion of labor, contractions become stronger and longer, and they begin coming closer together. For example, contractions may last fifty seconds, and a three-minute interval may separate the beginning of one contraction and the beginning of the next (contractions are measured by the time interval between the start of one contraction and the start of the next, and *not* by the time between the end of one and the start of the other). The cervix continues to efface during this period, and dilation of the cervix may be from 4 to 8 centimeters.

Women often become more serious about their experience during this period. Since the pain of the contractions increases during this period, this is often the time within which medication may be administered.

Transition. This is the last portion of the first stage of labor. It is the shortest phase of all of the labor period, typically lasting for no more than an hour or two. The cervix is usually completely effaced during this period, and it dilates from 8 to 10 centimeters (usually the point when the attending physician or nurse-midwife will allow the woman to begin to try to push the baby out). Contractions during this period can be quite long and strong and often may seem continuous.

The woman may experience an urge to push during this phase. Often this is experienced as an extreme pressure in the rectum. However, if the cervix is not completely dilated (i.e., 10 centimeters), the woman is asked to counteract this urge to push (e.g., by rapidly blowing or puffing air out of her mouth until the urge passes). If the woman should push before she is completely dilated, the cervix may become swollen as the baby's head pushes against it. If so, this will decrease the cervical opening and prolong labor.

In addition to the contractions other physical characteristics may typify the transition phase. Women often experience one or more of the following: belching, hiccups, nausea, vomiting, perspiration, trembling legs, leg cramps, feeling hot and cold simultaneously, or restlessness. Furthermore, many women become frustrated by the intensity of the contractions. They are unable to relax, are often quite irritable, and may make inappropriate requests (e.g., "Take me home; I've had enough of this."). Obviously, this is the most difficult, uncomfortable period of labor. Indeed, many women comment that the presence of this stage is why the process is called labor!

Second-Stage Labor

In this stage of labor the baby is born. Contractions during this period may temporarily stop and become less frequent. But although they may be further apart, they are still quite strong. The woman feels pressure in her back and rectal areas. With the cervix completely effaced and dilated, however, the woman may push during this period. Thus the baby's head "crowns"; that is, the top of the head protrudes out of the vaginal opening.

At this point, the woman may experience a warm, burning, and/or stretching sensation as the head is born. If necessary, an *episiotomy* is performed at this point in order to ensure that the tissue surrounding the vaginal opening will not be torn. An episiotomy involves surgical cutting of the skin and underlying tissue under the vaginal opening. After the head is born, the shoulders are delivered, and the rest of the infant's body slides out.

The woman may respond in several ways to her experiences during this stage of labor. Some women act very relieved and satisfied when they are allowed to push. Others find pushing uncomfortable or even painful. In turn, when the baby is born, the mother's reaction can range from excitement to relief or calm.

Third-Stage Labor

Less intense, but still rhythmic, contractions occur in this stage, as the placenta becomes detached from the uterine wall and is delivered. The uterus, which has been greatly expanded over the course of pregnancy, begins to shrink in size during this period and takes a globular shape the size of a grapefruit.

The mother may experience a feeling of fullness in her vagina and may react in many ways to the birth of her child; both laughing and crying have been reported. If the mother plans to breast-feed, the first such feeding may occur at this time.

Types of Delivery

As we have noted, the vast majority of infants are delivered in the same way: through the vaginal (or birth) canal. However, even through this conventional route into the world there is some variation. In addition, a large number of women in the United States do not have vaginal deliveries. Their babies are delivered through cesarean sections. Since, especially for the first baby, it is virtually impossible to be completely certain about the type of delivery one will experience, it is important to know about, and be prepared for, all types of delivery. In this section we discuss the several variations that occur in regard to vaginal deliveries and, in addition, we detail the nature of the cesarean delivery.

Vaginal Deliveries. More than 90 percent of all vaginal deliveries occur with the baby born in a headfirst position. Typically, the baby's head is facing down, toward the ground. However, in about 4 percent of all vaginal births, the baby's buttocks or feet are born first. This is called a *breech birth*.

In both normal headfirst or breech positions, the obstetrician or nurse-midwife will typically not hesitate to proceed with a vaginal delivery. However, some babies are not positioned in either of these two vertical (to the cervix) positions; they are positioned horizontally (or crosswise), which is called a transverse presentation. Today, obstetricians or nurse-midwives will not vaginally deliver a baby lying in such a position; they will try to turn the baby into one of the other two positions. If such an attempt is not successful, the baby will have to be delivered by cesarean section.

There are other variations that may occur with vaginal deliveries. If the health of the mother or of the fetus is in danger if the pregnancy is not terminated (e.g., due to fetal postmaturity) and/or if the labor is not progressing rapidly enough (e.g., the cervix is not dilating rapidly enough, or contractions are not strong enough), the attending medical staff may attempt to *induce* labor. Such induction may involve several procedures. One is to rupture the amniotic membranes. This is done with a long, thin plastic instrument which has a small hook at its end. The physician places the instrument between his or her fingers, placing them up the birth canal, and gently pricks the membranes, causing them to rupture. This procedure is not an especially painful one for the woman and takes only a few moments to perform.

Another way that labor may be induced is by administering a hormone (oxytocin). This hormone acts to initiate contractions of the uterus and/or strengthen relatively weak contractions that may have already begun. If such induction does not lead to a vaginal delivery, there are yet other ways the attending medical staff may achieve one. For example, a forceps delivery may be undertaken. Often, a forceps delivery is done if the fetus is in some sort of danger (e.g., if it is getting too little oxygen or is experiencing too much pressure on the head). Forceps are two pieces of metal, shaped like two large, oblong, and interlocking spoons. The forceps is inserted up the birth canal and fit around the head of the fetus. The physician then pulls the head down and out of the birth canal.

In the past, there was some variation in the types of forceps deliveries that were attempted. There were both high and low forceps deliveries. High deliveries occurred when the head of the fetus had not descended very far down the birth canal; thus the forceps had to be placed high into the birth canal. However, all forceps deliveries are dangerous, because damage can occur to the head and result in permanent brain damage to the infant. High forceps deliveries are especially dangerous in this regard, and as a consequence, today they are not frequently attempted. Instead, the physician may opt for a cesarean delivery. Although involving abdominal surgery, this formerly quite risky operation has today progressed in its safety and efficiency due to modern surgical advances. As such, in cases where there is danger to the mother or fetus if pregnancy or labor continue, and/or when vaginal delivery is problematic, a cesarean delivery is often safe and advisable.

Cesarean Deliveries. A cesarean delivery involves surgery. As such, many of the features of this type of delivery are those associated with general operating procedures. We will discuss the procedures the medical staff undertakes before, during, and after the delivery. Although the procedures involved in cesarean deliveries may seem more uniform than in other deliveries, there are many normal variations.

For example, if a spinal, as opposed to a general, anesthesia, is used, then the mother, numb only from the midsection of her body down, may be awake for the delivery. Thus, as opposed to what happens when a cesarean section is performed under general anesthesia, the mother can see the baby as soon as it is delivered and checked by the physician. In addition, if spinal anesthesia is used, the hospital may allow the father to be present during the entire procedure. Other variations in cesarean section procedures exist. For instance, as we will see below, the type of incision that is made may differ. We turn to the presurgical preparations that are made for delivering by cesarean section.

Preparations for Cesarean Delivery. Once a decision is made to deliver a baby by cesarean section, preparations for surgery take a relatively short time. As little as thirty minutes may be needed (of course, in an emergency things can proceed much more rapidly). Preparation for surgery usually involves shaving the hair from the abdomen and pubic area, taking a blood sample, administering an injection to dry up bodily secretions, and inserting a catheter into the bladder and an intravenous needle (usually into the arm). In addition a consent form for the surgery has to be signed.

The catheter is inserted to keep the bladder empty during surgery and for the period of time after the surgery when the woman may not be able to walk to a toilet.

The intravenous connection is made to provide fluids and a quick route for any medication or blood that might be necessary during the surgery. After the surgery, the intravenous connection delivers nourishment to the woman and, typically, oxytocin to help the uterus contract and shrink. As soon as the anesthetic takes effect, the operation to deliver the baby begins.

The Delivery. The entire surgical procedure normally takes about thirty to forty-five minutes. However, the baby is typically delivered within five to fifteen minutes after the surgery begins. If the mother has had a spinal anesthetic and is awake during the surgery, she will, nevertheless, not be able to see the actual surgical procedures. She is draped, with only her abdomen exposed, and a screen is placed at her chest level (below her neck). If the father is present, he is typically seated at the mother's head; he has a fairly complete view of the entire procedure. After the woman's abdomen is bathed with an antiseptic fluid, the actual surgery begins. The first step in the surgery is to make a skin incision. Two types of incisions are possible. One is a transverse incision, which is a cut made from side to side at, or just below, the pubic hairline. Often this type of incision is preferred for cosmetic reasons. Indeed, it is popularly known as the bikini incision. When the incision heals, the scar is almost invisible.

A second type of incision is the vertical skin incision. It is made at the midline of the abdomen and runs down from a point just below the navel to a point close to the pubic bone. When this incision heals, the scar will always be somewhat visible. Because the vertical incision is the quickest to make, it is often used when time is essential.

After the skin is incised, incisions are made in each of the underlying layers of subcutaneous fat and fascia. The abdominal muscles are separated and the peritoneum is opened. This exposes the uterus. The bladder is carefully pushed away, and then, typically, an incision is made in the lower portion of the uterus. Any amniotic fluid in the uterus is suctioned away.

At this point the surgeon lifts the baby out of the uterus, and the baby is born. The baby's nose and mouth are carefully suctioned to remove mucus, and then, typically, the baby emits his or her first cry.

If the mother is awake, she may have experienced a sensation of pressure or weight being lifted as the baby is delivered. If awake, she is shown the baby to admire. After the baby is placed under a warming lamp and examined, he or she may then be given to the father to hold if he is present.

As this is occurring, the rest of the surgery is being completed. The placenta is delivered, and the incision made in each layer of tissue is stitched with absorbable stitches. Nondissolvable stitches, clamps, or dissolvable stitches are used to close the skin incision. A dressing is applied and the mother is taken to the recovery room.

After the Delivery. In the recovery room the mother's pulse, respiration, temperature, and blood pressure are monitored. If awake, the baby is brought to the mother. If breast feeding is to be undertaken, it may begin at this time. After a relatively short time in the recovery room (usually no more than a few hours), the mother is moved to a regular hospital room.

Although pain and discomfort occur as a consequence of any surgery, the medical staff will invariably take steps to ease the discomfort. The mother who has had a cesarean delivery will have to stay in the hospital for a few days more than the mother who had a typical vaginal delivery. However, the mother who has delivered

BOX 4.4

ASSESSING THE HEALTH OF THE FETUS

Modern medical techniques enable physicians to assess the health of a to-be-born child. Although still encased in the amniotic sac within the mother's uterine walls, the fetus can be evaluated for the presence of various diseases. In addition, as labor is progressing, the health status of the fetus can be monitored.

A major technique currently used to appraise the health status of the fetus, even at times relatively early in the pregnancy, is *amniocentesis*. This technique involves extracting a small amount of amniotic fluid, surrounding the fetus, in order for chemical analyses to be conducted. The extraction of fluid is done by placing a syringe needle through the abdomen and the amniotic membranes and into the amniotic fluid. The placement of the needle is based on information obtained from ultrasound recordings and is done in a manner so that the fetus is not touched. The woman experiences little discomfort.

The amniotic fluid contains cells from the fetus. Analysis of these cells can reveal whether the infant is suffering from various disorders, e.g., from Down's syndrome. Analysis can also reveal if the fetus is male or female.

During labor, the health or level of distress of the fetus can be evaluated by use of medical equipment falling under the general heading of a *fetal monitor*. Depending on the specific fetal monitor used, several types of important information about the labor and the fetus can be obtained.

A belt, strapped around the woman's back and having a pressure sensitive device on it which is placed on the abdomen, can record the strength and periodicity of contractions. Contractions can also be measured by electrodes hooked to other parts of the woman's body. Knowledge of the strength and periodicity of the contractions can aid the obstetrician in decisions about whether oxytocin is needed to strengthen contractions to a level sufficient for delivery and/or whether a vaginal delivery is possible given the strength and

types of contractions being shown. For instance, at particular stages of labor if a contraction begins before a preceding one has actually terminated —a phenomenon known as coupling—the uterine muscles may be too fatigued to allow the woman to safely proceed with a vaginal delivery. Information derived from recordings of contractions on the fetal monitor can give the physician a basis for deciding whether coupling is present and, if so, what should be done.

Often, a decision about what should be done is made on the basis of additional information about the labor, information about the stress being experienced by the fetus. Such information is also obtained from the fetal monitor recordings. Through use of an ultrasound recording device placed on the woman's abdomen, the physician can record and hear the fetal heartbeat. After the amniotic membranes have broken, electrodes can be placed on the scalp of the fetus if the fetal head is engaged for a normal delivery; the fetal heartbeat can also be recorded in this way. Information about the strength and regularity of the fetal heartbeat can provide the physician with key information about the presence of fetal stress. If the heart rate drops to too low a level (e.g., 60 beats per minute—a low rate for a fetus) for too long a time, this may indicate some problem (e.g., with the umbilical cord) that requires immediate intervention. Thus use of the fetal monitor can help the physician decide whether progression toward a normal vaginal delivery can be made safely or whether a cesarean section or a forceps delivery needs to be attempted.

Fetal monitoring can save many lives. However, the safety of all aspects of fetal monitoring, e.g., the use of ultrasound, has not been definitely established. In addition, the attachment of numerous recording devices to the woman during her labor can enhance her feelings of discomfort. Nevertheless, in cases where there is good reason to suspect high risk or extreme fetal distress, fetal monitoring has strong advantages.

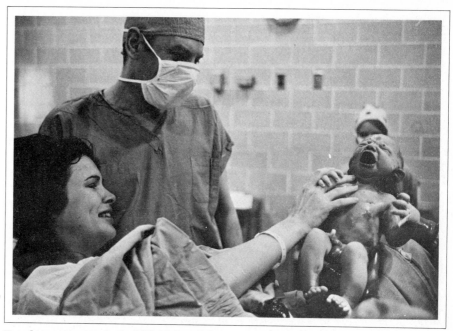

Proud parents meet their child: Today, fathers as well as mothers often meet
the newborn immediately following delivery.
(Mary Ellen Mark/Archive Pictures, Inc.)

her baby through cesarean delivery can use the extra recovery time for spending
more time with her new baby in an environment where both her and her baby's
needs are receiving the best attention possible.

In sum, there are several ways in which healthy babies can be delivered. As
noted, there is really no way of being precisely certain about what variations will
occur within a particular birth experience. In part, what sort of delivery the woman
experiences depends on the health or level of stress the fetus is experiencing during
the labor period. Physicians have available to them several ways of monitoring the
health of the fetus during pregnancy and, especially, during labor. Some of these
techniques are discussed in Box 4.4.

In sum, we have seen that when a new person emerges from the mother's
womb, he or she has already traversed a long and complex developmental path. It
already has a history of interactions with its context. Thus the neonate is not an incompetent organism; the newborn has actively developing physical and behavioral
functions. These characteristics continue to develop throughout the period of infancy. In the following two chapters we discuss these developments.

CHAPTER SUMMARY

There are three periods of prenatal development. The period of the ovum begins at
fertilization and lasts until the zygote becomes attached to the uterine wall. This period usually lasts from ten to fourteen days. The period of the embryo lasts from
about the end of the second week after conception through to the end of the tenth

week. The various bodily tissues, organs, and systems emerge and begin to develop in this period. The period of the fetus lasts from the end of the second month after conception until birth. Further growth and differentiation of the bodily organs and systems developed in the embryonic period continue here. In addition, these organs all begin to function.

There are several influences on prenatal development. Poor maternal nutrition, maternal alcohol and tobacco consumption, and maternal drug use can result in negative developments both prenatally and postnatally. Contraction of diseases by the mother and marked emotional stress in the mother during her pregnancy are also associated with such problems. There is some relation between maternal age and prenatal and postnatal problems, and there is also some evidence that teratogens, acting through males, can affect the developing embryo and fetus.

At about the end of 280 days, labor begins. Obstetrical medication delivered during pregnancy is often necessary and useful. However, there do seem to be some effects of such drugs on the infant, especially when the dosage is high. In addition, both prematurity and postmaturity are associated with problems for the infant.

Although there is great variation in labors and deliveries, there does seem to be a fairly generalizable series of stages in labor. The onset of labor is often signaled by the beginning of rhythmic contractions, by bloody shows, and/or by the rupturing of the amniotic sac. The first stage of labor is the longest, usually lasting from ten to fourteen hours. Regularly spaced contractions become established during this period (early first-stage labor), then become stronger, longer, and closer together (middle first-stage labor), and finally can be quite long and strong and may even seem continuous (transition). In the second stage of labor, the baby is born. Contractions during this period may stop temporarily and then become less frequent. In the third stage of labor, the placenta is delivered.

There are various types of delivery. Most infants are delivered through the vaginal (or birth) canal, typically in a headfirst position. However, some babies are born buttocks or feet first (i.e., breech). When the position of the fetus precludes a vaginal delivery, a cesarean delivery is done. This type of delivery involves surgery.

chapter 5

Infancy: physical growth and cognitive development

CHAPTER OVERVIEW

The neonate is not a neophyte. The infant has sensory, motor, and cognitive capacities that are identifiable at or shortly after birth. As the infant grows, these capabilities develop further. In this chapter we discuss these changes. In particular we focus on several ways of studying cognitive development: the learning approach, the psychometric approach, and the Piagetian approach.

CHAPTER OUTLINE

PHYSICAL GROWTH
SENSORY AND PERCEPTUAL CHANGES

Taste and Touch
Hearing
Vision

INFANT REFLEXES AND MOTOR DEVELOPMENT
COMPLEX BEHAVIORAL SEQUENCES AND INFANT STATES
APPROACHES TO THE STUDY OF COGNITION
LEARNING

What Is Learning?
Types of Learning

CLASSICAL CONDITIONING

Characteristics of Classical Conditioning

OPERANT CONDITIONING

Characteristics of Operant Conditioning
Secondary Reinforcement and Chaining

CONDITIONING IN INFANCY
OBSERVATIONAL LEARNING: MODELING AND IMITATION

Observational Learning Processes
Developmental Changes in Imitation in Infancy

PSYCHOMETRIC INTELLIGENCE

A Definition of Cognition
Quantitative Dimensions of Cognitive Development
Intelligence Tests
Measures of Infant Intelligence

THE PIAGETIAN APPROACH TO THE STUDY OF COGNITION
PIAGET'S THEORY OF COGNITIVE DEVELOPMENT

Stage-Independent Conceptions of Cognitive Development
Stages of Cognitive Development
The Sensorimotor Stage

CHAPTER SUMMARY

ISSUES TO CONSIDER

What is the rate of physical growth during infancy?

What is sensation and what is perception?

What is the nature of the infant's perceptual capacities in regard to the modalities of taste, touch, hearing, and vision?

What is the nature of the infant's motor repertoire?

How do infant states change?

What is learning, and what are the major characteristics of classical and operant conditioning?

Can infants learn?

What is the psychometric approach to intelligence, and what are the major uses and limitations of infant intelligence tests?

What is the Piagetian, or cognitive developmental, approach to cognition?

What are the major stage-independent conceptions in Piaget's theory?

What are the major developments that occur in the sensorimotor stage?

*A*s explained in Chapter 1, life-span developmentalists focus on behavior change processes in order to understand development. They recognize, however, that such intraindividual changes proceed at different rates for different people. That is, they know there are interindividual differences in intraindividual change. Indeed, goals of the life-span approach to studying human development are to describe, explain, and optimize both intraindividual changes and interindividual differences in such changes.

However, because of their focus on changes and differences, life-span developmentalists see a strict focus on age limits for any period of life as quite limited. Age is only a very rough guideline, or marker, for change. At any one age people will differ in regard to how far any particular process (e.g., regarding personality or cognitive changes) has advanced. Thus in considering the several physical and behavioral change processes that characterize the infancy period—or any other age period we will discuss—age limits are only quite general guidelines. They serve to indicate when, for one or several processes, changes typically exist within a particular range. For example, Mussen et al. (1979) argue that infancy spans the first 18 to 24 months of life because beyond this age range most people show changes in their cognitive functioning characteristic of a distinctive, qualitative difference in thought; as Piaget (1950, 1970) also argues, Mussen et al. note that the person at about 2 years of age shows representational cognitive functioning, and can deal with other people and with objects through use of internal, symbolic activity. When this change occurs, Mussen et al. contend it is significant enough to argue that the changes of infancy are past and the person is now in his or her childhood.

Many developmentalists, for similar reasons regarding other processes, also use the 18-month to 2-year point as the upper limit of the infancy period. As such, most of our discussion of the physical and behavioral changes characterizing infancy will pertain to developments within the first 2 years of life after birth.

PHYSICAL GROWTH

Upon birth the average infant weighs 7.5 pounds and is about 20 inches in length. However, the next 24 months of life typically involve considerable changes in both these characteristics. For example, by 2 years of age, most boys have achieved 49.5 percent of what will be their final adult height and most girls have achieved 52.8 percent of their final mature height (Bayley, 1956). In fact, in the first year of life alone, boys achieve 42.4 percent of their final height and girls reach 44.7 percent of their final height (Bayley, 1956).

However, during the first 2 years and, in fact, across the life span, different parts of the body grow at different rates. Before birth the head is the fastest growing part of the body, attaining almost three-quarters of its final adult size by the time the person is born. In the first year of life the trunk grows more than do other bodily areas; from the end of the first year of life until puberty the legs grow the most. Exhibit 5.1 illustrates that at different times of life a body part may represent a different proportion of total body size.

SENSORY AND PERCEPTUAL CHANGES

Sensation refers to the reception of stimulation. When our sense organs—cells in our eyes, ears, or fingers, for example—are presented with appropriate stimulation, sensation occurs. However, perception is *sensation with meaning* (Schneirla, 1957). When associations are made to sensory stimulation, for instance, when we interpret or attach a value or a label to a sensation, perception occurs.

144

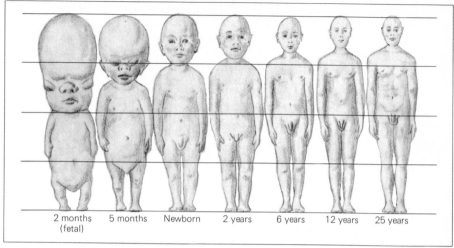

EXHIBIT 5.1 Across the life span a body part represents a different proportion of total body size. (Source: Jackson, 1929. By permission.)

For example, if a young infant, a monkey, or a chimp were to touch or pick up a hot coffee cup, we would expect that the sensation of pain would be associated with the dropping of the cup. Even if the cup were an extremely valuable antique one, we would expect the same response. The meaning of the cup, its value, would not be associated with the sensation. However, at some point in the human life span such meaning can be attached to the sensory stimulation. We have suggested that this occurs in early childhood, with the advent of symbolic, or representational, ability; indeed, this development is what we see as the major distinction between infancy and childhood. Thus when an association can be made to a sensation, when meaning is attached, perception takes place. If the person did not drop the cup because of its value, despite the fact that the sensation of pain was received, then this would be an example of perception.

This example is probably not representative of associations in infancy. The association of meaning to sensation probably does not involve the use of verbal labels or of values in the early portions of infancy. Nevertheless, there is a lot of sensory functioning in infancy. Moreover, as the person advances toward childhood, initial perceptual activity takes place, and this involves the role of associations acquired, for instance, through principles of learning and through preverbal instances of cognitive (e.g., memory) functioning. (Learning and cognitive processes in infancy are discussed in later portions of this chapter.)

Taste and Touch

Newborns have been found to suck at different rates on a nipple in relation to whether the nipple contained glucose or sucrose (two sweet-tasting substances) or water (Engen, Lipsitt, & Peck, 1974). In fact, in one study of 1- to 3-day-old infants (Nowlis & Kessen, 1976), increases in strength of the sucking response were directly associated with increases in concentrations of these sweet-tasting substances.

Although not extensively investigated, there is evidence that tactile (touch) perception exists in infants. Soroka, Corter, and Abramovitch (1979) studied forty 10-month-old infants by first giving them two minutes to tactually explore an object in a totally darkened room. Subsequently, during a two-minute test trial in the dark, half the infants were given the same object and half were given a novel shape. Infants presented with the novel shape spent a significantly longer time manipulating the object than did infants presented with familiar forms. This difference was taken as evidence for the ability to *discriminate* (to respond differently to different stimuli). Thus in the absence of visual information, these young infants showed the capability of tactually discriminating novel and familiar shapes (by showing a preference for novelty within the tactual sensory modality). Similar findings, using a similar procedure with twenty-five 1-year-olds, were reported by Gottfried and Rose (1980).

Hearing

Infants' hearing is well developed. Sounds can be localized, and sounds of differing loudness and duration will be responded to differently (Brackbill, 1970). For example, Muir and Field (1979) reported that a majority of newborn infants turned their heads toward a continuous sound source presented 90° from the midline of their heads. Similarly, at birth and at 1 month of age infants are capable of turning their heads to an off-centered sound (Field, Muir, Pilon, Sinclair, & Dodwell, 1980). Moreover, by 16 weeks of age binaural cues can be detected (Bundy, 1980).

Newborns respond differently to different sounds (Kinney & Kagan, 1976; Leventhal & Lipsitt, 1964), and they are especially sensitive to the sound of the normal human voice (Barrett-Goldfarb & Whitehurst, 1973). For example, infants (aged 6 to 12 weeks) learned to suck more when their sucking produced human speech than when there was no relation between sucking and speech (Williams & Golenski, 1978). Other research by Williams and Golenski indicates that such results may depend on young infants' state of alertness. Infants in more alert states show higher rates of sucking, following a change in sound, than do infants in less alert states.

Other research also shows that infants attend to human speech. Infants, 4.5 to 5.5 months of age, continued to attend to speech syllables which changed from test trial to test trial but stopped showing attention to a speechlike stimulus which did not change (i.e., it remained constant) from trial to trial (Trehub & Curran, 1979). In another, quite interesting study, DeCasper and Fifer (1980) arranged a situation wherein newborn infants could suck on a nipple through which no nutrients were delivered. However, by sucking on this *nonnutritive nipple* in different ways, the newborn could produce either the voice of his or her mother or the voice of another female. The infants learned to suck in the way which produced their mothers' voices, and they produced it more often than the other voice.

Finally, there is evidence that infants can interrelate information from two sense modalities, i.e., they can show *intersensory integration*. Spelke (1979) studied such functioning in regard to the auditory and the visual modes of perception. Spelke found that 4-month-old infants respond to relationships between visual and auditory stimulation that carry information about an object. Apparently, infants detected the temporal synchrony of an object's sounds and its visually specified impacts. Thus we see that another key mode of sensing and perceiving the world—the visual—is functional in infancy. We now discuss this modality.

Vision

Considerable research has been conducted on vision in infancy. It has been suggested that while newborns cannot focus their eyes well at all distances (Fantz, Ordy, & Udelf, 1962)—that is, they do not show good accommodation—they do respond differently to light and dark (Fantz et al., 1962), see color (Bornstein, Kessen, & Weiskopf, 1976; Fagan, 1974), can discriminate separations between stimuli (that is, they show visual acuity; Frantz et al., 1962), and can follow a moving stimulus (Bower & Paterson, 1973; McKenzie & Day, 1976). For example, in one study of color vision in infancy Bornstein (1979) showed that 4-month-old babies who were *habituated* to a single hue (that is, who were repeatedly shown the color until they stopped responding to it) *discriminated* a change in hue; that is, they dishabituated, or started responding to the new stimulus. In turn 4-month-old babies who were habituated to several hues also showed habituation to a novel hue. In other words, they *generalized* their response; they responded to a new stimulus in a manner akin to the way they had responded to a previous one.

From early in their lives infants respond more to patterned stimuli than to non-patterned stimuli; in this sense they are held to show a preference. Fantz (1961), for example, found that infants showed more *attention* to—as measured by length of time their eyes fixated upon—a picture of a bull's eye than to a plain red, white, or yellow stimulus. Moreover, he found that more patterned stimuli than the bull's-eye were given even greater attention by the infants. A stimulus having newspaper print was fixated longer than the bull's-eye. Similarly, in other research (Milewski, 1979) 3-month-old infants were found to be sensitive to the configuration of a visual pattern (e.g., arrangements of three dots) and to have the ability to detect invariance in patterns.

Most interestingly in the work of Fantz (1961) and others a stimulus pattern that looked like a face was shown the most attention of all. Hainline (1978), for instance, also found that more attention is paid to a face stimulus than to a nonface stimulus; and both she and Maurer and Salapatek (1976) report developmental changes, involving increased and more complex fixations on faces, across the first few months of infancy.

In fact, infants tend to fixate longer on more complex stimuli than on less complex ones (Martin, 1975; Sigman & Parmelee, 1974). However, developmental changes can be expected in which complex patterns may be attended to. Wetherford and Cohen (1973) found that while younger infants prefer familiar patterns, older infants prefer novel ones. Similarly, Leahy (1976) reported that features of simple geometric figures that initially attracted the most fixations received fewer fixations with repeated exposures, and Rubenstein (1974) found that infants look at novel objects more than familiar ones. These findings about infant developmental changes in regard to pattern perception have led some developmentalists to assert that infant perception is qualitatively similar to that of the adult (Bond, 1972).

Other aspects of vision have been investigated as well. When visual sensation is received, the part of our eye stimulated is termed our retina. A visual stimulus far from us "takes up" less space on our retina (i.e., it stimulates a smaller area of it) than does a stimulus close to us. For example, it we watch an airplane taxi down a runway, take off, and then disappear from sight, the airplane takes up a smaller and smaller amount of space on our retina. However, we *know*, as we watch the airplane disappear, that it is not shrinking. Although it is casting a smaller image on our retina, we know that its actual size is constant. Bower (1966) has demonstrated that infants show such size constancy—in his experiments in regard to the perception of cubes. In another study of size constancy McKenzie, Tootell, and Day (1980) used 4-, 6-, and 8-month-old infants as subjects. The results showed that the presence of size constancy depends on the distance the object is from the infant and the age of the infant. That is, at each age level the infants were repeatedly shown a three-dimensional model of a human head until the infants stopped attending to the stimulus (i.e., until they *habituated* it). At 6 and 8 months size constancy occurred for the head model up to but not beyond a distance of 70 centimeters. At 4 months of age, and even at a short distance (30 to 60 centimeters), size constancy was not apparent.

Other forms of perceptual constancy have been studied in infancy. Caron, Caron, and Carlson (1979) found evidence for the perception of shape constancy in 12-week-old infants, even when the object is rotated to be at different angles. Similarly, Schwartz and Day (1979) studied infants between 8 and 17 weeks. They report that these young infants were able to recognize shapes and were able to use relational information in this performance. However, there appear to be limits to the young infants' ability to discriminate among solid forms; Cook, Field, and Griffiths (1978) found that 12-week-old infants could distinguish a cube from a photograph of a cube and a concave solid (an L-form). But these infants could not discriminate between the cube and either a wedge or a pyramid type of solid. Similarly, Ruff (1978) found that 9-month-old, but not 6-month-old, infants were capable of recognizing the invariant form of the objects she presented to them.

Infants also show depth perception. Gibson and Walk (1960) placed young infants in the middle of a table covered with a transparent piece of glass. On one side of the table a checkered pattern was attached to the glass; on the other side the same pattern was several feet below the glass—this gave the appearance of a sharp drop-off, or of a visual "cliff." Infants, 6 to 14 months of age, crawled to their mothers when the mothers were standing by the "shallow" side; they did not do so when their mothers were standing near the deep side. Other studies have found results consistent with these, studying infants from as young as 6 weeks old to 9 months old (Schwartz, Campos, & Baisel, 1973). In a study of the predictors of such perceptual ability, Rader, Bausano, and Richards (1980) assessed twenty-two infants, aged 6.7 to 12.3 months, on the visual cliff apparatus. The infants were tested while in a walker and when crawling. Age and experience factors were evaluated as predictors

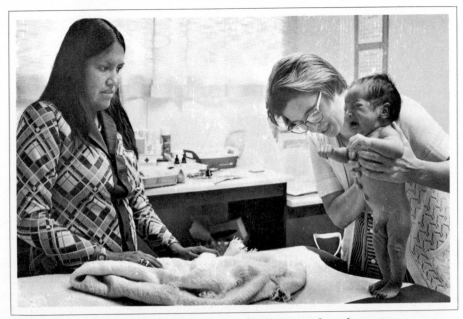

Part of infants' health checkups involves testing for their muscle and motor development. (Mimi Forsyth/Monkmeyer Press Photo Service.)

of avoidance (of the drop-off) or nonavoidance in the crawling condition. "Age when first crawled" was the best predictor of avoidance behavior. The earlier a child crawled, the better the avoidance and, by inference, the better the depth perception.

Studies using techniques other than the visual cliff have also been used to assess depth perception in infancy. In one such attempt Yonas, Oberg, and Narcia (1978) found evidence for sensitivity to binocular depth information among 20-week-old infants but not among 16-week-old infants.

Thus infants have the visual perception ability to cope appropriately with numerous aspects of their stimulus world. Several researchers have sought to discover whether these abilities are applied in the infants' perception of their broader physical context. One line of research that has developed is the investigation of spatial perception (see Liben, Patterson, & Newcombe, 1981), that is, the infant's perception of the physical space within which he or she exists.

Research done by Acredolo and her associates exemplifies this line of research. In one study (Acredolo, 1978) the ability of infants at 6, 11, and 16 months to keep track of their relationship to a place in space was assessed. Infants were trained to expect an event to occur to their right or to their left. The infants were then moved so that their view of space was reversed. The direction in which they turned in anticipation of the event indicated whether they were perceiving the location objectively (that is, in terms of its *actual* place in space) or "egocentrically" (that is, whether the place was "where" a left turn of the child's head put it). Following twenty-four infants longitudinally, Acredolo found an age-associated shift, from egocentric responding at 6 and 11 months to objective responding at 16 months. Similarly, using a procedure like that of Acredolo (1978), Rieser (1979) found that although some 6-month-olds can orient toward a location in space by use of a physical landmark, in many cases their visual search is egocentric, and this is associated with poor performance.

Another aspect of the infants' perception of his or her real-world context that has recently begun to be investigated is the perception of television among infants. Hollenbeck and Slaby (1979) found that 6-month-old infants looked longer at a television set when both the sound and picture were on than when only the picture was on. In turn, they looked longer at these two patterned types of visual stimulation than they did when unpatterned visual stimulation was displayed on the television.

Thus as we have seen with other sensory modalities in infancy, the infant's visual mode of perception is quite functional. In fact, as we saw in regard to hearing, there is considerable evidence that infants use their sensory modalities in concert, as opposed to independently from each other. That is, there is a liaison between the senses in infancy. Infants can transduce, or integrate, information from one sense modality to another; in other words, they show *intersensory integration*.

One study demonstrating this ability was reported by Rose, Gottfried, and Bridger (1978). These researchers compared full-term middle-class, full-term lower-class, and preterm infants on cross-modal and visual-intramodal functioning. All three groups demonstrated comparable intramodal, visual-visual functioning. That is, they showed equal levels of recognition memory of a visually presented object that had been previously presented visually. However, only the full-term middle-class group was able to transfer information across modalities, for example, in recognizing a stimulus, now presented visually, that previously had been presented tactually.

Similarly, Gottfried, Rose, and Bridger (1978) presented 6-, 9-, and 12-month-old infants with objects either visually or by combining visual with forms of touch stimulation. Memory for objects presented in these ways was assessed by measuring the infants' differential preferences for novel and familiar stimuli—a topic that we have examined earlier in this section. Gottfried et al. found that the oldest infants showed evidence of memory in all conditions, but the younger infants showed evidence for memory only in the visual condition.

In sum, infants begin life with their sensory apparatus functioning very well, and they progress rapidly toward the environment of elaborate perceptual functioning. Infants' reflex and motor developments show similar trends.

INFANT REFLEXES AND MOTOR DEVELOPMENT

The infant does more than just receive stimulation or attach meaning to stimulation. The infant can act on its world from the first moments of birth onward. The newborn infant has a large repertoire of *reflexes*. Reflexes are *relatively invariant motor outputs (i.e., muscular movements) in response to particular sensory inputs (e.g., stimulation)*. For example, neonates show the blink reflex. If a flash of light is presented to the eyes, the eyelids will close. This reflex, present in the newborn, will continue to be present across the life span. Throughout it serves the purpose of protecting the eyes from harsh visual stimulation.

Other reflexes also serve adaptive functions but, of course, may differ in regard to the particular sensory and motor elements involved. For example, in response to light tactile (touch) stimulation of the cheek, the infant will show the *rooting reflex;* that is, the infant turns its head in the direction of the stimulation. Although this reflex disappears by 3 or 4 months of age, it is clear that such a reflex also has adaptive significance. (i.e., it aids survival). In human evolution it was most probably the mother's nipple that stimulated the cheek of the infant. Turning the head in the direction of stimulation thus enhances the closeness of the infant's mouth to the source of food.

Exhibit 5.2 presents a list of newborn reflexes, how they may be tested for, the motor response involved in the reflex, any developmental changes associated with the reflex (e.g., whether or not the reflex stays with the person after infancy), and the adaptive significance of the reflex. It should be noted that the ages associated with the developmental changes presented in this exhibit represent only normative data, and as such, there is a good deal of interindividual variability associated with these norms.

Of course, neonatal reflexes are not the only motor behavior shown by the infant. As the person moves through his or her infancy, considerable nonreflexive, complicated motor behavior is seen; often, this requires the coordination of several muscles from several parts of the body. Norms for the attainment of particular motor behaviors exist (Bayley, 1935; Gesell & Amatruda, 1941; Shirley, 1933). Exhibit 5.3 lists some of these. As with Exhibit 5.2, the ages listed in this exhibit are norms and, as such, have considerable interindividual variability associated with them.

There has been research interest in the development of motor behaviors in infancy, particularly concerning the relation between reaching and visual stimulation and concerning the implications of such development for other facets of development. Ruff and Halton (1978) did not find evidence for directed reaching (at a visually presented ball) among nine alert infants aged 7 to 15 days. Extensions of their arms and hands were as frequent in the absence as in the presence of the ball. However, by 2 months of age better-developed eye-hand coordination ability was present.

Provine and Westerman (1979) studied the development of the ability to extend the hand across the body midline to contact a visually presented object among forty-eight normal, full-term 9- to 20-week-old infants. One of the infant's arms was restrained while the actions of the opposite-side (the *contralateral*) arm was observed. Infants first contacted objects placed in front of the same-side (the *ipsilateral*) shoulder, then at the body midline, and later in front of the contralateral shoulder. Between 9 and 17 weeks, success at contacting objects at the body midline progressed from 33 to 93 percent. During this same interval, success in contacting objects presented in front of the contralateral shoulder increased from 0 to 71 percent. By 18 to 20 weeks, all infants contacted objects in all three positions.

The characteristics of the visual stimulus that may promote reaching have been investigated. McGuire and Turkewitz (1978) assessed the relationship between visual stimulus intensity and higher movements in infants from 10 to 15 weeks of age and from 20 to 25 weeks of age. These infants were shown a cone that varied in size, brightness, and distance from them. Older infants tended to extend their fingers more than younger ones. However, as intensity increased, the younger infants' responses showed decreases in extension and increases in withdrawal from the stimulus. Among the older infants there was no clear relation between stimulus intensity and either finger approach or withdrawal.

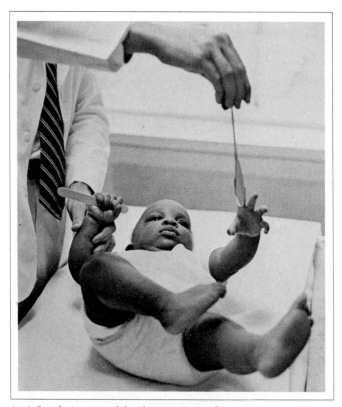

An infant being tested for the grasping reflex.
(Herb Levart/Photo Researchers, Inc.)

EXHIBIT 5.2

Reflexes of the Neonate

Name of Reflex	Test of the Reflex	Response	Developmental Changes	Adaptive Significance
Blink	Light flash.	Closing of both eyelids.	Permanent.	Protection of eyes from strong stimuli.
Biceps reflex	Tap on the tendon of the biceps.	Short contraction of the biceps.	In the first few days it is brisker than in later days.	Absent in depressed infants or in cases of congenital muscular disease.
Knee jerk, or patellar tendon	Tap on the tendon below patella, or kneecap.	Quick extension or kick of the knee.	More pronounced in the first 2 days than later.	Absent or difficult to obtain in depressed infants or infants with muscular disease; exaggerated in hyperexcitable infants.
Babinski	Gentle stroke on the side of the infant's foot from heel to toes.	Dorsal flexion of the big toe; extension of the other toes.	Usually disappears near the end of the first year; replaced by plantar flexion of great toe as in the normal adult.	Absent in defects of the lower spine.
Withdrawal reflex	Pinprick is applied to the sole of the infant's foot.	Leg flexion.	Constantly present during the first 10 days; present but less intense later.	Absent with sciatic nerve damage.
Plantar, or toe grasp	Pressure is applied with finger against the balls of the infant's feet.	Plantar flexion of all toes.	Disappears between 8 and 12 months.	Absent in defects of the lower spinal cord.
Palmar, or automatic hand grasp	A rod or finger is pressed against the infant's palm.	Infant grasps the object.	Disappears at 3 to 4 months; increases during the first month and then gradually declines; replaced by voluntary grasp between 4 and 5 months.	Response is weak or absent in depressed babies; sucking movements facilitate grasping.

As noted, some investigators have been concerned with the relationship between motor development and other aspects of development. Although measures of infant motor, or psychomotor, functioning do not appear to relate to measures of cognitive functioning in later life (e.g., Lewis & McGurk, 1972), there is some evidence that motor development is related to other physical developments *within* the infancy period. For example, Black, Steinschneider, and Sheehe (1979) studied 122 full-term healthy infants over the course of the first 4 weeks of their lives. The researchers recorded periods of sleep *apnea* (periods during which there is a pause in

EXHIBIT **5.2** (*Continued*)

Name of Reflex	Test of the Reflex	Response	Developmental Changes	Adaptive Significance
Moro reflex	Sudden loud sound or jarring (for example, banging on the examination table); or head drop—head is dropped a few inches; or baby drop—baby is suspended horizontally and the examiner lowers his or her hands rapidly about 6 inches and stops abruptly.	Arms are thrown out in extension, and then brought toward each other in a convulsive manner; hands are fanned out at first and then clenched tightly.	Disappears in 6 to 7 months.	Absent or constantly weak moro indicates serious disturbance of the central nervous system.
Stepping	Baby is supported in upright position; examiner moves the infant forward and tilts him or her slightly to one side.	Rhythmic stepping movements.	Disappears in 3 to 4 months.	Absent in depressed infants.
Rooting response	Cheek of infant is stimulated by light pressure of the finger.	Baby turns head toward finger, opens mouth, and tries to suck finger.	Disappears at approximately 3 to 4 months.	Absent in depressed infants; appears in adults only in severe cerebral palsy diseases.
Sucking response	Index finger is inserted about 3 to 4 centimeters into the mouth.	Rhythmical sucking.	Sucking is often less intensive and less regular during the first 3 to 4 days.	Poor sucking (weak, slow, and short periods) is found in apathetic babies; maternal medication during childbirth may depress sucking.
Babkin, or palmarmental reflex	Pressure is applied on both of baby's palms when lying back.	Mouth opens, eyes close, and head returns to midline.	Disappears in 3 to 4 months.	General depression of central nervous system inhibits this response.

Source: Adapted from Hetherington and Parke (1979).

breathing during sleep) among the infants. Prolonged sleep apnea has been hypothesized to be part of a pathological process resulting in the Sudden Infant Death Syndrome (SIDS), a syndrome discussed in Chapter 4 of this book. At 9 months of age these infants were given a standardized test of their psychomotor development (the Bayley Scales of Infant Development, to be discussed in a later section of this chapter). Those infants who had the highest scores for measures of sleep apnea, taken in the first weeks of life, averaged significantly lower in psychomotor development at 9 months.

EXHIBIT 5.3   Some Norms of Infant Motor Development

Motor Behavior of Infant	Normative Age
Raises chin while lying on stomach	1 month
Raises chest while lying on stomach	2 months
Sits with support	4 months
Sits without support	7 months
Stands with help	8 months
Stands by holding furniture	9 months
Creeps	10 months
Walks when led	11 months
Pulls self up to stand	12 months
Climbs stairs	13 months
Stands alone	14 months
Walks alone	15 months
Goes up and down stairs without help	18 months
Can run and walk backward	24 months

Source: Adapted from Gesell & Amatruda, 1941; Shirley, 1933.

COMPLEX BEHAVIORAL SEQUENCES AND INFANT STATES

The development of the infant's reflexive and motor functioning is involved in an array of more complex sequences of behavior. These behaviors are those often involved in allowing the infant's survival as a biological organism; they involve infants' sleep and waking patterns, their toileting behavior, and their eating and drinking behaviors. Exhibit 5.4 summarizes some developmental changes in these functions.

Different infant states may be related. For example, Gaensbauer and Emde (1973) found that the frequency of alert, wakeful periods in newborns could be altered by different types of feeding schedules, and Harper, Hoppenbrouwers, Bannett, Hodgman, Sterman, and McGinty (1977) found a relation between feeding periods and infants' sleep cycles. Thus the sequences and categories of infant states, such as sleep-wakefulness cycles, studied by several researchers (Campos & Brack-

EXHIBIT 5.4   Changes in Sleep and Waking Cycles, Toileting, Eating, and Drinking Behaviors in Infancy

Cycle or Behavior	General Characteristics
Sleep and waking cycles	Neonatal period—sleep 80% of time per day
	6 to 7 months—sleep through the night without awakening
	12 months—sleep 50% of time per day
Toileting behaviors	Neonatal period—involuntary release of waste products
	2 months—usually two bowel movements per day
	4 months—predictable interval of time between feeding and bowel movement
Eating and drinking behaviors	Neonatal period—seven to eight feedings a day
	1 month—five to six feedings a day
	2 months—solid foods introduced
	12 months—three meals a day

bill, 1973; Thoman, 1975), may be interactive with the other complex behaviors and states of the infant. Indeed, there is evidence that infant states interact with infant perceptual and motor abilities to produce behavioral functioning.

Williams and Golenski (1979) investigated the interaction between infant state and speech sound discriminations. Infants in more alert states (e.g., waking) were found to show higher rates of sucking following a change in sound than was the case for infants in less alert (e.g., sleeping) states.

Other studies have investigated whether different responsivity to stimuli exist within a particular state. For example, Rose, Schmidt, and Bridger (1978) studied infants while they were in different portions of their sleep state; that is, infants were studied while they slept quietly and while they actively moved while sleeping. Rose et al. were interested in whether infants show a different *threshold* to tactile stimulation in these sleep periods. A threshold is the minimal amount of stimulation needed to evoke a response. If a lot of stimulation is needed, then a high threshold exists; if little stimulation is needed, then a low threshold exists. In the earliest portion of the infant's sleep period, Rose et al. found no threshold differences to stimulation between active and quiet sleep phases, in regard to either behavioral or heartrate responses. In a subsequent portion of the sleep period, however, infants became unresponsive behaviorally in both sleep states, but in the quiet sleep phase, their heart rate accelerated in response to strong tactile stimulation. Moreover, throughout their sleep period the threshold of stimulation necessary to evoke a heart response was lower than for a behavioral response. The authors summarized their findings by indicating that sleep state is not the major factor per se in determining infants' tactile threshold; rather, the time elapsed after beginning a sleep period seems most important in neonates.

However, that such a conclusion may not be generalizable for other types of stimulation is a conclusion one may draw from other research by these same authors (Schmidt, Rose, & Bridger, 1980). Here, the authors presented tactile stimulation either alone *or* in combination with auditory (heartbeat) stimulation. When the infants were in active sleep, the sound improved the behavioral responses of the infants, but during quiet sleep, the sound had no effect on either heart or motor responses. In addition, the sound seemed to act to increase the length of time the infants slept. Other researchers, for example, Field, Dempsey, Hatch, Ting, and Clifton (1979), have also used combinations of auditory and tactile stimulation in studying both full-term and preterm infants; these researchers find that differences between these two groups in response to this combined stimulation may be understood by reference to state differences that characterize full-term and preterm infants.

In sum, the infant has well-developed sensory and perceptual functioning. In addition, well-developed reflexive and other motor behaviors are present throughout this period of life. Most importantly, however, from the first hours of life onward the infant has the capactiy to integrate his or her sensory input with his or her motor output. The acquisition of connections between stimulation and responses as a function of experience is an obviously necessary part of adaptive functioning. In a brief way we can term functioning to establish such connections *learning*. Indeed, many of the assessments of infant perceptual ability that we have just discussed capitalized on infants' learning abilities. For instance, some authors (e.g., see Kimble, 1961) consider habituation an elementary form of learning. In addition, discrimination and generalization are processes of learning (Kimble, 1961), and we have seen various researchers assess infants' perceptual abilities by determining whether they could, for instance, acquire a discrimination between two colors (e.g., Bornstein, 1979).

The presence of learning in infancy underscores the fact that the human infant is a capable, competent organism, one that is cognitively competent. Indeed, learning is one instance of cognitive functioning in infancy. In the next section of this chapter we discuss several of the instances of cognition in infancy. We begin by considering the topic of "learning." We discuss the types of learning that exist and also discuss their presence and development in infancy.

APPROACHES TO THE STUDY OF COGNITION

During infancy, the person's intellectual, or cognitive, capacities undergo marked development. These processes have been studied in at least three different general ways by human developmentalists. These approaches may be labeled as the learning approach, the psychometric approach, and the structural, or cognitive developmental, approach. This last approach is most identified with the work of Jean Piaget (introduced in Chapter 2). In this chapter we will present details of each of these three approaches and then indicate what research derived from each approach tells us about cognitive development in infancy.

LEARNING

Psychologists have long debated about, first, what learning is and, second, how many types of learning exist (Kimble, 1961). Before we can discuss the presence and development of learning in infancy, we have to deal with both of these debates. Accordingly, we will first define learning. Second, we will indicate the basic types of learning that are thought to exist.

What Is Learning?

As with any complex psychological phenomenon, there is no general theoretical agreement about what constitutes learning. Although it is fair to say that in general more attention has been paid by psychologists to learning than to any other psychological process, different workers in this area maintain markedly different conceptualizations about what variables are actually involved in learning and what, in fact, constitutes learning (see, for example, Gewirtz & Stingle, 1968; Bolles, 1972). Despite this controversy there is, to some extent, a general agreement about the empirical components of learning. However, we must recognize that even this empirical analysis might raise some controversial points; nevertheless, there is relatively more agreement about the empirical components of the learning process than there is about its theoretical characteristics.

Accordingly, we may focus on an empirical definition of learning offered by Kimble (1961). *Learning* is a relatively permanent change in behavior potentiality which occurs as a result of reinforced practice, excluding changes due to maturation, fatigue, and/or injury to the nervous system. This is obviously a complex definition, including many important component concepts. In order to understand this empirical definition, let us consider the various key words in this definition.

1. *Relatively Permanent.* Learning is defined as a relatively permanent change in behavior. By this we mean that the changes that comprise learning are changes that tend to remain with the person. When a person learns, he or she acquires a behavior; that is, the behavior in the person's repertoire has been changed, and this addition tends to remain. Although you might question this component of the definition, there is considerable empirical support for it. Let us make an extreme example for the purpose of illustration. Many people study a particular subject early in their educational careers (e.g., a foreign language). As adults, they may believe that they no longer remember anything they learned. Yet if they attempt to relearn the foreign language, they usually find that it takes them a shorter time to reach their previous level of competence. Such a *savings effect*, which has frequently been experimentally verified, indicates that previously acquired changes in the person tend to remain relatively permanent. Although the person did not retain everything he or she had previously acquired when beginning to relearn, the fact that he or she relearned the material more easily than had been the case originally suggests that some of the learning had been saved; it was relatively permanent.

2. Change. Learning constitutes a modification of behavior. The person has a repertoire of behavioral responses. With learning, however, this repertoire is altered; more responses are added to the repertoire, and thus a change has occurred. This is illustrated in Exhibit 5.5.

3. Behavioral Potentiality. This term refers to a most important concept we must deal with when attempting to understand learning. One sees a child *perform* a task. Thus on a basis of relatively permanent changes in such performance, one may infer that learning has taken place. In other words, the term *learning* may be used as a summary term, describing an empirical process involving changes in performance. Hence the term *behavior potentiality* refers to the *learning-performance distinction.* We do not see learning per se; rather, we see performance changes, and we summarize these changes by the term learning.

For example, a student may be given a pretest in mathematics. After a certain score on that pretest is achieved, the student is exposed to a given type of instruction designed to improve the score (performance) on a posttest. If the student's score increases, if the performance is enhanced through exposure to this instructional technique, we say that the student has learned. Yet we do not actually see this phenomenon called learning. We only see an alteration in performance. Still, we infer that learning has taken place on the basis of the observed change in performance. Simply, then, learning is not a variable we directly observe. It has the status of an *intervening variable* in psychology; that is, it summarizes observed changes in performance.

4. Reinforced Practice. This term, too, is particularly important for our understanding of learning. Practice per se of a behavior might lead only to fatigue. Yet many learning psychologists would suggest that for *some* of the major forms of learning if

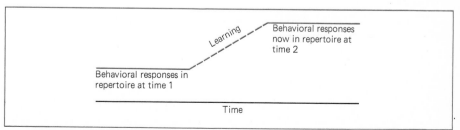

EXHIBIT 5.5   Learning involves a change in the behavioral-response repertoire of an organism. (Source: Lerner, 1976, p. 256.)

practice is combined with reinforcement, learning will occur. In other words, when a behavior is reinforced, a relatively permanent change in behavior potentiality will be obtained. Reinforcement, therefore, is an important component in our empirical conceptualization of the learning process. The application of reinforcement will make the acquisition of a relatively permanent change in behavior potentiality more probable. The exclusion of reinforcement, on the other hand, will not lead to such incremental changes. (Nevertheless, it should be noted that at least for one form of learning—observational learning—reinforcement may not be necessary for learning per se, but only for performance; Bandura, 1965, 1971. This form of learning will be discussed below.)

What, then, is reinforcement? A *reinforcement,* or a reinforcing stimulus, may be defined as any stimulus that produces or maintains behavior. For instance, salivation is a form of behavior. Given the appropriate stimulus conditions, our salivary glands will excrete a liquid substance termed saliva. One way of eliciting this salivary behavior is to show a hungry person his or her favorite food; the person will probably salivate, and the production of this salivary behavior is under the control of the food stimulus. In other words, the food is a reinforcing stimulus because it produces the salivary behavior.

Another form of behavior involves the muscular responses involved in digging a ditch. It is obvious that each time we shovel dirt up from the ground, we are behaving. Now, if someone comes up to us and simply gives us a shovel and tells us to start digging, we might readily decline this request. However, if the person says that for each shovel of dirt we dig, we will be given one dollar, most of us will now engage in shoveling behavior quite readily. What leads to the emission of shoveling behavior in this instance? Clearly, it is that the attainment of the money is made contingent upon our performing the behavior. The money that follows our behavior maintains it. If no money were to follow the behavior, we would probably not shovel. Yet since our attainment of the money is contingent upon our performance, the money maintains our behavior. Hence in this example money is a reinforcing stimulus; it leads us to emit an appropriate response. Our response emission is maintained because a reinforcing stimulus follows (is contingent upon) our behavior.

In sum, a reinforcing stimulus is a stimulus that will lead either to the elicitation of a response (as in the case of the food producing the salivation) or to the emission of a response in order to obtain the reinforcing stimulus (as in the case of digging). In either case the relation between behavior and reinforcement is that a reinforcing stimulus makes a behavior more probable. It is more likely that behavior will occur when it is reinforced. In general, reinforcement is needed for learning because *a reinforcing stimulus is a stimulus that increases the probability of a behavior* (cf. Skinner, 1950).

5. *Excluding Changes Due to Maturation, Fatigue, and/or Injury to the Nervous System.* Not all changes in behavior involve alterations due to the learning process. Only those relatively permanent changes in behavioral potentiality that occur as a result of reinforced practice may be attributable to the learning process. Changes that may not be reasonably attributed to learning are accounted for by variables involved in such things as maturation, development, fatigue, or nervous system injury. We may have learned, for example, that cookies are kept in a jar on top of the refrigerator. Yet because we are not tall enough (and because there are no chairs around), we will not perform the behavior of reaching into the cookie jar; however, when the appropriate maturation has occurred (e.g., we have grown 10 inches), such behavior may now be very typical. Fatigue may also lead to changes in behavior. Even if we are digging a ditch for one dollar per shovelful of dirt, our rate of shoveling will, after a time, decrease. Although our behavior might still be maintained by the monetary reward, it will change; it will slow down simply because we experience muscular fatigue. Finally, it is obvious that after some injury to our nervous system (e.g., serious brain damage following a car accident), our behavior will change, but such changes are attributable to the injury rather than to learning.

Types of Learning

We have seen some of the essential components of an empirical conceptualization of the learning process, but we have not considered the ways in which these components may interrelate in order to produce learning. What is the way, or ways, in which learning takes place? How does learning occur? How do reinforcing stimuli interrelate with behavior to produce learning? Generally, learning psychologists maintain that there are three types of learning. Two types of learning—classical and operant conditioning—represent the paradigms (models) of the learning process that are most involved with reinforcement as a key element in the acquisition of responses. As noted earlier, a third type of learning—observational learning—treats reinforcement as important in performing a learned response, but not necessarily as crucial in the acquisition of the behavior. We will now consider the essential components of each of these models.

CLASSICAL CONDITIONING

Classical conditioning involves behavior that is elicited by preceding reinforcing stimuli, while operant conditioning involves behavior that is controlled by succeeding reinforcing stimuli (Bijou & Baer, 1961). In this section we discuss the characteristics of classical conditioning. Many of us are probably familiar with what we have here termed the *classical learning paradigm* through knowledge of the work of Ivan P. Pavlov, the Russian physiologist who discovered this type of learning. We know that through happenstance Pavlov discovered that the dogs he was using as experimental subjects would salivate to stimuli other than food (e.g., his white lab coat or a bell). His studies of the variables involved in such associations led to an understanding of the nature of one of two basic types of learning we have noted, classical learning (or conditioning).

To understand the functioning of the classical paradigm we must first focus on the initial associations with which the organism enters the learning situation. Let us first posit that there exists a stimulus that reliably (i.e., regularly) elicits a response of strong magnitude. For example, we may assume that food is a stimulus that will reliably elicit a salivary response. Now, let us assume further that no prior association, or learning, is necessary to establish this relation between the stimulus and the response, between the food and the salivation. Given this assumption, we may term the food stimulus an *unconditional,* or *unconditioned, stimulus* (UCS). A UCS, then, is a stimulus that reliably elicits a response of strong magnitude without previous learning. Hence a UCS is an unlearned stimulus. Moreover, since no previous association, or learning, is necessary in order for the UCS to elicit the salivation, this salivation response is an *unconditioned response* (UCR), an unlearned response. Thus, to begin with, the classical paradigm may be represented as

$$\text{UCS (food)} \rightarrow \text{UCR (salivation)} \tag{1}$$

Ivan Pavlov in his laboratory. (The Bettmann Archive, Inc.)

Now, there exist numerous stimuli in the organism's environment which, of course, do not elicit this particular UCR. That is, they are neutral in respect to the above UCS → UCR relation; such neutral stimuli do *not* elicit the UCR. However, if one takes such a neutral stimulus (for example, a bell) and repeatedly pairs this stimulus with the UCS, then a different state of affairs will soon exist. For instance, suppose one rings the bell (that is, presents this neutral stimulus) and then presents the UCS; after such repeated pairings, the previously neutral stimulus will come to elicit a response markedly similar to the UCR. Because of the repeated pairings of the bell and the food, the bell begins to acquire some of the properties of the food: it, too, leads to salivation. We say, then, that the bell, previously a neutral stimulus, has become a *conditioned stimulus* (CS), a learned stimulus. Through the repeated pairings of the bell and the food, the bell becomes a CS and comes to elicit a learned, or a *conditioned, response* (CR). In this case the bell comes to elicit salivation. However, this salivation is not a UCR; rather, it is a response to a stimulus that obtained its control over behavior through its association with another stimulus. Since the organism learned this CS–UCS association, the response to the CS is a learned, or a conditioned, response. Hence because of

$$\text{CS (bell) –UCS (food)} \tag{2}$$

a learned response is acquired. A learned association, one that did not previously exist, between a previously neutral stimulus and a response to this stimulus is acquired. This association may be represented as

$$\text{CS (bell)} \rightarrow \text{CR (salivation)} \tag{3}$$

This is the classical conditioning, or learning, paradigm. A previously neutral stimulus precedes an unconditioned stimulus, and after repeated pairings of this CS and UCS, the CS comes to elicit a response markedly similar to the original unconditioned response. A summary of this paradigm may be represented as:

$$CS-UCS \rightarrow UCR-CR$$
$$\text{(bell)}-\text{(food)} \rightarrow \text{(salivation)}-\text{(salivation)}.$$

(4)

Characteristics of Classical Conditioning

We see that in classical learning, or conditioning, the CS or the UCS is said to elicit the response. The term elicit is used to signify the fact that classical conditioning involves the autonomic nervous system and the body's involuntary musculature. Hence we have neither to learn to salivate to food nor to learn to blink in response to a puff of air to our eyes. Rather, such responses are reflexive in nature; they are involuntary responses—hence classical conditioning involves the conditioning of our involuntary, reflexive responses. Such responses are almost literally pulled out of us by the stimulus in question (e.g., the food or the puff of air). Thus we say that classical responses are elicited. Moreover, since classical conditioning involves our involuntary nervous system, many of our emotional responses are acquired through classical conditioning. For example, a rat will show unconditioned emotional behavior in response to a strong electrical shock delivered to the floor of its cage. The shock is a UCS and the emotional response is a UCR. If a buzzer is always sounded two seconds before the shock begins, however, then after some time the rat will show a similar emotional response to the buzzer. The buzzer will become a CS and the emotional response will become a CR, a *conditioned emotional response* (CER). Similarly, if a human pedestrian witnesses a particularly bloody auto accident at a certain street corner, then he or she might experience unpleasant emotional reactions in response to such a sight (e.g., nausea). This sight of blood will be a UCS, the emotional response a UCR. If at some later time, however, the person is walking by that same street corner and is again overcome with nausea, this might indicate that the particular corner was established, in just one trial, as a CS and the emotional response was a CR.

Thus we see that classically learned responses can be considered reflexive in nature and that emotional responses are particularly susceptible to classical learning. Reinforcement, of course, plays an essential role in both these characteristics of classical learning. In the case of classical conditioning the role of the reinforcing stimulus is to produce behavior. That is, classical responses are elicited—they are produced—by the UCS. Since the UCS (e.g., food) produces the response (e.g., salivation), it provides, or represents, the reinforcing stimulus within the classical conditioning paradigm. In classical conditioning the UCS (the reinforcement) elicits (produces) the response, and thus we see that in this paradigm reinforcement produces the response. However, after classical learning has been accomplished, that is, after the CS has become able to produce (elicit) the CR, there is obviously another stimulus that now produces responding. Obviously, the CS also produces a response. However, because the CS obtained its reinforcing power (or efficacy) through its association with the UCS, we must distinguish between these two levels of reinforcers. Accordingly, we may term the UCS a primary reinforcing stimulus and the CS a secondary reinforcing stimulus. However, in the case of both the UCS and the CS, reinforcement always produces a response within the classical conditioning paradigm. In classical conditioning the reinforcing stimulus always leads to the response.

Although we have termed this type of learning classical conditioning, other names are used by learning psychologists to represent this process. For example, some psychologists refer to classical learning, or conditioning, as *Pavlovian conditioning*. Others refer to it as *associative shifting*, in recognition of the shift in association between the UCS and the response to the CS and the response. Other psychologists use the term *stimulus substitution* for similar reasons. Finally, another major term representing this type of learning is *respondent conditioning;* this term is used to highlight the fact that this type of learning involves responses to previous stimulation (Bijou & Baer, 1961; Skinner, 1938). However, all terms refer to a type of learning involving responses elicited by preceding, reinforcing stimulation. On the other hand, another major type of learning that exists—operant learning—works in quite another way. Let us now turn our attention to a consideration of the operant learning paradigm.

OPERANT CONDITIONING

Many of us have had the experience of being in a bus or subway station on a hot day. Suppose we are in such a situation and that we are hot and tired. Looking around the station, we might see two machines, a telephone on the wall and a soda pop machine. Being hot, tired, and perhaps a bit thirsty, we might put a coin in the soda pop machine and get a glass of soda pop. Certainly, however, we would not put the coin in the phone if we wanted something to drink, for such a response would not be followed by the appropriate (needed) stimulus—soda pop. In other words, only if we emit the appropriate response (putting a coin in a slot) in the presence of the correct stimulus (the soda machine) will we get the "needed" stimulus. If in the presence of certain stimuli we emit a certain response, this response will be followed by an appropriate stimulus. Emission of the same response, but in the presence of an inappropriate stimulus (in this case the telephone), will not be followed by the appropriate stimulus (the soda pop).

The above description represents an analogy of the operant conditioning paradigm. Certain stimuli in the environment are discriminated (responded to differentially) from other stimuli in the environment on the basis of the fact that responses emitted in the presence of some stimuli are followed by a reinforcing stimulus, while responses in the presence of other stimuli are not followed by a reinforcement. We know from our above discussion of reinforcement that if a response is followed by a reinforcing stimulus, the future occurrence of that response will be more probable. That is, if the occurrence of a reinforcing stimulus is made contingent upon the emission of a response, then the probability of that response occurring will increase. A response of putting a coin into a slot will be followed by a reinforcing stimulus (soda pop) if that slot works the dispensing mechanism of a soda pop machine. Since a response to that stimulus (the soda pop machine) will be followed by a reinforcement (soda pop), while responses to another stimulus (e.g., the telephone) will not be followed by the reinforcement, a discrimination between these two stimuli is established; responses in the presence of the soda machine are quite probable and responses in the presence of the telephone are not. This is the case because the former responses are reinforced, while the latter are not. Thus stimuli are discriminated on the basis of the consequences of responses made in their presence.

Hence operant conditioning involves learning to emit a response (R) in the presence of an appropriate, i.e., discriminative, stimulus (S^D); response in the presence of such discriminative stimuli is followed by a reinforcing stimulus (S^R). Simply, in terms of the above analogy, the presence of soda pop is contingent upon the emission of a correct response—a response to a soda pop machine and not a telephone.

A discriminative stimulus (S^D), then, is a stimulus that cues the occasion for a response. This is the case because a response (R) in the presence of an S^D will lead to the attainment of a reinforcing stimulus (S^R). Since the attainment of reinforcement is contingent upon the emission of responses in the presence of only certain stimuli, operant learning involves the acquisition of responses in the presence of S^D's and the absence of responses in the absence of S^D's. We may represent the operant paradigm as:

$$S^D\text{--}R \rightarrow S^R \tag{5}$$

The work of Skinner (1938, 1950) is, of course, most associated with operant conditioning. Most of his work has dealt with animals, and an example of operant conditioning with animals may be used to further illustrate this paradigm. Suppose you place a rat into a small experimental chamber (what has been termed by some a "Skinner box") containing a small lever (or bar) protruding from the wall and a magazine capable of delivering food into a cup also protruding from the wall. The animal when placed in the chamber will assuredly move around in it and eventually press the bar. If, when this happens, food is delivered, the animal will continue to press the bar if and as long as it is hungry. That is, if the animal has been deprived of food and if the bar press leads to food, the animal will bar press until satiated. However, if there is also a light in the experimental chamber which can be turned on or off by the experimenter, this light can soon be established as an S^D. As long as the rat is not satiated, it will continue to bar press. However, if the light is turned on and off at random intervals and bar presses occurring only when the light is on are followed by food, then responses when the light is off will soon diminish (or drop out entirely). Responses only in the presence of the light will remain. This example of operant conditioning may be represented as:

$$S^D \text{ (light on)--R (bar press)} \rightarrow S^R \text{ (food)} \tag{6}$$

Thus operant conditioning involves the acquisition of responses in the presence of certain (i.e., discriminative) stimuli, stimuli cueing the occasion for a response to be followed by a reinforcement. Responses emitted in the presence of such stimuli will be followed by a reinforcement; responses emitted when such stimuli are not present will not be followed by a reinforcement. The attainment of reinforcement is contingent upon the emission of a response in the presence of a discriminative stimulus.

Characteristics of Operant Conditioning

We see that within the operant conditioning paradigm, responses are emitted in the presence of a discriminative stimulus, and in fact a discriminative stimulus is established as such because such responses are followed by a reinforcing stimulus. An organism emits a response in the presence of an S^D because such responses are associated with succeeding reinforcement. In other words, responses are maintained in the presence of an S^D, and an S^D is, in turn, established as such because the occurrence of a reinforcement is contingent upon such responses. Thus responses are maintained because they are followed by reinforcing stimuli, or, conversely, the stimulus following the response in the operant conditioning paradigm maintains the production of that response.

In operant conditioning, the response produces the reinforcement (as opposed to classical conditioning, wherein the reinforcement produces the response). The organism must respond—in the presence of the appropriate stimulus (the S^D)—in order to obtain reinforcement. In other words, the organism is instrumental in obtaining reinforcement; it must operate in its environment in order to obtain reinforcement. In the presence of the appropriate, cueing, discriminative stimulus, the organism must itself respond (through pressing a bar, pecking a key, turning a handle, putting a foot on a peddle, etc.) in order for a reinforcement to be obtained. As opposed to classical learning, wherein responses are elicited by the UCS (the reinforcing stimulus), the organism must learn to emit the appropriate response, the response that will lead to the production of an S^R. The animal must behave in order to obtain a reinforcement. Thus since the organism itself is instrumental in obtaining an S^R, another name typically applied to operant conditioning is instrumental learning.

The term *emit* is used within the context of this paradigm to signify the fact that operant responses involve the body's voluntary musculature. Thus operant conditioning involves learning to use the voluntary musculature to emit a response that will be followed by a reinforcement. However, this is not to say that operant responses are not controlled. Rather, such responses are under the control of the stimulus environment. Since operant responses are maintained by succeeding, reinforcing stimuli, their probability of occurrence is determined by the degree of absence or presence of reinforcing stimuli; operant responses are also set by discriminative stimuli, since such responses are reinforced only when emitted in the presence of the appropriate discriminative stimuli. We see, then, that although operant learning involves the voluntary musculature, such responses are in the final analysis maintained and controlled by the reinforcing stimuli following them.

However, because operant responses do involve the voluntary muscles of the body, the major portion of motor behavior is thought to be acquired through operant conditioning. Such broad involvement in the majority of behaviors emitted by organisms may be illustrated by various kinds of seemingly disparate motor behaviors. If, while driving our car, we see a stop sign at the end of the road, we take our foot off the accelerator pedal and place it on the brake pedal. Soon after this response, our car comes to a halt. This motor behavior series may be interpreted as an instance of an operant sequence. The stop sign represents an S^D, cueing the occasion for a motor response (R) of lifting the foot off one pedal and placing it on another. The S^R in this instance is the halting of the car. Operant conditioning may also be involved in the exploratory behavior of young children; for example, while exploring the kitchen of his or her home when a roast is cooking in the oven, a young child puts a hand on the oven and then, of course, rapidly removes it. In further explorations the child continues to move away from the oven. Here we say that the oven is an S^D, which is associated with the R of moving away. The reinforcing stimulus in this instance is represented by the heat from the oven, and as we will soon see, this heat represents a negatively reinforcing stimulus. The point now is that the occurrence of most motor behaviors may be construed as being consistent with the operant paradigm.

Yet it is obvious that most of our daily motor behavior is much more involved than the behaviors described above. People typically emit long sequences, or chains, of behaviors mediated by the voluntary musculature, rather than single, discrete responses. Still, in principle, such complex chains of responses can be completely accounted for within the operant conditioning paradigm. However, such an accounting requires reference again to the principle of secondary reinforcement.

Secondary Reinforcement and Chaining

In our discussion of classical conditioning we saw that after repeated pairing with the UCS, the CS, too, obtains reinforcing power. Through its association with the primary reinforcing UCS, the CS thus becomes a secondary reinforcement—that is, a stimulus that acquires its ability to reinforce behavior through association with another reinforcer. Similarly, within the operant conditioning paradigm the S^D, through its association with the S^R, becomes a secondary reinforcing stimulus (S^r). The S^D cues the occasion when an R will be followed by an S^R, and this predictability between the occurrence of the discriminative stimulus and that of the reinforcing stimulus establishes the S^D as a secondary reinforcement. Because of the nature of its association with the S^R, reinforcing efficacy accrues to the S^D, establishing it as an S^r—a secondary reinforcement. Thus simply,

$$S^D = S^r \tag{7}$$

Because the S^D is also an S^r, this means that other responses may be acquired with the S^D used as the stimulus maintaining those responses. In turn, another S^D, which may be used to signal the occasion in which a certain response will now lead to the occurrence of the first S^D, will also, through its association with the latter discriminative stimulus, acquire reinforcing properties. In turn then this new S^D may now be used to reinforce yet another response. In this way, long sequences of behaviors (chains) may be established.

Some important conceptual implications of this chaining model may be pointed out. Our presentation of this model may enable one to see certain basic similarities between it and the mechanistic model discussed in Chapter 2. That is, the notion of chaining implies that to understand complex (e.g., adult) behavior all one must do is reduce this behavior to its constituent elements, these stimulus-response-stimulus connections. Hence from this view the development of behavior only involves building up these chains; as the child develops, she or he continually adds more of these identically constituted links.

CONDITIONING IN INFANCY

Can infants be classically conditioned? If so, at what age? What are the answers to these questions in regard to operant conditioning? It is fair to say that the absence or presence of conditioning in early infancy is a controversial topic (Fitzgerald & Brackbill, 1976; Sameroff, 1971, 1972; Sameroff & Cavanaugh, 1978).

Insofar as classical conditioning is concerned, Sameroff (1971, 1972) has taken the position that evidence for it in early infancy is far from unequivocal. However, some responses—such as the heart rate (Stamps & Porges, 1975; Crowell, Blurton, Kobayashi, McFarland, & Yang, 1976)—can be classically conditioned even in the neonatal period. Although evidence is less clear regarding the classical conditioning of behavior not so directly linked to the autonomic nervous system—as is heart rate—there have been some reports of successful attempts to classically condition motor responses such as sucking in the neonatal period. The results of one such study are presented in Exhibit 5.6. Using neonates as subjects, Kaye (1967) established conditioned sucking in an experimental group over baseline levels and beyond the levels shown in a control group (e.g., a group that did not have the appropriate pairing of the CS and the UCS).

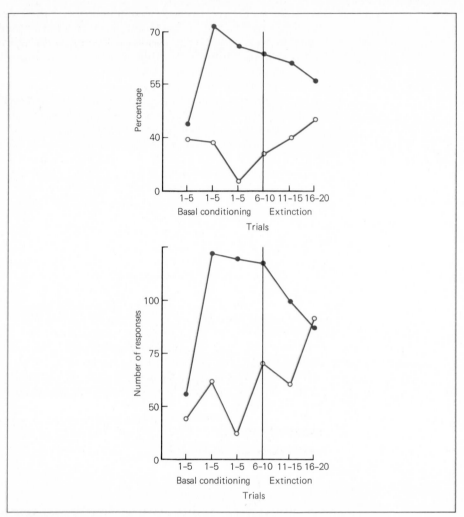

EXHIBIT 5.6   Classical conditioning of anticipatory sucking response. Closed circles, experimental group; open circles, control group. Top panel: percentage of trials on which at least one response occurred. Bottom panel: absolute number of CRs. (From Kaye, 1967, Fig. 12, p. 39. Reprinted by permission of the author and Academic Press, Inc.)

As previously noted, there is no unanimity as to whether *habituation* (a decrement in strength of a response to a stimulus, seen after repeated presentations of the stimulus) is actually a form of learning; and there is also no agreement on whether habituation is a simple or precursory form of classical conditioning (see Kimble, 1961). Nevertheless, many researchers use the phenomenon of habituation to study the infant's cognitive and/or learning competencies (see McCall, 1979). For example, in a study previously discussed in regard to visual perception, it was noted that Bornstein (1979) reported that 4-month-old babies who were habituated to a single color discriminated a change in color. In turn, he reported that 4-month-old infants who were habituated to a variety of colors generalized habituation to a novel color. In another study using a habituation procedure, Cohen and Strauss (1979) studied the ability of 18-, 24-, and 30-week-old infants to learn "categories" about adult female faces. At 30 weeks, infants learned to respond to a specific female face, regardless of its orientation (that is, how it was facing). However, such categorization (e.g., familiar, as opposed to not familiar) was not seen among the younger groups of infants.

There seem to be considerable individual differences—among infants and within an infant across his or her development—in patterns of habituation. McCall (1979) longitudinally studied infants between 5 and 10 months of age. Patterns of habituation in fixation time and in cardiac change to visual and auditory stimulation were assessed. Almost half the sample, when studied at 5 months, and almost 90 percent of the group, when studied at 10 months, did not show a regularly decreasing habituation pattern. Indeed, at 10 months of age, habituation was much more irregular and individually different than at 5 months of age. McCall (1979) did not find evidence for individual consistency in type of habituation, for any of the measures, across the two times of measurement.

Thus, despite the fact that habituation may be successfully used to investigate different types of perceptual and conceptual behavior among infants (see Cohen & Gelber, 1975), given the nature of individual differences in habituation, such as McCall (1979) reports, and also given the fact that some types of stimuli may be too complex to use in conjunction with a habituation paradigm (Lasky, 1979), infant-learning research using habituation must proceed with caution. For example, failure to dishabituate may not indicate a cognitive limitation in an infant. It may be that with a different stimulus and/or an infant or infants with different habituation patterns, alternative findings could have resulted.

In regard to operant conditioning there is less debate about the ability of even very young infants to learn in this way. In fact, Sameroff (1968) has demonstrated that the sucking response of the neonate can be operantly conditioned. Moreover, other researchers have demonstrated that young infants can learn behaviors that combine both classical and operant conditioning. Papousek (1967), for example, used the rooting reflex, discussed earlier, to demonstrate the combined influence of classical and operant conditioning; in response to a tactile UCS, the infant turned its head in the direction of the stimulation. A bell was paired with the UCS and represented the CS. However, upon turning the head, the infant received milk to suck on. Eventually head turning became an R eliciting the milk, which was the S^R.

Other researchers have shown that infants can be operantly conditioned through use of social stimulation. A classic example of this role of social stimulation is found in a study by Rheingold, Gerwirtz, and Ross (1959). After ascertaining the baseline (operant) level of vocalizations made by infants, the researchers made the administration of social stimulation to the infant (e.g., "cooing" to the infant, patting the infant's stomach) contingent on prior infant vocalizations. If the social stimulation provided to the infant served as a reinforcement, then increased vocalization should have been seen when the above contingency was operating; a decrease in vocalization, to baseline levels, should have been seen when the contingency was eliminated. As shown in Exhibit 5.7 the results of the Rheingold et al. (1959) study were consistent with these expectations. Although alternative explanations for these particular results have been offered, other studies using social stimulation as reinforcement have confirmed its role in operant conditioning in infancy (Finkelstein & Ramey, 1977).

For example, in a study already discussed in regard to auditory perception, it will be recalled that Williams and Golenski (1978) studied speech sound discrimination in infants aged 6 weeks to 12 weeks. The researchers presented contingent

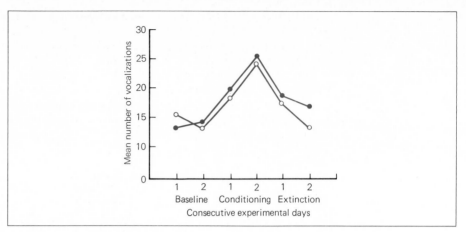

EXHIBIT 5.7   The mean number of vocalizations made by the two groups of subjects on the six days in the Rheingold, Gerwirtz, and Ross (1959) study of conditioning of infants' vocalizations. (Source: Rheingold et al., 1959.)

and noncontingent sound to the infants. That is, in the contingent situation high-amplitude nonnutritive sucking was reinforced with the presentation of a speech sound. There were two noncontingent conditions. Speech sounds were presented at either random or periodic intervals, but always independent of infant sucking. In all three conditions, the authors waited for the infants' rate of sucking to drop off. Then they either presented a new speech sound or continued to present the same speech sound. A significant increase in sucking rate occurred among those infants who were presented with the new speech sound *but* only if they had experienced contingent (reinforced) sucking.

Another, quite interesting, instance of the use of operant methodology to study infants' responsiveness to social stimulation is provided in the results of a study by DeCasper and Fifer (1980). By sucking on a nonnutritive nipple in different ways, a newborn infant could produce either the voice of its mother or the voice of another female. Infants learned how to suck in the manner that produced their mothers' voices, and they produced their mothers' voices more often than the other voice.

Other researchers have used operant techniques to assess whether differences in presentation of reinforcement affect infant learning. For example, Millar and Watson (1979) assessed whether delaying the presentation of contingent reinforcement influenced learning in 6- to 8-month-old infants. Infants were reinforced either immediately after emitting a response or only after a delay of three, six, or ten seconds. Infants showed reliable learning in the immediate reinforcement condition; however, in all delay conditions such evidence for learning was not found.

Finally, it should be noted that assessments of the learning capabilities of infants are used to explore the facets of the stimulus world about which the infant is presumably learning. For example, questions asked in recent research have been whether infants learn more about the form (shape) of objects in their world or the color of the objects they encounter. Casey (1979) reported a comparison of form and color discrimination learning in 1-year-old infants. The infants had to learn to discriminate either between two forms or between two colors. None of the infants given the color discrimination "problem" learned it, but about 50 percent of the infants given the form discrimination problem did learn it. Moreover, in contrast to infants who were given color problems, infants who were given form problems showed both an initial curiosity about changes in form and an ability to solve form problems.

In sum, although not flawless, and although there are great interindividual differences (qualifications which are, of course, applicable at any age level), infants show evidence of being able to interact with their world through their classical and operant learning capacities. Another way of studying the learning abilities of infants exists. It involves the assessment of the infant's imitative abilities and the study of observational learning, the third type of learning we have noted. We consider this form of learning next.

OBSERVATIONAL LEARNING: MODELING AND IMITATION

There is a class of learning phenomena that does not fall into either a classical conditioning or an operant conditioning category. This third type of learning may be termed *observational learning*. As discussed by Liebert and Wicks-Nelson (1981) this type of learning "occurs through exposure to the behaviors of others, the others being presented either live or symbolically in literature, films, television, and the like" (p. 170). As evidence that observational learning has occurred, the observer produces behaviors which are an *imitation* of that which was observed. Thus this form of learning is often labeled *imitative learning* and the actor that produced the to-be-copied or to-be-imitated learning is labeled the *model*.

Because one learns from observing others in one's social milieu, this form of learning is often treated under the heading of "social learning theory." In other words, theories which see basic learning phenomena as involving the imitation of behaviors observed in a model are social learning theories. Over the course of history, developmental psychology has seen several, somewhat different versions of social learning theory advanced (e.g., A. Davis, 1944; Miller & Dollard, 1941; Gewirtz, 1972; McCandless, 1970). Today, the position of Albert Bandura (1971, 1977, 1978, 1980a, 1980b) is most often thought of when people discuss principles of either observational learning or social learning theory.

Both classical and operant conditioning stress the role of reinforcement in the acquisition of behavior. In addition, insofar as operant learning is concerned, a response must first occur (and then be reinforced) in order for learning to occur (Shaffer, 1979). However, Bandura (1971) contends that neither classical nor operant learning principles are sufficient to explain the rich, complex, and often rapid development of behaviors in the repertoire of even very young children. In his view, learning can occur *vicariously*, i.e., through observation of others' behaviors and its consequences for them (Bandura, 1971). That is, children need not receive reinforcement, or even respond, in order to learn from a model. A classic experiment by Bandura (1965) illustrates this view. Bandura presented a short film to nursery school children, studied individually. In the film an adult was shown making a series of aggressive behaviors, for example, hitting a large, inflated plastic doll with a hammer or yelling at the doll. The children watching this film were divided in one of three groups, differentiated on the basis of the ending of the film which they saw. Group one was a "model rewarded" group. Here, at the end of the film a second adult gave the first one some candy for his "excellent" performance. Group two was a "model punished" one; a second adult scolded and spanked the first one. Children in group three saw neither reward nor punishment given to the model; here there were "neutral" consequences shown to the model.

After the film ended, children in all groups were individually placed in a room having many of the same toys that had been available to the model. Hidden observ-ers recorded each child's behavior. Children in both the "model rewarded" and the "neutral" groups imitated more than did children in the "model punished" group.

However, despite these performance differences there was evidence that children in all three groups equivalently learned the model's actions. Bandura (1965) offered children in all three groups rewards (e.g., a glass of fruit juice) if they would demonstrate what they had seen the model do in the film. This inducement canceled out the above-described performance differences, and children from all groups showed comparable amounts of imitative learning.

Thus from these data it is possible to see why Bandura (1965) makes a distinction between learning and performance. It appears that reward or punishment influences whether or not children will perform observed behavior; it does not appear to influence their learning of the behavior (Shaffer, 1979). In other words, the response consequences to the model affects whether an observationally learned behavior will be imitated. However, the mere observation of the model may be sufficient for learning to occur.

Observational Learning Processes

What are the requisite processes that determine the acquisition and performance of actions observed in a model? Processes both intrinsic and extrinsic to the child seem to play a role (Shaffer, 1979). However, because modern social learning theorists like Bandura (1977, 1978, 1980a, 1980b) emphasize intrinsic, cognitive processes in the acquisition of behaviors observed in a model, the theoretical approach is today often termed *cognitive social learning theory* (e.g., see Bandura, 1980a, 1980b). Shaffer (1979) has presented an excellent summary of both the cognitive and behavioral processes thought to be involved in observational learning. Among these are:

1 *Attentional processes.* The child must attend to the model in order to observe him, her, or it. What sort of variables negate the child's attention? The status of the model (e.g., his or her age) and the response consequences to the model (e.g., whether he or she is punished or rewarded) are features of the model and his or her social interactions to which the child attends (Bandura & Walters, 1963). A child's past learning history is also important for attention. For example, whether the model is a novel person or figure plays a role in evoking attention (Bandura & Walters, 1963). In addition, whether the model is presented in a medium of "intrinsic" interest to the children of a particular age, for instance, a cartoon figure presented to a young child, is an important variable in evoking attention (Shaffer, 1979). Finally, current situational influences, such as a teacher's instructions, can evoke attention.

2 *Memory processes.* To retain the information that has been observed, the child must use the cognitive processes involved in encoding, storage, and retrieval of the information.

3 *Motor processes.* The child must have the ability to translate the attended to and remembered behaviors into overt actions. He or she must have the *skill* to enact the behavior.

4 *Self-monitoring processes.* People are often aware that their performance may not conform to their image (memory) of the behavior they observed in others. For example, one of your authors often observes (on television) professional basketball players making graceful drives and jump shots. As he is in the process of trying to imitate these observations on the basketball court he has built in his driveway, he is quite aware of how his actions do not correspond—in either grace or outcome

(sinking the ball in the hoop)—with those he has witnessed. Often his lack of correspondence leaves your author with a feeling of great frustration—especially after repeated attempts at performance. As this example suggests, people also may have ideals of performance that they use as a standard against which to compare the behaviors they observe in themselves. These standards can be acquired through observational learning and/or through direct instruction, for instance, through a verbal statement, independent of observation (Bandura, 1978; Shaffer, 1979). Through applying these memories and standards, children can self-regulate and self-reward or self-punish their performance (Bandura, 1978).

5 Motivational processes. Here we return to reward and punishment. As noted earlier, rewards and punishments may not affect the acquisition of behaviors observed in a model (Bandura, 1965); however, they are the variables that appear to affect performance.

In sum, observational learning processes are quite distinct from those involved in classical and operant conditioning. Moreover, their distinction is today thought to lie to a great extent in the role that cognitive processes play in observational learning (Bandura, 1978, 1980a, 1980b). However, since cognitive processes show quite significant developmental changes throughout infancy and childhood, it is important to discuss the nature and development of observational learning through the infancy period. As such, we now turn to a discussion of the development of imitation in infancy.

Developmental Changes in Imitation in Infancy

There is some evidence that even in the first days, and certainly after the first few weeks and months, of life infants can imitate some behaviors. For example, data reported by Meltzoff and Moore (1977) indicate that 12- to 21-day-old infants can exhibit tongue protrusion in response to seeing an adult display a similar action. Meltzoff and Moore (1977) believe that such tongue protrusion represents the earliest form of selective imitation. However, this view has been criticized because their research did not consider the possibility that the matching behavior can be elicited by events other than the modeled behavior. To appraise whether this criticism is applicable, Jacobson (1979) studied twenty-four infants at 6, 10, and 14 weeks of age. She found that a moving pen and ball were as effective as the tongue model in eliciting tongue protrusion at 6 weeks of age. In turn, a dangling ring elicited as much hand opening and closing behavior at 14 weeks as did a model showing this behavior.

Most infants can imitate some adult sounds by 12 weeks of age, and many can imitate motor movements of the face, hands, etc., by the second half of the first year of life. In turn, by the end of the period of life that most scholars identify with the end of infancy—that is, by the time the child is 24 months old—infants' imitative capacities have increased so dramatically that the imitation of new words is a typical occurrence.

There are numerous data sets that document the development of imitative capacity through the infancy period. A major report of this development has been provided by McCall, Parke, and Kavanaugh (1977). One portion of their work involved the study of 12- to 24-month-olds. Infants watched a live adult perform several different behaviors (e.g., vocalizing, making a wooden puzzle, playing with some toys). After each behavior was performed, McCall et al. noted whether the infant imitated

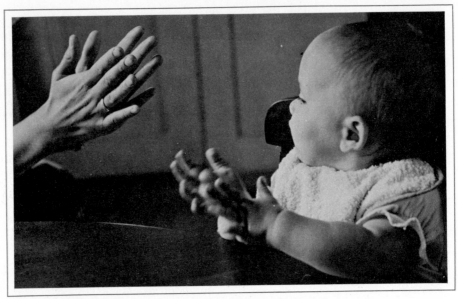

Infants in the first year of life can imitate the motor behavior of others.
(Suzanne Szasz/Photo Researchers, Inc.)

the adult's behavior. They found that the amount of imitation increased rapidly from 12 to 24 months of age.

Another portion of the McCall et al. report involved the presentation of televised as well as live models. As discussed in a subsequent chapter, there is great evidence that children beyond the age of infancy learn through observing televised models. Among infants, Slaby and Hollenbeck (1977) have found that 6-month-olds looked at a TV showing children's programs 49 percent of the available time. McCall et al. (1977) provide evidence that such observation is increasingly combined during infancy with imitative outcomes. Between 18 and 36 months of age children attended to a live model nearly all the time but showed an increasing tendency to attend to a televised model. At 18, 24, and 36 months of age the percentage of time spent watching the televised model was 68, 78, and 95 percent, respectively. Moreover, although both live and televised models were imitated at all age levels, at 18 and 24 months the live model was imitated more than the televised one. However, by 36 months of age imitation toward both models was virtually the same.

In sum, throughout infancy there is an increasing tendency to imitate behaviors observed in both live and televised models. Moreover, as the child makes the transition from infancy to childhood, there is evidence that either type of model is equally capable of evoking imitation. Given that imitative capacity begins at this time to involve language increasingly and that the child therefore becomes exposed to potential models through oral and written, as well as televised, media, it is clear that the potential richness of sources of observational learning experiences is quite great. Thus Bandura's (1971) point that much of the learning of the early years of life primarily involves observational learning—the imitation of models—seems an excellent one. Indeed, as we discuss many topics in subsequent chapters, for example, television effects, peer influences, and sex-role development, the important roles of imitation and of social learning theory will be seen. Here, however, we should note

Children often imitate what they see on television.
(Hella Hammid/Rapho/Photo Researchers, Inc.)

that other ways of studying the psychological abilities of infants exist. We consider another major one next.

PSYCHOMETRIC INTELLIGENCE

The study of cognitive development may be defined as the assessment of change processes involved with the acquisition and use of knowledge (Elkind, 1968). From this view, the terms *intelligence* and *knowledge* are just synonyms of cognition. However, there are many approaches to understanding cognition and what is involved in developing one's intelligence or one's knowledge. Yet despite the multiplicity of theories, definitions, and tests, two broad approaches to the study of cognitive development can be identified. Within these two major approaches, many of the major issues, findings, and problems in the study of human intellectual functioning and knowledge development over the life span exist. The two major approaches in this area of study are the quantitative ones, typically associated with the administration of intelligence tests and the derivation of IQ scores, and the qualitative ones, typically associated with the developmental theory of cognition of the Swiss scientist Jean Piaget (Elkind, 1968). We will treat both of these approaches in the remainder of this chapter. To do so, we will first offer a definition of intellectual-knowledge development that will be useful across both of these formats.

A Definition of Cognition

Within both the quantitative and the qualitative approaches to the study of intellectual-knowledge development, there is a basic concern with the same fundamental issue: What is the character of the person's knowledge? However, the two approaches differ in regard to the formulation of this concern. In the *quantitative, or psychometric, approach* there is an emphasis on *how much* the person knows relative to his or her age group. This emphasis leads to the derivation of a score (an intelligence quotient, or IQ) from a mental test used to represent the quantity of knowledge. Yet despite this emphasis on measuring whether a person knows more, as much, or less than others of his or her age group, proponents of this approach do not deny that the type of thinking or knowledge a person shows may differ at various points in development (Elkind, 1968). Indeed, different mental tests are often used to derive IQ scores at different developmental levels (Bloom, 1964).

The recognition of different modes, types, or qualities of thought is the emphasis of the second approach. Here, rather than stressing whether a person knows more or less information than his or her peers, the emphasis is on the way the person knows whatever he or she does know. Thus rather than emphasize the number of correct answers to test items, people using the *qualitative* approach to intelligence will consider the type of thinking a person uses to deal with information. Despite whether the answer to a question is correct, there will be a manner of thinking about the issue—for instance, in an abstract, or a hypothetical, way *or* in a concrete manner—which will be used by a person. It is this mode of thought—this quality—and not the correctness of the answer—that will be the focus in the second approach.

Thus both approaches are concerned with knowledge, with changes within people, and with differences between people in their knowledge. Those adhering to a quantitative approach will be concerned with how much knowledge a person has across his or her life and with differences between people in how much they know. Those looking at intelligence qualitatively, however, will be concerned with changes in the type of thinking shown by a person across his or her life and the differences among people (e.g., of different ages) in the quality of their thoughts. Hence although the kinds of changes and differences emphasized within the two approaches differ (Elkind, 1968), both approaches are concerned with characteristics of knowledge, with what and/or how a person "knows."

Accordingly, scientists specializing in this area of study are concerned most basically with knowing and with the psychological and behavioral processes involved in the development and use of knowledge (Elkind, 1968). Those working within the quantitative approach traditionally have used the term intelligence to reflect their area of study. Those working within the qualitative framework traditionally have used the term cognition most often, although they have considered the study of cognitive development and intellectual development as interchangeable endeavors (Neimark, 1975; Piaget, 1950, 1970). Thus one may find a common concern with knowledge and its development. Furthermore, there is a conceptual and an empirical (DeVries, 1974; Kuhn, 1976) compatibility between each approach. As a result, the term cognition may be used to represent the concern with intelligence and knowledge found within each of the two approaches identified.

Quantitative Dimensions of Cognitive Development

The quantitative approach to cognition emphasizes the measurement of mental functioning. For this reason it is also labeled the psychometric approach. The psychometric approach attempts to define and understand intellectual development by examining the interrelationships of various tests purported to measure intelligence.

Intelligence Tests

How is intelligence measured? What instruments have been developed to measure intelligence? The first intelligence test was devised in France in 1905 by Alfred Binet. His purpose was to devise a technique to screen children from the French public schools who were not mentally competent to profit from education. Intelligence tests were brought to the United States by Lewis Terman of Stanford University. Terman revised the scale of Binet by standardizing it on American samples. This new version was published as the Stanford-Binet in 1916. Since its initial introduction it has been restandardized and revised several times. The Stanford-Binet is an individual test of intelligence. Stated differently, the test is administered by a single administrator to a single person. But it has not remained the only intelligence test available. There are other individual intelligence tests, such as Wechsler's Intelligence Scale for Children (WISC) and Wechsler's Adult Intelligence Scale (WAIS).

Indeed, hundreds of tests have been developed to assess intelligence. Amazingly, these measures tend to correlate positively, although not necessarily highly, with one another. That is, relative to other people, an individual scoring high on a test of, say, arithmetic ability will tend to score high on a test of, say, vocabulary ability even though the tests appear to be measuring different things. In addition, performance on measures of intelligence tends to correlate positively with common-sense indexes of intellectual ability (e.g., scholastic achievement). Such relationships led early investigators to postulate the existence of a general intelligence which pervaded all cognitive tasks (Spearman, 1927). Let us now see what sort of tests are used to measure intelligence in infancy.

Measures of Infant Intelligence

Since infants do not have language, one cannot ask them to respond to the same sorts of items one can use with other children in tests of intelligence. One can assess babies' sensory and motor behaviors—and indeed this is what most infant tests of intelligence actually do—but such measures are not very similar to those found on tests of intelligence aimed at older children (Lewis & McGurk, 1972). Understandably, scores from infant intelligence tests are not highly related to intelligence measured at subsequent child, adolescent, or adult levels (Bloom, 1964). What this means is that a single measure of early sensorimotor behavior does not relate to later verbal, abstract reasoning, or spatial manipulation behaviors. Indeed, when adolescents or adults are measured at one point in time, their sensorimotor skills are also not related to their abstract mental abilities (Bloom, 1964).

In one major study illustrating the above points, Lewis and McGurk (1972) tested a group of infants at regular intervals during their first 24 months of life. The mental scale of the Bayley Scales of Infant Development was administered at 3, 6, 9, 12, 18, and 24 months, as was a test of whether the infant showed evidence of knowing the existence of an object was permanent even though he or she was not immediately perceiving it (i.e., a test of "object permanence" was administered). Finally, at 24 months the infants were given language comprehension and production tasks. All scores for all these tests were related to each other (through the calculation of correlation coefficients). In very few cases (in only 4 cases out of 30) were scores from one age related to scores of any other age—and it should be remembered that

BOX 5.1

DERIVATION OF AN IQ SCORE

Depending on the particular definition of intelligence being used, a person constructing a test of intelligence will devise items which presumably will provide measures of the particular cognitive abilities being evaluated. The validity of these presumptions typically will be ascertained with large groups of people. As a consequence, standards of performance will be developed.

The groups of people used in this test construction process will have particular demographic characteristics. For example, they may be white, middle-class, Protestant males and females between 3 and 25 years of age, living in urban areas in the midwestern United States. The performance of these groups will become the standard; that is, the average and range of responses seen with this group will become the norms of performance on the test. In sum, test norms are derived from the performances of people in the standardized population.

With such normative information we can know how a person having the characteristics of particular segments of the standardized population should perform. For instance, one can know what the typical 5-year-old from this population should score. Furthermore, one can appraise the performance of any given person relative to the performance of those in the standardized group. For example, one can see how the target person's actual score on the test items compares with the expected average score for his or her age group (Matarazzo, 1972). From such a comparison one can derive an IQ score.

There are several ways in which this score can be computed (Matarazzo, 1972). More recent methods use somewhat complicated statistical procedures, compared to the simpler methods associated with the technique originally employed. Although these more recent calculation procedures are necessary in construction and interpretation of test scores, the simpler, original formula for calculation will be used to illustrate the basic idea behind the derivation of the IQ score.

The purpose of the IQ score is to reflect in a single value how much knowledge, ability, or cognitive aptitude a person has relative to others in his or her group. The "group" corresponds to the standardized population, and typically in the study of human development, the person is compared to others of the same age. In short, one expresses with the IQ score a person's cognitive status relative to his or her age group.

Using the formula for the intelligence quotient (IQ) first developed by Stern (1912), one can express this relative standing as

$$IQ = \frac{MA}{CA} \times 100$$

where MA = mental age, or, in other words, one's actual score on the test, and CA = chronological age, or, in other words, the mean score that would be normatively expected for one's age. By dividing MA by CA, the formula allows for this relative evaluation, and by multiplying by 100 one avoids dealing with fractions and instead treats only whole numbers. Moreover, use of this formula ensures that the average IQ score for any age group will be 100.

Thus, if one's attained score is greater than the score expected for one's age, then one's IQ score would be better than the average person of one's own age (an example is a 10-year-old that knows what the average 12-year-old knows). One would always have in such a case an IQ above 100 (e.g., 120 in the above example). On the other hand, if one knew less than what was average for one's age, one would always have in such a case an IQ less than 100 (an example is a 12-year-old that knows what the average 10-year-old knows, and would have an 83 in this instance). If one's attained score at any age is the same as the average score for people of that age (if you are 10 years old and know as much as the average 10-year-old), then one's IQ would be 100. Said another way, at any age level, the average IQ equals 100.

This fact means that IQ scores are always relative appraisals. An IQ score expresses one's location in an age-based distribution of other IQ scores. As such, IQ is not an indication of any absolute level of knowledge. Indeed, it takes much more knowledge to attain an average score as a 12-year-old than it does as a 5-year-old and thus, for example, a 12-year-old with an IQ of 100 knows much more than does a 5-year-old with an IQ of 120. Yet one says the 5-year-old is "brighter" than the 12-year-old because, *relative* to their respective age groups, the 5-year-old scores higher.

all these scores were of performance *within* infancy. Thus, Lewis and McGurk (1972, p. 1176) conclude:

> Our results indicate that there is no reliable relation between successive measures of infant intelligence during the first 24 months of life. A similar picture emerges with respect to the measure of sensorimotor development—the object permanence scale—employed in our study. . . .
>
> All in all, these findings cast serious doubt on the notion that the concept of general intelligence is applicable to the period of infancy . . . our data tend to support the view, advanced by Bayley (1970), that at each stage of infant development intelligence comprises a set of relatively discrete abilities, or factors . . . therefore, there is no necessary continuity between intelligence as defined at one stage of development and as defined at another. (Lewis & McGurk, 1972, p. 1176)

Nevertheless, although infant "intelligence" tests are thus best thought of as measures of sensorimotor maturity rather than of abstract mental abilities, they may be useful—when used by an experienced and trained medical or psychological practitioner in conjunction with other assessments—for indicating when extreme variations from normative sensorimotor development exist *in the infancy period;* thus in such cases test results may alert pediatricians, psychologists, and parents to potential developmental problems with the child. In this regard, although Lewis and McGurk (1972, p. 1175) caution that "infant intelligence scales are invalid as measures of future potential" and that "the necessity for caution in this respect cannot be overstressed," they do note that:

> Despite these acknowledged limitations, infant intelligence scales are widely used in clinical situations, in the belief that, although lacking in predictive validity, these scales are valuable in assessing the overall health and developmental status of babies at the particular time of testing, relative to other babies of the same age. (Lewis & McGurk, 1972, p. 1175, emphasis of last phrase ours)

To amplify the points of Lewis and McGurk, it should be noted that the "items" on these infant tests relate to such behaviors as holding the head erect, eye-hand coordination, and eliciting various reflexes. Norms, which again are average responses for people of particular ages and socioeconomic backgrounds, have been established for when infants should show the behaviors appraised by these items. For example, on Bayley's (1968) Scale of Mental Development, 50 percent of the infants in the population with which the test was standardized, (i.e., the group with which norms were established) passed the following items (i.e., showed the following behaviors) at the following ages: blinks at shadow of hand—1.9 months; head follows vanishing spoon—3.2 months; picks up a cube—5.7 months; stirs with spoon in imitation—9.7 months; imitates words—12.5 months; and says two words—14.2 months.

From an infant's performance on such standardized items a developmental quotient, or DQ (derived from dividing developmental age by chronological age and multiplying by 100), may be calculated and interpreted in the manner we have seen associated with IQs. Thus if infants have very low DQs, this may be a sign of a major problem with their sensorimotor development

There are several infant tests that are currently used to appraise the sensorimotor status of the infant. Some of them are noted below.

1. *Bayley Scales of Mental and Motor Development* (Bayley, 1933, 1968). We have already introduced some information about this test. This test is one of the most often used measures of infant intelligence. Some of the items on this test are, (1) Can the infant follow a moving object with his or her eyes at about 2.5 months of age? (2) Can the infant reach for an object with one hand at about 13 months? (3) Can the infant put round blocks in round holes at about 17 months? Other items involve grasping objects, turning the head to follow objects, and, at older ages, imitating simple actions and drinking from a cup.

2. *The Gesell Developmental Schedules* (Gesell & Amatruda, 1947). This test is divided into four areas of development. It covers the age range from 4 weeks to 6 years, i.e., well beyond the infancy age level. Motor behavior is assessed through items relating to such behaviors as holding the head erect, sitting, standing, creeping, and walking. Adaptive behavior is assessed by such behaviors as eye-hand coordination and problem solving. In addition, language behaviors and social behaviors are assessed.

3. *The Cattell Infant Intelligence Scale* (Cattell, 1947). This test is designed to measure children within an age range of from 2 to 30 months. The test uses items from the Gesell schedules, from the Stanford-Binet, and from other tests as well. Some items involve following movement with the eyes, lifting the head, and using the finger.

4. *The Brazelton Neonatal Behavioral Assessment Scale* (Brazelton, 1973). The author of this test, a pediatrician, did not develop the test as a measure of infant intelligence; rather, as we have noted above, the test is used within the framework of the idea that a measure of early infant sensorimotor status may be useful in discriminating healthy infants from those with potential health problems. That is, as pointed out by St. Clair (1978), the view that behavior is closely related to underlying neurological mechanisms contributed to the idea to develop a behaviorally oriented infant assessment procedure.

The test is used with infants in the first 10 to 14 days of their lives, i.e., during their neonatal period. The test measures an infant's responses to the environment through taking neurological and behavioral assessments. Scores are based on the infant's best performance. Items involve behaviors relating to changing responses to a rattle and to a pinprick, alertness, and motor maturity. However, behavioral capabilities believed to reflect the quality of the parent-infant relationship are the focus of the assessments (St. Clair, 1978).

THE PIAGETIAN APPROACH TO THE STUDY OF COGNITION

The theory of Jean Piaget has become almost synonymous with the qualitative approach of the study of cognition (Elkind, 1968). Piaget's theory is a most complex one. As noted in Chapter 2, he divides the course of cognitive development into stages. Furthermore, to understand the whole range of stage progression, one must recognize that Piaget has concepts which cut across the stages he posits. He has concepts that are thus stage-independent and, in fact, are used to explain the basis of a person's development from one stage to another. Finally, however, in order to deal with all this complexity, one has to recognize that Piaget comes to the study of human cognitive development from a unique intellectual perspective. He was not trained as a social scientist and hence does not use a vocabulary typical of many psychologists or sociologists. Piaget's personal background is summarized in Box 5.2

PIAGET'S THEORY OF COGNITIVE DEVELOPMENT

Piaget proposes that cognitive development progresses through phases. He casts this proposal in the form of a stage theory of development. Within developmental psychology, a stage theory is one which holds that all people who develop pass through a series of qualitatively different (discontinuous) levels of organization in an invariant sequence. Thus stages are seen as universal levels of progressions. Within a stage theory, stages may not be skipped or reordered. The only interindividual differences possible are in the rate one passes through stages and the final level of development reached. Some people's development may not proceed, for instance, because of fixation; but *if* they did develop, they would have progressed in the specific sequence.

However, all stage theorists propose some mechanism by which a person changes from one level to another. Freud claimed, for example, that the movement of libido into specific bodily zones was the basis of a stage emerging. Piaget has his own ideas about general principles of development which apply at all stages and which, at the same time, account for progression from one level of functioning to the next.

Stage-Independent Conceptions of Cognitive Development

Piaget proposes principles of cognitive development that apply to all stages of development. These are general laws describing phenomena that continually function to provide a source of cognition throughout ontogeny. To understand them we must focus on the biological basis of Piaget's theory.

To Piaget (1950, 1970; Inhelder & Piaget, 1958; Piaget & Inhelder, 1969), cognition is an instance of a biological system, just like digestion, respiration, and circulation. As such, cognition is governed by laws that apply to any biological system and has functions and characteristics identical to any biological system

Like all biological systems, cognition has two components that are invariant: *organization* and *adaptation*. Cognition is always organized and it is always an adaptive system. Its functioning allows the person to adapt to his or her environment and, thus, to survive. Although he recognizes the fundamental importance of both these characteristics of cognitive functioning, Piaget devotes the major portion of his theorizing to adaptation. By understanding the adaptation involved in relations between person and environment, one may understand the bases of cognitive develop-

BOX 5.2

PIAGET: A BIOGRAPHICAL OVERVIEW

Jean Piaget was born in Switzerland in 1896. He died there in 1980. Piaget was intellectually precocious. He published his first scientific paper at the age of 10. During his own adolescence, he had published so many high-quality research papers on mollusks (sea creatures such as oysters and clams) that he was offered the position of curator of the mollusk collection in a Geneva museum until it was discovered that Piaget was only 14 years old (Flavell, 1963). To culminate these early achievements, Piaget received a doctorate in the natural sciences at the age of 21.

After receiving his doctorate in 1918, Piaget, maintaining a broad intellectual interest, became involved with work in psychology and maintained an active interest in *epistemology*, the philosophy of knowledge. Perhaps because it seemed that the best way to understand knowledge was to study how it develops, Piaget began studying the development of cognition in his own children. From these initial studies, his first book resulted, and in his books he began to devise a developmental theory of cognition. He viewed cognition as a developmental phenomenon rather than all development as a cognitive phenomenon.

Since Piaget came to his interest in cognitive development from his training in natural science and his interest in epistemology, his theory is colored by these intellectual roots. As a consequence, Piaget's theory uses many philosophically based terms and has a strong biological basis. This emphasis will become apparent as his theory is considered.

Jean Piaget. (The Bettmann Archive, Inc.)

ment. That is, the process of adaptation is divided into two *always* functioning complementary processes: *assimilation* and *accommodation*.

Assimilation. When a cell assimilates food, it takes the food in through its membrane, breaking it down to fit the needs of the cell. In other words, in assimilation an object taken in does not retain its original form or structure; it is altered to fit the existing structure. Hence, as Piaget has said, "from a biological point of view, assimilation is the integration of external elements into evolving or completed structures of an organism" (1970, pp. 706–707).

Cognitive assimilation involves changing an object, external to the subject (the person), to fit the already existing internal knowledge structure of the subject. For example, an infant may have knowledge of its mother's breast, gained through its *actions* on this external stimulus object. The infant has sucked on its mother's nipple and has come to "know" the mother's breast through the actions it performs in relation to it. The subject develops an internal cognitive structure from its actions on an external stimulus. Thus to Piaget the basis of knowledge lies in action.

Assimilation occurs when, for instance, the infant discovers its thumb and begins to suck on it. Knowledge of this external stimulus may be gained by the infant acting on the thumb as it did the nipple. The infant would integrate the thumb to the already existing cognitive structure pertaining to the mother's breast by altering actions on the thumb or, rather, fitting actions on the thumb, so as to incorporate the actions into an already existing cognitive structure. The infant would be changing the object to fit, or match, the structure of the subject, and so the infant would be assimilating.

Accommodation. The process that is the complement of assimilation is termed accommodation. Accommodation involves altering the subject to fit the object. Cognitive accommodation involves altering already existing cognitive structures in the subject to match new, external stimulus objects.

The infant could accommodate to its thumb, rather than assimilate it, by acting on the thumb not as it did on its mother's nipple, but rather by altering its actions. Such a change in action would modify the already existing cognitive structure. This structure is altered to include the different actions on this new object, and thus, there is a gain in knowledge. By the subject's altered actions on the different object, a corresponding alteration in the subject's cognitive structure occurs. Rather than matching the object to the subject, the subject—through differential actions—matches the object. Hence accommodation occurs.

Equilibration. Assimilation and accommodation are seen as complementary processes because Piaget postulates that there exists a fundamental factor in development, termed *equilibration.* Piaget proposes that a person's adaptation to the environment involves a balance—an equilibrium—between the activity of the person on the environment (assimilation) and the activity of the environment on the person (accommodation). In other words, when a person acts on the environment, he or she incorporates the external stimulus world into an already existing structure, and this is assimilation. At the same time, when the environment acts on the person, the person is altered in order to adjust to the external stimulus world, and this is accommodation.

Piaget hypothesizes an intrinsic orientation in people to equilibrate their actions on the environment with the environment's actions on them. Piaget proposes that neither of these two tendencies should always override the other if the person is to be adaptive. Thus equilibration is, to Piaget, the balance of interaction between subject and object (Piaget, 1952).

Piaget proposes that equilibration is the moving force behind cognitive development. Whenever the person alters the environment, incorporating it into an already existing internal structure, there must also be a compensatory alteration of the person's structure to match the objects in the external environment. There must be a balance in action—the basis of all knowledge—between person and environment.

Functional (or Reproductive) Assimilation. However, if cognitive development moves toward equilibration, then why, when such a balance is reached, does cognitive development not just stop there? It is not enough to argue that there are many things that impinge on the infant's world that necessitate further assimilations and accommodations. If the infant is in equilibrium, there would seem to be no reason to bother with other impinging stimulation.

To address this problem, Piaget specifies that there exist several other aspects of assimilation. Discussion of one of these—*functional* (or *reproductive*) *assimilation*—will illustrate how cognitive development continues to progress. The concept of reproductive assimilation refers to the fact that any cognitive structure brought about through assimilation continues to assimilate. Any biological system's equilibrium is necessarily temporary; the system must continue to function if adaptation is to be maintained. Ingested food may place the digestive system in equilibrium, but such balance is transitory since the food must necessarily be assimilated if digestion is to continue to subserve its adaptive function (to provide energy). Similarly, when a simple cognitive structure is developed on the basis of assimilation—such as that involved in our example of the infant's sucking on the mother's nipple—it continues to assimilate. It functions to reproduce itself. Such structures "apply themselves again and again to assimilate aspects of the environment" (Flavell, 1963, p. 35). Thus the concept of functional assimilation indicates that it is a basic property of assimilatory functioning to continue to assimilate.

Hence any equilibrium the infant establishes will be only transitory. The child would then assimilate other components of the environment, and this, in turn, would require a compensatory accommodation. When an equilibration is reached again, it, too, is short-lived, because a *disequilibrium* will inevitably result when the child continues to assimilate. Because of the disequilibrium resulting from continued functional assimilation, higher levels of cognitive development are reached when the child next equilibrates.

In summary, the occurrence of disequilibrium (through the process of reproductive assimilation) provides the source of cognitive development throughout all stages of life. This equilibration model provides a set of stage-independent concepts about cognitive development that account for a person's continued cognitive development. A consideration of Piaget's stage-dependent concepts follows.

Stages of Cognitive Development

As noted, Piaget (1950, 1970) describes four stages of cognitive development. People may differ in their rate of development through a stage, but some rough age boundaries for each stage can be indicated. The stage of cognitive development associated with infancy is the sensorimotor stage. We discuss this stage below; the other stages are discussed in later chapters.

The Sensorimotor Stage

Piaget labels the first stage sensorimotor and suggests that it lasts from birth through about 2 years. Changes in this stage involve the development of what Piaget terms *schemes*. A scheme is an organized sensorimotor action sequence. As such, the schemes existing throughout this first stage conveniently may be thought of as reflexive in nature. Similar to a reflex, a scheme is a *rigid* cognitive structure. The direction of the sequence involved in the scheme is always the same. Analogous to a reflex such as an eye blink, wherein a puff of air always precedes and leads to an eye blink, a scheme is also unidirectional; the motor component of the schematic sequence cannot be reversed.

For much of the early portions of the sensorimotor stage, the infant's cognitive development could be described as "out of sight, out of mind." The infant interacts with objects in the external world but acts as if their existence depended on his or her sensing them (Piaget, 1950). When objects are not in the infant's immediate sensory world, the infant acts as if they do not exist. The infant is *egocentric*; there is no differentiation between the existence of an object and sensory stimulation provided by that object (Elkind, 1967). There is no knowledge that objects exist permanently independently of the subject. Thus there is no scheme of object permanency.

Bremmer (1978) provides evidence for the presence of such egocentrism; 9-month-old infants were put in a situation wherein they saw an object hidden in one of two places but were prevented from reaching for it until after the spatial relationship between the infant and the object had been altered in some way. Bremmer found that, first, if infants had previous search experience at one place, they were likely to keep searching at this one place. This Bremmer saw as egocentric spatial coding; that is, the infants responded in terms of their past experience and not in terms of whether the object was placed in the place of their past experience. Second, Bremmer found that infants were better at locating the object after they had moved than they were after the object had moved. Finally, if specific differences between the covers of the two hiding places were made, less egocentric responding was seen.

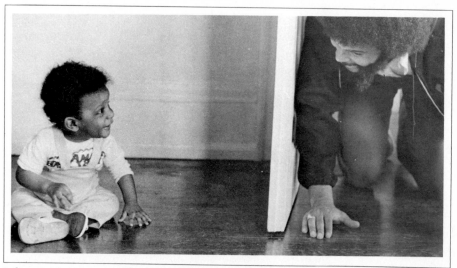

Infants in the sensorimotor stage enjoy playing peekaboo games.
(Photo Researchers, Inc. © Alice Kandell.)

All stages of cognitive development contain functioning that may be described as egocentric. Overcoming *sensorimotor egocentrism* involves the most important cognitive attainment of the child during this stage. All remaining portions of this stage contain schematic structures involving the infant's developing knowledge that there is object permanency in the world.

There are numerous instances of the infant's lack of a scheme of object permanency. For example, the child acts as if a person or object appears and disappears by virtue of going into and passing out of sight. Only as a consequence of repeated sensorimotor actions will a child develop an internalized representation of an object and come to know that it exists even though he or she is not perceiving it. When this occurs, the child has "conquered the object" (Elkind, 1967). The child's egocentrism has diminished enough to allow him or her to now know the difference between an object and the sensory impression it makes. This knowledge implies *representational ability*, an ability to represent internally an absent object and thus to act as if one knows it continues to exist. This major cognitive achievement enables the infant to progress to the next stage, the preoperational stage. We discuss this stage in Chapter 7.

CHAPTER SUMMARY

Upon birth the average infant weighs 7.5 pounds and is about 20 inches in length. By 2 years of age, most boys have achieved 49.5 percent of what will be their final adult height and most girls have achieved 52.8 percent of their final adult height. This instance of physical development in the infancy years (from birth to about age 2) is consistent with the elaborate developments in sensory, perceptual, and cognitive functioning in this period.

Sensation refers to the reception of stimulation. Perception is sensation with meaning. Sensory and perceptual abilities develop in infancy. Infants can discriminate among substances of different tastes and show a preference for touching novel, as opposed to familiar, stimulus objects. Infants' hearing is also well developed. They can localize sound and respond differentially to different sounds. Infants attend to human speech more so than to nonspeech sounds and show particular responsivity to the sound of their mothers' voices. Infants can also integrate information between auditory and visual sense modalities (i.e., they show intersensory integration).

In regard to vision, infants respond differentially to light and dark, show visual acuity, and can follow a moving stimulus. Even in early infancy, infants respond more to patterned than to unpatterned stimuli. Most attention seems to be shown to human facial or facelike patterns. Infants also show evidence of perceptual constancy, depth perception, and spatial perception.

In regard to motor functioning, there are numerous neonatal reflexes present in the newborns' behavioral repertoire. In addition, infants show a developing ability to coordinate their motor behavior (e.g., reaching) with their visual perception. The infant's developing motor abilities are part of the changing nature of the infant's complex behavior sequences and states, and the infant shows the ability to integrate the sensory stimulation impinging on him or her with this motor functioning.

This linkage suggests the presence of learning abilities in infancy, and, in fact, there is considerable evidence that infants learn. Empirically, learning is a relatively permanent change in behavior potentiality which occurs as a result of reinforced practice, excluding changes due to maturation, development, fatigue, and/or injury to the nervous system. Three major types of learning were discussed in this chapter: classical and operant conditioning and observational learning. In the first type of conditioning, a reinforcing stimulus elicits a response; in the latter, an emitted response is followed by a reinforcing stimulus. In observational learning, behaviors are imitated as a consequence of observing the actions of others; reinforcement appears to influence performance but not learning.

Although the data are far from unequivocal, there is evidence of classical learning in infancy. In regard to operant conditioning, there is relatively strong evidence that infants can learn in this way. In addition, there is good evidence that infants can learn by observing others. There is evidence that infants can and do imitate both live and filmed or televised models.

Another way in which the mental or psychological abilities of infants have been studied falls within the psychometric, or quantitative, tradition. Attempts have been made to develop psychometrically sound measures of infant intelligence. Most results indicate that measures of infant intelligence do not predict subsequent child, adolescent, or adult intelligence. Moreover, measures of intelligence taken at different times within the infancy period do not correlate well with each other.

Piaget's theory of cognitive development has been the most influential one in the qualitative approach to cognitive development. Piaget proposes that there exist four invariantly ordered, universal stages of cognitive development. People pass through these stages as a consequence of the functioning of an equilibration process which involves the functional invariants of assimilation and accommodation. The stage of cognitive development of Piaget's theory pertinent to infancy is the sensorimotor stage. This stage involves the elaboration of schemes, organized sensorimotor action sequences. The major cognitive achievement in this stage is the establishment of the scheme of object permanency.

chapter 6

Infancy: social and personality development

CHAPTER OVERVIEW

Infants are active participants in their social world. They are socialized by their caregivers, and they influence these caregivers as well. One basis of this reciprocal socialization between infants and their social world lies in infants' characteristics of individuality. We discuss various aspects of infants' individuality, or personality, and of their social development.

CHAPTER OUTLINE

SOCIALIZATION

The Social Nature of Human Functioning
Person—Social Context Interactions

COMPONENTS OF THE SOCIAL CONTEXT

The Role of the Family

FAMILY INTERACTION: PROCESS AND PRODUCTS OF BIDIRECTIONAL SOCIALIZATION

Infant-Caregiver Interactions

ATTACHMENT IN INFANCY

A Definition of Attachment
Contrasting Theories of, and Methods for Studying, Attachment
Description of Attachment Phenomena
Explanation of Attachment Phenomena
Attachment and Optimization of Development in Infancy

PERSONALITY DEVELOPMENT IN INFANCY

Defining Personality
The Role of Erikson's Theory
Changes in Adaptive Demands across Life
Erikson's Theory of Psychosocial Development in Infancy

INVESTIGATING THE IMPLICATIONS OF PERSONALITY DEVELOPMENT IN INFANCY

The Kagan and Moss Study of Birth to Maturity
The New York Longitudinal Study

EFFECTS OF SOCIAL ISOLATION AND DEPRIVATION

Social Isolation in Rhesus Monkeys
Social Isolation and Deprivation in Human Children
The Reversibility of Early Deprivation Effects

CHAPTER SUMMARY

ISSUES TO CONSIDER

What are the bases of the view that human behavior is both biological and social?

What is the adaptive basis of society?

What is socialization, and why is it thought to involve bidirectional, reciprocal interactions?

What is the role of the family in human development?

What evidence indicates that infants and their caregivers interact reciprocally?

What is attachment, and how does a life-span perspective aid in the description, explanation, and optimization of attachment phenomena?

What is the nature of the differences that exist in alternative definitions of personality? How is a life-span conception useful here?

How may Erikson's theory be of use in a life-span conception of infant personality development?

Are there any implications for later life of personality development in infancy?

What are the effects of social isolation and deprivation in infancy? Can these effects be modified?

SOCIALIZATION

*S*ocialization is a process by which members of one group (e.g., members of one generation) influence the behaviors and personalities of members of another group (e.g., members of another generation). People often think that socialization involves only the influence of parents or other adults on children. However, we will argue that socialization is a bidirectional process—infants and children as much influence their parents as their parents influence them. Moreover, in later chapters we will argue that socialization occurs across the life span—we continue to be socialized throughout our adult and aged years.

Our presentation in this chapter will indicate that humans are responsive to others around them across their life. In other words, human behavior is always influenced by the social context. In fact, the social character of human behavior seems to be the key feature of adaptive functioning.

The Social Nature of Human Functioning

Human behavior is both biological and social. Several lines of evidence support this view. First, biological adaptation requires meeting the demands of the environment. Typical environments are populated by other members of one's species. Thus adjustment to these other organisms is a requirement of survival (Tobach & Schneirla, 1968).

Human evolution has promoted this synthesis of biological and social functioning. The relative defenselessness of early humans, coupled with the dangers of living on the open African savanna, made group living essential for survival (Masters, 1978; Washburn, 1961). Therefore, the content of evolution was such that it was

Biological adaptation requires meeting the specific demands of one's setting. (Woodfin Camp and Associates.)

more adaptive to act in concert with the group than in isolation. Accordingly, processes supporting social relations (e.g., attachment, empathy) were selected over the course of human evolution (Hoffman, 1978; Hogan, Johnson, & Emler, 1978; Sahlins, 1978).

Finally, we may note that no form of life, as we know it, comes into existence independent of other life. No animal lives in total isolation from others of its species across its entire life span (Tobach & Schneirla, 1968). Thus, for several reasons, humans at all portions of their life spans may be seen as embedded in a social context.

Person—Social Context Interactions

There is reason to believe that a specific type of relation exists between people and their contexts: a multidirectional one. Adaptation to one's context involves bidirectional influences involving changes in the context to fit individual "needs" and changes in the individual to meet contextual "demands" (Dobzhansky, 1973; Harris, 1957). Piaget's (1970) concepts of assimilation and accommodation describe these alternatively directed processes, respectively. Similarly, Riegel (1975, 1976) sees human development as involving a synthesis of inner-biological, individual-psychological, outer-physical, and sociocultural processes; each level of process influences and is influenced by all others.

On the basis of views such as these, one should expect that the behavioral development of the infant should be influenced by variables that may exist in a particular social context and that, in addition, the character of the social context should be altered in relation to the individual characteristics of the infants interacting within it (Hartup, 1978; Lewis & Lee-Painter, 1974; Sameroff, 1975).

Such complex, multidirectional relations between an individual and his or her context have been labeled as *dynamic interactions* (Lerner, 1978, 1979; Lerner & Busch-Rossnagel, 1981; Lerner & Spanier, 1978a, 1980). In brief, this strong interactional view is that there are several sources of development and that changes in one source (e.g., the family or the individual) will promote changes in all others.

The study of interactions between individuals and their social contexts has been undertaken at several portions of the life span. Indeed, data supportive of the presence of such relations exist in infancy (Lewis & Rosenblum, 1974), childhood (Burgess & Conger, 1978), and adolescence (Bengtson & Troll, 1978). However, despite this support, it is the case that, at this writing, most of the empirical literature pertaining to any particular portion of the life span has not been derived on the basis of a conceptual framework emphasizing dynamic interactions. In turn, the research literature on social functioning at any portion of the life span has generally not been designed to demonstrate reciprocal relations between people and their contexts. In fact, the theoretical themes running through most existing research on social behavior emphasizes either biological or psychological variables as providing the primary impetus for social engagement or disengagement (e.g., Cumming & Henry, 1961; Erikson, 1959, 1968) or the unidirectional molding of social behavior by the external socializing environment (Hartup, 1978). Dynamic person-context interactions have thus received relatively little empirical attention.

Nevertheless, it is our view that as a consequence of the evolutionary and ontogenetic bases of individual—social context relations, noted above, it is both useful and necessary to consider human development within a contextual framework. In this chapter we consider research on socialization within the period of infancy. The usefulness of our interaction views about socialization will be illustrated by showing that ideas about bidirectional influences can help integrate this information. To demonstrate this use, we will have to discuss some key features of the social context.

COMPONENTS OF THE SOCIAL CONTEXT

What is there in the social context that facilitates our social responsiveness to it? The social context is composed of elements other than people of various age and cohort groupings, for it is also composed of social institutions. Much of the impact of the social context is influenced by social institutions. A *social institution* is an ongoing aspect of society around which many of life's most important activities are centered. Examples of social institutions are the family, religion, the legal system, the educational system, the mass media, and the political system. All social institutions change from time to time, and much of the social change is based on the continuing evolution that can be seen in a society's social institutions. Thus the social context of children is composed of both other people (e.g., parents) and social institutions. One social institution in particular—the family—is a major one throughout much of life in that it represents a key factor in the development of our social responsiveness.

The Role of the Family

Perhaps no social institution has as great an influence on development as the family. The family is the basic social unit of society in that it is the most common location for reproduction, childbearing, and childrearing. Virtually all children in all cultures

Social institutions such as the family and the church provide important influences on human development. (United Press International.)

are socialized by families, although the form of the family unit may vary slightly. Thus a *family* can be defined as a unit of related individuals in which children are produced and reared. But why are there families? Indeed, why are there social institutions or even societies?

As we have noted, anthropological information suggests that if humans could have survived in complete social isolation, social behavior would not have been promoted (Masters, 1978; Washburn, 1961). Social interaction provides a combination of behaviors which allows humanity to be perpetuated and social relations to be maintained. Thus because they function to aid survival, the structure of such social interactions became institutionalized. In this way society evolves. A society may be defined as any continuing system of social rules governing the behavior of people interacting within the system. But as we have just seen that individuals need society for their survival, we will now see that society needs individuals as well.

As the settings within which humans lived changed and became more complex, more differentiated and complex adaptational demands were placed on people in order for them to maintain and perpetuate themselves. One social unit could not, for example, produce all the resources necessary for survival. As such, different units took on different roles, some people becoming hunters, others cooks, shepherds, builders, or protectors. Thus role structure became more complex, more specialized, and more interdependent as society evolved.

But in order to ensure that the roles maintaining society would be performed by people having the skills and commitments necessary for social survival, people must engage in behavior to perpetuate the social order. Children born into society have to be instructed in the rules and tasks of that society in order to ensure their eventual contribution to society's maintenance. The process by which members of one generation shape the behaviors and personalities of members of another generation is termed *socialization;* and one function of socialization is to ensure that there will be members of society capable of meeting the adaptational (survival) demands of people in that society. Thus children are taught what is necessary to do in order to coexist with others and to survive. Although all societies teach their new members what to do, the precise attributes of what one has to do in order to survive will differ from one society to another.

Thus society needs new individuals born into a social context that maximizes the likelihood of socially approved socialization. In order for society to survive, the new members must be committed to its maintenance and perpetuation. As such, society must ensure that new members are efficiently and economically taught values that promote these goals. As a consequence, an institution is created in all societies which is the most efficient one for socializing new members. This core socialization unit is what we call the *family*, and it is given the function of bringing new members into the society within a context that ensures the stable continuation of the society. That is, if children are born outside of marriage, then not only are they born of parents who have behaved in defiance of basic and adaptive societal rules, but also, because of this situation, they are not in a context that seems the most likely to ensure societally approved socialization. In sum, for these reasons we have termed the family the basic unit of society and stated its role as the core socialization unit.

Given the prominence of the family, we can expect it to be a key interactive source of the person's development. This chapter will detail several aspects of this influence. We will review several sources of data suggesting that dynamic interactions within families provide a major basis of infant development. In turn, we will also often see that family interactions also interact with components of the broader social context.

FAMILY INTERACTION:
PROCESS AND PRODUCTS OF BIDIRECTIONAL SOCIALIZATION

The traditional social science approach to the study of the infant in the family has been to focus on the role of parental caregiving practices, such as rearing or disciplinary techniques, in the infant's behavioral and social development. As noted by Hartup (1978), the approach has often been based on a "social mold" conception of child-family relation. That is, the view that the infant's behavior is primarily shaped by how the parents treat the infant has been a popular one.

However, we have suggested an alternative view: While the parents shape the infant, the infant shapes them. That is, from our view, *socialization*, the processes by which members of one generation shape the behavior of members of other generations, is not seen as a unidirectional effect of parents on children. Rather, there is bidirectional socialization—infants and parents influence each other; and as such, one generation is not necessarily passively molded by the other. Instead, they interact. Moreover, the nature of these interactions does not exist in isolation from the changing social context. As social change proceeds, the child-family interactions that are involved in bidirectional socialization are altered.

Infant-Caregiver Interactions

Both parents and infants behave to maintain the other's behavior. For example, vocalizations by the infant are exchanged with parental vocalizations and perhaps also with touching. An example of this system is found in data reported by Lewis and Lee-Painter (1974). The looking behavior of an infant elicits more maternal vocalization than its touching behavior, but maternal touching and vocalization appear to evoke equal levels of infant vocalization.

Socialization in the family often involves the interaction of younger and older siblings. (United Nations/W. A. Graham.)

Although this information indicates that parental behaviors may indeed be contingent on infant behavior, is the nature of the relationship interactive? The data just cited seem to indicate that it is, and additional evidence reported by Lewis and Wilson (1972) supports this conclusion. Although they found no overall differences between middle-class and working-class mothers in the frequency of vocalization in the presence of their infants, a difference in the interactional use of vocalizations was obtained: middle-class mothers were more likely to respond to their infants' vocalizations with a vocalization than were working-class mothers, 78 to 43 percent, respectively. Similarly, Lewis (1972) reports that interactional sex differences also occur: female infants are more likely to respond with vocalization to a maternal behavior than are male infants, although no total amount of vocalization difference exists between the sexes.

Characteristics of the infant other than his or her sex contribute to interactions with their caregivers. Green, Gustafson, and West (1980) studied the social interactions of fourteen infants and their mothers, when the infants were 6, 8, and 12 months of age, in order to assess how changing social and motor capabilities of infants affect their daily social encounters. Green et al. found that with increasing age, infants more often initiated interactions and mothers performed fewer caretaking activities. The infants' locomotor abilities were found to be related to many of the changes in maternal caregiving, and thus Green et al. concluded that an infant's social environment is determined in part by the infants' developmental status.

Other characteristics of the infant that may affect his or her social context are suggested in the results of a study by Frodi, Lamb, Leavitt, Donovan, Neff, and Sherry (1978). Parents were shown videotapes of either a normal, full-term newborn or of a premature infant. Sound tracks were dubbed so that half of the normal and half of the premature infants emitted the cry of a normal infant, while the other half emitted the cry of a premature infant. The results indicated that the cry of the premature infant elicited greater physiological arousal among the parents and was viewed by the parents as more aversive. These effects were strongest when the premature cry was paired with the videotape picture of the premature infant.

These results suggest that an infant, by virtue of appearance and behavior (in this case, type of cry) can affect his or her own development. Babies who look and sound normal are not reacted to as negatively as are babies who have the look and/or sound of a premature infant. Indeed, if premature babies do represent such a relatively aversive stimulus to others in their social world, then an effect of such an infant on his or her own development may be to promote negative feedback from the caregivers. For example, such infants may be "at risk" for experiencing child abuse (Frodi et al., 1978). In fact, such a relation between prematurity and probability of child abuse exists (Lamb, 1978).

There are other lines of evidence which also demonstrate that infants can be producers of their own development (Busch-Rossnagel, 1981). If the infant is not merely reactive, he or she must be capable of initiating a flow of interactive behavior, maintaining it, and terminating it. Data indicate that infants do have such competencies. Bell (1974) indicates that in the first few weeks of life, crying elicits caregiver approach behavior, and such approach increases the likelihood that visual, olfactory, and tactile stimuli provided by the infant can elicit other caregiver behaviors. For example, in a study that illustrates how developing infant characteristics alter the interaction in the infant-mother dyad, Wolff (1966) reports that by a few weeks of age, infants' crying behavior comes to follow a predictable cycle. Infants move from a state of quiescence to soft whispering, then through gentle movements, rhythmic kicking, uncoordinated thrashing, and finally intense crying. As the pat-

Infants and their caregivers stimulate each other in their interactions.
(Charles Higgins/Photo Researchers, Inc.)

tern of changes emerges and becomes predictable to the mother, a response to one of the behaviors in the sequence might avert others. Kicking, for instance, might stimulate caregiving, which, in turn, would terminate an aversive infant behavior such as crying (Bell, 1974).

Such interactions are not automatic, however. The mother or father will not respond to every infant cry. There is some evidence that suggests that noncaregiving behavior is also influenced by infant behavior. Infant crying can become so excessive that it disrupts or terminates the flow of interactions. Robson and Moss (1970) found that mothers reported decreases in their feelings of attachment toward their 3-month-old infants if crying and other demands for physiological caregiving did not decrease, as they do in most infants.

Moreover, data reported by Thomas, Chess, and Birch (1970) suggest that developmental differences between children with easy-to-care-for versus difficult-to-care-for styles of behavior might arise as a consequence of the differential levels of energy their caregivers have to expend in order to maintain a relationship with them. Indeed, infants who have "easy-to-care-for" characteristics (e.g., a positive mood, a moderate level of activity, and regular, predictable eating, sleeping, and toileting behaviors) and, in addition, have a robust physique and good health, are often found to be "invulnerable" (Anthony, 1974; Heider, 1966) to the known negative effects of poor parenting, even when this parenting is done by a psychotic parent. On the other hand, children who lack the robust physique and good health and, especially, show the characteristics of a "difficult-to-care-for" child (e.g., a negative mood, a high activity level, and irregular behavior; Thomas & Chess, 1977) are more often associated with parental interactions that are not as favorable. One reason for this relation may be that the excessive aversive behavior by the child may require parents to expend too much of the caregiving or love resources, which leads to a change in the parent-child relationship.

Consistent with this possibility are data reported by Bell and Ainsworth (1972). Analyzing the relation between infant crying and maternal ignoring of crying in the four quarters of the infant's first year, they report that the more an infant cried in any one quarter, the more the mother ignored the cry in the subsequent quarter. These data suggest that for some infants their behavior was excessive enough to exceed their mothers' caregiving tolerance limits. After caregiving behavior failed to lead to a decrement in crying, the mother began to withdraw the more the infant cried; this led to increased crying and further maternal withdrawal (Bell, 1974).

The above data indicate that infant-parent relationships are interactional and that the infant is capable of initiating and maintaining a relationship if initiation behavior does not exceed caregiver limits. But what of the parent per se? The above data, in emphasizing the role of the child in family interaction, may serve to counter the stress on the parent as the main familial socialization agent. But parents contribute too.

Brazelton, Koslowski, and Main (1974) report that the maternal attributes most necessary for maintaining interactions with an infant are the development of a sensitivity to the infant's capacity for attention and understanding the infant's need for either complete or partial withdrawal after a period of attention. Here, too, the infant's contribution is again apparent. Cycles of attention and nonattention were found in all periods of prolonged interaction, and Brazelton et al. (1974) interpreted nonattentive, looking-away behaviors as reflecting the infant's need to maintain some control over the amount of stimulation taken in during an intense interaction period.

Additional evidence for the view that infants and their families are joined in a mutually interactive system comes from a study by Belsky (1979). Middle-class families with infants 15 months of age were observed in their own homes. Belsky found that there were more similarities than differences in maternal and paternal behavior, that parents had a slight preference for interaction with same-sex children, and that each parent showed more active parenting when alone with the infant than when the spouse was present. In turn, the infants were also influenced by the social situation, i.e., "infant with one parent" versus "infant with both parents." Infants directed more social behavior toward each parent when they were alone with them than when both parents were present.

Thus, infants' behaviors contribute to dynamic interactions between themselves and their caregivers. Not only do parents similarly contribute to such a balance in exchanges when infants' behaviors are too intense, but they also do so when behaviors are not intense enough. Bell (1974) indicates that extreme lethargy in an infant elicits various parental behaviors designed to intensify the infant's activity levels. Thus, parents intensify approach behavior when there is too little stimulation from the infant but substitute withdrawal behavior when there exists too much stimulation from the infant. For instance, Escalona (1968) reports that mothers stimulate sleepy infants in order to bring them into a state appropriate for interaction and then often must act to quiet the infant when a behavioral state too intense for interaction is obtained. In turn, Brazelton (1962) reports that if an infant's state is unalterable by the mother, markedly negative effects on the infant may ensue. In short, an infant behavior characteristic that provides a basis of parental caregiving behavior is state alterability (Bell, 1974).

Infants and their caregivers exist in a reciprocal social relationship. Not only is there evidence that the infant's skill in such social interactions increases throughout the infancy period, but also there is reason to believe that such development has important implications for other facets of the infant's functioning. Hay (1979), for instance, reports that between 12 and 24 months of age, infants' cooperative interchanges and sharing with their parents increased in frequency. Hay believes that such development helps the infant learn to behave prosocially. In addition, Farran and Ramey (1980) report that those infants who have higher scores on a measure of involvement with infant-mother interactions also show evidence of higher intellectual performance at 20 months of age.

Thus finding that involvement in reciprocal dyadic social relationships in infancy is associated with other positive personal and social functioning allows us to return to the point made earlier in the chapter. That is, dynamic, or reciprocal, interactions between infants and their social context have important implications for adaptive functioning.

In sum, social relationships between infants and their caregivers not only appear to be ubiquitous and reciprocal aspects of the infant's social world but also may be "necessary" (in the sense of having fundamental adaptive significance). Indeed, there are some scholars (e.g., Bowlby, 1969) who have suggested that as part of humans' evolutionary heritage there exists a propensity to form particular types of social relationships in infancy. The study of such relationships has arisen under the label of "attachment," and we now turn to a discussion of this topic.

ATTACHMENT IN INFANCY

The study of attachment in infancy is a useful topic with which to integrate previous discussions of infancy because several different processes appear to be involved in attachment, e.g., learning, cognitive, and socialization processes all seem to enter in. However, the inclusiveness of the topic has, very often, made for problems when trying to define attachment precisely.

To illustrate, we may note that Ainsworth (1973, p. 3) distinguishes, on the one hand, between attachment, dependency, and object relations and between attachment and attachment behavior, on the other (Ainsworth, 1973, p. 2). However, she stresses that the "hallmark of attachment is *behavior* that promotes proximity to or contact with the specific figure or figures to whom the person is attached" (Ainsworth, 1973, p. 2, italics added), and in so doing she offers a concept of attachment which appears very similar to what others call dependency (see Maccoby & Masters, 1970, p. 75). In fact, Maccoby and Masters (1970) believe that the term *dependency* has a technical meaning which includes the same classes of behavior that other authors (e.g., Bowlby, 1969) would call attachment.

A Definition of Attachment

We have integrated diverse conceptions of attachment to form a definition. The definition has five components, with each successive component representing an aspect of attachment having somewhat less consensus in the literature.

Attachment refers to (1) proximity and/or contact seeking and/or maintaining behavior; (2) shown to one or a few specific others; (3) which elicits reciprocal behaviors in, or the presence of, these others; (4) if these reciprocal behaviors are not seen, an aversive state exists for the person emitting the attachment behavior (e.g., as measured through distress behavior); and (5) this may lead the attached person to seek alternative attachment opportunities from among his or her broader social network (Lamb, 1977; Weinraub, Brooks, & Lewis, 1977).

Contrasting Theories of, and Methods for Studying, Attachment

A diverse array of theories has been offered to account for attachment phenomena. For example, Bowlby (1969) and Ainsworth (1973) have offered ethological-instinct views of attachment or dependency, and various social learning theories of attachment have been presented (Bijou & Baer, 1965; Gewirtz, 1972, 1976; Mischel, 1966; Sears, 1972). Since each theory differs in regard to its definitions of attachment, it also varies in the behaviors on which it focuses. Moreover, the study of attachment varies in relation to the different definitions. Methodological variations have involved, for instance, evaluating attachment behaviors in different situations, e.g., controlled laboratory manipulative studies, controlled observations in structured or unstructured play and/or separation situations, and in vivo naturalistic assessments.

Given such theoretical and methodological differences, one need not be surprised to find that the conclusions about attachment are equally varied. Major empirical issues addressed have been whether the mother or father is the major attachment figure, whether sex differences exist in infant attachment, whether there are age-associated differences in attachment, and whether there are differing developmental courses for various indexes of attachment (e.g., selective smiling or crying on separation). Conclusions derived from these inquiries have generally lent support to both sides of the issue. Some data have indicated, for instance, that the mother is the major attachment figure (e.g., Ban & Lewis, 1974; Fleener & Cairns, 1970), while other data suggest that the father is the major figure (Ban & Lewis, 1974). Still further data indicate no attachment differences toward the parents (Cohen & Campos, 1974; Lamb, 1975, 1977; Willemsen, Flaherty, Heaton, & Ritchey, 1974). Similarly equivocal data trends exist in regard to the other major empirical issues that have been addressed.

While such discrepancies in theories, methods, and findings may lead some to infer that attachment phenomena are too elusive and variable for comprehensive study, such an inference may be premature. Rather than regarding them as deficits, the current attributes of the attachment literature are, in fact, viewed as assets if integrated with a life-span orientation. We will attempt to do this by organizing our discussion around the general concepts of the description, explanation, and modification of development as outlined in Chapter 1.

Description of Attachment Phenomena

The precise behaviors defining attachment may be different at various points in development (Weinraub et al., 1977). Cohen (1974) notes that proximity-promoting behaviors are characteristic of attachment during infancy and early childhood but that they begin to diminish in occurrence thereafter. Cohen (1974) suggests that the behavior selectively directed to an attachment figure may be different at succeeding periods across the life course. Evidence for such transformation is derived from a report by Lewis, Weinraub, and Ban (1972). Proximity-seeking behaviors decreased as infants aged and were replaced with distal attachment behaviors such as looking and vocalizing.

Moreover, as stressed in life-span writings (Baltes & Schaie, 1973; Baltes & Willis, 1977), attachment behaviors may develop in multidirectional manners. For example, Maccoby and Feldman (1972) found that among 2- and 3-year-olds proximity seeking to mothers did not change with age, but if the concept of distal attachment was employed (as indexed, e.g., through speaking, smiling, or showing objects), then such attachment behavior did increase with age. Although distal attachment was not assessed, children (age 10, 14, and 18 months) were evaluated in a study by Coates, Anderson, and Hartup (1972) for their attachment through assessment of touching, proximity-seeking, and vocalization behaviors. While vocalization showed significant stability only from 10 to 14 months, touching and proximity seeking were significantly stable from both 10 to 14 months and from 14 to 18 months. Finally, decreases in some classes of attachment behaviors have also been found. Stayton, Ainsworth, and Main (1973) studied infants during their first year of life. It was found that smiles on greeting were more frequently reserved for mothers among the youngest infants studied (i.e., 3- to 6-month-olds); such a selective response decreased thereafter. In sum, it appears that depending on the behavior class being described, attachment behaviors vary in their developmental direction.

Contexts of Attachment. We have stressed that any developmental change is related to other alterations in the social context within which a given change is embedded. Attachment behaviors appear to change in relation both to other individual processes and to broader contextual variables, for example, those related to the social and cultural context.

To illustrate, relations may exist among attachment behaviors and physical or physiological variables. Fraiberg (1975), for instance, studied the development of attachment in ten blind but otherwise neurologically intact infants over the course of their first 2 years of life. Her results indicated that several attachment behaviors emerged within an age range comparable to normal, sighted children (e.g., a familiar voice irregularly elicited a smile in the range of from 1 to 3 months, and negative reactions to strangers emerged between 7 and 15 months). Thus visual processes appeared not to be necessary components or precursors of these attachment behaviors. On the other hand, separation anxiety among blind children was first seen between 11 and 12 months—a delay in comparison to sighted children, yet comparable to the former group's delay in attaining particular levels of object permanence. Leifer, Leiderman, Barnett, and Williams (1972) also provide data illustrating the use of physical status data for attachment research. Reciprocal mother-infant behaviors were studied with groups of premature and full-term infants. Mothers of full-term infants smile at them and hold them closer to their bodies more than do mothers of premature infants. Accordingly, maturity status appears to be associated with differential processes of mother-infant interaction.

Socioeconomic status may also be a useful marker for variables related to attachment. Tulkin (1973), for example, studied the attachment behaviors of middle- and working-class female infants in a play situation and after separation from their mothers. The two social-class groups did not differ in their play behaviors or in the separation condition (the working-class children cried as often as the middle-class children). Middle-class children did, however, cry sooner in the separation condition. In addition, the working-class infant of the working mother cried less frequently and crawled less often to the mother on her return after separation than did the working-class infant of the nonworking mother.

There are other indications of the influence of the social context on attachment. Several studies have used a standardized assessment procedure developed by Ainsworth and associates, the "Ainsworth strange-situation" procedure (Ainsworth, Blehar, Waters, & Wall, 1979). This is a structured, laboratory observation procedure, wherein infants are scored for their behaviors to their mothers and to a stranger in what for the infants is a strange (a laboratory) context. Vaughn, Egeland, Sroufe, and Waters (1979) studied 100 economically disadvantaged mothers and their infants by using the strange-situation procedure. The infants were studied at 12 months and again at 18 months. Infants were classified as either securely attached (e.g., they did not show distress behavior when apart from their mothers) or as showing various types of "anxious" (distress) behaviors. At both age periods, sixty-two infants had the same classification of attachment. However, an infant's being in the anxious attachment category was associated with his or her having experienced a less stable caretaking environment (e.g., in regard to family employment, financial, and health patterns) than was the case with an infant in the secure attachment category. Moreover, change from secure to anxious attachment between 12 and 18 months of age was associated with infants having families experiencing more stressful events than was the case for infants who remained in the secure attachment category across the 12- to 18-month age span. Similar evidence for impact of the context in accounting for individual differences in attachment is found in a study by Waters (1978).

Thus some changes in the family context of the infant can alter attachment behaviors. Other aspects of the social context can also affect attachment in infancy. Blanchard and Main (1979) observed infants, aged between 1 and 2 years, using both the above-described laboratory procedure as well as observations of the infants' behaviors in the day-care centers within which they were placed. There were individual differences among the children in their scores for avoidance of their parents upon reunion (in both the laboratory and the day-care observations) and in their scores and emotional adjustment in the day-care center. However, the relations between these scores suggested that experiencing the day-care context had a facilitative effect on these infants' attachment to parents and on social-emotional adjustment. That is, children who had spent a longer period of time in substitute-care facilities, such as provided by day care, had lower scores for avoidance and higher scores for social-emotional adjustment.

Because of historical changes in the political and economic nature of our social context, mothers of today are more often engaged in full-time employment outside of the home than was the case in earlier historical eras (Hoffman, 1979). We will discuss the issue of maternal employment, and of its effects on children, in Chapter 8. Here, however, we may note that some mothers may be concerned that because of their employment, and the need to place their children in day-care facilities, they may be adversely affecting their children's development; however, studies such as that by Blanchard and Main (1979) provide a basis for ameliorating such concerns. Indeed, a review of known effects of day-care placement (Belsky & Steinberg, 1978) suggests that not only does experiencing a quality day-care program have no negative effects on children, but also as suggested by Blanchard and Main (1979), some positive outcomes may occur. Indeed, in our view, the weight of current evidence supports the conclusions of Belsky and Steinberg (1978). Thus findings, like those of Schubert, Bradley-Johnson, and Nuttal (1980)—that employed mothers have vocal and nonvocal communication patterns with their 15- to 17-month-old infants which are not different from those of nonemployed mothers—are representative of those which portray the impact of maternal employment on the mother-infant relationship.

Finally, as a last instance of the impact of the social context on attachment, we may note that cross-cultural variables, as markers of sociocultural and historical processes, may be useful for evaluating the extensiveness of attachment change. First, despite cultural and methodological differences, Ainsworth's (1963, 1964) study of Ugandan children and Schaffer and Emerson's (1964) study of Scottish children found evidence for intercultural consistency in some aspects of attachment behavior. Such comparisons may provide evidence of the consistency of particular attachment processes. Cross-cultural comparative research can therefore show the robustness of a particular aspect of attachment, despite presumed cultural differences in such behaviors as childrearing. For instance, a comparison was made by Maccoby and Feldman (1972) between American home-reared children, age 2.5 years, and similarly aged Israeli kibbutz-reared children. Results showed a high similarity in reactions to mothers in separation distress and, as such, illustrate intercultural congruity despite known differences in the rearing conditions with infants from the two cultural groups.

In turn, however, some intercultural differences do exist. These may provide an initial basis for future investigations about the reason for such disparities. For example, Ugandan children develop specific attachments earlier than Scottish children, but they get intense fears of strangers later than do Scottish children (Ainsworth, 1973; Maccoby & Masters, 1970). These differences suggest that it would be profitable to explore familial interaction and childrearing practice differences between the two cultural groups.

Explanation of Attachment Phenomena

Application of life-span thinking to the study of developmental changes in attachment behavior may help clarify the explanation of attachment phenomena. The life-span view emphasizes that several explanatory perspectives may be of use in attempts to account for behavior-change processes across the entire life course (Lerner & Ryff, 1978). Such pluralism aids in explaining attachment processes as they occur during infancy. Ainsworth (1973), for instance, suggests that while the very initial emergence of attachment behaviors may be explained by genetically biased processes (Bowlby, 1969), subsequent attachment behavior can best be accounted for by nurture-based principles of social learning (e.g., as may be found in Gewirtz, 1972, 1976). Similarly, Huston-Stein and Baltes (1976) point out that while the first appearance of attachment-related behaviors (e.g., smiling) may be a function of maturation, the succeeding development associated with the behavior may primarily involve environmental stimulation and reinforcement.

The use of explanatory pluralism may be illustrated further. Yarrow (1963, 1964), for example, reports that the development of attachment is predicated on visual discrimination learning and on the development of the scheme of object permanency. Accordingly, an integration of ideas associated with social learning (Gewirtz, 1972) and cognitive developmental theory (Piaget, 1970), respectively, may be of use.

Attachment and Optimization of Development in Infancy

Why should one want to enhance attachment? Earlier in this chapter we noted that human behavior has evolved to be socially interdependent behavior, and, in fact, we saw that reciprocal relations between infants and their caregivers have adaptive implications for both groups. Attachment describes a set of behaviors linking the infant to his or her social context. As such, intervening to enhance attachment should have a positive influence on the infant's adaptive functioning. Indeed, there is evidence that attachment is an adaptive process and that particular qualities of attachment (e.g., as found in secure attachment) are related to adaptive functioning in diverse domains of infant development.

Matas, Arend, and Sroufe (1978) studied the relation between the quality of attachment in infancy and the quality of play and problem-solving behavior at 2 years of age. When tested at 2 years of age, infants who were securely attached at 18 months of age were found to be more enthusiastic, persistent, and cooperative than were infants who were insecurely attached at 18 months of age. Similarly, Waters, Wippman, and Sroufe (1979) found that secure attachment involves more than the absence of negative or maladaptive behavior directed toward a caregiver. Smiling and smiling combined with vocalizing and/or showing toys distinguished securely from anxiously attached 18-month-old infants during free-play periods. In addition, securely attached infants had better ratings for the quality of their affective sharing, at both 18 and 24 months of age, than did non-securely attached infants. Moreover, the quality of the infant-mother attachment related to both personal and interpersonal competence in the preschool play group at age 3.5 years. Results consistent with the above are found in a study by Easterbrooks and Lamb (1979). Studying 18-month-olds, these researchers found that the quality of the infant-mother attachment was related to infant peer competence. Securely attached infants engaged in more frequent and more sophisticated interaction with peers than did non-securely attached infants.

Thus the above findings indicate that attachment is a phenomenon that indeed has adaptive implications. As such, attempts to enhance attachment may be of use in interventions aimed at optimizing human functioning. As with descriptive and explanatory endeavors, the life-span view may be of use in optimizing attachment in infancy.

Attachment phenomena are excellent exemplars of processes amenable to the life-span view for developmental intervention, given the variability we have seen associated with attachment behaviors, as well as the fact that attachment is related to variables in the sociocultural context (Lamb, 1977; Weinraub et al., 1977).

Both infants and their caregivers obtain satisfaction from their attachments to each other. (Tim Davis Photography.)

For example, since attachment may not be expressed by a single behavior either within or between developmental periods but rather can be expected to show situational and life-course transformations (Weinraub et al., 1977), an almost infinite array of behaviors become potential targets of attempts to enhance attachment. Moreover, given the relation of attachment phenomena to the sociocultural context, different levels of analysis may be used in such attempts.

For example, American mothers are more likely than Japanese mothers to use distal behaviors to interact with their 3-month-old infants (Caudill & Weinstein, 1969). Japanese mothers, on the other hand, tend to use proximal behaviors for interaction with their infants (Weinraub et al., 1977). Similarly, at later portions of the life span it may be less likely that a father of Anglo-Saxon descent would express affection to a grown son through such proximal behaviors as kissing than would a father of Mediterranean descent (Weinraub et al., 1977, p. 35). If an interventionalist understands the expression of attachment behaviors across the life span, as they are embedded in particular sociocultural contexts, appropriate targets of intervention can be chosen.

In turn, given the explanatory pluralism we have discussed, mechanisms for producing change in attachment may range from alterations of the schedules of reinforcement for attachment behavior in one individual to modification, redesign, or invention of social institutions fostering attachment to numerous figures involved in a broad social network (e.g., day-care centers for young infants).

However, such mechanisms must be selected in recognition of the cultural meaning attached to particular target behaviors at different ontogenetic times. Attachment behaviors designed to be acquired or maintained at particular portions of life may have to be extinguished or redirected to alternative figures. In early infancy it may be appropriate, for example, to design intervention strategies to increase the probability of proximal attachment behaviors to one or a few specific others. However, as a transformation in attachment occurs during later infancy (cf. Weinraub et al., 1977) and distal attachment behaviors supercede proximal ones, the goal of intervention might be to extinquish proximal behaviors while ensuring acquisition of distal behaviors to the same figures.

In sum, attachment appears to be a process that may adaptively link an infant to his or her social context. A life-span perspective may be of use not only in describing and explaining variables relating to attachment but also in optimizing the intraindividual change processes involved in attachment phenomena. Of course, not all individuals will change in the same way. As emphasized in the life-span view, there are interindividual differences in intraindividual change. Accordingly, in order to fully understand adaptive linkages between infants and their social context, we need to understand such individuality. This concern leads us to a discussion of personality development in infancy.

PERSONALITY DEVELOPMENT IN INFANCY

Personality is a term in the vocabulary of most people. It is also a concept used by many social scientists. However, the way most people use the word differs from the manner in which it most typically is used by scientists. In its common, everyday use personality is often seen as something which helps one "win friends and influence people," something which can be turned on or off at will (as in, "turn on your personality when you go for a job interview"). Often, personality is also seen as something a person develops to make up for deficits in other areas (e.g., we hear people say that a person is not good looking but has a good personality). Yet despite this range of everyday "definitions" of personality, no conception within this range corresponds to the use of the term in science.

Defining Personality

Most social scientists use the term *personality* to refer to some *relatively* enduring characteristics of an individual, although in trying to describe and explain the basis and nature of these characteristics of individuality, there is as much diversity of opinion among social scientists as there is among nonscientists. Indeed, Wiggins (1968) wrote more than a decade ago that the primary scientific issue in the study of personality was one of defining the term. This problem of definition still exists today.

The issue is not one of asking whether people have particular patterns of thoughts, feelings, and behaviors. Rather, it is one of asking where the particular set of characteristics that typifies a person came from and what precisely are these characteristics. We may recognize the former concern as the nature-nurture issue. Some theorists, like Lorenz (1965) or Hall (1904), see the characteristics that comprise the person as lying in instincts, while other theorists, like Davis (1944), Mischel (1970), or McCandless (1970), see external stimuli as shaping the person. Thus while the former group of theorists emphasizes bases of personality lying within the person, the latter emphasizes bases of personality lying within the situation. Still another group of theorists stresses that a person's behavior derives from an interaction between person and situation (Bowers, 1973), and a range of opinion exists also about the type of interaction between person and context involved in individuality.

Not only do scientists disagree about the bases of personality, but they do not have similar ideas about what are the component thoughts, feelings, or behaviors of the individual or, in fact, whether all these things are necessary to study in the assessment of personality. One way of understanding this dimension of controversy is to focus on an often-cited observation by Kluckhohn and Murray (1948, p. 35) that in certain respects every person is (1) like all other people; (2) like some other people; and (3) like no other person (Gallatin, 1975). Most existing ideas of personality have tended to focus on only one of these three aspects of the individual and have diminished or ignored the other two.

For instance, the stage theory of Sigmund Freud (1949) emphasizes that all people pass through the specified psychosexual stages in an invariant sequence. In stressing such universality, Freud is indicating that any person is like all others. The constitutional typology of William Sheldon (1940, 1942) emphasizes that people exist in personality (or temperament) groups, depending on their possession of a physique that is either predominantly endomorphic (fat), mesomorphic (muscular), or ectomorphic (thin). In stressing that such types of people exist, Sheldon is indicating that a given person is like some, but not all, others. Finally, mechanistic social learning theories (Bijou, 1976; Bijou & Baer, 1961; Davis, 1944; McCandless, 1970) are conceptions which have the potential of speaking to the uniqueness of any person. Although the basis of behavior—the nurture laws of stimulus-response (S-R) learning—remains the same for all people, any given person may encounter any particular S-R reward or punishment history. As such, any person could have a unique repertoire of behavior. Indeed, Mischel (1973), a leading representative of this social learning position, has argued that in the real world, such differences in learning history invariably occur. As a consequence, a given person is like no other (Mischel, 1973).

In summary, because of differences in ideas about the basis of personality and about the characteristics that comprise personality, there is no single definition of personality that can be agreed on without exception. Yet a *useful* conception of personality can be advanced. In our view, such a conception would not focus on *just* universal, or group, or individual characteristics. This would be incomplete. For example, although the views of Freud, Sheldon, and Mischel are useful in depicting the universal, group, and individual characteristics of a person, respectively, their use is limited by the fact that they do not consider how these three types of characteristics interrelate. Every person has (1) universal characteristics, e.g., those relating to continual orientations to adaptation and survival; (2) group characteristics, e.g., those relating to society and culture membership; *and* (3) individual characteristics, e.g., those pertaining to the unique genotype-environment interaction of the person. Accordingly, it appears that a most useful definition of personality would be one which considers the range of individual (or, *idiographic*) *and* general (or, *nomothetic*) characteristics a person may possess and how these characteristics are related to the person's social world. This section attempts to offer such a definition.

A life-span view of development emphasizes that change is constant across life and that any one part of life is a consequence of preceding portions and an antecedent of following portions. Second, the context within which the person develops is constantly changing also. This indicates the need for a multidisciplinary orientation and suggests that a *personological* analysis of personality is limited. Such a view seeks to understand individual changes through exclusive recourse to individual phenomena.

Consistent with discussions about the life-span view presented in earlier chapters, personality can be assumed to have the following attributes:

1 Personality is a life-span process.

2 As such, any particular portion of the life span is understood in terms of how it is textured and shaped by the developments in life preceding it and, in turn, how it similarly contributes to the portions of the life span following it.

3 This life-span progression is understood best by examining processes lying at all levels of analysis. This leads to a view encompassing both general (for example, adaptational) and specific (for example, familial) aspects of the person throughout the life span.

The Role of Erikson's Theory

Erik Erikson's (1959, 1963, 1968) account of personality development *in part* meets the attributes of a life-span conception. Erikson *describes* personality development as a life-span process and uses multidisciplinary ideas to understand personality. For example, in adolescence, a person develops an aspect of the ego, an *identity*, based on attaining a role, which is a socioculturally based set of behavior prescriptions.

Erikson divides the life span into eight stages of development. Like Freud, on whose theory he bases his own ideas, Erikson sees these stages as proceeding in a fixed, universal sequence. Moreover, within each stage a person is supposed to develop a particular component of his or her personality—specifically, a part of the ego—*if* he or she is to proceed to develop adaptively. However, to Erikson, a person does not have an infinite amount of time in which to accomplish this development. In accordance with one's innately fixed maturational timetable, Erikson (1959) believes that stage progression will continue independently of whether one has devel-

Erik H. Erikson. (W. W. Norton/Jon Erikson.)

oped as expected. Since the focus of development shifts when a new stage emerges, it is *critical* that one develop appropriately in each stage. Since there is no environmental interaction that can alter the maturational timetable of progression, if one does not develop as expected in a critical period, then (1) that part will never be developed as adequately as it might have been and (2) the rest of development is altered unfavorably (Erikson, 1959).

Thus, while Erikson (1959, 1963, 1968) offers insightful and useful *descriptions* of phenomena associated with adolescence, his explanations, being based on weak interactional, universalistic, critical-period notions, suffer the shortcomings of such ideas. We have discussed these problems in Chapter 3. Yet it is possible to use Erikson's ideas at least as an initial framework for describing the phenomena associated with personality development.

Some Roots of Erikson's Ideas. Erik H. Erikson was born in Frankfurt, Germany, in 1902 and moved to the United States in the early 1930s. As a young man, Erikson served as a tutor to the children of some of the associates of Sigmund Freud. While working in that capacity, Erikson came under the influence of both Sigmund Freud and his daughter Anna. Accordingly, Erikson received training in psychoanalysis, and after moving to the United States and settling in the Boston area, he soon established his expertise in the area of childhood psychoanalytic practice.

Through his practice, as well as through the results of some empirical investigations (Erikson, 1963), Erikson began to evolve a theory of affective (or emotional) development which complemented the theory of Sigmund Freud. Erikson's theory altered the essential focus of past psychoanalytic theorizing, from a focus on the id to one on the ego.

When one changes focus from the id to the ego, one immediately recognizes the necessity of dealing with the society around the person. The function of the ego is survival, adjustment to the demands of reality. That reality is shaped, formed, and provided by the society that the person is a part of. An appropriate adjustment to reality in one society, allowing the person to survive, might be inefficient or even totally inappropriate in another society. Hence when one says that the child is adapting to reality, one is saying, in effect, that the child is adapting to the demands of his or her own particular society. To understand this societal shaping of adaptational demands, it is useful to recall our earlier discussion which indicated that society itself is an adaptive phenomenon. It evolves a structure (of roles and other institutions) to provide a means to adapt to its context. It socializes its members to fit into this structure.

Thus although it is held by Freud and Erikson that all infants pass through the same oral stage and need to deal with reality in order to obtain the appropriate oral stimulation, the way they obtain it may be different in different societies. In one society, for example, there may be prolonged breast-feeding by the mother. The infant need only seek the mother's breast, which may never be very far away, in order to obtain the needed oral stimulation. In another society, however, infants may be weaned relatively early. A few days after birth the mother might return to work and leave the infant in the care of a grandparent or older sibling (DuBois, 1944). Although the infant might also still need oral stimulation, adjustments to reality different from those involved with the former infant will have to be made. Thus the specifics of a child's society must be understood when one considers the implications for adaptation. In some societies we must learn to hunt, fish, and make arrowheads in order to survive. In other contexts such skills are not as useful as learning to read, to write, and to do arithmetic.

In sum, society, the roles it evolves, and the process of socialization within society are all components of adaptive individual and social functioning. To Erikson, the aspect of the person attaining the competency to perform these individual-social linkages is the ego, that aspect of the personality believed in psychoanalytic theory to be governed by the reality principle. Despite whether or not one chooses to talk of an ego as being involved in these linkages, the person must attain those skills requisite for survival in his or her society.

Yet it is clear that the demands placed on the person (or the ego) are not constant across life. Although society may expect certain behaviors of its adult members —behaviors which both maintain and perpetuate society—similar expectations are not maintained for infants, children, and, in some societies, adolescents. In other words, the adaptive demands of an infant are not the same as those of a child or an adolescent or adult.

Changes in Adaptive Demands across Life

Erikson believed that the implications of the ego for human psychological functioning were not given sufficient attention by Freud. When such attention was given, it seemed clear that humans are not only biological and psychological creatures but *social* creatures as well. To Erikson, a child's psychological development can be fully understood only when considered within the context of the society in which the child is growing up. Thus one can see why Erikson's most famous book is entitled *Childhood and Society* (1963).

By changing the focus of Freud's psychoanalytic theory by giving primary consideration to implications of the ego, rather than the id, and thus stressing the interrelation of the ego and the societal forces affecting it, Erikson is concerned with a person's *psychosocial* development throughout life. However, as psychosocial development proceeds, new adjustment demands are placed on the ego.

Society alters its specifications for adaptive behavior at different times in a person's life. As noted above, in infancy, society (specifically the family) expects "incorporation" from an infant. All that an infant must do in order to be deemed *socially* adaptive is to be stimulated and consume food from caregivers. We would not expect a person of this age to do much more than this. Certainly, we would not expect the infant to get a summer job or follow career goals. Yet we may expect such behavior from an 18-year-old. Indeed, if all one did on one's summer vacation from college were to take in stimulation (from the sun) and incorporate food (from the kitchen), this would not be considered very appropriate behavior by one's parents.

Thus a behavior deemed adaptive at one time in life is not similarly going to be seen as functional for the rest of life. Rather, new behaviors must emerge. Although we still have to be incorporative at age 18, we have to do more. Identity-related behaviors may have to come to predominate at this time of life. Unless one shows these behaviors, and shows them to sufficient degrees, one may not be judged as adaptive. One may not meet the demands of his or her society. In summary, the person must always meet the societally shaped demands of his or her world, but these demands are altered continually across the person's life span.

A similar conception has been advanced by Havighurst (1951, 1953, 1956), who believes that as people progress across their life span, there are certain tasks they must master at different portions of life. He terms such change-related requirements *developmental tasks* and notes that the specific tasks which occur at each particular portion of life arise out of particular combinations of pressures from inner-biological (e.g., physical maturation), psychological (e.g., aspirations in life), and sociocultural (e.g., cultural expectations) influences (Havighurst, 1956). These pressures require an adjustment on the part of the person. Since at different times in the life span the combination of pressures from each of the levels is different, then at each successive portion of life a distinct set of adjustment demands is placed on the developing person. Consequently, a developmental task "arises at or about a certain period in the life of an individual, successful achievement of which leads to his happiness and to success with later tasks, while failure leads to unhappiness in the individual, disapproval by society, and difficulty with later tasks" (Havighurst, 1953).

The particular combination of pressures from the inner-biological, psychological, and sociocultural levels of analysis that exist during any period of life will vary as a consequence of historical changes in the context of that life period (Thornburg, 1970). Tasks that may have been developmentally necessary for people at one period of history (e.g., selection of an occupation) may be delayed or accelerated in other eras.

Returning to Erikson's terminology, we may note that although the ego has to develop the capabilities of dealing with its society, the capabilities society expects are altered across life. Erikson believes that there are eight different capabilities that have to be developed. In other words, society places eight successive demands on the ego, and thus there are eight stages of psychosocial development. Within each stage, the ego must develop that capability necessary for adaptive functioning in order to proceed with the development of further ego capability.

Indeed, it has been seen that Erikson asserts that at each stage it is critical for an adequate capability to be attained. In any event, Erikson theorized that until the requisite demand within each stage is—or is not—met, the ego is in a state of *psychosocial crisis*. If the developing ego attains the appropriate abilities, the emotional crisis will be resolved successfully and, Erikson believes, healthy ego development can proceed. In turn, of course, if the appropriate attributes of the ego are not developed, negative emotional consequences will ensue. Of course, for this view to be of use, infants must be capable of manifesting a range of emotional reactions, and current research indicates that they can (Hiatt, Campos, & Emde, 1979; Izard, Huebner, Risser, McGinnes, & Dougherty, 1980; MacDonald & Silverman, 1978). For example, Izard et al. (1980) report that 1- to 9-month-old infants can be shown to reliably express at least eight different emotions (interest, joy, surprise, sadness, anger, disgust, contempt, and fear).

To assess the aptness of Erikson's ideas, as well as to gauge the descriptive utility of Erikson's stages for understanding personality development, we turn now to an overview of the first two psychosocial stages Erikson describes. These two stages pertain to the infancy period. We will discuss the other stages in later chapters. After this presentation we shall focus on research relevant to personality development in infancy.

Erikson's Theory of Psychosocial Development in Infancy

As a follower of Freudian psychoanalysis, Erikson sees the stages of psychosocial development as complementary to Freud's psychosexual ones. Accordingly, the id-based psychosexual stages exist along with the ego-based psychosocial ones. But although these psychosocial stages have some similarity to the psychosexual ones, they go beyond them in that they comprise stages in the ego's continual functioning.

Stage 1. The Oral-Sensory Stage. Freud's first stage of psychosexual development is termed the oral stage. In that stage the infant is concerned with obtaining appropriate stimulation in the oral zone. Erikson believes, however, that when one changes one's focus to the ego, one sees that the newborn infant is concerned not merely with oral stimulation. Rather, all the newborn infant's senses are being bombarded with stimulation—its eyes, ears, nose, and all other sense-receptor sites. Thus in order to begin to deal effectively with the social world, the infant must be able to incorporate all this sensory information effectively; hence Erikson terms this psychosocial state the oral-sensory stage. The ego must develop the capability of dealing with the wealth of sensory stimulation constantly impinging on it.

However, the necessity of dealing with this stimulation evokes an emotional crisis for the infant. If the infant experiences the sensory world as relatively pleasant or benign, one sort of emotion will result. Alternatively, if the child's sensory stimulation experiences are negative or harsh, then another type of feeling will result. If the infant has relatively pleasant sensory experiences, he or she will come to *feel* that the world is a relatively benign, supportive place and that it will not hurt or shock him or her. To Erikson, the infant will develop *a sense of basic trust*. If, however, the infant experiences pain or discomfort, he or she will feel that the world is not supportive but that there is pain and danger in the world. The infant will develop *a sense of mistrust*.

The infant thus faces an emotional crisis, precipitated by the nature and quality of the sensory world he or she attempts to incorporate. The emotional crisis is between *trust and mistrust*. (We may delete the phrase "a sense of" for the sake of brevity of presentation, with the understanding that this phrase is always to be applied to all the various stage-specific feelings.) Erikson thinks of these two feelings as being bipolar, alternative endpoints along a single dimension. Erikson would represent the continuum of trust versus mistrust as:

trust ————————————— mistrust

The ends of this bipolar continuum represent the alternative emotional outcomes of this stage of psychosocial development, but Erikson stresses that people do not, and should not, develop *either* complete trust *or* complete mistrust.

If a person develops complete trust, Erikson points out that this will be as un-adaptive as developing complete mistrust—the person will not recognize the real dangers that exist in the world (e.g., the person will never look when crossing the street because of the conviction that no driver would ever hurt him or her, or one might never strive to provide for oneself because one feels that the world will surely provide). On the other side, however, a person whose feeling falls on the far end of the mistrust side of the continuum will never attempt to interact with the world be-cause of feelings that assuredly the world will be hurtful. Erikson notes that such a person feels that there is no chance of anything but pain resulting from social inter-actions in the world. Thus we see that it is necessary to develop a feeling that lies somewhere between the two endpoints of the bipolar continuum. If one develops more trust than mistrust, then Erikson believes that healthy ego development will proceed. If, however, one develops greater mistrust than trust, then unhealthy, non-optimal ego development will proceed. One can develop a location at any point along this continuum. It is this "infinity of placement" which gives this view its idiographic component and, thus, gives it its relevance to universal, group, and indi-vidual aspects of the person.

Having a feeling located closer to the trust end of the continuum means that the ego has developed the appropriate capabilities, allowing it to deal effectively with the sensory input from the world. Having a feeling located closer to the other side of the continuum, however, means that the appropriate ego capabilities have not de-veloped. This location will affect the ego's functioning as the child enters the next stage of psychosocial development.

Stage 2. The Anal-Musculature Stage. Freud's second stage of psychosexual devel-opment is termed the anal stage. Here we may remember that the infant obtains gratification through the exercise of his or her anal musculature. To Erikson, how-ever, psychosocial development involves the other muscles of the body as well. Psy-chosocial development thus involves developing control over all of one's muscles. Analogous to use of the anal muscles, however, the infant must learn when to exer-cise and when not to exercise all his or her bodily muscles (Erikson, 1963).

Accordingly, if the child feels in control of his or her own body, the child will develop *a sense of autonomy.* On the other hand, if the child finds himself or herself unable to exert this independent control, if he or she finds that others have to do for him or her what one is expected to do for oneself, then one will develop *a sense of shame and doubt.* One will feel shame because one is not showing the ability to control one's own movements (e.g., bowel movements or movements involved in feeding oneself), and this may evoke disapproval from significant other persons (e.g., parents). Moreover, because one is experiencing this inability to control self, one will feel doubt about one's capabilities for so doing. Shame is felt because others do things for the person that both the individual and others feel should be done for oneself.

INVESTIGATING THE IMPLICATIONS OF PERSONALITY DEVELOPMENT IN INFANCY

Erikson's theory indicates that personality development across life should be characterized by both continuity and change. That is, he suggests that at *all* times of life the person must establish a means of adaptively relating to his or her society. But what the person must do to establish this relation, how the person meets particular social demands, and even what these demands are may change across life. Based on our particular set of individual and group characteristics, we should establish patterns of behaviors characteristic to us across life *and* special for us at particular portions of our life as well.

To best test these general expectations—of consistency and of change—it would be most useful to study people from different birth cohorts across their entire life span; however, for practical, economic, as well as other reasons, such research has rarely been attempted. Thus studies of people either at one or a few portions of the life cycle have been most often done. In addition to measuring people of different cohorts and of different age ranges, these studies often measure different aspects of personality. Thus the life-span characteristics of personality development from infancy onward are not easy to piece together.

However, one way to illustrate how such a major attempt might be done is to present data from a few key studies of personality development that *have* been relatively long-term, longitudinal ones. This focus will indicate that, consistent with Erikson's general notions, there are both continuous and changing aspects of the personality across the life span and that these features of personality have their roots in developments in infancy. Moreover, our review here will also lead to a second emphasis associated with Erikson's descriptions. That is, personality is an adaptive process linking the person with his or her context (indeed this *is* the general, continuous feature of personality in the present authors' view). We focus here on the results of two major studies of personality development from birth onward.

The Kagan and Moss Study of Birth to Maturity

In their 1962 book, *Birth to Maturity*, Jerome Kagan and Howard Moss report the results of a longitudinal study of psychological development. The study began in 1929 and continued through the late 1950s. Children enrolled in the Fels Research Institute's longitudinal population during the years between 1929 and 1939 were selected for repeated observations. In this way eighty-nine children (forty-four boys and forty-five girls) were selected for study. Subjects were mainly from the midwest, Protestant in their religious affiliations, white, of middle-class socioeconomic backgrounds, and the children of relatively well educated parents.

Obviously, not everyone who develops has the above characteristics. Thus one limitation of the Kagan and Moss study is in regard to external validity; that is, their sample is not necessarily representative of broader populations. Accordingly, while we may be unsure about the extent to which we may generalize the specific results of the Kagan and Moss study, we may at least expect the findings to provide us with some interesting (if tentative) suggestions about the course of personality-behavioral development from infancy onward.

Sources of Data. The initial information about the children pertained to their development from birth through early adolescence. The children were administered various tests of their intelligence and their personality, and these assessments were combined with live observations of them in their homes, nurseries, and, later, their schools and day camps. In addition, assessments of the mothers of the children were made, and teacher interviews were conducted. Through these procedures, data allowing for the measurement of many different variables across the first 14 years of life were obtained. Each variable was conceptualized as representing a behavioral dimension, with endpoints lying at 1 and 7. For example, for the variable "dependency," a score of 7 might indicate high dependency and a score of 1 might indicate low dependency, and a person's score could fall anywhere along this dimension.

Kagan and Moss reduced the first fourteen years of data into four consecutive and overlapping age periods: birth to 3 years (infancy and early childhood); 3 to 6 years (preschool years); 6 to 10 years (early school years); and 10 to 14 years (preadolescent and early adolescent years). Of their eighty-nine subjects studied in their first 14 years of life, seventy-one participated in the second phase of data gathering. When they returned, in the time between mid-1957 and late 1959, they were between 19 and 21 years of age. Thus a final age period—adulthood—was now introduced into the study.

Results of the Study. Perhaps the most consistent finding obtained by Kagan and Moss was that many of the children's behaviors shown in the third age period (the early school years, from 6 to 10 years of age) were fairly good predictors of similar early adulthood behaviors (Kagan & Moss, 1962, p. 266). Similarly, a few behaviors seen in the second age period (the preschool years, from 3 to 6 years of age) were also related to theoretically similar adult behaviors. However, in general, relatively little consistency between infancy and adult life was found. Nevertheless, as we will soon see, what consistency was found was quite provocative. Thus such adult behaviors as dependency (on the family) or anxiety (in social interactions) seemed to be related to analogous behavior-personality characteristics in the periods of early or middle childhood. Accordingly, Kagan and Moss conclude that such findings "offer strong support to the popular notion that aspects of adult personality begin to take form during early childhood" (1962, pp. 266–267).

Despite such overall continuity in personality development, an important qualification must be made. Despite the fact that changes in the variable of age period were often associated with continuity in the expression of various behavior-personality attributes, another variable—sex—affected this relation. Kagan and Moss found that age-period continuity in various behavioral characteristics was essentially dependent upon whether or not that behavior was consistent with traditional sex-role standards. For example, degrees of childhood passivity and childhood dependency remained continuous for adult women but were not similarly continuous for adult men. Kagan and Moss argue that traditional sex-role standards in our culture place negative sanctions on passive and dependent behaviors among males. Studies of stereotypes about the ideal masculine figure in our society find that the most positively evaluated male figure is one who is viewed as dominant, aggressive, and instrumentally effective (Lerner & Korn, 1972). Men who do not display such characteristics are negatively evaluated (Lerner & Korn, 1972). Kagan and Moss believe, however, that no corresponding negative sanctions about such behaviors exist for women in our society. Thus the authors found continuity between childhood passivity and dependency and adult passivity and dependency for females. A similar relation for males was not found.

On the other hand, through an analogous argument we might expect aggressive, angry, and sexual behaviors to be continuous for males but not continuous for females. In fact, such a finding was obtained. For example, "Childhood rage reactions and frequent dating during preadolescence predicted adult aggressive and sexual predispositions, respectively, for men but not for women" (Kagan & Moss, 1962, p. 268).

Of course, certain behaviors could be expected to remain similarly continuous for both sexes. Intellectually oriented behaviors (e.g., attempting to master school-work) and sex-appropriate interest behaviors (e.g., fishing for males, knitting for females) are consistent with traditional sex-role standards for either sex. Accordingly, Kagan and Moss found that such behaviors showed a marked degree of continuity for both sexes from their early school years through their early adulthood.

In sum, Kagan and Moss found overall age continuity for many personality-behavior characteristics. For many behaviors one could predict the adult's type of functioning through knowledge of his or her functioning in respect to conceptually consistent childhood variables. Yet whether or not such overall continuity was found depended on whether a particular behavior was consistent with traditional societal sex-role standards. Those behaviors that were consistent with sex-role standards remained stable; for those that were not, discontinuity was seen. Kagan and Moss summarize this portion of their results by stating:

> *It appears that when a childhood behavior is congruent with traditional sex-role characteristics, it is likely to be predictive of phenotypically similar behaviors in adulthood. When it conflicts with sex-role standards, the relevant motive is more likely to find expression in theoretically consistent substitute behaviors that are socially more acceptable than the original response. In sum, the individual's desire to mold his overt behavior in concordance with the culture's definition of sex-appropriate responses is a major determinant of the patterns of continuity and discontinuity in his development. (1962, p. 269)*

The Sleeper Effect. The above results present an important illustration of the potential empirical outcomes derived from longitudinal study. However, this technique provides the opportunity for finding other types of results; for instance, it may be the case that an important event occurs early in a person's life. The event will provide a cause for some of the person's behaviors, but this effect may not be seen right away. Simply, "there may be a lag between a cause and open manifestation of the effect" (Kagan & Moss, 1962, p. 277). In other words, one may see a *sleeper effect* in development.

Two such instances of a sleeper effect in development occurred in the Kagan and Moss study. It is here that some connection between personality in infancy and personality in later life was found. Among males, passivity and fear of bodily harm measured in the first age period (birth to 3 years) were found to be better predictors of a conceptually similar adult behavior (e.g., love-object dependency) than were the other measurements of the childhood behaviors. Thus males who were passive and feared bodily harm in their first 3 years of life (and thus may be surmised to have been dependent on their mothers for support and protection) were found to be similarly dependent on their love object (e.g., their wives) when they reached adulthood. Yet measurements of passivity and fear of bodily harm in the other three age periods did not predict adult male dependency. Although one may attempt to account for this finding through reference to possible problems in the measurement of dependency (perhaps the measures during the intervening periods were not sensitive enough to adequately measure dependency), one may also speculate that this sleeper effect is an instance of the same underlying personality characteristic being expressed in these two measurements.

A similarly interesting sleeper effect was found with females, when certain measures of the mother's behavior toward the child during the first 3 years of the child's life were related to various aspects of the child's own adolescent and adult functioning. A mother having a critical attitude toward her daughter during the first age period (birth to 3 years) was highly predictive of adult achievement behavior on the part of the daughter. However, a similar attitude in the other three age periods was not so related to adult female behavior. Similarly, maternal protection of the female child during the child's first 3 years of life was related to a conceptually consistent adult female behavior on the part of the daughter (e.g., withdrawal from stress), while similar maternal behaviors during the child's later age periods were not so related to adult female behaviors.

Through these findings, we see that Kagan and Moss were able to discover events and/or behaviors that occurred early in an infant's life that, while not relating to similar behaviors in immediately succeeding age periods, did highly relate to later, adult functioning. In the next study we will consider, further evidence will be presented that personality in infancy may have implications for later life.

The New York Longitudinal Study

The New York Longitudinal Study (NYLS) began in 1956 in New York City and continues through this writing. The study involves a longitudinal assessment of a relatively large (i.e., for a longitudinal study) group of children from their first days of life onward. There were 133 children in the initial sample and remarkably, at this writing more than 99 percent remain in the study. The purpose of the NYLS is to ascertain how a person's characteristics of individuality provide a basis of his or her development across the course of a person's life span.

Why Are Children Different? Any developmental psychologist would admit that children are different. Thus debates about the development of individuality do not focus on whether or not children are different; rather, they concern either the ways in which children differ or the sources of differences. The leaders of the New York Longitudinal Study—Alexander Thomas, Stella Chess, Sam Korn, Herbert Birch, and Margaret Hertzig—thought that the basis of a child's individuality lies in an interaction between his or her heredity and environment (see Chapter 3 for a discussion of nature-nurture interaction). The Thomas group sought to ascertain how the individuality that arose out of such an interaction contributed to the person's development. Specifically, the Thomas, Chess, Birch, Hertzig, and Korn (1963, p. 1) study was "concerned with identifying characteristics of individuality in behavior during the first months of life and with exploring the degree to which these characteristics are persistent and influence the development of later psychological organization."

Because of their theoretical point of view, Thomas et al. felt it necessary to focus on intraindividual consistencies and changes in the characteristics of a child over the course of development. Such an assessment of the contributions and implications of a child's individually different *style of behavioral reactivity*—the child's style of responding or reacting to the world—might supply important information bearing on an important topic in child development; i.e., how the child— as a consequence of his or her individuality—could affect his or her own development. Thus the researchers had to ascertain how the children with different styles of reactivity interact differently with the world over the course of their development and how such differential interactions continue to provide a source of the child's development. Accordingly, Thomas et al. had to study their subjects longitudinally. However, at the time of the study's inception, there was no acceptable definition of the dimensions of a child's characteristics of reactivity, or temperament. Although other psychologists (e.g., Sheldon, 1940, 1942) had provided definitions of temperament (or behavioral style), these were unacceptable to the Thomas group. Although the Thomas group knew that by *temperament* they meant only a general term representing the various aspects of how a child individually reacted to the world, they had no preconceived ideas about what sort of different reactive characteristics comprised a child's temperament.

This problem was complicated by the fact that the researchers felt it crucial to obtain measures of the child's temperament in all situations the child engaged in. If the researchers wanted to accurately ascertain how the child did whatever he or she did, and not just how the child went about doing certain things (for instance, those things that might be involved in a once-a-month experimental assessment session), they would have to obtain observations of the child's temperamental style continuously. The Thomas group decided to use each child's parent as the observer of the child's temperament; through appropriately designed interviews with each parent (i.e., interviews structured to get behavior descriptions and not interpretations), the researchers were able to obtain the needed information about the child's style of reactivity in all daily interactions with the world. At times, when parents insisted on interpreting rather than describing, the interviewer carefully reworded, rephrased, or repeated the question so as to obtain in each case a step-by-step description of the behavior in question. This insistence on description, rather than interpretation, was illustrated by Thomas et al. (1963, p. 25) in the following example of a segment of a parental interview:

INTERVIEWER: *"What did the baby do the first time he was given cereal?"*
PARENT: *"He couldn't stand it."*
INTERVIEWER: *"What makes you think he disliked it? What did he do?"*
PARENT: *"He spit it out and when another spoonful was offered he turned his head to the side."*

The Thomas group conducted interviews at three-month intervals during the child's first year of life. After the first year, subsequent interviews were conducted at six-month intervals.

The Attributes of Temperament. As noted, the Thomas group did not have any predetermined theoretical notions concerning what the attributes of temperament were. However, by performing an inductive content analysis of their data, they were able to generate nine reliably scored attributes of temperament into which the various behavioral descriptions could be placed. These categories comprise the nine attributes of temperament discovered by the Thomas group. The attributes are listed in Exhibit 6.1. Note that the nine categories refer to descriptions of the "how," or the style, of behavior.

Results of the NYLS. The NYLS is an ongoing, longitudinal project. The data from the study are still being analyzed, and the children of the project are still being studied. Thus we may speak only of those results of the investigation that have already been published (e.g., in Chess, Thomas, & Birch, 1965; Thomas & Chess, 1970, 1977, 1980; Thomas, Chess, & Birch, 1968, 1970; Thomas et al., 1963).

The first major finding was that children do show individually different temperamental repertoires and interrelations; moreover, these individual differences in temperament do become distinct—they can be discerned—even in the first few weeks of the child's life. Although some children tend to be similar in their tempera-

EXHIBIT 6.1 The Nine Categories of Temperament Identified in the New York Longitudinal Study, and the Definitions and Ratings Associated with Each

Temperamental Category	Definition	Types of Ratings Given
1. Activity level	The proportion of active periods to inactive ones	High, moderate, low
2. Rhythmicity	Regularity of hunger, excretion, sleep, and wakefulness	Regular, variable, irregular
3. Approach-withdrawal	The response to a new object or person	Approach, variable, withdrawal
4. Adaptability	The ease with which a child adapts to changes in his or her environment	Adaptive, variable, nonadaptive
5. Intensity of reaction	The energy of response, regardless of its quality or direction	Intense, variable, mild
6. Threshold of responsiveness	The intensity of stimulation required to evoke a discernible response	High, moderate, low
7. Quality of mood	The amount of friendly, pleasant, joyful behavior as contrasted with unpleasant, unfriendly behavior	Positive, variable, negative
8. Distractibility	The degree to which extraneous stimuli alter behavior	Yes, variable, no
9. Attention span and persistence	The amount of time devoted to an activity, and the effect of distraction on the activity	High, variable, low

Source: Adapted from Thomas et al., 1963, 1970.

mental styles, different arrays of scores for each of the different attributes were found. Thus in particular for the attributes of activity level, threshold, intensity, mood, and distractability, marked individual differences in temperamental repertoires were evident (Thomas et al., 1963, p. 57).

Moreover, these individually different temperamental styles were not systematically related either to the parents' method of childrearing or to the parents' own personality styles (Thomas et al., 1970). This finding indicates not only that children are individually different but also that these characteristics of individuality are not simply related to what the parent does to the child or to the parents' own personality characteristics.

The second major finding of the NYLS is that these characteristics of individuality, first identified in the child's first 3 months of life, tend to continue to characterize the child over the course of later years. Exhibit 6.2 shows that for each of the nine categories of temperament there is consistency in behavioral style across infancy (i.e., at 2 months, 6 months, 1 year, and 2 years of age) and from infancy into middle and late childhood (i.e., at 5 years and 10 years of age, respectively).

In sum, the results of the Thomas et al. (1963, 1968, 1970) NYLS and of the Kagan and Moss (1962) study of "birth to maturity" provide some support for the view that personality developments in infancy have some implications for personality functioning in later life. Does this mean that our personality is set in our earliest years? Is there just consistency across the life span? Our focus on Erikson's (1959, 1968) ideas led to the view that both consistency and change may characterize the life span. In addition, there has been recent emphasis on the potential for change, or plastic development, at all points in life (Brim & Kagan, 1980).

Accordingly, it is appropriate to ask, "To what extent do patterns of behavior developed in infancy provide an immutable basis of functioning in later life?" Do particular experiences, for example, those in which the infant may be deprived of social stimulation, provide a necessary (in this case negative) outcome in later life? Or can interventions at later portions of the life span optimize the development of even formerly deprived infants? We address these issues in the next section of this chapter.

EFFECTS OF SOCIAL ISOLATION AND DEPRIVATION

Most people know of Harry F. Harlow's studies (e.g., Harlow & Zimmerman, 1959) of rhesus monkeys reared by surrogate cloth or wire monkeys. Harlow was interested in learning about the nature of mothering, and sought to find out if the ties infants showed to mothers, for example, were due more to the food they provided or to the warmth and comfort they delivered. In one study two "wire monkeys" were placed in the infant monkey's cage. One wire monkey delivered food through a nipple. The other was wrapped in a soft cloth but did not deliver food. All other features of the monkeys were identical. Harlow (Harlow & Harlow, 1962; Harlow & Zimmerman, 1959) found that except when feeding, the infant clung more often to the cloth-wrapped "monkey," the one, Harlow contended, that provided "contact comfort." In addition, in times of stress the infant was most likely to cling to the cloth mother.

As part of these investigations Harlow had to isolate the infant monkeys from their real mothers. In a sense, Harlow's studies of mothering were also studies of social isolation or of deprivation of social interaction with particular members of one's species. Of course, Harlow recognized this aspect of his work. Indeed he and other investigators began to do a series of studies on the effects of social isolation or deprivation on monkeys. Much of what is known about the effects of social isolation comes from these studies.

Social Isolation in Rhesus Monkeys

It is, of course, unethical to isolate or deprive human infants experimentally or intentionally in the ways that one can ethically study rhesus monkeys. Thus although we cannot necessarily generalize the results of studies done on nonhumans, we need to be aware of results of studies of nonhumans. Indeed, at the very least, such studies provide hypotheses about the role of social interaction in humans.

There have been numerous studies of the effects of early social deprivation in rhesus monkeys (e.g., Harlow & Harlow, 1962, 1966, 1970). Monkeys socially isolated for the first 6 months or longer show extremely abnormal patterns of behavior when removed from isolation. These problems often persist into adulthood. For example, after 3 months of social deprivation, an infant monkey removed from isolation avoids all social contact and assumes motor stances wherein it buries its head in its arms and crouches (Fuller & Clark, 1966a, 1966b). Monkeys isolated 6 months or longer develop abnormal social and sexual behavior during their adolescence and adulthood (Harlow & Harlow, 1962). Indeed, Suomi and Harlow (1972, p. 166) note that the longer monkeys are "denied the opportunity to interact with peers, the more gross are their social inadequacies."

However, the effects of such social isolation are not irreversible. In fact, following up on the observation of Suomi and Harlow (1972), several studies have shown

EXHIBIT 6.2

Behavior Illustrations for Ratings of the Various Temperamental Attributes at Various Ages

Temperamental Quality	Rating	2 Months	6 Months
Activity level	High	Moves often in sleep. Wriggles when diaper is changed	Tries to stand in tub. Bounces in crib. Crawls after dog
	Low	Does not move when being dressed or during sleep	Passive in bath. Plays quietly in crib and falls asleep
Rhythmicity	Regular	Has been on 4-hour feeding schedule since birth. Regular bowel movement	Is asleep at 6:30 every night. Awakes at 7 A.M. Constant food intake
	Irregular	Awakes at a different time each morning. Size of feeding varies	Variable length of nap. Variable food intake
Distractibility	Distractible	Will stop crying for food if rocked. Stops fussing if given pacifier when diaper is being changed	Stops crying when mother sings. Will remain still while clothing is changed if given a toy
	Not Distractible	Will not stop crying when diaper is changed. Fusses after eating even if rocked	Stops crying only after dressing is finished. Cries until given bottle
Approach-withdrawal	Positive	Smiles and licks washcloth. Has always liked bottle	Likes new foods. Enjoyed first bath in a large tub. Smiles and gurgles
	Negative	Rejected cereal the first time. Cries when strangers appear	Cries at strangers. Delays playing with new toys.
Adaptability	Adaptive	Was passive during first bath; now enjoys bathing. Smiles at nurse	Used to dislike new foods; now accepts them well
	Not Adaptive	Still startled by sudden, sharp noise. Resists diapering	Does not cooperate with dressing. Fusses and cries when left with sitter

1 Year	2 Years	5 Years	10 Years
Walks rapidly. Eats eagerly. Climbs into everything	Climbs furniture. Explores. Gets in and out of bed while being put to sleep	Leaves table often during meals. Always runs	Plays ball and engages in other sports. Cannot sit still long enough to do homework
Finishes bottle slowly. Goes to sleep easily. Allows nail cutting without fussing	Enjoys quiet play with puzzles. Can listen to records for hours	Takes a long time to dress. Sits quietly on long automobile rides	Likes chess and reading. Eats very slowly
Naps after lunch each day. Always drinks bottle before bed	Eats a big lunch each day. Always has a snack before bedtime	Falls asleep when put to bed. Bowel movement regular	Eats only at mealtimes. Sleeps the same amount of time each night
Will not fall asleep for an hour or more. Moves bowels at a different time each day	Nap time changes from day to day. Toilet training difficult because of unpredictable bowel movement	Variable food intake. Variable time of bowel movement	Variable food intake. Falls asleep at a different time each night
Cries when face is washed unless it is made into a game	Will stop tantrum if another activity is suggested	Can be coaxed out of forbidden activity by being led into something else	Needs absolute silence for homework. Has a hard time choosing a shirt in a store because they all appeal to him or her
Cries when toy is taken away and rejects substitute	Screams if refused some desired object. Ignores mother's calling	Seems not to hear if involved in favorite activity. Cries for a long time when hurt	Can read a book while television set is at high volume. Does chores on schedule
Approaches strangers readily. Sleeps well in new surroundings	Slept well the first time he or she stayed overnight at grandparents' house	Entered school building unhesitatingly. Tries new foods	Went to camp happily. Loved to ski the first time
Stiffened when placed on sled. Will not sleep in strange bed	Avoids strange children in the playground. Whimpers first time at beach. Will not go into water	Hid behind mother when entering school	Severely homesick at camp during first days. Does not like new activities
Was afraid of toy animals at first; now plays with them happily	Obeys quickly. Stayed contentedly with grandparents for a week	Hesitated to go to nursery school at first; now goes eagerly. Slept well on camping trip	Likes camp, although homesick during first days. Learns enthusiastically
Continues to reject new foods each time they are offered	Cries and screams each time hair is cut. Disobeys persistently	Has to be hand led into classroom each day. Bounces on bed in spite of spankings	Does not adjust well to new school or new teacher; comes home late for dinner even when punished

EXHIBIT 6.2 (*Continued*)

Temperamental Quality	Rating	2 Months	6 Months
Attention span and persistence	Long	If soiled, continues to cry until changed. Repeatedly rejects water if wanting milk	Watches toy mobile over crib intently. "Coos" frequently
	Short	Cries when awakened but stops almost immediately. Objects only mildly if cereal precedes bottle	Sucks pacifier for only a few minutes and spits it out
Intensity of reaction	Intense	Cries when diaper is wet. Rejects food vigorously when satisfied	Cries loudly at the sound of thunder. Makes sucking movement when vitamins are administered
	Mild	Does not cry when diaper is wet. Whimpers instead of crying when hungry	Does not kick often in tub. Does not smile. Screams and kicks when temperature is taken
Threshold of responsiveness	Low	Stops sucking on bottle when approached	Refuses fruit he or she likes when vitamins are added. Hides head from bright light
	High	Is not startled by loud noises. Takes bottle and breast equally well	Eats everything. Does not object to diaper being wet or soiled
Quality of mood	Positive	Smacks lips when first tasting new food. Smiles at parents	Plays and splashes in bath. Smiles at everyone
	Negative	Fusses after nursing. Cries when carriage is rocked	Cries when taken from tub. Cries when given food she or he does not like

1 Year	2 Years	5 Years	10 Years
Plays by self in play-pen for more than an hour. Listens to singing for long periods	Works on a puzzle until completed. Watches when shown how to do something	Practiced riding a two-wheeled bicycle for hours until he or she mastered it. Spent over an hour reading a book	Reads for two hours before sleeping. Does homework carefully
Loses interest in a toy after a few minutes. Gives up easily if she or he falls while attempting to walk	Gives up easily if a toy is hard to use. Asks for help immediately if undressing becomes difficult	Still cannot tie his or her shoes because of giving up when not successful. Fidgets when parents read to him or her	Gets up frequently from homework for a snack. Never finishes a book
Laughs hard when father plays roughly. Screamed and kicked when temperature was taken	Yells if feeling excitement or delight. Cries loudly if a toy is taken away	Rushes to greet father. Gets hiccups from laughing hard	Tears up an entire page of homework if one mistake is made. Slams door of room when teased by younger sibling
Does not fuss much when clothing is pulled on over head	Looks surprised and does not hit back when another child hits her or him	Drops eyes and remains silent when given a firm parental "no."	When a mistake is made on a model airplane, corrects it quietly. Does not comment when reprimanded
Spits out food he or she does not like. Giggles when tickled	Runs to door when father comes home. Must always be tucked tightly into bed	Always notices when mother puts new dress on for first time. Refuses milk if it is not ice-cold	Rejects fatty foods. Adjusts shower until water is at exactly the right temperature
Eats foods he or she likes even if mixed with disliked food. Can be left easily with strangers	Can be left with anyone. Falls asleep easily on either back or stomach	Does not hear loud, sudden noises when reading. Does not object to injections	Never complains when sick. Eats all foods
Likes bottle; reaches for it and smiles. Laughs loudly when playing peekaboo	Plays with sibling; laughs and giggles. Smiles when he or she succeeds in putting shoes on	Laughs loudly while watching television cartoons. Smiles at everyone	Enjoys new accomplishments. Laughs when reading a funny passage aloud
Cries when given injections. Cries when left alone	Cries and squirms when given haircut. Cries when mother leaves	Objects to putting boots on. Cries when frustrated	Cries when unable to solve a homework problem. Very "weepy" if not getting enough sleep

that the effects of social isolation in rhesus monkeys can be revised by exposing iso-late-reared monkeys to peer "therapist" monkeys. Chamove (1978), for example, studied a group of rhesus monkey infants who were reared in three different conditions and then subdivided into three groups. That is, the infants were reared either as (1) socially "sophisticated" (i.e., nonisolated) monkeys for the first 9 months of their lives; (2) partially isolated monkeys, reared alone from birth for 9 months; and (3) socially naive 3-month-old infant monkeys. When groups of the 9-month-old partial isolates were paired with monkeys from their own and the other two rearing conditions, social play was greatest when the isolates were paired with "sophisticated" monkeys ("therapists") and was least when isolates were paired with isolates. In turn, fear responses were most frequent when monkeys from any of the three rearing conditions were paired with isolates. The 3-month-old monkeys were affected by pairing the most; they showed reduced play when paired with the isolated 9-month-olds but demonstrated play when with the sophisticated monkeys.

Even among somewhat older rhesus monkeys, effects of social isolation can be reversed. Novak (1979) demonstrated that 3-year-old "rehabilitated" isolates, who were still deficient when compared to normally reared age-mates in their social behaviors, could be made to exhibit appropriate behavior to age-mate monkeys. The isolate monkeys had been isolated for the first year of life. Although showing some recovery from the effects of isolation when "treated by" (exposed to) younger therapist monkeys, the isolates, now 3 years old, still had problems in their social behavior (e.g., they were attacked in social encounters with normal monkeys). Although the social behavior problems of the isolates did not change following 10 weeks of visual exposure to their age-mates, it did improve substantially after the isolates were housed permanently as a group of four rather than individually. Follow-up tests with new groups of normal age-mates, and with monkey infants, confirmed the newly acquired social competency of the isolates and their ability to interact appropriately and nonaggressively with young animals. Thus even after a long period of isolation, wherein problems associated with the isolation had become well established, there was plasticity sufficient in the potential behavioral repertoire of the isolated monkeys to overcome the adverse influences of isolation.

Thus we see the powerful, negative effects in monkeys of experiencing early social deprivation. In addition, we see the crucial role of peers in moderating these effects. Although, as noted, we do not have *experimental* evidence about the role of social deprivation in humans, we do, nevertheless, have important sources of information.

Social Isolation and Deprivation in Human Children

Children developing in certain institutional settings often experience a degree of social isolation. In the 1940s, several studies found that children who develop in the first year of life in understaffed institutions do not develop as adequately as do children reared at home by their parents (e.g., Spitz, 1945, 1946). For example, in one study by Spitz, infants growing up in a foundling-home institution were isolated from one another by sheets hung around most of their cribs; thus they received little physical or social visual stimulation. The children had few toys and were generally kept in their cribs for their first year of life. A second group of infants was reared in a nursery which provided a fairly stimulating environment. Spitz found that the two groups of children became increasingly dissimilar as they aged. The foundling-home children lagged further and further behind the other children. In addition, the foundling-home children developed severe behavioral and emotional problems. In fact, Spitz concluded that "while the children in 'Nursery' developed into normal healthy toddlers, a two-year observation of 'Foundling-home' showed that the emotionally starved children never learned to speak, to walk, to feed themselves. With one or two exceptions . . . those who survived were human wrecks who behaved in a manner of agitated or of apathetic idiots" (Spitz, 1949, p. 149) who "would lie or sit with wide open, expressionless eyes, frozen, immobile faces, and a far away expression as if in a daze, apparently not perceiving what went on in their environment" (Spitz & Wolf, 1946, p. 314).

In a more recent study, Provence & Lipton (1962) studied physically healthy infants who were living in an institution that provided appropriate nutritional and medical care but low levels of social stimulation. By the end of the first year of life, the children came to resemble those studied by Spitz—although these institutionalized infants had not differed from normal home-reared ones in the early months of life.

There are some studies which show that such early social deprivation can have long-lasting effects on humans. Goldfarb (1945, 1947) studied one group of children who left an institutional setting for rearing in a foster home during the first year of their lives and a second group of children who remained in an unstimulating orphanage for the first 3 years of their lives before leaving for foster homes. Goldfarb assessed each group when they were just more than 3, 6, 8, and 12 years old. On all measures, the second group was inferior to the first. Children in the second (i.e., institutional) group had difficulty in forming close personal relationships and tended to be socially withdrawn.

In sum, depriving infants of social interaction can have severely negative effects on their development. However, as emphasized in earlier chapters, the human organism is a plastic (i.e., changeable) organism. Consistent with our life-span interests in the optimization of human development, as well as this life-span focus on plasticity, we may ask if these severe effects of social isolation are available for change. We treat this issue in the next section of the chapter.

The Reversibility of Early Deprivation Effects

Several studies suggest that negative effects of being deprived of social interaction can be reduced. For example, in a study by Skeels (1966), mentally retarded women served as substitute mothers for a group of institutionalized infants, and the infants' social and intellectual development markedly improved. Similarly, reversibility of early deprivation has been found in other studies (e.g., Clarke & Clarke, 1976).

In one major study, in Guatemala, Kagan and Klein (1973) found severe retardation among infants living in an isolated farming area. Compared to their American peers, they were behind on a variety of cognitive and social skills. Kagan (1976) believes that the children's restricted environment may contribute to this slow development. Guatemalan infants are tightly clothed and restricted to the inside of a windowless hut. Adults interact minimally with the child, and the child has few conventional toys to play with. But during the second year of life, the child is allowed to leave the hut. Thus he or she can now interact with the diversified and complex world outside the unstimulating hut. By middle childhood the child, once minimally attended to and unstimulated, is now given a large array of complex and adultlike tasks—all of which require extensive interactions with the context. Although the interpretation of these data is not unequivocal (see Meacham, 1975), these changed experiences do seem to enhance cognitive development. By 10 years of age, despite the initial developmental slowness of Guatemalan children, they do seem as capable as American children on several cognitive tasks.

In sum, institutionalization, involving limited social interactions and/or severe early deprivation of stimulation, can have a negative effect on a child's social, intellectual, and motor development. However, research with monkeys (Harlow & Zimmerman, 1959; Harlow & Harlow, 1962), as well as with humans (Kagan & Klein, 1973), indicates that these effects of early deprivation can be reversed. The key for adequate development appears to be a responsive social environment, e.g., as provided by the peer group. Accordingly, as we now turn to a discussion of the childhood years, we will continue to see the importance of the social context and of peers (as well as other facets of the child's social world).

CHAPTER SUMMARY

This chapter stresses the link between social and individual processes during infancy. Socialization is a process by which members of one generation or group influence the behaviors and personalities of another generation or group. Human evolution has involved the elaboration of processes making people responsive to their social context; and with this linkage between biological and social functioning, socialization is facilitated.

People interact reciprocally with their social context, which includes other people and social institutions. These reciprocal relations exist in infancy and, in large part, involve interactions between infants and their caregivers (typically their parents). Considerable research supports the idea that the nature of infant-family relations is one of bidirectional socialization. The study of attachment behaviors illustrates these ideas. In addition, attachment behaviors may be usefully understood through application of concepts used in the life-span view of human development.

Attachment behaviors show multidirectional change across infancy (and later life), and they are moderated by variables associated with various levels of the context. Moreover, a pluralistic approach to the explanation of attachment development seems of use; for example, ideas associated with maturational, social learning, and cognitive developmental approaches all seem relevant. Attachment development is related to adaptive developments in areas of functioning other than social. As such, attempts to optimize attachment may be seen as relevant to enhancing human functioning. Because attachment is a contextually sensitive developmental phenomenon, several optimization strategies are available to interventionists.

The nature of one's links to the social context shows interindividual variation. Thus an understanding of individuality, of personality, is of use in an attempt to understand infant–social context reciprocities. There are many different definitions of personality. Our life-span conception stresses the combined role of universal, group, and individual aspects of personality.

Erikson's descriptions of personality development are seen as useful in understanding these features of personality. His ideas stress that personality development across life should be characterized by both continuity and change.

There is evidence that personality development in infancy does have implications for development in later life. The studies by Kagan and Moss (1962) and by Thomas et al. (1963, 1968, 1977, 1980) stress this. However, although there may be strong effects of early infancy experiences, there is also evidence that these impacts of infant development on later personal and interpersonal functioning can be ameliorated. Evidence exists that both on the nonhuman and human level, developments in infancy can be altered by social interventions.

Part Three

Child and Adolescent Development

Child and adolescent development, although not traditionally approached from a life-span perspective, has often been studied by scholars from multiple disciplines. For example, in the first several decades of this century child development was often studied within university institutes, e.g., at the University of Iowa, at the University of Minnesota, and at the University of California, Berkeley, which were designed to be multidisciplinary. However, this multidisciplinary perspective began to erode by the 1950s, and was replaced by a largely univocal psychological view of development. Similarly, adolescent development has been a concern of, and taught by, scholars working in departments of sociology, human biology, home economics, and education, as well as psychology. However, work in each of these disciplines has remained relatively independent.

Theoretical and empirical advancements in the study of child and adolescent development have been impressive across this history. Although it is of course the case that there are still great gaps in our knowledge, we know a good deal about cognitive developments, about the effects of peers on social development, about the nature of physical changes in childhood and adolescence, and about family relations in these portions of life—to cite just a few areas of knowledge. Moreover, as a consequence of recent conceptual and empirical advances in the social sciences, many of which are associated with the life-span perspective, there is a reintroduction of a multidisciplinary perspective in the study of child and adolescent development. Thus, today, we have almost come full circle: Work on basic processes of child and adolescent development by scholars from several disciplines is continuing, but at the same time there is a growing recognition of the importance of casting this knowledge in an integrative, often multidisciplinary, framework.

Issues pertinent to reciprocal socialization, to children's effects on their caregivers, and to the role of historical change in influencing basic processes of development provide examples of the theoretical and empirical concerns which are today leading social scientists toward a renewed appreciation of a multidisciplinary, and often life-span, approach to child and adolescent development. In addition, major changes in the social ecology of child and adolescent life, and in the character of their families, are also acting as an impetus for these approaches. For example, a proportion of children greater than ever before will experience the divorce of their parents, will live for at least a portion of their youth in single-parent homes, and will have mothers who are engaged in full-time employment outside the home. More youth than ever before, both in proportion and in absolute numbers, will be sexually active, will be involved in a pregnancy out of wedlock, will contract venereal disease, and will use drugs and alcohol.

In sum, at the very time we are learning more and more that we need to know more and more about the basic processes of childhood and adolescence, we are discovering that the nature of life during these periods is showing important changes. These changes are what make, in our view, a multidisciplinary, life-span perspective so important in the study of these age periods. These changes are also what make such an approach an exciting and challenging one to adopt. Our goal is to convey both this importance and excitement in the chapters of this section.

chapter 7

Childhood: cognitive development

CHAPTER OVERVIEW

The cognitive changes first identified during the years of infancy burgeon in childhood. New cognitive achievements become elaborated; language is one such example. The child begins to apply these cognitive developments in new and broader contexts; school settings are a major instance. In this chapter we will consider some of the cognitive changes of childhood, placing particular emphasis on psychometric and Piagetian approaches. We also will see how cognitive changes affect the child's functioning in his or her broader social context by considering the topics of language, of moral development, and of the role of schools in child development.

CHAPTER OUTLINE

LEARNING IN CHILDHOOD
Developmental Changes in the "Rules" of Learning

PSYCHOMETRIC INTELLIGENCE IN CHILDHOOD
Stability and Change during Childhood
Variables That Moderate Intellectual Change
Improving Cognitive Functioning

PIAGETIAN IDEAS ABOUT COGNITIVE DEVELOPMENT IN CHILDHOOD
The Preoperational Stage
The Concrete Operational Stage

LANGUAGE DEVELOPMENT
Theories of Language Acquisition
Stages of Vocalization
Stages of Linguistic Speech
Functions of Language

MORAL DEVELOPMENT IN CHILDHOOD
DEFINITIONS OF MORAL DEVELOPMENT
Freud's Nature-Oriented Theory
Nurture-Oriented Social Learning Theories
Interactionist Cognitive Developmental Theories

KOHLBERG'S THEORY OF MORAL-REASONING DEVELOPMENT
Kohlberg's Method of Assessing Moral Reasoning

STAGES OF MORAL REASONING IN KOHLBERG'S THEORY:
FORMER AND CURRENT LEVEL AND STAGE DESCRIPTIONS
The Former Version of the Theory
The Current Version of the Theory

CHARACTERISTICS OF MORAL-REASONING STAGE DEVELOPMENT
A CRITIQUE OF KOHLBERG'S THEORY
Methodological Appraisals of Kohlberg's Theory
Empirical Appraisals of Kohlberg's Theory

VARIABLES RELATING TO MORAL DEVELOPMENT
Effects of Models
Family Interactions
Peer Interactions

THE ROLE OF SCHOOLS
Physical Characteristics of Schools
Effects of Early School Programs

CHAPTER SUMMARY

ISSUES TO CONSIDER

What are the major developmental changes that occur in childhood learning? Why do they show the limits of simple stimulus-response models?

What is the nature of stability and change in IQ scores during childhood?

What variables influence changes in IQ scores in childhood?

Is there evidence that cognitive functioning can be improved?

What are the major characteristics of the preoperational and the concrete operational stages of cognitive development in Piaget's theory?

What are the leading theories of language acquisition? What is the nature of the evidence in support of each?

What are the major stages of language development?

What are the functions of language?

Why is moral development an important topic to study in childhood?

What are the major theories of moral development that have been forwarded?

What is the major distinction between the former and current versions of Kohlberg's theory?

To what extent does current evidence support Kohlberg's ideas?

What variables seem to relate to moral-reasoning development?

What are the effects of physical characteristics of schools on cognitive development?

Are there any positive effects of early school programs on cognitive development?

LEARNING IN CHILDHOOD

*A*s recently as 1970, Stevenson, a leading scholar of learning in childhood, noted that "only recently has the study of children's learning received a great deal of attention" (p. 235). However, as he noted, the amount of attention paid to this topic increased greatly by 1970. Thus, today, a large amount of information has been collected. As such our review of learning in childhood will be quite selective, focusing on topics that illustrate developmental changes in children's learning abilities.

Developmental Changes in the "Rules" of Learning

In Chapter 5 we saw that there was evidence that the principles, or rules, of classical and operant conditioning were relevant to learning among infants. That is, research designed to demonstrate the role of operant and classical conditioning phenomena were often seen to be successful. McCall (1977) has noted, however, that the relevance of some rules, principles, or "laws" in one situation (or at any age level) does not mean they are necessarily relevant to all situations or age levels. A demonstration that procedures may elicit a particular behavior does not mean that the procedures (always) do elicit that behavior in the "real" world.

Analogously, we may suggest that although the rules of operant and classical conditioning, as involved in infancy, may be relevant in childhood, it does not mean that they suffice to account for all learning phenomena in childhood. Indeed, as stressed by Stevenson (1970, p. 339), "Some of the problems that a stimulus-response approach to learning faces are encountered in developmental studies, where response characteristics of children at one age are displaced by other quite different forms of behavior at later ages."

A classic study by Stevenson and one of his colleagues illustrates this point. Stevenson and Weir (1961) presented 3-, 5-, 7-, and 9-year-old children with a task wherein they were required to make a choice of three possible alternatives. Children could indicate their choices by pressing a button. Marbles, which could be exchanged for a prize, were used as reinforcements, but two of the choices presented to the children were never reinforced.

Stevenson and Weir assessed the proportion of subjects who, on a second trial, repeated their response after their first response had been either rewarded by a marble or not rewarded. According to principles of reinforcement, one might expect that for all children a choice reinforced by a marble should be repeated, while a choice not so reinforced should not be repeated. This expectation was supported by the responses of the 3-year-olds; more than 80 percent of these children repeated their first choice if it had been rewarded by a marble, and fewer than 50 percent did not when their choice was not rewarded. This means that most of the 3-year-olds who made the rewarded choice the first time were rewarded the second time, since only one choice was a rewarded one. Thus their "strategy" of "sticking with the rewarded choice" would lead to their *maximizing* their rewards.

We might expect that the 9-year-olds, who "know" much more than 3-year-olds, would do even better at maximizing their rewards. Certainly, the traditionally understood principle of reinforcement would predict that once this oldest group learned which choice was rewarded they would maintain their response to it. However, this is not what Stevenson and Weir (1961) found.

Almost 80 percent of the 9-year-olds *changed* their response after receiving a reward. In fact, among the 5- and 7-year-olds there was an increasing tendency to move in the direction of the 9-year-olds; that is, there was an increasing tendency to make a response which differed from the one already rewarded. Stevenson (1970) notes that the older children were using a "win-shift, lose-shift" strategy to direct their responses, a strategy not easily accounted for in traditional reinforcement terms; however, the youngest children's responses seemed to be able to be accounted for by traditional reinforcement rules. Thus in this situation the oldest children were not receiving as many rewards for their responses to the three-choice problems as were the 3-year-olds.

Does this mean that 3-year-olds are better problem solvers than 9-year-olds? Not at all. Other studies of developmental changes in children's learning to solve problems suggest that what is occurring is the emergence of a new developmental phenomenon in children. Studies of problem-solving behavior by Kendler and Kendler (1962) illustrate the presence of this development.

Kendler and Kendler (1962) devised a way to study problem-solving behavior in various species of organism (e.g., rats) as well as in humans of various ages (e.g., nursery school children and college students). In the procedure they devised, subjects are presented first with two large squares and two small squares. One of each type of square is black, one of each type white. Thus there is a large black and a large white square, and a small white and a small black square. The subjects' task is to learn to respond either to the color dimension (and thus ignore the size) or to the size dimension (and thus ignore the color). For example, a subject may be presented with a large black and a small white square on one trial and then perhaps a large white square and a small black square on another trial. Now, if size is the aspect of the stimuli that should be responded to and, further, a response toward the bigger of the two squares will always lead to a reward, the subject should choose the large stimulus in each trial, no matter what the color. In other words, the subject first learns to respond to the difference in size and to ignore (not respond to) differences in color of the squares.

Rats, nursery school children, and college students can all learn this first problem-solving task. The interesting thing about this type of problem solving is what happens when the rules about the relevant aspect of the stimuli are changed. In the first problem-solving task, size was the relevant dimension (the big squares were rewarded and the small squares were not). Now, without directly cueing the subject that this rule has changed, it is possible to still keep the size of the stimuli as the relevant dimension (and the color as the nonrelevant) but make choice of the small squares the response that will be rewarded. Thus the same dimension of the stimuli (size) is still relevant, but there has been a reversal in which aspect of the size (from large to small) will lead to a reward. Kendler and Kendler call this type of alteration a *reversal shift*; the same stimulus dimension is still related to reward, but which of the two stimuli within this same dimension is positive and which is negative are merely reversed. A second type of shift may occur, however, in the second problem-solving task. Instead of size being the reward-relevant dimension, color can be. Now response to the black squares (regardless of their size) will lead to a reward, and response to the white squares (regardless of their size) will not. This type of change involves a shift to the other dimension of the stimuli and is not within the same dimension. Hence the Kendlers term this second type of possible change a *nonreversal shift*. Exhibit 7.1 illustrates the reversal and the nonreversal shifts. In all cases the stimuli toward which a response will lead to a reward are marked +, while the stimuli toward which a response will not be rewarded are marked −.

Kendler and Kendler (1962) review the studies of reversal and nonreversal problem solving done with rats, nursery school children, and college students. After learning the first problem (for example, after making ten correct responses to the

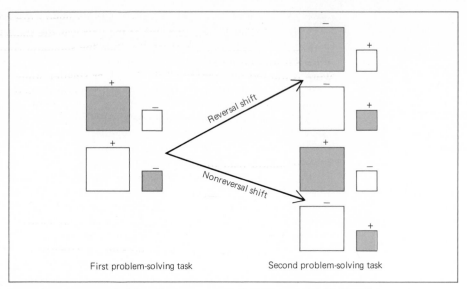

First problem-solving task　　　　　Second problem-solving task

EXHIBIT 7.1　Examples of a reversal shift and a nonreversal shift. (Source: H. H. Kendler and T. S. Kendler, "Vertical and Horizontal Processes in Problem Solving," *Psychological Review 69*, 1962. Copyright 1962 by the American Psychological Association. Reprinted by permission.)

large-sized stimuli), would it then be easier to learn a reversal shift or a nonreversal shift (again using the criterion of ten consecutive correct responses for learning)? The Kendlers' review indicates that rats learn a nonreversal shift easier than a reversal shift. Moreover, so do most nursery school children. Like rats, these human children reach the criterion for making a nonreversal shift faster than they reach the criterion for making a reversal shift. However, somewhat older children, as well as college students, find a reversal shift easier.

The Kendlers interpret these age changes by suggesting that in development there emerges a new mental process in children such that they move from responding ratlike to college student-like; this new mental process, not present at earlier ages (e.g., efficient language processes), alters children's problem-solving behavior such that a reversal shift becomes easier than a nonreversal shift. Hence while children's problem-solving behavior at the nursery school level can be accounted for by reference to processes apparently also identifiable in rats, their later behavior may be explained by the emergence of a new mental process. Certainly the processes present in the nursery school children provided a developmental basis for the processes seen among the older children. That is, it would be very unlikely to find children who now functioned like a college student but never functioned like a rat. Yet these former processes are not sufficient to account for the behavior of the older children. The problem-solving type of behavior reverses, and this alteration appears related to the emergence of a new mental function. Although other interpretations of these findings have been offered (see Esposito, 1975), the present point is that the work reported by Kendler and Kendler (1962) illustrates the developmental nature of learning in childhood. But what is involved in this development?

As noted by Stevenson (1970), a study by Graham, Ernhart, Craft, and Berman

(1964) may help answer this question. Children, aged 2 to 4.5 years, were presented with pairs of stimuli; one stimulus of the pair was the same for all children, while the other was either larger or smaller than the standard. Half of the children were always rewarded for choosing a particular stimulus—the standard one. Thus they were rewarded for "ignoring" the larger or smaller stimulus. However, as in the work of the Kendlers, a response to the relation between the stimuli was rewarded for the other children. That is, choice of the larger of the two stimuli (which in some children was the standard stimulus and in others was the other stimulus) was rewarded.

Principles of reinforcement would lead to the expectation that consistency of reward should lead to more rapid learning than should inconsistency of reward. Thus the group of children that was consistently rewarded for response to the same stimulus—the standard—should have learned more rapidly. From the work of Kendler and Kendler we may suspect that this did not occur. Indeed it did not. At all ages children learned the relational problem more easily.

What these children thus learned is that the bigger of the two stimuli was rewarded. As such, both Kendler and Kendler (1962) and Stevenson (1970) suggest that what is guiding the responses of these children is not an overt, mechanical S-R connection; rather, a mental (cognitive) process mediates between the stimuli presented to the children and their responses. The role of such *verbal mediation* (e.g., "Pick the big one") is of course dependent on the emergence of language processes and other forms of symbolic or representational ability. Later in this chapter we will directly explore symbolic and language developments in childhood. Here we should note that although there is still controversy surrounding the role of verbal mediation in accounting for developmental changes in the rules of learning in childhood (e.g., Cantor & Spiker, 1978), there are considerable data supporting the role of some such mediational process.

For example, Kemler (1978) studied elementary school children's problem-solving procedures in response to discrimination tasks. In the first two of three studies she conducted, Kemler studied the strategies used by children 7.5 and 10.5 years old, respectively. Not only did both age groups efficiently use the stimulus information presented to them, but, and especially in the older group, they tended to reject permanently strategies for response that had failed to be rewarded. They showed they could remember past strategies. In the third study Kemler (1978) conducted, 5-year-olds were studied. While, like the older children, they could successfully use information about the stimuli currently being presented to them to guide their responses, they did not show evidence of use of memories of past strategies. In fact, they showed a counterproductive tendency to reuse previously failed strategies.

In sum, as noted earlier, there are many topics that may be discussed under the rubric of learning in childhood, and a review of all of them is not possible in an introductory treatment. However, we will have reason at other points in the text to treat some of the more prominent topics; for example, observational learning will be discussed when we consider the effects among children of television viewing and of models of aggression or of prosocial behavior in the next chapter. Here, we may conclude by noting that the believed bases of developmental changes in children's learning—i.e., the role of cognitive mediational processes—have today made the study of childhood learning a far less stimulus-response, mechanistic concern than it was formerly (see Stevenson, 1970). Today both in childhood and across the rest of the life span as well, the study of learning and/or of social learning is a much more cognitively oriented concern (e.g., Bandura, 1978).

PSYCHOMETRIC INTELLIGENCE IN CHILDHOOD

Research within the psychometric tradition has focused on several important developmental issues. These include the degree of stability or change of intelligence during the childhood years and the implications of childhood intelligence for later functioning. In addition, psychometric research in childhood has tried to discover the variables that moderate intellectual change. Research has also been aimed at determining if intellectual functioning can be optimized. We discuss all these topics in this section.

Stability and Change during Childhood

By virtue of the ability to read and understand this book, the reader may be assumed to have at least an above-average level of intelligence. Did a correspondingly high level of intelligence exist prior to the late adolescent or early adulthood years? Said another way, would a measure of one's cognitive ability taken in the very early school years have served as a good basis for predicting one's current intellectual level?

In turn, does the level of intelligence a person displays in late adolescence or early adulthood tell anything about ability in later life? Will people, in their aged years, continue to display the above-average intellectual level they may now show? Or will they change, perhaps in the direction of decline? In short, what does a measure of intelligence taken at one level allow one to predict about intelligence at another age level?

At least for age levels from infancy through late adolescence, data derived from both cross-sectional and longitudinal research suggest a similar course of changes in IQ scores. The data in Exhibit 7.2 are adapted from a longitudinal investigation (Bayley, 1949). Illustrated here is a period of greatly increasing relations among scores from the early years of life through late childhood (about 9 to 10 years of age) with intelligence measured at 18 years; after this period, there is a slower increase in the relations among the scores to about age 16 or 17 years. From this portion of middle adolescence through the early adult years, the curve tends to flatten out, indicating no differences in scores between measured years. The data suggest that a person's IQ score becomes increasingly stable from birth through midadolescence and that at about age 18 years a maximal score is reached and maintained for some time.

Bayley (1949) studied a group of people from birth to age 18 years. Although Bayley had to employ different tests of intelligence at succeeding age levels in order to measure subjects appropriately, the number of tests she used (five) was relatively low for research of this kind (Bloom, 1964). Bayley related IQ scores at one age to those of other ages through use of a statistic termed the *correlation*. A correlation may vary from -1.0 to $+1.0$, with values closer to $+1.0$ indicating a strong, positive relation between two sets of scores. A positive correlation means that as scores in one set of measures increase, so, too, do scores in the other set. A negative correlation means that as scores in one set increase, scores in the other decrease. Thus a correlation "coefficient" of $+0.75$ between height and weight, for instance, would mean that greater height was associated with greater weight.

Bayley (1949) found that intelligence at age 1 year was not related to intelligence at age 18 years. The correlation between IQ scores from these two age levels was 0.0 (there was no relationship whatsoever). The correlation between IQ scores at 2 years and at 18 years was $+0.41$; between 4 years and 18 years, $+0.71$; and be-

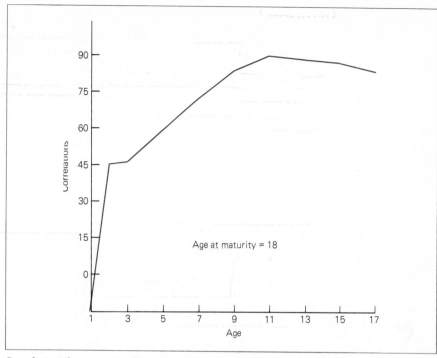

EXHIBIT 7.2   Correlations between intelligence measured at each age and intelligence at age 18. (Source: Longitudinal data collected by Bayley, 1949.)

tween 11 years and 18 years, +0.92. Thus there is a rapid movement toward stability of IQ scores across the early years of life. This movement toward stability slows during adolescence. By age 18 years, IQ scores have become stable for a time.

One reason for the low relation between early childhood intelligence and adolescent intelligence is that what is measured at age 1 is not akin to what is measured at age 18. At the earlier period, measures involve sensorimotor behavior, while at the later period, they involve such abilities as verbal reasoning and mathematical concepts. There is little reason to expect scores on these two types of tests to relate to each other across age, since even when such tests are administered at the same point in life (e.g., adulthood), they are not correlated (Bloom, 1964; Cronbach, 1960). However, as tests at age levels beyond 1 year come increasingly to include measures of abilities akin to those measured at age 18 years, there is a corresponding increase in the correlation. Thus between ages 1 and 18 years there seems to be an increasingly greater "overlap" (Anderson, 1939; Bloom, 1964) in IQ scores.

Variables That Moderate Intellectual Change

What are the variables that influence or moderate the stability or instability of psychometric changes across childhood? Researchers have begun to move beyond mere description of the course of intellectual change across life and have attempted to discover those biological, personal, social, and environmental variables which may moderate these changes. The potential number of variables which may be significant is enormous. Here we attempt to provide some illustrations of such influences.

1) (Achievement) From late childhood and continuing through adolescence, a relationship exists between scores on intelligence tests, grades earned in school, and scores on standardized *achievement tests* (Lavin, 1965). Such tests are designed to measure what has been learned already (Brody & Brody, 1976). One may distinguish between *aptitude tests*—such as intelligence tests—which are designed to measure one's *capacity* to learn, and achievement tests, which measure previous intellectual accomplishment.

The correlation between IQ scores and grades, for instance, is usually reported to be about ±0.5 (Brody & Brody, 1976), although this relation tends to be higher in high school than in college. Even when the correlation between aptitude (IQ) scores and achievement test scores has been found to be highest—about +0.7, in the late childhood and early adolescent portions of life (Crans, Kenny, & Campbell, 1972)—it is still true that more than half the difference in people's achievement scores is associated with factors other than IQ scores.

Nevertheless, there are some longitudinal data which suggest that among people who have *very* high IQ scores, there is a tendency to show especially high "achievement." Terman (1925b; Terman & Oden, 1959) longitudinally studied several hundred male and female persons from their late childhood and early adolescence through their adult years. At the time of their recruitment into the study, the average IQ score for this large group was 151, and very few had scores less than 140. The accomplishments of this group turned out to be impressive. Numerous men and women published novels, poetry, short stories, and plays, as well as scientific and scholarly papers and books. Many achieved national recognition, for example, by being listed in *Who's Who in America* and/or by becoming governmental, business, military, or educational leaders.

However, it should be emphasized that not *all* people in this high-IQ group showed such achievement. For instance, most of the women in the sample opted to be homemakers and not to pursue a career outside the home. Again it is seen that IQ and achievement do not stand in any one-to-one relation.

2) (Sex Differences.) Another popular variable that has been considered is sex. Researchers have been concerned with whether males or females tend to show similar patterns of development of intellectual abilities. Sex, like age, may be regarded as only a marker variable for processes of behavior change. However, studying sex differences may provide a reason for looking at how biological, sociocultural, psychological, or social processes may vary in relation to sex. Thus a search for the bases of cognitive changes might be made more efficient.

Unfortunately, the sexes do not differ consistently in tests of total or composite abilities (Maccoby & Jacklin, 1974). In fact, in a major review of the research literatures pertinent to sex differences, Maccoby and Jacklin (1974) can draw only a few, tentative generalizations about the presence and direction of sex differences. Although Block (1976) has shown that even these generalizations are quite problematic, she does agree with Maccoby and Jacklin that boys and men *tend* to score higher than girls and women in tests of spatial and quantitative ability and, in turn, girls and women score higher than boys and men on measures of verbal ability.

Yet even these generalizations are only first approximations of trends in the literature and need to be qualified. For instance, the generalization that females are better on verbal tasks must be qualified by noting that this advantage does not become marked until about grade 10 or 11. Although the difference is seen through the rest of adolescence and into young adulthood, it is not found in every study.

Similarly, the generalization that males show better quantitative ability must be qualified, since Maccoby and Jacklin (1974) find no evidence of such a sex difference in the preschool years. Moreover, there appears to be no sex difference in mastery of numerical operations and concepts in the early school years. In fact, the majority of studies using representative samples show no sex differences up to adolescence. However, when differences are found after this age level, they tend to favor males. Although Maccoby and Jacklin reach similar conclusions about the age-related emergence of sex differences insofar as visual-spatial abilities are concerned, we must again emphasize that the data are equivocal.

Not only did a study by Fitzgerald, Nesselroade, and Baltes (1973) report no sex differences in regard to adolescent number and space factors, but Waber (1977), studying forty male and forty female 10- to 16-year-olds, found no significant sex differences on either spatial or verbal ability measures.

Together, these studies of sex differences suggest that a search for processes that *consistently* may vary with sex may not be fruitful. In addition to methodological differences and problems in the various studies that assess sex differences (Block, 1976; Maccoby & Jacklin, 1974), it may be that these studies have contrasting findings because the contexts within which females and males develop are different. However, these context differences are often unassessed. Some of what is known about these effects is summarized next. We will have much more to say about sex differences in Chapter 11.

3) **Social Interaction Effects.** When one does consider those studies which evaluate the behavior-change processes involved in various settings, many variables that seem to moderate intellectual change are uncovered. Although there are various dimensions along which researchers have evaluated the influence of contextual variables on psychometric intelligence (Bijou, 1976), one dimension that has attracted considerable attention is the role of social interactions on IQ scores. Basically, the issue addressed here is whether people who develop in different contexts attain different IQ scores. Several research directions converge to indicate the answer is yes.

Wolf (1964) noted that although the correlation between socioeconomic status and IQ is generally reported to be +0.4, there is no indication as to what produces this relationship. Wolf hypothesized that parent-child interaction may be involved; children at different socioeconomic levels may interact with their parents in different ways. Wolf measured thirteen variables pertaining to parent-child interaction in a group of sixty mothers and their sixty fifth-grade children. The variables related to such categories as pressures for achievement and pressures for general learning. The combined (multiple) correlation between these thirteen parent-child interaction variables and IQ scores was +0.76, indicating that children differing in their social interaction context also differ in their psychometric intelligence.

These results suggest that if a person interacts in a milieu where pressures for certain behaviors are maintained about him or her, then performance in accordance with these pressures may occur. These pressures may be behavioral urging or prompting, as in Wolf's (1964) study. In other words, people's attitudes about a target person's behavior may create a social climate which channels the target's behavior in a way consistent with the attitudes (Lerner & Spanier, 1980; Rosenthal, 1966).

Moreover, a target person's own expectations about the social interactions involved in intellectual performance can contribute to such a self-fulfilling prophecy. Kagan (1969) notes that black children's poor performance on standard IQ tests may be due to social factors, as well as to the bias in test construction against blacks (the tests were standardized on white middle-class populations). Kagan says that blacks may not interact comfortably in an intellectual evaluation where they are appraised by white examiners. They may anticipate stress and in fact feel uneasy in such a situation, and this may interfere with performance.

Thus, one may speculate that if blacks could anticipate a favorable interaction with their white evaluator (or if the evaluator were black), their test performance would improve, at least insofar as attitudes about the interaction influenced behavior. Results of a study by Piersel, Brody, and Kratochwill (1977) support this idea. These investigators gave a standardized intelligence test to sixty-three disadvantaged, minority children (64 percent black, 25 percent Mexican-American, and 11 percent of mixed ethnicity). Children were divided into three groups: one group was administered the test by a white tester; a second group was given feedback for correct and incorrect responses; and a third group was allowed to view another minority child being given the test by a white tester. Presumably, the vicarious experience involved in this third condition should have reduced any apprehensions the minority children had about the situation and lead to better test performance. Indeed, not only did the children in this third group have higher test scores than those in the other two groups, but also, as amount of evaluation apprehension decreased, IQ scores increased.

Thus the attitudes of both teachers and students involved in an interactive situation can influence IQ scores. Although race is an easily discerned cue for eliciting attitudes about social interaction, there may often be cues that are more subtle. There may be aspects of either person involved in the parent-child or teacher-student interaction that lead to interactions which may facilitate or negatively influence IQ scores. These aspects need not be physical cues like race or sex. In fact, a study of Gordon and Thomas (1967) indicates that as far as teachers are concerned, a salient cue may be an aspect of the student's personality.

Gordon and Thomas (1967) noted that different children tend to show characteristic styles of behavior in the classroom. They were able to describe four such styles

Interactions between children and their parents can influence the development of abilities and interests. (Scott/Wakefield.)

of reactivity, or "temperament" types, among a group of kindergarten children. Students were rated as:

1 *Plungers*, who jump right into any new activity
2 *Go-alongers*, who participate after the activity is initiated
3 *Sideliners*, who participate only partially
4 *Nonparticipators*, who do not take part at all

Experienced teachers, who had taught the children, were asked to rate each child's temperament and to estimate his or her intelligence. Gordon and Thomas then took an actual measure of intelligence; however, the teachers did not know a child's score when making the intelligence estimates.

The results showed that although there was only a very low correlation between temperament and IQ (a correlation of about +0.2), the teachers estimated a higher relation to exist (about +0.6). Moreover, the teachers saw the plungers as brightest, the go-alongers as next brightest, the sideliners as next brightest, and the nonparticipators as least intelligent. Yet there was *no* actual IQ difference among these groups.

Thus differential teacher expectations based on a non-intelligence-related variable like temperament can lead to a self-fulfilling prophecy. A circular process is suggested by data from a study by Lerner and Miller (1971). These researchers repeated the procedure of Gordon and Thomas except that the subjects were all seventh-grade students in s small suburban midwestern junior high school. At this age level there was a high correlation between temperament and intelligence (about +0.6). Although cross-sectional in design, this latter study is consistent with the notion that variables that may relate to different processes of social interaction may influence a person's IQ score.

Children have different styles of participation in school activities.
(Constantine Manos/Magnum Photos, Inc.)

In summary, psychometric intelligence scores appear to be influenced by interactions in a particular social context and by broader sociocultural and historical influences associated with birth cohort membership and time of measurement. Although such plasticity has led some developmentalists to focus on the situational or cohort influences on IQ (Baltes, Cornelius, & Nesselroade, 1978), it has led others to reject psychometric intelligence as a useful topic of inquiry (McClelland, 1973). Finally, because of their belief in plasticity, it has lead some scientists to try to intervene to optimize intellectual performance.

Improving Cognitive Functioning

Although there is no evidence of major differences in intelligence among infants of different races or social classes, poor white and poor black children have, on the average, significantly lower IQ scores than do middle-class children by the time the school years begin (Mussen, Conger, Kagan, & Geiwitz, 1979, p. 192). Moreover, as these children from the lower social classes continue through school, they do progressively worse and worse. Major attempts at intervening to enhance the cognitive development of these children have occurred. While those efforts have neither met with unequivocal success nor avoided criticism (see Jensen, 1969), the weight of current evidence indicates that intervention can succeed. However, such attempts have to involve major changes in the lives of the youths involved and are most successful if occurring early in their lives. Moreover, even in those intervention efforts wherein immediate enhancement of cognitive functioning is not found, there is some evidence that positive, although delayed, effects of intervention can occur. In a review of ten long-term follow-up studies of children involved in preschool intervention projects in the 1960s, Palmer (1977) notes that when these children reach later elementary or junior high school, they score higher on various achievement tests and on IQ tests than do children who did not experience early intervention programs.

There are numerous, specific examples of successful intervention efforts. In one classic study, Skeels (1966) found that the intelligence of seriously retarded babies being reared in an orphanage could be enhanced by assigning them, before 3 years of age, to "mothers" who themselves were mentally retarded older girls living in an institutional setting. This group was compared with a matched group of infants who remained in the orphanage. The infants who were reared by the mentally retarded mothers gained an average of 32 IQ points within 2.5 years, while the comparison group showed an average loss of 21 IQ points. Twenty years after leaving their respective institutions, both groups were again studied. The mothered children were all self-supporting and had completed an average of 12 years of school. The average grade completed by the comparison group was the fourth, none was really self-supporting, and many were in state institutions.

Other evidence of the long-term effects of interventions exists. Heber and his associates (e.g., Heber, Garber, Harrington, and Hoffman, 1971, 1972; Garber & Heber, 1973) have been involved in the Milwaukee Project, an intervention program involving both family and infant intervention aimed at preventing retardation by intervention early in life. The project used an experimental and a control group, each composed of black, ghetto mothers whose IQs were 75 or less and their infants. Although tested when the experimental group was tested, the control group received no treatment. The treatment that the infants in the experimental group received involved intensive, thoroughly planned educational regimens, continued over the course of several years, and presented by a specific teacher assigned to the child. Although at 6, 10, and 14 months of age the experimental and the control children did not differ, the experimental group began to show significantly superior cognitive abilities by 18 months of age. They continued to move away from the control-group children thereafter. By 5.5 years of age the control group's average IQ was 94, while the experimental group's average IQ was 124.

Finally, there is evidence that enhancement programs begun after infancy can also work. Gray and her colleagues (Gray & Klaus, 1965; Klaus & Gray, 1968), in the Early Training Project of Peabody College, worked with black children about 3 years of age who came from poverty-stricken families. Children assigned to an experimental group received special training in special school sessions. Children in a control group received no such training. After training, the children in the experimental group were better than those in the control group in various verbal abilities, and follow-up studies—done after twenty-seven months and when two years of public school had been completed—showed that the experimental-group children continued to perform better than those in the other group. Moreover, Gray, Ramsey, and Klaus (1979) report some evidence for a continuing effect of the intervention, albeit a small one, when the groups reached their late adolescent and early adulthood years.

PIAGETIAN IDEAS ABOUT COGNITIVE DEVELOPMENT IN CHILDHOOD

In Chapter 5 we discussed the structural, cognitive developmental approach to the study of cognition taken by Jean Piaget. In that chapter we presented some of the general characteristics of his theory, and we saw some of the features he associates with the first stage of cognitive development in his theory, the sensorimotor stage. That stage describes cognitive developments which occur during infancy. In this chapter we discuss the two stages in Piaget's theory that are associated with the childhood years: the preoperational stage and the concrete operational stage.

The Preoperational Stage

The age range associated with this second stage in Piaget's theory is usually 2 through 6 or 7 years of age. The major cognitive achievements in this stage involve the elaboration of the representational ability that enabled the child to move beyond the sensorimotor stage. In the preoperational stage, true systems of representation, or symbolic functioning, emerge. In fact, Elkind (1967) has termed this stage the period of the "conquest of the symbol."

The most obvious example of the development of representational systems is language; words are used to symbolize objects, events, and feelings. Other indications of representational ability in this stage involve the emergence of *symbolic play* (as when the child uses two crossed sticks to represent a jet plane) and the emergence of *delayed imitation* (as when a child sees daddy smoking a pipe and pacing across the room and repeats the act hours later).

Cognitive development in the preoperational stage has limitations. The child in this stage is also egocentric, but here the egocentrism takes a form different from that seen in the previous stage. As always, egocentrism involves a failure to differentiate adequately between subject and object; but in this stage egocentrism involves a failure to separate the symbol of the subject from the object in the real world. The child has the ability to symbolize objects with words but fails to differentiate be-

Delayed imitation often involves the child's copying an adult's behavior some time after having observed it. (Richard Frieman/Photo Researchers, Inc.)

tween the words and the things to which the words refer. For example, the child believes that the word representing an object is inherent in it, that an object cannot have more than one word to symbolize it (Elkind, 1967). The child does not know that an object and a word symbolizing the object are two independent things.

One consequence of this type of egocentrism is that the child acts as if words carry more meaning than they actually do (Elkind, 1967). A broader consequence of this inability to hold two aspects of a situation separately in mind at the same time may be a more general inability to take into account simultaneously two different aspects of a stimulus array. One indication of this inability may be the preoperational child's failure to show conservation.

Conservation is the ability to know that one aspect of a stimulus array has remained unchanged, although other aspects of the stimulus array have changed. To illustrate, imagine a 5-year-old is presented with a mommy doll and a daddy doll. Six marbles are placed beside the daddy doll in positions directly corresponding to the mommy doll's marbles. The materials would look like those in Exhibit 7.3(*a*). If the 5-year-old is shown this arrangement and asked, "Which doll has more marbles to play with—the mommy doll or the daddy doll, or do they both have the same?" the child would probably say that both dolls have the same. However, if the mommy doll's marbles are spread out so that an arrangement as in Exhibit 7.3(*b*) is achieved, and the same question is asked, the preoperational 5-year-old child will answer that the mommy doll has more!

This is an example of the inability to conserve number. The child does not know that one aspect of the stimulus array—the number of marbles—has remained unchanged, although another aspect of the array—the positioning of the marbles—has changed. It seems the child cannot appreciate these two dimensions of the stimulus array at the same time. The child fails to show knowledge that the actions with the mommy doll's marbles can be *reserved* and be put back to the original stimulus array. When the next stage of cognitive development—the concrete operational

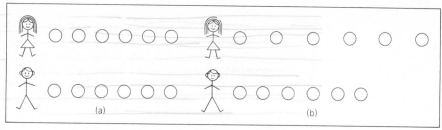

EXHIBIT 7.3   Examples of a test of number conservation.

stage—is discussed, it will be learned that the preoperational child cannot understand this *reversibility* of action yet. Thought is still dominated by schemes, and such structures are rigid and unidirectional. Thus even though one spreads out the marbles right in front of the child, the child still maintains that the altered array has more marbles. One might even return the array to its original form and again ask the same questions. The child would now probably say once again that they both have the same number, and even if one repeats these steps several times, the child's answer might correspondingly alternate between "same" and "more."

In sum, the preoperational child's inability to show conservation ability manifests itself in respect to several quantitative aspects of stimuli, such as mass, area, and volume. Although the preoperational child's thought has progressed well beyond that of the sensorimotor child—due mostly to extensive representational abilities—the preoperational child's cognition still has these limitations—due mostly to an inability to appreciate intellectually such things as the negation or reversibility of actions, as performed on the stimuli in the conservation tests. When one can mentally represent such reversible actions, one shows the ability that is the hallmark of the next stage.

The Concrete Operational Stage

Up to the point when a child enters the concrete operational stage (which spans from about 6 to 7 years of age through 11 or 12 years), the child's cognitive structure is composed predominantly of schemes. Being a unidirectional structure, there is no ability to mentally reverse various physical actions.

The emergence of *operational structures* gives the child this ability. An operation is an internalized action that is reversible. As opposed to schemes, operations allow the person to know that actions can be counteracted by reversing them. Moreover, operations *are* internalized actions. People in this stage do not have to see the action of rolling a clay sausage back into a ball to know that one can return the clay to its original shape. They can just think of this action. Their thought about concrete, physical actions does not require that they actually see these actions. They can reverse the actions in their heads and come to the same conclusion about the actions as if they actually concretely viewed the reshaping actions.

Operational cognitive ability extends the child's capacity to deal with the world. Because thought is now reversible, because the child can now appreciate the reciprocity in concrete actions with physical stimuli, the concrete operational stage is the period in which the child begins to show the conservation abilities lacking in the preoperational stage. Moreover, because operations are internalized actions, the child's cognitive abilities are also extended in that now one need not actually see actions performed in order to know about them. Simply, operations extend the scope of action by internalizing it.

Tomlinson-Keasey, Eisert, Kahle, Hardy-Brown, and Keasey (1979) conducted a four-year longitudinal study of the development of concrete operational thought in order to assess the structure of this cognitive level. Their subjects were fifty-six kindergarten children who were followed through three testing phases over two years; more than half these children were followed for an additional two years. Results showed that the children's level of ability in seriation and in numeration in kindergarten were excellent predictors of the acquisition of conservation of mass and volume in third grade. Other results indicated that classification skills and conservation were causally related.

Despite the accomplishments involved in the concrete operational stage, there are limitations. The label for this stage is concrete operational, and this denotes that thought is bound by concrete, physical reality. Although the child can deal with objects internally, actions and objects must have a concrete, real existence. Things or events that are *counterfactual*—that are not actually represented in the real world —cannot be understood by the concrete operational child.

An illustration of this point offered by Elkind (1967) is helpful. If someone asks you to imagine that coal is white and then asks you to indicate what color coal would be when burning at its hottest, you would probably have an answer to this counterfactual question. You might think that since coal is actually black and when burning at its hottest is white, then if it were white, it would be black when burning at its hottest. The point here is *not* the particular solution but the fact that you can deal with the counterfactual question. The concrete operational child, on the other hand, cannot do this. For example, the response might typically be, "But coal is black!" (Elkind, 1967). In essence, a major limitation of concrete operational thought is that it is limited to thinking about concrete, real things.

Other limitations of concrete thought exist. The concrete operational child is also egocentric (Elkind, 1967). For such children, egocentrism takes the form of an inability to differentiate between actions and objects *directly experienced* and the actions and objects one *thinks about*. Although the child can think independently of experience and can deal with an action whether it is experienced or just thought about, the child does not distinguish between knowledge gained through experience and knowledge gained from thought alone. If given some information about a physical situation (say a scientific problem) and asked to give a solution to the problem, the child will not have to see the actual physical objects in order to reach a solution. The child will think about it and form an answer. But the child will not recognize the answer as just one possible solution to the problem; rather, the child will think the answer is one and the same with the physical situation. The child will not see any difference between what he or she thinks and what is! Even if the child's ideas about experience are challenged and/or evidence is presented contradicting those ideas, the child will not alter the answer but will just reinterpret the opposing evidence to fit his or her ideas (Elkind, 1967). Rather than accommodate to the new data, the child will assimilate it to his or her already existing structure but will not recognize the hypothetical nature of this process. When the child attains the ability to think counterfactually, to see that reality and thoughts about reality are different, and can generate and recognize hypotheses about reality, the fourth stage in Piaget's theory is reached. This stage is the one associated specifically with adolescence (Elkind, 1967, 1978; Neimark, 1975). We will discuss it in Chapter 9.

LANGUAGE DEVELOPMENT

One of the most fascinating accomplishments for the young child is the acquisition of language. This seemingly effortless ability to acquire sounds, words, and grammatically correct sentences is nevertheless a complex and creative task. In the early stages of language development, the child's speech is not simply a poor copy of adult speech; rather, it is complex and governed by rules. By about the age of 5, the child will be using a language that in many ways is almost as sophisticated as that of an adult.

Theories of Language Acquisition

As with most psychological phenomena, there are a number of theories which seek to explain the why and how of language development. Today, there seem to be three major views or theories of language development incorporating the influence of heredity and environment: (1) *behaviorism,* or the learning theory approach, which argues that learning principles can account for language development; (2) *preformationism,* or a nativist view, which contends that language is innate; and (3) an *interactionist* view held by most current theorists, which recognizes the influences of both genetic and environmental factors on language development.

1). *The Learning Theory Approach.* From the beginning of life, the infant produces sounds that later will be the basis of the first words. As development progresses, new sounds appear in relatively systematic fashion. One attempt at explaining just how and why language development proceeds in this orderly fashion is through *learning theory principles.* Theorists who adhere to this view (e.g., Skinner, 1957) contend that parents *positively reinforce,* or reward, with praise and attention, any correct sounds emitted by the infant. These reinforced sounds will therefore occur more frequently in the child's speech and eventually will become an established part of his or her language. In addition, the learning theory view argues that sounds which are incorrect will not be reinforced and, therefore, such sounds will eventually disappear. Other learning theorists (e.g., Bijou, 1976) stress *imitation* and contend that the child picks up words, phrases, and sentences directly through imitation and then, through reinforcement, the child learns when to use them appropriately.

Learning theory accounts of language acquisition have not stood up well to empirical validation. For example, Leonard, Schwartz, Folger, Newhoff, and Wilcox (1979) studied the role of imitation in children's acquisition of certain lexical items (a lexical item is a word or phrase). They also assessed the factors influencing children's tendency to imitate those items and found that imitation did not appear to facilitate the subsequent spontaneous use of lexical items. In turn, children's tendency to imitate lexical items appeared to be influenced by such factors as the novelty of the lexical item and its referent and the informativeness of the referent. Similarly, Bloom, Hood, and Lightbown (1974) found that although there are both imitative and spontaneous types of words existing in a child's speech, an age-associated progression occurred wherein imitation of a particular lexical item decreased and the spontaneous use of that item increased.

Studies of parent-child interaction reveal that mothers are *not* more likely to reward sounds that occur in their language than they are to reward sounds that do not occur in their language (Wahler, 1969). Furthermore, linguists argue that the enormous degree of creativity in child language is not accounted for by learning theories. As Hetherington and Parke (1979, p. 263) state this criticism, the point is that "the number of specific stimulus-response connections that would be necessary to even begin to explain language is so enormous that there would not be enough time to acquire these connections in a whole lifetime, not to mention a few short years."

Learning theories also have not explained the regular sequence in which language develops regardless of culture. Despite the evidence which goes against learning theory accounts of the initial acquisition of language, learning principles have been shown to play a useful role in *modifying* already acquired language and in overcoming language deficits in some people. So although learning principles have not been shown to play a *critical* role in the normal acquisition of language by the young child, they can play an *important* and *useful* role later in the child's developing language.

2. *Preformationism.* Preformationist theories are popular among contemporary linguists. They stress that all individuals have an innate biological predisposition for language. They challenge behavioristic or learning theories because they view language as an abstract system of rules which cannot be learned through traditional learning principles. Preformationist theories can take several forms; for example, Chomsky (1968) argues that the human nervous system contains a mental structure which includes an innate concept of human language. Lenneberg (1967) holds that the human ability to speak language is an inherited species-specific characteristic of humans. Lenneberg (1967) contends that the systematic fashion by which language unfolds is tied to biological maturation in the same way as learning to walk is maturationally determined.

Lenneberg's theory has received considerable support, in part due to evidence that humans learn language more easily and quickly during one period of biological development, i.e., from infancy to puberty. Similarly, Lenneberg notes that particular language developments occur in a universal sequence and at about the same rate in *all* normal children, despite their native language and whatever cultural variation exists. Thus all children start babbling at about 6 months, say their first word at about 12 months, and start to use two-word combinations at about 24 months.

Further data lend credibility to Lenneberg's theory. That is, if language is an inherited species-specific characteristic, then all language must share some universal features. Researchers have concluded that a set of common principles indeed underlies all human languages (Greenberg, 1966).

However, some data do not support particular nativistic ideas. In Chapter 3 we discussed the nature-based "critical period" hypothesis. In regard to language this hypothesis holds that the acquisition of a first language must occur before particular brain developments (cerebral lateralization) is complete, i.e., at about the age of puberty. Snow and Hoefnagel-Höhle (1978) tested one prediction derived from this hypothesis; that is, that acquisition of a second language will be relatively fast, successful, and qualitatively similar to first-language acquisition only if it occurs before the critical period of prepuberty maturation is over.

This prediction was tested by studying longitudinally the naturalistic acquisition of Dutch by English speakers of different ages. The subjects were tested at three times during their first year in Holland, with an extensive test battery designed to assess several aspects of their second-language ability. The results were that the 12- and 15-year-old subjects and the adult subjects made the *fastest* progress during the first few months of learning Dutch and that at the end of the first year the 8- to 10- and the 12- to 15-year-olds had achieved the best control of Dutch. On the other hand, 3- to 5-year-old subjects scored lowest on all the tests employed. All these results ran exactly counter to what the nativistic, critical period hypothesis would lead us to expect.

In sum, although most theorists agree that humans are biologically prepared for the acquisition of language, they contend that nativistic biological principles cannot account for *all* aspects of language development. This is because critical interactions between the child and his or her social context appear evident.

3. *Interactionism.* The use of interactionist ideas to explain language development is recognized by most modern theorists today. They recognize that language develops through the interaction between heredity, maturation, and encounters with the environment. Humans are biologically prepared for learning language but experience with the spoken language is also considered a condition of language learning. In addition, this perspective emphasizes the *active* role the child plays in learning language, rather than the *passive* role he or she is thought to play by the learning theory approach. Furthermore, the role of others in the child's social environment in facilitating language acquisition is emphasized. Research has demonstrated that parents have a great influence on the ease with which their child develops language.

Parents and other adults modify their speech when talking to a child, and children are, in turn, more attentive to simplified speech (Nelson, 1973). By about 5 years of age, children themselves appear to have mastered this language interaction pattern, that is, to modify speech in regard to the age of the person to whom one is talking. Sachs and Devin (1976) recorded the speech of four 4- to 5-year-old children talking to different listeners (an adult, a peer, a baby, and a baby doll). The children's speech was different when talking to a baby or a doll as compared with the speech to a peer or an adult.

Other types of parent-child language interaction exist. Parents often prompt the child to use language or echo the child's speech. In addition, many parents use a technique termed *expansion*. If a child says "give mommy" and the adult then says "give it to mommy," this is an example of expansion. Middle-class parents use expansion very often with their children (Brown, 1973); children, in turn, often imitate such expansions.

Still other data support the role of interaction in language development. Moerk (1978) observed ten boys and ten girls between 1 and 5 years of age in interaction with their mothers in their homes. Mothers used several different types of techniques to motivate verbalization among their children. They also used several different types of techniques to instruct their children in the use of language. Although the frequencies of use of the various motivating techniques did not change with changes in the child's age or language level, the frequencies of use of the various instructional techniques declined. Finally, Moerk found evidence that mothers adapted their motivational techniques to the language capacities shown by their children; that is, mothers used particular groups of techniques in relation to particular child functions. This, Moerk (1978) contends, attests to the mutual interaction between mother and child.

This view is supported by results of a study by Ringler, Trause, Klaus, and Kennell (1978). As part of an ongoing study to determine the effects of early mother-infant contact on later maternal and child behavior, these researchers studied the relation between patterns of maternal speech while addressing their 2-year-olds and the children's speech and language comprehension at 5 years of age. Among those mother-child pairs who experienced extra contact in the immediate postpartum period, several components of maternal speech had a facilitative effect on children's later language performance. Among those pairs who did not have this early interaction, however, no such predictive effects were seen.

Hood and Bloom (1979) also show development of children's expressions about causality (i.e., about why an event occurs) can be understood as a feature of mutual influence between child and adult. Hood and Bloom examined 2- to 3-year-old children's expressions of causality in their naturally occurring conversations with adults. In regard to the content of their speech, the referential and functional uses of causal expressions for *both* children and adults were to ongoing or imminent situations; a speaker would comment on his or her intention to act or request the listener to act. In regard to the form of the child's speech, Hood and Bloom report: (1) an increasing use of connectives to link clauses, among all children, and (2) three main patterns of clause order, each used by different children (cause and effect; effect and cause; and equal use of both orders).

Finally, it should be noted that mothers and fathers may have somewhat different styles of verbal interaction with their children. Golinkoff and Ames (1979) studied the verbalization of twelve mothers and twelve fathers with their 19-month-old boys and girls. In one situation all three family members were together, while in another each parent was alone with the child. When all three family members were together, fathers spoke less and took fewer conversational turns than mothers. However, other than for this difference, fathers' speech and mothers' speech to their children did not tend to differ qualitatively or quantitatively.

Stages of Vocalization

Before uttering a first real word, the infant proceeds through a number of stages of early vocalization.

1) *Undifferentiated Crying (Birth to 1 Month).* During this stage of early vocalization, the newborn uses crying to signal all his or her needs. The crying is called undifferentiated because a listener will not be able to distinguish between the infant's cries of pain, hunger, and fear. Undifferentiated crying in the newborn is a reflexive action.

2) *Differentiated Crying (2 Months).* During this stage of early vocalization which develops at about 2 months of age, the crying is more distinguishable to the adult; therefore, it is termed differentiated crying. There are different patterns and pitches to signal hunger, pain, or distress, and it appears to be a better means of communication for the infant.

3) *Babbling (3 or 4 to 8 or 9 Months).* The next stage of early vocalization is *babbling*, which is the repetition of simple vowel and consonant sounds like "ma, ma, ma" and "da, da, da." It is at this stage that the infant produces *phonemes*, the basic sound units of the language. Babbling occurs most often when the baby is alone and contented, and it appears to be a universal phenomenon (Lenneberg, 1967).

4) *Lallation (6 to 8 Months).* This stage of early vocalization begins during a later stage of babbling, around 6 to 8 months. Unlike babbling, however, lallation involves the imperfect or accidental imitation of their own sounds and those of others, setting the stage for communication.

5) *Echolalia, or Imitation (9 or 10 Months).* This stage of early vocalization, or the pre-linguistic period, occurs when the infant consciously imitates the sounds he or she hears. The infant now begins to respond differently to adult speech, and by the end of the first year, infants can discriminate among the basic phonemes and reproduce adult language.

During these stages of prelinguistic development there is increasing evidence that the child can comprehend language although he or she cannot produce it. For example, Oviatt (1980) studied thirty 9- to 17-month-olds. In one experiment Oviatt gave these infants training on the name of a salient object (e.g., a live rabbit), and in a second experiment she trained them on the name of a simple action (e.g., pressing to activate a toy). In both experiments the infants were assessed for their ability to recognize the previously unfamiliar trained name. Responses were videotaped and scored for gaze, gesture, and vocalization. Both cross-sectional and longitudinal results from each experiment indicated great improvement in the children's ability to comprehend language.

6) *Patterned Speech (1 Year).* This final stage of early vocalization begins at about 1 year of age. At this time the child consciously produces adultlike intelligible sounds and, most importantly, uses them to communicate with those around him or her.

Stages of Linguistic Speech

Just like early vocalizations, once the child has acquired communicable speech, he or she proceeds to develop that speech through a series of stages.

One- and Two-Word Utterances. From ages 1 to 2, the child's speech is composed of one-word utterances, used to communicate and to function as sentences. At about 2 years of age, the child begins to combine words into two-word sentences, like "Where Mommy" and "Allgone shoe." These two-word combinations increase slowly, and seemingly suddenly there appear to be many of them. At this time, the child is also beginning to discover some of the simple rules of grammar.

Early Sentences. According to Roger Brown (1973), who conducted a longitudinal study of the development of language in three children (Adam, Sara, and Eve), there are five stages of linguistic development. Brown and his colleagues contend that the best index of language development in the early stages is through a measure called *mean length of utterance* (MLU), which is measured in terms of *morphemes*, the smallest units of speech that have meaning. Morphemes include the words *go* and *dog* as well as parts of words that have meaning such as *-ing, -ed, -s,* and other endings. Brown's first stage begins with the first two-word utterance, and lasts until MLU = 2.0 morphemes (stage 1), and continues until MLU = 4.0 morphemes (stage 5).

There is evidence that girls go through these stages earlier than do boys. For example, Schachter, Shore, Hodapp, Chalfin, and Bundy (1978) studied male and female young children, aged either about 2 or 2.5 years. The sexes were matched for age, class, and race. The girls in the youngest group were significantly advanced in both MLU and the length of their longest utterances; similar advancements were seen among girls in the older group.

Stage 1 Speech. During Stage 1, the earliest sentences are produced by the child. These sentences are sometimes termed *telegraphic* because they are abbreviated, or telegraphic, versions of adult speech. They are composed mainly of nouns and verbs. Thus the child is often leaving out small words (such as *of* or *the*) and word endings from his or her speech. For example, instead of saying "I'll give you the doll," the young child might say, "I give doll" or "give doll."

The child is following simple rules of grammar at this stage in his or her linguistic development, as evidenced by the fact that they put the words into the correct position and that they tend to make errors that are consistent with the rules they have learned. For example, one type of error is *overgeneralization* or *overregularization*. For example, when first learning that adding an *-s* or an *-es* pluralizes a noun, a child might say "fishes" for more than one fish or "foots" or "feets" for more than one foot. Or when the child has learned the rule that past tenses are formed by adding *-ed,* he or she may use this rule to make a past tense out of an irregular verb, e.g., "I goed."

A study by Kuczaj (1978) illustrates some of these features of speech. Kuczaj studied children's judgments of grammatical and ungrammatical irregular past-tense verb forms. He found that 3- and 4-year-old children are more likely to accept a base word with its *-ed* form (e.g., *eated*) than past tense plus *-ed* (e.g., *ated*), while 5- and 6-year-olds are about as likely to accept both forms. In another report by Kuczaj (1979) it was found that young children tend to learn suffixes better than prefixes (cf. Slobin, 1973).

Stage 2 Speech. According to Brown, Stage 2 starts with an MLU of 2.0 and extends to 2.5. At this time the children's sentences are becoming more and more complex; they learn prepositions, a few articles, irregular verbs, some verb tense inflection, plurals, etc. These are not acquired simultaneously, nor are they used perfectly. These *grammatical morphemes* seem to be acquired in a remarkably *invariant* order. Why this happens is still not entirely known, but there is a definite constancy with which these morphemes appear in children's speech.

Stages 3 and 4 Speech. These stages cover the range from an MLU of 2.5 to 3.5. During these stages, vocabulary increases and grammatical rules are used more consistently.

Stage 5 Speech. During this last stage, with an MLU of 3.5 to 4.0, all of Brown's subjects reached a major achievement: the construction of complex sentences in which two or more sentences are joined by the conjunction *and* ("You clap and I yell") or one sentence is embedded in another ("You hope you can go."). In addition, they begin to use complex sentences containing *wh-* clauses ("When I get older, I can go there").

In sum, in a relatively few short months from the appearance of the first word (at 12 months), there are rapid changes in the number of words in the child's vocabulary and in the complexity of the child's sentences. Children typically begin to use two words together at about 18 to 20 months of age. By 28 months children can use three-word sentences, and within another 10 months—by a little over the third birthday—sentences like "Why he going to have some seeds?" "The station will fix it," and "Who put dust on my hair?" are heard (Brown & Bellugi, 1964). In fact, after about 5 years of age, most children have become so expert in their native language that they can correctly differentiate between grammatical and ungrammatical sentences (e.g., deVilliers & deVilliers, 1974).

Functions of Language

Thus as language develops, it gives the child an increasingly more complex means for interacting with the social context. Indeed, Halliday (1975) has suggested that language has at least seven types of interactive functions:

1. *Instrumental*. The child uses language to express wishes ("I want . . . ").
2. *Regulation*. The child uses language to control others ("Do that!").
3. *Interpersonal*. The child uses language to interact with others (Let's talk about the movie").
4. *Personal*. The child uses language to express his or her individuality ("I am a good baseball player").
5. *Questioning*. The child uses language to explore the world ("Tell me why the sky is blue").
6. *Imagination*. The child uses language to pretend ("Let's make believe we're Mommy and Daddy").
7. *Information*. The child uses language to communicate to others ("Let me tell you what I did in school today").

In sum, language functions to allow the child to engage his or her social context in numerous, important types of social interactions. We have seen, throughout this book, that interaction with the social context is important in order for adaptive human development to proceed. Language, in facilitating the child's interaction with the social context, enables optimal development to occur.

MORAL DEVELOPMENT IN CHILDHOOD

New parents often proudly tell their friends and relatives that "she is a *good* baby." Parents of older children, when leaving their child with a babysitter, may caution him or her to "behave, be good, for the sitter." Similarly, parents are quite proud when they know their child is a "well-mannered" young person, a person who behaves appropriately.

From the earliest days children encounter social rules. Children come to understand quite early in their lives that their parents expect them to "be good," to follow these rules. Soon, most children come to know, and often believe, these same social rules. These changes in knowing and understanding rules of society—rules of how people *should* relate to their social world—involve what developmentalists term *moral development*.

In all societies children are expected to show moral development, and a great deal of attention has been paid to this topic by child and life-span developmentalists. We discuss their work in this chapter. We will see that the cognitive developmental approach to moral development is today the most influential approach to the study of moral development.

DEFINITIONS OF MORAL DEVELOPMENT

What is morality? How does one know whether a person is or is not moral? When and how does morality develop, and what are the changes that people go through as they show this development? Theories derived from the nature, nurture, and interactionist conceptions discussed in previous chapters provide different answers to these questions. Indeed, there are three major types of theories of moral development that are present in the current study of human development—i.e., theories which stress the role of nature, of nurture, or of interaction. Theorists whose ideas have already been discussed may be used to illustrate each of these types.

Freud's Nature-Oriented Theory

Sigmund Freud's (1949) psychoanalytic theory of psychosexual stage development takes a weak interactionist stance regarding nature and nurture and, as such, sees each stage emerging in an intrinsically determined, universalistic manner. Accordingly, all people experience an Oedipal conflict in their third psychosexual stage (the phallic stage). The successful resolution of this conflict will result in the formation of the structure of the personality Freud labeled as the superego. This structure has two components: the ego-ideal and the conscience. The latter represents the internalization, into one's mental life, of society's rules, laws, codes, ethics, and mores. In short, by about 5 years of age (with the maturationally determined termination of the phallic stage), superego development typically will be complete. When this occurs, the person's conscience will be formed as much as it ever can be, and this, in turn, means that by about 5 years of age, the person will have completed his or her moral development.

One may recognize two attributes of Freud's psychoanalytic view of moral development. First, Freud would identify a person as morally developed or not on the basis of whether that person showed *behavior* consistent with society's rules. Because of the internalization involved with conscience formation, Freud would only be able to know when a person had completed this formation on the basis of behavioral consistency with these external social rules. Accordingly, to Freud (1949), moral development involves increasing behavioral consistency with society's rules, and as soon as a person shows such behavioral congruence (at about 5 years of age), he or she is completely morally developed. This conception indicates then that as long as two people—say a 5-year-old and a 21-year-old—show an identical response in a moral situation, they are identically morally developed.

Second, Freud does not deal with the *content* of behavior. As long as the response conforms with the particular rules of the society, then that response shows internalization, conscience formation, and thus moral development. Hence because different societies can and do prescribe different sorts of rules for behaviors, Freud says there is no morally universalistic behavioral content; rather, what is seen as moral behavior is defined *relative* to a particular society.

Nurture-Oriented Social Learning Theories

Social learning theorists see behavior as a response to stimulation (Davis, 1944; McCandless, 1970). Such responses may arise either from external environmental sources, such as lights, sounds, or other people, or from internal bodily sources, such as drives (McCandless, 1970). Nevertheless, in either case, responses become linked to stimulation on the basis of whether reward or punishment is associated with a particular stimulus-response connection (Bijou & Baer, 1961). Those responses leading to reward stay in the person's behavioral repertoire, while those associated with punishment do not. The social environment determines which responses will or will not be rewarded and, as such, behavior development involves learning to emit those responses leading to reward and not to emit those responses leading to punishment.

Although social learning theorists differ in regard to the details of how such learning takes place (Bandura & Walters, 1963; Davis, 1944; McCandless, 1970; Sears, 1957), there is general consensus that development involves increasing coordination of behaviors to social rules. Thus the comparability of this position with that of Freud's is evident. Moreover, it is clear that a social learning conception of behavior development, in general, and of moral development, in particular, are virtually indistinct. There is nothing qualitatively different between behavior labeled as moral and behavior labeled as social, personal, or anything else for that matter. All behavior follows the principles of social learning, and as such, all behavior involves the conformity of the person's responses to the rules of society.

Although basing their views on quite distinct ideas about the *basis* of response conformity to societal rules, both psychoanalytic and social learning theorists would judge that if a young child and an adult emitted the same response in a moral situation, they would therefore be equally morally developed. Moreover, theorists from both persuasions would say that if in a particular society killing of certain other people was condoned and, in fact, rewarded (e.g., the murder of Jews in Nazi Germany or the institutionalized killing of some female infants in some primitive societies), this would be morally acceptable behavior insofar as that society was concerned. That is, because of moral relativism, any society may establish any behavior as moral.

Interactionist Cognitive Developmental Theories

Another view of moral development has become increasingly prominent in American social science since the late 1950s. This conception not only rejects the focus on responses as an index of moral development but also stresses that a *universalistic* view of moral development must be taken and, thus, rejects moral relativism. This third theory was initiated on the basis of the contribution of Jean Piaget. The position has, however, become more prominently advanced by theorists who, working from a cognitive developmental position like Piaget's, have expanded his initial conceptions.

Piaget (1965) became a major contributor to the topic of moral development by offering a theory that saw a child's morality as progressing through phases. However, he saw the child as having "two moralities," i.e., as progressing through a two-phase sequence. However, the target of concern in this sequence is not behavior that might require moral action; rather, it is *reasoning* about moral responses in such situations. Thus in his major statement of his views regarding moral development, Piaget (1965, p. 7) cautions readers that they will find "no direct analysis of child morality as it is practiced in home and school life or in children's societies. It is the moral judgment that we propose to investigate, not moral behavior."

Kohlberg (1958, 1963), Turiel (1969), and other followers of the cognitive developmental view believe that this focus on reasoning, and not on responses, derives from the view that the same moral response may be associated with two quite distinct reasons for behavior. Unless one understands *the reasons why* people believe an act is moral or not, one will not be able to see the complexity of moral development that actually exists (Turiel, 1969).

On the basis of his research, Piaget (1965) formulated two phases as characterizing children's moral-reasoning development. There are several dimensions distinguishing phase one—labeled *heteronomous morality*—from phase two—labeled *autonomous morality*. In phase one, the child is *objective* in his or her moral judgments. An act is judged right or wrong solely in terms of the consequences of the act.

If one breaks a vase, one would be judged by a child in this stage as morally culpable, whether or not the breaking was an accident. This type of judgment is based, Piaget believes, on the *moral realism* of the child. Rules are seen as unchangeable, externally (i.e., societally) imposed requirements for behavior; these rules are imposed by adults on the child and require unyielding acceptance. Such a "relationship of constraint" is seen as necessary because punishment for disobedience to rules is seen as an automatic consequence of the behavior. In short, acts are judged objectively as good or bad, and if a bad act is emitted, there will be *imminent justice,* i.e., immediate punishment that occurs automatically.

At phase two, however, children become *subjective* in their moral judgments. This means they take the *intentions* of the person into account when judging the moral rightness or wrongness of an act. If one breaks a vase out of spite or anger, one would be judged morally wrong. But if one broke the vase because of clumsiness, then no moral culpability would be seen. This second type of judgment, Piaget believes, is based on the child's *moral rationality.* Rules are seen as outcomes of agreements between people who are *not* in a relation of social constraint but, rather, of cooperation and autonomy. That is, each person is equal in such a relation, and as such, rules are made in relation to the *mutual* interest of those involved. Thus acts are judged good or bad in terms of the principles of this "contract." When punishment is to be associated with violation of contract rules is determined by humans and is not a consequence of some reflexive, automatic punisher.

Accordingly, although a 5-year-old and a 21-year-old might behave in similar ways in a moral situation, e.g., neither would cheat on an exam or steal from a friend, such response congruence would not mean there exists similarity in the reasons underlying the responses. The younger child might not act in an unacceptable moral manner due to a belief that because imminent physical punishment was associated with cheating or stealing, one had better obey rules against inappropriate behavior. Simply, one should not steal because one gets punished for it.

However, the 21-year-old might see such reasoning as "immature." Here the reason for not stealing might involve an *implicit* agreement among friends to respect each other's rights and property. The fact that there may be physical punishment associated with stealing would be *irrelevant* to a reason based on a conception of such mutual trust. Thus, to Piaget, because of the presence of such different types of moral reasonings, the 5- and the 21-year-old would not have similar levels of moral development, despite their response comparability.

In summary, Piaget (1965) would say that all people pass through these two phases of moral reasoning. In other words, he would suggest a sequence first involving an objective and concrete morality based on constraints imposed by the powerful (e.g., adults) on the nonpowerful (e.g., children). Second, a subjective morality follows, based on an abstract understanding of the implicit contracts involved in cooperation and autonomy relationships.

Piaget's denial of the importance of focusing *just* on the moral response and his stress on universal orderings to morality represented an approach to the study of moral development that was quite distinct from the morally relativistic, response-centered approaches of psychoanalysis and social learning theory. As such, it stimulated considerable interest among developmental researchers, especially because it offered a provocative framework for assessing changes in morality beyond the level of early childhood. However, this interest it stimulated soon led to Piaget's theory being replaced as the focus of developmental research inquiry. Following Piaget's general cognitive developmental theoretical approach, Lawrence Kohlberg (1958, 1963a, 1963b) obtained evidence that Piaget's two-phase model was not sufficient to take into account all the types of changes in moral reasoning through which people progressed. Kohlberg saw it necessary to devise a theory involving several stages of moral-reasoning development in order to encompass all the qualitative changes he discerned. Interest in moral development in the 1960s and 1970s was centered on assessing the usefulness of Kohlberg's universalistic theory.

KOHLBERG'S THEORY OF MORAL-REASONING DEVELOPMENT

Kohlberg's theory of moral development, like Piaget's, is based on the notion that by focusing only on the response in a moral situation, one may ignore important distinctions in the moral reasoning of people at different points in their life span, reasoning differences that in fact may give different meaning to the exact same response at various developmental levels. Because the response alone does not necessarily give a clue as to underlying reasoning, "an individual's response must be examined in light of how he perceives the moral situation, what the meaning of the situation is to the person responding, and the relation of his choice to that meaning: the cognitive and emotional processes in making moral judgments" (Turiel, 1969, p. 95).

Because of these issues, Kohlberg rejected response-oriented approaches to understanding moral development and chose to investigate the reasons underlying moral responses (Kohlberg, 1958, 1963a). He devised a way to ascertain underlying reasoning through his construction of a moral development interview. Information derived from this interview provided the data for the theory he formulated.

Kohlberg's Method of Assessing Moral Reasoning

Kohlberg devised a series of stories, each presenting imaginary moral dilemmas. We will present one such story and then evelute the features it offers in providing a technique for assessing moral reasoning.

> One day air raid sirens began to sound. Everyone realized that a hydrogen bomb was going to be dropped on the city by the enemy, and that the only way to survive was to be in a bomb shelter. Not everyone had bomb shelters, but those who did ran quickly to them. Since Mr. and Mrs. Jones had built a shelter, they immediately went to it where they had enough air space inside to last them for exactly five days. They knew that after five days the fallout would have diminished to the point where they could safely leave the shelter. If they left before that, they would die. There was enough air for the Joneses only. Their next-door neighbors had not built a shelter and were trying to get in. The Joneses knew that they would not have enough air if they let the neighbors in, and that they would all die if they came inside. So they refused to let them in.
>
> So now the neighbors were trying to break the door down in order to get in. Mr. Jones took his rifle and told them to go away or else he would shoot. They would not go away. So he either had to shoot them or let them come into the shelter.

What are the features of this story that make it a moral dilemma? First, as is true of all of Kohlberg's moral dilemma stories (Turiel, 1969), the story presents a conflict to the listener. In this particular story the conflict involves the need for choice between two culturally unacceptable alternatives: killing others so that you might survive or allowing others and yourself and family to die. The story presents a dilemma because it puts the listener in a conflict situation such that any response is clearly not the only conceivable one acceptable. As such, the particular response is irrelevant. What is of concern is the reasoning used to resolve the conflict. Thus Kohlberg asks the listener not just to tell him what Mr. Jones should do but why Mr. Jones should do whatever the listener decides.

Kohlberg would first ask, "What should Mr. Jones do?" Next he would ask, "Does he have the right to shoot his neighbors if he feels that they would all die if he let them in since there would not be enough air to last very long? Why?" Then, "Does he have the right to keep his neighbors out of his shelter even though he knows they will die if he keeps them out? Why?" And finally Kohlberg would ask, "Does he have the right to let them in if he knows they will all die? Why?"

On the basis of an elaborate and complicated system for scoring the reasons people give to questions about this and other dilemmas in his interview (Kohlberg, 1958, 1963a; Kurtines & Greif, 1974), a system which is currently undergoing considerable revision (Colby, 1979), Kohlberg classifies people into different reasoning categories. This classification led him to formulate the idea that there existed a sequence in the types of reasons people offered about their responses to moral dilemmas. The types of moral reasonings people used passed through a series of qualitatively different stages. However, contrary to the two cognitive developmental phases Piaget (1965) proposed, Kohlberg first (1958, 1963a) argued that there are six stages in the development of moral reasoning and asserted that these stages were divided into three levels, with each level being associated with two stages. Currently, however, Kohlberg (1976, 1978) and his collaborators (Colby, 1978, 1979) have revised and refined the theory somewhat. It appears that evidence for only five stages exists (Colby, 1979; Kohlberg, 1978); that is, the last level has only one stage in it. In addition, with the stages redefined, new systems of scoring moral development are currently being developed. Nevertheless, in both the present and the former version of the theory, both the levels and the stages within and across them are seen to form a universal and invariant sequence of progression.

STAGES OF MORAL REASONING IN KOHLBERG'S THEORY: FORMER AND CURRENT LEVEL AND STAGE DESCRIPTIONS

In this section we present a description of the stages and levels in both the former and the present formulation of Kohlberg's theory. We present both versions for several *quite* important reasons. First, human development is an active scientific discipline. It is important for students of the science to realize that scientific theories are not "carved in stone" but are themselves developing sets of ideas; Kohlberg's theory is a case in point. Second, Kohlberg and his associates revised the theory because of research evidence and conceptual criticism that the theory, as it stood, could not adequately treat. Because this revision is a relatively recent one, and much of the existing data pertinent to moral development were collected in regard to the former version, we need to see what the former characteristics of the theory were in order to (1) understand a lot of the already collected data about moral development and (2) judge if appropriate revisions now exist in the theory. We will see that the theory may be able to handle more, but certainly not all, of the criticism advanced against it.

There are no age limits typically associated with Kohlberg's stages or levels. However, since in both versions of the theory the very first stage does seem to rest on some minimal representational ability, we may presume it does not emerge prior to the preoperational period in Piaget's theory, i.e., somewhere between 2 to 6 or 7 years of age. Moreover, many of the subsequent stages of moral reasoning seem to be dependent on formal operational thinking (Kohlberg, 1973). As such, they may be expected to be involved more typically with adolescence and adulthood, at least insofar as western culture is concerned (Simpson, 1974).

The Former Version of the Theory

Kohlberg's (1958, 1963) earlier formulation of his theory included three levels of moral-reasoning development; there were two stages within each level. Box 7.1 describes the levels and stages of the former version of Kohlberg's theory.

BOX 7.1

LEVELS AND STAGES IN THE FORMER VERSION OF KOHLBERG'S THEORY OF MORAL-REASONING DEVELOPMENT

Level 1: Preconventional Moral Reasoning
Within the first level, the first two stages of moral reasoning emerge. Although these two stages involve qualitatively different thought processes about moral conflicts, they do have a general similarity. For both stages a person's moral reasoning involves reference to *external* and *physical* events and objects, as opposed to such things as society's standards, as the source for decisions about moral rightness or wrongness.

Stage 1. *Obedience and punishment orientation*. The reference to external, physical things is well illustrated in this first stage of moral development. Kohlberg sees this stage as being dominated by moral reasoning involving reference merely to obedience or punishment by powerful figures. Thus an act is judged wrong or right if it is or is not associated with punishment. Reasoning here is similar to what we have seen involved in Piaget's first stage of moral reasoning. In stage 1, a person reasons that one must be obedient to powerful authority because that authority *is* powerful; it can punish you. Acts, then, are judged as not moral only because they are associated with these external, physical sanctions.

Stage 2. Naively egotistic orientation. Reference to external, physical events is also made in this stage. However, an act is judged right if it is involved with an external event that satisfies the needs of the person or, sometimes, the needs of someone very close to the person (e.g., a father or

a wife). Thus even though stealing is wrong—because it is associated with punishment—reasoning at this level might lead to the assertion that stealing is right *if* the act of stealing is instrumental in satisfying a need of the person. For example, if the person was very hungry, then stealing food would be seen as a moral act in this instance.

Although this second stage also involves major reference to external, physical events as the source of rightness or wrongness, the perspective of self-needs (or, sometimes, the needs of significant others) is also brought into consideration (albeit egocentrically). Thus the development in this second stage gradually brings about a transition of perspective, a perspective involving people. This transition then leads to the next level of moral reasoning.

Level 2: Conventional Moral Reasoning
In this second level of moral reasoning, the person's thinking involves reference to acting as others expect. Acts are judged right if they conform to roles that others (i.e., society) think a person should play. An act is seen as moral if it accords with the established order of society.

Stage 3. Good-person orientation. Here the person is oriented toward being seen as a good boy or a good girl by others. The person sees society as providing certain general, or stereotyped, roles for people. If you act in accord with these role prescriptions, you will win the approval of other people, and hence you will be labeled a good person. Thus acts that help others, that lead to the approval of others, or that simply *should*—given certain role expectations by society—lead to the approval of others will be judged as moral.

Stage 4. Authority and social-order-maintenance orientation. A more formal view of society's rules and institutions emerges in this stage. Rather than just acting in accord with the rules and institutions of society to earn approval, the person comes to see these rules and institutions of society as ends in themselves. That is, acts that are in accord with the maintenance of the rules of society and that allow the institutions of social order (e.g., the government) to continue functioning are seen as moral. The social order and institutions of society must be maintained for their own sake; they are ends in themselves. A moral person is one who "does his or her duty" and maintains established authority, social order, and institutions of society. A person is simply not moral if his or her acts are counter to these goals.

Reasoning at this level involves a consideration of a person's role in reference to society. In addition, at stage 4, in contrast with stage 3, moral thinking involves viewing the social order to do one's duty and be moral; however, this thinking may lead the person to consider the alternative, or reverse side of the issue. The person may begin to think about what society must do in order for it to be judged as moral. If and when such considerations begin to emerge, the person will gradually make a transition into the next level of moral reasoning.

Level 3. Postconventional Moral Reasoning

This is the last level in the development of moral reasoning. Moral judgments are made in reference to the view that there are arbitrary, subjective elements in social rules. The rules and institutions of society are not absolute, but relative. Other rules, equally as reasonable, may have been established. Thus the rules and institutions of society are no longer viewed as ends in themselves, but as subjective. Such postconventional reasoning, which is related to formal operational thinking and thus to adolescence as well, also develops through two stages.

Stage 5. Contractual legalistic orientation. In this stage, similar to Piaget's (1965) second phase, the person recognizes that a reciprocity, an implicit contract, exists between self and society. One must conform to society's rules and institutions (do one's duty) because society, in turn, will do its duty and provide one with certain protections. Thus the institutions of society are seen not as ends in themselves but as part of a contract. From this view a person would not steal because this would violate the implicit social contract, which includes mutual respect for the rights of other members of the society.

Thus the person sees any specific set of rules in society as somewhat arbitrary. But one's duty is to fulfill one's part of the contract (e.g., not to steal from others), just as it is necessary for society to fulfill its part of the contract (e.g., it will provide institutions and laws protecting one's property from being stolen). The person sees an element of subjectivism in the rules of society, and this recognition may lead into the last stage of moral-reasoning development.

Stage 6. Conscience, or principle, orientation. Here there is more formal recognition that societal rules are arbitrary. One sees not only that a given, implicit contract between a person and society is a somewhat arbitrary, subjective phenomenon but also that one's interpretation of the meaning and boundaries of this contract is subjective. One person may give one interpretation to these rules, while another person may give a different interpretation. From this perspective, the ultimate appeal in making moral judgments must be to one's own conscience.

The person comes to believe that there may be rules that transcend those of specific, given social contracts. Since a person's own subjective view of this contract must be seen as legitimate, a person's own views must be the ultimate source of moral judgments. One's conscience, one's set of personal *principles*, must be appealed to as the ultimate source of moral decisions. To summarize, stage 6 reasoning involves an appeal to transcendent *universal* principles of morality, rules that find their source in the person's own conscience.

The Current Version of the Theory

In the present version of the theory (Colby, 1978, 1979; Kohlberg, 1976, 1978), there are again three levels of moral-reasoning development, generally labeled as in the former version. The last level, however, includes only one stage. The first two levels each have two stages. The major change in the theory is in the definition of these five stages. They focus on the social perspective of the person, *moving toward increasingly greater scope* (i.e., including more people and their institutions) *and greater abstraction* (i.e., one moves from physicalistic reasoning to reasoning about values, rights, and implicit contracts). As shown in Box 7.2, the levels are seen in essentially the same way as in the former version; however, the characteristics of each stage within each level have been changed in the ways we have described.

BOX 7.2

LEVELS AND STAGES IN THE REVISED VERSION OF KOHLBERG'S THEORY OF MORAL-REASONING DEVELOPMENT

Level 1: Preconventional

Stage 1. Heteronomous morality. "Egocentric point of view." The person doesn't consider the interests of others or recognize that they differ from the actor's; the person doesn't relate two points of view. Actions are considered physically rather than in terms of psychological interests of others. There is a confusion of authority's perspective with one's own.

Stage 2. Individualism, instrumental purpose, and exchange. "Concrete individualistic perspective." The person is aware that everybody has interests to pursue and that these can conflict; right is relative (in the concrete individualistic sense).

Level 2. Conventional

Stage 3. Mutual interpersonal expectations, relationships, and interpersonal conformity. "Perspective of the individual in relationships with other individuals." The person is aware of shared feelings, agreements, and expectations—which take primacy over individual interests—and relates points of view through the concrete "golden rule," putting oneself "in the other guy's shoes." The person does not yet consider generalized system perspective.

Stage 4. Social system and conscience. "Differentiates societal point of view from interpersonal agreement or motives." At this stage, the person takes the point of view of the system that defines roles and rules and considers individual relations in terms of play in the system.

Level 3. Postconventional, or Principled

Stage 5. Social contract or utility and individual rights. "Prior-to-society perspective." The rational individual is aware of values and rights prior to social attachments and contracts. Such a person integrates perspectives by formal mechanisms of agreement, contract, objective impartiality, and due process; considers moral and legal points of view; and recognizes that they sometimes conflict and finds it difficult to integrate them.

Adapted from Kohlberg (1976).

CHARACTERISTICS OF MORAL-REASONING STAGE DEVELOPMENT

Kohlberg and his associates (e.g., Colby, 1978, 1979; Turiel, 1969) have done more than just describe the ordering and nature of the above stages. They also have attempted to describe the nature of intraindividual *change* from one stage to another. Turiel (1969), for example, notes that development through the stages of moral reasoning is a gradual process. Transition from one stage to another is not abrupt; rather, movement is characterized by gradual shifts in the most frequent type of reasoning given by a person over the course of development. Thus at any given point in life, a person will be functioning at more than one stage at the same time.

Data collected by Kohlberg and his associates (Colby, 1979) support this point. Kohlberg (1958) studied a group of males who, since original testing, have been followed longitudinally. When first tested, the people ranged in age level from late childhood to middle adolescence. Today they are all in their adult years. Exhibit 7.4 shows the percentage of reasoning at each of five stages of development for the various age levels the people progressed through over the course of this continuing study. For example, at the 10-year-old age level, most moral reasoning was at stage 2 but there were still several instances of reasoning at other stages. In turn, at the 36-year-old age level, most reasoning is at stage 4, but reasoning at several other stages is evident.

Because of such stage mixture, one must have a large sample of a person's moral reasonings in order to accurately determine that person's stage of moral reasoning. Only such a large sample will allow one to discover the most frequently occurring type of reasoning the person uses to make moral decisions. Stage mixture, then, is not only an everpresent but a necessary component of moral reasoning development. As Turiel has said, "Stage mixture serves to facilitate the perception of contra-

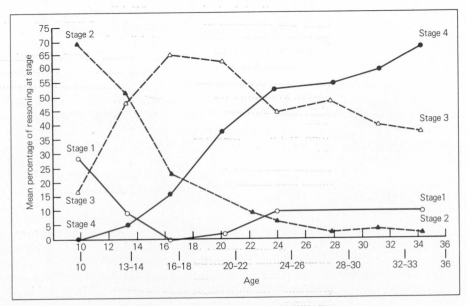

EXHIBIT 7.4   Mean percentage of reasoning at each stage for each age group. (Source: Colby, 1979.)

dictions, making the individual more susceptible to disequilibrium and consequently more likely to progress developmentally" (1969, p. 130).

However, people may experience differing degrees of cognitive conflict—and therefore disequilibrium—in their lives. Accordingly, they may pass through the stages of moral reasoning at different rates if, in fact, they pass through them at all. Thus different people are likely to reach different levels of moral thinking at any one time in their lives. In fact, Turiel (1969) has said that his research indicates that most Americans are modally at stage 4 in their moral-reasoning development. That is, for the majority of today's Americans, moral correctness is evaluted in reference to the maintenance of established social order and the institutions of that order.

A CRITIQUE OF KOHLBERG'S THEORY

Evaluators of Kohlberg's work can be classified into two areas: (1) those who have considered Kohlberg's method of evaluating moral reasoning development and (2) those who have tried to ascertain if moral reasoning follows the stagelike sequence Kohlberg formulated. Consideration of the first area of evaluation will make us cautious about the generalizability of information about moral reasoning derived from use of Kohlberg's interview. Consideration of the second area will support the idea that there are indeed qualitative changes across life in moral reasoning (although not necessarily completely consistent with the order Kohlberg suggests) and that these changes are related to behavior-change processes ranging from the psychological to the sociocultural and historical.

Methodological Appraisals of Kohlberg's Theory

A major methodological critique of Kohlberg's approach to studying moral development was presented by Kurtines and Greif (1974). They identified several major problems with how scores from Kohlberg's interview are derived. The scores represent the basic data for evaluating a person's level of moral reasoning. Kurtines and Greif contend that because there are many areas of methodological concern associated with his interview, it is most difficult to unequivocally evaluate the usefulness of Kohlberg's theory. Attempts have been made to answer several of the major objections Kurtines and Greif raised about Kohlberg's method (e.g., see Colby, 1978; Kuhn, 1976). As noted, a major revision in the theory has evolved (Colby, 1979; Kohlberg, 1976, 1978), and a new system of scoring the responses to the moral dilemmas has been developed (Colby, 1978, 1979). Nevertheless, it is important to remain appropriately cautious, since a lot of new data have not yet been generated about how successful these answers to the criticisms have been. Thus it is appropriate to review at least some of the major problems that have been previously identified.

Kurtines and Greif (1974) noted problems relating to the reliability of the moral-reasoning interview. A measure is *reliable* when consistent scores are repeatedly obtained from administration of the measure.

In addition to administration and scoring problems, Kurtines and Greif identify problems with the *content* of the dilemmas. These problems relate to the *validity* of the interview measure. A measure is valid if it assesses what it purports to assess. Does the interview provide, as Kohlberg claims (1958, 1963a, 1971), a measure of *universal sequences* of moral reasoning (i.e., sequences that apply equally to all people of all cultures at all times of measurement)? Kurtines and Greif (1974) note that the main characters in the dilemmas are male. If a person recognizes the differential role expectations for males and females in traditional western culture, then the gender of the main character in the dilemma may influence that person's moral reasonings. In light of this, it is not surprising that some research using Kohlberg's scale shows females to be less morally mature than males (Kohlberg & Kramer, 1969), although it must be noted that most studies using Kohlberg's measures show males and females to be quite similar in their reasonings (Maccoby & Jacklin, 1974, pp. 114–117). Yet Kurtines and Greif's criticism of sex bias in the content of the dilemmas is consistent with other objections to the content of the interview.

For example, Simpson (1974) sees Kohlberg's dilemmas and his scoring system to be culturally biased. Eisenberg-Berg (1976) notes that all dilemmas pertain only to constraining situations, i.e., to a person's being pressured by two moral values affecting himself or herself. However, prosocial issues, e.g., risking one's own life to save someone else's, are never evaluated. Thus there is a value bias in Kohlberg's interviews.

In sum, because of the various biases in the content and scoring of his dilemmas, there is reason to be wary about whether Kohlberg has offered a technique to validly assess universal sequences in moral-reasoning development. Yet despite these limitations, Kohlberg's approach continues to be an important one in studying moral development. Furthermore, there is *some* empirical support for the specifics of his theory and for the more general notion that there are indeed qualitative changes in moral reasoning across life. Keeping the methodological limitations of this work in mind, we now turn to research relevant to Kohlberg's theory.

Empirical Appraisals of Kohlberg's Theory

Two major, interrelated issues have been involved in the empirical assessment of Kohlberg's theory (1958, 1963a, 1963b, 1971, 1976). First, people have investigated whether moral reasoning progresses through a sequence akin to that suggested by Kohlberg. Second, if such a sequence does exist, is it universal and irreversible in sequence (as is a requirement of all stage theories of development like Kohlberg's; Lerner, 1976)? Finally, a third issue has arisen as a consequence of interest in the first two. People have concerned themselves with the variables that may moderate moral-reasoning stage sequencing.

Sequences in Moral Judgments. Most research relevant to Kohlberg's ideas has dealt with whether reasoning proceeds in a sequence consistent with his theory. Although many studies have not used his exact measures, there is strong evidence that there exists an age-associated development toward principled reasoning. However, although people at older ages are more likely than younger people to offer principle-based moral judgments that take into account the intentions—rather than the mere actions—of a person, the sequence does not appear as inevitable, as unfluctuating, or as smooth as Kohlberg might predict.

Data collected in the longitudinal study being conducted by Kohlberg and his colleagues (Colby, 1979), discussed earlier, provide the strongest support of the view that people, from late childhood to the early part of the middle-adult years, go through the stages of moral reasoning in the manner Kohlberg specifies. Exhibit 7.5 shows a smooth, continuous increase, from age 10 to 36 years, in the average moral maturity score derived from people's responses to the dilemmas of the interview. In

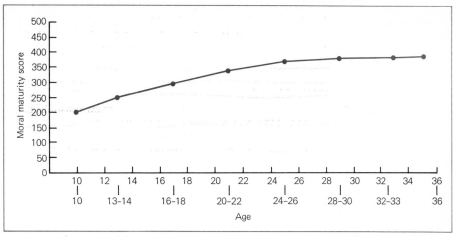

EXHIBIT 7.5   Mean moral maturity score for each group. (Source: Colby, 1979.)

turn, the data in Exhibit 7.6 show that there is an age-associated increase in the percentage of people reasoning at each of the succeeding stages.

Data other than those provided by Kohlberg and his colleagues also suggest an age-associated progression toward principle- or intention-based, as opposed to consequence-based, moral reasoning. Eisenberg-Berg and Neal (1979) studied a small group of 48- to 63-month-old preschoolers. The children were observed and questioned by a familiar researcher about their spontaneous prosocial behavior (helping, sharing, or comforting behaviors) over a twelve-week period. The children justified their behavior primarily with references to others' needs and pragmatic considerations. They used little punishment and authority-oriented, stereotyped, approval-oriented, or hedonistic reasoning. There were no age or sex differences in this sample. Thus these very young children showed moral reasoning that appears associated with the first level of reasoning in Kohlberg's theory.

In turn, Eisenberg-Berg (1979) examined the development of older children's reasoning about prosocial moral dilemmas. She studied 125 elementary and high school students. Elementary school children's reasoning tended to be hedonistic, stereotyped, approval-oriented, and interpersonally oriented and/or tended to involve labeling of others' needs. Stereotyped images of persons and interpersonally

EXHIBIT 7.6   **Percentage of Subjects at Each Stage within Each Age Group**

	Age, in Years							
Stage	**10**	**13−14**	**16−18**	**20−22**	**24−26**	**28−30**	**32−33**	**36**
1/2	71	21	7	0	0	0	0	0
2	39	33	28	0	0	0	0	0
2/3	8	57	22	8	5	0	0	0
3	2	12	42	21	6	13	4	0
3/4	0	1	16	18	17	27	15	6
4	0	0	0	21	14	25	25	14
4/5	0	0	0	0	31	39	23	8

Source: Colby, 1979.

Children's prosocial behaviors involve providing help or care
to other children. (Richard Frieman/Photo Researchers, Inc.)

oriented or approval-oriented reasoning decreased with age, and alternatively, the
use of empathic considerations and judgments reflecting internalized values increased with age. Again consistent with a general progression toward more principled reasoning, associated with the highest level in Kohlberg's theory, people at
older ages had more abstract bases of their moral reasonings. Similar progressions
are found in other studies.

Weiner and Peter (1973) studied 300 people, age 4 to 18 years. In making moral
judgments, subjects were asked to take the intentions and abilities of the actor and
the outcome of the act into account. At higher age levels the main determinant of
moral evaluation was the subjective intentions of the actor, rather than the objective
outcome of his or her act. Similarly, Hewitt (1975) read Dutch male children and
early adolescents, age 8 to 12 years, stories about a harm-doer. The harm-doer's intentions were either good or bad, and the results of the harm were either minor or
serious. The older males differentiated the harm-doer's behavior as being naughty
or not on the basis of intentions and provocations for the harmful act. The younger
males did not make these distinctions. This developmental trend to a predomination
of abstract reasons for judging moral acts is supported with data reported by Keasey
(1974). Preadolescent and late-adolescent (i.e., college) females rated responses to
Kohlberg's dilemmas. Their ratings were given in the context of being presented
with (1) an *opinion* as to how the dilemma should be resolved and (2) a *moral reason*, at various stages, that supported the opinion. There was a greater relative influ-

ence of opinion agreement, rather than stage of moral reasoning, in shaping the ratings of the preadolescents as opposed to the college students.

Similarly, evidence for development toward a more abstract conceptualization of morality is provided in a report by Edwards (1974). Seven hundred 7- through 15-year-olds were asked to define terms like "right" and "wrong." Older groups had higher moral development scores, and although there were no socioeconomic status differences, females did better on these evaluations than did males. This finding is consistent with the higher verbal ability noted for females of the age range studied (Maccoby & Jacklin, 1974) and suggests that depending on how moral thinking is measured, there may be either no sex differences or differences favoring females (which is counter to Freud's views, of course).

Further evidence exists for a sequential movement to stages involving more abstract, principle-based, and intentionality-oriented forms of judgment. Davison, Robbins, and Swanson (1978) studied 160 people ranging in school level from junior high school through graduate school. To measure moral reasoning, they used a scale developed by Rest, Cooper, Coder, Masanz, and Anderson (1974). Using this device, Davison et al. found that a subject's pattern of stage scores was generally consistent with the ordering of stages in Kohlberg's theory and that as one instance in support of this, reasoning at adjacent stages (e.g., stages 3 and 4) was more similar than reasoning at nonadjacent stages (e.g., stages 3 and 5).

Thus there is some evidence to indicate that although people generally seem to progress toward intentional, principled reasonings, they do not *necessarily* conform to a stagelike progression. Data reported by Surber (1977) support this view. Surber studied kindergarten through adult subjects. At all these age levels, i.e., *not* just at the older ones, the intentions of the actor and the consequences of the act influenced judgments. As Kohlberg might predict, the major age trend was a decrease in the importance of consequences in making moral judgment. Yet the fact that intentions were used continuously from early childhood through adulthood suggests that moral reasoning is less stagelike than suggested by Kohlberg's universalistic theory.

Moreover, there are other reasons why Kohlberg's universalistic stage divisions —of consequence-based moral reasoning being associated with early stages and intention-based reasoning being associated with only the latter or last stages—are not compelling ones. Hill and Enzle (1977) found that first-, third-, and fifth-grade children could be trained to consider intentions, rather than consequences, when making moral judgments. Moreover, even without any explicit training, kindergartners, second-, and fourth-graders have been found to take the intentions of an actor in a story into account when making moral judgments about that actor (Elkind & Dabek, 1977). Furthermore, in a study (Darley, Klosson, & Zanna, 1978) that involved the measurement of people ranging in age from 5 to 44 years, it was found that mitigating circumstances involved in a harmful act in a story were taken into account by subjects at *all* age levels. That is, mitigating circumstances led to less recommended punishment for the harm-doer across the entire age range sampled.

In turn, how one measures moral reasoning seems to affect whether people make consequence- or intention-based judgments. Chandler, Greenspan, and Barenboim (1973) found that when 7-year-olds were asked to make moral judgments in response to a verbally presented dilemma, their judgments were largely based on consequences. However, when responding to videotaped dilemmas, their judgments were largely based on intentions. Similarly, Feldman, Klosson, Parsons, Rholes, and Ruble (1976) report that whether 4- to 5- and 8- to 9-year-olds make intentions- or consequence-based judgments depends on the order in which questions about such reasonings are presented to them. Moreover, in a review of the literature, Karniol (1978) concludes that children do use intentions and ignore consequences when the acts they are asked to judge are explicitly specified. Finally, Kohlberg might suggest that the absence or presence of reciprocity between actors in a moral situation should be information taken into account only at the highest stages of moral reasoning (e.g., stage 4 or 5). However, there are several studies which show that children in kindergarten and elementary school make judgments, moral and otherwise, based on the reciprocity between actors and, thus, reason like those adults assessed in the same studies (Berndt, 1977; Brickman & Bryan, 1976; Peterson, Hartmann, & Gelfand, 1977).

In sum, there is much data suggesting that reciprocity and intentions information can be and are used from early childhood through the adult years in making moral judgments. As noted above, this indicates that moral reasoning does not have the universalistic stage features Kohlberg suggests. Further evidence of the lack of universalism of Kohlberg's ideas is derived from cross-cultural studies. We now turn to them.

The Universality of Kohlberg's Stages. Most cross-cultural studies of moral development have been cross-sectional in design. Although they provide data somewhat consistent with the notions of stage progression outlined by Kohlberg for the first four stages, they do not tend to show invariant progression into the highest level of moral reasoning, i.e., the principled one. This in itself is not necessarily contrary to a stage theory. Such a theory does not say that people will necessarily reach the highest levels but only that if they do, they will go through the specified sequence. Yet since cross-sectional studies confound age and cohort effects, by testing at only one measurement time, they cannot be regarded as providing unequivocal support for Kohlberg's theory. Indeed, in the one cross-cultural study that used a sequential design (White, Bushnell, & Regnemer, 1978), time of testing effects were seen to contribute most to moral-reasoning scores.

To be specific, all cross-cultural research before 1978 had been cross-sectional in design. For instance, studies done in British Honduras (Gorsuch & Barnes, 1973), Canada (Kohlberg & Kramer, 1969), Kenya (Edwards, 1974), Great Britian (Kohlberg & Kramer, 1969), the Bahamas (White, 1975), and Taiwan (Kohlberg & Kramer, 1969) had consistently indicated age differences in moral reasoning through Kohlberg's stages 1 through 4. However, reasoning beyond these first four stages appeared so infrequently in these studies that their age-related development in populations other than those derived from the United States remains uncertain.

This cultural difference may be due to the absence of particular, but largely unknown, variables which may moderate reasoning development in cultures other than the United States. Some support for this idea is found in the data of Edwards (1975), who reported that people in highly industrialized settings move through lower stages more rapidly, and achieve higher stages more often, than people in less industrialized and less urban settings (Salili, Maehr, & Gillmore, 1976). In turn, the cultural difference may be due to the cultural bias in Kohlberg's theory and scoring system, identified by Simpson (1974). Other cultures may not include as morally relevant the types of situations or values used by Kohlberg to score the higher stages of moral reasoning.

Although current data cannot decide between these two alternatives, the study by White et al. (1978) suggests that any changes in moral-reasoning scores are more attributable to events occurring within the cultural milieu when people are tested than to age- or stage-related phenomena. White et al. studied 426 Bahamian males and females, aged 8 through 17 years. All subjects were interviewed repeatedly over a three-year period using Kohlberg's interview; that is, their design was sequential in that they repeated testing of their initial cross-sectional sample over the course of the three-year period. There was an upward stage movement within and between birth cohort groups. In other words, the cross-sectional data at each year showed age-group differences; older age groups showed higher scores. Moreover, *within* each cohort there was an upward movement in reasoning scores from initial to final testing.

However, not only did none of the subjects in the sample show reasoning above stage 3, but the age differences in reasoning scores were associated much more with time of testing than age or cohort. The time of testing effect was 3.5 times greater than the age effect. This was the major effect found in the study; e.g., there were no sex differences found. Thus the results of the within-cohort longitudinal comparison, which involved a confounding of age and time effects, suggest that variables in the milieu of the subjects at the time they were tested, rather than those variables that may covary with age, account for the age differences found in the study.

It is clear that with such time-related effects playing the major role in contributing to changes in moral development, the universality of Kohlberg's stages is in doubt. Similar doubt about the adequacy of his stage theory is derived from the few studies directly assessing the irreversibility of the stages he posits.

The Irreversibility of Kohlberg's Stages. Kuhn (1976) used a short-term longitudinal design to assess the continuity in moral reasoning of 5- to 8-year-olds as measured by Kohlberg's interview. Across each of 2 six-month intervals, subjects were as likely to show progressive change (an increase in scores) as regressive change (a decrease in scores). However, over the longer period (one year) encompassed by the study, subjects tended to show progressive change, most of which involved *slight* advancement toward the next stage in Kohlberg's sequence. Although these data bear only on the sequence irreversibility of the first three stages of Kohlberg's theory—since these were the stages subjects reasoned at—they do provide partial support for his theory.

The considerable short-term (i.e., six months) fluctuation could have been due to several sources. Other studies (Bandura & McDonald, 1963) have shown that children can be *induced* to move forward or backward in their moral reasoning. Since subjects have reasoning structures at several stages available to them, such inducement would be expected to be obtained on the basis of stage mixture (Turiel, 1969). However, the subjects in Kuhn's (1976) study were not prompted to show either continuity or discontinuity in their reasoning. Thus the observed fluctuation could be due either to measurement unreliability associated with Kohlberg's interview or to genuine fluctuation in judgment (Kuhn, 1976), fluctuation perhaps naturally moderated by variables that happened to be acting at the particular times at which Kuhn's subjects were longitudinally tested. Indeed, such a testing effect in a longitudinal study is likely, given the data of White et al. (1978).

However, despite what variables may be operating to produce fluctuation or continuity in reasoning, there are data suggesting that when the irreversible sequencing of Kohlberg's stages is evaluated at age levels beyond childhood, their universal ordering does not hold. Holstein (1976) studied the individual developmental sequences of fifty-two middle-class adolescents and of forty-eight of their fathers and forty-nine of their mothers. These family groups were followed longitudinally for a three-year period. Over this sequence there was a stepwise progression in moral-reasoning development *but not from stage to stage*. Rather, people progressed from level to level. However, even this held only for levels 1 and 2 of Kohlberg's theory; that is, people only progressed sequentially from preconventional to conventional morality. Moreover, insofar as irreversibility is concerned, regression in reasoning was found in the higher (postconventional) stages. Furthermore, since the subjects in Holstein's study were of different ages and from different cohorts and since all these groups nevertheless showed the progression limitations and reversibility characteristics, we may infer that time of testing effects again are most associated with whatever age differences were obtained.

In conclusion, there do not seem to be characteristics of universality and irreversibility associated with Kohlberg's stages. Rather, variables acting in the particular historical context of people at particular times seem to be most related to whether one sees qualitative changes in moral reasoning and what direction those changes take. In other words, although evidence exists for a general progression from objective morality to subjective, intentional, and principled reasoning, it appears that the nature and direction of this sequence is a changeable phenomenon; it is dependent on other than stage-related variables. In the next section, we consider evidence pertaining to those non-stage-related variables which may moderate the shift in moral thinking from objective to principled reasoning.

VARIABLES RELATING TO MORAL DEVELOPMENT

Moral development involves an orientation of the person toward others in his or her world. This relation suggests the social relational character of morality. There are data suggesting that children who do have different levels of moral reasoning also have different types of social interactional experiences. Most of these data relate to interactions with a model, with family members, or with peers.

Effects of Models

Several studies show that exposure to a person modeling behaviors in a moral situation can lead to people, across a wide age range, showing comparable behaviors. Walker and Richards (1976) exposed first- and second-grade children either to a model who made moral judgments on the basis of the motives underlying an act (intentions), or on the basis of the consequences of an act, or to no model at all. While children in the last group showed no changes in the type of moral judgments they offered, children in the other two groups showed change in their moral reasoning in the direction of the particular model they were exposed to. In turn, Eisenberg-Berg and Geisheker (1979) exposed third- and fourth-graders to emphatic, normative, or neutral speeches, or "preachings," in their terms, delivered by a principal or teacher. Empathic preachings significantly enhanced children's generosity behavior (donating a portion of their earnings anonymously). Similarly, Brody and Henderson (1977) report that children exposed to adult and peer models who consistently displayed advanced moral judgments themselves produced advanced moral judgments.

One reason that researchers study the effect of models is that they believe that they are simulating the actual influences that other people—for example, peers or family members—may have on the developing person. Other researchers directly study the influence of such groups. We now turn to those studies.

Family Interactions

Mussen, Harris, Rutherford, and Keasey (1970) studied honesty and altruism among preadolescents, ranging in age fom 11.6 to 12.6 years. Girls who showed high levels of honesty and altruism were found to have warm, intimate interactions with their mothers and high self-esteem. However, for boys, honesty was negatively related to gratifying relationships with parents and peers and with self-esteem, but altruism was associated with good personal ego strength.

Hoffman (1975), in a study of fifth- and seventh-grade white, middle-class children and their parents, concluded that differences in the moral orientation of children (e.g., in their consideration for others, feelings of fear or of guilt upon transgression) are at least in part due to different discipline and affection patterns of parents.

Data reported by Santrock (1975) provide direct support for this notion. Subjects were 120 six- to ten-year-old, predominantly lower-class boys from either an intact home environment or a family where the father was absent (due to separation, divorce, or death). Based on reports by the subjects, the sons of divorced women experienced more power assertiveness disciplinary action (as opposed, for instance, to love withdrawal discipline) than did the sons of widows. Such interaction differences influenced moral behavior. According to teachers' ratings, the sons of divorced women have more social deviation, but more advanced moral judgment, than do the sons of widows. Moreover, in relation to interaction differences associated with being in a father-absent or a father-present home, there were also differences in moral functioning. Teachers rated the former group of boys as less advanced in moral development than the latter group.

Thus social interactions involving parents and children do seem to provide a basis of moral development differences. Yet such familial social interactions do not seem to involve children's replicating the moral-reasoning orientations of their parents. Haan, Langer, and Kohlberg (1976) found that in a large sample of children, ranging in age from 10 to 30 years, and their parents, there was little relation between moral stages among family members. Although husbands' and wives' moral stages were correlated at a low level, there was no relation among siblings' moral levels. Parents' and daughters' stages were also unrelated, and parents' and sons' stages were related only among younger sons. Thus although family interactions do contribute to moral functioning, they do not seem to shape it totally.

Peer Interactions

Social interactions outside the family may be more likely to advance moral development because by avoiding the inevitable power differences between parents and children, they more readily may promote the reciprocal and mutual interactions involved in decentered, morally principled thinking. It may be that children's greater interaction with their peers, who by definition are equal to them, provides them with the precise context necessary to facilitate moral development

Findings reported by Gerson and Damon (1978) and by Haan (1978) support this position. Gerson and Damon studied children aged 4 to 10 years old. The longer the children took part with their peers in a group discussion of how to distribute candy, the more likely they were to agree to an equal distribution of candy. This was true at all ages but most markedly among the older children. That is, 77 percent of all children in the study made an equal distribution by their final choice, but 100 percent of the 8- to 10-year-olds did so.

In turn, in the study reported by Haan (1978) six adolescent friendship groups (sexually mixed but either all black or all white) participated in a series of moral "games" to see whether behavior was accounted for better by Kohlberg's formal, abstract reasoning ideas or by an interpersonal formulation stressing that "moral solutions are achieved through dialogues that strive for balanced agreements among participants" (Haan, 1978, p. 286). The games presented to subjects, aged 13 to 17 years, included having to role-play life in two cultures, competing or cooperating as teams in a game having only one winner, constructing a society, and role playing being the last survivors on earth and deciding what to do.

Haan scored the subjects on the Kohlberg and "interpersonal" measures of morality across games. Although these scores fluctuated between games, the interpersonal scores were more stable than the Kohlberg scores, particularly in games involving stress (e.g., the competition one). Moreover, in all situations requiring action, all subjects used interpersonal morality—which required a balance of positions among all those who were interacting—more than the measures of moral reasoning associated with Kohlberg's theory. Thus insofar as moral behavior was concerned, moral reasoning involved establishment of social interaction agreements among the adolescents and was not primarily based on principled thoughts independent of the group consensus.

Yet the use of such interaction-based, social reciprocity morality facilitated development of both the interpersonal and Kohlberg-related measures of morality. After the games, the levels of scores of both these explanations of morality increased for all six adolescent groups, as compared to a control group that did not play any moral games. This suggests that any explanation of moral development is enhanced by considering the interaction between abstract reasoning and the demands of the particular social situation.

In sum, social interactions appear to foster the cognitive developments involved in changes in moral reasoning. There is one particular institution in the social context which, other than the family, may be the most important arena for moral reasoning as well as all other aspects of cognitive development. This institution is, of course, the educational institution, the "school." We close our discussion of cognitive development in childhood with a discussion of schools.

THE ROLE OF SCHOOLS

Over the course of the last century the role of social institutions in socializing the child has altered substantially. Former eras were characterized by the family's being virtually the sole socialization institution (Goldberg & Deutsch, 1977). However, today many of the former functions of the family have been given over to other institutions (see Bowman & Spanier, 1978). Certainly, the educational function of social institutions has been taken over by schools. Indeed, other than the family, schools represent the social institution having the major opportunity to influence the developing child (cf. Hetherington and Parke, 1979).

Of course, the family's influence on the child often interacts with that of the school on the child. For instance, Hess and Shipman (1967) found that lower-class mothers have different attitudes toward their children's involvement in school than do middle-class mothers. For example, the former group stressed that obeying school rules was an end in itself; the latter group stressed that obeying the rules was a means by which the child could better learn his or her lessons. Similarly, Lytton (1976) reports that better-educated parents use prohibitions less and reasoning more than less educated parents. Nevertheless, despite the potential relation between family and school directives, the fact that most children now go to school an average of 180 days a year suggests that school per se can have a powerful influence on the child. Data indicate that indeed this is the case.

Two recent areas of research that have involved the effects of school on child development are (1) the effects of the physical characteristics of schools and (2) the effects of early schooling programs, such as preschools, nurserys, and day-care centers. We will treat each of these separately.

Physical Characteristics of Schools

The impact of the physical characteristics of schools on a child's development in an educational setting has received continued attention by scholars (e.g., see Gump, 1975, 1980). Two areas of research within this broad concern have been how the physical construction or setting of classrooms affects children and how the number of other children physically present in a classroom (i.e., classroom crowding) affects children.

Allen (1974), for example, studied language development among students learning in a traditionally constructed classroom—that is, within rooms enclosed by walls, ceilings, doors, etc.—as compared to students learning in open-area classrooms—that is, areas not divided from others by walls, etc. Although the students did not differ in regard to self-esteem or attitudes toward their school experience, the students in the open classroom environment achieved better scores on tests of language development.

Similarly, Ramey and Piper (1974) assessed various aspects of creative thinking among first-, fourth-, and eighth-grade students enrolled in either an open or a traditional classroom. Children in the traditional setting showed more verbal activity than children in the open setting; however, more figural creativity (for example, in

Education in open classrooms may encourage creativity in children.
(Erich Hartmann/Magnum Photos, Inc.)

regard to drawings) was shown by the children in the open classroom than in the traditional one.

Another way of assessing the impact of the physical setting of the school experience on the child's development is through conducting cross-cultural research. Different cultures may have differences in the actual construction of schools; and/or cultures may differ sufficiently in the levels of economic wealth and/or urbanization such that studying two cultures simulates an experiment wherein physical characteristics of the school are varied (see Baltes, Reese, & Nesselroade, 1977, for a discussion of cross-cultural research as simulational research).

In one cross-cultural study, conducted by Wilkinson, Parker, and Stevenson (1979), 943 five- and six-year-old children living in jungle villages near Lamas, Peru, and in slum settlements in Lima, Peru, were studied. Some 6-year-olds attended school and others did not. The children were tested for their cognitive abilities regarding memory. Many differences were found between children who did and who did not attend school. While age and place of residence did not relate to memory in and of themselves, particular combinations of place of residence and schooling experience in that location did. The effects of school were smaller in Lamas than in Lima; i.e., schooling in the jungle setting was not as facilitative as schooling in other settings.

In another cross-cultural study, conducted by Irwin, Engle, Yarbrough, Klein, and Townsend (1978), the relationships among intellectual ability prior to schooling opportunities, characteristics of family and home environment, and elementary school attendance (i.e., just being in a school setting) and performance were studied in three rural Guatemalan communities. Preschool mental test performances, family socioeconomic status level, and indexes of parental values concerning education were all associated with attending or not attending school. Length of school attendance was related to preschooling mental test scores for girls and to family socioeconomic level and parental values for boys. Based on findings such as these, Irwin et al. concluded that schooled and nonschooled peers in most semiliterate communities are unlikely to be originally comparable.

In sum, the physical setting within which one goes to school may affect one's educational attainment. However, the precise nature of the influence appears to vary in relation to the types of attainment being assessed.

Alternatively, however, the effects of crowding on school attainment can be summarized less equivocally. Increased class size tends to decrease valued educationally relevant activities. For example, studies spanning over three decades have shown that class size influences the types of interactions in the classroom. Increased class size is associated with decreased individual participation in class activities (Dawe, 1934). Moreover, attention to the teacher and to the lesson material is less in crowded, as compared to uncrowded, classrooms (Krantz & Risley, 1977).

Effects of Early School Programs

More and more families are placing their children in various types of educational settings before the children are old enough (and/or required by state law) to enroll in kindergarten or first grade. Often the reason for this early enrollment is a desire to give the children an early exposure to educational settings in order to prepare them personally, socially, and intellectually for what will be a major force in their lives *for the next twelve to sixteen or twenty years.* More recently, financial and political reasons have also arisen. As women, for both political and economic reasons, more often seek careers and full-time employment, they often place their children in such settings. Indeed, about one-third of all mothers of children less than 3 years of age who live in intact homes (i.e., with their husbands) work full-time, and the figure for mothers of 3- to 5-year-olds in intact marriages is about 50 percent (Hetherington, 1979; Hoffman, 1979). These figures are higher in single-parent families.

Thus increasing proportions of children are experiencing a preschool program, either in the form of a nursery, a day-care center, or some analogous sort of arrangement. For example, in 1965, 27 percent of all 3- to 5-year-olds were in such a setting and by 1973 this percentage had risen to 41 percent (Bane, 1976). Similarly, from 1970 to 1976 the number of children in nursery schools arose from a total of 1.1 million to 1.5 million, an increase of more than 36 percent in just six years (U.S. Bureau of the Census, 1977a). These percentages are likely to increase in future years.

Controversy has existed about the effects of such trends for the child's development. This has been especially the case in regard to day-care placements, wherein the child is typically away from the parents for the entire working day. Evidence can be cited that day care has a disruptive effect on the child's relations with parents, especially his or her attachment to the mother, and on the child's intellectual and emotional development (e.g., Blehar, 1974; Schwartz, Strickland, & Krolick, 1974). On the other hand, evidence can be cited that indicates that such placement either has no effect on personal and social development or, in turn, enhances such functioning (e.g., Cornelius & Denney, 1975; Maccoby & Feldman, 1972; Portnoy & Simmons, 1978).

To determine exactly what conclusions could be drawn about the effects of day care on a child's development, Belsky and Steinberg (1978) performed a thorough review of the existing literature. Their conclusions provide virtually no support for those who claim that day care will have an inevitably negative influence on child development. Indeed, they conclude that the experience of high-quality day care (1) has neither salutary nor deleterious effects on the intellectual development of the child; (2) is not disruptive of the child's emotional bond with the mother; and (3) increases the degree to which the child interacts, both positively and negatively, with peers.

Evidence emerging since the Belsky and Steinberg (1978) review tends to bolster their conclusion that day care does not adversely affect mother-child interaction. For example, Portnoy and Simmons (1978) studied the attachment behavior of thirty-five white, middle-class 3.5- to 4-year-olds, who had experienced different rearing histories, through use of a series of standardized episodes involving separations and reunions with the mother and a stranger (the Ainsworth strange-situation test, discussed in Chapter 6). Group 1 children had been cared for continuously at home by their mothers. Children in group 2 were cared for at home by their mothers until age 3, when they were enrolled in a group day-care center. Group 3 children were enrolled in family day care at 1 year of age and entered a group day-care setting approximately two years later. No significant differences in attachment patterns were found for children with different rearing histories or as a function of the interaction between rearing history and sex. Moreover, Cummings (1980) reports that while children he studied became distressed when their mothers left them in an experimental situation, at a day-care center most children did not become distressed or attempt to follow their mothers when left with a caregiver. Similarly, Ramey, Farran, and Campbell (1979) compared two groups of mothers, all of whom had children who were at risk for intellectual retardation due to sociocultural factors. One group of mothers had their children enrolled in a day-care program; the other group's children were not enrolled. The two groups of mothers interacted with their children in comparable ways.

Finally, Rubenstein and Howes (1979) compared social interaction and play behavior in a day-care center and at home for two matched groups of 18-month-olds. Adult-infant, infant-peer, and infant-toy interaction were studied. More adult-infant play, tactile contact, and reciprocal smiling were found in day care. More infant verbal responsiveness to maternal talking, more infant crying, and more maternal restrictiveness were found at home. Developmental level of play with toys was higher in day care, a difference associated with interaction with peers. No adverse effects of daily mother-infant separation were noted in the daily social and play behavior of the day-care group. Finally, the importance of peers as social objects for the toddler emerged from this study. Peers seem to contribute to the high levels of play and to the positive effect noted in day care and also seem to facilitate separation from adult caregivers. (We have stressed the role of peers in Chapter 6; we will focus directly on the role of peers in childhood in Chapter 8.)

Thus these above studies all confirm the Belsky and Steinberg (1978) contention that day care does not adversely affect mother-child interaction. However, another conclusion they reached—that preschool programs, like day-care ones, do not necessarily facilitate intellectual development—may have to be revised in light of recent evidence. First, the New York State Education Department (1980) released the results of a four-year longitudinal study of experimental prekindergarten programs operated in New York State. The findings are that the programs have a strong, positive effect on children in both cognitive and noncognitive development *while they are enrolled in the program.* The study suggests that there are two important

Attendance at a quality day-care center may have facilitative effects on children's development. (Burk Uzzle/Magnum Photos, Inc.)

components of such programs, each necessary if such programs are to have a long-term effect on participating children. First, extensive parental involvement was important, especially in facilitating the child's cognitive development. Second, continuity between the prekindergarten experience and subsequent school experience also was found necessary to ensure long-term effects of the preschool program.

These results seem to speak against those of other studies which suggest that preschool programs do not have lasting cognitive effects (see Jensen, 1969). Another study, published by the Consortium for Longitudinal Studies (1980), also suggests that preschool experiences can facilitate later cognitive functioning. In this study the original data on low-income children who participated in preschool programs in the 1960s were reexamined. These children, now in their late childhood and early adolescent years, were also retested, using direct measures of actual academic performance. The results indicated that (1) early education programs significantly reduced the number of children assigned to special education classes and (2) early education programs significantly increased children's scores on fourth-grade mathematics achievement tests. Moreover, there was tentative evidence for an increase in fourth-grade reading test scores as well.

In sum, both the physical and social environment of the school impacts on the child's development. In turn, although more and more children are experiencing a school context at ages earlier than was the case for former cohorts of children, such social changes do not appear to have any deleterious effect on child development. However, there is another instance of the social context—television—that is known to have effects, both positive and negative, on the child. Since it is almost the case that when today's child is not playing with peers, or in school, or sleeping, he or she is likely to be viewing television, we will have reason to discuss the influence of television on the child as we turn, in Chapter 8, to a consideration of personality and social development in childhood.

CHAPTER SUMMARY

There are several approaches to the study of cognitive development in childhood. In addition, numerous topics relating to cognition (e.g., language) are studied by researchers from different perspectives.

The study of children's learning has expanded greatly in the last two decades. Current evidence indicates that there are developmental changes in the rules of or variables involved in children's learning. Cognitive processes, such as verbal mediation, take on a greater role as children develop.

Psychometric studies of children's intelligence exist also. There are data indicating that a person's IQ score becomes increasingly stable from birth through the childhood years and up to midadolescence. Several variables seem to moderate intellectual change. Achievement is related to psychometric intelligence, and there is evidence for at least some sex differences in mental abilities. Social interaction differences also relate to IQ score differences.

A great deal of attention to childhood cognitive development has been paid by those taking a Piagetian approach. Piaget describes two stages of cognitive development pertinent to childhood. In the preoperational stage, systems of representation (e.g., language) develop. However, the child shows cognitive limitations, as evidenced by his or her inability to conserve various dimensions of stimuli. In the concrete operational stage, operations develop; that is, internalized, reversible thought emerges. But a limitation of thought is that it is tied to the concrete, real world.

Language is one topic in cognition that has been approached by scientists working from different theoretical perspectives. Moral development is another such topic. We discussed both these topics.

There are nature, nurture, and interactionist theories of language development. Current evidence lends most support to interactionist ideas. Language development appears to proceed through stages, and as it does, it can be seen to provide numerous personal and social functions.

Moral development has been approached from a Freudian psychoanalytic perspective, from social learning perspectives, and from cognitive developmental perspectives. Although both Piaget and Kohlberg offer examples of the latter type of theory, it is possible to summarize much of the current theoretical, methodological, and empirical work in the area by reference to Kohlberg's theory.

By presenting standard moral dilemma stories to subjects he interviews, Kohlberg has obtained data allowing him to develop a stage theory of moral-reasoning development. In the initial version of his theory, there were three levels of moral-reasoning development and two stages within each level. In the current version of the theory, there are again three levels (preconventional, conventional, and postconventional); however, while the first two levels each has two stages associated with it, the last level is made of only one stage. There are a lot of data indicating that moral reasoning does not necessarily develop in the universal stagelike manner described by Kohlberg. On the other hand, there is strong support for the view that people's morality does generally become more abstract and intentionality-based with development. As with language development, there is evidence that social interactions, with family and peers, facilitates moral development.

Other than the family, the school provides a major institution within which such interactions occur. Schools with different physical characteristics influence cognitive development in children differently. There are positive cognitive and social effects of early school programs, for example, as may occur with quality day-care settings.

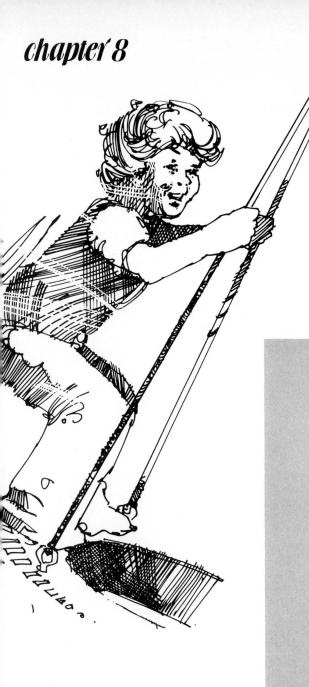

chapter 8

Childhood: social and personality development

CHAPTER OVERVIEW

In this chapter we consider many of the key features of a child's social world: family, peers, play, and the media (most importantly, television). As in infancy, children's individuality contributes to their interactions in their social world. As in previous chapters, the idea of reciprocal relations between children and their social world is a key one in our discussion.

CHAPTER OUTLINE

THE FAMILY

ERIKSON'S THEORY OF PERSONALITY DEVELOPMENT IN CHILDHOOD

Stage 3. The Genital-Locomotor Stage
Stage 4. Latency

THE ROLE OF THE FAMILY IN CHILD DEVELOPMENT

Parental Childrearing Practices
Child Effects
The Family in Contemporary Context

PEER RELATIONS IN CHILDHOOD

Functions of the Peer Group
Determinants of Peer Acceptance and Popularity
Physical (Facial) Attractiveness
Rate of Maturation

THE ROLE OF PLAY IN CHILD DEVELOPMENT

Functions of Play
The Development of Play

INFLUENCES OF TELEVISION ON CHILD DEVELOPMENT

Variables Related to Television Viewing
Effects of Viewing Televised Aggression
Effects of Viewing Prosocial Television
Effects of Television Commercials
Conclusions

CHAPTER SUMMARY

ISSUES TO CONSIDER

What are the major features of Erikson's theory in regard to the childhood years?
What is the role of the family in child development?
What types of parental childrearing practices exist? What are their effects on the child?
What effects do divorce and maternal employment have on child development?
What are the major characteristics of child-abusing families?
Why is it useful to view parent-child relations in a reciprocal manner?
What are the functions of the peer group in child development? What determines peer acceptance?
What are the roles of play in child development?
What are the potential positive and negative influences of television on child development?

THE FAMILY

*I*n Chapter 6 we discussed socialization and the nature of the relations between members of different generations. That is, we stressed that socialization processes involved bidirectional, reciprocal relations between infants and others in their social context, i.e., parents and other caregivers. In other words, we suggested that in order to understand the effect of the social context on the child, one needed to appreciate the child's contribution to this effect. This focus on individuality led to a concern with personality development and to a presentation of some of the descriptions of individual–social context relations provided in Erik Erikson's theory of personality development.

In this chapter we continue and extend these themes to a discussion of social and personality development in childhood. We first consider the nature of personality development in childhood as described in Erikson's theory; here we will see that, as in earlier stages of life, the adequacy of the person's development depends importantly on the nature of his or her interactions with his or her social context. As in infancy, a key element of the child's social world is the family. However, as was also the case to some extent in infancy, peers are important parts of the social context; they play increasingly more prominent roles in the development of children. Accordingly, we discuss the role of the parental-familial contexts for children's development, and we then turn to a discussion of peers, as well as still other aspects of the social context of childhood, e.g., the role of television.

ERIKSON'S THEORY OF PERSONALITY DEVELOPMENT IN CHILDHOOD

We have seen in earlier chapters that just as the child's individuality is shaped by his or her social context, the child's individuality influences the social context. Previously, we discussed the relevance of descriptions of personality development derived from Erik Erikson's theory for understanding these bidirectional relations from a life-span perspective. The first two stages in his theory, the oral-sensory and the anal-musculature, involved crises of trust versus mistrust and autonomy versus shame and doubt, respectively. Here we present Erikson's descriptions of the psychosocial tasks he sees in his theory associated with those stages of development occurring during the childhood years.

Stage 3. The Genital-Locomotor Stage

If the child has developed appropriately within the anal-musculature stage and gained the ability to control his or her own movements, the child will now have a chance to use these abilities. This third psychosocial stage corresponds to Freud's psychosexual phallic stage. Although Erikson does not dismiss the psychosexual implications of the Oedipal conflict, he specifies that such a development also has important psychosocial implications. If the child is to resolve successfully the Oedipal conflict, he or she must begin to move independently away from the parental figures. The child must begin to employ the previously developed self-control over the muscles, take his or her own steps into the world, and thereby break the Oedipal ties. What Erikson says is that society expects the child not to remain "tied to the mother's apron strings" but rather to locomote (walk off) by oneself and thereby eliminate such attachments. The child must be able to move freely in interaction with the environment.

Accordingly, if the child is able to step into the world without the parent being there to guide or prod, the child will develop *a sense of initiative.* He or she will feel that the self can decide when to use locomotor abilities to interact with the world. On the other hand, if the child does not move off on his or her own, if he or she remains tied to the parent for directives about the exercise of locomotor functioning, the child will not feel a sense of initiative. Rather, the child will feel *a sense of guilt.* To Erikson, this is because the child's Oedipal attachments remain relatively intact. To the extent that they continue to exist while society expects evidence of their being eliminated, the child will feel guilt.

Stage 4. Latency

Freud did not pay a good deal of attention to the psychosexual latency stage because of his belief that the libido is submerged. Consequently, the stage had little if any psychosexual importance to Freud. However, Erikson attaches a great deal of psychosocial importance to the latency years. Erikson believes that in all societies children begin at this stage to learn the requisite tasks necessary for being adult members of society. In our society this psychosocial directive, in part, takes the form of the child's being sent off to school. In other societies this same psychosocial orientation may take the form of teaching the child to farm, cook, hunt, or fish.

Accordingly, if the child learns these skills well, if he or she learns what to do and how to do it, he or she will develop *a sense of industry.* The child will feel that he or she knows what to do to be a capably functioning adult member of society. The child will feel that he or she can be industrious, that he or she has the capability *to do.* On the other side of the continuum lies the feeling associated with failures in these psychosocial developments. If the child feels that he or she has not learned to perform capably the requisite tasks of the society (while the child feels others around him or her have acquired this), he or she will feel *a sense of inferiority.*

In sum, the adequacy of the child's personality development is related to his or her interacting with the social context. In childhood, as in infancy, the family is a key component of this context. We discuss its influence next.

THE ROLE OF THE FAMILY IN CHILD DEVELOPMENT

As discussed in Chapter 6, families represent a necessary and ubiquitous part of a child's social world. In this section we will be concerned with how parents influence their children (e.g., through their styles of childrearing) and, of course, how children influence their parents. However, the family is only one institution in the social context. As the context changes across history, an impact on the family, and on parent-child relations, may be expected. As such we will also be concerned with how the family functions in the social context as it exists currently. This focus will allow us to stress how phenomena associated with particular features of the social context (e.g., economic changes) produce certain types of families (e.g., child-abusing ones) whose behavioral interactions can be understood as involving reciprocal socialization influences. We turn first, however, to a discussion of how parents can influence their children.

Parental Childrearing Practices

A prominent series of reports by Baumrind (1967, 1968, 1971, 1972) usefully summarize much of what is known about parental influence during the childhood years. In describing three general types of parental *caregiving practices*—childrearing and disciplinary behaviors used by a person toward a child in his or her care—Baumrind has identified clusters of parenting practices that may be seen in other independent studies. In addition, by showing that these different clusters are related differentially to child behavior, she has provided demonstrations of the role of the parent on a child's functioning.

One type of parent Baumrind identifies is labeled the *authoritarian parent*. This type of parent tries to shape, control, and evaluate the behavior and attitudes of the child in accordance with a set, typically absolute standard of behavior. This parent stresses the value of obedience to his or her authority. He or she favors punitive, forceful measures to curb "self-will" whenever the child's behaviors or beliefs conflict with what the parent thinks is correct. The parent's belief in respect for authority is combined with an orientation to have respect for work, preservation of order, and traditional social structure. The authoritarian parent does not encourage verbal give-and-take. Instead, he or she believes that a child should accept the word of the parent for what is correct conduct (Baumrind, 1968, p. 261).

A second type of parent Baumrind identifies is labeled the *permissive parent*. This type of parent attempts to behave toward the child's behaviors, desires, and impulses in a nonpunishing, accepting, and affirming manner. The parent consults with the child about decisions regarding family "policy" and offers the child rationales for family rules. But the permissive parent does *not* present himself or herself to the child as an active "agent" with the responsibility for shaping or modifying the child's present or future behavior. Rather, this type of parent presents himself or herself as a family "resource" for the child, someone to use as the child wishes. This parent largely allows the child to govern his or her own behavior. As such, the permissive parent avoids exercising control over the child and, in fact, often does not encourage the child to obey external (social) standards. Thus reason, but not overt power, is used by this parent in his or her attempts to rear the child (Baumrind, 1968, p. 256).

The third type of parent Baumrind identifies is labeled the *authoritative parent*. This type of parent tries to direct his or her child's activities through the use of a rational issue-oriented style. Through explanations to and reasoning with the child, the parent attempts to *induce* the desired behavior in the child. That is, through *induction*—direction through reasoning and explanation (Hoffman, 1970)—the authoritative parent tries to control the child. Therefore, such a parent encourages verbal give-and-take, because this allows him or her to share with the child the reasoning behind a policy. However, this parent exercises firm control over the child but not to the extent that the child is overburdened with restrictions. Rather, the child's interests and developmentally specific needs and behavioral capabilities are taken into account when deciding how to treat the child. While keeping his or her own parental and adult rights in mind, the authoritative parent, by combining power with induction, attempts to rear the child with rules that see the *rights and duties of parents and children* as complementary (Baumrind, 1968, p. 261; Gardner, 1978).

In Baumrind's (1967, 1968, 1971, 1972) view, extremely authoritarian or permissive parenting has equally bad consequences for children's development in that neither is as effective in leading to desired child behavior as is authoritative parenting. Most parents, whether they are authoritarian, permissive, or authoritative, want their children to show similar behaviors (Baumrind, 1972). These behaviors, which include friendliness, cooperation, orientation toward achievement, and interpersonal dominance, form a cluster of behaviors that describe a socially competent, responsible, and independent person. Authoritarian or permissive parents do not seem as successful as do authoritative ones in rearing children showing this gen-

The authoritarian parent attempts to instill obedience without encouraging verbal give-and-take with the child. (Erika/Peter Arnold, Inc.)

erally desired cluster of behaviors. As compared to the other two parenting types, authoritative parents are more likely to have children who score higher on measures of purposive, independent, and dominant behaviors and on measures of cooperation with adults and friendliness with children. These children are more instrumentally competent.

As noted earlier, data from other studies suggest the presence of the three parenting types identified by Baumrind and that the different parenting styles have contrasting roles in children's development. Becker, Peterson, Luria, Shoemaker, and Hellman (1962) interviewed the parents of young children. They found that parental use of induction was related to parental warmth and that, in turn, restrictiveness in parents was related to their use of power, punishment, and physical means to assert their authority. Coppersmith (1967) found that mothers who were accepting, supportive, caring, concerned, and loving, *and* who enforced established rules consistently, but sought the views of the child in a context of free and open discussion, had sons who were higher in self-esteem than did mothers who treated their sons harshly, gave little guidance, and enforced rules inconsistently. Similarly, Mussen, Harris, Rutherford, and Keasey (1970) found that girls who showed high levels of self-esteem, honesty, and altruism had warm, intimate interactions with their mothers. Although corresponding relations were not found with the boys Mussen et al. (1970) studied, Sears (1970) found that when at least one parent was warm and accepting, both male and female children were likely to have high self-esteem. In this regard, a recent report by Rutter (1979) indicates that a good relationship with one parent may "buffer" the child from the potentially detrimental effects of a poor relationship with the other parent. Clearly, this suggests that children with two parents may be at an advantage relative to those with just one, since they have an alternative relationship to fall back upon if one parent-child relationship turns sour.

Endsley, Hutcherson, Garner, and Martin (1979) assessed the relations between various maternal interaction patterns and children's curiosity. The frequency with which mothers (1) oriented their children to explore novel materials presented in a laboratory play situation, (2) responded positively to their children's exploration, (3) answered their children's questions, and (4) explored the novel materials themselves were highly related to high levels of child curiosity. However, authoritarian mothers were less likely to show these behaviors than were nonauthoritarian mothers; thus their children showed less curiosity. Similarly, Jones, Richel, and Smith (1980) found that maternal restrictiveness was associated with evasive problem-solving strategies in their preschool children; in turn, the more restrictive the mothers were, the less their children used personal appeal and negotiation problem-solving strategies. In turn, maternal nurturance was related to children's reliance on authority. Thus the Endsley et al. (1979) and the Jones et al. (1980) studies, as well as others (e.g., Schaffer & Crook, 1980), add to the evidence cited above which indicates that the different childrearing patterns identified by Baumrind are associated with differences in child development.

Hoffman (1970) identified several parental disciplinary strategies that are, in part, compatible with aspects of the parenting types described by Baumrind (1967, 1968, 1971, 1972). Parents who use *power assertive discipline* employ physical punishment, threats of punishment, and physical attempts to control their child's behavior. Parents who use *psychological discipline* may employ one or both of the following techniques: (1) *love withdrawal*, which involves the parent's temporarily taking away love from the child because the child's actions have made the parent feel disappointed or ashamed, and (2) *induction*, which involves parental use of rationality and explanation in attempts to influence the child's behavior. Thus at least insofar as power assertiveness and induction are concerned, there seems to be some consistency between Hoffman's descriptions and those pertaining to the authoritarian and the authoritative parent, respectively.

Hoffman (1970) has found that the frequent use of power assertion by a parent is associated with weak moral development in the school-age child. Indeed, children of parents who use power assertion techniques frequently tend themselves to show high levels of aggressive behavior (Anthony, 1970; Chwast, 1972; Feshbach & Feshbach, 1972). Similarly, inconsistent parental discipline or parental conflict over discipline is associated with aggressive or delinquent behavior in children (Hetherington, Cox, & Cox, 1979; Patterson, 1977, 1978). Although Hoffman (1970) did not find that love withdrawal was frequently related to moral functioning, he did find that inductive discipline combined with parental affection was associated with moral development. Similarly, Hoffman (1975), in a study of fifth- and seventh-grade white middle-class children and their parents, concluded that differences in the moral orientation of children (e.g., in their consideration for others, feelings of fear or of guilt upon transgression) are at least in part due to different discipline and affection patterns of parents. Again, induction and affection were associated with higher moral functioning.

Interestingly, in an entirely independent and longitudinal study, McCall (1974) has reported results remarkably similar to those of Hoffman and Baumrind. Parents whose children evidenced increasing IQ scores between 2 and 11 years of age, he discovered, were substantially rewarding of their children's behavior, clear in their disciplinary policies, medium to severe in their efforts to control the child, and highly encouraging of intellectual acceleration. Children whose IQ scores declined across the same developmental period, in contrast, had parents who provided either the most severe or the most lenient penalties when their children misbehaved and who rarely encouraged accelerated or mature behavior.

It would appear, then, that the same kind of parenting which Baumrind labeled authoritative, namely, that which is loving *and* controlling, and which she found to be associated with instrumental competence during the preschool years, also tends to promote moral and intellectual development once the child enters school.

Child Effects

Few doubt any longer that older children shape their parents' childrearing behavior. There is no shortage of studies documenting such child effects. One illustrative study is reported by Parke and Sawin (1975), who show that the punitiveness of parental discipline is likely to be a direct consequence of the child's response to discipline. The defiant child, for example, is likely to elicit from his or her caregiver increasingly severe punishment strategies. This suggests, of course, that the authoritarian parent whom we spoke of earlier might be more inclined to behave in an authoritative fashion if the child were inclined to cooperate with such rearing.

The reciprocal cause-effect cycle in the parent-child relationship is constant, so any analysis such as this makes it difficult to determine who is primarily responsible for a caregiver's behavior. Methodologies that tear apart these dynamic interactions remain relatively undeveloped at this time. One promising technique, which experimentally demonstrates child effects on adults' caregiving strategies, has been reported by Keller and Bell (1979). These researchers trained three 9-year-old girls to act as research confederates. The girls were trained to act either high or low in person orientation. Twenty-four female college students, unaware that the children were trained, participated in four 5-minute periods with one of the children. Although the four periods differed in the actual materials used, in all of them the adult was instructed to encourage the child to do something for another child (such as sew a pillow for a handicapped child). In the high person-orientation condition, the child attended to the adult's face and answered promptly; in the low condition, the child looked primarily at the toys and craft materials and delayed answering the adult. The experimentally produced variations in children's behavior affected the adults' socialization techniques. The verbal interaction of adults with children high in person orientation was characterized by reasoning about the consequences of acts, while that with children lower in person orientation largely involved bargaining with material rewards.

The child's gender and age have been favorite characteristics of study for investigators. In regard to the role of gender in determining some of the parental rearing practices we have discussed, a study of fifth-grade children and their mothers indicates that boys receive more power assertion and love withdrawal and less induction than do girls (Zussman, 1978). Similarly, Noller (1978) finds that parent-daughter dyads show higher levels of overall interaction and affectionate exchange than do father-son dyads. This latter finding documents what is known as a sex-of-parent by sex-of-child interaction.

With regard to the influence of age or developmental status, Reichle, Longhurst, and Stepanich (1976) found that the child can influence the quality of the mother's verbal interaction with him or her. More so than was the case with the mothers of 2-year-olds, mothers of 3-year-olds used more complex explanations. Similarly, the child's age has been found to influence the play behavior both mothers and fathers show to the child (Pakizegi, 1978). At older age levels, the status of the child's physical maturation, rather than age or reasoning abilities, has been found to be a key element in influencing patterns of verbal interaction within the family (Steinberg & Hill, 1978). What appears at first glance to be changes in parenting that are the result of age, this latter study shows, can turn out to be caused by other child characteristics that are merely correlated with age (e.g., physical maturation). This is why we say in developmental work that age is a "marker variable," that is, a summarizing index that subsumes a variety of salient characteristics (e.g., cognitive ability, social skill, physical status). It is not the child's age that always makes the difference, then, but the capabilities someone at a certain age possesses that stimulates parents to respond to a child in a way which may not be appropriate to a younger (and less skilled) or older (and more skilled) individual.

Probably the most dramatic effect of the child's developmental status on the care received is realized by considering the changing nature of what we have described as optimal parenting practices between infancy and the preschool- or school-age years. To document this effect, we must first consider the similarity in growth-promoting parenting practices across these ontogenic epochs.

Across childhood, parenting that is essentially attuned to the child's capabilities and to the developmental tasks he or she faces seems to promote the kinds of child outcomes that Americans hold dear: security, independence, social competence, and intellectual achievement. In infancy this sensitivity translates into being able to read the child's often-subtle cues and appropriately respond to his or her needs in reasonably brief periods of time. In childhood, sensitivity means continuing the warmth and affection provided in the early years but increasing the demands for age-appropriate behavior. The parent during this period must be willing, and able, to direct the child's behavior and activities without squelching his or her developing independence and industry. Thus the authoritative parent listens to the child and considers his or her point of view legitimate even when the parent disagrees.

The change in parenting that we have summarized no doubt results from the changing nature of the child; it is not merely an automatic response to the child's age. For example, as the child's cognitive skills develop, reasoning and demands for delayed gratification become effective disciplinary strategies. Similar efforts, we are not surprised to learn, would be useless with an 8-month-old. Similarly, as the child becomes increasingly able to regulate his or her own behavior, parental control provides a useful scaffold to support such emerging competence. Ultimately, the sensitive and thus competent parent must be willing to let go of this control and permit the child to test his or her own limits and discover his or her own boundaries. Indeed, by the time the child reaches adolescence, the competent parent has set the stage so that the child has the building blocks to encounter successfully the transition from childhood to adolescence. In sum, such a developmental approach to parenting not only increases the likelihood of highly valued child outcomes but also, from a scientific standpoint, illustrates the child's effect on his or her caregivers and, thus, the bidirectional nature of the parent-child relationship: Children's capacities influence their parents' behavior, which, in turn, influences subsequent child development.

Consideration of the child *in* the family leads us to recognize that the child does not only influence the parents' rearing strategy. Indeed, the child's effect extends to the marital relationship. Family sociologists have been interested in such an influence, particularly in how the age of the oldest child in the family affects the quality of marital satisfaction.

It has been suggested that a U-shaped curve, or function, characterizes marital quality across the family life cycle. Specifically, satisfaction levels were thought to be highest prior to the arrival of children, to decrease through the childbearing and early childrearing years, and to increase once again as children mature and eventually leave home (e.g., Rollins & Feldman, 1970; Rollins & Cannon, 1974; Rollins & Galligan, 1978; Deutscher, 1964; Miller, 1976). However, more recent research (Spanier, Lewis, & Cole, 1975) brings these earlier findings into question. The upturn in the marital satisfaction curve following the departure of children away from the home may itself be an artifact of selective attrition from study populations. As the years of marriage increase, dissatisfied couples are increasingly likely to divorce or separate, leaving mostly satisfied couples to be investigated. Additionally, those couples that do remain together may report high levels of satisfaction due to their own needs to justify the long-term maintenance of their relationships. Caution seems warranted, then, when interpreting cross-sectional life-cycle data in terms of child effects on marital relations. This should not make us lose sight of the fact, however, that the child's influence surely extends beyond parenting behavior (Belsky & Tolan, 1981).

In conclusion, we have seen in this analysis of family-child interaction how parents influence their children and how children can influence their parents. Additionally, we have noted that the effect of the child extends beyond the parenting role to the relationship that exists between parents.

The Family in Contemporary Context

The family is not a static institution. The family is altered in relation to changes in the broader social context. We here discuss three contemporary issues that arise as a consequence of changes in the social context—divorce, maternal employment, and child abuse. This discussion will allow us to illustrate the role of the child and the family in the wider context of human development. More specifically, our purpose is to suggest how child abuse, divorce, and maternal employment are issues that are most fruitfully considered from a vantage point that focuses upon what transpires simultaneously within and beyond the family.

Divorce. The divorce rate reached an all-time high by the end of the decade of the 1970s (National Center for Health Statistics, 1980). In the United States there are now more than 1,150,000 divorces each year, involving more than 2,300,000 adults and more than 1,100,000 children. About 2 of every 5 divorcing couples have at least one child. An average of about two children are involved in each divorce in which there are any children under the age of 18 (Spanier & Glick, 1980).

During the fifteen-year period from 1960 to 1975, the number of children under 18 whose parents divorced increased by 143 percent, while the number of divorces increased by 164 percent. This increase in marital disruption has resulted in a significant increase in the number of children living with only one parent. In the period from 1970 to 1978, for example, there was a 43 percent increase in the number of children under 18 living with their mothers only and a 32 percent increase in the number of children living with their fathers only.

However, it should be pointed out that fathers obtain custody of children following divorce only 10 percent of the time. Mothers have obtained custody about 90 percent of the time throughout the 1960s and 1970s, and there has been very little shift in this trend (Spanier & Glick, 1980). The apparent increase in custody for fathers that some claim to have observed is actually a reflection of the *numbers* of fathers that obtain custody, not of the *proportion* of all custody awards that go to fathers. The dramatic increase in the divorce rate has actually had a more pronounced effect

Most children live in families, and families today are composed of multiple generations. An infant, his mother, her father, his mother, and her mother are seen here. (Photo by R. M. Lerner.)

on the numbers of one-parent families headed by women, since they have received custody of children most often during this period of record high divorce rates. It has been estimated that about 45 percent of the children born in recent years can expect to spend at least a portion of childhood and adolescence in a home with only one parent (Glick & Norton, 1977).

Exhibit 8.1 shows the distribution of persons under 18 years old in the United States, with comparisons between 1970 and 1978. It can be seen that most American children live in families of some kind. However, less than four-fifths lived with two parents in 1978. About 1 in 6 children lives with his or her mother only, and only 1 in 60 children lives with his or her father only. In fact, there are about twice as many children living with neither parent (3 percent) as there are children living with their fathers (1.6 percent). Of the more than 10 million children living with their mothers, the largest number have mothers who are divorced, followed by those whose mothers are separated, who have never been married, or who have been widowed (U.S. Bureau of the Census, 1979).

Child Custody following Divorce. Divorce can be a difficult experience for both the parents and children involved. One of the most troublesome issues concerns who will obtain custody of the children. As noted above, fathers are unlikely to obtain custody of children in most cases, and this tends to be true regardless of the socioeconomic characteristics or marital history of the wife (Spanier & Glick, 1980). Fathers are somewhat more likely to obtain custody of all children if their former wives have remarried following divorce in a first marriage; but the differences are slight. Fathers are also more likely to obtain custody of children if the children are all boys; they are least likely to obtain custody if the children are all girls (Spanier & Glick, 1980).

It should also be noted that more than 75 percent of divorced persons remarry, and they do so relatively quickly. In fact, about 50 percent of all divorced persons have remarried within three years of the final decree. Thus many children obtain

EXHIBIT 8.1   **Living Arrangements of Persons under 18 Years Old: 1978 and 1970**

Living Arrangements and Marital Status of Mother	1978*	1970*	Percent Change, 1970 to 1978
Total persons under 18†	63,206	69,458	−9.0
In families	62,767	68,685	−8.6
Living with two parents	49,132	58,939	−16.6
Living with mother only	10,725	7,452	43.9
Separated	2,943	2,332	26.2
Other married, husband absent	567	902	−37.1
Widowed	1,250	1,395	−10.4
Divorced	4,335	2,296	88.8
Single	1,633	527	209.9
Living with father only	985	747	31.9
Living with neither parent	1,924	1,547	24.4
Not in families	439	773	−43.2
Percent	100.0	100.0	
In families	99.3	98.9	
Living with two parents	77.7	84.9	
Living with mother only	17.0	10.7	
Separated	4.7	3.4	
Other married, husband absent	0.9	1.3	
Widowed	2.0	2.0	
Divorced	6.9	3.3	
Single	2.6	0.8	
Living with father only	1.6	1.1	
Living with neither parent	3.0	2.2	
Not in families	0.7	1.1	

* Numbers in thousands.
† Excludes persons under 18 years old who were heads or wives of heads of families or sub-families. Figures for 1970 include inmates of institutions. Inmates were excluded from CPS coverage beginning in 1972.
Source: U.S. Bureau of the Census, 1979.

stepparents during their childhood. *Reconstituted families*—those formed following a remarriage—may provide a very different physical, emotional, and interpersonal context than the child was used to before the remarriage. In some cases, the adjustment to the new family arrangement is quite smooth; in other cases, the adjustment may be especially difficult. There is often the challenge of maintaining contacts with the two biological parents and their spouses and, potentially, with four sets of grandparents and an even larger extended family network. These special kinship networks which are created by divorce and remarriage are becoming more and more common and are now being studied by social scientists. Let us consider some of the effects of divorce on the parent-child relation.

Effects of Divorce. Hetherington (1979) notes that most children experience the transition from living in an intact family to one wherein a divorce has occurred as a painful event. She indicates that typical initial reactions to divorce are anger, fear, depression, and guilt. Typically, more than a year goes by after the divorce before children's tensions are reduced and more positive feelings begin to occur (Hetherington, 1979).

There is no one pattern of reaction to divorce characteristic of all children. Con-

In reconstituted families, children from each parent's former marriage, or marriages, must learn to live together. (United Press International, Inc.)

siderable interindividual differences in the quality and intensity of responses and adjustments to marital separation exist. Several variables have been found to be associated with this variation.

There are some indications that boys are more vulnerable to the adverse effects of divorce than are girls, although the reasons for this difference are not clear (Hetherington, 1979). Extrafamilial factors are also related to variation in children's response to divorce. The stress created within a family by a divorce can be increased or decreased by support (or lack of it) from other social institutions, from the family's social network of friends and relatives, by the quality of the family's housing and neighborhood, by the availability of child care, and by the economic status of the single-parent family (Colletta, 1978; Hetherington, 1979).

Whether or not a child adjusts successfully to a divorce is related to his or her level of cognitive and social development. Hetherington (1979) and others (e.g., Tessman, 1978) note that younger children are more likely to be dependent on parents than are older children; the former are less able to appraise accurately all the aspects of the divorce situation and of the parents' feelings. Indeed, because of their greater egocentrism, younger children may be more inclined to view the divorce as a result of their actions rather than as a consequence of marital difficulties.

Processes of Influence. Popular reports in the news media of the effects of divorce often leave the impression that there is something automatic about marital separation disrupting child development. Serious consideration of the social interactional processes by which divorce affects the child suggests that not only is this not the case but that when certain conditions are met, the separation of parents need *not* impair the children's social, emotional, and cognitive functioning. In fact, under certain circumstances, the dissolution of a marriage may be in the best interests of the child. This is most likely to be the case when children are reared in homes high in marital discord, since there is an abundance of evidence indicating that marital discord by itself is associated with impaired development.

Divorce is not necessary to disrupt child functioning. One common finding is that children from maritally conflicted homes are likely to engage in aggressive and antisocial behavior (Gibson, 1969). Johnson and Lobitz (1974), for example, discerned a consistent negative relationship between marital satisfaction and levels of observed child deviance. Similarly, Rutter (1971) reported in a large *epidemiological* study of 10-year-olds that those youngsters from families characterized by severe marital discord showed an increased rate of both behavioral deviance in school and psychiatric disorders. Others have found similar relationships.

These data and other data reported regarding the effects of divorce seem to result from a complex chain of events whereby stress in the marital relationship undermines parental functioning, which, in turn, fails to support child development. Hetherington, Cox, and Cox (1979) were able to document such a process upon discovering that it was not divorce per se that negatively affected children, but rather marital and postdivorce relations between parents that determined the influence of parental separation. When divorced parents maintained reasonably friendly relations, mothers with custody of their preschoolers tended to be more involved and supportive of their children than when continued strife characterized the relationship between two separated parents. And it was primarily when the quality of parenting was compromised by such stressful relations that Hetherington and her colleagues found that the quality of children's academic performance, peer relations, and sense of self suffered.

In fact, it appears that divorce did not so much have a direct as an indirect effect on the child: Stress in the marital relationship negatively influenced parental functioning and, thereby, impacted the child. This suggested, and indeed Hetherington et al. discovered, that if divorce led to an *improvement* in relations between parents, children benefited since their parents' caregiving was not undermined by the stress emanating from the marital relationship. Unfortunately, these investigators also found that when conflict characterized the marital relationship in intact two-parent households, parenting and child development tended to be negatively affected. Children from such families differed little from those raised by divorced parents still in conflict with their former spouses.

This analysis suggests that the marital relationship is an important, though generally indirect, influence on child development. This discussion highlights one way in which the family functions as a set of interdependent social relationships. Such a perspective on the family serves to enhance our understanding of child development by making us aware that what occurs within the parent-child relationship can be influenced importantly by forces beyond the relationship itself. In the case of divorce we discovered that it is marital, or ex-marital, relations that indirectly exert an influence on the child's development. This discovery should alert us to search for other complex pathways of influence and to look to family sociologists, as students of marriage, for assistance in understanding the child in the family. As students of child development become increasingly interested in the role that support systems (like informal social networks and day-care facilities) play in enhancing parental functioning (Bronfenbrenner, 1979; Cochran & Brassard, 1979; Garbarino, 1976), we must not overlook the most immediate parental support system of all. As our discussion of divorce has demonstrated, the support that a marriage provides or fails to provide can make a sizable difference in the way in which children develop in the family.

Maternal Employment. Considering all married women—not just those with young children—51 percent of those aged 18 to 34 were in the labor force in 1978. The proportion was 43 percent for those with at least one child. Moreover, in 1978, one-third of the married women who had given birth within the past year were currently in the labor force (U.S. Bureau of the Census, 1979). In addition, more than 40 percent of married mothers with a preschool child (2 to 4 years old) are employed (U.S. Bureau of the Census, 1979), and throughout the last decade, over 50 percent of married mothers with school-age children have been employed (Hoffman, 1979). All these rates are higher for mothers in single-parent families. Therefore, employment outside the home is now, and is likely to continue to be, a part of a woman's role, for a majority of American mothers. Such a trend requires some rethinking of the idea that a woman quits work following the birth of her first child.

The increase in working mothers is indicative of the desire of many women to establish careers in life. It may also indicate a willingness on the part of men and women to enhance a family's standard of living by having two incomes. Apparently, many couples see this as a worthwhile sacrifice; however, many women value work because of the rewards it brings. In a national sample of working women studied in 1976, 76 percent said they would continue to work even if they did not have to (Dubnoff, Veroff, & Kulka, 1978). This percentage represents a considerable increase over the percentage of working women responding similarly to the same question asked in a 1957 survey (Hoffman, 1979). It also demands that we recognize that ever-increasing numbers of children are being cared for by persons other than their parents. This care comes in a variety of forms: in-home care by relatives or babysitters, family day care in which small groups of children are cared for in the homes of others, and center-based day care in which larger groups of children are cared for in church basements, school facilities, or other formal establishments intentionally designed to provide care for the young (Belsky, Steinberg, & Walker, 1982).

Such changing attitudes of American women, and their changing behavior, illustrate the importance of the wider social context in understanding how children develop. This is because the recent and skyrocketing growth of mothers working outside the home while their children are still young can be traced to the women's movement and the national economy. While contemporary ideology regarding the role of women in society has encouraged many mothers to pursue careers and educational goals, economic forces have driven other women to work, as two incomes become necessary to maintain the standard of living in an inflationary economy. The rise in utilization of day-care services in the past decade, which is expected to increase into the 1980s, represents a social adaptation to the changing nature of the American family (Belsky, Steinberg, & Walker, 1982). How do such changes influence child development?

Effects on Children. There is clear evidence that the effect of maternal employment and/or day-care rearing need not be negative. Most preschool children of working mothers are cared for in their own homes and not in day-care centers (U.S. Department of Labor, 1977), and thus one need not fear that most face any effects of "institutionalization" (Hoffman, 1979). This is not to say that day-care placement is equivalent to institutionalization. Indeed, studies thus far have *not* demonstrated adverse effects of *quality* day care for infants and young children (Belsky & Steinberg, 1978; Belsky, Steinberg, & Walker, 1982). In addition, there are some data indicating that there is no difference in the amount of one-to-one mother-child contact occurring between working and nonworking mothers and their preschool children (Goldberg, 1977).

Effects of maternal employment on the child have been documented best in regard to the child's own vocational aspirations and expectations, and these effects are most pronounced in regard to the daughters of "working mothers." As summarized by Huston-Stein and Higgins-Trenk (1978, pp. 279–280):

> *The most consistent and well-documented correlate of career orientation and departure from traditional feminine roles is maternal employment during childhood and adolescence. Daughters of employed mothers (i.e., mothers who were employed during some period of the daughter's childhood or adolescence) more often aspire to a career outside the home (Almquist & Angrist, 1971; Hoffman, 1974; Stein, 1973), get better grades in college (Nichols & Schauffer, 1975), and aspire to more advanced education (Hoffman, 1974; Stein, 1973). College women who have chosen a traditionally masculine occupation more often had employed mothers than those preparing for feminine occupations. (Almquist, 1974; Tangri, 1972)*

Moreover, when females are raised within a family in which their mother is employed, it has also been shown that (1) they have less stereotyped views of female roles than do daughters of nonworking mothers; (2) they have a broader definition of the female role, often including attributes that are traditionally male ones; and (3) they are more likely to emulate their mothers; that is, they more often name their mother as the person they aspire to be than is the case with daughters of nonworking mothers (Huston-Stein & Higgins-Trenk, 1978).

These effects of developing in a family wherein the mother is employed outside the home can be identified in early childhood. Bacon and Lerner (1975) found that second-, fourth-, and sixth-grade females whose mothers were employed outside the home had societal vocational role perceptions that were more egalitarian than were the perceptions of grade-mate daughters of nonworking mothers. Gold and Andres (1978b) found that the sex-role concepts of both female *and* male nursery school children were more egalitarian if their mothers were employed. These children's perceptions of their mothers, along a negative-positive dimension, were not related to maternal employment status. However, fathers were perceived more negatively by their sons if the mothers were employed.

Thus maternal employment may relate to more than the child's vocational orientation. Gold and Andres (1978a) document this idea further. They assessed the sex-role concepts, personality adjustment, and academic achievement of 223 ten-year-old boys and girls, with either full-time employed or nonemployed mothers from working-class or middle-class families. Children with employed mothers had the most egalitarian sex-role concepts. However, this relation was primarily dependent not on maternal employment per se but rather on their mothers' greater satisfaction with their roles. There was also some relation between maternal employment status and the children's achievement. Middle-class boys with employed mothers had lower scores on language and mathematics achievement tests than the other middle-class children.

The data of Gold and Andres (1978a) suggest also that the effects of developing in a family wherein the mother is employed outside the home may involve family relations that differ from those found in families where the mother is not employed outside the home. They report that employed mothers and their husbands had more similar behavior patterns within the home and child-care attitudes than did nonemployed mothers and their husbands. Moreover, there are some data to suggest that mothers of achievement-oriented females take steps to promote independence, rather than dependence, in their daughters (Stein & Bailey, 1973). Since achievement orientations exist among daughters of working mothers, it may be that such interactions exist in these settings. Moreover, the father can promote the nontraditional vocational development of the daughter. The father having high occupational status more often promotes such achievement in his daughter, especially when the daughter is the oldest child or when there is no son (Huston-Stein & Higgins-Trenk, 1978). In summary, there are data to suggest that interaction in family settings having particular characteristics may promote the development of vocational role orientations and behaviors that are nontraditional.

Accordingly, the information reviewed in this section indicates that as a consequence of a particular social change, alteration in the nature of family-child interaction is likely to occur. Clearly these studies of maternal employment indicate that while the child develops in the family, both reside in, and are influenced by, the wider contexts of community, society, and culture.

Child Abuse. Child abuse is "any nonaccidental injury sustained by a child under 18 years of age resulting from acts of commission or omission by a parent, guardian, or caretaker" (Burgess, 1979, p. 1). Because of legal and moral sanctions, as well as lack of clarity of when appropriate physical punishment ends and *abuse* or *neglect* begins, it is difficult to know precisely how many children are abused in their families each year. For example, in 1973 estimates ranged from 41,000 (Light, 1973) to 1,500,000 (Fontana, 1973). More recent estimates continue to be within this wide range (Starr, 1979). Most experts agree, however, that whatever the exact number, the problem is large and quite serious (Garbarino, 1976, 1977; Parke & Collmer, 1975).

The Etiology of Child Abuse. Several explanations have been offered to account for the disturbing social phenomenon of child maltreatment. Probably the most common, and most popular, explanation is one that directs attention to psychological disturbances on the part of the parents, since parents who are abusers are often impulsive, immature, self-centered, and hypersensitive (Parke & Collmer, 1975). Frodi and Lamb (1980), however, offer provocative evidence that in addition to these factors, child-abusing parents may have different physiological reactions to children than do non-child-abusing parents. Frodi and Lamb had fourteen child-abusers and a matched group of nonabusers watch videotapes of crying and smiling infants. Their psychological responses were monitored throughout the session. After each videotape, the subjects described their emotional responses on a mood-adjective checklist. The crying infant elicited heart-rate acceleration and increases in skin electrical activity from both groups, although the abusers experienced greater increases in heart rate and reported more aversion and less sympathy. The nonabusers responded to the smiling infant with no change in or declines in physiological activation. The abusers, however, responded similarly to the smile and cry stimuli.

The fact that many parents who mistreat their children were themselves mistreated during their own childhood suggests further that abusive and neglectful parenting may have learned patterns of functioning (Belsky, 1980; Parke & Collmer, 1975). Indeed, it has been suggested that abusive parents are simply modeling the parenting they received or that because of years of insensitive care, they simply have not learned how to nurture and care for others. They lack the ability to take the point of view of their children and sometimes even expect their children to care for them.

The fact that many parents who were mistreated during their own childhoods do not mistreat their own children demonstrates that a sole focus on the individual parent will not adequately account for the occurrence of child abuse or neglect. Surprising as it may seem, there are reasons to believe that children are often the unintentional instigators of their own poor care. Evidence suggestive of this possibility comes from research indicating that premature and high-risk infants are more likely to be mistreated than are their full-term, healthy counterparts. The reason probably is that high-risk infants are more difficult to care for; they tend to be less responsive, less alert, and even less attractive (Belsky, 1980; Lamb, 1978).

Burgess (1979, p. 11) has noted that other characteristics of the child are associated with his or her abuse. Working from an evolutionary perspective, he suggests that these attributes of the abused child:

> . . . *usually involve either low reproductive potential for those children later in life, e.g., retardation and Down's Syndrome (Martin, Beezley, Conway, & Kempe, 1974; Sandgrund, Gaines, & Green, 1974) or they require effortful and costly care. For instance, Johnson and Morse (1968) reported that 70 percent of their child abuse cases showed some form of developmental problem ranging from poor speech to physical deformities and handicaps. Moreover, even the child welfare workers who dealt with these children found them hard to handle, fussy, demanding, stubborn, negativistic, and unsmiling. As I have noted elsewhere (Burgess, 1979, p. 160), "the child who represents an unwanted pregnancy, or who resembles a disliked or unfaithful spouse, the chronically sick child, the hyperactive child or the otherwise difficult-to-handle child may incite abusive behavior."*

This discussion focuses attention on the role of the parent and the role of the child in the occurrence of child abuse and strongly suggests that a reciprocal, interactive process is responsible for the mistreatment of children.

It is important to recognize that factors beyond the internal workings of families also play a role in child maltreatment. Especially important are sociological characteristics of families like unemployment (Light, 1973), frequent family moves (Gil, 1970), and social isolation from formal and informal support systems (e.g., friends, church). Indeed, there is evidence that indicates that child abuse is correlated with social class (Burgess, 1979; Starr, 1979). Gil (1970) reported that child abuse is more likely in families having a lower socioeconomic status; he noted that more than 48 percent of abusive families had annual incomes less than $5000 (the national percentage of families within this income was 25 percent at the time of Gil's 1970 report). Similarly, Burgess (1979) reviews data indicating that parents whose annual income is less than $6000 report abusing their children at a rate 62 percent higher than other parents; and in data reported by the American Humane Association (1978), insufficient income has been cited as a factor in almost 50 percent of child-abuse and child-neglect cases.

Finally, we must point out that cultural belief systems and values contribute to the child-abuse problem. America's exposure to violence (TV, sports) and its widespread acceptance of physical punishment as an appropriate means of disciplining children illustrate the role that the wider social context can play in creating environments in which child abuse can thrive. The denigration of the role of parent and the popular belief that children belong to parents, to be handled as the family chooses, also contribute to the incidence of child maltreatment (Belsky, 1980).

In summary, characteristics of the parent, the child, as well as of the family setting per se, and the cultural context of childhood seem to be related to the incidence of child abuse. Thus recent reviews of data on the etiology of child maltreatment reach conclusions consistent with the present authors' dynamic interactional model. Lamb (1978) notes, for example, the maladaptive interaction patterns found in families showing child abuse:

> . . . *are dependent on the multiplicative and complex interaction among (a) child characteristics and propensities (as in characteristics of difficult or premature infants); (b) parental personalities and styles; (c) the history of interaction between the infants and their parents; and (d) the nature (be it stressful or supportive) of the wider social context in which the family system is embedded. (Lamb, 1978, p. 156)*

Belsky (1980) offers an *ecological* framework capable of integrating what have often been viewed as divergent explanations of the etiology of child maltreatment. Parents, he argues, bring to the family and their roles as parents developmental histories that may predispose them to treat their offspring in an abusive and neglectful manner, while stress-promoting social forces both within the family (e.g., handicapped child, marital conflict) and beyond it (social isolation, unemployment) increase the likelihood that parent-child conflict will occur. The fact that a parent's response to such conflict takes the form of maltreatment is considered to be a consequence of both the parent's own prior experience as a child and the prevailing values and childrearing practices that characterize the society, subculture, or community in which the child and the family are embedded.

Once we recognize that child abuse is not simply a problem of a disturbed parent, we can move beyond traditional interventions that focus on parental disturbance alone, to propose efforts to attack the child-abuse problem at several levels of analysis (e.g., parent, child, family, community, society). Similarly, once we recognize that it is stress created by divorce or marital conflict that undermines parenting and thereby disrupts child development, we can set as our task the development of support systems to reduce this stress when marital separation threatens the functioning of a child. The rapid increase in the utilization of supplementary child-care services, which we have witnessed in recent years and which has accompanied the equally dramatic increase in the number of working mothers, indicates that society can successfully adapt to social change in a way that need not threaten the quality of persons reared in contemporary American society.

PEER RELATIONS IN CHILDHOOD

A *peer group* can be said to exist at any age level. Often such a social group is defined as all people who are social equals or who are similar on characteristics such as age or grade level (cf. Hetherington & Parke, 1975). Recent definitions of peers, however, have stressed behavioral and/or psychological similarity. Lewis and Rosenblum (1974), for example, suggest that peers are children who interact at about the same level of behavioral complexity.

Willard Hartup, who is one of the leading investigators of the role of peer interaction in child development, has noted that there are no well-formulated theories of peer influence (Hartup, 1970). Nevertheless, there is evidence that peers play an important role in child development.

Functions of the Peer Group

In our preceding discussion of the potential effects of social interaction deprivation, we noted evidence that being deprived of peer interaction can have severe negative effects on emotional and behavioral development. Indeed, Roff, Sells, and Golden (1972) found that poor peer relations in childhood are associated with severe maladjustment in later life. Similarly, Furman and Masters (1980) report that the more disliked a child is by his or her peers, the greater the incidence of social deprivation.

In turn, having exposure to peer interaction seems to aid in a person's socialization, in the person's learning to live by rules of the social groups within which he or she lives. An excellent example of this facilitative influence comes from a study by Furman, Rahe, and Hartup (1979); twenty-four socially withdrawn preschool children were assigned to one of three conditions: (1) socialization experience with a younger child during ten play sessions; (2) socialization experience with an age-mate during a similar series of sessions; and (3) no treatment. The socialization sessions, particularly those with the younger partner, increased the sociability of the withdrawn children in the classrooms.

Reinforcing Functions. These socialization functions of peers seem to occur because peers act as both reinforcing agents and as models for a child's development. In one major study of such functions, Charlesworth and Hartup (1967) studied the daily interactions of nursery school children over a long period. The investigators found that the children tended to reinforce each other's social interaction (for example, by a friendly gesture being met with another friendly gesture); however, at all ages girls tended to reinforce girls more than boys, and boys tended to reinforce boys more than girls. Finally, Charlesworth and Hartup found that reinforcement tended to get reinforcement; that is, they found reciprocity in the children's mutual reinforcements. They also noted that popular children appeared to be most involved in this reinforcement function, and other research has confirmed the relation between peer popularity and reinforcement. Gottman, Gonso, and Rasmussen (1975) studied 198 third- and fourth-graders; popular children distributed and received more positive reinforcements than unpopular children.

Peer reinforcement may foster the development of sex-role behavior. Lamb and Roopnarine (1979) observed the occurrence of sex-stereotyped activities and peer responses to them. Both traditionally sex-appropriate and sex-inappropriate activities occurred. Peer reinforcement occurred more frequently than did punishment, and it was usually administered in support of traditional sex-role behaviors. However, boys were more likely to administer reinforcements than were girls. Those activities that were positively reinforced continued longer than did punished activities. Positive reinforcement for male-typed activities affected boys more than girls; in turn, positive reinforcement for female-typed activities affected girls more than boys.

Age Differences in Peer Interaction. Other studies have confirmed and extended the Charlesworth and Hartup (1967) findings regarding age differences in peer interaction. Reuter and Yunik (1973) found that the amount of peer interaction and the duration of the average social interaction increased with age among 3-, 4-, and 5-year-old nursery school children. Gottfried and Sealy (1974) found analogous results in a study of 3- to 5-year-old mostly black children in a rural Head Start center. Again, older children engaged more frequently in peer social activity. In addition, a sex difference was found—males showed more peer interaction than females within the studied age range.

Findings consistent with the above are also reported by Finkelstein, Dent, Gallacher, and Ramey (1978). Infants and toddlers were observed in a day-care setting. With age, the frequency of teacher-child interaction decreased and peer interaction increased. Finkelstein et al. note that the increase in peer interaction appeared to be related to the toddlers' greater capacity for reciprocal social behaviors. Berndt (1979), studying developmental changes in conformity to parents and peers, reports analogous results. Studying children from the third through the twelfth grades, Berndt found that conformity to parents decreased steadily with age; but conformity behaviors to peers tended to increase until the sixth through the ninth grades, when it seemed to peak.

Finally, Lougee, Grueneich, and Hartup (1977) found age differences in preschool interaction, but here in relation to whether the peer group was composed of same- or mixed-age members. Social interaction and verbal communication were least frequent in younger same-age groups, intermediate in mixed-age groups, and most frequent in older same-age groups.

A child's development often involves interaction in a peer group composed of children of different ages. (Photo by J. V. Lerner.)

Other studies have investigated the nature of peer interaction when the peer group is composed of children of different ages. Results of such studies suggest that individual behavior is enhanced and that total performance of the group (e.g., on tasks of skill) is not decreased when peer groups of differentially aged members interact. For example, recall the study by Furman et al. (1979), discussed above. Another illustration is provided in a study by Graziano, French, Brownell, and Hartup (1976); here several types of groups of three children were studied. There were two types of same-age groups: one was made up of three first-graders (a 111 group), the other of three third-graders (a 333 group). There were two types of mixed-age groups: one was made up of one first-grader and two third-graders (a 133 group), the other of two first-graders and one third-grader (a 113 group). All groups were given a block-building task to perform, i.e., they were asked to build high towers out of the blocks. Overall, Graziano et al. found *no* differences among the four types of peer groups in their total productivity (that is, the total number of blocks stacked) or in their choice of strategies for building a high tower that would not easily fall. Interestingly, however, the third-graders who were in the groups with the two first-graders (i.e., the 113 groups) had higher performance scores than did third-graders in either the 133 or 333 groups. Thus for older children, interaction with younger peers may actually facilitate aspects of their behavioral functioning.

In turn, however, results of another study by Graziano (1978) suggest that different standards of fair play may be used by children in same-age and mixed-age peer groups. Studying first- and third-graders, Graziano (1978) found that younger children use cues associated with older children (i.e., physical size) as a basis for reward deservingness. However, older children base their reward distribution primarily on task performance. Thus first- and third-graders were asked to distribute rewards to two other children, when the relative age, size, and task performance of these two other children were independently manipulated. Graziano found that (1) first-graders were less likely to follow a task-based fair-play rule than were third-graders when one of the other children was larger and older but (2) when dealing with same-size age-mates, first-graders were no less likely to follow a task-based rule than were third-graders.

Reciprocity among Peers. Charlesworth and Hartup (1967) found that peers tended to reciprocate the social reinforcements given to them by their peers. In addition, they found that this mutuality was especially prevalent among children popular in the peer group. Gottman et al. (1975) found similar results. Thus there appears to be a tendency for children to want to reinforce and be reinforced by popular peers.

Other studies have investigated peer reciprocity. For example, Patterson, Littman, and Bricker (1967) studied peer interaction among thirty-six nursery school children. They were particularly interested in the nature of aggressive peer interaction. If a child's aggressive behavior was reinforced by a peer's crying or "giving in," then aggression tended to be strengthened. The child would be likely to aggress again (shove or hit, for example) toward the same peer. However, if the peer countered the child's aggression with reciprocal aggression, then the child's aggression tended to diminish—at least in respect to the reciprocating peer. Alternative, more pacific interactions were subsequently shown to the reciprocating peer; and these, in turn, were returned.

Thus mutual reinforcement and reciprocity serve a socialization function; as illustrated by the Patterson et al. (1967) study, we see that peer interaction can socialize the child into socially acceptable behavior.

Peers as Models. We have noted that peers also serve as models for behavior. Such a function can also socialize the child into socially desired behavior. Hartup and Coates (1967) found that children who observe the behavior of an altruistic classmate tend to imitate the behavior more so than do children who have not experienced this observation.

Other studies show that peers may be especially useful as models. Hicks (1965) assessed the role of peer and adult models of aggression in 3- to 6-year-old children. Children exposed to either model showed more aggression than did children not exposed to a model. Peer models were at least as effective as were adult ones in evoking aggressive responses. Indeed, the most effective model of all was a male peer one.

In sum, peers function to socialize children and may do so in either socially desirable (e.g., altruistic) or socially undesirable (e.g., aggressive) manners. Moreover, the socialization function of peers appears to be especially pronounced among *popular* peers. In the next section we consider the determinants of peer popularity.

Determinants of Peer Acceptance and Popularity

Not all children are equally liked by their peer group. Some children have more friends than others, and some children have a lot of difficulty in establishing and maintaining cordial relationships. We have noted the importance of peers for optimal functioning and have indicated that disruption of peer relations in childhood can have negative consequences for healthy psychosocial functioning in adolescence and adulthood (e.g., Roff et al., 1972). Thus it is important to have knowledge of the variables that may lead to peer acceptance or rejection. Although some studies have indicated that differences in peer popularity are associated with cognitive developmental variables, such as egocentrism (Rubin, 1972), many studies indicate that the variables related to peer popularity are essentially nondevelopmental ones.

Birth Order. Whether you are born as the first, second, or third child in your family is not something you have any control over. It is not a variable that changes with your development. Like the name your parents give you, it is a variable present at birth which is nondevelopmental in nature. Yet differences in peer popularity are associated with different names (Shaffer, 1979) and with one's birth position, or birth order.

There has been a continual interest in the relations of birth order to psychological and social development (see Zajonc & Markus, 1975). Differences between firstborns and later borns in intellectual achievement have been noted, for example. Recently, however, interest in birth order has extended to the study of peer-group popularity. In one major study (Miller & Maruyama, 1976) the relation between birth order and peer popularity was assessed by relating measures of school popularity in friendship, play, and schoolwork situations to birth order in a sample of 1750 grade school children. The children were of different racial-ethnic groups (black, Mexican-American, and white). For the measures of friendship and play popularity, *later-born children* were found to be more popular than their early-born peers.

Body Build. Although a variable more available to change than is birth order, one's body build, or physique, has been found to be related to peer popularity. Researchers have studied the positive and negative attitudes children show to pictures of peers who differ in body type. For example, pictures of a fat (endomorph), average or muscular (mesomorph), and thin (ectomorph) peer are presented and children are asked to "pick the boy (or girl) who best fits" phrases such as "picks the games to play; others like him (or her); most want as a friend; has many friends; be picked leader; kind; and neat"—which are positive characteristics—and "least want as a friend; gets teased; is left out of games; mean; sloppy; sad; has few friends; and not be picked as leader"—which are negative characteristics.

Results have consistently shown that from kindergarten through sixth grade, among males and females, and in cultural settings in the United States, Mexico, and Japan, children attribute mostly positive characteristics to the mesomorph and mostly negative characteristics to the endomorph and to the ectomorph (Lerner, 1969; Lerner & Korn, 1972; Lerner & Iwawaki, 1975; Lerner & Pool, 1972; Staffeiri, 1967, 1972). Moreover, children show differential interpersonal behavior to peers whose body builds differ in the above ways. *Personal space* is the amount of distance one tends to maintain toward another person when interacting with him or her. Evidence exists that the closer one comes to another, the greater one likes that person, whereas the further one stays from another, the greater the degree of dislike toward the person (Meisels & Guardo, 1969). Consistent with the findings about attitudes concerning body build, children show greatest personal space toward chubby males and females and least personal space toward average or muscular males or females—and, again, these findings hold across the kindergarten through sixth-grade age levels and in both United States and Japanese cultural settings (Lerner, 1973; Lerner, Karabenick, & Meisels, 1975; Lerner, Iwawaki, & Chihara, 1976).

Physical (Facial) Attractiveness

Differences in attitudes about and behavior toward body builds may be indicative of a more general set of attitudes and behavioral differences: those shown to children differing in physical attractiveness. Several studies have shown that children who differ in physical attractiveness—usually measured by the attractiveness of their faces—have different types of peer interactions and popularity (Berscheid & Walster, 1974). Indeed, a general summary of these data can be made by the phrase "Beauty is the Best" (Berscheid & Walster, 1974). This stereotype is held by preschool children regarding their peers and, interestingly, by mothers and fathers (and teachers) about some facets of their own children's (or students') behavior (Adams & Crane, 1980).

Interpersonal popularity data derived from nursery school children support this idea. By the end of the preschool term, unattractive children were liked less than were more attractive children (Dion & Berscheid, cited in Berscheid & Walster, 1974). The nursery school children believed that aggressive, antisocial behavior (e.g., hitting and yelling at the teacher) was more characteristic of the unattractive children, while attractive children were seen as more independent, appeared not afraid of anything, and were seen as self-sufficient.

Langlois and Downs (1979) also found some behavioral differences toward peers among physically attractive and physically unattractive children. Although no differences based on attractiveness were found among the 3-year-olds they studied, Langlois and Downs report that 5-year-old unattractive children aggressed against peers more often than did attractive children. Moreover, unattractive children tended to be more active than attractive children.

Similarly, Lerner and Lerner (1977) found that among fourth- and sixth-grade males and females, physical attractiveness was positively related to positive peer relations and was negatively related to negative peer relations. In addition, physically attractive children—that is, the ones who enjoyed this greater peer popularity —were found to be better adjusted than their physically unattractive peers.

As with relations between body build and attitudes, the impact of physical attractiveness on peer popularity appears to hold across ethnic background. Langlois and Stephen (1977) showed black, white, and Mexican-American children photographs of an attractive and unattractive child from each of the three ethnic groups. Each child was asked to rate the pictured child on several characteristics. Children from all three ethnic groups responded primarily on the basis of physical attractiveness. Attractive children were liked more; were seen as being smarter; and were rated higher on sharing and friendliness and lower on meanness and hitting other children than were unattractive children.

Finally, it should be noted that there are data indicating that teachers tend to rate their physically attractive pupils more favorably than their unattractive ones. They see the physically attractive child as having a better personality, as being more adjusted, and as having more academic capability (Adams & Crane, 1980; Clifford & Walster, 1973; Lerner & Lerner, 1977; Rich, 1975; Ross & Salvia, 1975). In fact, there is some evidence that teachers actually give physically attractive children better grades (Lerner & Lerner, 1977).

Rate of Maturation

A final physical variable that is well known to relate to peer popularity is rate of physical maturation. These relations will be discussed in some detail in Chapter 9. Here we need only note that early-maturing males enjoy more peer popularity than do their late-maturing age-mates (Jones & Bayley, 1950; Mussen & Jones, 1957) and that these differences are related to the ones pertaining to body build and physical attractiveness. Early maturers tend to be muscular, or mesomorphic, and this body build is rated as the most physically attractive one among males (Lerner & Brackney, 1978; Lerner & Korn, 1972). In turn, late-maturing males tend to be either thin (ectomorphic) or plump (endomorphic). These body types are not regarded as physically attractive among males—or females, for that matter (Lerner & Korn, 1972; Staffieri, 1972).

In sum, variables often not directly in control of the child are linked to his or her peer popularity. Although there are data indicating that greater peer familiarity or acceptance facilitates the personal (e.g., cognitive) and interpersonal (e.g., helping orientation) behaviors of children (e.g., Doyle, Connolly, & Rivest, 1980; Ladd & Oden, 1979), the point we are here making is that variables such as attractiveness or body type may provide key bases of acceptance and thus opportunity for familiarity. As we have seen, differences in peer popularity amount to differences in such things as which child will be included or excluded from games or from friendships. That is, differences will show themselves in who is chosen to be regarded as a friend, as someone whom the child will regard as a playmate. Differences in whom one will have as a playmate may be quite important in child development because, as we will now see, most of peer interaction occurs in the "arena" of the activity we label as play. That is, "peer interaction" and "play" are substantially overlapping activities in a child's development.

THE ROLE OF PLAY IN CHILD DEVELOPMENT

Although educators, psychologists, and sociologists have been concerned with the topic of play for numerous years (see Weisler & McCall, 1976, for a review), there is still no general consensus as to how to define play. Following other developmental-

Peer popularity can be affected when similarly aged peers mature
at different rates. (Alice Kandell/Rapho/Photo Researchers, Inc.)

ists (e.g., Hetherington & Parke, 1979), however, we will use Dearden's (1967) definition of play as nonserious and self-contained activity engaged in for the sheer satisfaction it brings.

It is known that most social interactions of peers occur in the activity labeled as
play. Indeed, children spend more of their time, outside of school, playing with
friends than involved in any other activity (Hetherington & Parke, 1979). As does
peer interaction, play serves several functions for the developing child.

Functions of Play

Play appears to facilitate the cognitive, social, and emotional development of children. Although these developments are interrelated, it is useful to separate them for
discussion.

Cognitive Development and Play. Both theory and data suggest that play enhances cognitive development. As discussed in Chapter 3, Piaget sees assimilation as the driving force of cognitive development. New stimuli the child encounters as he or she actively *explores* the environment are first assimilated (requiring, of course, an equilibrating accommodation). Piaget labels the dominance of assimilatory activity over accommodational activity as play and thus sees it as a prime component of cognitive growth. Similarly, Berlyne (1966) sees play as an activity which satisfies the child's exploratory drive to search through and discover his or her environment.

There are data which suggest that play does facilitate meeting the challenge of knowing one's environment. For example, in one study Rosen (1974) gave disadvantaged kindergarten children forty days of instruction and practice in a particular type of play. She trained them in sociodramatic play; i.e., she trained them to act out selected social problems. In comparison to children who did not experience this play training, the children whom she instructed showed improvements in problem-solving behavior, e.g., in their effectiveness in solving group problems requiring maximum cooperation and minimum competition. In addition, the trained children improved in their *role-taking* skills—another instance of advanced cognitive functioning.

Similarly, Rubin and Maioni (1975) found that dramatic play in preschooling was related to the ability to correctly classify objects and to perspective taking—other indexes of cognitive advancement. In turn, Burns and Brainerd (1979) found

Play is used to practice future roles. (Elaine M. Ward.)

that giving preschool children either constructive or dramatic play sessions produced substantial and equivalent improvement in several perspective-taking skills. Furthermore, Jennings, Harmon, Morgan, Gaiter, and Yarrow (1979) report that the quality (but not the quantity) of exploratory play among 1-year-olds is positively related to their cognitive development.

Social Development and Play. Rubin and Maioni (1975) also studied the relation between play and peer popularity. Not only was dramatic play associated with cognitive functioning, but it was also related to peer popularity. Those preschoolers who engaged in more dramatic play were also the most popular. Thus play not only is apparently related to those cognitive functions which are necessary if one is going to develop socially—taking the perspective of others, for example—but those children who engage in those play behaviors associated with such cognitive activities appear to be popular among their peers. Indeed, since play is in part imaginery, or assimilatory, activity, cognitive functions that allow the child to practice future roles (e.g., playing a parent), experience roles (e.g., be treated as a spouse by another child), and encounter the feelings of others should be prominent in play activities. Indeed, the Rubin and Maioni (1975) study provides evidence of these relations.

Further evidence for the relation between social development and play comes from a study by Vandell (1979). She compared the mother-son and father-son interactions of toddlers who were completely home-reared with the parent-child interactions of toddlers who were participants in a daily three-hour play group. Although there were no differences in the play-group and home-care toddlers before the play-group experience, significant differences were found after the play-group experience. The play-group children became more active in their parent-child interactions and more responsive to the interaction initiations of their parents. The parents of the play-group children seemed to change too; they became significantly less dominant toward their children.

Thus one implication of the Vandell study is that as a child's play behavior changes, so do the behaviors shown to the child by the parent. Another study (Crawley, Rogers, Friedman, Iacobbo, Criticos, Richardson, & Thompson, 1978) makes a similar point: As the child develops, the structure of mother-child play interactions also changes. Crawley et al. studied 4-, 6-, and 8-month-old infants. Mothers of the youngest group typically played games conducive to the direct stimulation of the infant's attentional and affective responses; however, mothers of the oldest group incorporated games possessing a motor behavior that could be readily assumed by the infant.

Emotional Development and Play. Hartup (1976) has noted that emotional problems in children have been associated with age-inappropriate play, with disruptions in play patterns, and with problems with peer popularity. In turn, of course, typical play behavior may be expected to facilitate both peer-group interaction and the solving of emotional problems. In fact, the relations of play to both cognitive and social development we have discussed would assure this role. In addition, we have reviewed both nonhuman and human data which indicate that when children are denied or do not have peer interaction, and, therefore, play experiences, they develop abnormal behavioral and emotional functions (Roff et al., 1972; Suomi & Harlow, 1972).

In sum, play serves several crucial functions in child development. However, since play is so highly related to other psychosocial developments, we might suspect that play itself undergoes developmental changes. In the next section we discuss evidence indicating that this is indeed the case.

The Development of Play

The types of play behaviors children typically engage in change across their development. Indeed, there seems to be a fairly general sequence in the development of play. A major basis for the identification of this sequence comes from a classic study by Parten and Newhall (1943).

Parten and Newhall studied the nature of social interactions among 2- to 5-year-olds. They discovered, first, that the number of social contacts children made increased with age. In addition, however, they found that the *type* of social interaction children showed also varied in relation to their age. The youngest children spent most of their time engaged in *solitary play*. This involved really no interaction between a child and another. However, older children more often showed a different type of play. They tended to engage in *parallel play*. Here, they played side by side, often engaging in similar behaviors (e.g., putting sand into a pail); but the children engaged in such behavior did not really interact with each other. The oldest children in the study tended to engage most often in truly interactive play. Labeled *cooperative play*, children here engaged in conversation and reciprocal, coordinated exchanges.

Other research has tended to confirm a sequence in the development of play behavior. For example, Eckerman, Whatley, and Kutz (1975) studied the growth of social play with peers during the second year of life. They found that by 2 years of age, social play exceeded solitary play and that the social partner was most often a peer as opposed to a parent. Similarly, Smith (1978) found that among forty-eight preschool children, longitudinally observed, group play increased and solitary play decreased over a nine-month period. However, parallel play did not vary much in overall occurrence, and not all children showed evidence of a solitary-parallel-group sequence. In turn, Rubin, Watson, and Jambor (1978) found that kindergarten children showed less unoccupied or solitary play and more group play and dramatic play than preschoolers. Preschoolers engaged in more solitary and parallel play and in less group play than their kindergarten counterparts.

Social-class status has been found to be related to the rate of development through this developmental sequence. Rubin, Maioni, and Hornung (1976) found that middle-class preschoolers engaged in less parallel play and more associative and cooperative play than did their lower-class age-mates.

In sum, play with peers serves many important developmental functions for the child, and play itself shows a developmental sequence at least in part related to variables (e.g., socioeconomic ones) in the child's broader social context. Thus the broader social context of a child cannot be dismissed in understanding the nature of his or her development; indeed, this is a point we have been emphasizing throughout this text. Accordingly, we now turn to a consideration of another quite prominent part of most children's social context and appraise its impact on child development. We will focus on one of those aspects of the social context that engages much of the child's interactions when the child is not interacting and/or playing with peers or not in school. That is, we consider the influence of television viewing on the child.

INFLUENCES OF TELEVISION ON CHILD DEVELOPMENT

Of all the dimensions of the social context influencing the development of the contemporary child, no dimension is associated so closely with children of the current historical era as is television. Children of other historical eras were exposed to the society outside their own community to some degree; instruments for such mass communications include the media of newspapers, radio, and film. However, only children living in the most recent decades in this society's history have been exposed to television.

This increasingly greater exposure to television may be illustrated by data on ownership of televisions. About 97 percent of all households have a television set and 61 percent of all households have a color set. Moreover, 45 percent of all households have two or more sets (U.S. Bureau of the Census, 1977a).

Variables Related to Television Viewing

Although it is the case that all children now watch more television than in previous decades, television-viewing habits are still related to socioeconomic, intellectual, personality, and developmental variables. In general, television viewing begins at age 2, increases rapidly until age 7, and continues to rise until it peaks at adolescence (Lyle & Hoffman, 1972). Fowles (1975) reports that children as young as 18 months are attentive to television and its material and visual qualities. Cereal companies have learned that 2-year-olds can identify cereal boxes with premiums on them (Choate, 1975). Between 3.5 and 4, children are able to recognize that the cereal is separate from the program.

Schramm, Lyle, and Parker (1961) found that upper-class children, children with higher IQs, and children of parents with higher education watched TV less than lower-class children, children with lower IQs, and children of parents with less education. In a more recent study (Lyle & Hoffman, 1972) similar relations were found. Some personality characteristics also appear to be of importance in television viewing. For example, Murray (1972) found heavy viewing to be related to interpersonal passivity. Schramm et al. (1961) found that in times of personal stress or frustration, television use increases.

In addition to research on the variables related to viewing television, there has been considerable research on the effects of viewing on child development. Research has fallen into three major categories. First, there have been studies of the effects of viewing violence and aggression on the development of similar behaviors. Second, there have been studies of the effects of viewing television shows displaying prosocial behavior (e.g., helping or administering of positive reinforcements) on the development of similar behaviors. Third, there have been studies evaluating the influence of commercial advertisement on children's behaviors. We will review each of these areas separately.

Effects of Viewing Televised Aggression

Two major strategies have characterized empirical attempts to assess whether television viewing of violence is related to violent or aggressive behavior in children. One strategy is an attempt to manipulate experimentally the viewing patterns of children by exposing them to regimens, or "diets," of violent or nonviolent programming in order to see the effects on behavior. A second strategy is to assess the actual viewing patterns of children in order to see whether naturally occurring amounts of violent viewing are associated with aggression. McCall (1977) has argued that although experimental studies show selected variables can influence behavior, they do not necessarily show whether these variables influence behavior in the "real world." Thus, we first review studies that show that viewing television violence indeed can influence the child to behave aggressively; in the next section we consider whether such viewing actually does have this influence.

Effects of Manipulated Television-Viewing Patterns. Although experimental stud-ies are useful—in that they indicate the presence of certain effects under controlled conditions—such studies can be limited in several ways (see Chapter 1). One im-portant limitation is that if samples are not representative, then effects identified through controlled manipulation cannot necessarily be generalized to broad popula-tions. This is a problem in the experimental studies of manipulated television view-ing on children's behavior (Stein & Friedrich, 1975). Most studies have used atypi-cal populations of children (for example, delinquents) and/or have only assessed males. As such, these limitations make it even more necessary to be cautious about generalizing results to children as a whole. Nevertheless, most short-term (for exam-ple, one experimental session) studies are consistent in their findings; they show an increase in aggression after exposure to aggressive films. For instance, Leifer and Roberts (1972) showed one of six television programs, varying in amount of vio-lence, to male and female kindergarten, third-, sixth-, ninth-, and twelfth-grade stu-dents. As the level of viewed violence increased, students indicated on a question-naire that they were more likely to use physical aggression in conflict situations. However, a second study by Leifer and Roberts (1972), and one by Collins (1973), studying fifth- through twelfth-grade males and females, found no difference in the incidence of answers about the likelihood of using violence between those students who saw a violent program and those who saw a neutral travelogue. However, three other studies, assessing only males, found that viewing violent programs led to more violent responses than did viewing neutral programs (Hartmann, 1969; Leifer & Roberts, 1972; Walters & Thomas, 1963).

Although the majority of short-term studies suggest that aggression can be in-creased by viewing violent programs, there are some other short-term studies which do not find this relationship. Similarly, studies which have involved longer-term manipulations (several days or weeks of manipulations) are divided on the presence or absence of such a relation. Feshbach and Singer (1971), Wells (1973), and Parke, Berkowitz, Leyens, West, and Sebastian (1977) studied various groups of children and adolescents, ranging from 8 to 18 years. All studies involved a repeated expo-sure, of from several days (Parke et al., 1977) to several weeks (Feshbach & Singer, 1971; Wells, 1973), to either violent programs or neutral ones. The aggression of the males in these studies was rated in several ways, for example, by peer or counselor behavior ratings or by questionnaire responses. For about half the measures of ag-gression in these studies, it was found that exposure to violent programs led to more aggression than did exposure to nonviolent ones. However, for about the remaining half of the measures, viewing of nonviolent programs led to more aggression than did exposure to violent ones. In fact, in the Wells (1973) study, a thorough investiga-tion involving males from ten schools in different geographical areas and from dif-ferent socioeconomic levels, there were no significant differences in aggression behaviors between males exposed to the two types of program contents.

Accordingly, the results of the above studies of the effects of manipulated view-ing show that children can, at least sometimes, be influenced under somewhat con-trolled conditions to show aggressive behavior in response to viewing violent pro-grams. A major issue (McCall, 1977) now is whether children in fact do exhibit violent or aggressive behavior in the real world in relation to their actual viewing of violent programs. Studies of effects of actual television-viewing patterns provide both indirect and direct evidence that such a real-world relation indeed exists.

Effects of Actual Television-Viewing Patterns. The amount of television actually viewed increases gradually from age 3 years to the beginning of adolescence, after which the total number of hours viewed decreases among high school–age adolescents (Lyle, 1972). Although the absolute number of violent programs watched also decreases during this period, the proportion of programs watched which are violent, and the preference to watch violent programs, increases throughout adolescence (Lyle & Hoffman, 1972).

Although in childhood, males and females watch television about equally, females watch more during adolescence (Lyle, 1972). However, adolescent males still watch more violent programs despite this difference (Stein & Friedrich, 1972, 1975). Although *most* recent research does not find any relation between amount of television watching and either intelligence or school achievement (Stein & Friedrich, 1975), when relationships are found, high amounts of television viewing are associated with low intelligence or poor achievement (Friedrich & Stein, 1973; Stein & Friedrich, 1972). Nevertheless, despite the equivocal nature of the relation between actual television viewing and academic-related functions, there is a far from equivocal connection between actual patterns of viewing violent television and violent and aggressive behaviors.

Children who view a lot of television violence are more likely to approve of it, and to consider it an effective means of conflict resolution, than are children who watch relatively little televised violence (Dominick & Greenburg, 1972). Moreover, children whose favorite programs are violent approve of aggression more than do those with nonviolent program favorites (McIntyre & Teevan, 1972). Since it is known that such aggressive attitudes are positively related to both self-reports and peer ratings of aggressive behavior (McLeod, Atkin, & Chaffee, 1972a, 1972b), this violent viewing may have a great effect on actual aggressive and violent behavior. This possibility is enhanced by the results of several studies which show that children who view a lot of television consider such violence more realistic than do those who watch only a little televised violence (Lefkowitz, Eron, Walder, & Huesmann, 1972; McLeod et al., 1972a, 1972b). Indeed, about one-third of the children studied by McIntyre and Teevan (1972) believed that their favorite television programs were "true to life," regardless of whether these programs were fictional or not. Similar appraisals of the reality of television programs—despite whether they are fictional —also have been found among adults (Gross, 1974).

Thus television viewing of aggression is related to aggressive attitudes and to the belief that the depiction of aggression on television programs reflects real-life behavior. Moreover, these attitudes and beliefs may be at least indirectly associated with aggressive behavior, since aggressive attitudes and aggressive behaviors have been found related. Direct evidence of a relation between violent television and commission of violent behaviors exists also, however.

McLeod et al. (1972a, 1972b) found that the frequency of viewing violence was related positively to self-reported aggression in two samples of male and female sixth- to tenth-graders. In one sample, peer and teacher ratings of aggression were related similarly to the frequency of violent viewing (although parent ratings were not related). Similarly, McIntyre and Teevan (1972) found that the amount of violence in the favorite television programs of junior high school and high school students was related both to self-reports of aggressive behavior and to reports of serious delinquent behavior (such as those requiring police attention).

Furthermore, in a ten-year longitudinal study (Eron, Lefkowitz, Huesmann, & Walder, 1972; Lefkowitz et al., 1972), relating television preference and aggressive behavior (as rated by peers), a positive relation was found between preference for violent television programs and aggressive behavior among third-grade males. More interestingly, however, these males' third-grade preferences for violent television were also related to aggressive behavior ten years later. However, aggressive behaviors at grade 3 and at the ten-year follow-up were not related to preferences for violent viewing at the time of the follow-up. Moreover, television-viewing preferences were not related to aggression among the females studied.

In summary, most of the evidence indicates that there is a relation between the actual viewing of televised violence and the incidence of aggressive behaviors (Stein & Friedrich, 1975). Although certainly not applicable to all children, the recurrence of this relationship in studies assessing effects of actual viewing patterns suggests that altering the incidence of violence on television can change the probability of violent behaviors among today's youth. Thus television—as a most prevalent mass media influence—has a demonstrated empirical effect on children's functioning. This effect can lead to the greater likelihood of either positive or negative social behaviors. Stein & Friedrich (1975, p. 247) have argued that "the responsibility rests heavily on researchers to inform the public that this ready form of relaxation is, in fact, a powerful teacher. . . . They must also be informed of the great potential inherent in the medium for enhancing the quality of life for individuals and for society as a whole." Similar conclusions will be drawn from our review of the other areas of research on the influences of television viewing.

Effects of Viewing Prosocial Television

Although there have been fewer empirical studies of the effects of viewing prosocial television programs than of viewing aggressive or violent television programs, the data that do exist demonstrate "that television and film programs can modify viewers' social behavior in a prosocial direction. Generosity, helping, cooperation, friendliness, adhering to rules, delaying gratification, and lack of fear can all be increased by television material" (Rushton, 1979, p. 345).

In one study having results consistent with Rushton's conclusion, for example, Coates, Pusser, and Goodman (1976) observed preschool children's behavior before, during, and after one week of exposure to two programs: *Sesame Street* and *Mister Rogers' Neighborhood*. Observations consisted of the frequency of the children's giving positive reinforcement and punishment to other children and to adults in the preschool. For children whose initial (previewing) scores were low, *Sesame Street* increased the giving of positive reinforcement and punishment to, and social contacts with, other children and with adults in the preschool. For children whose initial levels of giving reinforcement and punishments were high, *Sesame Street* had no effect on behavior. However, for *all* children *Mister Rogers' Neighborhood* increased the giving of positive reinforcement to other children and adults. On the other hand, other studies (Gorn, Goldberg, and Kanungo, 1976) have shown that being exposed to segments of *Sesame Street* which contain nonwhite children increased 3- to 5-year-old white children's preference for playing with nonwhites as opposed to whites.

Similarly, using first-grade children as subjects, Sprafkin, Liebert, and Poulous (1975) found that children who watched a program from the *Lassie* series, which included a dramatic example of a boy helping a dog, scored higher on a measure of helping than did children who watched other television shows that did not include scenes of helping. Finally, in their study of aggressive and prosocial television programs' effects on preschool children, Friedrich and Stein (1973) found effects of both types of programs. Insofar as the prosocial programs were concerned, children exposed to this type of content showed higher levels of task persistence and of rule obedience and tolerance for delay than children who viewed programs with neutral content. These differences were especially pronounced for children with above-average intelligence.

Friedrich-Cofer, Huston-Stein, Kipnis, Susman, and Clewett (1979) assessed the social, imaginative, and self-regulatory behaviors of 141 children in Head Start centers before and during one of four experimental treatments: (1) watching neutral films on TV; (2) watching prosocial TV; (3) watching prosocial TV, plus being exposed to prosocial play materials; and (4) watching prosocial TV, plus being exposed to prosocial play materials, *plus* having a teacher who was trained for inculcating behaviors related to prosocial functioning. Each treatment lasted for eight weeks.

Exposure to prosocial behavior by itself did not seem to influence behavior. Children who experienced prosocial TV and the related play materials had high levels of positive social interaction with peers and adults, had high levels of imaginative play, and had high levels of assertiveness and of aggressiveness. However, for those students who experienced the prosocial TV plus the prosocial play materials plus the trained teacher, i.e., students in group 4, there was no increase in aggression. But the positive effects seen with the group 3 children—that is, high levels of positive social interaction, of imaginative play, and of assertiveness—were seen. Thus the study suggests that when combined with other features of the social context, prosocial TV can facilitate socially positive behaviors in children.

In sum, these and other studies (e.g., McCall, Parke, & Kavanaugh, 1977) support the position of Rushton (1979, p. 345):

> *Television has the power to influence the social behavior of viewers in the direction of the content of the programs. If, on the one hand, prosocial helping and kindness make up the content of television programming, the audience will come to regard this conduct as appropriate, normative behavior. On the other hand, if antisocial behaviors and uncontrolled aggression are shown frequently, the viewers will think of this kind of action as the norm. This statement should not be surprising. Advertisers spend billions of dollars a year on United States television. They believe, correctly, that brief, 30-second exposures of their product, repeated over and over, will significantly modify the viewing public's behavior in regard to these products. . . . The message therefore is quite clear: Viewers learn from watching television and what they learn depends on what they watch.*

Thus Rushton's (1979) conclusions move us from the effects of television on antisocial or prosocial behavior to the question of the effects of televised advertisements on behavior. Do we learn what the advertiser's message is? Do we believe it? Do we buy or urge others to buy the product? More importantly, what are the answers to these questions in regard to children? Only some of these questions can be answered by existing research.

Effects of Television Commercials

Relative to the billions of dollars that have been spent in attempts to influence children through exposing them to television commercials, there have been few published studies documenting the effectiveness of these exposures. Most studies fall into one of two categories: (1) those assessing children's attitudes toward television commercials and (2) those assessing whether children attempt to buy—or actually influence their parents to buy—the products they see advertised. We will review each of these topics separately.

Children's Attitudes toward Commercials. Studies done on children's attitudes tend to show that younger children and/or children at lower levels of cognitive development tend to have more positive appraisals of commercials. Conversely, older and/or brighter (or more cognitively developed) children tend to dislike and/or disbelieve commercials.

For example, McNeal (1964) investigated the attitudes of children toward advertising; he interviewed 5-, 7-, and 9-year-old children. Subjects were very aware of television advertisements, and there appeared to be an increasing dislike and mistrust of the advertisements among older children. Older children believed that the ads were generally "annoying," "silly," and took too much time from the program in progress. The children who liked the advertisements thought them entertaining, particularly the musical jingles and animated cartoons.

Other studies have generally found that children gradually become more critical in regard to television advertising as they grow older. Garry (1967) reported that preschoolers and primary grade children take what they see on television as real, and in a study that investigated whether or not children judge television advertisements as real, Barcus (1969) found that the "reality" of the ads was negatively correlated with age and IQ but positively correlated with amount of viewing. Lewis and Lewis (1974) reported that, of 208 fifth- and sixth-grade students, 70 percent believed the commercials shown and 47 percent accepted all commercial messages as true.

Ferguson (1975) studied several hundred fourth- and sixth-grade students to assess the relation between cognitive development and the children's attitudes toward commercials. At high levels of cognitive development, attitudes toward commercials tended to be negative, while at low levels, they tended to be more positive. However, there was no consistency of the direction of attitudes among children in the middle levels.

Finally, it should be noted that there is some evidence that children do not particularly attend to television commercials. Zuckerman, Ziegler, and Stevenson (1978) videotaped second-, third-, and fourth-grade children while they were watching a standard fifteen-minute television presentation. The children's behavior was typified by generally low levels of attention to the presentation and by a decline in even this low level of attention during commercials. In addition, the children showed poor memory for the content of the commercials. Whatever attention they showed was more related to recognition of the visual than to the auditory portion of the programming. Finally, Zuckerman et al. found evidence for rapid habituation to the content of the commercials.

Children's Purchase Attempts. A few studies have investigated whether children attempt to persuade their parents to buy advertised products. Generally, the findings indicate that children do make such "purchase-influence attempts," and that such attempts increase in relation to the amount of television viewed by the child.

For example, Wells and LoSciuto (1966) unobtrusively observed shoppers in urban and suburban grocery stores. Their findings indicated that children exerted a strong influence on their parents' purchases, especially at the cereal and candy counters, but even to some degree at detergent displays. For example, 59 percent of all children made purchase-influence attempts for cereal, with 36 percent succeeding. Longstreet and Orme (1967) reported that 70 percent of the children tested asked their parents to buy the products advertised on TV and 89 percent of the parents responded by buying the foods. Ward and Wackman (1971), relying entirely on questionnaire data, found that food products, especially breakfast cereal, candy, and soft drinks, were requested very frequently by children in every age group. Ward and Wackman also found that younger children are more likely to try to influence parental purchases than older ones but, while they tried harder, they were less successful at influencing parents' purchases than older children. Child viewing habits have been correlated with food preferences, nutritional knowledge, and purchase-influence attempts. Sharaga (1974) and Dussere (1974) have both reported that children who watched more television ate more highly sugared cereals, snack products, and highly advertised foods and possessed poorer nutritional knowledge and information.

Galst and White (1976) investigated the relationship between a general measure of the reinforcement value of television advertisements for each child, the amount of television he or she viewed, and the number and fate of his or her purchase-influence attempts in the supermarket. They found that the number of purchase-influence attempts increased with age (from 3 years 11 months to 5 years 11 months) and that the more television the child viewed, the more purchase-influence attempts the child made. Their data also indicated that cereal and candy were the most frequently requested (and most heavily advertised) food products.

Conclusions

All the studies reviewed in this section show that television viewing has a powerful influence on a child's personal and social developments. Given the prominence of television in the social context of today's child, we must agree with Rushton (1979, p. 346):

> *Television is much more than mere entertainment. It is also a source of observational learning experiences, a setter of norms. It helps to determine what viewers will judge to be appropriate behavior in a variety of situations. Indeed, television may well have become one of the major agencies of socialization that our society possesses.*

CHAPTER SUMMARY

In this chapter we discussed key features of the child's social context. We indicated how the child, as a consequence of his or her individuality and activity, as much influences this context as the context influences him or her. We opened our discussion by focusing on personality development in childhood, as it is described by Erikson. As such, we considered the attributes of the genital-locomotor stage and of the latency stage.

The family is the key social institution in most children's development. Studies of families have shown that parents use different techniques to rear their children. Baumrind has identified these three techniques: authoritarian, authoritative, and permissive. These parenting styles are associated with differences in children's behavior development. In turn, as noted, children affect their parents, and in so doing, they provide a basis of feedback to themselves, feedback influencing their own development.

These reciprocal interactions within the family are influenced by events in the broader social context. Changes in this context impacting on the family were illustrated by discussion of the topics of divorce, of maternal employment, and of child abuse.

Peers provide another key feature of children's social context. Peer deprivation can have severe negative and emotional effects on a child's development. In turn, peers can act to reinforce behavior and can serve as models for behavior. Several variables determine peer acceptance and popularity; among these are birth order, body build, physical (facial) attractiveness, and rate of maturation.

A major activity children engage in with peers is play. Play has several functions in a child's development. For example, play is related to cognitive, social, and emotional development. Play itself seems to develop. Children often move from solitary play, to parallel, to cooperative play.

Another major feature of the social context for modern children is television. Viewing television can have both positive and negative effects on a child's development. Viewing of televised aggression can increase aggression in children. In turn, several instances of prosocial behavior (e.g., helping) can be enhanced by viewing television. Similarly, viewing of television commercials can affect children's purchase attempts.

chapter 9

Adolescence: physical and cognitive development

CHAPTER OVERVIEW

Adolescence is a portion of life which involves considerable quantitative and qualitative changes. In this chapter we discuss several of the key features of the most noticeable set of changes involved in adolescence: physical changes. These changes often have important implications for other processes of adolescent development. One of the key ones is cognition. As such, we discuss cognitive development in adolescence.

CHAPTER OUTLINE

A DEFINITION OF ADOLESCENCE

AN OVERVIEW OF THE ADOLESCENT'S TRANSITIONAL EXPERIENCES

PHYSICAL AND PHYSIOLOGICAL CHANGES IN ADOLESCENCE

Neuroendocrine Mechanisms of Adolescent Bodily Change
Prepubescence
Pubescence
Postpubescence
The Role of Sociocultural Variation
The Role of Historical Variation
Individual, Sociocultural, and Historical Implications of Adolescent Bodily Changes
Problems of Physical Maturation and Functioning

COGNITIVE DEVELOPMENT IN ADOLESCENCE

The Psychometric Study of Adolescent Cognition
The Piagetian Approach to the Study of Adolescent Cognition

IS THERE A FIFTH STAGE OF COGNITIVE DEVELOPMENT?

CHAPTER SUMMARY

ISSUES TO CONSIDER

Why is identity a core problem of adolescence?
What are the major periods of adolescent bodily change? What are their characteristics?
What implications do variations in rate of bodily change in adolescence have for personality and social development?
What are the major problems of adolescent physical growth?
Why are eating disorders so problematic in adolescence?
What is the age differentiation hypothesis regarding adolescent intelligence?
What are the major characteristics of formal operational thought?

A DEFINITION OF ADOLESCENCE

*A*n adolescent is not just someone who has attained a given age, reached reproductive maturity, or moved beyond the dependent roles of a child. An adolescent may be all of these things, or none of them, or much more. This is because adolescence involves changes among numerous processes involving the biological, psychological, sociological, cultural, and historical dimensions of existence. These processes change at varying rates, which means that some processes may be making transitions faster or slower than others. A person may have some processes (e.g., cognition) that are adultlike, others (e.g., emotions) which are still childish, and still others (e.g., physical makeup) which are somewhere in between these two. Therefore, to define adolescence, one must look across all the processes that are changing and make a statement about what are considered to be the most frequently occurring characteristics of adolescence.

If one looked at the many processes of a person—the physical, psychological, intellectual, emotional, sexual, educational, social, and moral—and found that most of these existed within the range of changes typically representative of a child, one would label the person a child. If most of these processes existed within the range typically representative of an adult, the person would be labeled an adult. If, however, most of them existed at neither the child nor the adult state but at the state of transition between these two periods, the person would be labeled an adolescent. Accordingly, *adolescent development is that period within the life span when most of the person's processes are in a state of transition from what typically is considered childhood to what typically is considered adulthood.*

Often, many of the specific processes of change in adolescence are associated with a particular disciplinary perspective (e.g., psychological, sociological, biological). Therefore, in this and the next chapter we discuss some of the most important perspectives and consider change processes within each of them. One of our purposes in this presentation is to indicate the role of each of these processes in adolescence and to show the limitations that could accrue if one defined adolescence on the basis of changes associated with any one of these processes *alone*. Another, more salient, purpose is to indicate how the *combination* of the changes involved in these processes makes adolescence a special time of life for the person, a time wherein unique issues confront the young person because of the combined effect of all these changes.

In the next section of this chapter we will try to depict some components of the adolescent experience that occur as a consequence of this combination. Indeed, we will try to explain how the person making the transition from childhood to adolescence is faced with a challenge of re-creating his or her self-definition—a developmental task some scientists regard as the core problem of adolescence—precisely because of the combination of changes involved in adolescence. In succeeding sections of this and the next chapter we will then indicate some of the issues and facts pertinent to many of these change processes. This discussion will result in our returning to the issue of self-definition—or identity—in order to synthesize the information in this and the next chapter and to provide an understanding of some key elements involved in the person's continuing his or her development from adolescence to young adulthood.

AN OVERVIEW OF THE ADOLESCENT'S TRANSITIONAL EXPERIENCES

The most obvious set of changes that the adolescent goes through are the anatomical and physiological ones. New hormones are being produced by the endocrine glands, and these produce alterations in the primary sexual characteristics and the emergence of secondary sexual characteristics. The person thus begins to look different and, as a consequence of the new hormones, feels different. There is also a change in the person's emotional functioning as a consequence of these physical and hormonal changes. New hormones induce sexual changes and produce new drives and new feelings. The new hormonal balance causes the person to feel things that he or she never really felt before. In combination with changing social influences, such as peer pressure, the mass media, and interest in members of the other sex, he or she becomes more sexually oriented.

All these physical, psychological, and emotional changes are complicated by the fact that the person is also undergoing cognitive changes. As noted in earlier chapters, new thought capabilities come to characterize the adolescent. Rather than being tied to the concrete physical reality of what is, the adolescent becomes capable of dealing with hypothetical and abstract aspects of reality. The way the world is organized is no longer seen as the only way it could be. That is, the system of government, the orders of parents, the adolescent's status in the peer group, and the rules imposed on him or her are no longer taken as immutable things. Rather, as the new thought capabilities that allow the person to think abstractly, hypothetically, and counterfactually come into existence, they allow the person to imagine how things could be. These imaginings could relate to government, self, parents' rules, or what he or she will do in life. In short, anything and everything becomes the focus of an adolescent's hypothetical, counterfactual, and imaginary thinking.

For both psychological and sociological reasons, the major focus of the adolescent's concern becomes the adolescent himself or herself. First, psychologically, the adolescent's inner processes are all going through changes, and the physical, physiological, emotional, and cognitive components of the person are undergoing major alterations. Now, with any object being able to be thought of in a new, different, hypothetical way, and with the individual so radically changing, it is appropriate for the person to focus on himself or herself to try to understand what is going on.

The person will ask: "What is the nature of this change? What will it do to me? What will I become? Am I the same person that I think I am?" As all these uncertainties are being introduced, another set of problems is introduced. That is, in regard to sociological functioning, at the very time that the adolescent is having all these concerns—at the very time that he or she may be least prepared to deal with further complications and uncertainties—other factors are imposed. The adolescent in our society typically is asked to make a choice, a decision about what he or she is going to do when grown. Society, perhaps in the form of parents or teachers, asks the adolescent to choose a role; in today's society, as soon as one gets into junior high or early in high school, one may be asked to choose whether a college preparatory program or a non-college preparatory program is desired. Thus at about age 13 years or so, one begins to put oneself on a path that will affect what will be done years and years later.

The point is that at precisely the time in life when adolescents are least ready to make a long-term choice, they often are asked to do so. In order to reconcile all the changes being experienced and to cope with all the demands put on them, American adolescents have to make these choices. To say what they will do, to commit themselves to a role, they have to know what their attributes and capabilities are. They have to know what they can do well and what they want to do. In short, they have to know themselves.

Many adolescents are in a dilemma. They cannot answer the social-role questions without settling other questions about themselves. The answers to one set of questions are interdependent with the answers to the others. A feeling of crisis may emerge, one which requires finding out precisely who one is now. This really is a question of self-definition, of self-identification. Erikson (1959) labels this dilemma the *identity crisis.*

To summarize, the adolescent—because of the impact of all the changes converging on him or her—may be described as in a state of crisis, a state of search for self-definition. Accordingly, the adolescent moves through his or her days attempting to find a place—or role—in society. Such definition will provide a set of rules for beliefs, attitudes, and values (an "ideology") and a prescription for behaviors (a role) that will enable the persons to know what they will "do with themselves" in the world. They will try to find out if they can begin to think like someone who engages in particular behaviors (e.g., being a doctor, a lawyer, a nurse, or a telephone operator). They will attempt to discover if they can adopt the ideology of and acquire the behaviors for particular roles.

The search for identity arises as a consequence of the combined influence of the changes the adolescent undergoes, and in turn, resolution of one's identity is a means of reconciling problems created by the combination of these changes. As noted, we will now consider some of the major facts and issues involved in several of these change processes. Our discussion will necessarily lead to a reconsideration of the issue of identity.

PHYSICAL AND PHYSIOLOGICAL CHANGES IN ADOLESCENCE

As a matter of convenience, the physical and physiological changes of adolescence may be labeled as "bodily" changes. These changes do not begin or end all at once; however, there is a general order to these changes that applies to most, but certainly not all, people (Katchadourian, 1977). Although divisions in this sequence are quite arbitrary because of the fluidity of bodily change (Schonfeld, 1969), it is convenient to speak of phases of bodily changes in adolescence in order to draw important distinctions among various degrees and types of change.

Bodily changes involve alterations in such components of the person as height, weight, fat, and muscle distribution, glandular secretions, and sexual characteristics. When changes in one or more of these characteristics have begun, but the majority of changes that will take place have not yet been initiated, the person may be labeled as being in the *prepubescent* (or prepuberty) phase. When most of those bodily changes that will eventually take place have been initiated, the person is in the *pubescent* (or puberty) phase. Finally, when most of those bodily changes have already occurred, the person is in the *postpubescent* (or postpuberty) phase; this phase ends when all changes are completed.

The bodily changes of adolescence involve a period of alteration of physical and physiological characteristics during which the person reaches an adult level of reproductive maturity. But puberty is not synonymous with all maturational changes. In fact, some authors (e.g., Schonfeld, 1969) see puberty as only one point within the pubescent phase. Puberty is the point at which the person is capable of reproducing (Schonfeld, 1969).

But this point is *not* synonymous with menarche (the first menstrual cycle) in females or with the first ejaculation in males. The initial menstrual cycles of females, for instance, typically are not accompanied by ovulation. Thus the early adolescent female usually is not able to conceive. Ovulation and the ability to conceive, which we have labeled puberty and others sometimes term *nubility,* occur one to three years after menarche (Hafez & Evans, 1973; Montagu, 1946). Even with puberty, however, regular ovulation typically is not established until several years after menarche (Hafez & Evans, 1973). Nevertheless, exceptions are increasingly evident. In one recent year (1975), for example, 30,000 females age 14 and under became pregnant in the United States (Alan Guttmacher Institute, 1976).

Similarly, for males there is a period of time between the first ejaculation, which usually occurs between 11 and 16 years of age (Kinsey, Pomeroy, & Martin, 1948) and the capability of fertilizing. At the time of first ejaculation, the male is usually sterile, and it is not for one to three years after this time that there are enough motile sperm in the ejaculate so that the male could be fertile (Schonfeld, 1969).

Thus neither menarche, the first seminal emission, nor puberty itself is commensurate with all bodily changes associated with the prepubescent, pubescent, and postpubescent phases of adolescence. In the following sections some of the major changes associated with each of these phases will be detailed. These changes will relate to both the *primary* and the *secondary* sexual characteristics of the person. Primary sexual characteristics are present at birth and involve the internal and external genitalia (e.g., the penis in males and the vagina in females). Secondary sexual characteristics are those which emerge to represent the two sexes during the prepubescent through postpubescent phases (e.g., breast development in females and pigmented facial hair in males). Moreover, these changes all involve nervous system and endocrine system mechanisms. After these general *neurohormonal* processes involved in all phases of adolescent bodily change are briefly described, some of their outcomes will be considered.

Neuroendocrine Mechanisms of Adolescent Bodily Change

There are two types of glandular systems. *Exocrine* glands have an opening (a duct) to the world outside the body. Examples of such glands are the salivary and the sweat glands. *Endocrine* glands, however, have no such external duct. Rather, they secrete a substance, *hormones*, directly into the bloodstream. The role of the endocrine glands is most important in understanding the changes associated with adolescent bodily changes.

The hormonal influences on adolescent bodily change begin to occur on the basis of stimulation from the central nervous system. On the basis of stimulation from some still unknown higher, cerebral cortex brain center, there is stimulation sent to a structure of the brain termed the *hypothalamus*. The hypothalamus in turn stimulates a major endocrine gland, the *pituitary*. More specifically, the front part of the pituitary (the *anterior*) is stimulated. In other words, the hypothalamus stimulates the anterior pituitary, and as a consequence, the anterior pituitary gland secretes certain hormones.

First, a *growth-stimulating hormone* (sometimes labeled as *GSH* or termed *somatotropin*) is secreted. This hormone acts on all body *(soma)* tissues to stimulate their rate of growth and nourishment *(trophe)*. A second type of hormone that is secreted, which is also a *trophic* hormone, stimulates specific tissues. In fact, the role of such hormones is to stimulate other endocrine glands to produce their own specific hormones. Although there are several trophic hormones released by the pituitary in order to initiate adolescence (Katchadourian, 1977; Schonfeld, 1969), two specific trophic hormones, *gonadotropic* hormones, act on these glands in males and females that are associated with each sex, respectively. These glands are the testes in males and the ovaries in females, and they are termed *gonads*. *Follicle-stimulating hormone (FSH)* encourages ovulation in the female and spermatogenesis in the male. In addition, it encourages release of estrogen in the female. *Luteinizing hormone (LH)* also encourages ovulation in the female and testicular development in the male. Furthermore, LH encourages the secretion of *progesterone* in the female and *testosterone* in the male. Thus gonadotrophic hormones produced by the anterior pituitary act specifically on the gonads of males and females, inducing the endocrine glands to produce particular hormones. These latter hormones provide a basis for the sexual maturation of males and females.

The hormones secreted by the gonads most generally are termed *androgens* (for males) and *estrogens* (for females). Their relative concentration in the body influences the nature of the changes we associate with prepubescence through postpubescence. The changes we typically find with males result from greater concentrations of androgens than estrogens, while more estrogens than androgens are necessary to see the changes we typically find with females. However, it should be noted that males and females have both types of hormones and that both play a role in fostering changes across the phases of adolescent sexual maturation. For instance, for both males and females, androgens initially stimulate growth and muscle development, but both androgens and estrogens are eventually needed to terminate growth (Schonfeld, 1969).

In sum, through the functioning of these hormones, the alterations associated with bodily change in adolescence occur; as such, sexual maturation proceeds. Interplays between gonadotrophins and androgens and estrogens lead to the emergence of mature, reproductively fertile sex cells in males and females: the sperm of the male and the ovum of the female.

Prepubescence

Because of the general orderliness of the sequence of bodily changes in adolescence, Schonfeld (1969) suggests that the beginning and end of prepubescence can be marked by the first evidence of sexual maturation and the appearance of pubic hair, respectively. In males, there is a progressive enlargement of the testicles, an enlargement and reddening of the scrotal sac, and an increase in the length and circumference of the penis (Schonfeld, 1969, p. 29). These changes all involve primary sexual characteristics. Insofar as secondary sexual characteristics are concerned, no true pubic hair is present, although the male may have downy hair.

In females, the changes marking prepubescence typically begin an average of two years earlier than with males. At the level of the cells and tissues of the body, the first sign of female development in this phase is the enlargement of the ovary and the ripening of the cells (termed *primary graffian follicles*) that will eventually evolve into mature reproductive cells *(ova)*. However, in contrast with males, the female changes in primary sexual characteristics typically are not visible for clinical or scientific inspection, and as such, changes in secondary sex characteristics are the indexes often used to mark prepubescent changes. In this phase there is a rounding of the hips and the first visible sign of breast development. The rounding of the hips occurs as a result of increased deposit of fat, while the first visible signs of breast development are marked by an elevation of the areola surrounding the nipple (Schonfeld, 1969). This raising produces a small conelike protuberance, known as the breast "bud." As with the male, there is no true pubic hair, although down may be present.

Pubescence

Schonfeld (1969) uses as an index of the onset of the pubescent phase the appearance of pubic hair and, in turn, defines the end of the major phase as when pubic hair development is complete. Whether or not one agrees with the appropriateness of these indexes, there is wide agreement that such pubic hair changes are important markers of the range of changes characteristic of this phase of major bodily changes (Katchadourian, 1977). Exhibit 9.1 shows the progressive changes involved in pubic hair development for adolescent females, and Exhibit 9.2 shows these changes for adolescent males.

In addition to pubic hair development, pubescence is marked by a growth spurt. The peak velocity of growth in height and weight occurs during this phase. As seen in Exhibit 9.3, there is a marked increase in the amount of height gained in the age range typically associated with pubescence for both males and females in the United States today. A corresponding velocity of weight gain in males and females is presented in Exhibit 9.4. In both Exhibits 9.3 and 9.4 it should be noted that the components of the adolescent growth spurt occur about two years earlier in females than in males. Exhibit 9.5 also shows the relative changes in growth velocity from infancy through the adolescent growth spurt and on to maturity. This figure is based on North American and western European populations. The growth rates for males and females are shown at regular intervals from infancy to maturity, and the relative appearance of each sex as a consequence of this growth is illustrated.

Other changes characterize people during pubescence. Menarche occurs in females, usually about 18 months after the maximum height increment of the growth

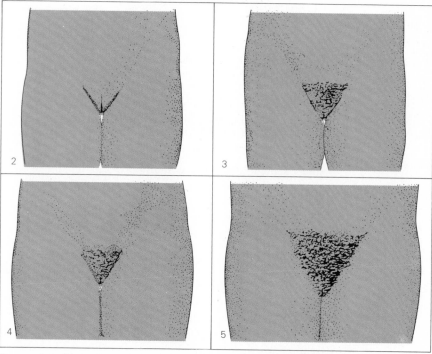

EXHIBIT 9.1   Stages of pubic hair development in adolescent girls: (1) early prepubescent (not shown) in which there is no true pubic hair; (2) late prepubescent–early pubescent in which sparse growth of downy hair is mainly at sides of labia; (3) pigmentation, coarsening, and curling with an increased amount of hair; (4) adult hair, but limited in area; (5) adult hair with horizontal upper border. (Source: Redrawn from M. M. Tanner, *Growth at Adolescence*, 2d ed., Oxford: Blackwell, 1962.)

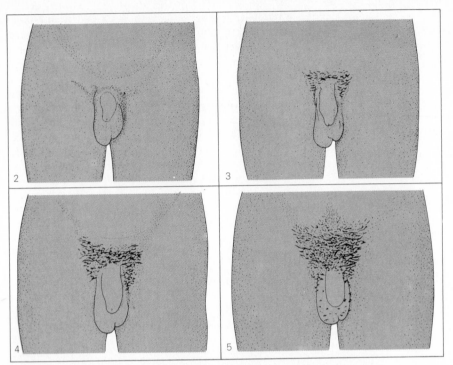

EXHIBIT 9.2   Stages of pubic hair development in adolescent boys: (1) early prepubescent
(not shown) in which there is no true pubic hair; (2) late prepubescent–early
pubescent in which sparse growth of downy hair is mainly at base of penis; (3)
pigmentation, coarsening, and curling with an increase in amount of hair; (4)
adult hair, but limited in area; (5) adult hair with horizontal upper border and
spreading to thighs. (Source: Redrawn from J. M. Tanner, *Growth at Adolescence*, 2d ed., Oxford: Blackwell, 1962.)

spurt (Schonfeld, 1969). In addition to this functional change, there are structural
alterations in the primary sexual characteristics of females. For instance, the vulva
and clitoris enlarge. Moreover, in regard to the secondary sexual characteristics of
females, there is increased breast development; there is a change from having breast
buds to having the areola and nipple elevated to form the "primary" breast (Schonfeld, 1969). This range of changes is described and illustrated in Exhibit 9.6.

The primary and secondary sexual characteristics of males also change during
pubescence. With regard to secondary characteristics, there is a deepening of the
voice, due to growth of the larynx. Both pigmented axillary and facial hair appear,
usually about two years after the emergence of pubic hair (Schonfeld, 1969). With
regard to primary sexual characteristics, the testes continue to enlarge, the scrotum
grows and becomes pigmented, and the penis becomes longer and increases in circumference (Schonfeld, 1969). These changes are described and illustrated in Exhibit 9.7

In summary, the pubescent phase of adolescent development involves a myriad
of structural and functional changes in the primary and secondary sexual characteristics of both sexes. Although not all such changes have been described, what most
authorities see as the major ones have been depicted (Katchadourian, 1977; Schonfeld, 1969; Tanner, 1973).

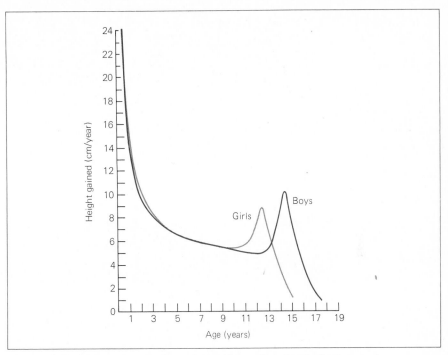

EXHIBIT 9.3   Typical individual curves showing velocity of growth in *height* for boys and girls. (Source: Redrawn from J. M. Tanner, R. H. Whitehouse, and M. Takaishi, *Archives of Diseases in Childhood*, 41:466, 1966.)

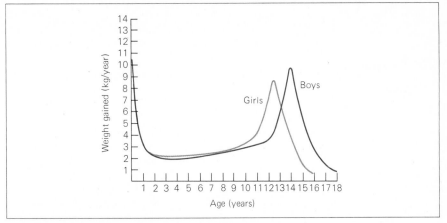

EXHIBIT 9.4   Typical individual curves showing velocity of *weight* gain in boys and girls. (Source: Redrawn from J. M. Tanner, R. H. Whitehouse, and M. Takaishi, in *Archives of Diseases in Childhood*, 41:466, 1966.)

Postpubescence

According to Schonfeld (1969, p. 28), the postpubescent phase of adolescence may be marked as starting when pubic hair growth is complete, when there is a deceleration of growth in height, and when there is completion of changes in the primary and

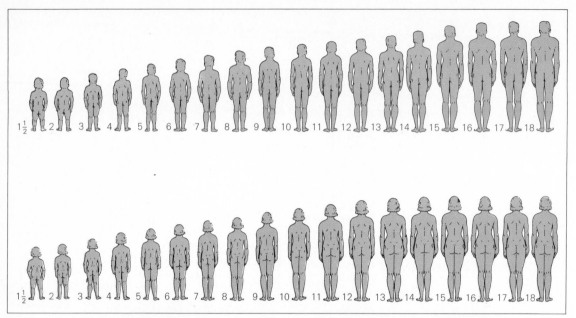

EXHIBIT 9.5   Rates of growth in development of boys and girls, shown at regular intervals
from infancy to maturity. Note the change in the form of the body as well as
the increase in height. (Source: Redrawn from J. M. Tanner, *Growing Up*.
Copyright 1973 by Scientific American, Inc. All rights reserved.)

secondary sexual characteristics and the person is fertile. Despite the extent of the
completed changes, there continue to be changes in the primary and secondary
sexual characteristics. In males, the beard usually starts to grow in this phase; for
females, there is further growth of axillary hair and breast development.

Thus the physical and physiological changes characterizing adolescence involve much more than just puberty; they include alterations spanning perhaps an
entire decade of the life span and involve at least three phases of bodily change.
Exhibit 9.8 summarizes the normative sequence of bodily changes for males in current United States society, and Exhibit 9.9 presents corresponding information for
females. While the information in these two exhibits represents norms generalizable across most segments of current United States society, there exist important
qualifications of these data. There is in some cases considerable variation in the rate
and quality of changes involved in all phases of adolescent bodily change, and this
variation is associated with both sociocultural and historical differences. These
sources of variation are considered in the succeeding two sections.

The Role of Sociocultural Variation

There are abundant data indicating that the body changes associated with adolescence vary in relation to their sociocultural environment. To illustrate, data pertinent to one aspect of bodily change—menarche—will be considered.

As depicted in Exhibit 9.10, there is a vast difference in the age of menarche in
different cultures and, further, in different social strata within a culture. Thus the

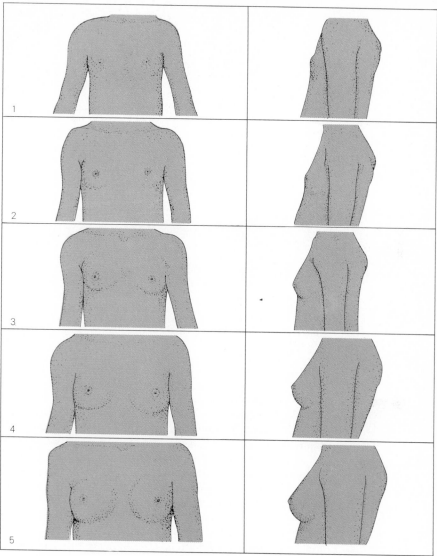

EXHIBIT 9.6   Stages of breast development in adolescent girls: (1) prepubescent flat appearance; (2) small, raised breast bud; (3) general enlargement and raising of breast and areola; (4) areola and nipple form contour separate from that of breast. (Source: Redrawn from J. M. Tanner, *Growth at Adolescence*, 2d ed., Oxford: Blackwell, 1962.)

median age at menarche for Cuban blacks or whites is 12.4 years—the lowest age listed in Exhibit 9.10—while the median age for the Bundi tribe (people of New Guinea) is 18.8 years—the highest age shown. Not only do the data of Exhibit 9.10 underscore that age per se is an inappropriate focus in an attempt to understand development, but also these data indicate that sociocultural differences are not related to race. As observed by Katchadourian (1977), for example, blacks were observed to have both early (in Cuba) *and* late (in New Guinea) maturation.

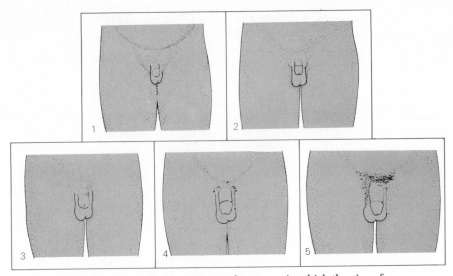

EXHIBIT 9.7   Stages of male genital development: (1) prepubescence, in which the size of
the testes and penis is like that in early childhood; (2) testes become larger
and scrotal skin reddens and coarsens; (3) continuation of stage 2, with lengthening of penis; (4) penis enlarges in general size and scrotal skin becomes pigmented; (5) adult genitalia. (Source: Redrawn from J. M. Tanner, *Growth at
Adolescence*, 2d ed., Oxford: Blackwell, 1962.)

  In addition, a study by Halbrecht, Sklorowski, and Tsafriv (1971) shows that significant differences exist within a given cultural group. In Israel, the age of menarche was found to be different for females born in Europe or the United States, *or*
whose father had been born in these different settings.

  The variables accounting for these differences might relate to nutritional or
health-care differences associated with various sociocultural settings. There is evidence for this view. Differences in age of menarche within the same country are
often seen between urban and poor rural areas. Tanner (1970) reports that in Ru-
mania the average age of menarche is 13.5 years in towns and 14.6 years in villages.
Corresponding differences between town-reared and rural-reared children have
been found in the Soviet Union (13.0 and 14.3 years, respectively) and in India (12.8
and 14.2 years, respectively). In all these contrasts, nutritional resources are less
abundant in the rural setting. That such scarcity generally may delay menarche is
possible in light of the fact that females who have many siblings (and thus a higher
probability of scarce resources) also have delayed menarche.

  Differential availability of resources in different socioeconomic strata within a
society is also associated with variation in age of menarche. In Hong Kong the age at
menarche is 12.5 years for children of the rich, 12.8 years for those of average income levels, and 13.3 years for the children of poor families (Tanner, 1970). However, different studies of the relation between age of menarche and social class define social class differently. Thus there is often a situation where the "poor" in one
country may be starving, while the poor in another society may have relatively adequate nutrition (Katchadourian, 1977). Nevertheless, a relation between socioeconomic status and menarche typically is found. Thus in European countries, the
range of socioeconomic status from rich to poor is associated with a range of about
two to four months in age of menarche (Katchadourian, 1977).

EXHIBIT 9.8 **Normative Sequence of Bodily Changes in Males**

Phase	Appearance of Sexual Characteristics	Average Ages	Age Range*
Childhood through pre-pubescence	Testes and penis have not grown since infancy; no pubic hair; growth in height constant; no spurt.		
Prepubescence	Testes begin to increase in size; scrotum grows, skin reddens and becomes coarser; penis follows with growth in length and circumference; no true pubic hair; may have down.	12–13 years	10–15 years
Pubescence	Pubic hair is pigmented, coarse and straight at base of penis, becoming progressively more curled and profuse, forming at first an inverse triangle and subsequently extending up to umbilicus; axillary hair starts after pubic hair; penis and testes continue growing; scrotum becomes larger, pigmented, and sculptured; marked spurt of growth in height with maximum increment about time pubic hair first develops and decelerates by time fully established; spontaneous or induced emissions follow but spermatozoa inadequate in number and motility (adolescent sterility); voice beginning to change as larynx enlarges.	13–16 years	11–18 years
Postpubescence	Facial and body hair appear and spread; pubic and axillary hair become denser; voice deepens; testes and penis continue to grow; emission has adequate number of motile spermatozoa for fertility; growth in height gradually decelerates, 98 percent of mature stature by 17.75 years ± 10 months; indentation of frontal hair line.	16–18 years	14–20 years
Postpubescence to adult	Mature, full development of primary and secondary sex characteristics; muscles and hairiness may continue increasing.	onset 18–20 years	onset 16–21 years

* This age range includes 80 percent of cases.
Source: Adapted from Schonfeld (1969, p. 30).

The Role of Historical Variation

Adolescent bodily changes are also related to history. This relation, often termed the *secular trend* (Katchadourian, 1977; Muuss, 1975b), can be illustrated by considerable data. In addition, it may be noted that since 1900, children of preschool age have been taller on an average of 1 centimeter and heavier on an average of 0.5 kilo-

EXHIBIT 9.9   **Normative Sequence of Bodily Changes in Females**

Phase	Appearance of Sexual Characteristics	Average Age	Age Range*
Childhood through pre-pubescence	No pubic hair; breasts are flat; growth in height is constant; no spurt.		
Prepubescence	Rounding of hips; breasts and nipples are elevated to form "bud" stage; no true pubic hair; may have down.	10–11 years	9–14 years
Pubescence	Pubic hair is pigmented, coarse, and straight primarily along labia, but progressively curled, and spreads over mons and becomes profuse with an inverse triangular pattern; axillary hair starts after pubic hair; marked growth spurt with maximum height increment 18 months before menarche; menarche: labia becomes enlarged, vaginal secretion becomes acid, breast development begins with areola and nipple elevated to form "primary" breast.	11–14 years	10–16 years
Postpubescence	Axillary hair in moderate quantity; pubic hair fully developed; breasts fill out, forming adult-type configuration; menstruation well established; growth in height decelerates, ceases at 16.25 ± 13 months.	14–16 years	13–18 years
Postpubescence to adulthood	Further growth of axillary hair; breasts fully developed.	onset 16–18 years	onset 15–19 years

* This age range includes 80 percent of cases.
Source: Adapted from Schonfeld (1969, p. 33).

gram per decade (Katchadourian, 1977). Furthermore, the changes in height and weight occurring during the adolescent growth spurt have involved increased gains of 2.5 centimeters and 2.5 kilograms, respectively (Falkner, 1972; Katachadourian, 1977). It has been noted that these changes occur in the context of increasingly earlier ages for the start of adolescent bodily changes (Muuss, 1975b; Schonfeld, 1969). Thus these alterations happen among present birth cohorts during a pubescent period, which both begins and ends earlier in life than was the case with past birth cohorts. Because adolescent bodily change stops earlier and earlier as history progresses, the absolute increases in final adult height and weight across history are not great (Bakwin & McLaughlin, 1964; Maresh, 1972). Yet there is no final decision about whether the historical trend for increasingly earlier adolescent bodily changes is continuing (Katchadourian, 1977; Schonfeld, 1969; Tanner, 1962).

Plasticity in aspects of bodily changes continued to characterize pubescent adolescents in some sociocultural settings, at least through the early 1970s, as illus-

EXHIBIT 9.10   **Median Age at Menarche in Several Populations**

Population or Location*	Median Age (Years)	Population or Location	Median Age (Years)*
Cuba		Tel Aviv, Israel‡	13.0
Negro	12.4	London, U.K.	13.1
White	12.4	Assam, India (city dwellers)	13.2
Mulatto	12.6	Burma (city dwellers)	13.2
Cuba†		Uganda (wealthy Kampala)	13.4
Negro	12.9	Oslo, Norway	13.5
White	13.0	France	13.5
Mulatto	13.0	Nigeria (wealthy Ibo)	14.1
Hong Kong (wealthy Chinese)	12.5	U.S.S.R. (rural Buriats)	15.0
Florence, Italy	12.5	South Africa (Transkei Banfu)	15.0
Wroclaw, Poland	12.6	Rwanda§	
Budapest, Hungary	12.8	Tutsi	16.5
California, U.S.A.	12.8	Huru	17.1
Colombo, Ceylon	12.8	New Guinea (Bundi)	18.8
Moscow, U.S.S.R.	13.0		

* Data from Tanner (1966b) unless otherwise indicated.
† Pospisilova-Zuzakova, Stukovsky, and Valsik (1965).
‡ Ber and Brociner (1964).
§ Hiernaux (1965).
Source: Jean Hiernaux, "Ethnic Differences in Growth and Development," *Eugenics Quarterly*, 1968, 15:12–21. Copyright 1968 American Eugenics Society.

trated by the data in Exhibit 9.11. Here, the declining age of menarche across history in different countries is illustrated. In Norway, for example, it may be noted that females of the 1840 birth cohort reached menarche at about 17 years, while Norwegian females in more current cohorts reach this point about four years earlier in life. The rate of decreased age of menarche in this country has been about four months per decade (Muuss, 1975b). Similarly, it may be noted in the exhibit that in the United States, women of a birth cohort around 1890 had menarche at about age 14 years. Today, however, females in the United States reach this point between one and two years earlier.

In sum, the bodily changes of adolescence are changeable phenomena, apparently quite responsive to influence by variables associated with sociocultural and historical variation. These bodily changes in turn may reciprocally impact on these other factors which influence them, and it may be that the *nature and rate* of adolescent bodily change have important implications for the individual, the society, and history. This idea is evaluated in the following sections of this chapter.

Individual, Sociocultural, and Historical Implications of Adolescent Bodily Changes

The quality and rate of any aspect of individual development influence the person. The type of thinking ability one possesses (e.g., whether it is concrete or abstract), and how quickly one changes from one type of thought to another (e.g., whether one attains the ability to think abstractly sooner than peers do), may be expected to influence a person's functioning. Similarly, the type of body characteristics one attains as a consequence of the prepubescent through postpubescent changes (e.g., whether

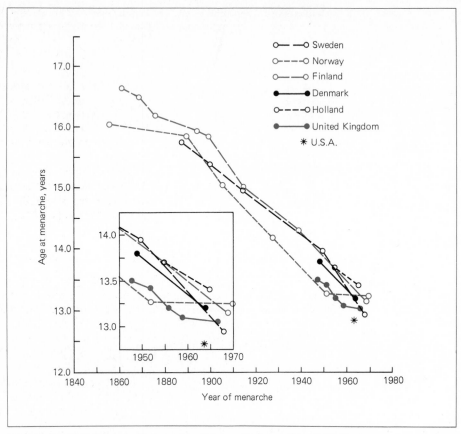

EXHIBIT 9.11   Historically declining age of menarche. (Source: Redrawn from J. M. Tanner, *Foetus into Man: Physical Growth from Conception to Maturity.* Cambridge, Mass.: Harvard University Press, 1978.)

the body is fat or muscular), and how fast one proceeds through these phases of adolescent bodily change (e.g., whether one has slower or faster breast development than one's peers), will influence one's personal and social functioning. Thus the inference that the nature and rate of bodily changes should influence psychological and social levels of analysis is just one instance of a more general view that the types and rates of all developmental changes are important parameters of human functioning.

This general view has a long tradition of study within adolescence, insofar as it has been raised in the context of bodily changes. Generally, the study of the issue has fallen under the label of the assessment of the implications of *early versus late maturation*. Adolescent bodily changes involve, in a functional sense, an alteration toward attainment of reproductive capability; this is an adaptive capability which obviously allows the species to survive. When an organism is capable of reproductive function, it is termed *mature*. Thus the physical and physiological changes associated with the prepubescent through postpubescent phases are labeled *maturational*, in that they are involved with reproductive capability.

However, this derivation of the term is stressed to indicate that when one speaks of maturational changes, one is not necessarily implying a solely nature-

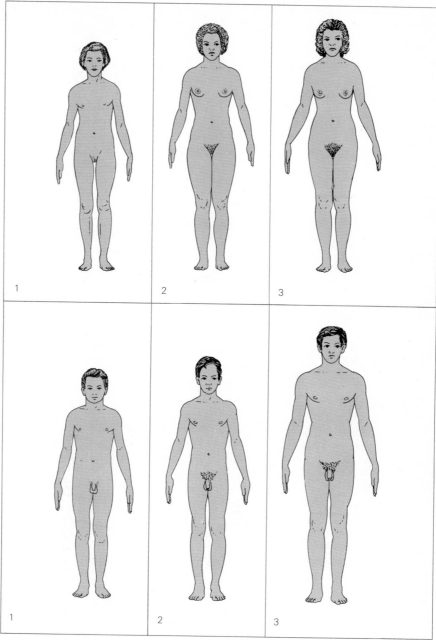

EXHIBIT 9.12   Variations in pubescent development. All three females are 12.75 years old and all three males are 14.75 years old, but they are in different phases of change. (Source: Redrawn from J. M. Tanner, *Growing Up*. Copyright 1973 by Scientific American, Inc. All rights reserved.)

based phenomenon. In an earlier chapter it was suggested that many people mistakenly have considered that any change, or *growth* (a term *usually* associated with changes in structure, as opposed to function), which could be labeled as maturational was a nature phenomenon. Not only are there arguments and data that suggest the limitations of such a view; but the data reviewed in preceding sections of this chapter—regarding the sociocultural and historical influences on bodily changes— also speak against the view.

Simply, it is our position that although prepubescent through postpubescent bodily changes may be labeled as maturational in order to indicate their association with the attainment of reproductive capability, the use of the term maturation does in no way imply a solely nature-based phenomenon. Indeed, it has been argued that all changes, be they associated with adolescent maturations or with maturation at any time in life, are products of interactions among nature and nurture processes.

Thus, the study of *early maturers* focuses on people whose nature-nurture interactional history has led them to go through the phases of adolescent bodily changes faster than usual. They show instability in that their bodily changes relative to their same-age peers are more rapid. In turn, the study of *late maturers* assesses people whose interactional history has resulted in passage through the phases of adolescent bodily change slower than usual. Later maturers also show instability; but here it is because their bodily changes relative to their same-age peers are slower. Although of the same age, the early maturer is accelerated in bodily growth (he or she is in a more advanced change phase), while the late maturer is delayed (he or she is in a less advanced phase, although of the same age).

Such differences in rate of development lead to quite distinct differences in bodily appearance. As seen in Exhibit 9.12, all the females are of the same age, but 1 has the characteristics of prepubescence, 2 of pubescence, and 3 of postpubescence. Similar differences are seen with the three identically aged males shown in the exhibit. Although the adolescents depicted in Exhibit 9.12 are shown at one point in their development, the contrasts in bodily appearance between early and late maturers become increasingly more pronounced over time. These developmental differences are shown for males in Exhibit 9.13 and for females in Exhibit 9.14.

It should be noted that at any given age in adolescence, it is possible to describe in different ways the *types* of bodies early and late maturers possess as a consequence of their contrasting rates. For example, the early-maturing male at age 16.5 has a body build which is muscular and which appears strong and athletic looking (see the first 16.5-year-old male in Exhibit 9.13). However, the most extreme late-maturing male at this time has a body type which is not muscular (see the second 16.5-year-old male in Exhibit 9.13). Indeed, such extreme late maturers often have a thin body type and a frail appearance (Jones & Bayley, 1950), often seen as "not having filled out." In addition, other late maturers—but not to the extreme extent of the first group—have a plump body (Tanner, 1962) and are regarded in these cases as "not having lost their baby fat." With females, early maturation typically results in a bodily appearance that combines both muscularity with some fatness, while late maturing is associated with thinness, as in males (see the two 14.5-year-olds, respectively, of Exhibit 9.14). These contrasts in the physical appearance of the body, in the type of build resulting from bodily changes, will be important when the inference that both the types and rates of bodily change in adolescence affect individual and social functioning is considered.

To evaluate this inference it should first be noted that there has not been a great amount of research on the psychosocial implications of early versus late maturation. What data do exist come mostly from longitudinal studies conducted at the Univer-

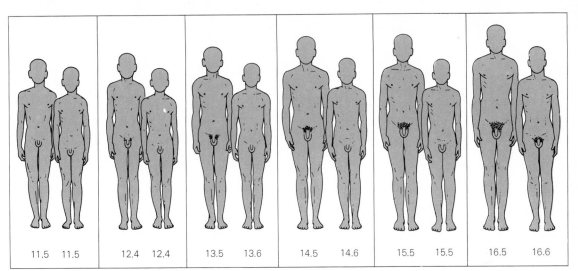

EXHIBIT 9.13   Contrasts between early- and late-maturing boys, aged 11.5 to 16.6 years.
(Source: Drawing based on photograph in F. H. Shuttleworth, *The Adolescent
Period: A Pictorial Atlas*, monograph of the Society for Research in Child De-
velopment, 1951, 14, Serial No. 50, 1949, p. 24.)

sity of California at Berkeley. These studies, labeled the California Growth Studies,
have the limitations of all studies with longitudinal designs. They involve a confounding of age and time effects, they may have biased samples, and there may be
effects associated with repeated testing.

Moreover, the data about early and late maturers from the California Growth
Study involve extremely small samples of subjects in most cases. Further, these
samples consist mostly of middle-class persons. Thus the data from these studies
can, at best, serve as a tentative basis for generalizations. Because there are important differences in the nature of the findings associated with early and late maturation in males and females, the data for each sex will be separately considered.

Implications of Early versus Late Maturation in Males. In one of the first studies of
early and late maturation derived from the California Growth Study, Jones and Bayley classified a group of ninety males on the basis of their rate of maturation. To
judge this rate, x-rays were taken of bones in various parts of the body. This index
reliably indicates the phase of adolescent bodily change and corresponds to other
measures of such change, including phase of pubic hair growth or testicular enlargement (McCandless, 1970). It is interesting to note that the use of this x-ray index of
maturation level has varied across history. Researchers in current cohorts recognize
the possible dangers of x-ray use more so than those who worked in the 1930s, when
the subjects of the California Growth Study were children and adolescents
(McCandless, 1970), and, therefore, use less exact measures of rate of maturation.
For example, indexes relating to pubic hair growth and even just general physical
appearance are used.

Nevertheless, on the basis of their x-ray indexes, Jones and Bayley (1950) were
able to find two groups within the original sample of ninety males who most consistently were accelerated (those whose maturation rate was in the top 20 percent of the
distribution) and who most consistently were delayed (and thus in the lowest 20

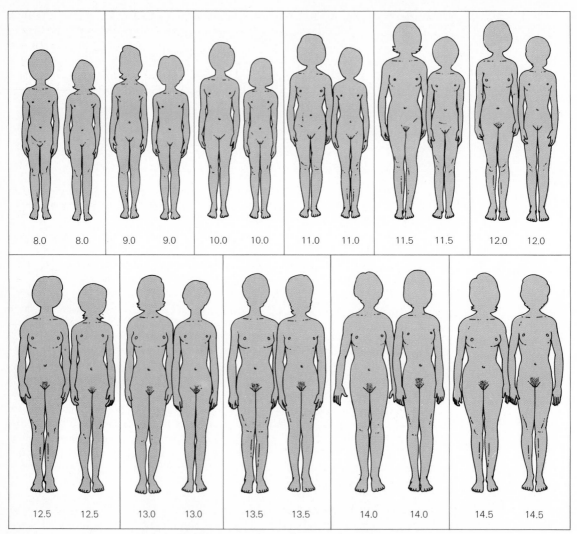

| 8.0 | 8.0 | 9.0 | 9.0 | 10.0 | 10.0 | 11.0 | 11.0 | 11.5 | 11.5 | 12.0 | 12.0 |
| 12.5 | 12.5 | 13.0 | 13.0 | 13.5 | 13.5 | 14.0 | 14.0 | 14.5 | 14.5 |

EXHIBIT 9.14   Contrasts between early- and late-maturing girls, aged 8 to 14.5 years. (Source: Drawings based on photographs in F. H. Shuttleworth, *The Adolescent Period: A Pictorial Atlas*, monograph of the Society for Research in Child Development, 1951, 14, Serial No. 50, 1949, pp. 13–14.)

percent of the distribution) in their respective maturational changes. There were sixteen males in each of these two groups. It was these groups of early and late maturers that were studied extensively regarding the psychosocial implications of their maturational rates. Unfortunately (because of limitations), the data from these thirty-two males provide much of what we know of these implications.

Jones and Bayley obtained ratings of the members of each group from two sources. First, a group of adults judged the males in terms of their social and personal characteristics. Second, some of the peers of these early- and late-maturing males made the same type of judgments. The early maturers—those whom it may be remembered had the athletic-looking type of body build—were judged more positively by the adults than were the late maturers—those who had primarily a thin body type. The early maturers were seen as more physically attractive, more masculine, better groomed, and less needful of status striving than the late maturers. On the other hand, the delayed males were regarded as more tense, more seeking of attention, more affected, and more childish than their early-maturing age-mates (McCandless, 1970). In addition, the late maturers were seen by the adults as less mature in heterosexual relations than their accelerated peers.

However, the adults did not judge there to be a difference between the two groups insofar as popularity, leadership, cheerfulness, and assurance were concerned. However, the peers of the males in these two groups did not quite agree with the adults here. In fact, they were quite a bit more negative about the late maturers than were the adults.

Their peers judged the late maturers to be more restless, more bossy, more attention seeking, less assured, less likely to have a sense of humor about themselves, less physically attractive, and less likely to have older friends than the early maturers. In addition, the peers saw the late-maturers as less popular and as less likely to be leaders than the early maturers. This means that members of the former group were *in fact* less popular and less often leaders than were members of the latter group, because such attributes are based on peer-group appraisals (McCandless, 1970). Independent data support this view. Latham (1951) found that junior high school boys who were more physically mature, and were therefore taller, heavier, and more muscular than their peers, were more likely to be chosen sport-team captains. In addition, these more mature boys were the dominant people in the leading social groups in school.

In summary, members of their adult and peer society afforded adolescent boys differential social and personal evaluations in accordance with whether such boys reached puberty before or after their group norm. If social interactions and appraisals contribute to individual functioning, then it may be expected that these contrasting social evaluations differentially would affect the two groups' psychological development.

The results of a second study, done by Mussen and Jones (1957), confirms this expectation. These researchers studied the same males assessed by Jones and Bayley. Mussen and Jones, however, administered the Thematic Apperception Test (TAT) to the males. This test is a *projective* personality test. In such a test, a relatively ambiguous stimulus is presented to a person (in the case of the TAT, pictures are presented). It is *assumed* that in describing the stimulus, the person will reveal his or her underlying personality "dynamics." In other words, his or her current conflicts, fears, desires, feelings, and thoughts about self and others will be revealed, or "projected," into the descriptions.

Although many of the interpretations of the subjects' projections did not show distinctions between the two groups, there was one major and statistically significant set of differences between early and late maturers. More often than was the case with the early maturers, the late maturers gave responses suggestive of feelings of inadequacy, weakness, rejection, rebelliousness toward their families (often seen as a desire to get free of parental restraint), and feelings of disapproval by parents and authorities. At the same time, the late maturers were more likely than the early maturers to give responses suggestive of dependency. Since a behavioral rejection by the peers of late maturers exists, it is to the parents that the late-maturing male must turn if anyone will meet his dependency needs. Thus the late maturer is dependent on the very social objects—the parents—toward whom he holds negative attitudes. There is a need to approach and a concomitant desire to avoid the parents.

Similarly aged adolescents often vary in their level of physical maturation. (Mimi Forsyth/Monkmeyer, Press Photo Service.)

In short, this means the late-maturing male exists in a state of conflict. In sum, very much more so than is the case with early-maturing males, late-maturing males have negative self-concepts. Indeed, although not statistically significant, the late maturers' tendency to feel they had a less attractive body, and their feelings of being less relaxed and assured and more restless, is consistent with this interpretation.

Early maturers tended to be characterized as having positive self-concepts. They lacked rebellious feelings toward their parents and gave responses indicative of self-confidence, independence, and the ability to play an adult role in interpersonal relationships. Although these males evidenced more aggressive themes in their responses than did the late-maturing males, such a feeling is quite consistent with the traditional masculine stereotype in our society of a dominant, aggressive, instrumentally effective person (Kagan & Moss, 1962; Lerner & Korn, 1972).

In summary, the adult and peer social appraisals of the two maturational groups were related to their functioning on the individual-psychological level. Information independent of the California Growth Study data indicates that early and late maturers have personal characteristics consistent with those reported by Mussen and Jones (1957). On the basis of an essentially qualitative analysis of their data, Kinsey et al. (1948) report that early and late maturers formed two quite distinct groups. Rate of maturation was indexed by the age at first ejaculation, as reported by the respondent through recall. Although this event is a questionable index of maturational rate and the recall may be faulty, the Kinsey et al. data are consistent with other information. The early maturers in the Kinsey et al. sample were seen to be perhaps the most alert, energetic, spontaneous, active, extroverted, and aggressive males studied. The late maturers, however, were appraised as slow, quiet, mild, unforceful, reserved, timid, introverted, and socially inept.

It should be noted that these differences were discerned among males who were several years postpubescence. Most were in late adolescence or early adulthood, if not well beyond. This fact suggests that the individual and social advantages of the early maturer and the problems of the late maturer may have implications for later development across the life span. There is ample evidence supporting this inference.

Weatherly (1964) reported data, independent of those from the California Growth Study, comparing early-, average-, and late-maturing college males on several measures of personality. The late maturers were found to have more responses indicative of feelings of guilt, inferiority, and depression and more responses showing needs for being encouraged and obtaining sympathy and understanding than males in other groups. On the other hand, members of the late-maturing group were less oriented toward leading, dominating, or controlling than were males in the two other maturational groups. However, Weatherly found that despite their need for support from others, late maturers were more rebellious and unconventional than were average or early maturers. Thus the conflict between dependency and independency, seem among the late maturers in the Mussen and Jones (1957) study, is identifiable in this independent and older sample of late maturers.

Still other evidence, from the California Growth Study data, indicates that at least for the early adult years, the differences between early- and late-maturing males remain stable. Ames (1957) interviewed early- and late-maturing males when they were in their early thirties. The major trend in the results of the study was that rate of maturation in adolescence was a better predictor of social and vocational behaviors in the males' thirties than were any *social* measures of such behaviors taken during childhood or adolescence. Basically, the direction of Ames's findings was the earlier the age of maturation (the more mature the male in adolescence), the higher the social and personal success in adulthood. The earlier the maturation was in adolescence, the higher was adult occupational status, informal and formal social activity, and business social life. In addition, of the forty males Ames studied, fifteen had supervisory or leadership business roles. Twelve of these fifteen were early maturers. Similarly, eight of the subjects held some office in a club or a civic group. All of these eight were early maturers.

In sum, although there were no known intellectual or social-class differences between the early and late maturational groups (McCandless, 1970), the early maturers led personally and socially more positive lives than the late maturers. Indeed, there was only one hint of disadvantage for the early-maturing males during this period of their life. Ames (1957) found a very slight tendency for this group to report they were less happy in their marriage than was the case with the late maturers.

Also studying these males when they were in their thirties, but through administration of several psychological tests, Jones (1957) did not find such a marital-quality difference. Jones also reported that physical attractiveness differences between the two groups were no longer present. The groups did not differ in attractiveness, grooming, marital status, family size, or educational level. However, the early maturers showed, as Ames also reported, earlier patterns of job success. In addition, on a personal level, these males scored higher on a test designed to measure their orientation for and ability toward making a good impression on others.

Interviews revealed that the late maturers seemed to be less settled at age 33, and on the basis of tests, they showed themselves to be less self-controlled, dominant, and responsible and more rebellious, touchy, impulsive, self-centered, and dependent on others than were the early maturers. However, Jones (1957) did find the late maturers to be more insightful and assertive than the early maturers.

Thus there is marked stability in the personality and social implications of rate of maturation from adolescence into the fourth decade of the life span, but there is evidence that by the end of this decade and the beginning of the next, these contrasts become unstable. In a follow-up study of California Growth Study males (Jones, 1965), done when they were in their forties, the advantages for the early maturers seemed to diminish. Possibly due to the fact that the bodily assets of the early maturers also tended to diminish relative to the late maturers at this time of life (middle age), the social and personal superiority of the early maturers, which seemed based on bodily appraisals, also decreased to a point where the psychosocial functioning of early- versus late-maturing males was almost identical. Although the psychological implications of early versus late maturation may disappear in the middle adulthood years, the above data still indicate clear, and *relatively* long-term, effects of maturational status on the psychosocial development of adolescent males. Before we attempt to interpret this relation between maturational rate and individual and social functioning, however, the nature of this relation for females is considered.

Implications of Early versus Late Maturation in Females. The data for females are not as complete as those for males. Early- and late-maturing females have not been followed over a similar duration of their lives as is the case with males. Moreover, what data do exist are somewhat more complicated.

Jones (1949) studied groups of early- and late-maturing females. Jones did not find the psychological and social advantages of early maturation with females that were found with males. In fact, the reverse seemed to be the case. In females, late maturation seemed to be associated with positive social appraisals and personality, and early maturation had correspondingly negative consequences. However, as was seen in Exhibit 9.14, early-maturing girls tend to have a body type which is more chubby and, hence, *less* attractive in American society (Staffieri, 1972) than have late-maturing girls, whose thinner, less plump bodies are more attractive (Staffieri, 1972). Since it was the late-maturing boys who had a more unattractive male physique (Lerner & Korn, 1972) and it was early-maturing girls who had a more unattractive female physique, relative to their respectively sexed but alternatively paced maturing peers, it may be that it is the socioculturally judged attractiveness of the body one derives from one's maturational retardation or advancement, rather than this retardation or advancement per se, which is most important in providing a source of the relationship between personality, social status, and the body. How one's body is appraised by one's society, and not the rate or stage of maturation that gives one this appearance per se, may be the basis of body-behavior relations.

It should be pointed out that although the above differential relations between early versus late maturation and social and personality development exist in young adolescent females, such differences seem to disappear by 17 years of age. Jones and Mussen (1958) studied the TAT responses of early- and late-maturing females when they were 17 years old, and no major differences in projective personality responses were seen as a function of their different maturational backgrounds.

Together, the Jones (1949) and Jones and Mussen (1958) studies suggest that whatever differences may exist initially between early- and late-maturing females, these contrasts tend to diminish much sooner in life than is the case with early- and late-maturing males. Other studies of implications of early and late maturation in females present data generally consistent with those reported by Jones (1949) about the relative advantages of a rapid maturation rate for adolescent psychosocial functioning. The data also indicate that these differences in early adolescence tend to disappear by late adolescence or early adulthood.

Stolz and Stolz (1944), in a longitudinal study, found that early-maturing females tended to be embarrassed more than their peers. Reasons for their greater embarrassment centered around their being bigger and taller than others—including their male peers, who matured from six months to two years later. Further, the females were embarrassed about their menstrual cycle. Thus like those of Jones (1949), the Stolz and Stolz data indicate disadvantages for early-maturing females. Again, one reason for the disadvantage of the early maturers may be that as a consequence of their maturational rate, they possess a physique type (an overly large, somewhat heavier one relative to peers) that is negatively evaluated by others in their social world (Lerner & Gellert, 1969; Lerner & Schroeder, 1971; Staffieri, 1972). However, as noted, these maturation-related differences among females tend to disappear.

Faust (1960) studied sixth- through ninth-grade females cross-sectionally. The females' maturational status was indexed by age of menarche, and the peers of the subjects were asked to make attributions of positive characteristics about them. In the sixth grade, the early-maturing females received fewer positive peer attributions; they received more peer rejection than did the premenarche sixth-grade females. However, in the seventh-grade cohort, and markedly so in the older two grade cohorts, there was a tendency for positive peer attributions to be associated with postmenarche females. The difference suggests that, as reported by Jones (1949) and Stolz and Stolz (1944), females who mature at an average or even late rate (e.g., the seventh, and the eighth and ninth grades, respectively) are more likely to have a psychosocial advantage in their early adolescence than are those females who mature early (e.g., the sixth-graders).

Again, that these contrasts disappear by middle to late adolescence is supported by information independent of that derived from the California Growth Study data reported by Jones and Mussen (1958). Nisbet, Illsley, Sutherland, and Douse (1964), although finding (contrary to the above) that females who matured earlier than others were in fact superior at age 13 years on some intellectual ability measures, noted these differences had largely disappeared by 16 years of age. Similarly, Shipman (1964) found that age at menarche was unrelated to a measure of timidity or of emotional maturity among adult females.

In sum, the effects of early versus late maturation seem to be more transitory in females than in males. Yet for either sex it may be the type of body one develops as a consequence of rate of maturation that has the major implication for social and personal functioning. If one develops a type of body that is viewed as attractive by members of one's society, as is the case with early maturation in males and late maturation in females, then this development may promote correspondingly favorable personality and social developments. Alternatively, development of a physique type considered unattractive by members of one's society, as in the case of late-maturing males and early-maturing females, may be associated with negative personality and social development.

Problems of Physical Maturation and Functioning

Some problems in adolescence involve hormone imbalances associated with the bodily changes of adolescence. These hormonal irregularities take many forms and vary in their incidence in the adolescent population. Three major instances of diseases involving hormonal irregularities that vary in this way are disturbances of growth, menstrual dysfunction, and acne. Other physical problems of adolescents involve their body but do not necessarily stem from hormone imbalances. Rather, adolescents behave in ways that promote these problems. They may overeat and become obese or undereat and experience a disorder known as anorexia nervosa. We consider each of these problems successively.

Disturbances of Growth. According to Katchadourian (1977), most disturbances in adolescent growth can be detected before puberty. As such, while they are labeled more properly as problems of childhood, they are included in discussion of adolescent disease, since the effects persist beyond childhood. Two major types of these diseases involve insufficient and excessive growth. There exist criteria for determining if either of these types of growth occurs.

If a person's height differs by more than 20 percent from the norm for people of his or her age and sex group, the person may be labeled as either "short" or "tall," depending, of course, on the direction of deviation. If the deviation exceeds 40 percent, then the person is labeled a "dwarf," if below the norm, or a "giant," if above it (Katchadourian, 1977). Given current adult height norms, men shorter than 4 feet 8 inches or taller than 6 feet 7 inches would be termed dwarfs and giants, respectively; the corresponding heights for women are 4 feet 5 inches and 6 feet 1 inch (Prader, 1974).

Short stature is typically caused by an insufficient production of hormones by the anterior pituitary gland. Sometimes, when this insufficiency is coupled with insufficient production of the sex hormone, a condition known as *delayed puberty* occurs. The person goes through the bodily changes associated with pubescence much later than usual.

In turn, a condition typically caused by the early production of sex hormones is termed *precocious puberty*. This condition exists when the changes of pubescence occur in females at age 8 years or earlier and in males at age 9 years or earlier (Katchadourian, 1977). This disorder is twice as common in females as in males, but extreme cases of the disease have been reported for both sexes. For example, in females, menarche has been observed in the first year of life and pregnancy has occurred as early as age 5.5 years; in males, penis development has occurred in 5-month-olds and spermatogenesis has been observed in 5-year-olds (Katchadourian, 1977).

Menstrual Dysfunction. Menstruation is a natural, normal function which produces varying degrees of discomfort among adolescent females. It may also be accompanied by a brief period of depression, fatigue, or irritability. Painful menstruation (*dysmenorrhea*) may be the result of hypersensitivity of the lining of the uterus, a too tightly closed cervix, unusual flexion of the uterus backward or forward, atrophy of the uterus, tumors, inflammation of organs adjacent to the uterus, infection, congestion due to constant standing, disorders of the endocrine glands, allergies, constipation, and other similar contributing factors. It may also be due to subtle psychological factors (Bowman & Spanier, 1978).

Many cases of dysmenorrhea may be relieved by adequate medical treatment. In some instances, dysmenorrhea may be sufficiently disabling that the adolescent female will be unable to function at her usual levels for a day or two each menstrual cycle. With modern medical remedies, however, only a very small number of females are severely disabled during their menstrual periods. On the other hand, many females do experience sufficient discomfort during their menstrual periods that they restrict their activity and take medication to help relieve the symptoms.

Other problems that may be associated with the menstrual cycle are failure to menstruate (*amenorrhea*), excessive menstrual bleeding, and premenstrual tension. Some females begin their menstrual periods later than others, and much of the menstrual activity in early adolescence is erratic. Some females who have experienced more or less regular menstruation cease having a period altogether. Amenorrhea may be caused by a number of medical conditions, and among adolescents especially, it often is caused by changes in emotions.

One common example of how amenorrhea is tied to the emotions relates to the fear of pregnancy. Some adolescent females who have had sexual intercourse without contraceptive protection fear that they may have become pregnant. They become very upset and anxious while they await their period. In extreme cases, this emotional upset interferes with the onset of the period. The delayed period continues to create anxiety, and a vicious circle of tension, anxiety, and delayed menstruation results. This phenomenon is why females who have a delayed period should always have a pregnancy test if pregnancy is suspected, rather than relying on the "missed" period itself as evidence of pregnancy.

Amenorrhea, then, may be characteristic of emotional disorders or temporary psychological distress among adolescent females, or it may be a symptom of underlying physical disorders or disease. In either case, medical and/or psychological treatment may be warranted.

For some young women there is a brief period during the menstrual cycle, usually just prior to the onset of menstruation but sometimes earlier and at times persisting for a day or two after menstruation begins, when they exhibit one or more of a cluster of symptoms. These symptoms together are termed the *premenstrual syndrome,* sometimes referred to as *premenstrual tension,* although tension is only one of the possible symptoms (Bowman & Spanier, 1978). Perhaps 60 percent of women experience the premenstrual syndrome. There are many possible symptoms, including headache, anxiety, inability to concentrate, depression, emotional outbursts, crying spells, hypersensitivity, unexplainable fears, and insomnia. Most women who experience premenstrual tension are able to control the expression of these symptoms, and the phenomenon is not a readily noticed problem for most adolescent females.

Acne. Acne is the most common disorder of medical significance in adolescence. Zeller (1970) reports that 85 percent of all adolescents have acne at least sometime during adolescence. The data presented in Exhibit 9.15 attest to this. Similar data have been reported by Burgoon (1975), who notes that female adolescents are most affected by acne in the age range from 14 to 17 years, while the corresponding age range for males is 16 to 19 years.

Because the presence of acne is enhanced by the presence of the hormone androgen, which is commonly present in greater concentrations in males, acne is more

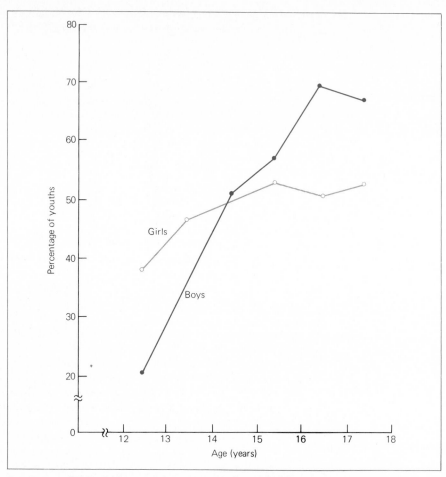

EXHIBIT 9.15   Percentage of U.S. youths reporting acne, pimples, or blackheads, by age and sex. (Source: Adapted from U.S. National Center for Health Statistics, *Vital Health Statistics*, Ser. 11, no. 147, 1975.)

common among males than females and infrequently occurs among prepubescent children of either sex (Katchadourian, 1977). The presence of acne represents a major concern for adolescents, despite its great prevalence. American adolescents spend more than $40 million a year for nonprescription acne "cures"; yet, according to Katchadourian (1977), there exists no convincing evidence that even medical treatment shortens the course of or cures acne.

Obesity. Although most American adolescents judge their own health to be very good to excellent (see Exhibit 9.16)—and, in fact, for most adolescents this is an accurate self-appraisal—many adolescents take actions which produce self-induced problems. Although many different adolescent behaviors can induce biological or physical problems, those associated with *eating disorders* are especially prevalent among American adolescents. Eating disorders may take the form of either too great or too little an intake of calories. In the former case a disorder termed *obesity* is a common occurrence; in the latter situation disorders such as anorexia nervosa occur.

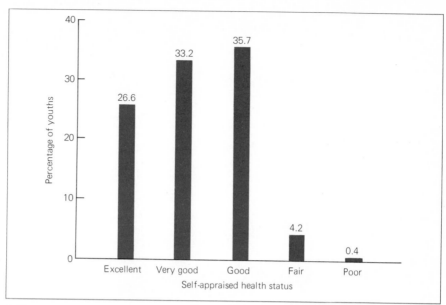

EXHIBIT 9.16   Self-appraised health status of U.S. youths between 12 and 17 years of age. (Source: U.S. National Center for Health Statistics, *Vital Health Statistics*, Ser. 11, no. 147, 1975.)

Obesity is the most prevalent eating disorder at all ages among Americans, and this of course includes adolescents (Katchadourian, 1977). About 10 percent of grade school children are obese; among adolescents estimates of the frequency of obesity range from 10 to 30 percent (Paulsen, 1972, p. 215). Furthermore, people who were overweight as children or adolescents have an 80 percent chance of remaining this way as adults; in fact, 50 percent of the markedly obese adult population is composed of people who were overweight children. Moreover, although between childhood and adulthood there is a general increase in the prevalence of obesity in the American population, between the ages of 20 and 50 years this increase is threefold (Moore, Stunkard, & Srole, 1952).

Criteria for obesity exist (Katchadourian, 1977). When a person is 10 percent above the average weight associated with others of his or her height, he or she can be classified as overweight. When a person is 20 percent above the average weight associated with others of his or her height, he or she can be classified as obese. However, it is important to make sure that this percentage of weight over the average is contributed by an excess of body fat, and not body muscle or bone. Thus a weight lifter might be heavier than average, but if this excess weight is due to more body muscle than average, the weight lifter would not be considered obese.

Obesity is 6 times more frequent among males and females of low socioeconomic status than those of high socioeconomic status (Goldblatt, Moore, & Stunkard, 1965); and there is some evidence (see Goodman, Dornbusch, Richardson, & Hastorf, 1963) that the obese are more likely to come from low socioeconomic backgrounds associated with particular ethnicities (e.g., Jewish and Italian youth).

Obesity is a physical handicap which has well-established behavioral implications. Stunkard and Mendelson (1967) found that obese adolescents viewed their bodies as loathsome and grotesque and that these views of their bodies persisted

Obesity is a common problem among today's adolescents. (United Press International, Inc.)

into later life, even if they lost weight. Moreover, Stunkard and Mendelson (1967) reported that others viewed the bodies of the obese with hostility and contempt. Such prejudiced reactions seem to affect the obese. Monello and Mayer (1963) found that obese adolescent females had personality characteristics similar to people from social groups that are victims of prejudice. Indeed, there is some evidence that there is social discrimination toward the obese. Canning and Mayer (1966) found that although neither application rates to high-level colleges nor academic qualifications differed for obese and nonobese high school seniors, the obese students, and particularly the obese females, were not admitted as frequently as were the nonobese students.

Anorexia nervosa. Because of such negative implications of obesity, adolescents often make concerted efforts to reduce their caloric intakes. At times, this effort can be seriously overdone, and in these instances, undernutrition can result. While common causes of undernutrition—the ingestion of too few calories for optimal functioning—are poverty and drug or alcohol use, an overconcern with caloric intake due to the presence of body fat can lead to a sustained refusal to eat and/or a serious loss of appetite (Katchadourian, 1977).

In such cases an emotional disorder termed *anorexia nervosa* can occur, which predominantly affects adolescent females. Indeed, only 4 to 6 percent of those having this disorder are males. Estimates are that as many as 1 in 250 females, within the high-risk age range of from 12 to 18 years, may have this disease (Crisp, 1970, 1974; Sours, 1969; Vigersky, 1977). Although the onset of this disorder can occur in the prepubescent period (or rarely even in adulthood), females who are affected usually first develop this disorder in their early to late adolescence.

There are several distinct features of this disorder (Andersen, 1977; Bruch, 1973, 1977; Crisp, 1970, 1974; Sours, 1969; Vigersky, 1977):

1 *Behavior directed toward losing weight.* Anorectics make drastic reductions in their total food intake, especially of foods high in carbohydrate and fat content. There are daily intakes of as little as 80 to 100 calories, and anorectics often induce vomiting after ingestion of food and/or make extensive use of laxatives (Bruch, 1973; Crisp, 1970, 1974; Vigersky, 1977).

2 *Peculiar patterns of handling food.* Although limiting themselves to a few low-calorie foods, anorectics often prepare elaborate meals for others, collect recipes, and become preoccupied with thoughts of food and the calories contained in foods (Bruch, 1973, 1977; Sours, 1969). In addition, anorectics frequently have been known to hoard, conceal, and crumble food.

3 *Weight loss.* Anorectics are characterized by a loss of at least 25 percent of original body weight, but frequently as much as 50 percent of original body weight is lost (Bruch, 1973, 1977; Crisp, 1970, 1974; Sours, 1969; Vigersky, 1977).

4 *Intense fear of gaining weight.* Anorectics fear they will become obese. This fear does not lessen with increased weight loss; indeed, they become preoccupied with the size and appearance of their body, often spending long periods of time gazing in the mirror (Bruch, 1973, 1977; Vigersky, 1977).

5 *Disturbance of body image.* Although preoccupied with their body, anorectics do not perceive it accurately. They misjudge body size, often believing they are overweight—despite increasing thinness—or, alternatively, that they look quite good, despite their poor physical state (Andersen, 1977; Bruch, 1973, 1977; Vigersky, 1977).

6 *Other medical and psychological problems.* Since anorexia nervosa predominately affects adolescent females, amenorrhea is a common complication of this disorder. However, this is an outcome of the anorexia, and not a cause of it. Indeed, anorexia nervosa occurs when there are no known medical problems that would account for the weight loss (Bruch, 1973, 1977; Crisp, 1970, 1974; Sours, 1969; Vigersky, 1977). However, obesity often precedes anorexia nervosa, and some estimates are that about one-third of all anorectics have been at least slightly overweight prior to the onset of the disorder (Crisp, 1970, 1974; Sours, 1969; Vigersky, 1977).

It is generally believed that family problems and withdrawal from normal social relationships are the most common underlying causes of anorexia nervosa. For example, Bruch (1977) finds that two-thirds of anorectics come from families that only have daughters and/or stress that "only outstanding is good enough." Treatment of the anorectics' psychosocial and medical problems is difficult, since most anorectics deny their problem and resist therapy (Bruch, 1973, 1977; Crisp, 1970, 1974).

With few calories ingested and no body fat to use for calorie consumption, an anorectic's muscle tissue is soon metabolized. Hence organ failures and heart attacks are frequently the "final" complications of anorexia nervosa. As such, there is a high mortality rate (15 to 21 percent) among those adolescents having this disorder (Crisp, 1970, 1974; Katchadourian, 1977; Vigersky, 1977). Literally, such adolescents starve themselves to death.

COGNITIVE DEVELOPMENT IN ADOLESCENCE

Much of what we have said in this chapter emphasizes that the adolescent exists in a "new" body and that the physical and physiological changes providing this newness appear to have implications for the person's psychosocial functioning. One set of processes contributing to the impact that bodily changes have for behavioral development are cognitive processes. There is evidence that there are both quantitative and, especially, qualitative changes in adolescent cognition. The adolescent thinks about his or her new body and about his or her new self in new ways.

Accordingly, because of the importance of cognitive development for these and other changes in adolescence, we now turn to a discussion of quantitative and qualitative changes in adolescent intellectual functioning. First, we briefly note some issues and data in the psychometric study of adolescent cognition. Then we focus in some depth on that dimension of adolescent cognition, the qualitative, that has received most recent theoretical and empirical attention (Elkind, 1967; Neimark, 1975). As in previous chapters, the theory of Jean Piaget is most important in our concern with qualitative changes in adolescent thought.

The Psychometric Study of Adolescent Cognition

It has been argued that early infant or childhood measures of intelligence do not measure what is assessed as intelligence in adolescence. Are those types of abilities that are present in adolescence first present in this period of life? In other words, is there continuity or discontinuity in the type of cognitive abilities present from childhood to adolescence? This question raises the issue of the age differentiation hypothesis. It is one of the key issues in the psychometric study of adolescent cognition.

The Differentiation of Cognitive Abilities. The number of different abilities comprising psychometric intelligence has been a continuing source of controversy. Estimates have ranged from 1 (Spearman, 1927) to over 100 (Guilford, 1967). From a developmental perspective, the number of abilities that characterize the person at different points in the life span also has been controversial, especially insofar as adolescence is concerned. Burt (1954, 1955) and Garrett (1946) formulated the age differentiation hypothesis of intelligence to address this controversy.

As originally put, this idea suggests that a global ability structure emerges in childhood. That is, there is a general sort of intelligence, with all measures of cognitive ability highly related. Furthermore, it is argued that although all measures of cognitive ability remain correlated across the life span, there are groupings of abilities that form (as we saw in Schaie's sequential research). In other words, there is a differentiation from one to several groups of abilities, and it is held that this change characterizes the adolescent period.

According to Fitzgerald, Nesselroade, and Baltes (1973), there are data which suggest a revision of this general idea. A sequence of *integration* (early childhood), *differentiation* (childhood through adolescence into adulthood), and *dedifferentiation* or *reintegration* (late adulthood) is suggested. This ordering involves increasingly higher interrelations in childhood among whatever mental abilities are present, and as such, the presence of one global or general ability dimension is hypothesized. After this, by late childhood and into adolescence, there is differentiation into several ability dimensions. In late adulthood, a return to the less differentiated structure is suggested.

Fitzgerald et al. (1973) note that the existence of this life-span process is uncertain, since there have been studies both finding and not finding it. In fact, reviews of this research (Anastasi, 1970; Reinert, 1970) show that studies divide about equally between supporting and rejecting the age differentiation hypothesis. However, the reviews have indicated that several methodological problems exist in this research (e.g., use of small homogeneous samples, use of tests that are not sensitive to changes in the number of abilities present, and failure to include control groups). As such, to decide whether the age differentiation hypothesis is a useful notion, as least insofar as adolescent development is concerned, Fitzgerald et al. conducted a study which avoided such methodological problems.

From a stratified probability sample of about 2100 junior high and high school students in West Virginia, a group of 1891 volunteers was studied. Students were divided into three grade-level groups: grades 7 to 8 (sample size, N, was 783), grades 9 to 10 ($N = 630$), and grades 11 to 12 ($N = 478$). Within each group, subjects were divided about equally by sex.

The age differentiation hypothesis suggests a gradual change through adolescence in the structure of intelligence from the global nature of childhood to the differentiation of adulthood. Consequently, Fitzgerald et al. used the Thurstone and Thurstone (1962) conception of the primary mental abilities of adulthood to define the structure of adult intelligence. This conception of adult intelligence sees cognitive abilities as grouped into four categories (verbal, number, reasoning, and space factors). Using this structure of abilities as a target, Fitzgerald et al. assessed whether there was an increasingly greater correspondence across the grade groups with the target, adult structure, than would be expected on the basis of the age differentiation hypothesis.

The results of this matching failed to find evidence for differentiation. At all grade levels, the same number of factors (four) were found. Thus from early through middle adolescence, the same level of differentiation was present. In addition, the four factors present in each group were found to be clearly the same as those that Thurstone had indicated comprised adult intellectual structure. Data within and across groups indicated that the factors comprising adult ability were already present among adolescents within the seventh grade. In short, from early adolescence to adulthood, people have a stable ability structure, and if there is differentiation of cognitive abilities, it emerges prior to adolescence.

In sum, the rapid increases in the stability of ability scores from early to late childhood and the cohort-related stability or instability of the adolescent through adult years of life do not seem to involve marked alterations in the psychometric abilities possessed by the person after the period of rapid increases in childhood test scores.

The Piagetian Approach to the Study of Adolescent Cognition

As we have noted, Piaget's approach to the study of adolescent cognition is the leading one when a structural, qualitative approach toward cognitive development is taken (Elkind, 1967; Neimark, 1975). The infancy and childhood portions of life were seen in previous chapters to be associated with the first three stages of Piaget's theory, the sensorimotor, the preoperational, and the concrete operational. When the child attains the ability to think counterfactually, to see that reality and thoughts about reality are different, and can generate and recognize hypotheses about reality, the fourth stage in Piaget's theory is reached. This stage is one associated specifically with adolescence (Elkind, 1967, 1968; Neimark, 1975).

The Formal Operational Stage. The last stage of cognitive development in Piaget's theory is termed the formal operational stage. It begins at about 11 or 12 years of age and continues for the rest of life, according to Piaget (1972). It is because of the lower age limit typically associated with this stage, in both theory and research (see Neimark, 1975), that the study of the qualitative aspect of adolescent cognition is associated with the evaluation of formal operational development.

In this stage, thought becomes hypothetical in emphasis. Now discriminating between thoughts about reality and actual reality, the child comes to recognize that his or her thoughts about reality have an element of arbitrariness about them, that they may not actually be real representations about the true nature of experience. Thus the child's thoughts about reality take on a hypothetical "if . . . then" characteristic: *"if* something *were* the case, *then* something else would follow."* In forming such hypotheses about the world, the child's thought can be seen to correspond to formal, scientific, logical thinking. This emergence accounts for the label applied to this stage—the formal operational stage.

Another perspective on the quality of this stage is provided by Neimark (1975). She explains the distinctive quality of adolescent thought by noting:

> *Although the properties and relations at issue during the concrete operational stage are abstract in the sense of being derived from objects and events, they are still dependent upon specifics of the objects and events from which they derive; that is, they are empirically based abstractions rather than pure abstractions. In this sense the elements of concrete operational thought are "concrete" rather than "abstract" or "formal." On the other hand, propositions, the elements of formal operational thought, are abstract in the sense that the truth value of a statement can be freed from a dependence upon the evidence of experience and, instead, determined logically from the truth values of other propositions to which it bears a formal, logical relationship. This type of reasoning, deriving from the form of propositions rather than their content, is new in the development of the child: deductive rather than inductive thought. (Neimark, 1975, pp. 547–548)*

In other words, because the concrete operational child can only form abstractions relevant to phenomena or problems that exist, thoughts about a given topic cannot be integrated with *potentially* relevant but nonempirical (i.e., hypothetical) aspects of the problem. In this sense, mental operations are not coordinated, and the child cannot reach solutions by means of general theories or the postulation of all possible solutions to a problem (Wadsworth, 1971). In turn, the formal operational person shows these attributes of thought.

To be able to deal with all potentially relevant aspects of a problem, the person has to be able to *transform* (i.e., alter, or rearrange) the problem so as to contend with all its possible forms. The cognitive structure that characterizes formal operations allows such complete transformations. This structure is termed the *INRC group* by Piaget (Inhelder & Piaget, 1958). That is, all transformations of a problem may be obtained through the application of the components of this group: identity, negation, reciprocal, and correlative transformations. Simply, one can think of all aspects of a problem by, for instance, recognizing the problem in terms of its singular attributes (an identity transformation), canceling the existence of the problem (a negation operation), taking its opposite (a reciprocal transformation), or relating it to other problems (a correlative transformation).

Since only the coordinated use of all these transformations allows all potentially applicable aspects of a problem to be dealt with, it is not until the INRC group is established that the person possesses a cognitive structure appropriate for dealing with pure abstractions, with the set of all propositions pertinent to a problem that is the "object" of thought in the formal operational stage. To indicate the attributes of thought characteristic of people in the formal operational stage, some examples of tasks typically used to measure such cognitive functioning can be offered.

The Measurement of Formal Operational Thought. In discussing preoperational thinking, some of the various conservation tasks used to index the child's level of functioning were noted. Different types of problems are presented to children and adolescents in order to assess whether they show the types of mental functions representative of formal operations. Inhelder and Piaget (1958) and Piaget and Inhelder (1969) describe several such formal operational task problems.

One of these tasks involves assessing the person for the presence of *combinatorial thought*. To explain the meaning of this term, Piaget and Inhelder (1969) describe a task wherein a subject is presented with five jars, each containing colorless liquids. Combining the liquids from three particular jars will produce a color, while any use of the liquids from either of the other jars will not produce a color. The subject is shown that a color can be produced but is not shown which combination will do this.

Concrete operational children typically try to solve this problem by combining liquids two at a time, but after combining all pairs, or possibly trying to mix all five liquids together, their search for the workable combination usually stops. However, the formal operational child will explore all possible solutions and, typically, tests all possible combinations of two and three liquids until a color is produced.

Tasks involving certain types of *verbal problems* are usually unable to be solved unless formal operational ability is present (Piaget & Inhelder, 1969; Wadsworth, 1971). One such verbal problem is represented by the question, "If Jane is taller than Doris and is shorter than Francine, who is the shortest of the three?" (Wadsworth, 1971). Although concrete operational children may be able to solve an analogous problem, e.g., one dealing with sticks of various lengths when the elements of the problem (the sticks) are physically present, abstract verbal problems are usually not solved until formal operations have emerged.

Although formal operational thought can be illustrated by particular solutions to problems presented by many other tasks (Inhelder & Piaget, 1958), we may mention one other task, *the pendulum problem*, to illustrate the quality of thought at this period of functioning. A pendulum can be made to swing faster by shortening the string holding it. Conversely, it can be made to swing slower by lengthening the string. Concrete operational children typically adjust the *weight* of a pendulum when asked to alter its speed (Wadsworth, 1971). Alternatively, they may adjust string length and weight simultaneously and attribute any change in speed to the weight alteration (Wadsworth, 1971). However, in the formal operational period, subjects separate weight and string length, deal with them separately, and show knowledge that it is string length which is the variable relevant to the speed of swinging (Inhelder & Piaget, 1958).

In summary, with formal operations the child's thought is completely free from any dependence on concrete reality. Now the child can and does think not only in the "if . . . then" or "as if" manner, but counterfactually and completely abstractly as well. In her review of research on adolescent intellectual development, Neimark (1975) notes that there exists strong research evidence for the validity of formal operational thought as an empirical phenomenon distinct from concrete operational thought. There have been repeated demonstrations of age-related improvements in formal thought during adolescence (e.g., Martorano, 1977). Indeed, presentations by Moshman (1977) and Flavell and Wohlwill (1969) support Piaget's ideas about the structure of adolescent thought, as well as the way this stage is formed. Similarly, Roberge (1976) found that the formal operational structures necessary to solve complex problems and deal with conditional-reasoning arguments emerge in early adolescence, and a report by Strauss, Danziger, and Ramati (1977) shows that formal thoughts, which dominated college students' thinking, cannot easily be changed back to concrete operational. Thus formal operations emerge as a distinct stage in adolescence and continue to characterize thought thereafter. However, several qualifications of these conclusions are necessary.

Issues in the Assessment of Formal Operations. As will be noted below, the emergence of formal operations in adolescence does not say anything about whether another stage of cognition follows. Moreover, there is evidence that the emergence of a formal operational thought structure is not a characteristic of all people. As reviewed by Neimark (1975), studies done with older adolescents and adults in western cultures show that not all individuals attain the level of formal operations. For instance, Jackson (1965) found that only seven of sixteen persons 13 through 15 years old reached a conventional criterion of formal thought, and less than 75 percent of the 15-year-olds that Dale (1970) studied reached such a criterion. In fact, on the basis of results from two studies (Elkind, 1962; Towler & Wheatley, 1971), Neimark (1975) notes that there is evidence that only about 60 percent of samples of college students show appropriate performance on tasks requiring the conservation of volume. Similarly, Martorano (1977) found that in a sample of sixth- through twelfth-grade females, there was a grade-associated improvement on average scores for ten different tests of formal operations. However, not even the oldest group of subjects showed formal operations across all ten tasks.

Kuhn, Ho, and Adams (1979), however, provide data that suggest that the absence of a formal operational level of performance on the part of many older adolescents or adults may to a large extent reflect cognitive-processing difficulties in dealing with the format of problems, rather than the absence of underlying reasoning competencies. Kuhn et al. found a group of preadolescents and a group of first-year college students who showed no evidence of formal reasoning on particular measures of it. However, the two groups were administered additional kinds of assessments of formal operations, and when problems were repeated over a period of months, most of the college students showed immediate and substantial formal reasoning, whereas the preadolescents made only gradual and modest gains. Similarly, Roberge and Flexer (1979) showed that adults scored significantly higher than eighth-graders on tests of combinations and proportionality, although not on a test of propositional logic.

Moreover, although not all people may attain the formal operations level, once attained in adolescence it may not be continuous throughout the rest of life. Papalia (1972) found that less than 60 percent of her 65- to 82-year-old sample conserved volume, and Tomlinson-Keasey (1972) found evidence of formal operations in just about 50 percent of her middle-aged female sample. These latter two studies could be interpreted as showing that formal operations are not present in the postadolescent life periods of people of some (older) cohorts. This may be due either to their never reaching this stage or, after having attained it, losing their ability (or to assessment problems). However, without ontogenetic-historical data on subjects' lifespan performance, this issue presently cannot be decided.

Not only are generalizations about the course of formal operations across the life span within a culture tentative, but so, too, are such statements when made from a cross-cultural perspective. As summarized by Neimark (1975), people in many cultural settings do not attain formal operational abilities at the average early adolescent time that it occurs within western cultural settings. For example, Douglas and Wong (1977) reported that American adolescents (13- and 15-year-olds) were more advanced in formal operations than Hong Kong–Chinese youth of corresponding ages. In fact, in some nonwestern groups there is a failure ever to attain such thinking ability. Piaget (1969, 1972) himself notes such failures. Reasons for these differences have been suggested to lie in experiential contrasts found between rural and urban milieus (Peluffo, 1962; Youniss & Dean, 1974) and the kind of schooling experiences encountered (Goodnow, 1962; Goodnow & Bethon, 1966; Peluffo, 1962, 1967).

Although these explanations might lead one to expect socioeconomic or educational differences to be associated with formal operational attainment, the data do not support such a view. Neimark (1975) finds that socioeconomic status has no known effect on the development of formal thought, and in addition, she notes that only *very* profound differences in education seem to be associated with the development of formal thought. Moreover, variables such as sex and psychometric intelligence, which might be thought to mark experiential differences associated with formal thought development, do not seem useful. Simply, some studies find sex differences (e.g., Dale, 1970; Elkind, 1962) and some do not (e.g., Jackson, 1965; Lovell, 1961). Moreover, IQ-score differences do not relate to differences in formal operational development (Kuhn, 1976). Indeed, it seems that after some minimal level of psychometric intelligence is reached, variables other than those associated with IQ scores contribute to formal operational development (Neimark, 1975).

It has just been suggested that the variables readily thought of as possible facilitators of such progression do not seem consistently functional. It might be appropriate to try *directly* to train formal operational thought, but few studies have attempted this. Although Tomlinson-Keasey (1972) facilitated formal thought through training on one task, subjects did not generalize formal-thinking ability to other tasks. Schwebel's (1972) attempt at training also met with limited success, and Lathey (1970) had no success at all in attempting to produce volume conservation in 11-year-olds. Kuhn and Angelev (1976) did succeed in their intervention attempt, however. Fourth- and fifth-graders were given a fifteen-week program where they confronted problems requiring formal operational thought. The frequency of exposure to the problems (once per two weeks, once a week, or twice a week) was related directly to amount of advancement on three formal operational problems.

In sum, formal operations does not represent a level of thought reached universally by all people in all cultures. Even though uncertainty remains about the variables which provide a basis of transition to this level of thinking, there is strong evidence that formal operations do represent a distinct level of thought beyond concrete operations. Moreover, for most adolescents living in contemporary western culture—and for all those people now reading these words—formal operations represent an *attained* stage of thought. Although we have seen the assets of this type of thought, it must be noted that, as is the case with preceding levels of thought, formal operational thinking also has limitations. These are now considered.

Adolescent Egocentrism. Because anything and everything can become the object of the adolescent's newly developed abstract and hypothetical cognitive ability, the person not only may recognize his or her own thoughts as only one possible interpretation of reality but also may come to view reality as only one possible instance of a potentially unlimited number of possible realities. The concrete predomination of what *is* real is replaced by the abstract and hypothetical predomination of what *can be* real. All things in experience are thought about hypothetically, and even the adolescent's own thoughts can become objects of his or her hypothesizing.

In other words, one can now think about one's own thinking. Since the young person spends a good deal of time using these new thought capabilities, the person's own thought processes thereby become a major object of cognitive concern. This preoccupation, or *centration,* leads to a limitation of the newly developed formal operational thought; it leads to egocentrism within the formal operational stage. Elkind (1967) has labeled the egocentrism of this stage *adolescent egocentrism,* and sees it as having two components.

First, we have seen how the adolescent's own thoughts come to predominate his or her thinking. Because of this preoccupation, the adolescent fails to distinguish, or discriminate, between his or her own thinking and what others are thinking about. Being preoccupied with self, and not making the above discrimination, the adolescent comes to believe that others are as preoccupied with his or her appearance and behavior as he or she is (Elkind, 1967). Thus the adolescent constructs an *imaginary audience.*

An illustration of the functioning of the imaginary audience and of some emotional concomitants of this cognitive development may be seen if one thinks back to his or her days of early adolescence. Assuredly, some new fad, perhaps in regard to a particular style of clothing, sprang up among peers. Some adolescents perhaps were stuck with wearing the outdated style and were literally afraid to be seen in public. They may have been sure that as soon as they walked about without the appropriate clothes, everyone would notice the absence. Because they were so aware of the absence of this evidence of fitting in with everyone else, they were equally sure that everyone immediately would be aware of the shortcoming.

A second component of adolescent egocentrism also exists. The adolescent's thoughts and feelings are experienced as new and unique. Although to the adolescent they are, in fact, new and unique, the young person comes to believe that they are *historically* new and unique. That is, the adolescent constructs a *personal fable,* the belief that he or she is a unique, one-of-a-kind individual—a person having singular feelings and thoughts.

Here, too, it is easy to think of an illustration of the personal fable. We can think back to our early adolescent years and our first "love affair." No one ever *loved* as deeply, as totally . . . no one had ever felt the intense compassion, the devotion, the longing, the overwhelming fulfillment that we felt for this, our one true love. . . . Then remember a few days or weeks later. When it was over! The pain,

Because of the adolescent's egocentric belief in an imaginary audience, there is a tendency to conform to fads. (Nancy J. Pierce/Photo Researchers, Inc.)

the depression, the agony . . . no one had ever *suffered* as deeply, no one had ever been so wrongfully abused, so thoroughly tortured, so spitefully crushed by unrequited love. . . . We sat in our rooms, unmoving, and our mothers would say, "What's wrong with you? Come and eat." Our inevitable answer: "You don't understand. What do you know about love!"

Although the formal operational stage is the last stage of cognitive development in Piaget's theory, the egocentrism of this stage diminishes over the course of the person's subsequent cognitive functioning. According to Piaget (Inhelder & Piaget, 1958), the adolescent decenters through interaction with peers, elders, and—most importantly—with the assumption of adult roles and responsibilities: "The focal point of the decentering process is the entrance into the occupational world or the beginning of serious professional training. The adolescent becomes an adult when he undertakes a real job" (Inhelder & Piaget, 1958, p. 346).

Quite interestingly, it may be noted that Piaget sees the end point of adolescence as the adoption of a role in society. Earlier discussions also have stressed the centrality of role attainment as the major psychosocial task of adolescence; this emphasis predominates the views of adolescent development of Erik Erikson. In short, for many theoretical perspectives—including those stressing cognition and personality—the linkage of the adolescent with his or her society through the processes of role search and attainment is a core component of development.

In sum, with the attainment of formal operations, Piaget (1950, 1970, 1972) claims the person has reached the last stage of cognitive development. To him and his followers, no new cognitive structures emerge over the course of life (Flavell, 1970; Piaget, 1972), although the person may change through a differentiation or specialization of abilities within the common, formal structure (Neimark, 1975; Piaget, 1972). As a way of summary, Exhibit 9.17 presents the four stages of Piaget's theory and shows the cognitive achievements and limitations involved in each

EXHIBIT 9.17   **Piaget's Stages of Cognitive Development**

Stage	Approximate Age Range	Major Cognitive Achievements	Major Cognitive Limitations
Sensorimotor	0–2	Scheme of object permanency	Egocentrism: lack of ability to differentiate between self and external stimulus world
Preoperational	2–6 or 7	Systems of representation. Symbolic functioning (e.g., language, symbolic play, delayed imitation)	Lack of conservation ability. Egocentrism: lack of ability to differentiate between symbol and object
Concrete operational	6 or 7–12	Ability to show experience-independent thought (reversible, internalized actions). Conservation ability	Egocentrism: lack of ability to differentiate between thoughts about reality and actual experience of reality
Formal operational	12–	Ability to think hypothetically, counterfactually, and propositionally	Egocentrism: imaginary audience, personal fable

Source: Adapted from Lerner (1976).

stage. However, to close this chapter, it is important to note that Piaget's is not the only view about the character of qualitative change in adolescence (and later life).

IS THERE A FIFTH STAGE OF COGNITIVE DEVELOPMENT?

Although unquestionably seen as one of the greatest contributions to the study of human development, Piaget's work has not escaped criticism. These critiques have involved many dimensions of inquiry. For example, criticism has centered on measurement and other methodological problems of Piaget's work and the issue of whether the idea of a "stage" is useful, at least insofar as Piaget employs it (Brainerd, 1978). Insofar as adolescent development per se is concerned, the major issue of criticism has involved whether formal operations is indeed the last stage of cognitive development. Is there a fifth, and qualitatively distinct, level of thought that may follow formal operations?

Theorists like Riegel (1973, 1976b) and Arlin (1975) have pointed out that formal operational thought involves finding solutions to problems—of being able to generate and evaluate all possible combinations of hypotheses about an issue. However, they do not believe that such thinking allows one to recognize new problems and issues. Arlin (1975) proposed that a fifth stage of development, involving a new level of thought, may follow formal operations for some people and presented data showing that such a "problem-finding" stage existed. Arlin's (1975) ideas have been criticized (Fakouri, 1976), and as is the case with formal operations, there are some data speaking against the presence of such a fifth stage in some late adolescents (Cropper, Meck, & Ash, 1977). But there is some reason to believe that future research moving beyond the concerns emphasized by Piaget may show the use of a "fifth-stage" conception.

Riegel (1976b) suggests that if the impact of a dynamically changing environment is considered, then the sequence and end point of cognitive development that Piaget specifies would have to be altered. Riegel (1973, 1976b) proposes then that a fifth stage of cognitive development—termed *dialectical operations*—may follow the formal operational period. By attending to the person's dynamic interaction within a changing environment—a concern largely omitted by Piaget—Reigel (1973, 1976b) suggests new information may be provided about the qualitative nature of cognitive development in and after adolescence. It is for future research, however, to ascertain the appropriateness of this notion of a fifth stage.

CHAPTER SUMMARY

Adolescence is that period within the life span when most of the person's processes (such as the biological, psychological, and social) are in transition from what typically is considered childhood to what typically is considered adulthood. Because of the numerous changes that occur during adolescence, the person often is pressed into finding a coordinated way to reconcile these changes. Some scientists have depicted this endeavor as the adolescent's search for identity, or self-definition.

The bodily changes of adolescence may be divided into three successive periods. Changes in all these periods are promoted by the release of chemical substances, termed hormones, from ductless glands of the body, termed endocrine glands.

The first period of bodily change in adolescence is termed prepubescence. This period begins with the first evidence of sexual maturation and ends with the appearance of pubic hair. Signs of this stage in males are the progressive enlargement of the testicles and an increase in the length and circumference of the penis. In females signs are a rounding of the hips and the first visible indications of breast development.

In pubescence there is the growth of pubic hair and growth spurts in height and weight that occur for both sexes. However, as is the case for most bodily changes across all three periods, these changes occur about two years earlier in females than in males. For females the first menstruation (menarche) occurs and breast development continues. For males there is continued growth of the penis and testes and a change in the coloration of the scrotum. Postpubescence begins when pubic hair growth is complete and ends when all changes in primary and secondary sexual characteristics are complete.

There also exist sociocultural and historical differences, presumably due to variables relating to health and nutritional statuses of children and adolescents. The age of menarche, for example, varies today in relation to the cultural context within and differences between countries. Moreover, there has been a historical trend downward in the average age of menarche. A decrease of several months per decade from about 1840 to the present has been found.

Variation in the rate of bodily change in adolescence is related to psychological and social developments. Studies of early- and late-maturing adolescent males indicate that, in comparison to the latter group, the former group is better adjusted and has more favorable interactions with peers and adults. These advantages of early maturation, and the possession of a body type that has strong muscles and bones (a mesomorph), and disadvantages of late maturation, and either a plump (endomorph) or thin (ectomorph) body type, tend to continue through the middle adult years for males.

For females, however, somewhat different relations exist. Early maturation is associated with more psychosocial disadvantages than is late maturation. These relations for female adolescent rates of physical maturation, bodily appearance, and personality and social functioning have not been determined in later adult life.

It has been found for males that mesomorphy is the most positively regarded body type and that endomorphy and ectomorphy are the least favorably evaluated. For females, however, mesomorphy is not seen as the most favorable body type; instead, a build closer to ectomorphy is seen as more desirable.

Some adolescents have developmental problems relating to physical growth. Delayed puberty and precocious puberty are related to the insufficient production and the early production of hormones, respectively. Some physical problems may be known only to the adolescent initially. Painful menstruation, known as dysmenorrhea, and the failure to menstruate, known as amenorrhea, affect females and may have both medical and emotional causes. Other physical disorders are readily noticeable. Acne is an example of a problem which causes great concern to adolescents, since it is visible to others and is the most common disorder of medical significance during this period of life. About 85 percent of adolescents have acne at least sometime during adolescence. Eating disorders can be especially problematic in adolescence. Obesity affects a significant minority of adolescents. Less frequent, but even more severe in its effects, is anorexia nervosa, a disorder found almost exclusively among adolescent females. This condition involves severe loss of weight and is a sign of an emotional problem requiring professional attention.

Intelligence is a differentiated phenomenon. Prior to the adolescent years, a global set of intellectual abilities exists. Through childhood, adolescence, and early adulthood, a differentiation of mental capacities occurs. The variables that appear to moderate such changes are difficult to assess.

The study of cognition from a qualitative point of view is most associated with the theory and research of Jean Piaget. Knowledge about the reversibility of actions occurs in the concrete operational stage. Operations—internalized actions which are reversible—now exist, but thought is limited in that the person can only think about objects which have a concrete, real existence. The person does not have the ability to deal adequately with the arbitrary nature of thought.

Such recognition characterizes the formal operational stage, the stage most representative of adolescents in modern western society. Here the person can think of all possible combinations of elements of a problem to find a solution—the real and the imaginary. Because the person centers so much on this newly emerged thinking ability, some scientists believe the adolescent is characterized by a particular type of egocentrism. An adolescent may believe others are as preoccupied with the object of his or her own thoughts—himself or herself—as he or she is (this is termed imaginary audience). Because of the attention, the adolescent comes to believe he or she is a special, unique person (this is termed personal fable).

Research pertaining to Piaget's theory and adolescence has taken several directions. Data indicate that not all people within and across cultures attain formal operations or attain it completely. Other data suggest that in later life, formal operational abilities may be lost. However, since there are data showing that formal operational ability may be achieved through intervention, such loss may suggest the presence of nonoptimal contexts. Finally, both research and theory suggest that cognitive changes proceed beyond adolescence and that this may involve the emergence of a fifth stage of cognitive functioning.

chapter 10

Adolescence: social and personality development

CHAPTER OVERVIEW

The social world of the adolescent is one of transition from childhood to adulthood. Building on earlier development, the adolescent must find a way to contribute to the adult world he or she is about to enter. As in earlier portions of the life span, the adolescent's personality is involved in the person's social development. In this chapter we discuss several of the key features of the social and personality developments of adolescents. In particular, issues of adolescent sexuality and of adolescent identity are discussed.

CHAPTER OUTLINE

THE SOCIAL CONTEXT OF ADOLESCENCE

Peer Relationships
The Emergence of Peer Groups
Family Relationships

THE RELATIVE INFLUENCES OF PARENTS AND PEERS

THE GENERATION GAP

Studies of Intergenerational Relations

EDUCATION DURING ADOLESCENCE

Functions of the Educational System
Adolescents' Views about the Functions of Education
Family and Peer Influences on Education
Sociocultural and Historical Influences on Education

HIGH SCHOOL DROPOUTS

Magnitude of the Problem
Why Students Leave School
Consequences of Dropping Out

ADOLESCENTS AND THE POLITICAL SYSTEM

Legal Conceptions of Adolescence
Law and Adolescents
Adolescents as Voters
Political Socialization in Adolescence

SEXUALITY IN ADOLESCENCE

Sexual Behaviors of Adolescents

ADOLESCENT SEXUALITY TODAY

Incidence of Sexual Activity
The Initiation of Sex
Contraceptive Use
Reasons for Inadequate Protection
Sexual Standards

SOME CONSEQUENCES OF PREGNANCY OUTSIDE OF MARRIAGE IN ADOLESCENCE

Extent of the Problem
Outcomes of Premarital Pregnancy
Outcomes for Males

ISSUES TO CONSIDER

What is the nature of peer and family relationships in adolescence?

What evidence exists for a "generation gap"?

What are the functions of education in adolescence? Are peers or parents more influential in affecting adolescents' educational plans?

Why do adolescents drop out of school? What are the consequences of dropping out?

Are adolescents politically aware and active?

What are the forms of sexual behavior among adolescents? What is the incidence of sexual activity among today's adolescents?

Do sexually active adolescents use contraception regularly?

What are the major problems resulting from adolescent sexuality?

Why does Erikson claim that the identity crisis is the core problem of adolescence?

Does current research evidence provide support for Erikson's theory?

*T*hroughout our discussions of social and individual (personality) functioning in infancy and in childhood, we have emphasized that the person interacts reciprocally with his or her social context. Just as the person is affected by the context, the person affects the social world within which he or she lives. The social world of the adolescent is quite different from that of the infant or child. In part, the adolescent creates this change; as a consequence of his or her new cognitive functioning, new physical status, capabilities, and drives, the adolescent seeks out "broader horizons."

As such, in this chapter we will first depict some essential features of the adolescent's social context; peer, familial, educational, and other institutions in the context will be discussed. Next, we will consider one of the key changes brought about by the combined impact of physical and physiological changes, cognitive changes, and social context changes; we will discuss the emerging and changing nature of the adolescent's sexuality. We will stress how the adolescent uses his or her new body, and new conception of self, with elements of the social context (peers) in order to express and "test out" his or her individual sexuality. This sexual expression is, however, only one instance of the adolescent's expression of his or her new individuality. Like the other expressions, it may be understood as a part of the adolescent's search for a new definition of self—an identity. This new definition will, hopefully, "tie up" (integrate) all the processes of change that we have seen in the last chapter (and will see in this chapter) that typify the adolescent period. Thus we will end our discussion of adolescent development where it began in the last chapter—with a focus on Erikson's descriptions of the search for identity by the adolescent.

THE SOCIAL CONTEXT OF ADOLESCENCE

As we have noted, the adolescent's social context is broader and more complex than that of the infant and the child. Nevertheless, the components of the social context that were prominent in these earlier periods of the life span are still key ones in adolescence. Peers, family, and school are among the key elements of the adolescent's social world.

Peer Relationships

Perhaps the most notable social phenomenon of adolescence is the emergence of the marked salience of peer groups. Peers have the ability to make an adolescent feel on top of the world or at the bottom of the social ladder. Peers hold the key to adolescent popularity or rejection. Peers informally instruct the adolescent on how to talk, how to dress, and how to eat. And it is often the intensity of the way one conforms to peer-group norms that serves as a basis of parent-adolescent conflicts.

The study of peer relations is sometimes referred to as the study of *intragenerational* relationships. The study of parent-adolescent relationships, which crosses generations, is the study of *intergenerational* relationships. A *generation*, like a birth cohort, is a group of people born during one period of history. Usually, if one refers to all people born in just one year, the term birth cohort is used. The length of a generation is the span of time between one's birth and the birth of one's children. In the United States, this period averages about twenty-five years. The term generation, however, sometimes is used in a more general way to refer to spans of time varying from several years to three or four decades, and this latter use is applied in this book. Thus peer relations are sometimes called intragenerational relations, because peers are born within the same generation.

The Emergence of Peer Groups

It is sometimes said that there is strength in numbers. This theme seems to characterize much of peer-group interaction. Peer groups certainly exist in childhood, but they become more central to the individual in adolescence. The adolescent comes to rely heavily on the peer group for support, security, and guidance. Marked peer-group importance probably emerges because there is a great need for such support, security, and guidance during these years of transition, and it is easiest to find such things among others who are undergoing the same transition and with whom much time is spent.

From the standpoint of the adolescent, the peer group has much to offer. It is a sounding board; it gives constant feedback that all is well (or not so well) as adolescence progresses; it answers some of the important questions of adolescence (for example, about sexual relations and drug and alcohol use); and, by virtue of its size, it helps legitimize any behaviors or activities in which its members engage.

All peer groups change over time. Moreover, an adolescent may move from one peer group to another, and he or she may belong to more than one peer group. Furthermore, all adolescent peer groups in a given culture comprise the adolescent subculture. Thus when one speaks of peer groups, it is important to recognize the great variability in the existence of peer groups. However, despite such variability, there is much in common among adolescent peer groups.

During adolescence, persons begin new types of interaction with members of the opposite sex. They move into a new school environment (junior high school or high school), and they engage in a multitude of activities that may be new to them (for example, organized sports, dances, after-school clubs). All these new experiences can be mildly threatening if approached alone. The peer group often provides group support to its individual members so that these transitions become less formidable. Thus it can be argued that influential peer groups emerge during adolescence to help the individual through some new transitions and to provide support, security, and guidance to the persons involved.

In addition to the peer group's influence there is another part of the social context that is quite influential for adolescents. This is the family, and we explore its role next.

Family Relationships

Perhaps no social institution has as great an influence on development as the family. As we have emphasized in earlier chapters, the family is the basic social unit of society and is the most common location for reproduction, childbearing, and childrearing. Virtually all children in all cultures are socialized by families, although the form of the family unit may vary slightly.

Children and adolescents may find themselves in more than one family during their socialization. About 1 out of 5 children under 18 in the United States does not live with two parents (U.S. Bureau of the Census, 1978). There are a substantial number of single-parent families, and many children and adolescents live in families with two adults, but only one biological parent. These variations in living arrangements and family status come about, of course, as a result of the high rates of marital dissolution found in the United States. Divorce and remarriage are common in the United States today, with about 2 out of every 5 marriages involving recently married persons expected to end in divorce (U.S. Bureau of the Census, 1976). Moreover, about three-fourths of the persons involved in divorce eventually remarry—and those who remarry do so, on the average, about three years after the divorce. Thus many children and adolescents—perhaps 45 percent—can expect to live in a family setting without one of their natural parents for a period of time before they reach adulthood.

In summary, both peers and the family are components of the social context influencing adolescents. Further contributions of these components, as well as an appraisal of their relative impacts, are discussed below.

THE RELATIVE INFLUENCES OF PARENTS AND PEERS

Adolescents exist simultaneously within both family and peer groups, and one may ask how such dual commitments influence the adolescent's behavior and socialization. Do the family and peer relationships that adolescents have contribute in similar ways to development, or is one set of relationships more influential? There are data to suggest that both parents and peers are important; however, depending on the context and meaning of the social relationship, either parents or peers may be shown to be *more* influential.

Douvan and Adelson (1966) indicate that among 14- to 18-year-old male and female adolescents, there are few, if any, *serious* disagreements with parents. In fact, they report that in choosing their peers, adolescents are oriented toward those who have attitudes and values consistent with those maintained by the parents and adopted by the adolescents themselves. Similarly, Smith (1976), in a study of over 1000 sixth- through twelfth-grade urban and suburban black and white adolescents, found that the family, more so than the peer group, influences the adolescent to seek the advice and consider the opinions of parents. Similar findings have been reported by Kandel and Lesser (1972).

Thus although there are data indicating that in adolescence the person spends more time with peers than with parents (Bandura, 1964; Douvan & Adelson, 1966), this shift in time commitments does not necessarily indicate a corresponding alteration from parental to peer influence. For instance, Costanzo and Shaw (1966) found that conformity to peer-group norms increased between 7 and 12 years of age for males and for females but declined thereafter. Floyd and South (1972) studied sixth-, eighth-, tenth-, and twelfth-grade males' and females' orientation to parents and peers. They found less orientation to parents and more orientation to peers in older age groups, but there was a trend showing a mixed orientation—to both parents and peers—in these groups. Not only are the influences of parents and peers compatible in the values and behavioral orientations they direct to adolescents (Douvan & Adelson, 1966), but at older ages adolescents show an orientation to be influenced simultaneously by both these generational groups (Floyd & South, 1972).

Indeed, not only does it appear that this dual influence exists, but it also seems that which generational group is more influential at any particular time is dependent on the issue adolescents are confronting. Floyd and South (1972) found that when parents were seen as the better source about a particular issue, adolescents were more parent- than peer-oriented. Larson (1972) also found that the demands of the particular choice situation determined adolescents' choice, regardless of the direction of parent or peer pressures. This was true despite the fact that the fourth-, ninth-, and twelfth-graders Larson (1972) studied more often complied with parental than with peer desires. Similarly, Brittain (1963, 1969) found that both parents and peers influence adolescents, depending on the issue at hand; adolescent females were more likely to accept the advice of parents concerning the future and the advice of peers concerning school-related issues. Consistent with Brittain's data, Kandel and Lesser (1969) reported that 85 percent of middle-class adolescents and 82 percent of lower-class adolescents were influenced directly by parents in formulating future goals (in this case concerning educational plans).

Other studies also show an orientation to parents and peers, depending on the issue of concern. Chand, Crider, and Willets (1975) found agreement between adolescents and parents on issues related to religion and marriage but not on issues related to sex and drugs. Similarly, Kelley (1972) found high parent-adolescent similarity on moral issues but not on issues pertinent to style of dress, hair length, and hours of sleep.

Moreover, several studies indicate that although groups of adolescents and parents have somewhat different attitudes about issues of contemporary concern (e.g., war, drug use, and sexuality), most of these differences reflect contrasts in attitude intensity rather than attitude direction (Lerner, Karson, Meisels, & Knapp, 1975; Lerner & Knapp, 1975; Lerner, Schroeder, Rewitzer, & Weinstock, 1972; Weinstock & Lerner, 1972). That is, rather than one generational group agreeing with an issue while the other group disagrees (a directional difference), most generational group differences involved just different levels of agreement (or disagreement). For example, in regard to a statement such as, "Birth control devices and information should be made available to all who desire them," one study found that adolescents showed strong agreement with the item, while their mothers showed moderate agreement with the item (Lerner & Knapp, 1975).

Finally, consistent with the above data indicating influences by both peers and parents, there are data suggesting that adolescents perceive their own attitudes as lying between these two generational groups. In one study, adolescents were asked to rate their own attitudes toward a list of thirty-six statements pertaining to the issues of contemporary social concern noted above; in addition, they rated these same statements in terms of how they thought their peers would respond to them; last, they responded in terms of how they thought their parents would answer. The adolescents tended to see their own attitudes as lying *between* those of others of their own generation and those of their parents' generation (with twenty-seven of thirty-six—or 75 percent—of the items). Adolescents tended to place their own positions between the "conservative" end of the continuum, where they tended to put their parents, and the "liberal" end, where they tended to place peers. Interestingly, adolescents think their peers are more liberal than they actually are. In essence, this means that adolescents think their friends are, for example, using more drugs and having more sex than is actually the case.

In sum, the above data suggest that adolescents and their parents do not have many major differences in attitudes and values. Apparently, the impact of the intragenerational and intergenerational social contexts is often compatible. Indeed, adolescents tend to perceive that their values lie between those of their parents and peers.

THE GENERATION GAP

Despite the character of the influence of parents and peers that actually exists for the adolescent, there are recurring reports in the media and elsewhere that suggest significant disparities exist between the generations. The term *generation gap* is often used. The presence of such a gap would suggest a basis for conflict between adolescents and their parents.

Studies of Intergenerational Relations

Data reviewed previously suggest that when the *actual* attitudes of adolescents and their parents are compared, few major differences in attitudes can be found, although intensity differences often do exist. However, most studies find even this type of intergenerational disparity to occur in only a minority of attitude comparisons between the generations (Lerner et al., 1975; Lerner & Knapp, 1975; Lerner et al., 1972). These data suggest that the purported generation gap may not be real after all (Adelson, 1970).

Nevertheless, there are both empirical and theoretical reasons to expect that a generation gap might exist to some degree. There are data showing that adolescents and parents do not perceive the influence of social relationships accurately. Adolescents perceive their parents to be less influential than they actually are, while parents perceive that they are more influential than they actually are (Bengtson & Troll, 1978; Lerner, 1975; Lerner & Knapp, 1975). For example, two studies compared the actual and the perceived attitudes of adolescents and parents (Lerner et al., 1975; Lerner & Knapp, 1975). Actual intergenerational differences in attitudes about issues of contemporary societal concern occurred with fewer than 30 percent of the comparisons made in either study by Lerner and his colleagues. However, in both investigations, adolescents *overestimated* the magnitude of the differences that existed between themselves and their parents; they saw their parents as having attitudes less congruent with their own than was actually the case. In both studies, parents *underestimated* the extent of differences between themselves and their children; they saw their children and themselves as having attitudes very consistent with their own. Thus although only a small and selective generation gap can be said to actually exist, parents underestimate this division, while adolescents overestimate it.

Both psychological and sociological theorists have suggested reasons for the existence of these different perceptions regarding the generation gap. Erikson (1959, 1963) believes adolescence is a period in life which involves the establishment of a sense of personal identity, of self-definition. Other theorists (e.g., Elkind, 1967; Piaget, 1950, 1970) believe that adolescence is, as well, a period involving the development of new thought capabilities; these capabilities lead adolescents to believe their ideas are not only new in their own lives but in general as well. Together, then, adolescents need to establish their own identities and their own beliefs in the uniqueness of their thoughts. These orientations may lead them to believe they are quite different from those around them—especially their parents. This might result in their overemphasizing and magnifying whatever differences actually exist (Lerner, 1975).

A sociological approach suggests why parents might minimize the differences between themselves and their children. Bengtson and Kuypers (1971) suggest that members of the parental generational group have a stake in maximizing consistency between themselves and members of their children's generation. The parents have "invested" in society, for example, by pursuing their careers and accumulating society's resources. Because they want to protect their investment, they want to rear their children—the new members of society—in ways that will maintain the society in which they have invested. It may be that, as a consequence of such a "generational stake" (Bengtson & Kuypers, 1971; Bengtson & Troll, 1978), parents are oriented to believing they have produced children who—because they agree with parental attitudes—will protect their investment. This orientation is consistent with one Erikson (1959, 1963) describes; he believes that adults have a psychosocial need to feel they have generated children who will perpetuate society. This idea, as well as the generational stake idea, suggests that parents may be oriented to minimizing whatever differences exist between themselves and their adolescent children.

In summary, both theory and research combine to suggest that the adolescent's social context is composed of not one but several generation gaps. First, there exists a relatively minor and selective set of differences between adolescents and their parents. However, in addition to this actual gap, there exist two perceived gaps: the overestimated one of the adolescents and the underestimated one of the parents. The potential presence and function of these gaps make the social context of the adolescent a complex, diverse setting; we continue to see this complexity as we examine other dimensions of the social context. In addition, we will have continued reason to explore the relative influence of parents and peers on the other adolescent-institution interactions.

EDUCATION DURING ADOLESCENCE

Enrollment in school is a nearly universal characteristic of adolescence. At the beginning of adolescence, virtually all males and females attend school. About 99 percent of individuals aged 10 to 13 are enrolled in school in the United States, and this figure remains as high as 98 percent for 14- and 15-year-olds (U.S. Bureau of the Census, 1978). This high rate of school enrollment can be attributed to two factors: the high value Americans place on education and the compulsory laws of school attendance throughout the country.

In most states, however, a student is no longer compelled to attend school upon reaching his or her sixteenth or seventeenth birthday; consequently, only 89 percent of 16- and 17-year-olds are in school. About 85 percent of all males and females finish high school, and about half of all individuals who graduate from high school now go on to college (U.S. Bureau of the Census, 1978).

Contrary to what many may think, females drop out of high school as often as do males. The high dropout rate for females may be due to early marriage and the high incidence of adolescent pregnancy, as well as many reasons that typically apply to males. These reasons include attraction to employment more so than attraction to school, poor school performance, and peer-group pressures. Despite the notable proportion of individuals who drop out of high school, and the significant portion of the population who choose not to go on to college, education is a major influence in the lives of adolescents. For most of the years comprising the adolescent period, the preponderant majority of youths are enrolled in schools. With such a commitment of time, money, and attention by society to education, one may expect significant functions to be performed by the educational institution.

Functions of the Educational System

Schools perform many functions. Ideas about the multiple functions of the educational system have been expressed by several specialists in adolescent development. Ausubel, Montemayor, and Svajian (1977) see education basically as a training institution. They view the school as an agent of cultural transmission designed to perpetuate and improve a given way of life. Schools create a context for the transmission and attainment of basic knowledge. McCandless (1970) also considers schools to have a function of skills training and of cultural transmission of knowledge and values. However, he believes schools have at least one other major function, a "maintenance-actualization" role. McCandless believes the educational system creates a setting in which the adolescent can be happy and yet challenged; schools are a place to develop optimal personal and interpersonal attributes and, as such, maximize the person's ability to contribute to society.

Others have also noted the personal and interpersonal functions of schools. Ausubel et al. (1977) state that schools provide a context for social interactions and relationship development. Moreover, they facilitate the adolescent's emancipation from parents. School, then, affords an opportunity for the adolescent to earn his or her own social status. Status may be earned concurrently with school attendance by demonstrating a mastery of the curriculum, by attaining high class standing, and by nonacademic interactions with the peer group in school activities such as organized athletics or clubs (Ausubel et al., 1977).

Social status may be earned in the future through the training and education attained in school. Johnston and Bachman (1976) point out, for instance, that although far from perfect, there is a positive relation between an individual's advancement in education and the greater likelihood of success, particularly in work. They report educational advancement is positively related to lifetime income, high-status jobs, attractive working conditions, and opportunities for personal development. Johnston and Bachman (1976) also point out that the school serves a custodial role in society, in that a system of compulsory education, such as that found in the United States, highly structures the time and activity of students.

Coleman (1965) maintains that the major function of school to today's adolescent is as an interpersonal social milieu. Thornburg (1969, 1971) found that even during the height of the student activism in the late 1960s and early 1970s associated with the Vietnamese war, the draft, and civil rights, the dominant concern of college-age youths was their education. In summary, there are individual, interpersonal, and societal functions served by today's educational institutions.

Adolescents' Views about the Functions of Education

Several studies show that adolescents place greater emphasis on the school's role in interpersonal development than on the attainment and cultural transmission of skills and knowledge. In a study of urban and rural high schools having student bodies ranging from 100 to 2000 students, Coleman (1961) found that academic accomplishment was not a predominant basis of peer popularity. For instance, only 31 percent of the males studied wanted to be remembered by their peers as excellent students, but 45 percent wanted to be remembered as outstanding athletes. Among females, 28 percent wanted to be remembered as excellent students, but 72 percent wanted to be recalled as being popular. Similarly, Snyder (1972) found that among high school juniors, the most important criterion selected by both males and females for giving someone peer recognition and status involved personal qualities. Next in importance were material possessions, social activities, and athletics. Academic achievement followed all these qualities in their rankings.

Moreover, Johnston and Bachman (1976), in a study of a national probability sample of 2100 high school teachers, found that both teachers and students had almost identical views regarding what were the actual functions of current high schools. Both groups believed that athletics was the area receiving most emphasis in their schools. Functions related to skill attainment receive less emphasis, and those related to cultural transmission of norms and values still less.

However, other data reported by Johnston and Bachman and other researchers indicate greater salience for these latter functions. Frieson (1968) studied about 15,000 students in nineteen Canadian high schools. He found that students saw athletics and popularity as more important for current successful functioning, but they believed academic achievement was more important than either for future successful functioning. In addition to seeing a future utility for the skill attainment and cultural transmission roles, data by Johnston and Bachman (1976) suggest that adolescents hope that such functions will become the ones of primary emphasis in the schools of the future. Instead of athletics, the students and teachers in the study agreed that "increasing students' motivation and desire to learn" should be the most important function and that several issues relating to academic achievement were of utmost priority.

Among a national sampling of college seniors, there was an even greater stress on skill attainment and cultural transmission functions. Hadden (1969) found that 75 percent of the students studied saw college as a "symbol of hope in a troubled world" and fewer than 20 percent felt that what they were learning was "silly, wrong, and useless." Individual and interpersonal enhancement functions were also emphasized. About two-thirds of the sample agreed that "college has changed my whole view of myself."

It appears that all of the roles of education mentioned are recognized by today's students as important aspects of educational institutions. Moreover, data from a national sample of students show that more than 75 percent believe that schools are doing a good-to-excellent job in providing an appropriate context for students (Harris, 1969).

Family and Peer Influences on Education

The family and peers of adolescents influence both aspirations about and actual outcomes of the adolescents' accomplishments in school. According to information from the U.S. Bureau of the Census (1978), the college aspirations of high school seniors tend to be correlated with the educational attainment of the head of the household in which they live. About 70 percent of students who were living in households in which the head had completed at least one year of college themselves had definite college plans. When the head of the household had completed high school but not any college, only 45 percent of students had definite college plans. Of those living in families having a head of household who had not completed high school, only 35 percent had definite college plans.

Sewell and Shah (1968a, 1968b) found that parents' educational attainment is highly related to their adolescent children's educational aspirations and also to the actual success of the adolescents in the school setting. High educational attainment of parents, and particularly fathers, was found to predict similarly high (e.g., college) attainment by students.

There are particular types of parent-adolescent interactions that appear to facilitate successful school functioning. Morrow and Wilson (1961) found that high-achieving adolescents, as compared to a group of low achievers, tended to come from families where they were involved in family decisions, where ideas and activities were shared by family members, and where parents were likely to give approval and praise of the adolescents' performance and show trust in the adolescents' competence. In turn, low-achieving adolescents came from families marked by parental dominance and restrictiveness (Morrow & Wilson, 1961). Both Morrow & Wilson (1961) and Shaw and White (1965) found that high-achieving adolescents tend to identify with their parents, while low-achieving adolescents do not.

Still other data indicate that the types of parental behaviors found by Morrow and Wilson (1961) relate to high adolescent school achievement. Both Swift (1967) and Rehberg and Westby (1967) report that parental encouragement and rewards are associated with better adolescent school performance. Wolf (1964) reports that parent-child interactions that involve encouragement to achieve and development of language skills are highly correlated with intelligence.

Peers also influence adolescents' aspirations and educational performance; in most cases, there is convergence between family and peer influences. Rigsby and McDill (1972), in a study of over 20,000 adolescents, found that there was a positive relation between the proportion of peers perceived to have college plans, the actual proportion with college plans, and adolescents' own likelihood of planning for college.

Similarly, Kandel and Lesser (1969) found that if adolescent peer relationships were characterized by closeness and intimacy, then there was a great deal of correspondence between the educational aspirations of the peers and of the adolescent; however, most adolescents (57 percent) had educational plans that agreed with peers and parents. In turn, among those adolescents who disagreed with their parents, there was also a great likelihood (50 percent) that they would disagree with peers as well. In those cases when there was a discrepancy between parent and peer orientations, it was most likely that the parental orientation would prevail (Kandel & Lesser, 1969).

Sociocultural and Historical Influences on Education

The political, economic, and environmental changes in society, which are part of the sociocultural context, influence the nature of the educational institution. For instance, changes in political and economic pressures have resulted in a greater number of minority youths being a part of the student body of higher educational institutions. For example, only about 26 percent of blacks who were in their early twenties in the late 1960s had completed a year or more of college. This figure increased to 32 percent for blacks who were in their early twenties in the late 1970s. Whites, in contrast, attended college in similar proportions over the decade. About 4 in 10 whites completed a year or more of college during both time periods (U.S. Bureau of the Census, 1977b, 1978).

The choice of a major also may be influenced by sociocultural changes across history. Economic recessions, difficulty in finding employment, and environmental concerns were examples of social issues in the late 1960s and early 1970s. Changes in the popularity of different college major fields of study paralleled changes in these social issues. An analysis of the changes of majors found between 1966 and 1974, for example, indicates unpredictable enrollment shifts.

For instance, social science majors increased from about 640,000 to 950,000 between 1966 and 1972 but declined to 770,000 by 1974. The initial increase undoubtedly was related to heightened interest in social services, social change, and perhaps to a desire for more relevance in the lives of college students of that time. The decline that followed may have occurred partly because of a resulting oversupply of social science majors, given the level of available jobs (U.S. Bureau of the Census, 1978).

HIGH SCHOOL DROPOUTS

Only 85 percent of Americans finish high school. Approximately equal proportions of males and females drop out of school, mostly during the junior and senior years. Since all states have some form of compulsory attendance law requiring school enrollment until age 16 or 17, in the typical case adolescents, parents, and school personnel usually do not have to contend with the problem until midadolescence. But the seeds of a student's unrest which may lead to dropping out may develop at any time. In this section we explore the magnitude of the problem of school dropouts, some of the causes, and some of the consequences.

Magnitude of the Problem

Educational attainment of the American population continues to increase. Decade by decade since the turn of the century, an increasing proportion of the population has graduated from high school. The proportion of adolescents graduating from high school reached an all-time high in the 1970s, and is now stabilizing. Nevertheless, with about 15 percent of the population not completing high school, a sizable portion of adolescents still must contend with the problems associated with dropping out of school.

More than 800,000 adolescents dropped out of school during the 1975–1976 school year. Most dropouts do not return to high school after leaving it, although some obtain their high school degree through night school or by taking a high school equivalency examination. In 1975, 3 percent of persons graduated from high school by equivalency examination, and about 1 percent graduated by attending night school or by attending day school part-time (U.S. Bureau of the Census, 1977, 1978).

Blacks and persons of Spanish origin have very low rates of high school completion. About two-thirds of black adolescents finish high school, and about 55 percent of persons of Spanish origin finish high school. The completion rates for male and female members of these minority groups do not differ significantly (U.S. Bureau of the Census, 1978).

Why Students Leave School

Many studies of school dropouts are unable to point to a single reason for the problem. Each adolescent who decides not to finish school has his or her own reason, and the causes found in the dropout population as a whole are numerous. One national study, the longitudinal *Youth in Transition* project, conducted at the Institute for Social Research of the University of Michigan, found that family background factors and ability were related to educational attainment (Bachman, 1970). Those most likely to finish high school and enter college were from families with higher socio-economic standing, were more likely to come from intact homes, and had smaller family sizes. Individuals who dropped out of school had lower intelligence, vocabulary, and reading abilities (Bachman, Green, & Wirtenen, 1971). As might be expected, then, adolescents who find themselves struggling academically through school are most likely to drop out. Indeed, Bachman et al. found that reasons most often offered by adolescents for why they dropped out pertained to schoolwork and school authority.

There are other nonacademic reasons for leaving school. Dropouts may find little intellectual stimulation at home. They may encounter a negative evaluation of education by their parents or pressures from the family to obtain work to help support the family financially. Realistically, however, many adolescents who report that their parents encouraged them to drop out of school may be looking for someone else to blame. Even parents with little education themselves tend to respect the

Adolescents may drop out of school in order to work and earn money.
(E. and F. Bernstein/Peter Arnold, Inc.)

value of education and see educational attainment as a means of improving one's social standing. Probably only a small proportion of parents actually encourage their adolescent children to leave school. Bachman et al. (1971) note that some other reasons that adolescents give for dropping out are a desire for freedom, getting married, a need to earn money, wanting to get a job, and need to help the family financially.

Other nonacademic reasons may be more influential, however. Dropouts have lower levels of self-esteem and are more often depressed than those who finish high school, while those who did not drop out had higher needs for self-development. These and other personality factors may be involved in leaving school (Bachman et al., 1971).

It is not difficult to understand high school dropouts when one thinks of the adolescent who is doing poorly in school, receives little encouragement at home, perhaps is in trouble from time to time with teachers and administrators, and has to get out of bed every weekday morning for nine months every year to do something that he or she cannot tolerate. Poor school attendance, behavior problems in the classroom and on school grounds, vandalism, and delinquency are consequences which might be expected when one considers the frustration that some adolescents experience in what they consider to be an oppressive environment.

One study suggests that individuals who drop out of school, particularly those from lower-class backgrounds, have failed to achieve middle-class values, which emphasize education (Namenwirth, 1969). Success in school requires acceptance of the traditional middle-class values, and this may be an elusive goal for persons from lower-class families. Indeed, delinquency and other forms of socially deviant behavior are more likely to precede than to follow dropping out (Bachman et al., 1971). Another explanation of failure to finish high school suggests that the school experience simply fails to meet the personal, social, and vocational needs of many youths,

Adolescent delinquency and dropping out of school are often linked.
(Charles Gatewood.)

particularly those in urban areas and those from lower-class backgrounds (Mussen, Conger, & Kagan, 1974).

It is likely that adolescents who drop out of school very early are different from those who drop out later. There is some evidence that early dropouts have lower-than-average intelligence and may find it difficult to compete in school, even if they were motivated. Those who drop out later tend to have higher intelligence than those who are early dropouts, although both groups score lower on intelligence tests than do students who finish school (Voss, Wendling, & Elliott, 1966).

One ambitious study of high school dropouts was based on a national sample of 440,000 students attending 1300 public and private high schools throughout the country (Combs & Cooley, 1968). This study compared dropouts to high school graduates who did not go on to college. Males who dropped out of school were lower on all measures of ability, although the scores were similar for females. Males who finished school had greater interests in science, engineering, math, and related areas, whereas males who dropped out of school tended to have interests in skilled trades, labor, and music. Females showed similar differences. Male and female graduates perceived themselves to be more sociable, vigorous, calm, tidy, cultured, self-confident, and mature than did those who dropped out (Combs & Cooley, 1968).

Thus it appears that there are significant differences between adolescents who leave high school and those who finish. Certainly in contemporary society, the high school dropout is somewhat stigmatized, since he or she is violating a social norm and rejecting a social status which has come to be highly valued by virtually all segments of society. What, if any, are the consequences of such a decision?

Consequences of Dropping Out

The consequences of not finishing high school are difficult to assess. One form of evaluation is an economic one. High school dropouts can expect to have a lifetime earnings of about $575,000, on the average, based on the 1980 value of dollars. High school graduates are projected to earn substantially more—about $855,000—in their lifetime. Those who obtain college degrees may expect lifetime earnings of about $1,120,000 (U.S. Bureau of the Census, 1977a).

Of course, money is just one measure of the impact of differential levels of educational attainment. High school dropouts have higher unemployment rates than do persons with more education, are more likely to have blue-collar jobs, and tend to be employed in positions requiring minimal skills (U.S. Bureau of the Census, 1977a).

Thus one's educational level will usually determine one's employment possibilities. Many jobs require a high school degree, even if specific skills learned in high school are not required for the job. Many employers, then, expect employees to have high school degrees regardless of the job description. Furthermore, some employers require the degree for advancement to higher positions.

There are numerous other circumstances which are related to lower educational attainment. Although these events cannot be directly attributed to an incomplete high school education, they undoubtedly are products of a lifestyle which often includes dropping out of school. Poverty, early marriage, high fertility, high rates of marital disruption, higher mortality rates, and earlier death are all examples of variables related to educational attainment (U.S. Bureau of the Census, 1977a; Spanier & Glick, 1979). Finally, however, it may be noted that Bachman et al. (1971) report that more often than not those who dropped out regretted this action.

ADOLESCENTS AND THE POLITICAL SYSTEM

The political and governmental system existing in any social context provides a set of institutions linking the person to his or her society. Adolescence, in modern western culture, is a period within the life span wherein especially important social changes regarding the political system occur.

Today's adolescents invariably are influenced by the systems of government they encounter in their local communities and their state and nation. Often they are constrained in their behaviors by these institutions; for example, there may be laws governing use of alcohol. At other times they are encouraged to become involved in the political system. By an amendment to the U.S. Constitution, for example, 18-year-olds now have the right to vote in national elections. In this section we review the laws pertinent to adolescents and how adolescents behave in regard to their political system.

Legal Conceptions of Adolescence

There is no legal definition of adolescence. The law in most states recognizes, however, ages of minority (birth until the eighteenth birthday, for instance) and ages of majority (typically 18 years and older). As with a chronological definition of adolescence, such legal definitions of minors (children and "youths") and of adults are arbitrary. Different states or different countries might define children and youths and adults differently. Moreover, within the same state we may find variations in ages at which "adult" behaviors such as voting, drinking, or signing a legal contract may be permitted. But such variation per se has rarely been the focus of concern of those interested in the transitions of adolescence. Although changeable, a law is relatively static and, as such, generally has not been regarded as a dynamic component of a developmental view of adolescence.

Law typically is not a traditional component of conceptions of adolescence. Yet the laws about children and youths certainly reflect society's conception about the status, obligations, and responsibilities of these people. Moreover, consideration of how such laws have changed across the history of a society, or now exist differently in different cultures, will reveal the evolving and contrasting meanings attached to children and youths across time and geographical area. For example, a review of the philosophical and historical evolution of the notion of development, and of adolescent development in particular, suggests that for a great span of history, no distinctions in physiological or social responsibility were made between children and adults. Accordingly, being just "miniature adults," children were expected to work as long and in some of the same tasks as full-grown adults. If they shirked these or other social and moral responsibilities, they could be punished by the same laws as could adults.

Today, of course, there are laws protecting children with regard to work and legal culpability. In fact, children and youths are not allowed to work at certain jobs in some states, and in all states there are limits imposed on how long or hard they may be permitted to work or below what age work may not be permitted at all. Indeed, in many states youths have to have a "license" (i.e., apply for "working papers") in order to be employed in some positions. Similarly, the legal system for youthful offenders is not the same one an adult encounters. Up until the age of majority, youths in many states are not considered totally responsible for criminal acts they commit. Parents, for example, may be required to pay for damages or legal costs incurred by their children.

Accordingly, the nature of the legal discriminations between children and youths, on one hand, and adults, on the other, reflects the current psychosocial differences society sees or does not see regarding people at different points in their life spans. Furthermore, changes in these perspectives reflect the evolving philosophy about and attitudes toward children, youths, and adults that exist in a culture over time. Such changes may be seen as a product of other biological, cultural, legal, and historical changes that also exist. For instance, the number of persons under age 18 who were arrested more than doubled from 1960 to 1975 (to a total of more than 2 million). The rate of serious crimes almost doubled during this same period (Uniform Crime Reports, 1975). One may ask how the altering conceptions of youths, reflected in laws for them, have contributed to this increase. Furthermore, how may this increase influence changes in present laws?

Law and Adolescents

Adolescence has been described as the transitionary period between childhood and adulthood. This definition is particularly appropriate when we consider the legal status of the adolescent. In most states, 18 is defined as the age at which an individual is no longer legally responsible to parents nor are the parents responsible for the children beyond this time. The adolescent is legally *emancipated*.

Many things change for the adolescent upon reaching this age of majority (five states have older ages, either 19 or 21). Although the laws vary from state to state, it is typically at this time when the adolescent would be tried in adult court for commission of a crime. Before this age, he or she would have reported to juvenile court for adjudication. After this age, adolescents become responsible for their own debts;

they can sign legal contracts, buy property, leave home without being defined as runaways; and they can give consent for surgery or other medical procedures. These and a host of other rights and responsibilities come with legal emancipation.

For many adolescents, reaching the legal age of adulthood comes before they have reached adulthood socially, psychologically, or economically. In other words, the 18-year-old may still depend on parents but has the legal right to be considered totally apart from parents and family. This poses a conflict in some families, where parents wish to continue to exert influence over their children—influence which may be rejected by the children. The reverse is also true in some cases. Some parents want their children to become self-supporting and independent, but the adolescent may continue to wish to be supported by the parents, both economically and psychologically.

Most parents and adolescents realize that a balance between legal emancipation and psychological emancipation is necessary. College educations can rarely be financed without some form of parental support, and for non-college-bound adolescents, starting out on one's own can be difficult without some support from one's family. Although most families are able to conduct this transition in a smooth way, others find that it is a constant struggle to untie the knot that held the adolescent and parents together, whether or not the individuals involved were on good terms when the transition began.

Exhibit 10.1 shows the state-by-state breakdown of ages of majority as of the end of 1975. Also shown are some of the legal conditions applying to the individual's right to consent to medical care, obtain contraceptive services, obtain abortions, be treated for venereal disease, and receive pregnancy-connected care. It can be seen that there is great variability from state to state and that the intricacies of the law can be complex. The reader is also cautioned that some of the specific ages and rules which apply have assuredly changed since this exhibit was prepared. Furthermore, some of the information may be subject to legal interpretation.

It can be seen from Exhibit 10.1 that some states have also made determinations about what legal rights *minors* have. Whereas adolescents are often inclined to think about what rights they may gain upon emancipation, there are some rights which they have while still minors. In many states, for example, an adolescent may obtain pregnancy-related care, be treated for venereal disease, or even receive an abortion, regardless of age and without parental consent. Furthermore, most states will allow the adolescent to consent for medical treatment in an emergency when parental permission may not be readily available.

Upon legal emancipation, parents and adolescents need to investigate the status of health insurance policies, automobile insurance coverage, and any other plans, policies, or services which may no longer include the adolescent. In some cases, the adolescent will need to arrange for his or her own coverage as a result of this new legal status.

Adolescents as Voters

Before 1972, only persons 21 and over were permitted to vote in national elections. The right to vote was given to adolescents aged 18 to 21 after a turbulent time in the 1960s, when America's youths became quite politically active and vocal. Thus beginning with the 1972 presidential elections, persons 18 years old and over have been permitted to vote in national elections. Yet the voter turnout among this newly enfranchised group was low in the 1972 presidential election and even lower in the 1976 presidential election; 38 percent of persons 18 to 20 years old reported that they voted in 1976, whereas 63 percent of persons 25 and over voted in the 1976 presidential election (U.S. Bureau of the Census, 1978).

Thus, older adolescents who have the right to vote do not seem to exercise this right as often as persons who are older. Several reasons may account for this. In follow-up surveys, some youths simply reported that they were not interested in the election, and many reported that they were unable to register. The mobility of youths may affect their voter participation, since they are often away from home at the time of elections (U.S. Bureau of the Census, 1978).

Political Socialization in Adolescence

Earlier in this chapter, we discussed the alleged generation gap between adolescents and their parents in regard to attributes and values about contemporary social values. In that presentation we noted that although adolescents often may believe they have beliefs that contrast with those of their parents, there is, in fact, considerable similarity; it appears that parents do fairly well in transmitting their own beliefs about social issues to their children. A similar conclusion may be reached in regard to the political socialization of children and adolescents.

Although they often believe otherwise, adolescents tend to hold political attitudes consistent with those of their parents (Niemi, 1973). Indeed, there is evidence indicating that such socialization of political attitudes is evident, and quite resistant to change, by *early* adolescence. Gallatin (1975) notes that high school students' exposure to various types of social studies curricula does not appear to have much impact on political attitudes and behavior. In fact, Jennings and Niemi (1968) and Langton and Jennings (1968) found that high school students who had taken several civics courses did not answer factual questions better than those who had taken only one course. In addition, high school students in issues-oriented political courses did not differ greatly from those students in relatively conventional social studies courses, in regard to their concerns about social matters and politics. The researchers suggested that by the time people reach high school, they may have already been so politically socialized that course work cannot affect their attitudes. This view is supported by Hess and Torney (1967), who suggest that by the very beginning of adolescence people have acquired their basic orientations to the political system and are unlikely to alter their views thereafter.

However, there do seem to be changes in the *reasons* adolescents offer for their political views. Adelson (1971), Adelson and O'Neil (1966), and Gallatin (1972, 1975) report that in response to questions pertaining to government, law, and political parties, younger subjects were more concrete, authoritarian, and categorical in their thoughts and views, while older adolescents gave more abstract, humanitarian, qualified, and informed responses. Interestingly, Gallatin (1972) found that the older adolescents offered reasons for their answers that incorporated many of the political principles involved in the Bill of Rights of the U.S. Constitution.

In sum, the adolescent's social context is made of numerous elements, each with a different set of pressures to which the adolescent must adapt. Yet these pressures are put on the adolescent at just the time when he or she is changing in several ways—physically and cognitively, for example. The adolescent must deal with these changes as he or she deals with his or her changing world.

One of the key changes the adolescent must deal with is alterations in sexual and reproductive capacity. These changes, bringing about a "new" sexual drive (A. Freud, 1969), impel the adolescent to deal with his or her sexuality as he or she is adapting to a changing social context. The study of adolescent sexuality is then not

EXHIBIT 10.1

Age of Majority, and Ages at which State Legislation, Court Action, or Attorneys-General Opinions Have Specifically Affirmed the Right of Individuals to Consent for Medical Care in General, for Contraceptive Services, for Examination and Treatment of Pregnancy and VD, and for Abortion; as of December 31, 1975 (X = Any Age)*

| State | Age of Majority | May Consent for Medical Care in General | | | May Consent for: | | | |
		No Limitation	If Married (M) or Emancipated (E)	In Emergency	Contraception	Pregnancy-Connected Care	VD Care	Abortion
Ala.	19	14	E[9], M	X	14	X	X	14
Alaska	19, MF	19	E[6,7,8]	X[5,28]	X	X	X	18
Ariz.	18	18	E, M	X[10]	18	18	X	18
Ark.	18	X[2,4]	E, M	X	X[2,4,12]	X[14]	X	18
Calif.	18	18	15E[6], M	X	X[3]	X	12	X
Colo.	18[1]	18	15E[6], M	18	X[2]	18	X	X
Conn.	18	18	E, M	18	18	18	X[18]	18
Del.	18	18	E, M	18	12[11]	12	12	18
Fla.	18	18	E, M	X	X[20]	18	X	X
Ga.	18	18[3]	M[3]	X	XF[3]	X	X	X
Hawaii	18	18	18	18	18	14[27]	14[27]	14[27]
Idaho	18	18	18	18	X[4]	18	14	18
Ill.	18[1]	18	M[7]	X	X[15]	X	12	X[24]
Ind.	18	18	E, M	X	18	18	X	X[24]
Iowa	18, M	18	E, M	X	18[12]	18	X	18
Kans.	18	X[4], 16[5]	18	16	X[4]	X	X	X[4]
Ky.	18	18	E, M[8]	X	X[3]	X[3]	X	X
La.	18, M	X[21,3]	M	X	18[12]	18	X	18
Maine	18	18	E	X	X[23]	18	X	X
Md.	18	18	M[8]	X	X[3]	X	X	X
Mass.	18	18	E[3,6], M[5]	X	18	X	X[19]	X
Mich.	18	X[4]	E, M	X	X[4,12]	X[4]	X	X[4]
Minn.	18	18	E[6], M[8]	X	X[13]	X	X	X[29]
Miss.	21	X[4]	E, M	X	X[16]	X	X	X[4]
Mo.	18[1]	21	E, M[8]	X	21	X[14]	X	18
Mont.	18	X[23,3,14]	E[3,14] M[7,3,14]	X	18	X[14]	X	18[27]
Neb.	19, M	19	M	19	19	19	X	X[24]
Nev.	18	18	E[3,4,14] M[3,4,14]	X[3,4,14]	18	16	X	18
N.H.	18	X[4]	E, M	X[4]	X[4]	X[4]	14	X[4]
N.J.	18	18	E, M[7]	18	18	X	X	X
N. Mex.	18	18	E, M	X[10]	18[12]	X[17]	X	18
N.Y.	18	18	E, M[8]	X	X	X[4]	X	X[4,25]
N.C.	18	18	E, M	X	18	18	X	18
N. Dak.	18	18	E, M	18	18	18	14	18
Ohio	18	X[4]	18	X[4]	X[12,4]	X[4]	X	18

EXHIBIT 10.1 *(Continued)*

State	Age of Majority	No Limitation	If Married (M) or Emancipated (E)	In Emergency	Contraception	Pregnancy-Connected Care	VD Care	Abortion
		May Consent for Medical Care in General				May Consent for:		
Okla.	18	$X^{23,3,14}$	$E^{3,14}$ $M^{8,3,14}$	X	18^{12}	$X^{3,14}$	X	18
Oreg.	18^1, M	15	M	15	15^{14}	15^{14}	12	18
Pa.	21	18	E^9, M	X	X^{26}	X	X	X
R.I.	18	18	E	16, M	18	18	X	18
S.C.	18	16^{22}	E, M	X	16	16	X	16
S. Dak.	18	18	E, M	18	18	18	X	18
Tenn.	18	18	18	18	X^3	18	X	18
Tex.	18	18	16E, M	X	18	X	X	X
Utah	18, M	18	M	X	X	X	X	X^{27}
Vt.	18	18	E, M	18	18	18	12	18
Va.	18	18^3	E	18	$X^{3,14}$	X^{14}	X	18
Wash.	18	18	E	18	18	18	14	X
W. Va.	18	18	18	X	18^{12}	18	X	18
Wis.	18	18	E, M	18	18	18	X	18
Wyo.	19	19	19	19	19^{12}	19	X	19
D.C.	18	18^3	E, M^3	X	X^3	X^3	X	X
Total At 18	45	35	6	13	21	18	0	24
<18		12	44	36	27	31	51	26

* The fact that no affirmative legislation, court decision, or attorney-general's opinion has been found in a particular state does not mean that some or even all categories of minors below the ages shown in the table do not have the right to obtain some or all medical services on their own consent.

Note: Because of reporting lags, the table probably does not include all applicable legislation, cases, and attorneys-general opinions for 1975. M = married; F = female; E = emancipated.

1. For purposes of signing contracts. **2.** Excluding voluntary sterilization if under 18 and unmarried. **3.** Excluding voluntary sterilization. **4.** If mature enough to understand the nature and consequences of the treatment. **5.** If parent not immediately available. **6.** Emancipated defined as living apart from parents and managing own financial affairs. **7.** And/or pregnant. **8.** Or parent. **9.** Emancipated defined as a high school graduate, a parent, or pregnant. **10.** If no parent available, others may consent in loco parentis. **11.** If sexually active. **12.** Comprehensive family planning law permits (or does not exclude) services to minors without parental consent. **13.** Unless parent has previously notified treating agency of objection. **14.** Excluding abortion. **15.** If referred by clergy, physician, or Planned Parenthood or if "failure to provide such services would create a serious health hazard." **16.** If referred by clergy, physician, family planning clinic, school or institution of higher learning, or any state or local government agency. **17.** Examination only. **18.** In public health agencies, public or private hospitals, or clinics. **19.** In publicly maintained facilities. **20.** If married or pregnant or "may suffer, in the opinion of the physician, probable health hazards if such services are not provided." Surgical services excluded. **21.** If minor "is or believes himself to be afflicted with an illness or disease." **22.** Except for operation essential to health or life. **23.** If physician finds probable health hazard. **24.** Parental consent requirement temporarily enjoined by court. **25.** In New York City, municipal hospitals perform abortions on minors without parental consent if married, emancipated, or at least 17 years old or if seeking parental consent would endanger the physical or mental health of the patient. **26.** Minors are being served under a state law which permits doctors to serve minors of any age if delay in treatment "would increase the risk to the minor's life or health." **27.** Parent notification, but not consent, is required, where possible. **28.** If parent refuses to grant or withhold consent. **29.** County attorney stated that legislature did not intend to include abortion as pregnancy-related treatment.

Source: Paul, Pilpel, and Wechsler, 1976.

only an important topic in and of itself; in addition, it is illustrative of the relation between a changing person and a changing world.

SEXUALITY IN ADOLESCENCE

Sex is one of the greatest concerns of adolescents and of adults who worry about them and it is not difficult to imagine why this is so. Although sexual development begins in childhood, it is not until adolescence that sexuality becomes an important focus in the lives of most individuals.

As we have noted, the adolescent's hormonal balance changes throughout this period. The genitals develop to their nearly adult form. Both males and females begin to experience new feelings, including new bodily functions such as menstruation for females and ejaculation of seminal fluid for males. Perhaps more important than these physical changes are the social and psychological changes which accompany adolescence. Sexual *behavior* and sexual *interaction* take on meaning as the adolescent begins to think about social events in school, parties outside of school, love, and dating.

Society has taken a special interest in the topic of adolescent sexuality in recent years, undoubtedly because of the vast changes that have taken place in the extent of sexual behavior among young persons. In addition, there is growing concern about the potential consequences of sexual involvement during adolescence. Not only is sexuality sometimes a source of anxiety for the individuals who must negotiate adolescence, but it is also the source of some of our society's most troublesome social problems. Pregnancy outside of marriage is an increasingly prominent feature of the adolescent years for a sizable number of American youths. The consequences of pregnancy—abortion, illegitimacy, and early marriage—are all significant social issues. Moreover, many topics pertaining to sexual development, such as homosexuality, are of considerable interest and controversy. Finally, some of our most fervent religious and moral debates center around issues such as the acceptability of adolescent sexual behavior.

Sexual Behaviors of Adolescents

Sexual expression may take on many different forms. When we think of adolescent sexuality, we often think first of heterosexual intercourse. However, many forms of sexual activity short of intercourse are important during adolescence.

For males, the beginning of adolescent sexuality is often a *nocturnal emission* (or "wet dream"), a discharge of semen during sleep. The experience is accompanied by an orgasm and is often associated with sexually colored dreams. Although females may also have sexual dreams, and are capable of experiencing orgasm during sleep, they do not have a similar discharge.

Masturbation is another normal adolescent sexual activity (Kinsey et al., 1948, 1953; McCary, 1978). Masturbation for some begins in childhood, but it is a widespread phenomenon during adolescence. It is nearly universal among males, and a majority of adolescent females report masturbating at some time. Whereas it tends to become a regular activity for males once they begin, it ranges from a rather sporadic event for some females to a regular occurrence for others.

Contrary to common assumption, there is no evidence that masturbation is a harmful activity. Some adolescents develop an acute sense of guilt about masturbation, feeling ashamed or embarrassed. Parental threats to their children to avoid such sexual release will generally exacerbate feelings of guilt and are unlikely to prevent the practice. Adolescents have sexual tension which can be relieved through masturbation, and it is unlikely that an adolescent would easily abandon the activity. Furthermore, some women are able to achieve orgasm only by masturbation, and others report more intense orgasm from masturbation than from coitus (Ellis, 1958; Masters & Johnson, 1966; Hite, 1976).

Research shows that heterosexual activity progresses through a sequence of steps for the typical adolescent (Spanier, 1976). Kissing, light petting, heavy petting, and coitus are encountered in sequence for most persons. There is great variation as to when a person has his or her first romantic kiss, how quickly one progresses from slight involvement to complete sexual involvement, and when in the life cycle the sequence is completed. Indeed, some individuals have already had sexual intercourse by the time they reach puberty; others have never kissed someone (or had a date) by the conclusion of the adolescent years; and still others may go through the entire sequence of sexual involvement on the first date (Spanier, 1977; Zelnik & Kantner, 1977).

Interaction with members of the opposite sex increases dramatically in adolescence. (Jon Vincent Veltri/Photo Researchers, Inc.)

Petting is rarely discussed, but it is an important topic. For married persons or adolescents who are having sexual intercourse regularly, petting may only be a form of foreplay—a prelude to sexual intercourse. But for many adolescents, petting is the sexual focus of a relationship. We have many terms in our vocabulary to describe the phenomenon—making out, feeling up, going to third base, Russian hands and Roman fingers, etc. Many adolescents readily engage in petting since they find it a pleasurable and stimulating activity. In many cases, it is an indication of the affection and interest a male and female have for each other.

However, petting may lead to a level of arousal which makes it difficult not to advance further; consequently, many adolescents, particularly males (Kanin, 1970), find it difficult to be satisfied with petting as an end in itself. Indeed, it is not uncommon for the sexual tension generated in petting to lead to orgasm for males and females, without intercourse. Unrelieved sexual tension may sometimes result in pain in the region of the testes in the male, a condition known as *orchialgia*, or *testalgia*. It is pain probably due to congestion in the prostate gland (Cawood, 1971). Such pain is uncomfortable but disappears in a short time after the stimulation. Under similar circumstances of stimulation without sexual release, some females also have discomfort due to congestion and tension in the pelvic region; this condition is known as *vasocongestion*. As with males, it can be relieved through orgasm or will dissipate soon after the stimulation ceases.

Traditionally, males have been put in the position of aggressors in social and sexual relationships, and females have had the role of holding the males off. Much evidence suggests that this traditional stereotype is still applicable. Men are more active in every form of sexual behavior and still tend to assume that females are the ones who must set the limits in a given sexual relationship. Although the generalization still holds, it is clear that the times are changing. Females seem to be narrowing the gap in all forms of sexual behavior, and there seems to be more willingness on the part of the females to play a more nearly equal role in sexual involvement. The extent of this change is difficult to monitor, since all such changes have been overshadowed in recent years by the great increase in adolescent coitus. We now consider the extent of adolescent coital involvement in America today.

ADOLESCENT SEXUALITY TODAY

There can be little doubt that the nature and extent of adolescent sexual activity have changed in recent years. Although the definition of what constitutes a sexual revolution varies from person to person, most researchers who study adolescent sexuality would now agree on at least one fact—that a greater number of adolescents have sexual intercourse before reaching adulthood than at any previous time in this country's history.

Incidence of Sexual Activity

A recent national survey of never-married females aged 15 to 19 found that by age 19, 55 percent of the young women interviewed had already had sexual intercourse. Among all females aged 15 to 19, 35 percent were sexually experienced by the time of the interview (Zelnik & Kantner, 1977). In the five-year period between an earlier survey which used a comparable national sample and the more recent study, there was an increase of 30 percent in coital experience for all females between ages 15 and 19. By age 15, 18 percent of American females have already had sexual intercourse, suggesting that many of the concerns we have about pregnancy outside of marriage are justified for younger, as well as older, adolescents.

Exhibit 10.2 displays the steady increase in sexual involvement through the adolescent years. Figures for 1976 and 1971 are shown to indicate the magnitude of change which resulted in just five years. These figures are the best we have to date, since they are from national probability samples of almost 4000 young females in 1971 and 2000 young females in 1976. Actually, the estimates are conservative, since they do not include adolescents who were married by the time of the interview, females who are likely to have had an even higher incidence of coital experience.

This study interviewed females only, and there are no comparable data for males. However, virtually every earlier study of adolescent sexuality conducted over the past fifty years has shown the corresponding figures to be higher for males (Bowman & Spanier, 1978). Thus it can be reasonably assumed that this generalization still holds true for today's adolescents.

We need to be clear about what these data say and what they do not say. These figures reflect *cumulative incidence*, which refers to the total number of persons who have engaged in sexual intercourse at least once by the time of the interview. Cumulative incidence figures do not tell us how often, with how many partners, or under what circumstances such sexual involvement took place. Cumulative incidence figures are sometimes misleading to the unwary reader, because they may erroneously be interpreted as *active incidence* figures, which indicate the number of persons who are sexually active at a given time. Of course, active incidence is always lower than cumulative incidence. Let us now consider some of the other aspects of adolescent sexuality today.

About half the females who were sexually experienced indicated that they had not had intercourse during the month prior to the interview, and fewer than 3 in 10 had intercourse as many as three times in the month (Zelnik & Kantner, 1977). Thus we see that a good deal of adolescent sexual activity is rather sporadic. In fact, 15 percent of all sexually experienced adolescent females report they had sexual intercourse *only once*. Thus only a portion of the 35 percent of all adolescents aged 15 to 19 who are sexually experienced are having coitus on a regular basis.

The findings that more than a third of all middle and older adolescents are sex-

EXHIBIT 10.2 **Percent of Never-Married Women Aged 15 to 19 Who Have Ever Had Intercourse, by Age and Race, 1976 and 1971.**

	Study Year and Race									
	1976					1971				
	All	White		Black		All	White		Black	
Age	%	%	N	%	N	%	%	N	%	N
15–19	34.9	30.8	1232	62.7	654	26.8	21.4	2633	51.2	1339
15	18.0	13.8	276	38.4	133	13.8	10.9	642	30.5	344
16	25.4	22.6	301	52.6	135	21.2	16.9	662	46.2	320
17	40.9	36.1	277	68.4	139	26.6	21.8	646	58.8	296
18	45.2	43.6	220	74.1	143	36.8	32.3	396	62.7	228
19	55.2	48.7	158	83.6	104	46.8	39.4	287	76.2	151

Source: Adapted from Zelnik and Kantner, 1977.

ually experienced have caused great concern for many parents, teachers, school administrators, community agencies, and government policymakers. One fear is that many adolescents are indiscriminately having sexual relations with very little thought about the consequences and about the personal meaning for them. Actually, the data suggest that this is not what is happening. About half of all sexually experienced adolescents have had sexual intercourse with only one partner; less than one-third have had two or three partners; and only 10 percent have had six or more partners. Thus if we consider sexually experienced and sexually inexperienced adolescents together, less than 4 percent have engaged in coitus with six or more partners.

However, we also observe, in Exhibit 10.3, that the number of partners with whom adolescents had had sexual intercourse increased during the five years between the two surveys of American females. Furthermore, as we mentioned before, other research has shown that the figures are higher for males in all forms of sexual activity (Kinsey et al., 1948, 1953; Carns, 1973; Spanier, 1973). It can be concluded that the number of partners for males of corresponding ages will likely be higher on the average. We should also note that as we examine the individual age groups across adolescence, the number of partners increases. Thus among 19-year-old sexually experienced females, 14 percent had had six or more partners.

There are some notable racial differences in sexual activity (see Exhibit 10.2). By age 15, 38 percent of black females have had sexual intercourse, and by age 19, the figure is 84 percent. For a substantial number of black females, sexual activity actually begins before the female is even capable of conception. By the conclusion of their adolescence, sexual experience is quite prevalent among blacks, although our same caution about the differences between cumulative and active incidence is applicable in this discussion.

However, unless we think that these higher cumulative incidence figures suggest that blacks are any more promiscuous than whites, we point to the figures on number of partners. Although blacks are more likely to have had more than one sexual partner than are whites, they are less likely than whites to have had six or more sexual partners. Only 6 percent of sexually experienced black adolescent females have had six or more partners, compared to more than 11 percent for whites.

EXHIBIT 10.3   **Percent Distribution of Sexually Experienced Never-Married Women Aged 15 to 19, according to Number of Partners Ever, by Age, 1976 and 1971.**

Year	Number of Partners	Age		
		15–19	15–17	18–19
1976	1	50.1	54.0	45.3
	2–3	31.4	31.5	31.3
	4–5	8.7	8.4	9.1
	≥6	9.8	6.1	14.3
	Total	100.0	100.0	100.0
1971	1	61.5	66.5	56.1
	2–3	25.1	22.7	27.7
	4–5	7.8	5.9	9.9
	≥6	5.6	4.9	6.3
	Total	100.0	100.0	100.0

Source: Adapted from Zelnik and Kantner, 1977.

The Initiation of Sex

When and under what circumstances does sexual activity begin? We have already indicated that nocturnal emissions and masturbation typically are the earliest sexual experiences for males. At some time during the adolescent years, males and females usually begin their progression of heterosexual involvement in conjunction with dating relationships. In the typical case, sexual intercourse first occurs with a steady dating partner. For most adolescents, intercourse usually follows a period in which kissing and petting were the primary forms of sexual expression. Of course, there is considerable variation, with some adolescents having their first sexual experience very early in a dating relationship or outside of a dating relationship altogether.

Exhibit 10.4 shows that among sexually active adolescent females, the median age at which coitus first took place was 16.2. This finding is very important when we consider sex education and the availability of contraception. Sex education and contraceptive information and availability are too late to be of utmost benefit if provided only in the last years of high school. It is clear from these recent data that many adolescents need detailed information about reproduction and contraception during junior high school at the very latest.

Research shows that for whites, there is no relationship between first coitus and the age at menarche, but there is a correlation for blacks (Zelnik & Kantner, 1977). In other words, for blacks, the earlier the age at menarche, the earlier sex tends to begin. There are also some interesting seasonal variations. Nearly two-fifths of American females report that they first experienced sexual intercourse during the summer months (Zelnik & Kantner, 1977).

Parents of adolescents have always been curious about where sexual activity among adolescents takes place. Some parents encourage their children to bring dates home, presumably because the couple is less likely to become sexually involved if they are in a more supervised setting. Other parents, though, encourage their children to go out, thinking that places of greater social activity will discourage sexual activity. The data in Exhibit 10.5 tell us that the male's residence is the most common location for the first coital experience. It is likely, however, that in a large proportion of the cases, no one other than the young couple is at home at the time. A relative's or friend's home is a distant second, and the female's home is the third most common place. Fifty-eight percent of initial coital experiences, then, occur in the homes of the male or female, and almost 80 percent take place in someone's home. Only 5 percent occur in motels or hotels, less than 10 percent in automobiles,

EXHIBIT 10.4 **Median Age at First Intercourse of Sexually Experienced Never-Married Women Aged 15 to 19, by Age, 1976.**

Age at Survey	Median Age at First Intercourse
15–19	16.2
15	14.7
16	15.5
17	16.4
18	16.8
19	17.1

Source: Adapted from Zelnik and Kantner, 1977.

EXHIBIT 10.5   **Percent Distribution of Sexually Experienced
Never-Married Women Aged 15 to 19, according
to Place of Occurrence of First Intercourse, 1976**

Place of Occurrence	Percent
Respondent's home	16.3
Partner's home	41.6
Relative's or friend's home	21.4
Motel or hotel	5.2
Automobile	9.5
Other	6.0
Total	100.0

Source: Adapted from Zelnik and Kantner, 1977.

and the remaining 6 percent in other places, such as in the "great outdoors" (Zelnik & Kantner, 1977).

Data from an earlier national survey of college students provide us with additional information about the nature of the relationships between males and females at the time they initiate sexual activity (Carns, 1973). Males tend to have their first sexual experience at earlier ages than do females, and their first partner is more likely to be a pickup or casual date than is a female's first partner. A female's first partner is more likely to be someone to whom she is seriously committed or with whom she is in love. The first intercourse is less likely to be planned by males and less likely to be discussed if planned ahead of time. Males are more likely to report that they enjoyed their first sexual experience than are females. They are less likely to do it again with their first partner, they talk about it sooner, tell more people, and are more likely to receive an approving reaction from the persons they tell. Males tell their parents more than females do. They have a greater frequency, have more partners before they are married, and report less guilt about having had sexual intercourse than females.

These differences persist throughout the remainder of sexual activity in the adolescent years. It is not until marriage that males and females become more nearly equal in frequency of sexual intercourse. Then, of course, marital coitus would by definition provide equal experience for men and women. However, data on extramarital sexuality (Kinsey et al., 1948, 1953; Hunt, 1974) show that male-female differences in total sexual outlet still persist throughout the remainder of the life span, since extramarital coitus is more common for males and masturbation, which continues to some extent for many persons during marriage, is also more common for males.

These data on male-female differences from first sexual intercourse through adolescence and into adulthood further persuade us to consider modern sexual change as evolutionary rather than revolutionary. Adolescent heterosexual activity is still dominated by males. Sex is much more important in the adolescent male subculture than the adolescent female subculture. Males talk about sex more openly with each other, joke about it more, and experience greater peer pressure to perform sexually. Females traditionally have not discussed sex very much with their friends and have been quite secretive about their own activities. Although adolescent males and females are converging on these dimensions, the narrowing of the gap can probably be attributed more to changes in the female subculture than in the male.

But until males and females have a more nearly equal role in the initiation of sexual relationships or in the task of postponing sexual relations, a sexual revolution cannot be said to have occurred. We believe that a true sexual revolution would be characterized by greater equality among males and females in the *nature* of sexual interaction. Similarity in the numbers of sexually experienced persons will not signal a sexual revolution until the content and dynamics of the social *relationships* in which the sexual acts occur are more equal. In short, social and sexual relationships must change *qualitatively*, not just quantitatively.

Contraceptive Use

It should be evident from the figures presented above that researchers no longer need to concern themselves with the question of whether adolescents do or do not engage in sexual intercourse. The answer is "yes." Many do. The research topic which is now probably more important than any other in the study of adolescent sexuality is contraceptive use. Why do some adolescents use contraception, while others do not? What leads an adolescent to be contraceptively prepared and protected? What can be done to help adolescents who have decided to have sexual relations to avoid the unfortunate consequences of venereal disease, unwanted pregnancy, and perhaps abortion?

In 1976, one-fourth of sexually experienced adolescent females had *never* used contraception, an additional 45 percent had used contraception only sometimes, and only 30 percent reported that they had always used it. These figures represent a significant improvement in contraceptive use since 1971, but they still point to the serious problem of nonuse among adolescents. Almost 2 out of 3 adolescents reported that they had not used any form of contraception during their last sexual experience. Even by age 19, only a little more than two-thirds of the respondents report that they were contraceptively protected at the last intercourse.

Among adolescent female contraceptive users, oral contraceptives (birth control pills) are the most commonly used method. Pills were reported as the method used at last intercourse by almost one-third of the females. Condoms and withdrawal were next in popularity, with intrauterine devices (IUDs), douches, and other methods used by a small percentage. Less than half the adolescent women are using effective methods of contraception (pills, condoms, IUDs). Withdrawal, douches, and some of the methods included in the other category (rhythm, creams, foams, jellies) have relatively high failure rates, and many of the adolescents using these methods are actually at risk for pregnancy (Hatcher, Stewart, Guest, Finkelstein, & Goodwin, 1976). And almost 37 percent of the adolescents used no contraception at all at last intercourse.

Zelnik and Kantner (1977) report that the older a female is at the time of first intercourse, the more likely that she will commence contraception at the same time she begins to have sex. However, there is no evidence that the gap between age at first intercourse and age at first contraceptive use has declined for females as a whole in recent years. A study of the relative influences on males and females to use contraception demonstrated that partner influence was most important, parental influence least important, with peer influence in between (Thompson & Spanier, 1978).

Much has been said about oral contraceptives contributing to the increased incidence of adolescent sexuality. The fact that only a minority of sexually experienced adolescent females use oral contraceptives should suggest to us that pill use is not a major or even contributing factor in the *initiation* of sexual activity for most American adolescents. It may be true, however, that once sexual activity has begun, the availability of the pill makes sexual involvement easier and less worrisome.

About half the unmarried teenage women who have used oral contraceptives got their first prescription from a clinic rather than a private physician. According to Zelnik and Kantner, this contrasts with the practice of older, married women, who rely much more on the private physician for contraception.

Reasons for Inadequate Protection

Studies have brought to light a variety of reasons why people have premarital intercourse without adequate contraceptive protection (Lehfeldt, 1971; Sandberg & Jacobs, 1972; Shah, Zelnik, & Kantner, 1975). Among them are ignorance of which contraceptive methods are effective and where to get them; rejection of a method prescribed by a physician because the patient thinks it unsafe; objection to contraception on religious or moral grounds; denial that contraception works; irresponsibility; immaturity; willingness to take risks; availability of abortion; rebellion against society or parents; hostility toward the other sex; equation of love with self-sacrifice; a belief that intercourse is sinful and pregnancy is the punishment; a feeling that pregnancy is a gift of love; the belief that sex is for procreation only; unwillingness to deny oneself or to delay intercourse; a desire of the female to become pregnant; the feeling that "it can't or won't happen to me"; the belief that intercourse is a demonstration of love; the belief that the girl was too young to become pregnant; and the belief that intercourse was too infrequent or occurred at the wrong time of the month.

In future years, it will be increasingly important for parents and persons working with adolescents to pay attention to education about contraception and motivation for contraceptive use. The widespread lack of contraceptive use among adolescents comes at a time in history when the most effective methods of protection ever known to exist are readily available, when state laws restricting the distribution of contraception to minors have finally been eliminated (Paul et al., 1976), when there are family planning clinics and other health services within driving or commuting distance for virtually all American adolescent females, and when adolescent knowledge about contraception is at an all-time high.

Sexual Standards

Every person has a sexual standard. This standard may change during a person's life. Individuals often begin with a more restrictive standard—one which instructs us that sex is to be reserved for marriage. Yet one may then develop a new standard, perhaps after falling in love or coming to college, which permits him or her to have sexual intercourse before marriage in certain circumstances. Reiss (1960) differentiates between two types of sex: *Body-centered* sex has its emphasis on the physical nature of sex and *person-centered* sex has its emphasis on the emotional relationship between the individuals who are engaged in the sex act. Furthermore, *individuals* may be classified according to one of the four sexual standards:

1 *Abstinence.* Premarital intercourse is wrong for both men and women, regardless of circumstances.

2 *Permissiveness with affection.* Premarital intercourse is right for both men and women, as long as there is emotional attachment, love, or strong affection.

3 *Permissiveness without affection.* Premarital intercourse is right for both men and women whenever there is physical attraction, regardless of whether affection is present.

4 *Double standard.* Premarital intercourse is considered right for men but wrong for women.

Social scientists have debated whether the double standard or the abstinence standard was dominant throughout modern western history. Even though it has always been stated, since the emergence of Christianity, that abstinence was the norm, it is certainly true that it was never fully practiced. It is probably correct to say that, until recently, abstinence has been the culturally approved standard, whereas the double standard existed in reality for a significant portion, perhaps a majority, of the population. Today, most young persons adhere to a standard of permissiveness with affection, although it is likely that the commitment necessary for "affection" has been relaxed somewhat in recent years. The double standard is on its way out, although remnants of it must surely still be around, since male-female differences continue to be evident. The abstinence standard is reported by many young persons, although it appears that this standard is replaced by one of the permissiveness standards sometime before marriage for a majority of American youths. A minority of persons adhere to the abstinence standard, particularly those of strong religious conviction and those under strict parental influence.

The standard of permissiveness without affection has never been dominant. It probably involves only a small portion of the population but is undoubtedly increasing. There have been many times in history when young people were accused of advocating "free love" and sexual promiscuity. In the 1950s, some such persons were called "beatniks" in the mass media; in the 1960s, they were called "hippies"; and in the 1970s, some were known as "freaks." Whatever the term applied, it is correct to conclude that young people have never held one universal standard. And the standard of permissiveness without affection has never run rampant. There are a variety of sexual standards among American youths, and our society is becoming increasingly willing to allow each individual to choose his or her own standard.

SOME CONSEQUENCES OF PREGNANCY OUTSIDE OF MARRIAGE IN ADOLESCENCE

Pregnancy outside of marriage, as well as its possible outcomes—abortion, illegitimacy, forced marriage—is a prominent social problem in America today. Venereal disease is another possible consequence (which we discussed in Chapter 3).

Life entails risk. But apart from the circumstance of rape, the risk of pregnancy outside of marriage is entirely avoidable. Yet when abstinence is rejected as a sexual standard, an element of risk is always a consideration in sexual intercourse. The failure to use adequate contraception or any contraception at all drastically increases the risk of pregnancy, and our earlier discussion documents the magnitude of the potential problem. It should not be surprising to learn that pregnancy and childbirth among adolescents have come to be described in the last decade as a problem of epidemic proportions.

Extent of the Problem

There is no way at present to get a complete and accurate picture of premarital pregnancies, but there are useful statistics giving some indication of frequency. If we assume a reasonable degree of validity in available statistics and we (1) combine figures for births recorded as illegitimate; (2) add children born within marriage but conceived before the wedding; (3) assume the addition of an unknown number of premarital pregnancies terminated by abortion before or after the wedding; and (4) realize there are an unknown number of such pregnancies ending with miscarriage or a stillborn child, we could conservatively estimate that there are at least 1 million *premarital* pregnancies each year in the United States. Certainly the major proportion of these pregnancies is likely to be among adolescent females. Zelnik and Kantner (1978) estimate that almost 800,000 premarital pregnancies occur among females in the 15 to 19 age group each year. Contraceptive use is carefully considered in their analysis. They estimate that almost 700,000 additional adolescents would become pregnant each year if they did not use contraception. Furthermore, an additional 300,000 pregnancies could be prevented each year if contraception were used consistently by the sexually active females or their partners in this age group. They estimate that 9.3 percent of adolescent females aged 15 to 19 who have not yet married have been pregnant at least once.

Some adolescent females do not feel that the risk of pregnancy is very great for them. Exhibit 10.6 dramatically shows how mistaken this belief is. Indeed, 58 percent of sexually active adolescents who had never used contraception had already been pregnant at least once by the date of the survey, 42 percent of the sometimes users had been pregnant, and even 11 percent of those always using contraception had been pregnant. The lowest pregnancy rates—about 6 percent—were for females who had always used an effective medical method, such as birth control pills. Another analysis of adolescent pregnancy considered females below the age of 15. Jaffe and Dryfoos (1976) estimated that 30,000 young girls under age 15 become pregnant each year in the United States.

Most recent studies of pregnancy among adolescents indicate a trend suggest-

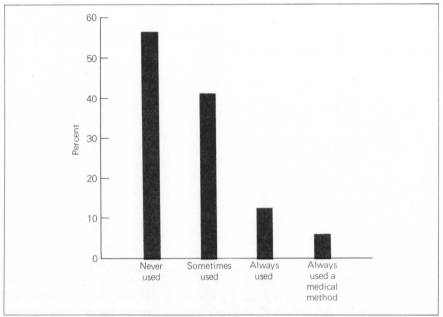

EXHIBIT 10.6   Percentage of sexually active young women who ever had a premarital pregnancy, by contraceptive-use status. (Source: Zelnik and Kantner, 1978.)

ing that females are becoming more regular and effective contraceptive users. Thus there can be some optimism for the future with regard to adolescent pregnancies. With better-accepted and more widespread sex education in the public schools, greater availability and use of effective contraception, and increasing social concern about the problem of adolescent sexuality and pregnancy, we can expect a gradual decrease in out-of-wedlock pregnancies among adolescent females. Nevertheless, the magnitude of the problem at present is still very great.

Outcomes of Premarital Pregnancy

Females facing a premarital pregnancy have several options:

1 They may marry.
2 They may have an abortion.
3 They may give birth to the child and keep it.
4 They may give birth to the child and give it up for adoption.

There is also a fifth possibility that does not involve a decision on the part of the female—*miscarriage* (the technical name is *spontaneous abortion*). The third and fourth options lead to what is known as an *illegitimate* birth, a term referring to any birth outside of marriage.

What do American adolescents do when faced with a pregnancy? Of the females who had ever been pregnant by the time they were interviewed in 1976 (Zelnik & Kantner, 1978), 28 percent married before the outcome of the pregnancy; an additional 9 percent married after the outcome of the pregnancy; and the remaining 63 percent never married by the time of the interview. Eleven percent of the females who had ever been pregnant were pregnant at the time of the interview, and they were included in the above figures.

In this national study of females aged 15 to 19, the researchers found that among white females, about half of all first pregnancies terminated in abortion. About 3 out of 10 pregnancies culminated in live births, about 1 out of 6 pregnancies ended in miscarriage, and 1.6 percent of the females had stillbirths.

There are some significant racial differences in how the outcome of a pregnancy is managed. Whites are about 4 times as likely as blacks to marry after discovering that they are pregnant. Data on abortion among blacks are not as reliable as those available for whites, but Zelnik and Kantner (1978) estimate that blacks are much more likely to carry a premarital pregnancy to term, while whites are much more likely to have an abortion. About 5 out of every 6 black adolescent females who report ever being pregnant state that they continued the pregnancy and had the child. Considering this tendency to continue a pregnancy and their lesser likelihood of marriage following a pregnancy, it is easy to understand why young black females have an illegitimacy rate which is substantially higher than the rate for whites.

Outcomes for Males

Since females get pregnant and males do not, we tend to think most about the consequences of pregnancy for the female; however, the male and his feelings should not be neglected. Sexual relationships during the adolescent years have potential rewards and potential pitfalls for the male as well. Whether or not a pregnancy results, the consequences can be troublesome. Let us review those which are possible:

An Unwanted Pregnancy. The greatest risk that a man faces is a possible pregnancy for the female. He may disclaim any involvement with her, but in the typical case, a man acknowledges his involvement and is affected by the pregnancy. The female carries the greater burden because, after all, she is pregnant and will usually be considered legally responsible for the child. The man, however, may be faced with the expenses of an abortion, a forced marriage, or a wedding which is quickly moved up. Above all, perhaps, he must at a more personal or emotional level deal with the problem that he has had a role in creating. But the consequences for the male may be even more severe.

Accusation of Rape by Force. Most cases of rape are violent acts committed by troubled men. However, cases are not unknown in which a woman voluntarily has sexual intercourse, and afterward, because she feels guilty, or because the fact is discovered, or for some other reason, she claims rape (see MacDonald, 1973). Usually there are no witnesses to intercourse. Hence it is the man's word against the woman's. She may not succeed in making her accusation hold up in court. But even after the fire is put out, so to speak, the odor of smoke remains in the air, and there may be people who still wonder whether the man was guilty and justice miscarried.

Accusation of Statutory Rape. Typically, state law specifies an "age of consent." This is the minimum age an unmarried girl must have reached in order legally to give consent to sexual intercourse. If below the age of consent, she cannot legally consent to having intercourse even though she voluntarily says the words agreeing to it. A comparable situation is found in an individual's inability to sign a valid legal contract before age 18 or 21 even though his or her name is on the current contract form.

Let us assume that in a given state the age of consent is 16. A physically well developed 15-year-old girl claims to be 17. A 21-year-old man takes her at her word. She voluntarily has sexual intercourse. Her parents discover this fact and make known her correct age. The man with whom the girl had intercourse is liable to prosecution for statutory rape, which constitutes a felony.

Accusation of Paternity. If a woman becomes pregnant, any man who has had intercourse with her within a given time span is a possible candidate for an accusation of paternity.

Contrary to popular assumption, paternity cannot be unequivocably ascertained by means of blood tests. It is true that blood types are hereditary. Blood tests made upon a man or men, mother, and child can ascertain whether a given man could or could not be the father. That is not equivalent, however, to determining that a given man *is* the father. For example, suppose Ms. X has a child. She admits having had intercourse with three men, namely, A, B, and C. Blood tests are made on all five persons. The tests indicate that Ms. X could not have had that child, with its particular blood type, with A or B. She could have had that child with C. That still does not prove beyond all doubt that C is the father. Ms. X may not have mentioned D, with whom she also had intercourse and whose blood type is the same as C's.

Blood tests can establish *nonpaternity* as a certainty in some cases but paternity only as a possibility. The scientific definition of "fact" and the legal definition of "fact" are not the same. The former implies certainty, while the latter implies only "beyond a reasonable doubt." The former is based on proof, the latter on evidence. Therefore, "proof" of paternity may reach a level of probability that is acceptable in a court (Krause, 1971).

Pressure to Marry in Case of Acknowledged Paternity. Pressure to marry in such cases may be external or internal. It may come from parents, fear of gossip, or concern for reputation. Or it may come from a sense of responsibility.

If the couple is engaged, there may be no pressure in the usual sense of the term because they plan to marry anyway. Their problem is relatively simpler, involving only an actual or asserted change of wedding date. In some cases, however, marrying earlier than they had expected is a serious upset to a couple's families and to its educational and occupational plans.

If, on the other hand, the couple had not planned to marry, would not be good choices as marriage partners for one another, do not know each other well enough, or are not in love, marriage to camouflage a pregnancy or to give a child a legal father and a name may be damaging to all three persons.

In sum, there are many key changes that typify adolescent sexuality, and there are several problems that may be associated with expressions of adolescent sexual behavior. Indeed, the topic of adolescent sexuality illustrates the issues involved for the adolescent in dealing with a changing self while dealing with a changing world. One may suggest that the adolescent must integrate all these inner and outer changes if he or she is to be adaptive. The adolescent must find a means to integrate all the alterations in his or her life. This achievement may be attained by forming a (new) sense of self.

Accordingly, to appraise the adequacy of this reasoning and as a means of seeing how one might combine the study of all change processes of adolescence under a common topic, we will conclude our discussion in this chapter by focusing on the adolescent's development of a sense of identity. This focus will require our being concerned once again with the ideas of Erik Erikson.

ERIKSON'S VIEWS OF PERSONALITY DEVELOPMENT IN ADOLESCENCE

In previous chapters we have discussed the usefulness of the descriptive components of Erik Erikson's (1959, 1963, 1968) theory of psychosocial development. His ideas have been presented as useful for understanding the nature of person–social context interrelations. The stages that Erikson sees as characterizing the infancy and childhood portions of life are the oral-sensory, the anal-musculature, the genital-locomotor, and the latency stages. The fifth stage in Erikson's theory is labeled *puberty and adolescence.*

Erikson's descriptions of development in this stage suggest that the events and outcomes of adolescence are central. The processes of behavior change existing prior to adolescence converge to create a crisis of self-definition. That is, the adolescent must establish an identity to maintain—or re-create—an integrated sense of self in the face of these alterations. This must be done at a time when social demands are also changing. This process involves a crisis resolution through role commitment. Hence he sees identity achievement as involving the adoption of a set of behaviors and an ideology which will have a potential for allowing the person to meet the psychosocial demands of the rest of the life span. Let us turn to a more detailed discussion of this stage.

Stage 5. Puberty and Adolescence

This stage of development corresponds to the genital stage of psychosexual development in Freud's theory. Erikson also is concerned with the implications of the emergence of a genital sex drive occurring at puberty. But as with the previous stages of psychosocial development, Erikson here looks at the broader psychosocial implications of all the physical, physiological, and psychological changes that emerge at puberty.

Erikson sees all the changes occurring at puberty as presenting the adolescent with serious psychosocial problems. The person has lived for about twelve years and has developed a sense of who he or she is and of what he or she is and is not capable of doing. If the person has developed successfully, he or she will have developed more trust than mistrust, more autonomy than shame and doubt, more initiative than guilt, and more industry than inferiority. In any event, all the feelings developed have gone into giving him or her a feeling about who one is and what one can do. Now, however, this knowledge is challenged. The adolescent finds himself or herself in a body that looks and feels different and, further, finds that he or she is thinking about this and all things in a new way. Thus all the associations the adolescent has had about the self in earlier stages may not now be relevant to this new person one finds oneself to be. Because of the need to have a coherent sense of self in order to be adaptive, the adolescent asks a crucial psychosocial question: "Who am I?"

Moreover, at precisely the time when the adolescent feels unsure about this, society begins to ask related questions about the adolescent. For instance, in our society the adolescent must now begin to make the first definite steps toward career objectives. For example, one has to make a decision about whether or not to enter college preparatory courses. Society asks adolescents what role they will play in society and wants to know how soon these soon-to-be-adult persons will contribute to its maintenance. Society wants to know what *socially prescribed set of behaviors*, functioning for the adaptive maintenance of society, will be adopted. Such a set of behaviors is a *role*, and thus the key aspect of this adolescent dilemma is one of finding a role. Yet how can one know what one can do and wants to do to contribute to society, and meet its demands, if one does not know who one is?

In summary, this question—"Who am I?"—is basically a question of self-definition, necessitated by the emergence of all the new feelings and capabilities arising during adolescence (e.g., the sex drive and formal thought), as well as by the demands placed on the adolescent by society. The adaptive challenge to find a role one can be committed to, and thus to achieve an identity, is the most important psychosocial task of adolescence. The emotional upheaval provoked by this crisis is termed by Erikson the *identity crisis*. To resolve this crisis and achieve a sense of identity, Erikson (1959) sees it necessary to attain a complex synthesis between psychological processes and societal goals and directives:

> *At one time, it will appear to refer to a conscious sense of individual identity; at another to an unconscious striving for a continuity of personal character; at a third, as a criterion for the silent doings of ego synthesis; and finally, as a maintenance of an inner solidarity with a group's ideals and identity. (Erikson, 1959, p. 57)*

To achieve identity, then, the adolescent must find an orientation to life that fulfills the attributes of the self while being consistent with what society expects of a person. As such, this orientation must be both individually and socially adaptive. That is, such a role cannot be something which is self-destructive (e.g., sustained fasting) or socially disapproved (e.g., criminal behavior). Indeed, Erikson terms the adoption of a role such as the latter *negative identity formation* and notes that although such roles exist in most societies, they are by definition ones which have severe sanctions associated with them. In trying to find an orientation to life that meets one's individual and societal demands, the adolescent is searching for a set of behavioral prescriptions—a role—that fulfills the biological, psychological, and social demands of life. Said another way, a role represents a synthesis of biological, psychological, and social adaptive demands. This is why Erikson (1959, 1963) sees the ego identity as having these three components.

To find such an identity, the adolescent must discover what he or she believes in, what his or her attitudes and ideals are. These factors, which can be said to define one's *ideology,* provide an important component of one's role. When we know who we are, we know what we do, and when we know what we do, we know our role in society.

Along with any role (e.g., wife, father, student, teacher) goes a set of orientations toward the world which serves to define that role. These attitudes, beliefs, and values give us some idea of what a person engaged in a particular role in society thinks of and does. Thus there is an ideology that serves to define a societal role. We know fairly well what the ideology of a Catholic priest is and how it is similar to and different from the ideology associated with a military general, or a professional artist, or a professional politician. The point is that along with any role goes a role-defining ideology. To solve one's identity crisis, one must be committed to a role, which, in turn, means showing commitment toward an ideology. Erikson (1963) terms such an emotional orientation *fidelity.*

If the adolescent finds his or her role in society, if he or she can show commitment to an ideology, he or she will have achieved a sense of identity. Alternatively, if the adolescent does not find a role to play in society, he or she will remain in the identity crisis. He or she typically might complain that he or she "does not know where I'm at" or that one cannot "get one's head together." In an attempt to resolve this crisis, the adolescent might try one role one day and another the next, perhaps successfully, but only temporarily, investing the self in many different things. Accordingly, Erikson maintains that if the adolescent does not resolve the identity crisis, he or she will feel a sense of *role confusion,* or *identity diffusion.* These two terms denote the adolescent's feelings associated with being unable to show commitment to a role and, hence, to achieve a crisis-resolving identity.

In summary, it may be seen that the identity crisis of adolescence is provoked by individual and societal changes and can only be resolved through commitment to a role balancing the individual and social demands raised by these changes. The terms *crisis* and *commitment* become hallmarks of the fifth stage of psychosocial development. Yet the adaptive struggle in this stage is not only preceded by events in earlier ones but also is influenced in its outcome by them. As Constantinople (1969, p. 358) points out:

> *In order to achieve a positive resolution of the identity crisis, the adolescent must sift through all of the attitudes toward himself and the world which have occurred over the years with the resolution of earlier crises, and he must fashion for himself a sense of who he is that will remain constant across situations and that can be shared by others when they interact with him.*

Furthermore, the identity the adolescent attains as a consequence of the psychosocial crises preceding and during adolescence will influence the rest of the life span. To Erikson, self-esteem is a feeling about the self which tends to remain constant across life and, thus, gives the person a coherent psychological basis for dealing with the demands of social reality. In one essay (1959) in which he cast the notion of identity in terms of self-esteem, Erikson says:

> *Self-esteem, confirmed at the end of a major crisis, grows to be a conviction that one is learning effective steps toward a tangible future, that one is developing a defined personality within a social reality which one understands (Erikson, 1959, p. 89).*

Constantinople (1969, p. 358) elaborates that in adolescence:

> *This self-esteem is the end product of successful resolutions of each crisis; the fewer or the less satisfactory the successful resolutions, the less self-esteem on which to build at this stage of development, and the greater the likelihood of a prolonged sense of identity diffusion, of not being sure of who one is and where one is going.*

Where one is going is on to the early portion of one's adulthood. There yet another psychosocial crisis will be faced. As implied, successful resolution of it, as well as of the remaining crises of the adult years, will rest on the attainment of an adequate identity. While later chapters will focus on the adult years of life, let us here conclude our discussion of Erikson's ideas about adolescence and then turn to an evaluation of his ideas for the empirical study of adolescent identity.

Conclusions

We have argued that Erikson's descriptions not only allow adolescent personality to be studied from a life-span framework but also allow this study to be focused on what Erikson and others believe to be the core aspect of personality in this portion of life: identity. Not only does our summary in the preceding section support this view about the centrality of identity in the study of adolescent personality, but the history of the scientific study of adolescence reflects a corresponding emphasis. Although the first edition of a major work presenting the main theories of adolescence (Muuss, 1966) barely made mention of Erikson's ideas, a more recent edition (Muuss, 1975c) devotes more discussion to the ideas of Erikson than to any of those of several other theorists mentioned.

Thus Erikson's ideas about adolescent development have become quite influential, and there has been considerable research attention to the study of identity, its development, correlates, and bases. This research has, however, extended far beyond the issues raised by Erikson himself. As noted about other major theorists, like Piaget and Kohlberg, a major use of their ideas lies in the works stimulated in others. As such, the next sections consider the nature of empirical findings about adolescent identity development.

IDENTITY DEVELOPMENT IN ADOLESCENCE: RESEARCH DIRECTIONS

Research about adolescent identity processes has fallen into three interrelated categories. First, there has been research assessing whether ego identity occurs in a stagelike progression such as Erikson specifies. Do issues of industry versus inferiority invariably precede those of identity versus role confusion, and, in turn, do these issues always become of concern prior to problems of intimacy versus isolation? If such universality of sequencing were found, then Erikson's (1959) ideas about the critical importance of successful development in a prior stage for subse-

CHAPTER 10 *Adolescence: Social and Personality Development*

quent stage functioning would be supported. If such universality were not found, then a search would be appropriate for those variables which provide a basis of the various sequences of ego development that could occur.

A second direction that research about adolescent identity processes has taken has been to focus on what changes, if any, occur in a person's identity status over time. Erikson describes the identity crisis (as he does the basic crisis of each stage) as a bipolar continuum ranging from identity to role confusion. Thus a person may occupy any position along this continuum or, in other words, have any one of a number of statuses along this dimension. A person may have a location along this continuum close to the identity end, have a status close to the confusion or diffusion end, or be located at any one of several points in between. Researchers (e.g., Marcia, 1964, 1966) have tried to describe the array of different statuses a person may have along this dimension and how, across life, these statuses may change.

Because Erikson's sequences have not been found to be universal, and because one's identity status within the adolescent stage of life also has been found to change, a third area of research pertinent to ego identity has arisen. People have been concerned with what variables are related to ego development and identity status changes. Although such work has been largely descriptive, we shall see that it provides ideas about the explanation of these changes.

Sequences in Ego Development

Does the emergence of an identity crisis in adolescence occur in a stagelike manner? Is the adolescent crisis, as Erikson described it, inevitably preceded by the specified childhood crisis and followed by the hypothesized adult ones? Although much research supports the idea of adolescence as a time of crisis in self-definition, the universal stagelike characteristics of this challenge are in doubt.

Several studies, independent of Erikson's framework, do show adolescence to be a period of change in self-definition. Montemayor and Eisen (1977) found that self-concept development from childhood to adolescence followed a sequence from concrete to abstract. This change was assessed by analyzing responses to the question "Who am I?" a question we have seen to be of central import in the identity crisis. Significant increases from grades 4 through 12 were seen in self-definitions relating to occupational role, individuality of one's existence, and ideology. All these are concerns of the adolescent role search as Erikson (1959, 1963) describes it. Decreases in self-definitions pertaining to territoriality, citizenship, possessions, resources, and physical self were also seen across this grade range.

Additional support for the view that adolescence is a time of personal reorganization, and one involving ego development, is found in studies by Haan (1974) and by Martin and Redmore (1978). In the former study, involving a longitudinal assessment of ninety-nine adolescents, it was found that being able to cope with adaptive demands in adulthood was apparently preceded by progressive reorganization of personality characteristics during adolescence. In the latter study, of thirty-two black children studied longitudinally from the sixth to the twelfth grade, thirty showed an increase in level of ego development, and these increases showed intraindividual stability (the correlation in ego scores between the two grades was +0.5).

Thus people's egos do develop, their personalities reorganize, and their self-definitions come to include occupational and ideological concerns, but such alterations do not necessarily correspond to stagelike progression. Research aimed directly at assessing such a conception has not provided complete support for Erikson's ideas. In the largest study involving an assessment of the presence of stagelike qualities in adolescent development, Constantinople (1969) tested more than 900 male and female college students from the University of Rochester. Her study was complex, involving both cross-sectional and longitudinal comparisons. In 1965 she tested members of the first-year through senior classes, and then in 1966 and again in 1967 she retested portions of these original groups.

To study Erikson's stages of ego development, she devised a test containing sixty items. She wrote five items for each of the first six stages in Erikson's theory to reflect the successful, or positive, end of the bipolar continuum and five items to reflect the negative end. Although there was some evidence that people's answers were somewhat influenced by how socially desirable a particular response to an item seemed, Constantinople (1969) nevertheless concluded that high scores on positive ends (e.g., industry, identity) *and* low scores on negative continuum ends (e.g., inferiority, diffusion) indicated successful resolution to a stage's crisis. Because our concern is adolescence, we focus on Constantinople's (1969) findings regarding the crises of industry versus inferiority, identity versus identity diffusion, and intimacy versus isolation.

Insofar as the scores for these crises are concerned, it appeared that, in general, scores on the positive continuum ends increased, while those for the negative ends decreased across groups. There was increasingly more successful stage resolution among males and females having higher college standing. Moreover, since between cohort differences (people studied at different times but in the same college class) did not appear great, it seems that the age-group differences may reflect age changes. Another characteristic of the cross-sectional data was a trend involving stage resolution being less evident for the intimacy versus isolation crisis, a finding consistent with Erikson's idea that identity issues must be solved first before intimacy ones can be dealt with. These college students may have just resolved their identities and made career plans by the senior years and just begun to focus on intimacy. But although seniors scored higher than first-year students on the positive industry and identity crisis ends, and lower on the negative inferiority end, there were *no* college class differences for diffusion or isolation—negative crisis ends that should have been lowered if successful development in accordance with Erikson's stage theory had occurred. This failure to show the expected developmental trend was particularly evident for females (Constantinople, 1969).

Constantinople (1969) undertook longitudinal follow-ups of the subjects in order to provide further information about the differences in successful resolution of stages in the cross-sectional data. These three-year repeated measurement studies gave unqualified support to the suggestion of developmental changes in the cross-sectional data insofar as alterations in identity and identity diffusion were concerned. There were consistent increases in the successful resolution of identity from the first year to the senior year across subjects *and* from one year to the next within subject groups. Even here, however, there are problems for Erikson's theory. Only males showed consistent decreases in the scores for identity diffusion. Changes in scores for the other crises did not always decrease or increase in accordance with Erikson's theory, and furthermore, the changes that did occur were often accounted for by time and cohort effects. Exhibit 10.7 shows the increase in identity scores for males and females from their first year through senior year, as well as the degrees of decrease in the diffusion scores for these subjects during this period.

Constantinople's (1969) data provide, at best, only partial support for the stagelike character of ego development. Similarly, Ciaccio (1971), studying ego development in male 5-, 8-, and 11-year-olds through use of a projective test, found only par-

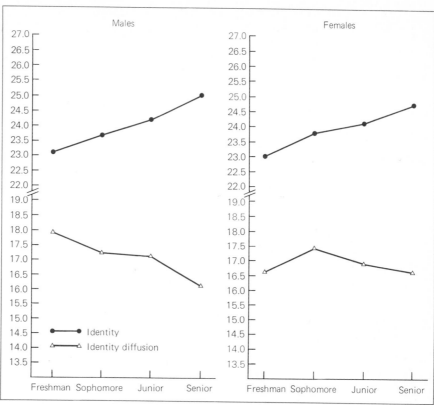

Males Females

EXHIBIT 10.7 Changes in identity and identity diffusion scores for male (upper graph) and female (lower graph) college subjects. (Source: Adapted from Constantinople, 1969, pp. 364–366.)

tial support for Erikson's theory. In support of Erikson, the youngest group showed the most interest with stage 2 crisis issues (autonomy versus shame and doubt), while the older two groups showed the most interest for the crises of stages 3 and 4. But support for Erikson's idea that the ego develops as it meets the different crises of succeeding stages was not found, since despite their varying interests, all groups showed most conflict for the stage 2 crisis.

Similarly, LaVoie (1976) studied sophomore, junior, and senior male and female high school students with Constantinople's (1969) measure of stage crisis resolution. LaVoie classified these adolescents on the basis of their degree of successful stage 5 crisis resolution (identity achievement) and found that although those who were high on identity scored higher on the positive crisis ends for stages 1 and 4, there were no differences between high- and low-identity achievers for the positive scores for stages 2, 3, and 6. Moreover, although LaVoie and Adams (in press) found that positive resolution of industry, identity, and intimacy crises, as measured by Constantinople's (1969) test, is related to adult attachment (i.e., liking and love) scores, negative crisis scores were not so related.

In summary, it appears that ego development does not proceed in the stagelike manner that Erikson suggests. Thus the ego identity crisis does not necessarily arise after the crises typically associated with earlier portions of life or precede those typically associated with later portions. Instead, a more plastic ordering of crises seems to exist between individuals (Douvan & Adelson, 1966; Gallatin, 1975). There seems to be a progressive increment in resolution of issues dealing with identity. For instance, the major consistency between cross-sectional and longitudinal data that Constantinople (1969) reported was the general movement across life toward a change in location along the identity–identity diffusion continuum. Most students changed in the direction of increased identity and decreased diffusion scores.

Hence there do seem to be developmental progressions involved in the resolution of the identity crisis of adolescence. Adolescents' changing status along the crisis continuum has thus become the second major focus of research on adolescent identity development.

The Differentiation of Identity Status

What are the major types of ego changes adolescents go through in meeting the challenge of the identity crisis? Are there adolescents who occupy locations along the identity–identity diffusion continuum other than at or near these two extremes? If so, what is their location or status? These were the major issues that James Marcia (1964, 1966) confronted as he began a series of provocative studies of identity development in adolescence. Rather than viewing identity as a global phenomenon, Marcia (1964, 1966) hypothesized that more differentiation existed. Erikson (1959) suggested this possibility, and in this idea Marcia found the basis of his own notions about identity status.

Erikson noted that the adolescent period involves a crisis in self-definition and in order to resolve this crisis one must commit oneself to a role. As discussed earlier, such commitment means adoption of an ideology (attitudes, values, beliefs) that coincides with the behavioral prescriptions for one's adopted role or occupation. Accordingly, to ascertain the adolescent's identity status, one should appraise his or her degrees of crisis and commitment. Using a semistructured open-ended interview, Marcia (1966) evaluated adolescents' level of crisis and commitment insofar as issues of occupational choice, religion, and political ideology were concerned.

Using eighty-six male college students as subjects, Marcia (1966) found evidence for four identity statuses. As Erikson (1959, 1963, 1968) said, two of these statuses were held by adolescents who had achieved identity and those who were in a state of identity diffusion. The former group had had a crisis period but now showed commitment to an occupation and to an ideology. The latter group may or may not have had a crisis; however, their defining characteristic was their lack of commitment. Moreover, they were not concerned about this lack of commitment.

The first of the two other statuses Marcia identified he labeled as *Moratorium*. Students here were in a crisis and had, at best, vague commitment to an occupation or to an ideology. However, Moratorium status adolescents were actively trying to make commitments. They were in a state of search.

The last identity status Marcia identified was termed *Foreclosure*. Adolescents in this status had never experienced a crisis. Yet they were highly committed. Their interviews showed they had adopted the identities their parents had wanted them to take and that they had done so with little or no question and with no crisis.

Marcia (1966) found some evidence that there were differences among these four groups that were consistent with Erikson's theory. As such, his research attests to the validity of each status. *Identity Achievers* were found to have the highest scores on an independent test of ego identity and did not follow authoritarian values (e.g., the belief that one should conform, a stress on obedience to authority) as much as members of some other statuses (e.g., Foreclosure adolescents). In addition, the Achievers maintained their feelings of self-esteem (i.e., positive self-regard) more than did members of other groups in the face of experimental manipulations having negative information about them. Foreclosure adolescents endorsed authoritarian values more than did other groups and had a self-esteem more vulnerable to negative information than had the Identity Achievers.

Evidence regarding the validity of the other two statuses was not as compelling. In a concept attainment task involving experimentally induced stress, the Moratorium subjects showed more variability than others but were otherwise not distinguishable from the Identity Achievers. Furthermore, the *Identity Diffusion* subjects performed lower than did the Identity Achievement subjects on the stress task, but in no other way did their behavior conform to theoretical expectations (Marcia, 1966).

However, other studies have provided more evidence that these four identity statuses exist. Schacter (1968) found that even among emotionally disturbed adolescent males, resolution of the identity crisis was positively related to attainment of occupational commitment. Marcia (1967) found that Identity Achievement and Moratorium adolescent males were less vulnerable to attempts to manipulate their self-esteem than were Foreclosed or Diffusion males, and Marcia and Friedman (1970) found that Identity Achievement adolescent females chose more difficult college majors than did Identity Diffusion Females. In addition, Foreclosure females, although like Identity Achievers in many respects, were higher in authoritarianism. However, they also had higher self-esteem and lower anxiety than did the Achievers, a finding not seemingly consistent with those of other studies.

Similarly, partial support for the fourfold differentiation of identity status comes from a study by Toder and Marcia (1973). Attempts were made to induce experimentally conformity to group demands among sixty-four college females. Identity Achievers conformed less than those having unstable statuses (the Moratorium and Diffusion subjects). However, the Foreclosure subjects also conformed less than those in the other two groups. Although this lack of conformity to group pressure would be expected on the basis of the Foreclosure subjects' stable identity status, it would not be expected by virtue of their authoritarian, and hence conforming and obedient, values.

Thus these data are far from unequivocal in showing distinctions among the four different statuses. Other data, however, provide more compelling support. Several studies have shown that when adolescents are studied longitudinally, they progress through the four identity statuses in ways consistent with theoretical expectations. For instance, Waterman and Waterman (1971) longitudinally studied ninety-two male adolescents through the course of their first college year. Using Marcia's (1966) interview, changes in occupational and ideological status were assessed. In regard to occupational commitment, there was a significant increase in the number of Moratorium subjects, i.e., those people actively searching, and a significant decrease in the number of Identity Diffusion subjects, i.e., those people not caring to search for commitment. Hence there was a group movement toward trying to make commitments and, hence, play an adaptive, functional role in society.

Insofar as ideology was concerned, there was a significant increase in the number of Identity Diffusion subjects. Indeed, about 44 percent of the subjects changed identity status for occupation over the course of their first year, and about 51 percent did so in regard to ideology; *but* people who had an Identity Achiever status at the beginning of the first-year period were just as likely to change as those who initially had other statuses.

These one-year longitudinal data do not provide compelling support for general progression toward more adaptive identity statuses in adolescence. However, the first year of college may involve demands on the person that produce instability in identity. Progressively more stability in identity achievement might be found if adolescents were followed beyond the first year. Longer-term longitudinal assessments support this expectation.

Waterman, Geary, and Waterman (1974) studied fifty-three male college seniors, all of whom had been subjects in the Waterman and Waterman (1971) study of identity development in the first year of college. Over this longer time period there were significant increases in the frequency of subjects falling in the Achievement status, for *both* occupation and ideology ratings. Moreover, although about 50 percent of these subjects stayed in the same status from the first to the senior year, the Achiever status was the most stable across this period. In turn, the Moratorium status was the least stable. Similarly, Marcia (1976), in a reinterview of thirty males who were given identity status interviews six to seven years earlier, found the Moratorium status to show a 100 percent rate of change. Moreover, Waterman and Goldman (1976), reinterviewing eighteen seniors studied as first-year students in 1970 and forty-one seniors studied as first-year students in 1971 found with both cohorts that there were significant increases in the frequency of Identity Achievement and decreases in Moratorium and Diffusion statuses. In summary, there was a very high probability for resolution of the identity crisis. About 75 percent of those youths studied over the entire period of their college experience reached a status of Identity Achievement.

We conclude, therefore, that there is some evidence for differentiation among ego identity statuses within adolescence. Most people move adaptively from other statuses toward Identity Achievement, and hence crisis resolution, by virtue of occupational and ideological role commitment. However, it is clear that there is variability in this pattern. Not all people go through this sequence. For instance, some begin and end adolescence as Foreclosure youths. People differ in their rates of development, and some never seem to attain an identity of either Achievement or Foreclosure. For example, Waterman et al. (1974) found that a substantial proportion of subjects completed their college years in the Identity Diffusion status. In their study, 13 percent of seniors were Diffuse on both occupational and ideology ratings, and an additional 33 percent were Diffuse in one or the other of these areas.

In summary, not only is there considerable plasticity of when in the sequence of other psychosocial crises the identity crisis occurs, but there also is plasticity in developments within this adolescent crisis. Because of this variability, the third area of research pertaining to adolescent identity has arisen. Researchers have become concerned with what variables may interrelate with those pertaining to identity processes. Through describing how identity may be moderated by those other variables to which it is related, this research provides an explanation of the basis of adolescent identity and, as such, its changing character.

Identity: Psychological Dimensions

Processes of cognitive and moral development are intertwined with those of identity. Podd (1972) found that principled moral reasoning, formal operational thought, and identity achievement were positively related. Those whose identity status was diffuse tended to show preconventional moral thought (Podd, 1972). Furthermore, independent data show the interrelation between advances in identity development and other theoretically relevant psychological functions.

Advances in formal operations involve progressions in dealing with abstract thought. If such development is indeed related to identity formation, then one should expect to see those who engage in abstraction processes tending to have the Identity Achievement status. If one construes engaging in poetry writing as an activity at least in part based on abstract thought, then suppport for this relation is found in a report by Waterman, Kohutis, and Pulone (1977). In two studies of college students, these investigators found that those people who wrote poetry were more likely to be in the Achievement status than those who did not write poetry. Moreover, poetry writers were less frequently found in the Foreclosure and Diffusion statuses than people who did not write poetry. Yet writing per se was not related to identity status, since there were no identity differences among students who did or who did not keep a personal journal or diary.

Similarly, Waterman and Goldman (1976) found that among the two cohorts they studied longitudinally from the first year of college to the senior year, an interest in literary and art forms was predictive of becoming Identity Achievers by the end of college for those who were not in this status as first-year students. Moreover, that identity status is related to school attainments is a finding reported by Jones and Strowig (1968). A group of 150 female and 167 male rural Wisconsin high school seniors were asked the "Who am I?" question, and answers were scored for identity development. Higher scores were positively related to achievement.

Thus advanced cognitive performance is associated with an advanced identity status. This relation suggests that a generally greater level of adaptive functioning is associated with identity attainment. Data reported by LaVoie (1976) support this idea. High school male and female students who were measured as having high identity had lower scores on measures of defensiveness, general adjustment, and neurosis than did students measured as low in identity. Moreover, the self-concepts of the high-identity scorers were more positive than those of the low-identity scorers. Similarly, Matteson (1977) found that among ninety-nine Danish youths, aged 17 to 18 years, more advanced identity status was related to the ability to control the expression of impulses among males, and to rejection of compliance to authority among both sex groups. It may be that in rejecting authority, high-identity youths see themselves as more capable of controlling their own lives.

Two studies support these interpretations. Waterman, Beubel, and Waterman (1970) found that Identity Achievers (and also Moratorium subjects) saw their own behavior more controlled by phenomena internal to and thus dependent on them than by events external to and thus independent of their control (e.g., luck or "fate"). In turn, the reverse ideas about the locus of control for one's behavior were held by subjects in the Foreclosure and Diffusion statuses. Similarly, Schenkel (1975) found that Identity Achievers were least dependent on extraneous cues from the environment in performing a perceptual task, while the reverse was the case with Identity Diffusion adolescents.

Thus the development of identity is associated with cognitive, adjustment, perpetual, and other psychological processes. However, the character of these psychological interactions may be moderated by the reciprocal relations they bear to interindividual social processes.

Identity: Social Dimensions

Since achieving an identity denotes finding a role meeting society's demands, identity processes are basically interpersonal ones. They link the person to society in a way that facilitates both individual and social maintenance and survival. One might expect, then, that people who have achieved identity should engage in interpersonal relationships useful in advancing this individual and social functioning, i.e., intimate relations. Orlofsky, Marcia, and Lesser (1973) found evidence of just such a relation. In a study of fifty-three college males, it was found that those subjects who were in the Identity Achievement status were among those who had the greatest capacity for intimate interpersonal relationships. The interpersonal relationships of Foreclosed and Diffusion students were stereotyped, superficial, and, hence, not very intimate. The searching Moratorium students showed the most variability between these two extremes. Similarly, Kacerguis and Adams (in press) studied forty-four male and forty-four female college students and found that those people of either sex who were in the Identity Achievement category were more likely to be engaged in intimate relationships than were males or females in the Foreclosure, Moratorium, and Diffusion categories. People in these latter three groups were much more variable in their level of intimacy.

Because identity links the adolescent to his or her social world, a basis of the different interpersonal styles of adolescents differing in identity status may lie in their social interaction history; perhaps this involves the family, since it is the major social institution delivering those societal demands to which the person must adapt. Recalling Erikson's (1959) conception of identity as being in part composed of self-esteem, O'Donnell's (1976) finding that in eighth- and eleventh-grade adolescents, the degree of positive feelings toward parents was generally more closely related to self-esteem than degree of these feelings to friends may be taken as support for the saliency of family interaction in identity development. Other studies show that different family structures, for example, presence of a working or a nonworking mother (Nelson, 1971) or father absence (Santrock, 1970), are associated with contrasts in levels of adjustment in adolescence or in ego development prior to adolescence, respectively. However, neither these reports nor that of O'Donnell suggests what sort of parental or familial functions may facilitate ego identity development.

Several studies suggest that parental personal and interpersonal characteristics may be transmitted to their offspring, in the context of the family milieu the parents help create, to foster identity development. LaVoie (1976) reports that male high school students high in identity reported less regulation and control by their mothers and fathers and more frequent praise by their fathers than did males low in identity. Similarly, LaVoie found that high-identity high school females reported less maternal restrictiveness and greater freedom to discuss problems with their mothers and fathers than did low-identity females. Thus high-identity adolescents appear to be characterized by a family milieu involving less parental restrictiveness and better parent-child communication than do low-identity adolescents. Waterman and Waterman (1971) and Matteson (1974) provide further data to support this conclusion.

In their longitudinal study of first-year college students Waterman and Waterman (1971) found that those students who showed stable Identity Achievement status for the entire year—and it will be recalled that many did not—scored significantly higher on a measure of family independence than did those students who changed out of the Achievement status. In addition, those students who initially were Foreclosed and then left this status by the end of the first year were also significantly higher scorers on a measure of family independence than those students who did not change out of this status.

In the study by Matteson (1974), involving ninth-grade students, a measure of adolescent self-esteem and of communication with parents was taken. In addition, the parents of the adolescents completed questionnaires about parent-adolescent communication and their own marital communication. Matteson reported that adolescents with low self-esteem viewed communication with their parents as less facilitative than did adolescents with high self-esteem. Moreover, parents of adolescents with low self-esteem perceived their communication with their spouses as less facilitative and rated their marriages as less satisfying than did parents of adolescents with high self-esteem.

Thus family milieu variables relating to communication quality among *all* family members and to patterns of parental control appear to relate to identity development. A family milieu having open communication and low restrictions on the individual seems to be most facilitative in providing a context for successful resolution of role search. Of course, milieus having such characteristics need not be just conventional American familial ones. Although there have been few studies, other types of family structures, on the one hand, or other types of social milieus, on the other, can promote such development.

Long, Henderson, and Platt (1973) studied fifty-one Israeli male and female adolescents, aged 11 to 13 years, reared in a kibbutz and compared them to two groups of same-aged youths reared in more traditional family settings. That is, in the kibbutz family system, children are reared collectively by adults who are not necessarily their parents. In fact, in this system children may often spend at most a few hours a day with either biological parent. Yet in the kibbutz system, the individual's contribution to the group is emphasized, as is equality of all group members despite their age or role. The adolescents reared in this setting showed more social interest and higher self-esteem than those adolescents reared in other settings.

Moreover, the college environment may be seen as a nonfamilial social milieu where open communication of ideas and minimal restrictiveness of search for roles are involved. Sanford (1962) has speculated that because of these properties the college experience promotes movement toward Identity Achievement. The longitudinal data of Waterman and Waterman (1971), Waterman et al. (1974), and Waterman and Goldman (1976) support this view. Most college students experience an identity crisis during their college years, and of these, 75 percent reach the Achievement status.

Although future research is needed to evaluate the appropriateness of these interpretations, it does appear that if an adolescent is placed in a social setting involving openness of social communication and minimal restrictiveness on role search, an adaptive coordination between self and society will be attained. An identity will be achieved.

CHAPTER SUMMARY

Adolescent development occurs in a social context which includes the family, peer groups, the schools, and the society in which one lives. In society, there are influences such as the political system and the legal system. To understand adolescents and their development, it is important to consider each of these influences and the interrelationships among them.

The importance of the peer group is widely known and indisputable. However, there is evidence to suggest that the family is perhaps the most important determinant of adolescent attitudes, values, and political and religious beliefs. We see the impact of the peer group more often in the day-to-day behavior of adolescents. The importance of automobiles and athletics and the presence of cliques and group behavior are results of the powerful force of the peer group. Although conflict between parents and adolescents exists, evidence has been presented to suggest that for most adolescents such conflict actually is minimal. Adolescents actually overestimate the magnitude of the differences that exist between themselves and their parents. Parents underestimate the extent of these differences.

Education is nearly a universal characteristic of adolescence, and about 85 percent of individuals finish high school. However, among those who drop out of school, numerous problems can occur. Males and females have roughly equal dropout rates, although the reasons for leaving school are sometimes different. Females, for example, sometimes leave school because of pregnancy or childbirth. High school dropouts earn less income than those who finish school; obtain less desirable jobs; marry earlier; have higher rates of marital disruption, higher mortality rates, and higher fertility rates; and are more likely to live in poverty.

Americans can now vote at age 18 in national elections, and this right highlights the role of the adolescent in the political system. But adolescents have a sometimes ambiguous role. State laws regarding drinking, signing contracts, and criminal behavior differ, and adolescence actually straddles the legal statuses of adult and minor.

It is with adolescence that sexual development, sexual behavior, and sexual interaction take on special importance to the individual. Society, too, has a great interest in the sexuality of adolescents, since some of our greatest social problems—venereal disease, illegitimacy, and early marriage—are all related to adolescent sexuality.

Masturbation for both males and females and nocturnal emissions for males only are two forms of sexual behavior found normally in adolescence. Holding hands, kissing, and petting are forms of sexual interaction that become increasingly prevalent in adolescence. Sexual intercourse is now experienced by the majority of adolescents in the United States, although for many the event does not occur until late adolescence and for others the frequency of such involvement and the duration of the relationships involved may be very short. Although there is now more sexual behavior during adolescence than at any previous time in history, it is still debatable whether there has been a sexual revolution. It can be argued that the term sexual evolution is a more appropriate notion, since the changes that have occurred in American society have really been gradual and in many ways consistent with other social change.

Sexual intercourse begins in early adolescence for a small number of persons, but by late adolescence, about half of white females and five-sixths of black females (and even higher proportions for males) have become sexually active. Sexual intercourse usually takes place in someone's home, most often the home of the male. Contrary to what some might believe, most adolescents who are sexually active have had only a small number of partners, with half having had only one partner. Unfortunately, most adolescents are not contraceptively protected when they begin sexual activity, and a large proportion are inadequately protected throughout adolescence. As a result, there are a great number of pregnancies among adolescent females.

Adolescents have differing sexual standards. Those having an abstinence standard believe that sex before marriage is wrong for both males and females, regardless of feeling involved. The standard of permissiveness with affection says that sex before marriage is acceptable for both males and females as long as there is emotional attachment or love. The standard of permissiveness without affection proposes that premarital intercourse is right whenever there is physical attraction, regardless of the degree of affection. The double standard exists when sex is considered right for men but wrong for women. Other sexual standards also exist.

There are hundreds of thousands of premarital pregnancies each year in the United States. Adolescent females who become pregnant have four options: they may marry, have an abortion, give birth to the child and keep it, or give birth to the child and give it up for adoption. Abortion is the most common alternative, and the large majority of abortions performed in the United States are on unmarried females, mostly adolescents. There are many consequences of an unwanted pregnancy for both males and females.

In adolescence the person needs to adopt a role and thus adopt a set of behaviors that will contribute to the maintenance and perpetuation of society. Yet because of all the changes that characterize adolescence, the person is unsure of who he or she is; and unless the person feels he or she has such self-knowledge, any role adoption cannot be made with assurance. Thus the crisis of adolescence is between identity at one extreme and role confusion (identity diffusion) at the other.

Research on identity development in adolescence has been primarily derived from these ideas of Erikson. One direction that research has taken is to determine if the identity crisis emerges in adolescence in the stagelike manner that Erikson suggests. Although most research indicates that adolescence is indeed a time for personal definition and reorganization, the sequence that Erikson offers does not seem to hold. Some adolescents, for instance, attain intimacy—the characteristic supposedly associated with the succeeding stage of successful ego development—simultaneously with or prior to the development of identity.

Another direction that research has taken is to indicate the types of identity statuses a person may possess during adolescence. Marcia has indicated that in addition to the Identity Achievement and the Identity Diffusion statuses, some adolescents fell into a Moratorium category (they were experiencing an identity crisis but were actually striving to make a commitment to a role), while others were in a status Marcia labeled as Foreclosure (adolescents here had never experienced a crisis and yet they were committed). Data indicate that over the course of the college years people change in their identity status more toward Achievement and away from Moratorium. Indeed by the end of college most people are Identity Achievers. Moreover, over time the Achievement category tends to be the most stable ego status.

A third direction that research has taken is to explore the variables that may affect the development of particular ego statuses in adolescence. Data indicate that formal operational ability, principled moral reasoning, positive self-concept, and an internal center of control are related to the attainment of Identity Achievement. Furthermore, particular types of social interaction (for example, warm, facilitative ones with parents) and relationships (for example, intimate, friendly ones with peers) in particular contexts (such as college) appear to be positively related to the development of Identity Achievement.

chapter 11

Sex Differences in Cognitive, Personality, and Social Development in Childhood and Adolescence

CHAPTER OVERVIEW

Aside from their obvious physical and physological differences, in what other ways are males and females different? As they move from infancy through childhood and adolescence, are there necessary and/or inevitable intellectual, personality, or social behavior developments for each sex group? In this chapter we review history and research bearing on these important and provocative issues.

CHAPTER OUTLINE

THEORIES OF SEX DIFFERENCES IN PSYCHOSOCIAL FUNCTIONING

Nature-Based Identification Theories
Cognitive Developmental Theories
Nurture Theories
Social Learning Theories
Conclusions

SEX-ROLE STEREOTYPES

Implications of Sex-Role Stereotypes for Socialization

THE ADAPTIVE BASIS OF SEX DIFFERENCES IN ROLES

Cultural-Historical Change and Sex-Role Evolution
Conclusions

RESEARCH ON SEX DIFFERENCES IN COGNITIVE, PERSONALITY, AND SOCIAL DEVELOPMENT

Sex Differences in Cognitive Processes
Sex Differences in Personality and Social Development
Agency and Communion in Self-Concept and Self-Esteem

SEX DIFFERENCES IN VOCATIONAL ROLE DEVELOPMENT

Sex Differences in Vocational Role Behavior
Sex Differences in Vocational Role Orientation

INTERACTIVE BASES OF THE DEVELOPMENT OF INDIVIDUALITY

CHAPTER SUMMARY

ISSUES TO CONSIDER

What are the characteristics of identification views of sex differences in behavior?

What are the features of cognitive theories of sex differences?

What are the various nurture views of sex differences in behavior? What are the key features of social learning theories of sex differences?

What are sex-role stereotypes? Do they relate to the sex differences described in nature and/or nurture theories?

What is the adaptive significance of roles and of sex roles?

To what extent do data from current research studies suggest the presence of sex differences in cognitive, personality, and social behaviors?

What are the characteristics of vocational role behaviors and orientations among male and female children and adolescents?

*A*n old song asks what little boys and little girls are made of. The idea of the song is that boys show certain behaviors, girls others. Indeed, much of the lore of our culture pertains to differences in the behavior of male and female children and youths. Of course, structural differences in primary and secondary sexual characteristics and functional differences in reproduction are universal contrasts between males and females. Only women can *menstruate, gestate* (carry children), and *lactate* (breast-feed). Only men can *impregnate* (Money & Ehrhardt, 1972).

To social scientists interested in the total development of individuals, it is important to try to discern both the obvious and the subtle characteristics that give the person his or her distinctiveness. More importantly, it is necessary to explain the bases of any behavioral differences between the sexes. This chapter examines some major dimensions of psychosocial differences between males and females during their childhood and adolescent years; in addition, we discuss various accounts of the bases and meaning of such differences.

THEORIES OF SEX DIFFERENCES IN PSYCHOSOCIAL FUNCTIONING

At the present time no one theory adequately accounts for sex differences in psychosocial behaviors across the life span. According to Worell (1981), "it may be that a unified theory of sex-role development is not possible because the conceptual and behavioral domains are too broadly defined" (p. 325). Each theory has been limited to restricted domains of knowledge, or cognition, affect, or emotion, or behavior. Traditionally, these theories fall into three rather distinct categories: nature-based identification theories, cognitive developmental theories, and nurture-based learning theories (Worell, 1981).

In the evolution of this latter approach, social learning theories have emerged which stress both internal (e.g., cognitive) and external (e.g., modeling) influences and, *especially*, the reciprocal relation between the two (e.g., Bandura, 1978; Worell, 1981). As we review the various theoretical approaches to the study of development of sex differences in behavior, we will see that current cognitive social learning theories (see Chapter 5 for a further discussion of such theory) are quite consistent with the nature-nurture interactionist position favored by your authors throughout this book.

Nature-Based Identification Theories

Identification theories suggest that a child's adoption of the behaviors, attitudes, and values of the same-sex parent by the age of 5 or 6 years leads to a personality structure consistent across time and situations (Worell, 1981). Deviations from the stereotyped conception of sex-appropriate behaviors are considered failures in the adequacy of identification (Erikson, 1963, 1968; Freud, 1923, 1965).

One example of this type of theory has been presented by Erikson (1963, 1968). Here it is suggested that cognitive, personality, and social functioning are innately tied to the physical and physiological characteristics of males and females (e.g., Erikson, 1964, 1968; see also Freud, 1923). Such a *nature* view says that there are inevitable differences between the sexes, since there are universal biological contrasts between them. Just as it is biologically adaptive (for species survival) to engage in male or female reproductive functions, it is also "biologically imperative" (Freud, 1923, 1949) to engage in those behaviors which are linked innately to one's biological status as a male or female. It is in this sense that sex differences in cognitive, personality, and social relations are seen as universal. To Erikson (1964, 1968), *anatomy is destiny!*

For example, Erikson (1964, 1968) theorizes that the female's genitalia must be used for incorporation of the male's penis in order for her to function in a biologically appropriate (i.e., reproductive) manner. Women must develop roles which allow this *inner space* to be fulfilled. To Erikson, women are oriented to roles as wives and mothers, and because this presumably adaptive role behavior is dependent on men, women should develop self-concepts which are characterized by such traits as dependency, submissiveness, and passivity—all attributes which lead to the presence of a negative self-esteem. Moreover, because of the nature of their self-evaluations and roles, females would not be oriented to develop themselves in endeavors requiring mastery of, or achievement with, complex material; that is, to compete with others intellectually would require a combined orientation toward competition, mastery, and achievement—all activities that would take time away from attainment of roles compatible with females' biological processes. Thus females would, on the basis of these ideas, not be expected to attain levels of intellectual functioning comparable to those of males.

The male, on the other hand, needs to use his genitalia to intrude on objects outside his body. Such "objects" may be a female's body, or, symbolically, any external environmental object. Oriented to this *outer space*, men must develop roles and self-conceptions which allow an independent, manipulative, active, and dominant mastery over their environment. Erikson argues that the attainment of such roles and self-concept traits would be associated with a positive self-esteem. Moreover, males' cognitive development should also differ from that of females. To better mas-

Despite Erikson's beliefs about sex differences in social behavior, male and female adolescents often show quite similar behaviors.
(Bob S. Smith/Rapho/Photo Researchers, Inc.)

ter their environment, males should, from the perspective of this theory, show greater cognitive achievement than should females.

Another idea that stresses the primacy of nature over nurture is often expressed through the belief that there exist initial, biologically based behavioral differences between the sexes (cf. Maccoby & Jacklin, 1974). Such congenital sex differences make male babies different stimuli to their caregivers than female babies. These biologically based behavioral differences elicit different reactions from caregivers, and thus the infant's behavior provides a biological basis of its own socialization experiences.

Criticisms of Identification Theories. There have been numerous criticisms leveled against nature-based identification theories, such as those forwarded by Erikson (1964, 1968). Worell (1981) has summarized many of these. They include:

1 Such theories stress the early consolidation and, therefore, the inflexibility of sex-typed behavior. However, critics note that people change their "commitment" to sex-typed behaviors across life.

2 Such theories require the same-sex parent to be the major role model. However, critics note that often parents of the other sex are the major role model (e.g., see Bronfenbrenner, 1960).

3 Such theories assume that (1) moral and ethical developments are linked directly into successful sex typing and (2) there are necessary sex differences in moral development such that males are more moral. However, critics note that available data support neither of these two assumptions (Hoffman, 1980).

Because of these and other problems with such theories (e.g., the stereotyped view of the female role), other theoretical accounts have been forwarded.

Cognitive Developmental Theories

In earlier chapters we have discussed the cognitive developmental theories of Piaget (1970) and of Kohlberg (1978). Cognitive developmental theories of sex differences emphasize the structural, stagelike progression of one's knowledge about one's own gender; because one knows one is a male or a female and because one acquires the knowledge of what it means behaviorally and cognitively (e.g., in regard to attitudes and values) to be a male or a female in one's society, one therefore seeks to acquire these sex-specific defining behaviors and cognitions. In other words, as Kohlberg (1966) emphasizes in his theory of the development of sex differences in behavior, a cognitive developmental theory of sex-typed behavioral development stresses the stagelike development of knowledge about one's and others' gender and indicates that individuals are active in their attempts to acquire relevant sex-typed information.

Criticisms of Cognitive Developmental Theories. Worell (1981) notes that Kohlberg's (1966) theory of sex differences in behavior development has been criticized in several ways. Critics note that in Kohlberg's theory the male role is seen as the prototype of development and that individual differences in behavior development, within each sex, are either ignored or are considered as deviations from "normal."

Also, critics points out that Kohlberg's (1966) view assumes that the most cognitively developed position is one which is sex-typed and stereotyped. Such an idea, however, contradicts the cognitive developmental view that cognition develops toward a final structural level involving flexible and reciprocal (reversible) thinking.

Because of such problems in the position forwarded by Kohlberg, other theoretical approaches have been followed.

Nurture Theories

Nurture ideas about the possible basis of sex differences exist also. As we indicated throughout preceding chapters, such approaches emphasize learning and/or behavioral principles (Worell, 1981). Maccoby and Jacklin (1974) describe several psychological theories aimed at explaining sex differences by stressing nurture processes such as praise or discouragement, self-socialization, and imitation. Discussion of this last process—imitation—will allow us to focus on recent nurture attempts to account for sex differences—those associated with social learning theory.

Praise or Discouragement. Parents and other socializing agents reward and praise boys for behaving in ways that they think of as "boylike." They actively discourage boys when they show "girllike" behaviors. In turn, girls receive praise for showing girllike behaviors and discouragement for showing boylike behaviors.

One such social learning interpretation of behavior development is provided by McCandless (1970). Because any behavior is, from his point of view, a phenomenon shaped through socially mediated rewards and punishments, there are no necessarily universal sex differences. Nevertheless, because males and females are seen to be rewarded and punished for different sets of behaviors, proponents of this view still predict that males and females will show different personality, cognitive, and social behaviors. Interestingly, McCandless (1970) has ideas regarding sex differences in cognitive, personality, and social behavior similar to those of Erikson (1964, 1968).

Drive-reducing, and hence rewarding, behaviors are associated with instrumentally effective and competent behaviors among males and with interpersonal behaviors of warmth and expressiveness for females. As with Erikson's theory, McCandless's ideas lead to the views that (1) males develop cognitive orientations and self-concepts characterized by mastery, instrumental effectiveness, activity, assertiveness, dominance, and competence—all favorable attributes that would lead to a positive self-esteem—while females develop cognitive orientations and self-concepts characterized by lack of mastery, passivity, submissiveness, and instrumental ineffectiveness—all unfavorable attributes associated with a negative self-esteem—and (2) because of contrasting self-concepts and self-esteems, males would be oriented to adopt higher-status roles, while females would be oriented to adopt lower-status ones.

Self-Socialization. A second nurture theory noted by Maccoby and Jacklin (1974) describes a process which builds on the other two noted above: imitation and praise or discouragement. On the basis of imitation, praise, and discouragement, the child develops a concept of what it is to be male or female in society. Once understanding his or her own identity (e.g., "I am a female"), he or she attempts to fit his or her own behavior to the concept of what behavior is sex-appropriate.

Imitation. Children choose same-sex models, typically the parent. These models are used for adopting their own behaviors (at least more so than is the case with models of the opposite sex). Recent major theoretical attempts have been made to use the principles of imitation derived from the study of observational learning (see Chapter 5) to account for sex differences in behavior development. These attempts fall under the topic of social learning theories (Worell, 1981).

Social Learning Theories

As discussed in Chapter 5, social learning theories emphasize principles of observational learning and stress that symbolic and vicarious reinforcement and punishment are the major mediators of behavior development. The approach provides for multiple sources of modeling (e.g., live or televised models) and reinforcement (e.g., from observing the response consequences to a model, from self-appraisal, or from direct reward or punishment from others).

From models and reinforcement history the individual's sex-role behavior develops. The individual comes to select same-sex behaviors and develops standards for same-sex role behavior, standards that become increasingly self-monitored, as a consequence of exposure to models and to reinforcement patterns (Worell, 1981). Moreover, since social learning theory (e.g., as in Bandura, 1965; see Chapter 5) emphasizes the distinction between learning and performance, individuals have knowledge of many behaviors they opt not to display, e.g., because of rewards and punishments present in the particular context (Worell, 1981). Flexibility of behavior is therefore possible as a consequence of changes in the contexts, in the patterns of rewards and punishments, and in the models the person observes (see Mischel, 1966, 1970).

Criticisms of Social Learning Theories. As noted by Worell (1981), past critics of social learning theory have indicated that the theory (e.g., as represented by McCandless, 1970) does not account for active cognitive monitoring by the individual and it does not pay adequate attention to the cognitive and motivational (reinforcement) functions of self-appraisals on the subsequent imitation of the person. We may recognize that these criticisms do not apply to the more recent versions of social learning theory. We have seen these recent formulations often labeled as cognitive social learning theory, because of its stress on cognitive processes as crucial to understanding the link between observations of a model and behavior (see Chapter 5 for details). Let us note the features of these more current views.

Reformulations of Social Learning Theories. Today, social learning theory attempts to integrate both cognitive and social learning variables into a cognitive social learning theory (Bandura, 1977, 1978; Mischel, 1973, 1977; Rosenthal & Zimmerman, 1978; Worell, 1981; Worell & Stilwell, 1981) in order to account for behavioral development. The emphasis in such accounts is on a continuous, reciprocal interaction between environmental variables (e.g., cultural and historical influences), observed behavior (e.g., the characteristics seen in available models), and covert (internal, e.g., cognitive) person variables. Such covert variables include self-generated expectancies, standards, attributions, and evaluations (Worell, 1981), and in interaction with external and objective variables that may control behavior (e.g., the direct reinforcement given for imitation of a particular model in a particular setting), they allow a cognitive social learning model of "reciprocal determinism" to be advanced (Worell, 1981).

Conclusions

Theories that stress *either* nature *or* nurture do not appear to be useful ones in accounting for sex differences in behavior development. As emphasized in recent social learning formulations (Worell, 1981), models which combine both internal cognitive *and* observational learning principles seem more useful. Stressing the reciprocal relation among internal and external processes (e.g., Bandura, 1978), such views can be expanded to include all aspects of the context of human life. Indeed, other relatively recent ideas have been advanced (e.g., Bem, 1975, 1979; Block, 1973; Sorell, 1979; Spence & Helmreich, 1978, 1979; Worell, 1981) which provide the essence of an interactionist view about the basis of sex differences in human development. Such views describe an individual's behavior development as a product of biological, psychological, social, and cultural interactions across history (Block 1973; Sarason, 1973; Sorell, 1979; Worell, 1981). It will be useful to present more details of such views in the context of our succeeding presentation, which focuses on research studies that allow us to assess the extent to which sex differences exist in cognitive, personality, and social behaviors (and if their degree of presence allows one to find support for nature, nurture, or interactionist theories).

Since the topic of sex differences in cognitive, personality, and social development is a controversial and complex one and since it is an area with hundreds of recent research studies and theoretical papers which are relevant (Block, 1973, 1976; Maccoby, 1966; Maccoby & Jacklin, 1974), not all studies of sex differences can be reviewed precisely in one chapter. Thus we will focus on those research literatures which allow some statements to be made about whether existing nature, nurture, or person-context interaction theories seem more useful in understanding the character of sex differences in cognitive, personality, and social processes. Simply, we will ask whether available research evidence shows males and females to differ in regard to cognitive, personality, and social development, and if so, then we will ask the extent to which they do differ. Are such differences consistent with a nature, a nurture, or an interactionist view of development? For example, do males and females differ in respect to self-concepts, self-esteems, and vocational role orientations in the ways Erikson's nature and McCandless's nurture ideas predict? Answering these questions is a complex task for at least two reasons.

First, the research pertaining to this question is voluminous and often contradictory. Second, despite the degree to which the available evidence *actually* shows sex differences consistent with various nature or nurture ideas, such as those of Erikson and McCandless, there is another, related body of research which shows that *people believe sex differences do exist*. Because such social beliefs can indeed provide a basis for behavior development—through the creation of self-fulfilling prophecies (Anthony, 1969; Rosenthal, 1966)—we turn first to a consideration of research pertaining to the social expectations people maintain about sex differences in human behavior.

SEX-ROLE STEREOTYPES

A social *stereotype* is an overgeneralized belief, i.e., some combination of cognition and feeling—some *attitude*—which invariantly characterizes a stimulus object (e.g., a person). A stereotype allows for little exception (Allport, 1954). Because of this rigidity, a social stereotype is relatively resistant to change and, as such, may become accepted as always true in a given society. The sex differences that Erikson and McCandless describe are consistent with traditional social stereotypes regarding males and females in American society. Such stereotypes are widely held in America today.

A *sex role* may be defined as a socially defined set of prescriptions for behavior for people of a particular sex group; *sex-role behavior* may be defined as behavioral functioning in accordance with the prescriptions, and *sex-role stereotypes* are the generalized beliefs that particular behaviors are characteristic of one sex group as opposed to the other (Worell, 1978). Broverman, Vogel, Broverman, Clarkson, and Rosenkrantz (1972) report a series of studies they conducted using a questionnaire that assessed perceptions of "typical masculine and feminine behavior." In order to study sex-role stereotypes, Broverman et al. gave a group of college males and females a long list (122 items), with each trait presented in a bipolar manner (e.g., "not at all aggressive" to "very aggressive"). They conceptualized sex roles as "the degree to which men and women are perceived to possess any particular trait" (Broverman et al., 1972). They found forty-one items with which at least 75 percent agreement existed among males and females as to which end of the bipolar dimension was more descriptive of the average man or of the average woman. These were the items, presented in Exhibit 11.1, they concluded formed the sex-role stereotypes for these students.

From this exhibit, it can be noted that those items associated with males, as judged by *both* males and females, are markedly consistent with the expectations derived from Erikson and McCandless. For instance, we find males described as very aggressive, very independent, very dominant, very active, very skilled in business, and not at all dependent. These items form what Broverman et al. (1972) term a competency cluster; and this is indeed identical to the competency-effectiveness set of behaviors that McCandless and Erikson predict. Not only are females judged by both males and females to be at the opposite (low) ends of these competency-effectiveness dimensions, but also they are judged to be high on the warmth-expressiveness items. For example, they are seen to be very gentle, very aware of feelings of others, very interested in appearance, and having a very strong need for security.

Moreover, although Exhibit 11.1 shows that for competency items the masculine end of the trait dimension is more desirable and for the warmth-expressiveness cluster the feminine end is more desirable, we see also that there are more competency items than warmth-expressiveness items. Thus there are more positively evaluated traits stereotypically associated with males than with females. Broverman et al. (1972) found evidence that these attitudes are quite pervasive in society. They report that their questionnaire has been given to 599 men and 383 women, who vary in age (from 17 to 60 years), educational level (from elementary school completed to an advanced graduate degree), religious orientation, and marital status. Among all the respondents there was considerable consensus about the different characteristics of males and females; the degree of consensus was not dependent on one's age, sex, religion, educational level, or marital status (Broverman et al., 1972).

In fact, one of the most striking instances of the high consensus regarding sex-role stereotypes was derived from a special sample that the researchers studied. They administered the questionnaire to seventy-nine practicing mental health workers (clinical psychologists, psychiatrists, and psychiatric social workers). Of the forty-six men in this sample, thirty-one held a Ph.D. or an M.D., while of the thirty-three women studied, eighteen held one of these degrees. The range of clinical experience in this group was from an internship to extensive professional practice. These professionals were asked to respond to the questionnaire three times, first to describe "a mature, healthy, socially competent adult male," second to give a corresponding description for an adult woman, and third for an adult (with no sex specified).

First, it was found that these male and female clinicians did not differ significantly from each other in their descriptions of men, women, and adults, respectively. Second, for each of these three categories there was high agreement about which end of the trait dimension reflected healthier behavior; thus these clinicians

EXHIBIT 11.1   **Stereotyped Sex-Role Items**

Competency Cluster (Masculine Pole Is More Desirable)

Feminine	Masculine
Not at all aggressive	Very aggressive
Not at all independent	Very independent
Very emotional	Not at all emotional
Does not hide emotions at all	Almost always hides emotions
Very subjective	Very objective
Very easily influenced	Not at all easily influenced
Very submissive	Very dominant
Dislikes math and science very much	Likes math and science very much
Very excitable in a minor crisis	Not at all excitable in a minor crisis
Very passive	Very active
Not at all competitive	Very competitive
Very illogical	Very logical
Very home oriented	Very worldly
Not at all skilled in business	Very skilled in business
Very sneaky	Very direct
Does not know the way of the world	Knows the way of the world
Feelings easily hurt	Feelings not easily hurt
Not at all adventurous	Very adventurous
Has difficulty making decisions	Can make decisions easily
Cries very easily	Never cries
Almost never acts as a leader	Almost always acts as a leader
Not at all self-confident	Very self-confident
Very uncomfortable about being aggressive	Not at all uncomfortable about being aggressive
Not at all ambitious	Very ambitious
Unable to separate feelings from ideas	Easily able to separate feelings from ideas
Very dependent	Not at all dependent
Very conceited about appearance	Never conceited about appearance
Thinks women are always superior to men	Thinks men are always superior to women
Does not talk freely about sex with men	Talks freely about sex with men

Warmth-Expressiveness Cluster (Feminine Pole Is More Desirable)

Doesn't use harsh language at all	Uses very harsh language
Very talkative	Not at all talkative
Very tactful	Very blunt
Very gentle	Very rough
Very aware of feelings of others	Not at all aware of feelings of others
Very religious	Not at all religious
Very interested in own appearance	Not at all interested in own appearance
Very neat in habits	Very sloppy in habits
Very quiet	Very loud
Very strong need for security	Very little need for security
Enjoys art and literature	Does not enjoy art and literature at all
Easily expresses tender feelings	Does not express tender feelings at all easily

Note: These results are based on the responses of 74 college men and 80 college women.
Source: Adapted from Broverman et al., 1972, p. 63.

Despite consistency in sex-role stereotypes, children of all ages often engage
in behaviors incompatible with traditional expectations.
(Raimondo Borea/Photo Researchers, Inc.)

had a generalized belief about what constituted good mental health. Third, it was
found that these clinical judgments regarding mental health were highly consistent
with college students' corresponding depictions.

Finally, and most importantly in terms of sex-role stereotypes, these professionals' judgments regarding what constituted the healthy male and the healthy female were consistent with sex-role stereotypes. The desirable masculine end of the
competency cluster was attributed to the healthy man, rather than to the healthy
woman, more than 90 percent of the time, while the desirable feminine end of the
warmth-expressiveness cluster was attributed to the healthy woman, rather than to
the healthy man, more than 60 percent of the time (Broverman et al., 1972, p. 70). If
one considers the content of the items attributed by these mental health workers to
males and females, respectively,

> . . . clinicians are suggesting that healthy women differ from healthy men by
> being more submissive, less independent, less adventurous, less objective, more
> easily influenced, less aggressive, less competitive, more excitable in minor cri-
> ses, more emotional, more conceited about their appearance, and having their
> feelings more easily hurt. (Broverman et al., 1972, p. 70)

Moreover, the mental health workers' judgments about what constitutes a
healthy adult and what constitutes a healthy man did not differ. However, there was
a difference when their view of the healthy adult was compared to their view of the
healthy woman. That is, among both the male and female professionals studied, the
general idea of mental health from a sex-unspecified adult is "actually applied to
men only, while healthy women are perceived as significantly *less* healthy by adult
standards" (Broverman et al., 1972, p. 71).

Not only is there evidence that sex-role stereotypes are fairly consistent across
the sex, age, and educational levels within society, but there is also evidence for

considerable cross-cultural consistency in sex-role stereotypes. In fact, Block (1973), in a study of six different countries (Norway, Sweden, Denmark, Finland, England, and the United States), not only found marked cross-cultural congruence but also found empirical verification of the differential emphases on competence-effectiveness and warmth-expressiveness for the two sexes that McCandless and Erikson predicted to exist and Broverman et al. (1972) found people to *believe* to exist.

Block's (1973) term for the type of behaviors we have labeled competence-effectiveness is *agency*. As shown in Exhibits 11.2 and 11.3, the items she sees as characteristic of agency (e.g., assertive, dominating, competitive, and independent) correspond to those in the Broverman et al. (1972) competency cluster. Block's (1973) term for the type of behavior we have labeled warmth-expressiveness is *communion*. The items she sees as comprising communion (e.g., loving, affectionate, sympathetic, and considerate) correspond to those in the Broverman et al. warmth-expressiveness cluster.

Block (1973) had four psychologists classify all these items into either the agency or the communion category. Those items that showed high agreement among the judges are shown by category in the first column of Exhibits 11.2 and 11.3. Exhibit 11.2 shows that among the university students in the samples, there were sixteen items on which males were more stereotypically associated than were females. In this group, *all* items receiving a classification were agency (competence) items. Moreover, although there are cross-cultural differences, the implications of which will be discussed below, we see that in at least four of the six cultural groups, both males and females within and across cultures agreed that males are higher than females in regard to being practical, shrewd, assertive, dominating, competitive, critical, and self-controlled.

EXHIBIT 11.2

Adjective Attributions among Students in Six Countries: Items on Which Males Are Stereotypically Associated

Agency-Communion Classification	Adjective	Country					
		United States	England	Sweden	Denmark	Finland	Norway
Agency	Practical, shrewd	X	X	X	X	X	X
Agency	Assertive	X	X		X	X	X
Agency	Dominating	X	X			X	X
Agency	Competitive	X	X				
Agency	Critical	X					X
	Self-controlled	X				X	
Agency	Rational, reasonable	X				X	
Agency	Ambitious						
	Feels guilty						X
	Moody		X				
Agency	Self-centered						
	Sense of humor				X		
	Responsible					X	
	Fair, just					X	
Agency	Independent						X
Agency	Adventurous						X

Note: X = significant difference found in number of attributions to males and females.
Source: Adapted from Block, 1973, p. 518.

Exhibit 11.3 shows that there were seventeen items on which females were more stereotypically associated than males. Of those eight items receiving a classification by the judges, seven were communion (warmth-expressiveness) ones. Moreover, in at least four of the six cultural groups, males and females within and across cultures agreed that females are higher than males in regard to being loving, affectionate, impulsive, sympathetic, and generous.

In summary, across groups in American society and in comparisons among samples from different societies, there is clear evidence that stereotypes exist which specify that different sets of behaviors are expected from males and females. This evidence shows that the male role is associated with individual effectiveness, independent competence, or agency. On the other hand, the evidence shows that the female role is associated with interpersonal warmth and expressiveness, or communion. Although the existence of these stereotypes means that people believe males and females differ in these ways, the existence of the stereotype does not *necessarily* mean that males and females actually behave differently along these dimensions. Thus although people's stereotypic beliefs do correspond to the theoretical expectations of Erikson and McCandless, this correspondence does not necessarily confirm their views because, as noted, stereotypic differences need not correspond to behavioral differences. Even if a correspondence between the stereotypes and behavior does exist, this relation would not provide unequivocal support for either Erikson or McCandless. Other processes, unspecified by these theorists, could be involved. Before we evaluate data pertinent to the relations between sex-role stereotypes and sex differences in cognition, personality, and social behavior, we consider some implications of the character of existing sex-role stereotypes.

EXHIBIT 11.3

Adjective Attributions among Students in Six Countries: Items on Which Females Are Stereotypically Associated

Agency-Communion Classification	Adjective	Country					
		United States	England	Sweden	Denmark	Finland	Norway
Communion	Loving, affectionate	X	X		X	X	X
	Impulsive	X	X	X		X	X
Communion	Sympathetic		X			X	
Communion	Generous			X			X
Agency	Vital, active	X		X			X
	Perceptive, aware						X
Communion	Sensitive	X	X				
	Reserved, shy	X	X				
Communion	Artistic					X	
	Curious						X
	Uncertain, indecisive					X	
	Talkative					X	
Communion	Helpful		X				
	Sense of humor		X				
	Idealistic				X		
	Cheerful						X
Communion	Considerate						X

Note: X = significant difference found in number of attributions to males and females.
Source: Adapted from Block, 1973, p. 519.

Implications of Sex-Role Stereotypes for Socialization

Lerner and Spanier (1980) presented some ideas about the potential role that stereotypes could have for behavioral and social development. On the basis of initial stereotypic appraisals of people categorized in a particular group (e.g., "adolescents," "endomorphs," "blacks," "women"), behavior is channeled in directions congruent with the stereotype. As a consequence, stereotype-consistent behavior often is developed. Once this self-fulfilling prophecy is created, behavior maintains the stereotype and a circular function is thus perpetuated. The applicability of this explanation for data pertinent to the implications of bodily changes in childhood and adolescence and to some cognitive developments in adolescence has been shown (Lerner & Spanier, 1980). Here we may note that there exist some reasons to believe in the presence of a similar self-fulfilling prophecy process being involved in the creation of sex differences in cognitive, personality, and social behavior.

It has often been posited that the socialization experiences of males and females differ in ways consistent with traditional sex-role stereotypes and, hence, with the existence of a self-fulfilling prophecy. For example, Maccoby and Jacklin (1974) detail the widely assumed belief that parents do a great deal during the child's early life to define for him or her what sex-appropriate behavior is *and* to guide the child toward the adoption of this behavior. Indeed, not only do most parents obviously dress boys and girls differently, but there are some data indicating that they urge children to develop sex-typed interests, e.g., by giving them sex-typed toys and by discouraging them (particularly boys) from engaging in activities they consider appropriate only for the opposite sex (Maccoby & Jacklin, 1974). For example, Miller and Swanson (1958) found that a majority of urban midwestern mothers who were studied channeled the behaviors of their children in ways congruent with traditional notions about divisions of labor (e.g., regarding "women's work" such as dishwashing). Brun-Gulbrandsen (1958) found in Norway results similar to those of Miller and Swanson (1958) and, in addition, found that mothers put more pressure on girls than on boys to conform to societal norms. Similarly, in a series of investigations involving the mothers and fathers of boys and girls ranging in age level from early childhood through late adolescence, Block (1973) found further evidence pertaining to stereotype-related differences in the ways males and females are socialized. The parents were asked to described their childrearing attitudes and behaviors regarding one of their own children. Block assumed that parents of boys are not (at least beforehand) intrinsically different from parents of girls. Therefore, differences in the way parents socialize males or females should reflect sex-role stereotyping imposed by parents on the children. Moreover, in comparing the parents of boys with the parents of girls, Block (1973) reports that the socialization practices for boys across the age range studied reflected an emphasis on achievement and competition, an insistence on control of feelings, and a concern for rule conformity (e.g., to parental authority). However, for girls of the age range studied, the socialization emphasis was placed—particularly by their fathers—on developing and maintaining close interpersonal relationships; the girls were encouraged to talk about their problems and were given comfort, reassurance, protection, and support (Block, 1973, p. 517).

While there does appear to be at least some evidence that parents strive to socialize their children in accordance with sex-role stereotypes, Maccoby & Jacklin (1974) conclude that, on the basis of their review of several hundred relevant studies, the weight of the evidence supports neither a socialization process of social learning based on sex-role stereotypes nor a nature-based conception stressing differential childrearing practices based on congenital biological-behavioral differences between the sexes. According to Maccoby and Jacklin (1974), most studies indicate that parents neither reward boys more for an aggressive or competitive behavior nor punish girls more for these behaviors. They note, "There is no evidence that parents are systematically reinforcing sons, more than daughters, for aggressive behavior" (Maccoby & Jacklin, 1974, p. 340). Similarly, there is little evidence that females are reinforced more for modesty or punished more for sexual exploration. Moreover, they conclude that parental childrearing differences for sons and daughters do not exist in regard to encouragement of independence or exploratory behavior or in restricting such behaviors.

Accordingly, there are considerable data indicating that parents treat boys and girls very much alike. Not only does this situation interfere with a social learning interpretation of sex differences which stresses socialization based on sex-role stereotypes, but Maccoby and Jacklin (1974, p. 343) assert that this suggests "there are probably not very many initial biologically-based behavior differences, at least not many that are strong enough to elicit clear, differential reactions from caregivers." Thus we have a situation supporting neither extant nature nor nurture views of the bases of sex differences. Instead, we see that people believe that sex differences

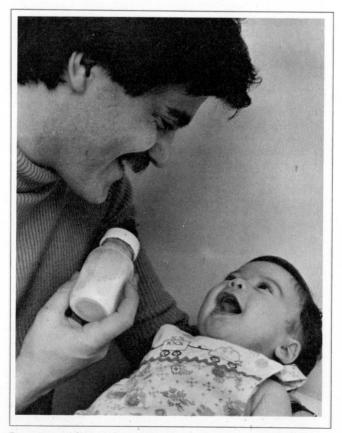

Parents treat their male and female children very similarly.
(Erika Stone/Peter Arnold, Inc.)

generally exist that parents sometimes seem to socialize children in accordance with these social stereotypes, but that most of the time they do not. This leads to an expectation that sex differences may not characterize human development most of the time, although they certainly can (and do) occur; in addition, however, it leaves us with the view that despite the expected paucity of sex differences in development, we may also expect that people believe (have stereotypes that) considerable sex differences exist. Why might people believe that sex differences generally exist despite evidence that they are not generally characteristic of human development? By addressing the issue of why particular behaviors comprise the (stereotyped) definition of the roles for each sex, we will not only answer this question but also provide a basis for elaborating our interactionist view about the bases of sex differences in human development.

THE ADAPTIVE BASIS OF SEX DIFFERENCES IN ROLES

Why are there traditional prescriptions of stereotypic sex roles, and why does their content exist as it does? Roles have adaptive significance. The function of roles is to allow society to maintain and perpetuate itself. Indeed, we may speculate that the anatomical and physiological constitution of the sexes resulted in the first social roles being differentiated on the basis of sex. Although there is no way to test this idea directly, sex roles, like all other social roles, should have some basis in the functions of people and their society. From this reasoning, it follows that sex differences in role behavior, at least initially, arose from the different tasks males and females performed for survival, including reproduction.

Although maintaining these sex differences—that is to say, making them traditional—could continue to serve this survival function (if the survival demands on people remained the same), it is also possible that sex roles could become traditionalized despite a change in the adaptive demands facing society. Indeed, not only is there cross-cultural empirical evidence consistent with our speculations about the basis of sex differences in roles, but there is at least some indirect support for the possibility that sex roles may not be evolving apace with social changes promoting new adaptive demands.

Barry, Bacon, and Child (1957) reviewed anthropological (ethnographic) material which described patterns of socialization in mostly nonliterate cultures. They reported a general trend of greater socialization pressures toward nurturance, responsibility, and obedience for females and greater socialization pressures toward self-reliance and achievement behaviors for males. In these relatively primitive societies, these different socialization pressures were seen to be associated with the contrasting biological and socioeconomic functions each sex had to assume when adult. As argued by Block (1973, p. 518), such findings suggest that:

> *When hunting or conquest is required for societal survival, the task naturally and functionally falls upon the male because of his intrinsically superior physical strength. So, boys more than girls receive training in self-reliance, achievement, and the agentic corollaries. Child-bearing is biologically assigned to women, and because, in marginally surviving societies, men must be out foraging for food, child rearing, with its requirement of continuous responsibility, is assigned to women. Thus, girls more than boys are socialized toward nurturance, responsibility, and other qualities of communion.*

Cultural-Historical Change and Sex-Role Evolution

In modern countries, the demands of day-to-day life are considerably different from those of primitive, marginally surviving societies. Accordingly, the meaning and function of these traditional sex roles may be different. As Block (1973, p. 519) states the issue:

> *The heritage and functional requiredness of sex typing in early and marginal cultures seem clear. The question for our times, however, is to what extent past socialization requirements must or should control current socialization emphases in our complex, technological, affluent society where, for example, physical strength is no longer especially important and where procreation is under some control. Under present conditions, and for the future, we might ask: What is necessary? What is "natural" in regard to sex typing?*

One way of addressing this issue is to reconsider the data on cross-cultural analyses of sex-role stereotypes presented in Exhibits 11.2 and 11.3. *If* sex roles do reflect to some extent the requirements placed on men and women in their particular societal setting, then differences in these settings could—to some extent—relate to differences in sex-role prescriptions in the different cultures. But despite the fact that all countries studied by Block (1973) are western societies and despite the fact that we have seen general trends in *all* cultures toward agency stereotypes for males and communion stereotypes for females, there are nevertheless differences in the socioeconomic, political, and physical environmental pressures of people in the respective societal settings.

Bakan (1966) found a relationship to exist between socialization pressures for agency (competence and effectiveness) and the presence of a capitalistic social and economic system, which he believed required an intensification of the agentic orientation. Consistent with this view, Block (1973) reported that in the two countries in her research that had long and widespread commitments to social welfare—Sweden and Denmark—there were fewer sex differences and less emphasis on agency than in the United States. In fact, Block (1973) found that American males were significantly different from the males of the other countries studied; they placed greater emphasis in their ratings on depictions of the male role as being adventurous, self-confident, assertive, restless, ambitious, self-centered, shrewd, and competitive. Their emphasis on these characteristics reflects their greater orientation to agency.

Moreover, also consistent with Bakan's (1966) idea of a relation between capitalism and agency, and our more general notion that behaviors associated with the roles of each sex necessarily reflect the sociocultural (e.g., economic, political, and environmental) demands placed on people at a particular time in their society's history, are Block's findings regarding the females. Despite being more oriented to communion than to agency, American women nevertheless placed greater emphasis in their ratings on agency terms than did women in the other countries studied (Block, 1973). To a significantly greater extent than did the women from the other cultural settings, the American women gave higher ratings to such traits as practical, adventurous, assertive, ambitious, self-centered, shrewd, and self-confident in their responses about a women's role (Block, 1973). These characteristics are also all agency ones.

The United States is the most capitalistic of the countries studied by Block (1973); and if we can assume that this characteristic is a salient one differentiating the countries, then the differences between the sex-role stereotypes of males and females in the United States and those of the other countries may be understood better. Thus differences in the behavior expected from males and females of a particular society are understandable on the basis of the sociocultural forces acting over time on the people. Moreover, not only should these different sociocultural-historical pressures influence behavioral expectations, but they should also influence childrearing (socialization) practices. If the adaptive demands placed on people in a society are different, the socialization of people to meet these demands should be different.

In support of this idea, Block (1973) reported that students' ratings of parental childrearing practices differed between the United States sample and those of the five other countries. In the United States, it was found that significantly greater emphasis was placed on early and clear sex typing and on competitive achievement and less importance was placed on the control of aggression in males (Block, 1973).

Thus these differences between American and European childrearing orientations and sex-role stereotypes not only show that there can be and are contrasts between different cultures (Block, 1973)—although they are highly similar, western ones—but also that these contrasts are related to familial, sociocultural, and historical differences. Moreover, the character of the sex differences in some cultural settings shows that because of its long-term nature, a presumably adaptive narrowing of some of the agency-communion differences found traditionally in other cultural settings (e.g., American ones) can occur; this suggests that humans can, insofar as sex-role stereotypes and socialization practices are concerned, overcome the divisions between the sexes that remain as perhaps less-than-optimal remnants from earlier times (Block, 1973).

Conclusions

The above evidence and arguments suggest that a biological notion, which stresses that anatomy is destiny, or a nurture notion, which emphasizes only social rewards and punishments, is a limited view. Biology certainly does exert pressures on psychosocial development; however, this influence does not occur independently of the demands of the cultural and historical milieu. The biological basis of one's psychosocial functioning relates to the adaptive orientation for survival. Hence although it may be adaptive at some time in a given society to perform roles highly associated with anatomical and physiological differences, these same roles may not be adaptive in other societies or at other times in history. The agency-communion differences that were functional in primitive times may no longer be so in a modern society having greater leisure time, nearly equal opportunity for employment, and almost universal formal education. A similar argument has been advanced by Self (1975).

Block (1973) shows that at least one measure of more developed psychosocial functioning—the presence of a principled level of moral reasoning—is associated with American college-age males and females who have less traditional sex-role definitions of themselves. In a series of studies by Bem (1974, 1975, 1977), it has been found that among men and women who have an orientation to *both* traditional male *and* traditional female behaviors, there is evidence for adaptive psychosocial functioning. Bem (1974, 1975) argued that internalization of the culturally stereotypic sex role inhibits the development of a fully adaptive and maximally satisfying behavioral repertoire. Instead, a male or female who identifies with both desirable masculine and desirable feminine characteristics—an *androgynous* person—is not only free from the limitations of stereotypic sex roles but should also be able to engage more effectively in both traditional male and traditional female behaviors across a variety of social situations, presumably because of his or her flexibility, than should a nonandrogynous person (Jones, Chernovetz, & Hansson, 1978). Several studies (Bem, 1974, 1975, 1977; Spence & Helmreich, 1978; Worell, 1978) have provided data validating this general idea, i.e., that the greater vocational and sex-role flexibility which defines androgynous men and women is associated with greater social competence and psychological health than is the case with males and females oriented to traditional sex-role adoption.

If one views individual development as reciprocally related to sociocultural change, one may be led to predict that the current historical context presses males and females to forgo the traditional vocational roles, perhaps adaptive in earlier epochs; instead, today's cohorts of males and females may be encouraged to adopt sex-role orientations showing flexibility and independence from traditional sex-role prescriptions in order to be adaptive in the current social context. Thus anatomy is not destiny. Rather, it is just one component of biology that *may be* of relevance for the adaptive roles in a particular sociocultural-historical milieu (cf. Self, 1975).

As such, a person's behavior in his or her cultural setting appears to involve much more than obtaining rewards for emitting behaviors that are drive-reducing. At any point in history, such behaviors involve (1) a coordination of the cultural, political, and economic values of one's society (2) as that society exists in a physical environmental setting placing changing survival demands on the people embedded within it (3) in the context of familial childrearing pressures and psychological (e.g., moral) developments.

We now consider cognitive, personality, and social domains of functioning, predicted to show sex differences by nature (e.g., Erikson) and nurture (e.g., McCandless) theorists, in order to ascertain whether data directly relevant to these predictions support our interactionist conclusions.

RESEARCH ON SEX DIFFERENCES IN COGNITIVE, PERSONALITY, AND SOCIAL DEVELOPMENT

The first issue to address is whether the existing research literature indicates that males and females differ in ways consistent with Erikson's, McCandless's, or the present authors' interactionist views. As noted, although there are literally hundreds of studies that may be considered in order to address this issue, there have been several recent attempts to integrate these studies (Block, 1976; Maccoby & Jacklin, 1974; O'Leary, 1977). One of these (Maccoby & Jacklin, 1974) was sufficently encompassing that it may serve as a basis for our presentation.

Maccoby and Jacklin (1974) evaluated the results of about 1600 research reports, published for the most part between 1966 and 1973. Maccoby and Jacklin derived these studies from those professional research journals that frequently include information about psychological sex differences, as well as from other sources (e.g., review chapters and theoretical papers). The studies they reviewed were classified into eight major topical areas (e.g., perceptual abilities, intellectual abilities, and achievement motivation). Moreover, within each of these areas, studies were sorted on the basis of their relevance to particular behaviors or constructs, e.g., aggression, dependency, helping, or anxiety. Although the number of studies dealing with the more than eighty behaviors or constructs evaluated by Maccoby and Jacklin differed from topic to topic, for each of these behaviors or constructs, tables were formed. These tables included information about the authors of the study, the ages and numbers of the people studied, and whether or not statistically significant differences between males and females had been found. Depending on the proportion of the studies done on a topic for which significant sex differences occurred, Maccoby and Jacklin drew conclusions about whether sex differences were or were not well established.

There are some problems with the conclusions they drew. First, the proportion of findings used to decide whether a sex difference was well established varied for different domains of behavior. Moreover, the number of studies done in different areas varied. In addition, not all studies pertinent to a particular behavior were of the same quality. For example, some studies assessed very small samples. In such cases it is difficult for statistics to confidently confirm a difference between the sexes. At these times it is said that the statistic lacks *power,* and in deciding whether a sex difference was well established, Maccoby and Jacklin (1974) did not consider "how the poor statistical power of certain studies may have influenced adversely the trend of the findings they are attempting to integrate" (Block, 1976, p. 287).

Another problem with the conclusions that Maccoby and Jacklin (1974) drew was an unevenness in the representation of various age groups in the research they reviewed. Seventy-five percent of the studies on which Maccoby and Jacklin based their conclusions involved people 12 years of age or *younger,* and about 40 percent studied preschool children (Block, 1976). There are very few studies of sex differences in the adulthood and aged years. This differential representation of age is important because there is some evidence (Terman & Tyler, 1954) that sex differences increase in frequency during adolescence. Thus Maccoby and Jacklin (1974) may have underestimated the proportion of sex differences that actually exists for a particular behavior by their review of studies of preadolescent samples. Moreover, sampling techniques varied from study to study, and there is little consideration of the possibility that studies claiming to measure the same phenomenon actually do so.

Because of such problems with the conclusions that Maccoby and Jacklin reached, Block (1976) attempted to draw her own conclusions by tallying the number of studies reviewed by Maccoby and Jacklin which pertained to various domains of behavior. In other words, Block took those studies reviewed by Maccoby and Jacklin for each of several behaviors and calculated the ratio of significant differences favoring one or the other sex in each set of studies. We may here adapt Block's (1976) tallying method and, for each of several behaviors, discuss the percentage of studies in which females were significantly higher, the percentage in which males were significantly higher, and the percentage not showing a significant difference.

Sex Differences in Cognitive Processes

First, we summarize the extent of sex differences found in the studies of various cognitive processes reviewed by Maccoby and Jacklin (1974); however, such a summary is complicated by several features of Maccoby and Jacklin's (1974) presentation. Their work attempted to integrate large areas of research under common topic headings. For example, as is the case with the several topics Maccoby and Jacklin (1974) surveyed, the studies of cognitive processes may be grouped into several categories. In regard to cognition these categories are sensation and perception; learning and memory; intelligence; and cognitive development. Moreover, within each category there are several subcategories of topics that have been studied. For example, in the sensation and perception category there have been studies of five topics: tactile sensitivity; tactile perception; audition (hearing); vision; psychomotor abilities. Still further complexity occurs because for each topic not all age levels have been equally studied; i.e., it is difficult to get a clear idea about the ontogenetic progression of sex differences because all studies of all topics do not have a common age range. Some, for example, study only 2-, 3-, and 4-year-olds. Finally, we should note that not all topics have been equally studied. For instance, there are only thirteen studies reported of tactile sensitivity in the first year of life; however, more than sixty studies of vision in the first year are reported.

Nevertheless, despite these limitations there are several themes that can be identified. First, in regard to those studies of cognitive processes which dealt with sensation and perception, 162 comparisons between males and females were made and in the vast majority of these—73.5 percent—there were no significant sex differences. Females scored higher than males on 17.9 percent of the comparisons, and males were higher scorers than females on 8.6 percent of the studies. Moreover, for all age groups and for all topics the sexes did not differ in the majority of the surveyed studies. Thus even though there are some areas (e.g., tactile sensitivity) where, at some age levels (e.g., among newborns), one sex group (e.g., here females) showed higher scores than the other more of the time, it is still the case that for most of the topics studied, most investigations *did not* report sex differences at most age levels.

This theme of lack of sex differences in the majority of investigations of most topics at most age levels is seen, too, in regard to the other categories classified under the heading of cognitive processes. Thus in regard to learning and memory, of the 214 comparisons surveyed by Maccoby and Jacklin (1974), there is no case (i.e., age grouping) wherein males exceed females or females exceed males in the majority of the studies. Indeed, it is again the case that across the total of 214 comparisons surveyed under this heading, females exceeded males on only 14.9 percent, males exceed females on only 7.5 percent, and the sexes did not differ on the vast majority of comparisons—77.6 percent.

Similarly, in regard to the third category surveyed under the cognitive processes heading (intelligence), general confirmation occurred for this trend of a lack of sex differences for most studies of most topics at most age levels. Indeed, of the 350 comparisons between the sexes surveyed by Maccoby and Jacklin (1974) only for studies of spontaneous verbal and vocal behavior, involving eleven comparisons within a 2- to 10-year-old age range, and for studies of tested verbal abilities, involving only three comparisons within a 6- to 18-month-old age range, do the sexes differ on most studies. Thus for fifteen of the seventeen tallies of topic by age grouping listed in the intelligence category, the sexes do not differ most of the time. In fact, although there are some topics—for instance, spatial ability—wherein when sex differences *do* occur, they tend to occur predominantly in one direction (and in the case of spatial abilities males appear to generally get higher scores when differences are found), the point is that even here the sexes do not differ in most cases. On only 19.4 percent of the comparisons did females score higher than males, on only 19.4 percent of the comparisons did males score higher than females, and on 61.2 percent of the comparisons the sexes did not differ significantly.

Finally, the last category surveyed by Maccoby and Jacklin under the topic of cognitive processes—cognitive development—again supports these trends. Indeed, there is only one combination of topic by age grouping (conceptual level at age 33 years), involving only two comparisons, where the sexes differ 50 percent of the time. Thus for the 123 comparisons surveyed by Maccoby and Jacklin (1974), females exceed males on only 13.8 percent, males exceed females on only 17.1 percent, and the sexes did not differ on the preponderant majority of comparisons—69.1 percent.

Insofar as the various cognitive functions reviewed by Maccoby and Jacklin

(1974) are concerned, we may conclude that differences between the sexes do not generally exist. Indeed, only for some topics studied at some age levels, and then in only some (and, infrequently, most) of the instances of measurement, do sex differences in cognitive processes occur. Thus any nature notion that specifies that the genes which determine one's sex also determine one's particular cognitive attributes can find no support in such data. In turn, a social learning notion that speaks of general socialization environment rules promoting specific behaviors for males and other specific behaviors for females cannot be seen as very plausible on the basis of these data. Clearly, sometimes males and females differ on some cognitive variables, but most often and for most variables they do not.

The job for science, then, in our view, would seem to be to discover what combinations of factors associated with the organism and with the environment lead to the absence and/or presence of sex differences for selected variables at successive times of life. In short, the study of people in context across their life spans seems to be the avenue of investigation promoted by the weight of the literature on sex differences in cognitive processes. Similar recommendations are to be found in our analysis of sex differences in personality and social development.

Sex Differences in Personality and Social Development

We may now turn to a summary of the extent of sex differences found in the studies of various personality and social developments reviewed by Maccoby and Jacklin (1974). We may organize our presentation in exactly the same way as our preceding one of cognitive processes with, however, one additional feature. The present authors have classified each of the behaviors we will discuss into either an agency or a communion category, based on the criteria for such behavior presented by Block (1973), Bakan (1966), and Broverman et al. (1972). The behaviors we have classified as communion ones include fifteen behaviors surveyed by Maccoby and Jacklin (1974): sensitivity to social reinforcement; touch and proximity to parent and resistance to separation from parent; touching and proximity to nonfamily adult; proximity and orientation to friends; positive social behavior toward nonfamily adult; positive social interactions with peers; self-report of liking for others; trust in others; sensitivity to social cues ("empathy"); helping; sharing; cooperation; compliance with adult requests and demands; conformity, compliance with fears, susceptibility to influence; and spontaneous imitation. The behaviors we have classified as agency ones include nine behaviors surveyed by Maccoby and Jacklin (1974): achievement striving; task persistence and involvement; curiosity and exploration; confidence in task performance; internal locus of control; activity level; aggression; competition; and dominance. As with the studies of cognitive processes summarized above, the study of most communion behaviors is not generally associated with sex differences. Indeed, for the 419 comparisons of male and female communion behaviors surveyed by Maccoby and Jacklin (1974), females scored higher than males in 19.8 percent of the cases, males scored higher than females in 13.4 percent of the cases, and the sexes did not differ significantly in more than two-thirds of the cases: 66.8 percent. Indeed, if one considered each communion behavior in relation to studies done at successive age levels, then, for the combinations of behavior by age grouping associated with thirteen of the fifteen communion behaviors, the sum total of the studies showing sex differences was *50 percent or less*. Thus more often than there were sex differences in a communion behavior, there were no such differences when the same behavior was assessed in males and females of the same age levels.

Similarly, most agency behaviors are not generally associated with sex differences. Of the 289 comparisons of male and female agency behaviors surveyed by Maccoby and Jacklin (1974), females scored higher than males in 10.4 percent of the cases, males scored higher than females in 40.1 percent of the cases, and the sexes did not differ significantly 49.5 percent of the time. Moreover, if one considered each agency behavior separately, then with seven of the nine agency behaviors the sum total of studies showing sex differences was 50 percent or less. Specifically, except for 15 studies of confidence in task performance and 94 studies of aggression, a general trend in the 180 (62.3 percent) additional comparisons of agency behaviors surveyed by Maccoby and Jacklin (1974) is that sex differences are much more likely to be absent than they are to be present. That is, of the fifteen comparisons made by Maccoby and Jacklin (1974) regarding confidence in task performance, sex differences favoring males were found in twelve and no comparison found a sex difference favoring females; similarly, of the ninety-four comparisons regarding aggression, fifty-two comparisons showed males scoring higher than females, five comparisons showed females scoring higher than males, and the remaining comparisons showed no sex differences. However, if one considers the remaining majority (77.8 percent) of agency behaviors, then of the 180 comparisons across these latter behaviors, females scored higher than males in 13.9 percent of the cases, males scored higher in 24.4 percent of the cases, and the two sexes did not differ significantly in the majority of comparisons: 61.7 percent.

Although the data regarding communion and agency behaviors suggest that *when* sex differences are found in communion behaviors, the differences most often "favor" females and, in turn, *when* sex differences are found in agency behaviors, the differences most often favor males, the point is that these differences are not seen in most studies.

It can be concluded then that most studies of either agency or communion behaviors *do not* show that the sexes differ. However, when they do differ, females tend to score higher on communion behaviors, such as dependency, social desirability, compliance, general anxiety, and staying in the proximity of friends (Block, 1976), and males tend to score higher on agency behaviors, such as aggression, confidence in task performance, dominance, and activity level (Block, 1976). Thus, only a minority of the studies are consistent with agency and communion differences in males and females posited by McCandless and by Erikson. As such, researchers must search for those biological-through-historical processes that produce such wide variations (such plasticity) in the presence and quality of communion and agency behaviors in *both* males and females.

In order to best discuss information pertinent to such interactive processes, we consider data relevant to the question of whether the self-concepts and self-esteems of males and females differ in the ways predicted by Erikson and McCandless.

Agency and Communion in Self-Concept and Self-Esteem

As with the nature (Erikson) and nurture (McCandless) predictions about behavioral differences between males and females, self-concepts should reflect orientation toward agency for males and toward communion for females if these ideas are useful. Moreover, because agency is composed of more positively evaluated traits than is communion (Broverman et al., 1972), the self-esteems of males would, from these perspectives, be predicted as higher (more positive) than those of females. Of course, the interactionist view we favor leads us to expect more plasticity.

Data relevant to self-concept and self-esteem differences are numerous and have been integrated also by Maccoby and Jacklin (1974) and evaluated by Block

(1976). A major methodological problem involved with these data, identified by Maccoby and Jacklin (1974), Block (1976), and Wylie (1974), however, serves to limit the usefulness of the data. Most studies assessing self-concept and/or self-esteem have used the person's own ratings of these constructs as the basis for measurement. Not only are such judgments obviously subjective, and probably considerably biased, but if only self-ratings are used, there is no way to check on the validity of the appraisals. In addition, Broverman et al. (1972) have shown that both males and females share the same sex-role stereotypes and tend to apply these stereotypes to themselves when characterizing their own behaviors. This stereotype-consistent appraisal may bias the self-concept and self-esteem scores in the direction of the predicted agency-communion differences. Other methodological problems exist. Researchers often do not distinguish between self-concept and self-esteem, they often define either of these terms in quite different ways, and they use different instruments to measure these constructs (Wylie, 1974). Because extensive information does not exist about how these different measures of self-esteem and/or self-concept relate to each other, important issues of interpretation are raised. For instance, one often does not know when a study using a particular measure shows that groups of males and females differ in their self-concepts whether these differences reflect (1) true differences between the groups; (2) differences due to the fact that a particular measure was used; or (3) some combination of 1 and 2. These methodological problems make it difficult to assess unequivocally whether males and females differ in their self-concepts and self-esteems in the ways that McCandless and Erikson predict.

Moreover, this problem is complicated by a related one. Some data can be found to support their views that males' self-concepts are characterized by individual competency, independent effectiveness (i.e., agency), and high, positive self-esteem, while females' self-concepts are characterized by warmth expressiveness and dependent passivity (i.e., communion) and a lower, more negative self-esteem than that found with males. However, data can also be found to contradict this view.

For example, the longitudinal study reported by Kagan and Moss (1962) showed that males' scores relating to aggression and dominance (both agency characteristics) tended to be stable and continuous from the early years of life through young adulthood but their scores relating to dependency and passivity (communion characteristics) tended to be unstable and discontinuous. In turn, communion-type characteristics were found to be stable and continuous for females during the age span, whereas agency-related characteristics tended to be unstable and discontinuous (Kagan & Moss, 1962). Similarly, in a study of the "self-images" of over 2000 children and adolescents in the late 1960s, Rosenberg and Simmons (1975) report that adolescent females have images of themselves that are more "people"-oriented, while the view adolescent males have of themselves stresses achievement and competence. Indeed, more females than males view their own sex with displeasure, and this is related to females having a more negative self-image than is the case with males (Simmons & Rosenberg, 1975). Moreover, cross-sectional data derived from 139 male and 142 female high school students, reported by Hakstian and Cattell (1975), are consistent with the above data, in that males were shown to be more tough-minded and realistic, while females were seen to be more tender-minded, dependent, and sensitive.

Yet other data are not consistent with these findings. For instance, although, as a consequence of their presumably greater dependency and domination by others, females would be more prone than males to conform to group pressures, studies of male and female 5- to 19-year-olds (Collins & Thomas, 1972) and of male and female

13- to 14-year-olds and 18- to 21-year-olds (Landsbaum & Willis, 1971) found no differences between the sexes in their incidence of conformity.

Thus the issue is not whether there are absolute, nonoverlapping differences between males and females but one of seeing the extent of the differences between the sexes in their self-concepts and their self-esteems. Despite the methodological biases and the equivocal nature of the findings, there is, *at best*, only partial support for self-concept and self-esteem sex differences.

Seven of the eight studies that Maccoby and Jacklin (1974) summarize for levels of strength and potency of the self-concept (i.e., agentic self-concept) showed males scoring higher than females. In the one remaining study, there was no sex difference. Five of these eight studies involved 5- to 17-year-olds and three involved 18-year-olds. Of nine studies of social self-concept (i.e., communion self-concept) done among 8- to 17-year-olds and of an additional three studies of social self-concept done with 18- to 45-year-olds, females scored higher than males on four of the nineteen and on two of the three studies within the two age groupings, respectively. In turn, males scored higher on one of the nine and on one of the three studies within the two age groupings, respectively. However, the more agency-oriented self-concepts of males and the more communion-oriented self-concepts of females do not appear to translate into self-esteem differences between the sexes.

Among those studies reviewed by Maccoby and Jacklin (1974) that looked at self-esteem sex differences, in the majority (61 percent) of the thirty-nine studies the sexes did not differ. Age level did not seem to moderate these results. There were two studies done among 3- to 4-year-olds, eighteen studies done among 6- to 12-year-olds, one study done among 11- to 17-year-olds, and eighteen studies done among 18- to 88-year-olds. There were no sex differences reported in the studies done at the first and the third age-level groupings. Females scored higher than males on five and on four of the studies within the second and fourth age-level groupings, respectively; males scored higher than females on two and on four of the studies within these two age-level groupings, respectively. Thus on the majority of studies within each of these last two groupings, the sexes did not differ.

Although the sexes may differ in regard to the items that they use to define themselves (Maccoby & Jacklin, 1974; O'Leary, 1977), they do not evaluate the items differentially when applying them to themselves. This means that males and females are likely to have comparable levels of self-esteem. In fact, in a series of studies of self-esteem (Lerner, Karabenick, & Stuart, 1973; Lerner, Orlos, & Knapp, 1976; Lerner & Karabenick, 1974; Lerner & Brackney, 1978), virtually identical levels of self-esteem were found in four independent cohorts of male and female late adolescents. In summary, it is clear that agency and communion sex differences are not characteristic of most adolescents studied and are not necessarily translated into the self-esteems of the sexes in any event. Accordingly, the question we are faced with is, What are the conditions when such sex differences do occur? Aid in answering this question is derived from a consideration of data relevant to sex differences in vocational role development (and to the nature and nurture predictions of Erikson and of McCandless).

SEX DIFFERENCES IN VOCATIONAL ROLE DEVELOPMENT

Both McCandless and Erikson predict that males and females should expect and aspire to play roles in society which are traditional sex-role ones. Such role orientations, it is held, are adaptive (for either biological or social reward reasons). The data to be reviewed provide both some support for and some refutation of these ideas.

Sex Differences in Vocational Role Behavior

Independent of the expectations and aspirations of males and females or of their implications for adaptive functioning, it is clear that the vocational roles of men and women in today's society are different. For instance, in the data reported in the 1970 U.S. Census, it was found that over 80 percent of the people engaged in the vocations of doctor, lawyer, dentist, truck driver, and farmer were males. Over 80 percent of the people engaged in the vocations of nurse, secretary, librarian, telephone operator, and elementary school teacher were females.

Moreover, in society, the primary role assigned to women is that of wife (O'Leary, 1977). In this country, only 5 percent of women never marry, and the average age of becoming a wife is 21 years (Spanier & Glick, 1979). In addition, the role of wife is associated with that of mother; most people entering marriage expect to have children. Another reason is that wives who are voluntarily childless are viewed negatively (Veevers, 1973) and are often characterized as neurotic or selfish (Bardwick, 1971). In fact, Russo (1976) notes that the number of children a woman has is sometimes used as a measure of her success in her mothering role. Although men are expected to marry and be fathers, there are role expectations for them outside of the marital union, and their success in these outside roles is of more import than their success as husbands and fathers (Block, 1973; O'Leary, 1977).

It is not certain that these role behavior divisions are adaptive, despite their traditionality. There is a higher incidence of mental illness among married women than among single women (O'Leary, 1977), and Gove and Tudor (1973) suggest that this difference may reflect the difficulties involved in engaging in traditional female roles which are not highly valued.

Findings by Block (1973) support these ideas. Block notes that the traditional socialization process widens the sex-role definitions and behavior options of men; we have just noted that in addition to being husbands and fathers, men are expected to (successfully) play roles outside of marriage. However, the traditional socialization process narrows the sex-role definitions and behavior options of women—wifing and mothering are the major roles some women play. About half the married American women are not employed outside the home (U.S. Bureau of the Census, 1978). Thus in her research about sex differences in ego development, Block (1973) concluded that it is more difficult for women to achieve higher levels of ego functioning because it involves conflict with prevailing cultural norms. As a consequence, few women of the cohort she studied had sex-role definitions that combined agency and communion orientations. Block (1973, p. 526) concludes that "it was simply too difficult and too lonely to oppose the cultural tide." Yet she notes that some balance of agency and communion was apparently necessary for advanced ego functioning. Highly socialized men had this adaptive status, but because of restricted socialization, highly socialized women did not (Block, 1973).

There is some evidence that adaptive behavior is associated with women who do engage in somewhat nontraditional role behavior. Although there are differences found in relation to socioeconomic status (Nye, 1974; Shappell, Hall, & Tarrier, 1971), employed women have been found to be more satisfied with their lives than are homemakers (Hoffman & Nye, 1974). Similarly, Birnbaum (1975) reports that middle-aged career women, whether single or married, had higher self-esteems than homemakers and even felt that they were better mothers. Moreover, Traeldal (1973) found that Norwegian women who were homemakers had stable feelings of life satisfaction across age but that the life satisfaction of women who were employed increased with age.

Contrary to expectations derived from Erikson's and McCandless's notions,

Today, males and females are increasingly engaging in nontraditional roles.
(Russ Kinne/Photo Researchers, Inc.)

then, traditional role behavior does not appear maximally adaptive for either males or females. Rather, the available evidence suggests that combinations of agency and communion behaviors can facilitate psychological and social processes related to ego development and life satisfaction (Block, 1973; O'Leary, 1974, 1977). However, Block (1973) suggests that most women do not achieve this; instead, they remain oriented toward traditional female vocational patterns.

Sex Differences in Vocational Role Orientation

Most data indicate that men and women remain oriented to traditional sex differences in vocational roles, despite the fact that the complexion of the American work force continues to change (Block, 1976). From 1900 to 1978 the percentage of all adult women (aged 20 to 64) who were in the labor force rose from 20 to 58 percent. The percentage of married women who were employed rose from 5.5 percent in 1900 to 48 percent in 1978 (Bureau of Labor Statistics, 1978).

The female work force shows uneven representation in relation to educational level and race. In 1977, 62.3 percent of women with college degrees and 71.5 percent of those with at least some graduate work were employed, while only 23 percent of women with eight or less years of education were employed. Black women were more likely than white women to work outside the home, and this difference is maintained across various educational levels (Bureau of Labor Statistics, 1978).

Furthermore, since the number of women receiving college degrees and the proportion of women in professional graduate schools are increasing rapidly, the future complexion of the American work force will certainly be altered. In fact, Kreps (1976) estimates that the average woman's work life will be only about ten years shorter than that of the average man.

Despite the continuing changes in the proportion of women in the work force, the vocational role orientations of females—and of males—remain quite traditional. In fact, stereotypes of females as vocationally incompetent, emotional, and unable to handle high-level jobs persist (Huston-Stein & Higgins-Trenk, 1978). Indeed, in a review of the literature by Huston-Stein and Higgins-Trenk, it was concluded that most studies show that females accept this stereotyped view of themselves, particularly in vocations that are traditionally "male."

Given the apparent acceptance of these stereotypes by males and females, it might be expected that vocational aspirations and expectations will be traditional. Data support this inference.

Looft (1971) asked first- and second-grade males and females to indicate their *personal* vocational role orientations, i.e., the vocations that they themselves expected to engage in as adults. The aspirations of both sexes were traditional. Boys commonly named football player and police officer as personal role orientations, while girls often named nurse or teacher. No girl mentioned vocations such as politician, lawyer, or scientist, although these were frequently noted by the boys.

In a related study, Bacon and Lerner (1975) interviewed second-, fourth-, and sixth-grade females about their personal vocational role orientation and in addition assessed the females' *societal* vocational role orientation, i.e., their conception of the roles in which the sexes could engage in society. At the higher two grade levels, females were more egalitarian (i.e., nontraditional) in their societal response than were females in the second grade; however, at all grade levels, most females had personal vocational orientations that were traditional. Thus, for most females, there was a self-other discrepancy between the nature of the vocational orientations associated with others (males *and* females) and the nature of the vocational orientation they associated with themselves. Others could have egalitarian vocations, but they aspired and expected to be traditional.

Lerner, Vincent, and Benson (1976) retested most of the second- and fourth-graders when they were in the third and fifth grades, respectively, and found this self-other discrepancy still evident. Furthermore, in an independent sample of fourth-, fifth-, and sixth-grade females and males, the self-other discrepancy was confirmed (Lerner, Benson, & Vincent, 1976). Both males and females at all grade levels showed similar and high levels of societal egalitarianism, but insofar as personal vocational orientations were concerned, both groups were traditional. Thus males associated themselves with the highly evaluated traditional male roles while associating others with egalitarian possibilities; females had personal associations with the less favorable traditional female roles and associated others (i.e., females) with more favorable egalitarian opportunities.

In our view, however, the fact that the vocational role orientations of both sexes are, at least in part, egalitarian and nontraditional in character is both encouraging—if one favors social equality for the sexes—and, at the same time, a basis for suggesting that the third prediction derived from Erikson's and McCandless's writings is not supported.

Both theorists would hold that there should be comparability between personal and societal orientations, since for Erikson, this relation would be necessary for biological adaptation, while for McCandless, it would be a product of learning and conforming to cultural norms for rewards. Moreover, not only would there be difficulty for these theorists' views when attempting to integrate the data about the dually directed personal-societal vocational orientations of both males and females, but also, the fact that there is increasing evidence that some females, especially college-educated ones, are expecting to combine agency-type careers with marriage and family goals (and hence communion) or, in turn, are placing emphasis on the former and not on the latter, would be problematic.

The percentage of college women who obtain graduate degrees, pursue careers, and yet engage in marriage has increased in recent years. Furthermore, fewer young women expect to be solely homemakers and mothers. There has been a decrease in recent years in women's involvement in marriage and childbearing. Birthrates reached an all-time low in the 1970s, while female employment rates reached an all-time high. Only about 1 in 4 married women in the 18- to 24-year-old age range expects to have three or more children, although the proportion of young married women who desire to remain childless (1 in 20) has not changed much in recent years (U.S. Bureau of the Census, 1978). In addition, it appears that there are the beginnings of some acceptance among males of this revised vocational role orientation among women. Although there are some data to indicate that middle-class males, especially those with high IQs, are the most traditional regarding vocational roles for males and females (Entwisle & Greenberger, 1972), other studies show that from 40 to 60 percent of college males would favor an interrupted career pattern for their wives; however, most define this pattern as meaning that wives should not work until children have completed school (Huston-Stein & Higgins-Trenk, 1978).

In conclusion, the vocational role predictions of McCandless and Erikson do not appear very consistent with existing data. Nontraditional vocational role behavior appears to be adaptive for both sexes and seems to involve a combination of agency and communion orientations. Moreover, the vocational role ideologies of males and females involve the perception—at least societally—of equal opportunity for males and females in endeavors traditionally defined as agency or communion. Not only may the self-other discrepancy in the personal vocational orientations of females facilitate social change, but there is also some tentative evidence that at least among people in a college setting, some increasing orientation toward combining agency and communion behaviors is occurring.

Thus in again rejecting the utility of either the McCandless or the Erikson position, we find evidence that complex alterations in personal development of males and females are occurring and these developments involve changes associated with particular cohorts who are embedded in particular social settings (e.g., college).

INTERACTIVE BASES OF THE DEVELOPMENT OF INDIVIDUALITY

Because of the particular combination of forces acting on a person, the individual character of development stands out. Block (1973, p. 513) notes that one's characterization of the attributes of the sexes "represents a synthesis of biological and cultural forces." We have noted that variables associated with many processes play a role in the social and personality development of males and females, and one may conclude that few sex differences must necessarily apply across time, context, and age. Indeed, the major implication of the analysis of this chapter is that individual differences are dependent on the person's developmental context. One component of this context is composed of phenomena associated with time of testing. In support of this view, recall the Nesselroade and Baltes (1974) sequential longitudinal study, discussed in Chapter 1. The basis of personality changes in adolescence was most related to the type of social change patterns which comprised the environmental milieu for *all* adolescents over time.

Perhaps the best example of how the changing social context provides a basis of individual development is derived from a study by Elder (1974), who presented longitudinal data about the development of people who were children and adolescents during the Great Depression in the United States (from 1929 to 1941). Elder reports that among a group of eighty-four males and eighty-three females born in 1920 to 1921, characteristics of the historical era produced alterations in the influence that education had on achievement, affected later adult health for youths from working-class families suffering deprivation during this era, and enhanced the importance of children in later adult marriages for youths who suffered hardships during the depression.

Other components of a person's context can be the physical and social characteristics of his or her school environment. Simmons, Rosenberg, and Rosenberg (1973) found that changes in the school context may influence personality. In a study of about 2000 children and adolescents, they found that in comparison to 8- to 11-year-old children, early adolescents—and particularly those 12 and 13 years of age—showed more self-consciousness, greater instability of self-image, and slightly lower self-esteem. However, they discovered that contextual, rather than age-associated effects, seemed to account for these findings. Upon completion of the sixth grade, one portion of the early-adolescent group had moved to a *new* school, i.e., a local junior high school, while the remaining portion of the early adolescents stayed in the same school (which offered seventh- and eighth-grade classes). The group of early adolescents who changed their school setting showed a much greater incidence of the personality changes than did the group that remained in elementary school. Thus variables related to the school context seem to influence the personality of young people. This idea finds further support in a study of 184 male and female black early adolescents (Eato, 1979). It has found that perceptions of the social environment of the school were a significant influence on the females' self-esteems and that perceptions of both the physical and social environment of the school setting significantly influenced the males' self-esteems.

Still another component of a person's context is provided by his or her family setting. Just as family interaction differences seem to provide a basis of different vocational role orientations among females, there seems to be a role that such interactions play in other personality development. For instance, Matteson (1974) found that adolescents with low self-esteems viewed communication with their parents as less facilitative than did adolescents with high self-esteems. Moreover, parents of low self-esteem adolescents perceived their communication with spouses as less facilitative and rated their marriages as less satisfactory than did parents of high self-esteem youths. Similarly, Scheck, Emmerick, and El-Assal (1974) found that feelings of internal (personal) control over one's life, as opposed to believing that fate or luck was in control, was associated with adolescent males who perceived parental support for their actions. Furthermore, interactive differences associated with different types of families promote individual differences in personality development. Long, Henderson, and Platt (1973) found that among 11- to 13-year-old Israeli children of both sexes, rearing in a kibbutz, as compared to more traditional familial rearing situations, was associated with higher self-esteem and social interaction among both sexes.

The cultural context can also be influential. Evidence suggests that development in different cultures is differentially related to the presence of sex differences. For instance, Offer and his colleagues (Offer & Howard, 1972; Offer, Ostrov, & Howard, 1977) report variation in sex differences from culture to culture. For example, the differences between the sexes in the United States are not as great as they are in Israeli and Irish cultural settings. However, there seem to be no differences in the types of sex differences found between American and Australian adolescent samples (Offer & Howard, 1972). Similarly, cultural context is related to the absence or presence of sex differences. Ramos (1974) found that among Brazilian adolescents of Japanese origin, there are no sex differences in self-esteem.

In sum, it is our view that the nature of individual differences between the sexes is dependent on interactions among biological, psychological, sociocultural, and historical influences. In other words, we stress the implications of all aspects of the person's context in attempts to understand his or her individual development.

Individual differences in cognitive, personality, and social behavior are influenced by one's social context. Differences in independence, superego control, achievement, importance placed on children, self-consciousness, and self-esteem exist in relation to one's context—for example, factors associated with particular times in history, birth cohort membership, school milieu, structure and function of the family unit, and cultural setting.

CHAPTER SUMMARY

Many sex differences exist between males and females. Erikson theorizes that the female's genitalia must be used for incorporation of the male's penis in order for the female to function in her biologically appropriate (that is, reproductive) manner. Women must develop roles which allow this inner space to be fulfilled. To Erikson, women are oriented toward roles as wives and mothers, and because this presumably adaptive role behavior is dependent on men, women should develop self-concepts which are characterized by such traits as dependency, submissiveness, and passivity; all attributes which would lead to the presence of a negative self-esteem. In addition, women should not compete or achieve in intellectual (cognitive) arenas as well as should men.

The male, on the other hand, needs to use his genitalia to intrude on objects outside his body. Such objects may be women's bodies or, symbolically, any external environmental object. Oriented to this outer space, men must develop roles and self-conceptions which allow an independent, manipulative, active, and dominant mastery over their environment. Erikson argues that the attainment of such roles and self-concept traits would be associated with a positive self-esteem and with greater cognitive competence.

McCandless has similar ideas regarding sex differences in adolescent personality and social behavior. Drive-reducing, and hence rewarding, behaviors are associated with instrumentally effective and competent behaviors among males and with interpersonal behaviors of warmth and expressiveness for females. Thus there are both nurture (McCandless) and nature (Erikson) theories of the bases and implications of sex differences. In this chapter we offered an interaction conception as well.

Data show that male and female late adolescents have sex-role stereotypes that are consistent with the predictions of Erikson and McCandless. Males and females from American and western European university settings share a common stereotype about the behaviors they believe associated with males and females. Males are believed to possess a group of behaviors characterized by competency and instrumental effectiveness, or an agency orientation; females are believed to possess a set of behaviors characterized by social warmth and interpersonal expressiveness, or a communion orientation.

Although it may have been the case that in earlier historical times (for example, in hunting and foraging societies) agency-communion distinctions may have served an adaptive function, recent data indicate that males and females who combine high levels of both agency and communion orientations are the most adaptive. Indeed, the weight of most recent evidence does not lend great support to the predictions of Erikson and McCandless.

In addition, although most men and women expect to marry and most women expect to have children, recent data pertinent to the vocational development of male and female adolescents suggest that traditional vocational role distinctions between the sexes are diminishing. Many men and women still expect to engage in traditional vocations. However, the percentage of college women who obtain graduate degrees and who pursue careers in addition to marriage has increased in recent years. Indeed, such nontraditional role behavior for women has been found more adaptive, for example, in regard to feelings of life satisfaction and self-esteem, than traditional endeavors. Women who have careers both inside and outside the home thus are able to meet many needs in their lives.

Individual differences in personality and social behavior are influenced by one's social context, for example, factors associated with particular times in history and birth cohort membership.

Part Four

Adulthood and Aging

In terms of time, the periods labeled adulthood and aging comprise the majority of the human life span. Until recently, however, these periods were seen as less significant and interesting, developmentally, than infancy, childhood, and adolescence. Adulthood was viewed as a time of stability when what had been developed earlier was utilized. Aging was viewed as a time of decline when what had been developed earlier was lost. We now know that this view is too simplistic. Adulthood and aging are just as significant and interesting as any other periods of the life cycle. A life-span developmental perspective stresses multidimensional and multidirectional change. Thus adulthood and aging are characterized by both "growth" and "decline."

In the case of cognitive processes, for example, there is growing evidence for large intraindividual modifiability of intellectual performance. Relatively short-term interventions are often successful in increasing performance including its maintenance and transfer (Denney, 1979; Baltes, Reese, & Lipsitt, 1980). Such results suggest that a lifelong capacity for cognitive change exists. Cognitive processes, including learning, memory, and intelligence, are examined in Chapter 12.

Similarly, in the case of social and personality processes there has been increasing focus on the dynamic quality of the aging process. For instance, the notion of aging refers not only to changes in the individual from birth to death but also to the flow of cohorts through time. The successive cohorts form age strata of individuals who are young, middle-aged, and old and who age in different ways as society changes (Riley, 1979). In the area of personality development, recent efforts have been directed toward identifying stages or phases of adult development (Gould, 1979; Levinson, 1978; Vaillant, 1978). These theorists suggest that opportunities for personality growth occur repeatedly throughout adulthood. Social and personality processes are examined in Chapter 13.

The life-span perspective also emphasizes relations between processes and how these are influenced by the individual's context. This emphasis has led to a focus on significant events in adulthood (Hultsch & Plemons, 1979). Events such as choosing an occupation, getting married, becoming a parent, divorcing, being promoted, becoming a grandparent, becoming ill, retiring, and experiencing death of a loved one shape the lives of individuals and families. It is in examining such events that the full richness of the interaction between the individual and his or her context may be seen. Such life transitions and life events are examined in Chapters 14 and 15. In Chapter 14 we outline a life-event and life-transition framework which provides the basis for this integration. Then in Chapter 15 we examine some of the life events unique to young adulthood, middle age, and old age. The final section of this chapter is devoted to death and dying—the final transition of life.

chapter 12

Adulthood and Aging: cognitive development

CHAPTER OVERVIEW

The prevailing cultural stereotype suggests that cognitive performance declines with increasing age during adulthood. In this chapter we will examine the adequacy of this view by reviewing several major approaches to learning, memory, and intelligence during this portion of the life span. We will also examine multiple influences on cognitive functioning and the extent to which the performance of adults can be improved.

CHAPTER OUTLINE

THE SLOWING OF COGNITIVE BEHAVIOR

Brain Electrical Activity and Slowing of Cognitive Behavior
Cardiovascular Disease and Slowing of Cognitive Behavior

LEARNING AND MEMORY IN ADULTHOOD AND AGING

Associative Learning
Information Processing
A Contextual Approach
Age versus Cohort

INTELLIGENCE IN ADULTHOOD AND AGING

Psychometric Intelligence
The Piagetian Approach
Intelligence in Context

CHAPTER SUMMARY

ISSUES TO CONSIDER

What is the relationship between central nervous system functioning and the slowing of cognitive behavior with age?

How does the pacing of tasks affect adult learning performance?

How do noncognitive factors, such as arousal and anxiety, influence adult learning?

What are the characteristics of the three memory stores proposed by the information processing model?

Why might older adults have difficulty remembering a long list of items, but not a short list of items?

What appears to be the primary source of difficulty for the adult: encoding or retrieval?

What techniques could be used to facilitate the learning of middle-aged and older adults?

What types of memory tasks, if any, do older adults do well on?

How does fluid-crystallized theory define the concept of intelligence?

How do fluid and crystallized abilities change over the life span?

How does Piaget's theory characterize intellectual development during adulthood?

What adult age differences are observed on Piagetian tasks of cognitive functioning, and how may these be interpreted?

What role do cohort effects play in adult age differences in intellectual performance?

How do biological and health variables affect intellectual functioning during adulthood and old age?

In what ways, and to what extent, are adult intellectual abilities subject to modification through the process of learning?

How do social and cultural factors influence intellectual performance during the adult years?

An old man skillfully defeated a group of much younger men at a gymnasium in a variety of athletic sports, including swimming, running, and weight lifting. "You must be at least 70 years old," said one of the young men admiringly, "yet you beat us at every sport we tried. Are you that good at everything?"

"Not at all," said the old man. "I'm not what I used to be. For instance, when I went to bed last night I had intercourse with my wife. When I woke up this morning I also had intercourse with her. Then I got out of bed to take a shower and when I returned I had intercourse once again. You see, my memory is bad." (Richman, 1977, p. 211)

*T*he perception that cognitive processes—learning, memory, and intelligence—decline with increasing age is one of our most pervasive cultural stereotypes. The conclusion that there are age-related deficits in these processes is also reflected in the scientific community. For example, Arenberg (1977) in a paper entitled "Memory and Learning Do Decline Late in Life" concludes that while age-related decrements in these processes are not inevitable, they are widespread and substantial. Arenberg suggests that to deny the existence of this pattern of decline is to engage in wishful thinking. Are age-related deficits in cognitive processes characteristic of adults? Let us examine this question.

THE SLOWING OF COGNITIVE BEHAVIOR

There is relatively clear evidence which indicates that, with advancing age, individuals show a tendency toward increasing slowness of response (Welford, 1958; 1977). This is a gradual change occurring across the entire life span, which shows up in a variety of so-called speeded tasks. *Speeded tasks* are those where errors would be unlikely if the individual had an unlimited amount of time to complete the tasks. Typically, they involve relatively simple responses such as pushing buttons, sorting items, crossing out items, and so forth. The objective, of course, is to complete the task as rapidly as possible. Several brief examples will suffice.

Reaction time (RT) tasks involve a measure of the time elapsing between the appearance of a signal and the beginning of a responding movement. Reaction time is usually viewed as a measure of *central nervous system* processing; that is, it involves perceptual and decision-making processes. Reaction time tasks vary in complexity. For example, *simple RT tasks* involve only one signal and one response (e.g., pushing a button when a light goes on). *Disjunctive RT tasks* involve multiple signals and/or responses (e.g., pushing the right-hand button when the red light goes on and the left-hand button when the green light goes on; pushing the button only when the red light goes on). Hodgkins (1962) examined simple RT performance (subject released a key when a signal light was lit) in over 400 females aged 6 to 84 years. Hodgkins found that mean speed increased with age until the late teens, remained constant until the midtwenties, and then declined steadily throughout the remainder of the age range. The degree of change in RT was 25 percent between the twenties and the sixties and 43 percent between the twenties and the seventies. The slowing seen with age in simple RT tasks is magnified in the case of disjunctive RT tasks (Griew, 1958) or tasks requiring the subject to remember previous signals and responses (Kay, 1954).

Slowing is also seen with sorting tasks. Botwinick, Robbin, and Brinley (1960) asked younger and older persons to sort seventy-one ordinary playing cards (excluding face cards) into slots in a sorting bin. They were asked to sort the cards by matching the number of cards to the number of a stimulus card located above the bin. When no match was possible, the card was to be placed in the last slot. Five levels of difficulty were created by varying the number of stimulus cards involved: one, three, five, seven, and nine. The results indicated that the older adults were slower than the young adults and that this difference increased as the complexity of the task increased.

A third type of speeded task involves copying simple materials or performing various canceling operations. For example, Botwinick and Storandt (1974) asked individuals ranging in age from the twenties to the seventies to copy rows of numbers and to "cancel" rows of 1/4-inch horizontal lines by drawing a vertical line through them. The results showed a steady decline with age in the speed of carrying out these tasks for both men and women.

As we have implied, the slowing of behavior with age appears in a wide range of tasks. Indeed, it appears to be a general characteristic of older adults. That is, younger adults appear to be fast or slow depending on the characteristics of the task and situation, e.g., familiarity, motivation, etc. Older adults, however, seem to have a characteristic or general slowing of behavior independent of task and situational characteristics (Birren, Riegel, & Morrison, 1962). This general slowing does not appear to be primarily a function of *peripheral nervous system* factors (e.g., sensory acuity; Botwinick, 1971; speed of peripheral nerve conduction; Birren & Botwinick, 1955; speed of movement once a response is initiated; Botwinick & Thompson, 1966). Rather, it appears to reflect a basic change in the way the central nervous system processes information (Birren, 1974).

Brain Electrical Activity and Slowing of Cognitive Behavior

Considerable evidence has accumulated to link changes in brain electrical activity to the slowing of behavior (Marsh & Thompson, 1977; Woodruff, 1979). The *electroencephalogram (EEG)* represents a record of the brain's electrical activity. It is obtained by attaching electrodes to the scalp and amplifying the resulting electrical activity many times. The human EEG displays continuous rhythmic activity—differing wavelike patterns varying in frequency and amplitude. There are four basic patterns, each of which is associated with different behavioral states. The dominant rhythm is the *alpha* with a frequency of 8 to 13 cycles per second (cps). It is associated with a relaxed, awake state. *Beta* rhythm is faster (18 to 30 cps) and is associated with an attentive, alert state. *Theta* (5 to 7 cps) and *delta* (0.5 to 4 cps) are slow frequencies associated with drowsiness and sleep, respectively.

The bulk of the work on aging has focused on the alpha rhythm. Generally, this research has suggested a slowing of *alpha frequency* with increasing age (Busse & Obrist, 1963; Obrist, 1954, 1963). Most of these studies also show a decrease in the amount of time older adults produce alpha, which has been labeled a decrease in *alpha abundance.* This research indicates that alpha frequency reaches its maximum of between 10 to 11 cps during adolescence and begins to gradually slow after the age of 25 to 30. The average frequency decreases to about 9 cps by the sixties and to 8 to 8.5 cps after age 80. This slowing has been confirmed in longitudinal data as well (Obrist, Henry, & Justiss, 1961; Wang & Busse, 1969).

The exact mechanisms involved in the slowing of the EEG with age remain somewhat unclear. However, the most powerful explanation associates it with central nervous system pathology (Obrist, 1972). Basically, it is proposed that reduced cerebral blood flow, with consequent hypoxia (reduced levels of oxygen), results in neuronal loss. The initial reduction of cerebral blood flow appears to be related to arteriosclerotic deposits or other vascular problems. Thus comparisons of matched samples with and without arteriosclerotic diseases typically yield differences in cerebral blood flow as well as alpha frequency (Dastur, Lane, Hansen, Kety, Butler,

Perlin, & Sokoloff, 1963; Obrist & Bissell, 1955). While evidence for a disease-related explanation of alpha slowing is strong, other basic aging mechanisms may be involved as well. There is also evidence to suggest that slowing also occurs in essentially disease-free older adults (Obrist, 1963).

The slowing of alpha with age has also been related to the slowing of behavior with age. Woodruff (1979) notes that two basic hypotheses have been proposed. The more general hypothesis is based on general theories of activation and arousal. These theories suggest that if the central nervous system of older adults is less responsive, then older adults should show slower EEG rhythms as well as slower behavior as a result. In this case, slowing is actually a result of reduced cortical excitability—alpha is merely an index of this state. Reduced excitability could be a result of disease processes or other fundamental changes in brain physiology. Surwillo (1968) has suggested a more specific hypothesis in which he proposes that alpha constitutes a kind of internal biological clock. In this case, the slowing of alpha is seen as causally linked to the slowing of behavior rather than as an index of reduced central nervous system excitability.

Both of these hypotheses, however, suggest that the slowing of alpha should be related to the slowing of behavior, and this appears to be the case. For example, Surwillo (1963) reported a correlation of 0.72 between average reaction time and average alpha period (inverse of alpha frequency) for 100 individuals aged 28 to 99 years. In spite of this, there is little support for the hypothesis that slowing of behavior is a direct result of the slowing of alpha.

In a study designed to examine this issue, Woodruff (1975) used biofeedback techniques to manipulate alpha. Woodruff trained ten younger and ten older adults to increase the abundance of alpha activity at their modal level and at frequencies above and below their model level. At each of these levels, Woodruff tested the individuals' reaction times once they had reached an appropriate criterion of alpha production. Comparison with control subjects indicated that biofeedback itself did not affect reaction time. Further, reaction times were faster for nine of ten older subjects and seven of ten younger subjects when they produced fast brain waves than when they produced slow brain waves. Thus Woodruff's data suggest that a shift in brain-wave frequency is related to a shift in reaction time; however, correlations between brain-wave frequency and reaction time within persons were relatively small. As a result, Woodruff's data do not support Surwillo's hypothesis that the slowing of alpha causes the slowing of behavior. Rather, she suggests that the arousal hypothesis is the simplest explanation—in order to produce faster brain waves, the arousal of the subject's central nervous system was increased, resulting in faster reaction times.

Thus the older adult's central nervous system appears to be in a state of underarousal compared to that of younger adults, possibly as a result of vascular pathology. As a result, older individuals have slower modal brain-wave frequencies, and these are associated with slower mean reaction times.

Cardiovascular Disease and Slowing of Cognitive Behavior

The relationship between central nervous system function and the slowing of behavior is also apparent in the link between health status and behavior (Birren & Spieth, 1962; Spieth, 1965).

For example, Spieth (1965) examined speeded performance among young and middle-aged men who had mild to moderate degrees of cardiovascular disease. The subjects of Spieth's inquiry were present or former air pilots and air traffic controllers ranging in age from 35 to 59 years. The research was done in connection with a physical examination for renewal of medical certification to fly or control air traffic. The individuals were classified into a number of health-status groups including: (1) healthy; (2) mild-moderate congenital or rheumatic heart defects; (3) arteriosclerotic or coronary disease without hypertension; (4) essential hypertension or cardiovascular disease with hypertension; and (5) history of cerebrovascular disease or old cerebrovascular accident. These groups reflect a continuum of disease severity from mild to moderately severe. Note, however, that these men were not acutely ill; essentially, they were under no restriction of ordinary activity.

The men were administered a battery of speeded tasks including simple and complex RT tasks. An interesting feature of Spieth's study is that one may assume that such tasks were not highly unfamiliar to the subjects because of their experience as pilots and air traffic controllers. The tasks, then, were probably relatively high in ecological validity for these individuals.

Spieth's results indicate that speed of performance declined in a relatively linear fashion with increasing degrees of cardiovascular impairment. This effect occurred on all tasks, although it was greater on complex compared to simple tasks. Spieth's study, then, suggests that individuals suffering even mild to moderate levels of cardiovascular impairment perform more poorly than healthy individuals.

Thus there appears to be an important link between biological functioning, particularly central nervous system functioning, and the slowing of cognitive behavior with age. Other variables are important as well. For example, it has been demonstrated that older persons can increase their speed of response with practice (Hoyer, Labouvie, & Baltes, 1973). Similarly, it has been demonstrated that older adults appear to be more cautious, exhibiting a tendency to compensate for a loss of speed by increased accuracy (Welford, 1958; Rabbitt & Birren, 1967). These factors, however, seem only to explain a portion of the slowing (Welford, 1977). Fundamentally, the slowing of cognitive behavior with age appears to be a function of a basic change in the speed of central nervous system functioning.

Even mild levels of cardiovascular disease are associated with impairment on complex tasks, like controlling air traffic. (United Nations.)

LEARNING AND MEMORY IN ADULTHOOD AND AGING

Learning is usually defined as a relatively lasting change in behavior that is brought about by experience; memory, as the storage or retrieval of things learned earlier. However, these basic definitions have been elaborated within several different theoretical traditions, and within each of these traditions, researchers have used different tasks, materials, and procedures. Accordingly, we will examine each of these traditions and their perceptions of learning and memory in adulthood and aging separately.

Associative Learning

Associative approaches are rooted in the assumption that all learning and memory are based on the association of ideas or events which occur together in time. This view, originating in classical times, was elaborated in the seventeeth century by the British associationist philosophers Thomas Hobbes and John Locke. Current associative theories of learning and memory reflect this basic assumption.

From an associative perspective, learning involves the formation of *stimulus-response (S-R) bonds*, and the contents of memory are defined by such associations. The act of remembering involves emitting previously acquired responses under appropriate stimulus conditions. Changes in learning and memory are seen as quantitative rather than qualitative. Acquisition may occur as a function of increases in the number of S-R associations or as a function of strengthening existing associations through processes such as repetition. Forgetting is a function of the loss or weakening of associative bonds through processes such as *decay* or *interference*.

Experimental Procedures. While investigators following an associative approach have generated numerous tasks to study learning and memory, two procedures— paired-associate and serial learning—have been used extensively. What is involved in each of these tasks?

In a *paired-associate* learning task, the individual learns to associate pairs of items, typically unrelated words, such that he or she can provide the second word of the pair when presented with the first word of the pair. A list of such pairs usually is presented for learning. The sample list shown in Exhibit 12.1 depicts the *anticipation method* of presentation. In this method, the stimulus (S) word is presented followed by the S-R word pair. Each component is presented, usually visually, for a brief interval of several seconds until the entire list has been seen. As noted in the exhibit, presentation of the S-R pair constitutes a study phase of the task. Thus fol-

EXHIBIT 12.1   **Sample Paired-Associate List**

List Words	Component	Phase	Interval
ARROW	S	Test	Anticipation
ARROW-STORM	S-R	Study	Inspection
IRON	S	Test	Anticipation
IRON-HARP	S-R	Study	Inspection
PIPE	S	Test	Anticipation
PIPE-WHALE	S-R	Study	Inspection
TOAST	S	Test	Anticipation
TOAST-CHAIR	S-R	Study	Inspection
OVEN	S	Test	Anticipation
OVEN-TREE	S-R	Study	Inspection

lowing the first trial when the pairs of words are seen for the first time, the individual's job is to provide the response (R) word of each pair when the stimulus word is presented alone (test phase). Following this, presentation of the S-R pair provides feedback to the individual on the accuracy of his or her response and an opportunity to study the pair again (study phase). The time span of the test phase is labeled the *anticipation interval*, while the time span of the study phase is labeled the *inspection interval*.

Usually, paired-associate lists are repeated several times, although the pairs often are shown in a different order on each trial. Several performance measures such as the number of trials to a specific criterion (e.g., one perfect recall of the list) or the number of errors per trial are obtained.

In a *serial learning task* a list of single items is used, and the individual's job is to learn the list in the exact order in which it is presented. That is, the words are presented one at a time, and the individual is to respond by naming the next word on the list before it is presented. Thus rather than associating pairs of words as in the paired-associate task, the focus is on associating each word with the next word in the list.

Pacing. It has been mentioned that the paired-associate and serial learning procedures of the associative approach are rapidly paced. In addition, the slowing of cognitive behavior has been viewed by some as a major characteristic of the aging process. Thus it is logical that investigators interested in the learning and memory of adults have examined the pacing of the task as a major independent variable.

What effect does the pacing of a task have on adults? Reviews of research on this variable indicate that the acquisition of young adults is superior to that of older adults and that the faster the pace of the task, the greater the age differences (Arenberg & Robertson-Tchabo, 1977; Witte, 1975).

For example, in an initial and now classical study, Canestrari (1963) presented younger (17 to 35 years) and older (60 to 69 years) individuals with three paired-associate tasks, each of which was presented at a different rate: 1.5 seconds, 3.0 seconds, and self-paced. Canestrari's (1963) findings are shown in Exhibit 12.2. Fewer errors occurred for both age groups at the slower (3.0-second) pace. However, the older learners benefited more from the slowing of the pace of the task than the younger learners. When the individuals were allowed to regulate the pace of the task themselves (self-paced), the older learners exhibited a further improvement in performance, while the younger learners did not. An analysis of the time taken during the self-paced condition showed the older learners took more time than the younger learners. This extra time tended to be taken during the test phase, rather than during the study phase, of the task. Related to this, Canestrari found that the differences between the age groups were accounted for by *errors of omission* rather than *errors of commission*. In other words, the higher error rates of the older learners occurred because they did not provide a response during the test phase rather than because they provided an incorrect response.

Since Canestrari (1963) varied both the anticipation and inspection intervals simultaneously, he could not determine whether older learners require additional time to provide a response, to study the pair, or both. In order to investigate this question, Monge and Hultsch (1971) varied both intervals independently. Each interval could be 2.2, 4.4, or 6.6 seconds in length. Young and middle-aged men learned a paired-associate list under one of the nine possible combinations of anticipation and inspection intervals. The results indicated that middle-aged learners benefited from both longer anticipation and inspection intervals, but young learners

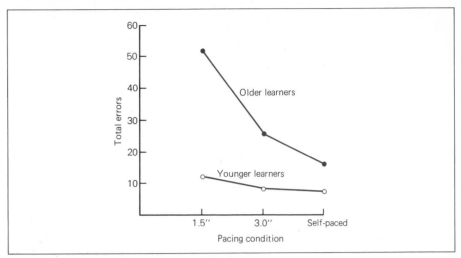

EXHIBIT 12.2   Mean total errors to criterion as a function of age group and pacing condition. (Source: Based on Canestrari, 1963, Table 2, p. 166.)

benefited only from longer inspection intervals. Therefore, longer inspection intervals benefit both age groups equally, while longer anticipation intervals benefit the middle-aged but not the young.

In general, then, younger adults appear to perform better on associative learning tasks than older adults, and the faster the pace of the task, the greater the age differences. But why should this be the case?

A major issue which has emerged from the verbal learning research we have been reviewing focuses on the question of whether the observed age-related deficits in acquisition are related to cognitive or noncognitive processes (Botwinick, 1967). In other words, is the older adult's difficulty related to learning processes (e.g., inefficiency in forming S-R bonds) or to noncognitive processes such as personality traits or states which are relevant to the learning situation (e.g., greater cautiousness or anxiety)?

For example, some researchers (e.g., Canestrari, 1963) suggested that the poorer performance of the older learners was simply due to an insufficient amount of time to make a response rather than to a learning disability. This conclusion was based largely on the fact that the age differences were linked to the length of the anticipation interval and were indexed by greater omission errors on the part of the older learners. This explanation proved to be an oversimplification (Arenberg, 1965). However, two other explanations have received some support. These suggest that the older adults' learning difficulty may be related, on the one hand, to physiological overarousal and anxiety and, on the other hand, to deficiencies in the use of mediational strategies.

Physiological Arousal and Anxiety. One explanation of age-related deficits in learning is based on the hypothesis that older adults are more aroused or anxious in learning situations than younger adults and, as a consequence, perform more poorly. This explanation suggests the other adults' problem is not learning per se, but performing adequately. Two sets of studies have examined this hypothesis. The first set of studies has focused on the *arousal* construct assessed by physiological measures, while the second set of studies has dealt with the *anxiety* construct assessed by self-report measures.

Eisdorfer (1968) and his colleagues have investigated the role of physiological arousal in age differences in learning performance. This work is based on a general theoretical formulation which characterized the relationship between arousal and performance as an inverted U-shaped curve. That is, an optimum level of arousal for performance is proposed. Up to this point, increases in arousal will be accompanied by increments in performance, and beyond this point, increases in arousal will be accompanied by decrements in performance.

Eisdorfer (1968) proposed that older adults are characterized by overarousal, rather than underarousal, in the learning situation. He investigated this hypothesis in several studies using serial learning tasks and a physiological measure of arousal. The latter consisted of the level of free fatty acid in the bloodstream—an index of autonomic nervous system activity in response to stress.

An initial study (Powell, Eisdorfer, & Bogdonoff, 1964) revealed substantial age differences in arousal during a serial learning task with older individuals showing higher levels than younger individuals. In particular, while the free fatty acid level of the younger age group plateaued during the learning task and declined following it, that of the older group increased throughout the task, reaching a peak approximately fifteen minutes following it. Thus older adults, rather than being less involved in the task, were more involved and under greater stress.

While this study suggests older learners exhibit higher levels of autonomic arousal, it does not establish a cause-effect linkage between arousal and learning performance. Such a linkage requires demonstrating that the manipulation of arousal levels affects learning performance. In an attempt to provide such evidence, Eisdorfer, Nowlin, and Wilkie (1970) varied arousal via administration of the drug propranolol, which blocks end-organ autonomic activity. A group of older men received either propranolol or a *placebo* (a substance having no physiological effect) during the learning task. An analysis of the free fatty acid levels of the two groups showed that, following administration of the drug and the placebo, the arousal level of the drug group decreased, while that of the placebo group increased. Furthermore, the drug group made significantly fewer errors than the placebo group; these data provide evidence of a causal relationship between arousal and learning performance. Unfortunately, no younger individuals were tested in this study. As a result, the question of whether age differences in performance are a function of a greater arousal on the part of the older learners is not definitively answered by these data.

Studies of physiological arousal and learning performance are complicated by several problems. For example, techniques such as drawing blood samples are themselves stressful and confound arousal produced by the learning task with arousal generated by the physiological measurement technique (Troyer, Eisdorfer, Bogdonoff, & Wilkie, 1967). Further, the use of a single measure of arousal is problematic because it appears that there are multiple arousal responses which may exhibit different rates of age-related change (Elias & Elias, 1977). Nevertheless, the research reviewed in this section suggests the possibility that part of the performance deficit exhibited by older learners may be related to overarousal.

A second set of studies has examined the stress-learning hypothesis by measuring self-reported anxiety. Indirect support for the hypothesis is suggested by studies indicating that supportive instructions, which presumably reduce anxiety, facilitate the performance of older adults to a greater extent than that of younger adults (e.g., Lair & Moon, 1972); however, these studies did not include an independent measure of anxiety. As a result, there is no way of determining the impact of the instructional conditions on age-related anxiety levels. However, Whitbourne (1976) used a separate measure of anxiety and found that older men were more anxious than younger men following a sentence memory task. Moreover, there was a negative relationship between text anxiety and memory performance; high anxiety scores were associated with low memory scores.

Additional evidence on the role of anxiety in the cognitive performance of adults comes from informal observations of older individuals in the research context. These observations suggest that older adults are more likely to refuse to participate in research than younger adults and, if they do participate, are more likely to withdraw prior to completion of the task. In the experimental setting, statements about declining abilities, dislike of tests, and fear of failure are frequently obtained from older adults but rarely obtained from younger adults. These informal data reinforce the conclusion that older adults find such experimental settings more stressful than younger adults. Researchers studying physiological arousal and state anxiety suggest the possibility that these states may account for part of the cognitive performance deficit typically observed with increasing adult age. Other states may affect learning and memory as well, as discussed in Box 12.1

BOX 12.1

MEMORY COMPLAINTS: ABILITY OR AFFECT?

Memory dysfunction is one of the most prominent stereotypes associated with aging. In fact, memory problems are one of the most frequently cited complaints reported by older adults. Yet there is evidence to suggest that the memory complaints of older adults may be related to degree of affect, particularly depression, rather than degree of ability.

In a study focused on this issue, individuals with varying degrees of depression and brain dysfunction were compared on memory complaints and on actual memory performance (Kahn, Zarit, Hilbert, & Niederehe, 1975). The sample of 153 adults aged 50 to 91 (average 65.1 years) was classified into two groups according to the presence or absence of signs of brain dysfunction. Measures of depression, memory complaint, and memory performance were then obtained from all individuals.

Within the normal brain function group, it was discovered that those who were more depressed had a greater number of memory complaints than those who were less depressed. Their actual memory performance, however, did not differ. Furthermore, individuals with normal brain function had as many memory complaints as those with altered brain function even through the latter group had far greater difficulty on the actual memory tasks. This work suggests that memory complaints among the elderly may be related, at least in part, to depression and other affective states rather than to changes in memory ability.

Memory dysfunction in adulthood may be related to depression. (Michael Weisbrot and Family Photography.)

Mediational Strategies. Another explanation of age-related deficits is based on the hypothesis that older adults are less adept at utilizing *mediational strategies*. Such strategies are seen as major mechanisms by which individuals form associations between stimuli and responses. This process has been conceptualized as the formation of a covert response which forms a link between S and R. These links may consist of verbal responses, visual images, or other covert responses. It is assumed that the overt stimulus (S) produces a covert response or mediator (r). The overt response (R) is then connected with the covert stimulus consequences (s) of this mediating response. Schematically, the process may be represented as follows:

$$S \rightarrow r-s \rightarrow R$$

For example, the pair FLOWER-MEADOW may be linked by visualizing a field of flowers; alternatively, the pair may be linked by the concept "nature." Mediational responses involve active processing on the part of the learner.

There is considerable evidence suggesting that older adults do not use mediators as extensively or as efficiently as younger adults. Several studies have requested younger and older learners to specify the strategies they used to learn items (Erber, 1976; Hulicka & Grossman, 1967; Rowe & Schnore, 1971). These studies indicate that younger learners are more likely than older learners to use mediational strategies spontaneously. Several investigators also reported that when mediators are used, older learners prefer verbal mediators to visual ones, while the reverse is true for younger learners (Hulicka, Sterns, & Grossman, 1967; Rowe & Schnore, 1971).

If older adults do not use mediators as extensively or effectively as younger adults, then instructions to use mediational strategies should benefit older learners more than younger learners. This appears to be the case (Hulicka & Grossman, 1967; Hulicka, Sterns, & Grossman, 1967; Treat & Reese, 1976). The use of mediators supplied by the experimenter also appears to benefit older learners (Canestrari, 1968), although this manipulation may not be as effective as self-generated mediators. While mediational instructions typically have reduced observed age-related differences in paired-associate learning performance, they have generally not eliminated such differences, since older learners require more time to develop and apply mediators (Treat & Reese, 1976).

These studies suggest, then, that older learners do not spontaneously use mediational strategies, particularly strategies based on imagery, as frequently or efficiently as younger learners. However, when instructed to use mediators or when mediators are supplied by the experimenter, older learners are able to use these devices and performance improves. In spite of this, their performance only rarely equals that of younger learners.

If we assume that younger and older adults are characterized by learning strategy differences, then these differences must still be explained. To date, little research has been done on the potential antecedents of these strategy differences. Two rather global hypotheses contrasting physiological and experiential antecedents have been offered (Witte, 1975). First, it has been suggested that physiological degeneration of the central nervous system with increasing age may impair the spontaneous application of such higher-order learning strategies. Second, it has been suggested that such strategies are likely to reach their optimum development in formal learning situations such as school. As a result, the higher-order learning strategies of older individuals may be inefficient because of disuse. Alternatively, the educational experience of older cohorts may have deemphasized the use of such strategies. There is weak evidence related to both of these hypotheses. For example, Rust (1965) reported differences in the use of mediational strategies between individuals with and without symptoms of arteriosclerosis. Gladis (1964) found that by presenting multiple paired-associate tasks, which presumably would provide practice in higher-order skills, the performance of middle-aged learners is improved to a greater extent than that of young learners. Overall, however, the antecedents of these age differences remain to be demonstrated.

Transfer and Interference. Transfer of training—the effect of learning one task on the learning or retention of another task—is a major concern in associative approaches to learning and memory. If learning the first task facilitates learning the second task, then transfer is positive. If learning the first task interferes with learning the second task, then transfer is negative. Transfer may occur specifically or non-specifically. *Specific transfer effects* are dependent on the similarity between tasks such as the degree to which the words of two lists are synonymous. *Nonspecific transfer effects* are the results of more general factors such as warm-up or learning to learn. One major hypothesis derived from this framework has been the suggestion that older adults may be more susceptible to interference than younger adults (Welford, 1958).

One approach to investigating the role of interference in adult learning has been to examine the effect of established verbal habits on the learning process by varying the associative strength of the word pairs to be learned. When one word frequently elicits another in a free-association task, the pair is said to have *high associative strength;* when one word infrequently elicits another in such a task, the pair is said to have *low associative strength.* Such free associations presumably reflect the individual's established verbal habits. For example, the pair DARK-LIGHT has high associative strength, while the pair DARK-FAST has low associative strength.

If the associative habits of older adults are more established than those of younger adults, it follows that age-related differences should be minimized with lists that are high in associative strength. This appears to be the case (Botwinick & Storandt, 1974; Kausler & Lair, 1966; Zaretsky & Halberstam, 1968). In the most comprehensive of these studies, Botwinick and Storandt (1974) presented adults ranging in age from 21 to 80 years with three paired-associate lists which varied in terms of their associative strength and difficulty: the low-difficulty list consisted of high–associative strength word pairs such as OCEAN-WATER; the moderate-difficulty list consisted of low–associative strength word pairs such as BOOK-HAIR; and the high-difficulty list consisted of consonant word pairs such as FP-WAGON. Botwinick and Storandt (1974) found no age-related performance differences on the easy list but marked age-related differences in performance on the moderate and difficult lists.

If the performance of older adults is facilitated by lists which are consistent with preestablished verbal habits, it follows that lists which are constructed specifically to conflict with these habits should interfere with their performance. In an attempt to test this hypothesis, Lair, Moon, and Kausler (1969) compared middle-aged and older individuals on lists which were high and low in response competition. The low-competition list consisted of pairs of relatively low associative strength. The high-competition list consisted of high–associative strength word pairs such as BLOSSOM-FLOWER and HOT-COLD. However, instead of being paired together, these words were re-paired so that for each stimulus word its highly associated response was present in the list but paired with another word (e.g., BLOSSOM-COLD, HOT-FLOWER). Such a list should produce a great deal of interference with prior verbal habits. Differences between the age groups were much larger for the high-competition list than for the low-competition list, thus supporting the interference hypothesis. The older adults were more susceptible to the effects of response competition as would be expected when strongly reinforced verbal habits competed with the formation of new associations.

Information Processing

Information processing approaches to learning and memory, pioneered by Broadbent (1958), are based on the principles underlying modern computers. For example, the computer contains an input unit which enters the information into the computer, a working unit which holds information being actively processed, and a core unit which stores information for later use. Incorporating these principles, information processing approaches to learning and memory are based on the concepts of storage structures and control operations. A generalized model is outlined in Exhibit 12.3. Typically, three types of *storage structures* are proposed: sensory stores (sensory memory), a short-term store (primary memory), and a long-term store (secondary memory). Information is retrieved from one store and entered into the next by *control operations* such as attention, elaboration, and organization which transform the information involved.

Sensory, Short-Term, and Long-Term Stores. Stimuli are received and registered in the modality-specific (i.e., visual and auditory) sensory stores. Sensory memory is part of the peripheral sensory system, and items are represented as literal, visual, or auditory copies. These representations persist only for a brief time, decaying in the absence of further processing. Information is retrieved from the sensory stores by attending to it, thereby entering it into a short-term store. Here items are coded in auditory or other physical fashion. The capacity of the short-term store is limited to about five units, with information being lost principally by displacement. The duration of primary memory from the short-term store can be extended by rehearsal, or the material can be transferred to long-term storage by processing the items in terms of their semantic content (meaning). Retrieval from the long-term store is dependent on the development of a retrieval plan based on elaboration or organization of

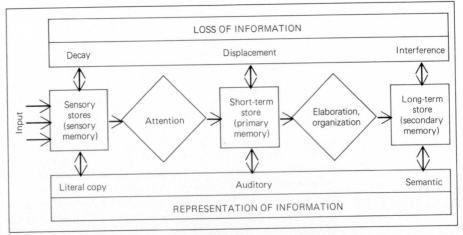

EXHIBIT 12.3   A generalized information processing model of learning and memory.

the information. The long-term store has unlimited capacity, and the duration of secondary memory is lengthy, if not permanent.

Since the division of memory into three components is one of convenience, we are not implying that items are placed into one of three separate memory systems as a secretary might place a letter into one of three separate files. Instead, there is a continuing elaboration of the memory trace from its initial perceptual processing to its integration into the individual's knowledge structure. This point was emphasized by Waugh and Norman (1965), who argued that the definition of the various stages of memory should be linked to processes rather than retention intervals. This is important because Waugh and Norman showed that long-term memory processes (e.g., elaboration) mediate performance, in part, even in tasks where the retention interval is extremely short. To avoid confusion, they suggested relabeling short-term and long-term memory *primary memory* and *secondary memory*, respectively. This terminology will be used in the following sections.

Sensory Memory. As summarized above, *sensory memory* is a part of the peripheral sensory system. Research on adult age differences in this system is very limited. Although several studies have suggested the possibility of age-related decrements in visual sensory memory (Schonfield & Wenger, 1975; Walsh, 1975), these appear of limited significance to overall age differences in memory performance (Craik, 1977).

Primary Memory. As we noted earlier, *primary memory* is a temporary maintenance system for conscious processing. As such, it serves an important control function for both storage and retrieval from the permanent maintenance system of secondary memory.

One relatively pure measure of primary memory is derived from the *free-recall* task. In free-recall, the individual is presented with a series of items during an input phase and is asked to recall as many of the items as possible in any order during an output phase. Presentation of the items may be simultaneous or successive but is usually successive. There may be just one input and one output phase, or several input and output phases may be combined in an alternating or other type of sequence. Single words are usually the items of concern, although other types of material such as syllables, letters, and geometric figures are used. What is "free" about free-recall is the order in which the individual may recall items.

Typically, one outcome of the free-recall procedure is that the last few items of the list are recalled first. This *recency effect* is considered to be a measure of primary memory. Both Craik (1968b) and Raymond (1971) report finding no age-related differences in recency using this task. These authors concluded that primary memory processing does not decline with increasing age.

Primary memory also may be assessed by immediate memory span tasks. The *immediate memory span* is defined as the longest string of items (digits, letters, words) that can be immediately reproduced in the order of presentation. This task probably involves both primary- and secondary-memory components. That is, the average immediate memory span is about five items for words and seven items for digits. The capacity of primary memory, however, is estimated to be smaller than this. Thus, immediate memory span tasks reflect a small secondary-memory component as well as a large primary-memory component.

If primary memory does not decline with age, then one would expect little decrement with immediate memory span tasks. This appears to be the case. For example, several studies have found no age-related decrements on digit span tasks (Craik, 1968a; Drachman & Leavitt, 1972; Talland, 1968). Other studies have found slight decrements (Botwinick & Storandt, 1974). This slight decline probably is related to the secondary-memory component of the task. While age-related differences in performance are not observed on the usual memory span task, this is not so when the task is modified to require division of attention or reorganization of the material (Craik, 1977; Talland, 1968). In this case, older adults find the task more difficult than younger adults.

The research reviewed in this section suggests that age differences in primary memory are minimal (Craik, 1977). Although there is evidence that the rate of search decreases with age, the capacity of this system appears to be unrelated to age. Furthermore, there is little evidence to suggest that older adults are more susceptible to interference effects in primary memory than younger adults.

Secondary Memory. Secondary memory is a permanent maintenance system characterized by semantic content. If sensory- and primary-memory processes are only minimally related to age, then observed age-related differences in performance should be in secondary memory.

We have noted that retrieval from primary memory and entry into secondary memory requires *elaboration* and *organization* of the material. For example, it has been suggested that memory depends on the individual's perceptual and cognitive analysis of the material: the more elaborate the analysis, the better the acquisition and retention of the material (Craik & Lockhart, 1972). More elaborate or deeper levels of processing are those involving semantic analyses. Thus it is possible to define a progression from relatively unelaborated or shallow levels of processing (e.g., physical characteristics of stimuli) to relatively elaborated or deeper levels of processing (e.g., semantic meaning characteristics of stimuli).

The role of elaboration in memory may be illustrated by an experiment conducted by Craik and Tulving (1975). Individuals were presented with a word list and asked to perform three orienting tasks involving different levels of analysis. They found that both recall and recognition of the words increased as the levels of processing increased. The generation of a rich, elaborate code for the to-be-remembered items, therefore, is central to secondary memory. The most effective encoding, however, is not restricted to a single item. Rather, the to-be-remembered items are organized into higher-order units. The formation of such units has been labeled *chunking* (Miller, 1956). Chunks are based largely on the principles of grouping and relating. The importance of organizational processes in secondary memory is illustrated by a series of studies completed by Mandler (1967), who asked individuals to sort large sets of unrelated words into categories of their own choosing prior to free-recall. Following this, free-recall of the words was requested. The findings revealed a strong relationship between the number of categories used during sorting and the number of words recalled during free-recall. The greater number of categories (chunks) the individual made, the better the recall.

Compared to younger adults, older adults are deficient in terms of the elaborative and organizational processes of secondary memory (Craik, 1977; Hultsch, 1969, 1971). Generally, the older adult's difficulty appears to be of the production deficiency or inefficiency variety. That is, older adults do not spontaneously use organizational strategies as extensively as younger adults, or if they do, they use them less effectively. However, when various organizational strategies are built into the situation, the performance of older adults improves significantly. The following two studies illustrate these conclusions.

Eysenck (1974) conducted a study in which he applied the orienting task procedure described earlier. Individuals performed one of four orienting tasks: (1) counting the numbers of letters in each word; (2) finding a word that rhymed with each word; (3) finding a suitable modifying adjective for each word; and (4) forming an image of each word. These conditions were presumed to reflect a continuum from shallow to deep processing. In addition, a fifth condition instructed the individuals to learn the words. All groups subsequently were asked to recall the words. Eysenck (1974) found that the differences between the younger and older individuals were greatest when the orienting task required deeper processing of the material. These results led Eysenck (1974) to suggest that older individuals exhibited a "processing deficit" at deeper semantic levels.

Hultsch (1971) used Mandler's (1967) procedure, summarized earlier, in which individuals are asked to categorize words to a criterion of two identical sorts prior to free-recall. Individuals from three age groups performed the task. In order to determine the impact of organizational processes on recall, the opportunity to organize the words was manipulated experimentally. Half the individuals at each age level were instructed to sort the words into from two to seven categories. These "nonsorting" individuals inspected the words one at a time for the same number of trials as taken by a randomly assigned "sorting" partner to reach criterion. Thus, the sorting and nonsorting conditions were designed to maximize and minimize the opportunity for the individual to organize the material in ways that were meaningful, while equating the number of input trials prior to recall. The results of this study are summarized in Exhibit 12.4, which indicates that middle-aged and older individuals exhibited less of a recall deficit under conditions that maximized the possibility for meaningful organization.

These studies have examined age-related differences in elaborative and organizational processes of secondary memory by manipulating conditions (e.g., instructions) presumed to influence these processes. It is also possible to measure organizational behavior itself and several indexes have been developed which measure the amount of organization in the individual's recall (Bousfield & Bousfield, 1966; Tulving, 1962). These measures are based on successive trials in spite of the fact the

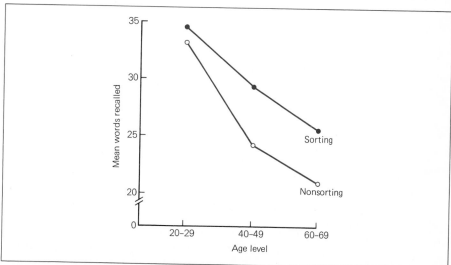

EXHIBIT 12.4   Mean number of words correctly recalled as a function of age and sorting condition. (Source: Hultsch, 1971.)

words are presented in a different order each time. Tulving (1962) has suggested that these pairings represent the formation of an organized unit by the learner. The tendency to recall words together as a unit increases systematically over trials and is correlated with the amount of correct recall. Consistent with our previous conclusions, research using these measures, appropriately computed, shows significant age-related differences in amount of organization in recall (Hultsch, 1974).

This sample of studies suggests that older adults are less effective or efficient at elaborating or organizing to-be-remembered items, although this deficit can be reduced by various manipulations which facilitate these processes. Studies other than those already mentioned are consistent with this conclusion (Earhard, 1977; Laurence & Trotter, 1971; Perlmutter, 1978).

Encoding versus Storage versus Retrieval. The studies reviewed above suggest age-related decrements in the elaborative and organizational processes of secondary memory. The question remains whether the problem reflects difficulty involving *encoding* (formation of a code at the time of input), *storage* (retention of the code until the time of output), or *retrieval* (utilization of the code at the time of output).

Of these processes, there is little evidence to support the hypotheses that age-related differences are the result of faulty code retention. Presumably such forgetting is a function of interference processes, and the available research provides little evidence for age differences in susceptibility to interference in secondary memory (Hultsch & Craig, 1976; Smith, 1974, 1975). This suggests the locus of the difficulty is in original encoding of the material or utilization of this code at the time of retrieval.

Recall versus Recognition. One approach to comparing storage and retrieval processes has been to compare *recall* and *recognition*. In a now classical study, Schonfield and Robertson (1966) compared the recall and recognition performance of individuals ranging in age from 20 to 75 years. As shown in Exhibit 12.5, there were systematic age-related declines in recall scores but not in recognition scores. Other studies have found similar results. These results led Schonfield and Robertson (1966), as well as other investigators, to conclude that the memory deficit of older individuals primarily reflects a retrieval, rather than a storage, problem.

More recent evidence (Smith, 1980) suggests that age differences in recall and recognition performance may be explained on the basis of encoding strategy. It has been shown, for instance, that while recall is particularly dependent on the grouping and relating of items (organization), recognition is particularly dependent on discriminating the items from one another (elaboration). In this sense, retrieval plans are more important for recall than for recognition. Smith (1980) has reported that younger individuals spontaneously tend to use organizational strategies. Therefore, on a recall task where organizational strategies are primary, younger individuals perform better than older individuals. However, on a recognition task where elaborational strategies are primary, the organizationally based strategies of the younger individuals are not as effective. As a result, age-related differences in performance typically are reduced. While older individuals tend to use elaborative strategies spontaneously, there is evidence of a processing deficit in this strategy as well, because conditions that facilitate elaborative strategies do not improve the performance of older individuals to the same extent they improve the performance of younger individuals (Eysenck, 1974; Mason, 1979). Therefore, the recent work of Smith and his colleagues (Smith, 1980) suggests that age-related differences in recall and recognition tasks are a function of differential encoding strategies.

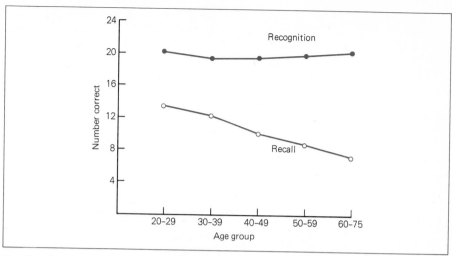

EXHIBIT 12.5   Recall and recognition scores as a function of age.
(Source: Based on Schonfield and Robertson, 1966, Table 3, p. 231.)

The findings summarized in the preceding sections suggest that the age-related deficits in secondary memory observed under multiple-testing conditions are due to the same basic cause—namely, a failure to engage in deep semantic elaborational and organizational processing of the material (Craik, 1977; Smith, 1980).

A Contextual Approach

In 1932, Bartlett published a book entitled *Remembering* in which he described a series of experiments on the retention of meaningful text materials. For example, in one study, participants attempted to recall an unusual Indian folk tale on several occasions over a period of months. Bartlett reported that there was a high proportion of inaccuracy in the participants' recall and that they appeared to be unaware of the extent of this inaccuracy. Further, the inaccuracy resulted not only from omission and condensation but also from transformation of the original material. These elaborations appeared to be efforts to recast the unusual tale used into a form compatible with the participant's cultural knowledge and social conventions. These results led Bartlett (1932) to suggest that we form schemata, or concepts, of the world based on past experience. During learning, new information is integrated with existing schemata. When the material to be remembered conflicts with existing schemata, as in the case of the unusual folk tale used by Bartlett, recall is distorted. Thus memory is viewed as an active process involving reconstruction and elaboration of the original information.

Although Bartlett's work was published almost fifty years ago, it was essentially ignored for many years. Recently, however, many of the issues raised by Bartlett have reemerged as researchers have begun to formulate a contextual approach to learning and memory (Jenkins, 1974; Meacham, 1977; Bransford, McCarrell, Franks, & Nitsch, 1977). Such an approach yields a different conceptualization of learning and memory than either the associative or information processing views. The contextual approach, for example, does not view learning and memory as involving associative bonds between stimuli and responses or storage structures and

How and what we remember depends on the cultural context. In a culture such as the one shown here, the storyteller could retain a complete tribal history. (Woodfin Camp & Associates.)

control processes. Rather, it focuses on the nature of the events the individual experiences. What is learned and remembered depends on the total context of the event; for example, the physical, psychological, and social context in which the event was experienced; the knowledge, abilities, and characteristics the individual brings to the context; the situation in which we ask for evidence of remembering; and so forth. This perspective, then, views learning and memory as a by-product of the transaction between the individual and the context. Because of this focus on cognition as a by-product, learning and memory are not seen as isolated processes. Rather, the emphasis is on the interaction of the various perceptual, inferential, linguistic, problem-solving, personality, and social processes that contribute to understanding events. In addition, memory is not seen as a static process; rather, remembering is a reconstruction of past events. This depends, in large measure, on the degree to which the material has been articulated with past experience during acquisition. Thus the individual continually constructs and reconstructs events as the context changes. Finally, because memory processes do not represent isolated behaviors, the contextual approach does not focus on circumscribed tasks of limited meaning. Rather, the emphasis is on tasks high in ecological validity, tasks that are meaningful to the individual and which relate to his or her everyday life.

Meaning and Context in Learning and Memory. Much of the work within the contextual framework has focused on knowledge actualization and comprehension and memory for meaningful sentence and text materials.

BOX 12.2

EDUCATING ADULTS

In recent years, various writers have suggested that education should be reconceptualized as a continuous lifelong process (Agruso, 1978; Birren & Woodruff, 1973). Rapid cultural change, increasing numbers of adults reaching old age, and increasing amounts of leisure time underscore the importance of such proposals. Of course, the implementation of life-span education programs involves many philosophical, economic, and technical considerations. From a purely technical standpoint, however, available research on adult learning and memory provides a number of suggestions for structuring educational interventions for adults. Some of these are summarized below.

1. *Pacing.* Allow individuals to set their own pace, if possible. Tasks or methods involving significant time pressure are likely to be difficult for adults. **2.** *Arousal and anxiety.* Some degree of arousal is necessary for learning. However, older adults may become too aroused or anxious in a learning situation. Allow individuals an opportunity to become familiar with the situation. Minimize the role of competition and evaluation. **3.** *Fatigue.* Some tasks may produce considerable mental or physical fatigue—a problem which is likely to particularly affect older adults. Shorten the instruction sessions or provide frequent rest breaks. **4.** *Difficulty.* Many tasks are quite complex. Arrange materials from the simple to the complex in order to build individuals' confidence and skills. **5.** *Errors.* Structure the task so errors are avoided and do not have to be unlearned. **6.** *Practice.* Provide an opportunity for practice on similar but different tasks. Such practice helps to develop generalizable higher-order skills. **7.** *Feedback.* Provide information on the adequacy of previous responses. **8.** *Cues.* Materials should be presented to compensate for the potential sensory problems of older adults. Direct attention toward the relevant aspects of the task. Reduce the level of irrelevant information to a minimum. **9.** *Organization.* Learning and remembering often require that information be grouped or related in some way. Instruct individuals in the use of various mnemonic techniques (e.g., mental images, verbal associations, etc.) which may be used to elaborate or organize the material. **10.** *Relevance and experience.* People learn and remember what is important to them. Attempt to make the task relevant to the individual's concerns. Performance is likely to be facilitated to the extent that the individuals are able to integrate the new information with known information.

Knowledge actualization (Lachman & Lachman, 1980) involves memory for "real-world" knowledge. This encompasses an enormous range of information such as the location of the nearest gas station, the name of the starting quarterback of the Pittsburgh Steelers, the knowledge that it is important to maintain a balanced diet, and the fact that gold has greatly increased and decreased in value over the last several years. Some of the information is salient to one's daily life, and some of it is not. However, all of it was acquired during a lifetime of formal education and everyday experience. In other words, it was not acquired for the express purpose of remembering it in the laboratory.

The comprehension of meaningful sentence or text material involves abstracting and organizing information from the text and integrating this information with what one already knows (Kintsch & van Dijk, 1978). This processing occurs at multiple levels including the *graphemic* (letter), *phonemic* (sounds), *lexical* (words), and *semantic* (meaning). Each of these analyses leaves traces in memory. Normally, however, a reader or listener is concerned primarily with the meaning of the text. Therefore, memory for the semantic level is likely to be stronger than memory for other levels. As a result, the knowledge one gains from processing the text goes beyond what is shown in verbatim recall or recognition. Memory for the meaning or gist of the material is likely to be far more complete and longer lasting than memory for the specific wording or other surface properties of the material.

Knowledge Actualization. We have mentioned that knowledge actualization involves memory for accumulated world knowledge which has been acquired through education and other real-world experience. Studies examining such memory have found either no age differences (Lachman, Lachman, & Thronesbery, 1979) or age differences favoring older adults (Botwinick & Storandt, 1974, 1980; Perlmutter, 1978). For example, Lachman et al. (1979) asked young (19 to 22 years), middle-aged (44 to 53 years), and elderly (65 to 74 years) adults to respond to 190 questions covering such topics as famous people, news events, history, geography, the Bible, literature, sports, mythology, and general information (e.g., "What was the former name of Muhammed Ali?"; "What is the capital of Cambodia?"). No evidence of age differences in retrieval of world knowledge was found in this study. The elderly group actually answered more questions correctly than the younger groups, although the differences were not statistically significant.

Dated information was purposely omitted from the Lachman et al. (1979) study. However, as one might expect, there are cohort differences in world knowledge. For example, Botwinick and Storandt (1980) examined recall and recognition of historical and entertainment facts from each of seven decades—1910 through 1970 (e.g., "What was the name of the plane in which Lindberg flew the Atlantic?"; "Who was 'the Sweater Girl?'"; "What was the name of the first man to set foot on the moon?"). Subjects from each decade from the twenties through the seventies performed the task. With the exception of those in their seventies, the older adults (forties, fifties, sixties) recalled and recognized more information than the younger adults (twenties, thirties) although the differences were not large. But to some extent, knowledge actualization was cohort specific. Items from earlier decades were recalled less well by most subjects. However, older adults recalled material from earlier decades better than younger adults. The reverse was true for material from recent decades.

In general, it appears that older adults are able to recall factual information as well as or better than younger adults, although some of this information is cohort

Age differences would not be expected to relate to performing tasks pertaining to memory for world knowledge. (NBC).

specific. Older adults also appear to have accurate knowledge of their own memory processes—knowledge which has been labeled *metamemory*. For example, Perlmutter (1978) found no age differences on measures of memory knowledge (e.g., "Is it easier to remember visual things than verbal things?"), memory strategy use (e.g., "How often do you write reminder notes?"), and memory monitoring (e.g., prediction of the number of items that would be recalled following various memory tasks).

Comprehension and Memory for Sentences and Text. Only a few studies have examined adult age differences in comprehension and memory for sentence and text materials. Several of these have required verbatim recall, and the typical finding in these instances is that older adults recall fewer words than younger adults. But what about the retention of the meaning or gist of the material as opposed to the exact wording of the material? In this case, the findings are more diverse.

Several studies have found age-related deficits in the gist recall of text materials similar to those found in the verbatim recall of text on list materials (Dixon, Simon, Nowak, & Hultsch, 1982; Gordon & Clark, 1974; Taub, 1975). For example, Dixon et al. (1982) asked younger, middle-aged, and older adults to read or listen to several short texts which dealt with recent news events (e.g., the nuclear accident at Three-Mile Island). The subjects were asked to remember as much as they could about the texts but not to recall them verbatim. It was found that, in general, the older adults recalled only half as much information as the younger adults. The middle-aged group performed in between these two groups. Furthermore, the differences between the age groups seemed to be most pronounced in the case of the main ideas, as opposed to the details, of the studies. Such a finding suggests that the older adults may have had difficulty discovering or utilizing the underlying organization of the text. As we have noted, this kind of organizational deficit appears to contribute to the older adults' difficulty in verbatim recall of word lists as well. Interestingly, however, other recent studies using similar materials and techniques have failed to find age-related differences in text processing (Harker, Kent, Hartley, Finkle, & Walsh, 1980; Meyer & Rice, 1981). For example, Meyer and Rice (1981) found no age differences in total recall, recall of main ideas, or use of the text's structure on a passage about parakeets. One significant difference between the two studies described above is reflected in the educational level of the participants. The subjects in the Dixon et al. study were mainly high school graduates while those in the Meyer and Rice study were mainly college graduates. Thus in the case of text processing, age-related differences may interact strongly with education, occupation, and other indexes of cognitive activity and lifestyle.

In this context, Perlmutter (1980) suggests that memory mechanisms (e.g., encoding, retrieval) may decline with age. However, the individuals' store of information about the world tends to increase with age. To the extent that the individual has remained cognitively active and the task requires such world knowledge, older adults may perform as well as younger adults in spite of less effective memory mechanisms.

Age versus Cohort

Virtually all the studies reviewed thus far have been based on the cross-sectional data collection strategy. As we saw in Chapter 1, this data collection strategy confounds age and cohort effects. In the absence of appropriate data collection strategies, it is not possible to describe these results as age-related changes. There is only one source of data which allows us to address age-related changes. These data are from the Baltimore longitudinal study of men reported by Arenberg and Robertson-Tschabo (1977). Analysis of their data led these investigators to conclude that learning performance actually does decline with increasing age, particularly after 60 years of age. However, additional data based on the sequential data collection strategies discussed in Chapter 1 are required to resolve this question.

INTELLIGENCE IN ADULTHOOD AND AGING

Early theorists and researchers maintained that universal decline in intelligence during adulthood is evident as a function of intrinsic, biologically based aging processes. Wechsler (1958), for example, portrays a bleak picture stating that "nearly all studies . . . have shown that most human abilities . . . decline progressively after reaching a peak somewhere between age 18 and 25" (p. 135).

This early descriptive work, however, suffers from a number of methodological problems. First, most of this research was based on measures of intelligence which were developed within an atheoretical framework. The most widely used instrument—the Wechsler Adult Intelligence Scale (WAIS)—consists of eleven subtests. Six subtests deal with verbal content and five subtests focus on perceptual-motor performance. The test produces three scores: a verbal score, a performance score, and a full-scale (composite) score. Considerable research demonstrates that older adults show more decline on performance scores than on verbal scores. However, it is difficult to attach meaning to these results in the absence of a theoretical framework. Second, much of the early work was based on either cross-sectional or longitudinal data collection strategies. As we explained in Chapter 1, these strategies have significant limitations because of their confounding with cohort, time of measurement, selective survival, and other variables. Indeed, because of these difficulties, early cross-sectional and longitudinal results tend to display somewhat dissimilar pictures of intellectual decline with age. Cross-sectional results reflect earlier and steeper declines, whereas longitudinal strategies point to later and less steep declines. Such differing results are not surprising given the different types of confounds produced in either data collection strategy. For example, in the case of the cross-sectional strategy, differences may be magnified because of the presence of cohort differences favoring later-born cohorts. With the longitudinal strategy, however, decline may be underestimated because of selective survival effects—those who are less able drop out or die, while those who are more able remain to be tested. In any event, despite these difficulties, early descriptive research firmly established the conclusion that intelligence declines with increasing chronological age during adulthood.

But does intelligence in fact decline with increasing age? Although extensive research has been conducted during the roughly twenty years since Wechsler answered "yes" to this question, there actually is no definitive answer. Rather than witnessing researchers coming closer to a precise "yes" or "no" response, we have witnessed increasing controversy over the timing, extent, and sources of intellectual change during adulthood. On the one hand, Baltes and Schaie (1974) have concluded that "general intellectual decline in old age is largely a myth" (p. 35), while, on the other hand, Botwinick (1977) has concluded that "decline in intellectual ability is clearly part of the aging picture" (p. 580). Since these disagreements reflect differing sets of assumptions which result in varying theoretical and methodological approaches to the phenomenon, we will review three areas of research on intelligence. As each research area is examined, the centrality of different intellectual patterns and sources of intellectual change during adulthood will be examined.

Psychometric Intelligence

One major attempt to define and understand intellectual development has involved examining the interrelationships of various tests purported to measure intelligence. A multitude of such tests have been developed. Amazingly, these measures tend to correlate positively, although not highly, with one another; that is, relative to other people, an individual scoring high on a test of, say, arithmetic ability will tend to score high on a test of, say, vocabulary ability even though the tests appear to be measuring different things. In addition, performance on measures of intelligence tends to correlate positively with commonsense indexes (e.g., scholastic achievement) of intellectual ability. Such relationships led early investigators to postulate the existence of a *general intelligence* which pervaded all cognitive tasks (Spearman, 1927).

Nevertheless, modern theories of intelligence clearly have suggested that it is more useful to conceptualize this domain as containing multiple abilities. As a consequence, investigators have attempted to find dimensions that describe what is common to various tests. These results suggest that the bulk of individual differences found among many ability tests may be accounted for by a relatively small number of *primary mental abilities*. Estimates of the number of such ability factors range from 7 to 120, although most investigators tend to focus on the lower end of this range.

Finding common dimensions to various tests of intelligence requires the use of *factor-analytic techniques* and has yielded several major theoretical formulations of intelligence (Cattell, 1971; Guilford, 1967; Horn, 1970, 1978; Thurstone & Thurstone, 1941; Vernon, 1961). However, few of these theorists have incorporated any developmental variables into their theories. An exception has been the work of Cattell (1963), later refined by Horn (1970, 1978). As a result, their theoretical effort has become the major psychometric approach within which adult intelligence has been examined.

Fluid and Crystallized Intelligence. Horn (1978) has argued that even a reduction of the multitude of intelligence tests to twenty or thirty primary mental abilities is too complex at this stage of our thinking. An alternative strategy is to examine what is common to the primary abilities by applying the same statistical techniques to these factors as was originally applied to the individual tests. Such analyses generate a set of second-order abilities, and within this framework, the work of Cattell and Horn suggests that the intelligence domain may be described by two basic second-order abilities: *fluid intelligence* and *crystallized intelligence*.

According to Horn (1978), both fluid and crystallized intelligence involve behaviors characteristic of the essence of human intelligence: perceiving relationships, abstracting, reasoning, concept formation, and problem solving. However, they reflect different processes of acquisition, are influenced by different antecedents, are reflected in different measures, and show different patterns of change over the course of adulthood. Let us examine these differences.

On the one hand, fluid intelligence reflects incidental learning processes—the degree to which the individual has developed unique qualities of thinking independent of culturally based content. Crystallized intelligence, on the other hand, reflects intentional learning processes—the degree to which the individual has been acculturated, that is, has incorporated the knowledge and skills of the culture into thinking and actions.

As one would expect, given this distinction between incidental and intentional processes of acquisition, fluid and crystallized intelligence are indexed by different types of tests. No single measure of fluid or crystallized intelligence exists, because each of these abilities is a conglomerate of several abilities indexed by many measures (Horn, 1978). Thus any given test may reflect both abilities, although some tests are relatively pure measures of one or the other. Regardless, fluid intelligence tends to be indexed by tests which minimize the role of cultural knowledge, while crystallized intelligence tends to be indexed by tests which maximize the role of such knowledge. Turn to the sample items illustrated in Exhibit 12.6 and see how well you fare. As illustrated, relatively little cultural knowledge, other than basic terms and relationships, is required to answer the fluid items. But in order to answer the crystallized items, considerable knowledge about the culture in which you live must have been acquired.

Developmental Patterns and Sources of Influence. Horn's theoretical model describing the development of fluid and crystallized intelligence over the life span and the major sources of influence on this development are shown in Exhibit 12.7. As indicated in this figure, the theory postulates that fluid intelligence declines during adulthood after a peak in early adulthood, while crystallized intelligence increases throughout adulthood.

We have noted that incidental learning produces fluid intelligence, and according to Horn and Donaldson (1980), incidental learning processes are particularly influenced by physiological and neurological functioning. As indicated in Exhibit 12.7, the physiological base deteriorates from birth through death as a result of negative maturational processes, injury, and illness. For example, in late adulthood there is considerable evidence of such deterioration indexed by loss of brain weight, reduction of cerebral blood flow and oxygen consumption, and increase in inert waste products in the neurons. Positive maturational influences are significant during early life but decrease in later life. These positive maturational influences mask the loss of the physiological base during childhood and adolescence, resulting in gains in fluid intelligence. Following biological maturity, however, the effects of physiological loss become more apparent and fluid intelligence declines.

We have noted that acculturation produces crystallized intelligence. As shown in Exhibit 12.7, the experiences which produce such acculturation accumulate over the life span. The individual continues to learn about the culture, and much of what is learned is not forgotten. As a result, information is added to the system, and this information also may be restructured as it is organized and related in different ways. Skills may be more finely tuned through practice. These processes occur throughout adult life, although in different contexts such as the family, school, and work. As a result, crystallized intelligence is hypothesized to increase during adulthood.

Research has tended to confirm the predictions of Horn's (1970, 1978) theory. For example, Horn and Cattell (1966, 1967) examined age differences in fluid and crystallized tests as well as tests which combined these two factors (omnibus tests). Their results are displayed in Exhibit 12.8. Fluid intelligence decreases steadily from adolescence through middle age, crystallized intelligence increases, and omnibus measures show few age-related differences as the two factors cancel each other out.

The Piagetian Approach

Another major approach concerned with the development of intellectual thought is based on the work of Piaget (Piaget, 1950, 1951, 1952, 1968). As we have seen in earlier chapters, classical Piagetian theory focuses on childhood and adolescence.

Once formal operations are achieved, qualitative change is assumed to cease. Piaget and Inhelder (1969) state: "Finally, after the age of eleven or twelve, nascent formal thought restructures the concrete operations by subordinating them to new structures *whose development will continue throughout adolescence and all of later life*" (pp. 152–153, italics added).

More recently, Piaget (1972) has suggested that while all normal individuals attain formal operations, at least between 15 and 20 years of age if not between 11 and 14 years of age, they do so in different areas according to their aptitudes and occupational specializations. For example, a carpenter may be capable of *hypothetical thought* (e.g., dissociating variables, performing combinational analyses, reasoning with propositions involving negations and reciprocities) within the context of constructing a house but not within the context of traditional Piagetian tasks which are based on logical-mathematical concepts. Similarly, a theoretical physicist may be capable of formal reasoning with logical-mathematical tasks but may perform at a concrete level when attempting to construct a shed for his or her garden tools. In either case, according to Piaget (1972), the individual's lack of knowledge or loss of previously acquired knowledge will hinder the application of formal operations and result in the application of concrete operations to the problem.

Thus Piaget (1972) suggests that all individuals progress through a series of four stages of cognitive thought culminating in formal operations attained sometime during adolescence. This latter stage will be applied or used differently by individuals according to their particular aptitudes or experiences. In spite of this differentiation of formal processes during adolescence, and presumably during adulthood, no new qualitative changes during the adult years are proposed. Other theorists (e.g., Riegel, 1973; Arlin, 1975), as noted in an earlier chapter, have proposed the existence of a fifth stage—noting that formal operational thought is confined to finding solutions to problems—the generation and evaluation of all possible combinations of hypotheses. Arlin (1975), for example, proposes a "problem-finding" stage focused on the discovery of new problems and ideas. Nevertheless, evidence for a fifth stage is limited at this time, and traditional Piagetian theory suggests that adult intelligence is characterized by stability. However, recently a number of researchers applying Piaget's theory to the latter portion of the life span have found evidence of age-related differences on a variety of Piagetian tasks (see Denney, 1974c; Hooper & Sheehan, 1977; and Papalia & Bielby, 1974, for review). We will examine several of these differences in the following sections.

Conservation Abilities. Several studies have investigated conservation abilities in adults and the elderly and generally have revealed that older adults do more poorly on such tasks than adolescents or young adults. For example, Papalia (1972) investigated conservation of number, substance, weight, and volume in individuals ranging in age from 6 to 82 years. With the exception of the 55- to 64-year-old group, performance tended to decrease with increasing chronological age during adulthood. Further, the elderly adults performed best on those conservation abilities which appear earlier during childhood (number, substance) and worst on those which appear later in childhood (weight, volume). This led Papalia (1972) to conclude that there is a regression with age to simpler modes of responding. She speculated that this regression was a function of increasing neurological decrement with increasing age. Other studies have supported Papalia's (1972) basic conclusion that older adults frequently exhibit poorer performance on the various conservation tasks than younger adults (Papalia, Kennedy, & Sheehan, 1973; Papalia, Salverson, & True, 1973; Rubin, Attewell, Tierney, & Tumolo, 1973).

EXHIBIT 12-6

Sample Test Items Marking Fluid and Crystallized Intelligence

Secondary Ability	Primary Ability	Directions/Items
Fluid	Induction	Each problem has five groups of letters with four letters in each group. Four of the groups of letters are alike in some way. You are to find the rule that makes these four groups alike. The fifth group is different from them and will not fit the rule.*
		1. NOPQ DEFL ABCD HIJK UVWX 2. NLIK PLIK QLIK THIK VLIK 3. VEBT XGDV ZIFX KXVH MZXJ
Fluid	Visualization	Below is a geometric figure. Beneath the figure are several problems. Each problem consists of a row of five shaded pieces. Your task is to decide which of the five shaded pieces when put together will make the complete figure. Any number of shaded pieces from 2 to 5 may be used to make the complete figure. Each piece may be turned around to any position but it cannot be turned over.†

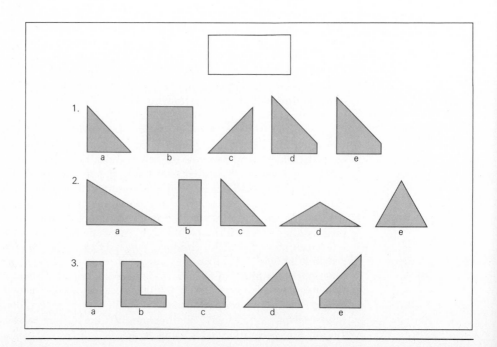

EXHIBIT 12-6

Sample Test Items Marking Fluid and Crystallized Intelligence (*Continued*)

Secondary Ability	Primary Ability	Directions/Items
Crystallized	Verbal meaning	Choose one of the four words in the right-hand box which has the same meaning or nearly the same meaning as the word in the left-hand box.‡

	run	try
attempt	hate	stop
pecuniary	involving money	esthetic
	trifling	unusual
germane	microbe	contagious
	relevant	different

Secondary Ability	Primary Ability	Directions/Items
Crystallized	Mechanical knowledge	Complete each of the statements by selecting the correct alternative or answer.§

1. The process of heating two pieces of heavy metal so hot that they will fuse (melt together) is known as:
   riveting
   soldering
   welding
   forging
2. A paint sprayer functions in exactly the same way as a
   centrifugal water pump
   carbon-dioxide fire extinguisher
   perfume atomizer
   vacuum cleaner
3. The tool used to rotate a cylindrical object such as a water pipe is a
   Stillson wrench
   open-end wrench
   box-end wrench
   socket wrench

Answers: Induction: 1. DEFL, 2. THIK, 3. VEBT
        Visualization: 1. a,c,d,e 2. a,d,e 3. b,c,e
        Verbal meaning: 1. try, 2. involving money, 3. relevant
        Mechanical knowledge: 1. welding, 2. perfume atomizer, 3. Stillson wrench
* Letter Sets Test, I-1; Educational Testing Service, 1962, 1976.
† Form Board Test, VZ-1; Educational Testing Service, 1962, 1976.
‡ Vocabulary Test, V-5; Educational Testing Service, 1962, 1976.
§ Mechanical Information Test, MK-2; Educational Testing Service, 1962 (test no longer in print).
Source: Ekstrom, French, Harman, & Dermen, 1976; French, Ekstrom, & Price, 1963.

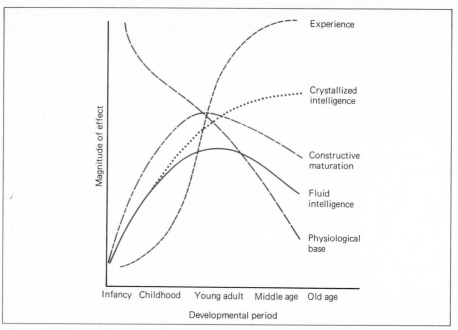

EXHIBIT 12.7   Life-span development of intelligence in relation to major influences.
(Source: Horn and Donaldson, 1980.)

Classification Abilities. A number of studies have also investigated classificatory behavior on a variety of tasks. These studies have generally shown poorer performance on the part of the elderly adults as compared to younger adults (Denney, 1974b). For example, several studies have examined adult age differences on *free-classification*

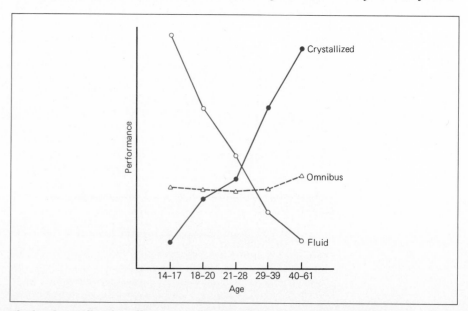

EXHIBIT 12.8   Fluid and crystallized intelligence as a function of age. (Source: Horn, 1970.)

tasks in which individuals are asked to group stimuli that are alike or that go together in some way. Many of these studies have noted that younger or older adults employ different classification criteria (Denney, 1974a; Denney & Lennon, 1972). On the one hand, younger adults tend to classify items on the basis of perceptual or conceptual similarity—that is, common attributes shared by all items. Thus a pick and a shovel may be classified together because they both have handles or because they are both tools. On the other hand, elderly adults tend to classify items on the basis of their functional interrelationships as defined by past experience or the testing situation. A shovel and a wheelbarrow, therefore, may be classified together because the shovel is used to put dirt into the wheelbarrow. Such functional classification is also characteristic of young children, leading some investigators to suggest the possibility of a structural regression during later life. However, Denney (1974b) suggested the use of complementary criteria may simply be a more natural way of organizing experience, rather than a structural regression caused by neurological degeneration.

Formal Operational Abilities. There have been relatively few studies of older adults' capacity to engage in the hypothetical thought characteristic of formal operations. Those which have been done suggest that older adults perform more poorly on such tasks than younger adults. For example, Tomlinson-Keasey (1972) presented sixth-grade, young adult, and middle-aged females with a series of formal operational tasks (e.g., pendulum, balance, and flexibility problems). In the case of the sixth-grade girls, only 32 percent of the girls' responses were at the formal level, with 4 percent at the most advanced stage. The percentages for the young adult and middle-aged women were 67 and 23 percent and 53 and 17 percent, respectively. Thus, the middle-aged women did not perform as well as the young women. Other studies using elderly adults show even more pronounced age differences (Clayton & Overton, 1976). Still other investigations examining different aspects of formal reasoning (e.g., volume conservation, combinatorial reasoning) support the conclusion of age-related differences on these tasks (Papalia, 1972).

However, the use of formal operations appears to depend heavily on the particular problem and the individual's past experience. Several recent studies have used everyday, as well as traditional, forms of formal operational problems (Sinnott, 1975; Sinnott & Guttman, 1978), and these studies suggest that adults solve everyday problems more easily than traditional problems. Further, the advantage provided by the use of everyday materials appears to be greater for older adults than for younger adults. For example, Sinnott (1975) found the performance of her younger adults increased by 10 percent with familiar materials, while that of her older adults increased by 25 percent. At a more general level, however, adults of all ages often perform poorly on formal operational problems whether their content is familiar or not. For instance, Capon and Kuhn (1979) found that few adult female shoppers were able to use a proportional reasoning strategy in order to determine which of two sizes of a common item sold in the supermarket was the better buy. In general, research suggests that the use of formal operations is far from universal among adults and appears to depend heavily on experience.

Competence versus Performance and Piagetian Tasks. As summarized in our brief review, there appears to be strong evidence to suggest that older adults do not perform as well as younger adults on a variety of Piagetian tasks. How is this fact to be interpreted? Two interpretations have dominated the literature.

Competence refers to the formal logical representation of cognitive structures.

Thus the older adult's poorer performance may reflect a loss of cognitive structures necessary for logical thought (Hooper, Fitzgerald, & Papalia, 1971; Papalia, 1972; Rubin et al., 1973). Generally, it has been assumed that this loss is a result of inevitable neurological degeneration accompanying the aging process. Further, since competencies appearing late in childhood are lost first and competencies appearing early in childhood are lost last, it has been suggested that the deficit in adulthood reflects a structural regression reversing the order of development in childhood.

Performance refers to processes by which available competence is assessed and applied in real situations. Thus the older adult's poorer performance may reflect a number of task or situational factors which interfere with performance even though the underlying structural competence required is unimpaired. Such factors may include "lack of familiarity with the testing situation, irrelevance of the tasks, disuse of relevant skills or strategies, memory limitations, and preferential modes of thinking" (Hornblum & Overton, 1976, pp. 68–69).

In many respects it is impossible to evaluate these two interpretations at the present time. For example, to date, all the studies examining adult performance on Piagetian tasks have employed a cross-sectional data collection strategy. Adequate evaluation of the structural regression hypothesis requires the application of longitudinal or sequential data collection strategies in order to examine change rather than differences (see Chapter 1). Thus at the present time the competence interpretation is speculative. Evidence relevant to the performance interpretation is also limited, since most investigators have simply compared the performance of different age groups under standard conditions. One alternative is to expose individuals to short-term training experiences. Improvements following training are assumed to reflect changes in performance factors (e.g., attention, strategy use), rather than changes in underlying competence (Bearison, 1974). In a study of this type, Hornblum and Overton (1976) were able to improve older adults' performance on several conservation tasks simply by providing verbal feedback on the accuracy of their responses during training. However, few studies of this type are available to illustrate the extent of training possible.

Intelligence in Context

The research reviewed in the last two sections emphasizes both gains and losses in intellectual functioning during adulthood. Gains are seen in abilities which reflect acculturation—measures of crystallized intelligence. Losses are seen in abilities which reflect incidental learning—measures of fluid intelligence and Piagetian operations. These latter abilities, as we discussed, are thought to be particularly affected by the degeneration of the physiological base with aging. Thus research derived from both psychometric theory and Piagetian theory suggests that while some intellectual functions remain stable or increase, others must be expected to decline as a logical consequence of the aging process.

In contrast to this view, a number of researchers have argued that aging does not necessarily imply inevitable, irreversible, and universal decrement—even in the case of fluid abilities (Baltes & Labouvie, 1973; Labouvie-Vief, 1977; Schaie, 1970, 1974, 1979). Although these researchers do not deny the reality of decrement in cognitive functioning in many elderly, they suggest that the past view of intellectual decrement is too pessimistic. Rather, the emphasis is on the relative plasticity and period-specific nature of intellectual performance functions.

Thus a contextual approach places heavy emphasis on the contextual determinants of intellectual functioning. Unquestionably, age-related performance differences do occur, and these tend to follow the pattern specified by fluid-crystallized theory. However, from a contextual perspective, it is questionable whether such functions represent irreversible and universal age changes. On the one hand, there is emphasis on the potential for change.

> *The routine cognitive performance of older individuals as they function in psychological research settings may be an extremely poor indicator of what they can do, and this competence-performance gap may be decreased by relatively benign interventions. It therefore appears that there may be a good deal more plasticity to intelligence in old age than has been acknowledged thus far. (Labouvie-Vief, 1977, p. 245)*

On the other hand, there is emphasis on the role of context, particularly the historical context within which individuals experience events.

> *It is in the nature of cohort effects and historical change that the data presented are restricted to the culture and generations studied. Cultural change over the last decade has been rapid. Accordingly, the relative deprivation of the current elderly and the relative contributions of cohort effects may be particularly pronounced at this point in historical time. (Barton, Plemons, Willis, & Baltes, 1975, p. 235)*

Thus the contextual perspective does not so much propose a different definition or theory of intelligence as it focuses on a different set of antecedents and methods.

BOX 12.3

THE INTELLIGENCE WAR

In a series of articles in the *American Psychologist,* John Horn and Gary Donaldson (Horn & Donaldson, 1976; 1977) and Paul Baltes and Warner Schaie (Baltes & Schaie, 1976; Schaie & Baltes, 1977) exchanged their views on the course of intellectual development during adulthood. Considerable disagreement was expressed over the extent and significance of intellectual decline during the latter part of the life cycle.

Horn and Donaldson criticized both the theoretical conception and data collection and analysis strategies of Baltes and Schaie. They concluded, "It is thus premature and incorrect to infer on the basis of existing evidence that there is no notable maturational decline in intellectual abilities in adulthood and that the major portion of such change as might be indicated represents only differences between the environments of persons of different generations" (Horn & Donaldson, 1976, p. 707).

In their replies, Baltes and Schaie argued that Horn and Donaldson misinterpreted both their theoretical and methodological efforts. They

concluded, "It seems fair to conclude that research on intelligence in adulthood and old age has pointed to large interindividual differences, multidimensionality, multidirectionality, and the import of generational differences. . . . We see these findings as suggesting much more plasticity in adult development than what has traditionally been assumed" (Baltes & Schaie, 1976, p. 721).

The somewhat heated debate between these researchers illustrates the degree of controversy surrounding the issue of intellectual decline in adulthood. In our view, despite their sharply expressed differences, neither Horn and Donaldson nor Baltes and Schaie are that far apart. Both appear to agree that intellectual decline does occur during adulthood, particularly late in life. However, both also appear to agree that adult intellectual functioning is modifiable and that different individuals exhibit different patterns of development. Nevertheless, the data on intellectual change during the adult years are open to many interpretations, and it is likely that this topic will continue to be a source of lively controversy.

Age versus Cohort. At the heart of the disagreement over intellectual change during adulthood is the issue of whether the observed differences reflect age-related change or cohort-related differences. From Chapter 1, you may recall that the cross-sectional and longitudinal data collection strategies typically used in developmental investigations have a number of limitations. In particular, the cross-sectional strategy confounds age with cohort, and the longitudinal strategy confounds age with time of measurement. To the extent that history-graded or life-event sources of variance are antecedents of the behavior in question, these designs will provide an erroneous description of the developmental functions involved. Differences documented by the cross-sectional strategy may reflect age-related changes or the impact of differential experiences on different birth cohorts. Changes documented by the longitudinal strategy may reflect true age-related changes or the impact of events occurring at a given point in historical time which affect individuals of all ages. Some resolution of this dilemma is possible, however. We mentioned, for example, that Schaie (1965, 1973) and Baltes (1968) proposed strategies involving the simultaneous and sequential application of cross-sectional and longitudinal strategies in order to partially unconfound age and historical-evolutionary effects.

A number of studies are now available which support the view that history-graded sources of variance contribute substantially to adult differences in intellectual performance (Nesselroade, Schaie, & Baltes, 1972; Schaie & Labouvie-Vief, 1974; Schaie, Labouvie, & Buech, 1973; Schaie & Parham, 1977; Schaie & Strother, 1968). These reports are based on a large sample of individuals first tested in 1956 and retested in 1963, 1970, and 1977. In addition, new independent samples of individuals were tested in 1963, 1970, and 1977 in order to control for retesting and sample attrition effects. Using this data set, the various researchers examined several arrangements of the data in order to investigate different issues. However, we will use only a fraction of these data to illustrate several central themes. Exhibit 12.9, for example, contains the results of an analysis on one primary mental ability reported by Schaie and Labouvie-Vief (1974). In this figure, the cross-sectional gradients (top right) support the traditional finding of age-related decline. Note the general downward trend occurring after a peak in early adulthood. Notice also that the peak tends to occur later with later times of measurement (1956, 1963, 1970). Thus availability of three cross-sectional gradients obtained at different points in time hint at the role of historical variables such as increased education.

When the same data are plotted in terms of seven- and fourteen-year longitudinal gradients (bottom right), a different picture emerges. Here increments are observed for the younger groups (age 25, 46, or 53 at first testing), while genuine decrements appear to occur no earlier than age 60.

Exhibit 12.9 also shows the combined cross-sectional and longitudinal gradients for 1956 to 1963 (top left) and 1963 to 1970 (bottom left). In addition to repeating the information in the right-hand portion of the figure, these gradients allow comparison of same-aged individuals who were born at different points in time. For example, the boxed areas of the figure compare individuals who are all 53 years of age but who were born at different points in time. Note the large differences. Exhibit 12.10 represents a more comprehensive example of such a comparison. This figure estimates cohort differences in composite measures of intelligence and educational aptitude for individuals born during different periods of history from 1899 to 1931. Clearly, substantial differences are associated with the cohort variable. These illustrations are confirmed in much more detail by other analyses (Schaie & Labouvie-Vief, 1974).

In general, these results suggest the need to consider the role of historical-evo-

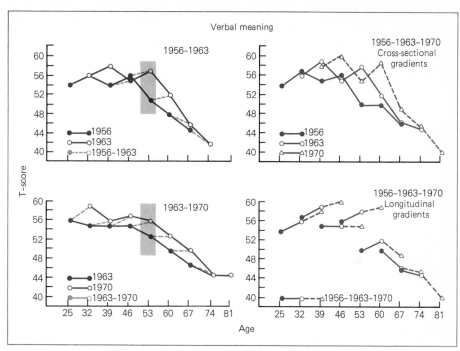

EXHIBIT 12.9   Cross-sectional and longitudinal age gradients for verbal meaning.
(Source: Schaie and Labouvie-Vief, 1974.)

lutionary change in intellectual functioning. For the most part, cohort differences are larger than age changes, particularly prior to age 65. Most of young adulthood, middle age, and early old age is characterized by stability or increases in intellectual performance. Decline, however, does occur relatively late in life (Schaie & Parham, 1977). This may be illustrated by the figures in Exhibit 12.11, which shows the age at which a reliable decrement over a seven-year period appears. The difference in the

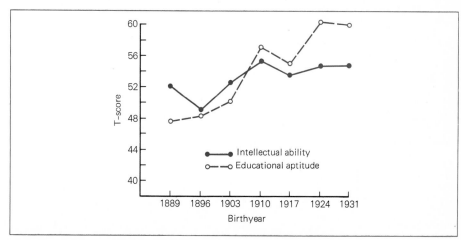

EXHIBIT 12.10   Cohort gradients for the composite measure of intellectual ability and educational aptitude. (Source: Schaie and Labouvie-Vief, 1974.)

EXHIBIT 12.11   **Ages at Which Reliable Decrement over a 7-Year Period Is First Shown**

	Age	
Variable	Repeated Measurement Study	Independent Random Samples Study
Verbal meaning		74
Space	74	67
Reasoning	74	
Number	74	60
Word fluency	53	39
Intellectual aptitude	67	67
Educational aptitude	74	74
Motor cognitive rigidity		
Personality-perceptual rigidity	74	
Psychomotor speed	67	67

Source: Schaie and Parham, 1977.

repeated measure and independent random samples reflects the positive bias of repeated measures due to sample attrition. With the exception of word fluency, most measures do not demonstrate a reliable decrement until the late sixties or middle seventies.

These data suggest that conclusions of an inevitable, universal, age-related decline are open to question. Rather, they emphasize the role of the environmental-historical context in which individuals develop. However, since cohort, like age, is a descriptive variable (see Chapter 1), identifying it as an important variable does not explain how it operates. Cohort reflects the operation of history-graded antecedents which differentially affect groups of individuals born at different points in time. Thus, for both age and cohort, the task is to identify these antecedents. Several sets of influences which may be useful in delineating a life-span perspective on adult intelligence will be examined in the next sections.

Biological Antecedents. In Chapter 3, we summarized some of the degenerative biological changes which occur during adulthood. The scope of these changes suggests the plausibility of Horn's (1970, 1978) view that the decline of intellectual abilities, particularly fluid abilities, is due to the degeneration of the physiological base, especially as it affects central nervous system functioning. However, it is important to distinguish whether biology-intelligence relationships are extrinsic, pathologically based processes related to stress or disease or are intrinsic, genetically based aging processes unrelated to pathology (Jarvik & Cohen, 1973). Although such a distinction is complex and can be made only tentatively, it is important. For example, from an optimization perspective, declines related to disease processes are more likely to be amenable to intervention than those which are related to the process of aging per se.

Within this context, evidence at the present time suggests that intellectual decline is not uniformly distributed in the population. Linkages between biological functioning and intellectual performance are observed primarily in populations with pathology such as cardiovascular disease. For example, in a now classical study, Birren, Butler, Greenhouse, Sokoloff, and Yarrow (1963) examined the relationship of health and psychological functioning in a group of community-dwelling men. Individuals were classified into two groups on the basis of health status. Group 1 consisted of "superhealthy" individuals with measurable evidence of trivial disease (e.g., partial deafness, varicose veins) or without any measurable evidence of disease at all. Group 2 consisted of "less healthy" individuals with measurable evidence of potentially serious disease (e.g., arteriosclerosis); however, these individuals were not acutely ill. That is, they showed only incipient signs of disease. Symptoms were not present at the behavior level, and the existence of the disease was unlikely to be detected by other than a rigorous medical examination. Nevertheless, in spite of the incipient nature of the disease involved, there were significant functional differences between groups 1 and 2. In particular, cerebral blood flow was approximately 16 percent lower in group 2 than in group 1. Cerebral oxygen consumption also tended to be lower. These differences were accounted for by the individuals in group 2 who showed incipient signs of arteriosclerosis. Thus groups 1 and 2 differed in terms of degree of pathology.

These groups also displayed significant differences in cognitive performance. Group 1 subjects scored higher than group 2 subjects on a battery of twenty-three tests of cognitive functioning. In particular, performance on the verbal scales of the WAIS was significantly related to health status—superhealthy group 1 individuals scored higher than less healthy group 2 individuals. The scores of group 1 elderly were likewise higher than those of a group of younger adults. The scores of group 2 elderly were equivalent to those of the younger adults. This led the authors to suggest that late-life illness such as that exhibited by group 2 results in a loss of ability. Finally, the correlations between physiological and cognitive indexes were more numerous and higher for group 2 than group 1. This latter result led Birren and his colleagues to formulate a "discontinuity hypothesis" concerning the relationship between biological functioning and cognitive functioning. That is, cognitive functioning is largely autonomous of biological functioning until certain limits are reached as a function of disease or trauma. At this point, a new set of relationships emerges. It therefore appears that intellectual functioning is built on a biological base. When the biological base is intact, there is little relationship between biology and intelligence. However, when the biological base is damaged by illness or injury, a relationship between biology and intellectual functioning emerges.

More recent studies, reviewed by Eisdorfer and Wilkie (1977), of the relationship of health to cognitive functioning have confirmed and expanded the findings of Birren and his colleagues.

Brain Function. A large number of studies have shown a significant relationship between indexes of central nervous system (CNS) pathology (electroencephalogram, blood flow, oxygen consumption) and intellectual impairment, with greater degrees of pathology related to poorer intellectual performance (Obrist, Busse, Eisdorfer, & Kleemeier, 1962; Wang, 1973; Wang, Obrist, & Busse, 1970). This relationship is seen both in institutionalized elderly with various brain disorders and in community-dwelling elderly with incipient signs of disease.

Cardiovascular Disease. Spieth (1965) found that airline pilots and air traffic controllers with cardiovascular disease did not perform as well as healthy individuals on a battery of cognitive tasks including subtests of the WAIS. Again, these individuals were not critically ill but were capable of normal activity. However, the more severe levels of disease were associated with greater cognitive impairment.

Hypertension. High blood pressure is related to decrements in intellectual functions. Wilkie and Eisdorfer (1971) found that individuals with high blood pressure showed significant intellectual decline over a ten-year period, while those with normal or slightly elevated blood pressure showed little change or increases.

Thus there is considerable evidence suggesting a relationship between biological and intellectual functioning in individuals with various types of pathology. Since such a relationship does not seem apparent in healthy adults, the relationship between biology and intelligence during adulthood appears to be related to pathology rather than to aging per se.

Terminal Drop. The significance of the relationship between biological functioning and intellectual functioning is underscored by an interesting set of research which has examined the relationship between intellectual decline and death (Jarvik & Falek, 1963; Kleemeier, 1962; Lieberman, 1965; Riegel & Riegel, 1972). Individuals' intellectual performance shows a marked decline up to several years prior to death. This phenomenon has been labeled *terminal drop* (Kleemeier, 1962). Terminal drop research suggests that there is little decline in intellectual functioning with age but that such decline occurs up to several years prior to death. However, since the incidence of mortality increases with age, an apparent decrement in intelligence is produced as larger and larger numbers of persons exhibit terminal drop.

While the terminal drop phenomenon is easily interpretable as a consequence of biological deterioration, it is important to realize that sociocultural factors may be involved as well—e.g., nutritional and health-care conditions. Thus it is conceivable that the relationship between death and intellectual decline may be cohort-specific.

Learning Antecedents. Since learning processes may be viewed as the antecedents or building blocks of intellectual abilities, one strategy for examining the role of learning processes in intellectual functioning is to determine the modifiability of intellectual performance via the manipulation of variables such as practice and strategies. Underlying this approach is the assumption that age-related performance declines may not reflect biologically based decline but rather experientially based variables such as lack of practice or appropriate strategies. A variety of intervention approaches has been applied in an attempt to modify the intellectual and problem-solving performance of older adults. These include such strategies as feedback, modeling, and strategy instruction.

Feedback. Several investigators have examined the role of feedback knowledge of the accuracy of one's response on intellectual performance. For example, Hornblum and Overton (1976) focused on the effect of feedback on conservation performance. An area conservation (surfaces) task was used for training. Individuals in the experimental group were provided with feedback contingent upon the correctness of their responses ("Yes, that's right; let's go on" versus "No, that's not right. There is [is not] the same amount of space remaining on the board"). Individuals in the control group received the same problems but were not given feedback. Following training, the participants received six posttests examining both area and volume conservation. Exposure to feedback increased conservation on posttests. These effects were apparent on both near (similar) and far (dissimilar) relations to the training task. This suggests that training activated existing operational structures.

Other studies have also provided evidence for the importance of feedback, although in some instances this variable has been combined with other training (Sanders, Sterns, Smith, & Sanders, 1975; Schultz & Hoyer, 1976).

Modeling. Denney and Denney (1974) used a modeling strategy to improve the performance of older adults on a concept identification task. The task was similar to the game Twenty Questions. The person was presented with a picture of a number of objects. The object of the task is to identify the object the experimenter is thinking of by asking questions that can be answered "yes" or "no." Younger adults tend to ask questions which exclude whole groups of items at a time (e.g., "Is it in the right half?") and, thus, solve the problem quickly. In contrast, older adults tend to ask questions that eliminate only one item at a time (e.g., "Is it the house?") and, thus, solve the problem more slowly (Denney & Denney, 1973). In the training study, the investigators were able to improve the performance of older adults by exposure to another person (model) using more efficient techniques. These included simply asking questions which eliminate more than one item at a time and asking such questions plus verbalizing the underlying strategy. Both techniques were effective in reducing the number of questions the elderly asked prior to solution.

Other studies have also shown facilitative effects of modeling on problem-solving performance (Denney, 1974a; Labouvie-Vief & Gonda, 1976; Meichenbaum, 1974).

Strategy Instruction. Several studies have attempted to facilitate older adults' problem-solving performance by verbally instructing them to use particular strategies. For example, Sanders, Sterns, Smith, and Sanders (1975) presented older adults with a concept identification task under four different conditions; participants were presented with a programmed learning sequence beginning with simple problems without irrelevant dimensions. They also were given strategy hints, memory cue cards, and verbal feedback after each response. In the reinforced training conditions, tokens for correct responses were given as well. In the practice condition, participants were given the same problems but no strategies, hints, or other training aids. Finally, in the control condition, participants were given only the pretest and posttest. Both training conditions led to improved performance compared with practice alone and the control condition. This research suggests that direct training can improve older adults' performance on problem-solving tasks. Other studies confirm this finding (Sanders, Sanders, Mayes, & Sielski, 1976). However, some techniques do not appear to facilitate performance (Heglin, 1956; Young, 1966). It appears that the usefulness of strategy instruction depends greatly on the particular task and training technique used.

New Directions. Thus several different training strategies have emphasized the relative modifiability of intellectual performance in adulthood. Learning, particularly the acquisition or sharpening of higher-order skills, seems to play a key role in intellectual development. However, with some exceptions, the research we have examined suffers from a number of deficiencies. The training is limited in scope and duration. Typically, training is provided on a specific task during a single session, and the effects of training are usually measured immediately following training but not at later points in time. As a result, it is not possible to assess the durability of the training effect. Also, no attempt is made to determine whether the training transfers to tasks similar to but different from the trained task. Finally, no attempt is made to relate the training program to any extant theories of learning or intelligence.

As an illustration of the type of research that is required, let us examine a recent study by Plemons, Willis, and Baltes (1978). These investigators attempted to examine the degree to which fluid intelligence can be modified in older adults (fluid abilities, you will recall, tend to show normative decline with increasing chronological age). The experimental group participated in an eight-session training program designed to facilitate understanding of the relational rules found in measures of figural relations—a primary ability reflective of fluid intelligence. The control group did not receive any training. The performance of these two groups was then compared on posttests administered at three points in time following training—after one week, after one month, and after six months. Four tests varying in their similarity to the training materials were used. One would expect significant training effects for tests similar to the training materials but insignificant training effects for tests dissimilar to the training materials.

The results are shown in Exhibit 12.12. In the case of the most direct measure (near-near transfer) of training, the group receiving training outperforms the group not receiving training (top-left panel). This effect was present even after six months. In the case of the less direct measure (near transfer) of fluid intelligence, immediate but not persistent training effects are obtained (top-right panel). Finally, in the case of measures dissimilar to the training items (far and far-far transfer), no training effects are obtained (bottom panels). Consequently, training may have a relatively long-term impact on intellectual performance, at least in the case of measures with a high degree of similarity to those involved in training.

Overall, the results of various training studies point to the importance of learning processes in intellectual development and suggest the performance of older adults may be relative plastic.

Socioenvironmental Antecedents. In the previous sections, we have emphasized sets of antecedents which are centered specifically *in* the aging individual, for example, biological processes and learning processes. However, from a life-span perspec-

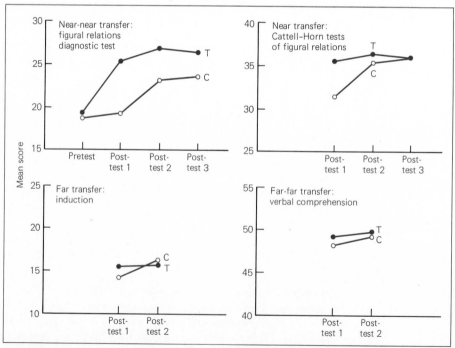

EXHIBIT 12-12   Mean score on all transfer tasks for the training (T) and control (C) on all occasions of measurement. (Source: Plemons, Willis, and Baltes, 1978.)

tive, the socioenvironmental context surrounding the individual may be a powerful source of antecedents affecting intellectual change during adulthood as well. Increasingly, researchers have begun to emphasize the role of these contingencies on the competence of adults (Baltes & Labouvie, 1973; Labouvie-Vief, 1977; Lawton & Nahemow, 1973).

Generally, it may be argued that the latter part of the life span, particularly the postretirement portion, is characterized by environmental contingencies which discourage the development of social and intellectual competence (Kuypers & Bengtson, 1973; Labouvie-Vief, 1977). For example, while young adulthood and middle age are marked by relatively well defined roles and expectations, this is less the case in older age (Bengtson, 1973; Rosow, 1974). If anything, the role is one of a "sick" role, and the expectations are ones of increasing dependence and incompetence. Further, it has been suggested that such expectations lead to a withdrawal of appropriate reinforcers for competent behavior. As a result, the process becomes one of a self-fulfilling prophecy in which the older individual expects to and actually does become less competent. In some settings, such as nursing homes, this modification of competencies actually results from reinforcement of incompetent behavior (e.g., performing self-care functions for the patient).

Kuypers and Bengtson (1973) apply a *social breakdown model* to suggest how the social environment interacting with self-concept and competence of the older individual creates a negative spiral of breakdown. Essentially, this involves (1) a socially vulnerable position in which the individual is (2) dependent on external sources of self-labeling, (3) negative social labeling of a group as incompetent, etc., (4) the socialization of the group into a dependent role, (5) learning of "skills" appropriate to this role, (6) an atrophy of previous skills, and (7) the individual's identification and self-labeling as sick, inadequate, and so forth. Many of the performance-related problems of older adults in cognitive testing situations such as anxiety and lack of confidence may be related to such factors.

CHAPTER SUMMARY

It has generally been concluded that cognitive processes decline with increasing age during adulthood. However, different pictures of this decline are painted by different theoretical approaches. The speed of the central nervous system decreases with increasing age. This conclusion has been based primarily on an examination of brain electrical activity measured by electroencephalogram recordings. The alpha rhythm of the EEG, for example, reaches its maximum frequency in adolescence and begins to slow gradually after young adulthood. This slowing may be related to disease processes (particularly vascular disease) as well as to basic aging processes. The older adult's central nervous system appears to be in a state of underarousal compared to that of younger adults. This slowing of central nervous system functioning appears to account for a large part of the pervasive behavioral slowing observed among older adults on a wide range of cognitive tasks.

Associative approaches define learning as the formation of S-R bonds, while forgetting involves the weakening or loss of such bonds. Research within this tradition has devoted considerable attention to the influence of the pacing of the task on adult learning and memory. Generally, the performance of older adults is impaired more by rapid pacing than the performance of young adults. The poorer performance of older adults on associative tasks, particularly those which are rapidly paced, has been attributed to both cognitive and noncognitive factors. On the one hand, it has been suggested that older adults are more cautious, aroused, or anxious in the learning situation, and that these noncognitive factors inhibit adequate performance. There is some support for these suggestions, particularly the arousal-anxiety hypothesis. On the other hand, it has been suggested that older adults' cognitive processing may be ineffective. In particular, it has been shown that older adults do not use mediators as extensively or as efficiently as younger adults. Similarly, older adults appear to be more susceptible to interference from previously acquired S-R bonds.

Information processing approaches propose that memory involves multiple storage structures and control processes and, typically, three storage structures (sensory, primary, secondary) are proposed. Information is retrieved from one store and entered into the next by various control processes (attention, elaboration, organization) which transform the information involved. Research suggests that there are relatively few age-related differences within the temporary sensory- or primary-memory systems. Large age differences, however, occur within the secondary-memory system. Secondary memory depends on the elaboration and organization of the information in terms of its semantic or meaning content. Compared to younger adults, older adults appear to be deficient in these processes. The older adult's difficulty, then, appears to involve primarily the encoding of the material rather than storing or retrieving it for later use.

A contextual approach to learning and memory suggests that what is learned and remembered depends on the total context of the event. Learning and memory are a by-product of the transaction between the individual and the context. Learning involves the integration of novel events with past experience, and remembering involves a reconstruction of past events. Much of the work within this tradition has focused on memory for world knowledge and comprehension and memory for meaningful sentence and text materials. Studies examining memory for world knowledge have generally found older adults retrieve such information as well as or better than younger adults. Similarly, older adults appear to have accurate knowledge about their own memory processes. Within some contexts, older adults appear to integrate and retain the meaning of sets of sentences and texts as well as younger adults.

Psychometric approaches attempt to define intellectual development by examining the interrelationships of various tests designed to measure intelligence. This work has suggested that intelligence may be conceptualized as containing multiple abilities. While several theories have been developed within this framework, that of Cattell and Horn has proved the most useful in viewing adult intelligence. These investigators propose that intelligence may be described by two basic factors—fluid and crystallized intelligence. Fluid intelligence, measured by tests that minimize the role of cultural knowledge, reflects the degree to which the individual has developed unique qualities of thinking through incidental learning. Crystallized intelligence, measured by tests that maximize the role of cultural knowledge, reflects the degree to which the individual has been acculturated through intentional learning. Research shows that fluid abilities tend to decrease, while crystallized abilities tend to increase over the adult life span.

Piaget proposes that cognitive development progresses through a sequence of four stages. According to this view, the final stage—formal operational thought—is reached during adolescence. Adult intelligence, therefore, is characterized by stability. In spite of this, research has found evidence of age-related differences on a variety of Piagetian tasks. Older adults generally have been found to perform more poorly than young adults on tasks reflective of both concrete (conservation, classification) and formal operational thought. Some researchers have suggested these results reflect structural regression resulting from neurological degeneration. Others have suggested that older adults' poorer performance may reflect situational factors such as motivation or lack of practice.

Research growing out of a contextual approach emphasizes the relative plasticity and cohort-specific nature of intellectual functioning. Using sequential data collection strategies, researchers have demonstrated that historical-evolutionary sources of variance contribute substantially to age differences in intellectual performance. For the most part cohort differences are larger than age changes prior to age 65. Much of young adulthood, middle age, and old age is characterized by stability or increases in intellectual performance. Decline, however, does occur late in life.

A variety of antecedents appear to influence intellectual development. Decline appears to be closely linked to pathology, particularly cardiovascular disease. Intellectual performance also shows a marked decline, labeled terminal drop, up to several years prior to death. The relative plasticity and importance of learning antecedents on intellectual development are demonstrated by research which has modified the intellectual and problem-solving performance of older adults. These studies have shown that various strategies such as feedback, modeling, and strategy instruction are effective in improving the performance of older adults. Finally, it is important to recognize that the socioenvironmental context may influence adults' cognitive functioning. It has been suggested that our culture tends to discourage the competence of older adults, thus leading to a self-fulfilling prophecy in which the older adult expects to and becomes less competent.

chapter 13

Adulthood and Aging: social and personality development

CHAPTER OVERVIEW

During adulthood, the social and personal world of the individual is broader and more complex than at any other point in the life cycle. People enter the major roles of life—worker, spouse, parent. In this chapter, we will examine the social processes which shape adult life: cohort succession, role allocation, role change, and role satisfaction. We will also examine the extent to which this changing social context affects the individual's personality. In particular, we will examine the issue of whether personality is shaped by events early in life and remains stable in adulthood or whether it undergoes major changes as the individual encounters new events and situations in adulthood.

CHAPTER OUTLINE

SOCIAL PROCESSES IN ADULTHOOD AND AGING

Age and Cohorts
Age and Roles
Role Satisfaction

PERSONALITY DEVELOPMENT IN ADULTHOOD AND AGING

Types of Personality Change
The Case for Discontinuity
The Case for Continuity and Stability
Life-Course Analysis

CHAPTER SUMMARY

ISSUES TO CONSIDER

How does the developmental process involve both aging and social change?
What is the process of cohort flow?
How many age grades are distinguished by our contemporary society?
How does age operate as a criterion for the allocation of roles?
How does age operate as a criterion for the evaluation of roles?
How do status and roles change over the life span?
How do the various age groups differ in the resources important for social exchanges?
How does social change affect the process of role allocation and evaluation?
What factors affect marital satisfaction during adulthood?
What factors affect occupational satisfaction during adulthood?
What factors affect life satisfaction during adulthood?
What types of changes must be distinguished in order to understand personality development in adulthood?
How does Erikson characterize personal development during adulthood?
How does Levinson characterize personal development during adulthood?
What is the evidence for personality change versus lack of change in adulthood?

SOCIAL PROCESSES IN ADULTHOOD AND AGING

As people grow older, they change. But development is not simply an individual process. It involves a dynamic interaction between individuals and society (Riley, 1976). As particular individuals are changing over the life cycle from birth to death, a related process is taking place: New cohorts of individuals are being born, replacing those who are dying off. Because they are born at a particular time in history and experience a unique sequence of events, individuals of different cohorts develop in different ways. Finally, as different cohorts experience these unique sequences of events and respond to them in unique ways, society itself (e.g., its institutions, roles, values) is changed. In turn, these changes provide a new context for the development of individuals and successive cohorts. Thus development involves the two interdependent processes of aging and social change. Let us examine the role of these processes in adult development.

Age and Cohorts

As we have noted previously, a cohort refers to a group of individuals who experience the same event within the same time interval (Ryder, 1965); one could talk of the 1942 birth cohort, the Great Depression cohort, the Vietnamese war cohort, or the women's liberation cohort. Generally, cohort effects are seen as having a historical base. That is, it is meaningful to talk of the women's liberation cohort but not of the married women's cohort. The most specific definition of cohort is birth cohort—the aggregate of persons born in a given year or period and who age together (Ryder, 1965).

Cohort Flow. The process of *cohort flow* reflects what happens to different cohorts over their life courses: their formation, modification through migration, and reduction and eventual dissolution through the death of their members (Riley, 1976). A cohort starts out with a given size which, except for additions from immigration, is the maximum size it can ever attain. Over the life course of the cohort, individual members die until the entire cohort is destroyed. A cohort also starts out with a given membership which has certain characteristics. Some of these are quite stable (e.g., sex, race), while others are not (e.g., socioeconomic status, intelligence). However, the composition of the cohort changes over time even with respect to the stable characteristics. For example, women outlive men and whites outlive blacks.

Age Strata. The succession of cohorts results in the formation of a series of *age strata* composed of persons of similar age. Within the population the size of the various age strata varies over time. Exhibit 13.1 illustrates the changes in the demographic profile of the United States from 1880 to 1970. In particular, the period has seen an increase in the size of older age strata and a decrease in the size of younger age strata. This trend is likely to continue as, for example, those individuals currently 20 to 35 years of age reach late life.

Age and Roles

What does it mean for the individual to be located in one age stratum rather than another? The age structure is linked to social processes because society uses age criteria for allocating roles and for evaluating performance in these roles (Eisenstadt, 1956).

A cohort refers to a group of people who experience the same event in a particular historical period. (United Press International, Inc.)

Age Grades. At a general level, this process is seen in the fact that all societies divide their members into groups according to age (Eisenstadt, 1956). For the most part, three age grades are distinguished: children, adults, and the aged. However, in our contemporary society, finer distinctions have been made; for example, adolescence has definitely been established as a stage of the life cycle (Muuss, 1975a), and youth (Keniston, 1970), middle age (Levinson, 1977), and multiple periods in old age (Neugarten, 1974) appear to be emerging.

In nonindustrial societies, the passage to adulthood is clearly marked by physical maturity. At this point, *rites de passage* ranging from physical changes (e.g., bodily mutilation) to social ceremonies (e.g., religious blessing) are performed. This results in a child or group of children emerging with adult status (Muuss, 1975a). In western societies, such rites have become rituals which no longer signify true adult status (e.g., bar mitzvahs). Generally, as a culture increases in technical and symbolic complexity, the period of time required to become an adult is prolonged. This

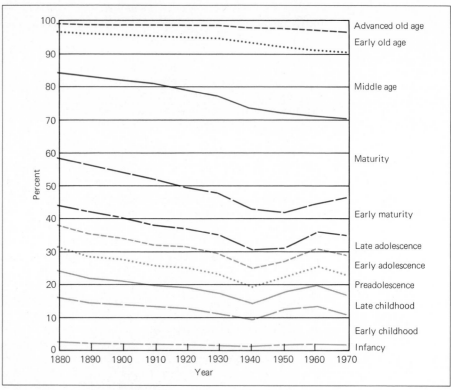

EXHIBIT 13.1   Distribution of U.S. population by age strata, 1880 to 1970.
(Source: Riley, Johnson, and Foner, 1972.)

lengthening period of preparation led to the emergence of adolescence as a stage of
the life cycle during the latter part of the nineteenth century (Demos & Demos,
1969). As a larger and larger proportion of adolescents have entered higher education following graduation from high school, Keniston (1970) identified another stage
labeled youth.

Within this context, the definition of an adult has become increasingly vague. In
some instances, the definition is specifically tied to age. However, even these criteria are imprecise. For example, in Pennsylvania an individual may vote, serve in
the armed forces, or marry without parental consent at age 18, but may not drink
alcoholic beverages until age 21. Our society, then, does not provide systematic
guidelines for adult status. In general, however, the beginning of adulthood is most
often defined as the point at which a person leaves school and becomes economically independent, usually in connection with full-time employment and/or marriage.

Adulthood is not a single phase. Middle age, for example, refers to a period
characterized by physical vigor, continued involvement in career, and diminished
family responsibilities. The recent emergence of this stage is the result not only of
increasing longevity but also of changes in the family life cycle. In particular, as
fewer children are born earlier in the lives of the parents, the length of time after the
children leave home and before retirement has increased significantly.

Old age is often defined chronologically, and age 65 is typically used as a marker. However, such criteria appear to be undergoing gradual change (e.g., the elimination of age 65 as a criterion for mandatory retirement). Distinctions other than chronological age may be more useful. For example, Neugarten (1974) has drawn a distinction between the young-old and the old-old. The former, drawn mainly from those aged 55 to 75, are characterized as relatively healthy, economically comfortable, and free from the traditional responsibilities of work and parenthood. In contrast, the old-old, typically over age 75, are more likely to suffer from physical disability and economic deprivation.

Age Structure of Roles and Age Norms. Within this general context of age grades, age is used as a criterion for entry into and exit from certain roles within the educational, economic, political, religious, and familial institutions of society. Age can operate directly for role entry and exit (e.g., one cannot become President unless one is at least 35 years of age; an individual of age 18 is no longer considered a juvenile by the justice system). Age can also operate as an indirect criterion because of its relationship to other factors (e.g., the necessity of completing high school tends to limit age of entry into college; biological changes tend to place an upper limit on the age of childbearing). Finally, age operates even more indirectly as social norms become established for major events and for behavior and as individuals are socialized to these norms (e.g., although individuals may marry for the first time at any point in adulthood, the normative expectation is for the first marriage to occur in young adulthood; individuals are expected to "act their age").

People appear to be aware of these *age norms* for role entry and exit and behavior. Neugarten and her colleagues, for instance, found adults possessed a high degree of consensus about the timing of major role transitions (Neugarten, Moore, & Lowe, 1965; Neugarten & Peterson, 1957). For example, most middle-class men and women agreed that the best age for a man to marry was from 20 to 25; that the best age for most people to finish school and go to work was from 20 to 22; that most men should be settled in a career by 24 to 26; that most men should hold their top jobs by 45 to 50; and that most men should be ready to retire by 60 to 65.

It appears that there is a normative *social clock* of the life cycle by which people anticipate the timing of major events. Such norms tell people whether their behavior is age-appropriate—*"on time"* or *"off time."* Indeed, Neugarten found that individuals readily report how their lives conform to this social clock (e.g., "I married early" or "I got a late start because of the Depression"; Neugarten & Hagestad, 1976, p. 44). These norms do show variations. For example, age-related expectations seem clearer for young adulthood than for middle age or old age. Similarly, respondents of higher socioeconomic status generally expect these transitions at later ages than respondents of lower socioeconomic status.

Neugarten, Moore, and Lowe (1965) also confirmed the common observation that there are age-related norms for specific behaviors (e.g., the appropriateness of a woman wearing a two-piece bathing suit at age 18, 30, or 45) They presented a questionnaire designed to measure these norms to 400 younger, middle-aged, and older middle-class adults. It was found that all age groups acknowledged the existence of age norms both in their own minds and in those of "most people." In general, the respondents saw themselves as less age-constrained than others. Similarly, younger adults were less age-constrained than middle-aged and older adults. That is, they were less likely to view age as an appropriate criterion for judging behavior. Neugarten has suggested that these results reflect a general weakening of age-related norms and sanctions in recent years (Hall, 1980).

Roles and the Life Cycle. What happens to people's roles in adulthood? How does status change with certain roles? According to Rosow (1976), "status is treated as a position in a social structure and role [is] the pattern of activity intrinsic to that position" (p. 458). From this conceptual framework, Rosow (1976) describes a typology of four role categories—Institutional, Tenuous, Informal, and Nonrole—in which status and role can vary independently. These categories may be described as follows:

Institutional. The Institutional role type refers to statuses with roles. They are roles in which normative expectations are associated with definite positions or attributes; e.g., men, women, professionals, manual workers, parents, children, Catholics, Jews, public officials, and so forth.

Tenuous. The Tenuous role type refers to statuses without roles. There are two sub-types—*titular* and *amorphous*. Titular positions include *honorific* (e.g., Nobel laureate) and *nominal* (e.g., token promotion) types. The former constitute "social promotions" in which the honor is symbolic, and no specific role activities beyond the most token are associated with the position. The latter constitute "social demotions" in which role functions are significantly limited. Amorphous positions include *de facto* types, in which objective circumstances prevent the individual from performing a role (e.g., chronically unemployed), and *role attrition* types, in which role responsibilities dwindle away or are lost (e.g., the elderly who have experienced the reduction or loss of institutional positions in the family, labor force, and so on).

Informal: The Informal role type refers to roles without statuses. These encompass a wide range of roles such as heroes, villains, playboys, heavies, blackmailers, prima donnas, gossips, and so on.

Nonrole: The Nonrole type refers to the absence of roles and statuses. This type simply permits the classification of idiosyncratic behavior and, for all practical purposes, is unimportant.

Rosow (1976) contends that the three role types (Institutional, Tenuous, Informal) exhibit different patterns of change over the life span as shown in Exhibit 13.2. Institutional roles increase through midlife, then decrease sharply during old age. Tenuous roles are relatively high during childhood, low during adulthood, and reach their highest point during old age. Finally, informal roles increase until adulthood, then level off. The relative imbalance between Institutional roles and Tenuous roles in old age (shaded portion of Exhibit 13.2) suggests that this portion of the life span becomes increasingly "nonnormative." According to Rosow (1976), this has several consequences:

> *First, the loss of roles excludes the aged from significant social participation and devalues them. It deprives them of vital functions that underlie their sense of worth, their self-conceptions and self-esteem. . . . Second, old age is the first stage of life with systematic status loss for an entire cohort. . . . Third, persons in our society are not socialized to the fate of aging. . . . Fourth, because society does not specify an aged role, the lives of the elderly are socially unstructured. . . . Finally, role loss deprives people of their social identity.* (pp. 466–467)

Rosow's taxonomy, therefore, provides a broad perspective within which the timing and ordering of various socializing events during adulthood may be considered.

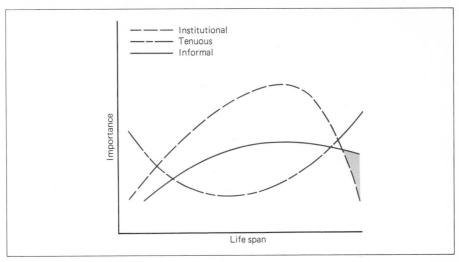

EXHIBIT 13.2 Relative importance of role types over the life span. (Source: Rosow, 1976.)

Age Status and Exchange. We have seen that society categorizes its members according to age and uses this categorization as a basis for allocating roles. As a result, the pattern of the individual's role involvement changes over the life cycle. However, society also differentially rewards various roles. Thus there are age-related inequalities in power, prestige, and resources within society, and in this sense, age operates as a basis of social stratification much like social class, race, and sex (Foner, 1975).

BOX 13.1

AGEISM AND SOCIETAL FACTORS

Several studies have described various forms of discriminatory inequality between age groups in our youth-oriented culture (Palmore & Whittington, 1970). In 1969, Butler called this inequality "ageism" and, at the White House Conference on Aging in 1971, ageism was condemned.

Many people, however, maintain that older adults still are not given meaningful societal roles and are not considered valued members of society (Butler, 1974; Kalish, 1975). Rosow (1976) contends that we are witnessing an increasing struggle for age rights as a result of rising economic costs, decreasing energy resources, and an increase in the numbers of adults, primarily the aged. Even the elderly, who view themselves slightly more positively than other age groups, feel being old is less desirable than being young (Colette-Pratt, 1976). A negative attitude toward the elderly is apparent by the time people reach adolescence (Hickey & Kalish, 1968).

Other investigators, however, paint a more positive picture of the situation of the elderly. Neugarten and Hagestad (1976) feel that increased consciousness about the elderly's plight is leading to a decrease in ageism. In fact, economic preferences are being conferred on the elderly, which is viewed as a status improvement. They cautiously admit, though, that such a trend could be reversed with continued inflation and economic stagnation.

Dowd (1975, 1980) has examined this issue within the framework of *exchange theory*. According to exchange theory, individuals enter into and maintain social interactions because they find them rewarding, and in the process of seeking such rewards, they incur costs. Costs refer either to unpleasant experiences or to the necessity of abandoning other rewarding activities in order to pursue the current activity. As in other economic exchanges, the difference between *rewards* and *costs* equals *profit* (or loss). The basic assumption underlying exchange theory is that social interactions between individuals may be characterized as an attempt to maximize profit (i.e., to increase reward and reduce cost).

Frequently, exchanges are not balanced. In this case, one person in the exchange has more *power* than the other. Specifically, the person who is less dependent on the exchange for the gratification he or she seeks has more power (Dowd, 1975). Ultimately, power depends greatly on *resources*, which are essentially anything an exchange partner considers rewarding (e.g., money, knowledge, social position). Such resources render the exchange partner susceptible to social influence. In general, the more resources a person possesses, the more power he or she is likely to have in social exchanges.

As shown in Exhibit 13.3, Dowd (1975) proposes a curvilinear relationship between age and degree of power resources. Power resources tend to be limited in young adulthood, reach their maximum in midlife, and decline sharply in old age. Cross-cultural research tends to support this hypothesis. For example, in an analysis of ethnographic data from forty-seven societies, Abarbanel (reported in Dowd, 1975) found control over most resources began immediately before or after marriage, peaked when the children were adolescents, and began to decline when the children married and left home. However, peak control was not reached until the late-family phase in preindustrial, nonurban societies. This latter finding is consistent with other research which reports an inverse relationship between the status of the aged and societal modernization (Bengtson, Dowd, & Smith, 1975; Cowgill & Holmes, 1972). The relationship between age and power resources is also modified by socioeconomic status (see Exhibit 13.3) and race (see Box 13.2).

From an exchange theory perspective, the problem of aging is essentially a problem of decreasing power resources (Dowd, 1975). The aged have little of instrumental value to exchange—their skills are obsolete or can be provided more effi-

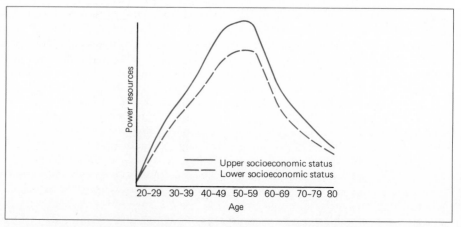

EXHIBIT 13.3   Hypothesized relationship between control of power resources and age within categories of socioeconomic status. (Source: Dowd, 1975.)

BOX 13.2

THE DOUBLE JEOPARDY HYPOTHESIS

What does it mean to be an elderly member of a racial minority in contemporary society? One perspective suggests that the minority aged suffer from a situation of "double jeopardy." That is, the impact of race and age discrimination combines to make the relative status of such groups weaker than that of other groups whose members are of like age but are not minorities. Another perspective suggests that advancing age acts to "level" racial inequalities that are present during middle life.

Dowd and Bengtson (1978) tested these two hypotheses by examining the variables of health, income, life satisfaction, and social participation (reported interaction with family, kin, neighbors, and friends) in a sample of over 1200 middle-aged and older blacks, Mexican-Americans, and whites in Los Angeles County. Differences between the three groups were particularly apparent in the areas of income and self-assessed health and offered support for the double jeopardy hypothesis. However, life satisfaction and social participation declined among all groups, supporting the "age-as-leveler" hypothesis. In some areas, then, minority aged do appear to be exposed to double jeopardy, while in other areas they do not.

ciently by others. As a result, Dowd argues, they must exchange generalized power resources which are universally experienced as rewarding: money, approval, respect, and compliance. Of these, approval and respect are less costly. Ultimately, however, the older person is often reduced to offering compliance as a medium of exchange.

The relationship between age and power resources proposed by Dowd is specific to the current historical and societal context. Several trends, if continued, could result in a less drastic decline in power resources in late life. For example, the higher levels of education of future cohorts may constitute a significant power resource. In other words, few of the differences in power resources between age groups are a direct result of developmental processes. Most can be traced to social norms, social expectations, and historical opportunity.

In this context, Kuypers and Bengtson (1973) have argued that at least part of the behavioral incompetence of older adults is the result of a self-fulfilling prophecy. They argue that the social reorganization which occurs in later life creates a negative cycle of events which leads, ultimately, to both incompetent behavior and negative self-identity on the part of the elderly. The events in the cycle, labeled the *social breakdown syndrome*, include:

1 A precondition of susceptibility
2 Dependence on external labeling
3 Social labeling as incompetent
4 Induction into a sick, dependent role
5 Learning of "skills" appropriate to new role
6 Atrophy of previous skills
7 Identification and self-labeling as "sick" or incompetent

These steps are presented graphically in Exhibit 13.4.

Kuypers and Bengtson (1973) argue that the role-related changes which occur in late life render the individual particularly susceptible to this cycle of breakdown.

(a)

(b)

The aged may show an array of emotional reactions to their social status.
The social breakdown syndrome is not inevitable.
(*a.* Gerhard E. Gscheidle/Peter Arnold, Inc.
b. Bettye Lane/Photo Researchers, Inc.)

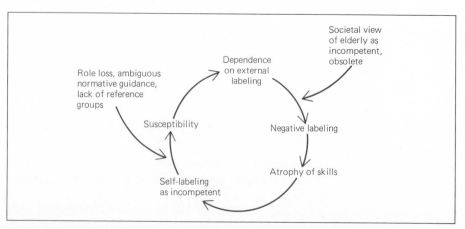

EXHIBIT 13.4   The social breakdown syndrome. (Source: Kuypers and Bengtson, 1973.)

Specifically, the familiar roles of adulthood are lost and not replaced by others of equal specificity. Related to this, the individual experiences a lack of age-specific norms to guide behavior. Finally, the individual is increasingly left without groups of defined others with whom to identify. These changes leave the older person vulnerable to and dependent on external sources of labeling. Within western societies, these sources tend to label the elderly adult as obsolete, incompetent, and worthless. Such negative labeling leads to social exchanges with negative characteristics. The individual is induced into a "sick role." Behaviors appropriate to the new role are learned and there is an atrophy of previous skills which are inconsistent with the new role. Finally, the individual identifies and labels himself or herself as incompetent. Kuypers and Bengtson suggest that this cycle of breakdown could be minimized by several interventions. In particular, they suggest that efforts should be made to liberate the person from the dominant social view that worth is contingent on performance of "productive" social roles.

Roles and Social Change. The age differentiation of people and roles constitutes two major structural components of the social system. However, it is important to recognize that these components interact with social change. In some respects, the age-related allocation of roles is uneventful. First-year students become sophomores, sophomores become juniors, and juniors become seniors; students become workers, workers become managers, and managers become retirees; men become husbands, husbands become fathers, and fathers become grandfathers. But sometimes there is an imbalance of people and roles within an age grade. Such imbalances typically occur when there is a discrepancy between the size or role preferences of a cohort's membership and the number and nature of the roles open in the role structure. Waring (1975) has labeled this situation *disordered cohort flow.*

Disordered cohort flow may be the result of several mechanisms. It may occur because of abrupt shifts in fertility, migration, or mortality. For example, baby booms, sudden influxes of individuals from other systems (refugees, immigrants), or sudden decreases in the mortality rate may create large cohorts which cannot be absorbed by the existing role system. Similarly, wars, epidemics, or emigrations may deplete a cohort, leaving it too small to fill the roles assigned to it.

The balance between people and roles may also be disturbed by economic, political, and social events which influence the availability of roles. For example, economic depressions may create a surplus of people even in small cohorts. In contrast, during periods of economic expansion or war, there may be too few people from all age grades to fill the role vacancies which occur.

Disordered cohort flow affects the behavior of both individuals and social institutions. Historically, one response has been large-scale transfers of people—emigration or immigration. For example, the massive immigration of individuals from Europe to the United States during the late nineteenth century was largely a function of economic conditions. Individuals who could not be absorbed by the stagnant economies of the old world found role vacancies in the rapidly expanding economy of this country. However, solutions based on massive emigration or immigration have become less feasible with increases in the population and decreases in the resources of the world.

A second response to disordered cohort flow involves the expansion or reduction of roles to conform to the size of the cohort. For example, increased numbers of school-age children after World War II resulted in an increased need for teachers. Similarly, as this cohort ages more and more, adult service providers will be required. In contrast, the current decrease in the number of school-age children is now resulting in decreased demand for teachers.

Disordered cohort flow affects the availability of various roles which, in turn, affects the composition of future cohorts. For example, Easterlin (reported in Collins, 1979) notes that in large cohorts, competition for jobs is intense and the resultant subjective feeling that "times are tough" curtails the process of family formation. In smaller cohorts, however, fewer people compete for jobs, and people feel optimistic about their opportunities. This stimulates the family formation process, producing an increase in the birthrate for that cohort. Easterlin theorizes that a self-generating mechanism produces cyclic swings in birthrate pattern and hence cohort structure. Examination of the birthrates twenty to twenty-five years ago suggests that during the 1980s, we will witness a growing scarcity of younger adults. Easterlin speculates that this will result in an increase in the birthrate and a decrease in the divorce and suicide rates.

Role Satisfaction

Once adulthood is achieved, individuals play out their lives striving for satisfaction in the roles they occupy. What factors influence role satisfaction? We will examine this issue within the context of three psychosocial dimensions: marital satisfaction, occupational satisfaction, and life satisfaction.

Marital Satisfaction. How does the quality of the marital relationship change over the family cycle? Duvall (1971) has divided the family life cycle into eight stages as follows:

Stage 1. Beginning families—married less than five years, no children
Stage 2. Childbearing families—oldest child, birth to 2 years 11 months
Stage 3. Families with preschoolers—oldest child, 3 years to 5 years 11 months
Stage 4. Families with school-age children—oldest child, 6 years to 12 years 11 months
Stage 5. Families with adolescents—oldest child, 13 years to 20 years 11 months
Stage 6. Launching families—first child gone to last child's leaving home
Stage 7. Families in the middle years—empty nest to retirement
Stage 8. Aging families—retirement of first spouse to death of first spouse

These stages may be combined to define four major periods of the marital relationship: the newlywed marriage (stage 1), the parental marriage (stages 2 to 6), the middle-age marriage (stage 7), and the aging marriage (stage 8). Each stage is accompanied by different tasks and goals.

BOX 13.3

ALTERNATIVE LIFESTYLES

Singleness
In the United States between 1970 and 1975, the percentage of 20- to 24-year-old unmarried females increased from 28 to 40 percent. Although

Glick (1977) feels that part of this increase reflects the postponement of marriage, an estimated 6 to 7 percent of these women will never marry.

Why would people want to remain single?

Several factors—including changing sex-role expectations, particularly for women, decreased cultural emphasis on childbearing, and the availability of special facilities and services (e.g., housing, newspapers and magazines, entertainment spots)—have probably contributed to the growth of this lifestyle. Evidence suggests that single women of all ages are happier single than single men (Campbell, 1975) and have lower rates of depression and neurosis than married women (Bernard, 1972). Moreover, Gubrium (1975) found single elders to constitute a distinct type, one that is different from other old people. In a study based upon interview data, he reported that the social world of single elders is relatively isolated but not perceived as lonely. In fact, they viewed their state of singleness as just an extension of their past.

Cohabitation: The Arrangement

The college years seem to be the time when young cohorts experiment with cohabitation. In a typical arrangement a couple shares expenses and living arrangements and engages in sexual activity. According to Skolnick (1973), the tension between commitment and freedom helps explain the advantages and disadvantages of such an alternative lifestyle. Most couples cohabit because they feel this arrangement will lead to greater emotional maturity and help them acquire ways to live with and relate to others (Stinnett & Walters, 1977). In a recent study (Stafford, Backman, & Dibona, 1977), married and cohabiting couples were compared. Data indicated that the same conflicts over the division of labor and role behavior exist for both couples, but cohabiting men do help with the laundry and dishes more often than married men and cohabiting women do more home repair and gardening than married women.

Dual-Career Couples

The advantages of a dual-career marriage are financial and personal. Although these marriages can be characterized by pride, companionship, role sharing, and feelings of self-expression and growth for women, they also have unique strains. As the wife assumes a full-time career and the hus-

band develops relationships with the children, time limitations and an increase in responsibilities may produce "role overload." There appears to be an optimal way to balance demands at home with the demands of careers, but there is often a problem in reaching this balance. The difficulty seems to lie with timing, or "role cycling." When one partner must spend more time in career behavior, the other must spend more time in home activities. However, at some point someone's "give" needs to become "take" in order to minimize stress and feelings of being taken advantage of (Rapoport & Rapoport, 1971). With changes that affect work patterns such as reducing the hours and/or days of a workweek, better child-care services, and efficient transportation, the dual-career pattern of marriage should increase. Moreover, with increasing inflation and financial stress, this alternative lifestyle may become a necessity.

Communes

In 1969, there were only 500 communes in the United States, but from 1975 to 1977, there were approximately 3000. Membership usually consists of middle- or upper-class adults between the ages of 20 and 28 (Stinnett & Walters, 1977). A shared ideology is present, although communes exhibit diversity in structure and functioning. Members want to create the intimacy of an extended family while striving for spiritual rebirth, personal growth, a "natural life," and greater freedom (Zablocki, 1971).

Homosexual Marriage

Since the obligations and responsibilities of partners in a homosexual marriage are neither institutionalized nor legally defined, these relationships are often unstable. Many homosexuals, however, do wish to develop marriages, whether or not legally sanctioned, and do not wish to change their sexual orientation. With psychologists and psychiatrists no longer regarding homosexuality as a mental illness, homosexuality became much more visible in the 1970s than it was in the 1950s. Hunt (1974), however, demonstrated that homosexual behavior has not increased since the time of the Kinsey report.

What happens to marital satisfaction as a family moves through this cycle? Most researchers agree that there is an initial decline in marital satisfaction during the early years of marriage, particularly after the birth of the first child (Blood & Wolfe, 1960; Pineo, 1961; Luckey, 1966). However, there is less agreement about what happens in the middle and later stages of the cycle. Some studies suggest a continued decline (Blood & Wolfe, 1960; Pineo, 1961). In contrast, other studies have pointed to a curvilinear relationship with continued decline during the parental stages followed by increases during the latter stages (Rollins & Cannon, 1974; Rollins & Feldman, 1970).

In a large-scale study designed to examine this issue, Spanier, Lewis, and Coles (1975) measured the marital satisfaction of 1584 men and women from three different areas of the country—Georgia, Ohio, and Iowa.

As shown in Exhibit 13.5, the Ohio sample displayed significant curvilinearity, the Georgia sample displayed a tendency toward curvilinearity, but the Iowa sample displayed a tendency toward stability. Support for the curvilinearity hypothesis, then, was suggestive but not conclusive.

There are several plausible explanations for such a curvilinear trend in marital satisfaction. First, the launching of the last child and the subsequent empty nest may not be a negative experience, particularly for women. For example, Livson (1974) contrasted the experience of two types of middle-aged, middle-class married women—"traditionals" and "independents." On the one hand, the traditional women found continued satisfaction in interpersonal satisfaction after the children left home. In particular, when leisure time was shared with their husbands, marital satisfaction increased. On the other hand, the independent women found new ways to express ambitions and desires which had been limited because of childrearing responsibilities. Thus for middle-class women the departure of the children from the home may provide new opportunities for personal growth. For men, the parental years may be particularly difficult because of role strain—conflicting demands between the roles of worker, spouse, and parent. In this context, Nydegger (1973) found that late fathers (over age 40) were more comfortable than early or on-time fathers. Men who became fathers early or on time felt the demands of their careers interfered with their performance as fathers.

Second, there is evidence which shows that women increasingly display assertive and instrumental behaviors in later life, while men increasingly display emotional and affiliative behaviors during this period (Neugarten & Guttman, 1958). This merging of interpersonal styles may be particularly important for marital satisfaction in middle and old age. For example, Stinnett, Carter, and Montgomery (1972) found that in stages 7 and 8, the affective aspects of the marital relationship are perceived as most rewarding. The most often cited areas were companionship and expression of "true" feelings. Consistent with the curvilinear hypothesis, more than 90 percent of these couples described their relationship as "very happy" or "happy" and half stated their marriage had improved with time. Exhibit 13.6 indicates the most rewarding and troublesome aspects of marriage during the later years.

Thus there appears to be some support for the hypothesis that marital satisfaction decreases from a high point during the newlywed period to a low point during the parental period and then increases during the middle-age and aging periods. This conclusion must be accepted cautiously, however. Almost all the studies examining this issue have used the cross-sectional data collection strategy. As we have noted, this strategy confounds age (family stage) differences with cohort differences. To the extent that different marital cohorts vary in marital satisfaction, the curvilinear pattern observed may not reflect developmental processes.

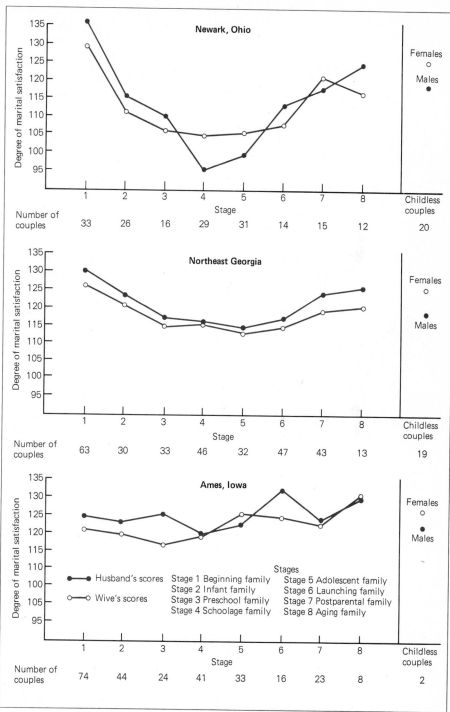

EXHIBIT 13.5   Spouses' marital satisfaction at eight stages of the family life cycle in three samples. (Source: Spanier, Lewis, and Coles, 1975.)

(a)

(b)

Marital satisfaction may change in relation to the stage of the family life cycle.
(*a*. Charles Gatewood; *b*. Alice Kandell/Photo Researchers, Inc.)

Occupational Satisfaction. The occupational role provides an identity for an individual that is both personal and social and occupies much of the adult's time and energy. What promotes occupational satisfaction? In general, occupational satisfaction appears to result when there is a good fit between the individual's abilities and interests and the characteristics and demands of the job (Healy, 1973; Quinn, Staines, & McCullough, 1974). These factors can include (1) familiarity with the occupation as a result of earlier experience (e.g., Rauner, 1962); (2) similarity between person-environment characteristics (e.g., Holland, 1973); (3) good occupational role models (e.g., Bell, 1970); (4) challenging, but not threatening, job requirements and expectations; (5) reduced concern with prestige; (6) match between personal values and expressed work values; and (7) the socialization context of the work environment (e.g., Henry, Sims, & Spray, 1971). Thus people work for different reasons, such as money, status, prestige, service, companionship, satisfaction, etc. (Terkel, 1972), and occupational satisfaction depends upon the fit between a person's reasons for working and the characteristics of the work situation. An individual's perception of work and associated reasons for working are determined largely by previous work experiences and occupational aspirations.

A complicated matching process between people and occupations goes on all the time. Employers want the people who can best fill the positions, and people want the positions that are best for them. Thus despite job discrimination related to

EXHIBIT 13.6   **Perceptions of Older Husbands and Wives Concerning the Most Troublesome and Most Rewarding Aspects of Marriage Relationships During the Later Years**

Aspect	Percentage
Rewarding aspects	
Companionship	18.4
Mutual expression of true feelings	17.8
Economic security	16.2
Being needed by mate	12.0
Affectionate relationship with mate	11.2
Sharing of common interests	9.3
Having physical needs cared for	7.6
Standing in the community	7.0
Troublesome aspects	
Different values and life philosophies	13.8
Lack of mutual interests	12.5
Mutual inability to express true feelings	8.6
Unsatisfactory affectional relationships	8.5
Frequent disagreements	8.5
Lack of companionship	7.7
Other	8.5
Nothing troublesome	36.2

Source: Stinnett, Carter, and Montgomery, 1972.

age, sex, race, and so forth, people who have aspiration, preparation, and ability tend to occupy jobs that produce the highest levels of satisfaction (Terkel, 1972).

With the focus of occupational satisfaction on "matching" person and position, two different types of work groups emerge. The first group of people derive little satisfaction from the work process itself, either in terms of accomplishments or feelings of self-fulfillment, but gain satisfaction from direct status benefits related to salary. These people feel work is a necessary evil one must suffer in order to do what one wants. The end of a workday marks the end of any work-related concerns. Satisfaction with activities (e.g., family and friends) outside of work is quite high; it is related to occupational satisfaction in that work provides the means to engage in those activities. The second group of people have a strong work orientation. For these people a major source of satisfaction comes from the process of work and the related accomplishments. Through work they feel useful, enjoy opportunities for self-expression, enjoy the companionship of coworkers, and derive a sense of status and self-esteem from their accomplishments on the job. Moreover, work concerns are continued during evenings and on weekends. Thus the satisfaction from leisure activities is relatively low because the satisfaction from occupation is so high (e.g., Friedmann & Havighurst, 1954; Havighurst, 1957; Meltzer, 1965).

Both orientations—income oriented and work oriented—exist across all fields of work, although there is a higher proportion of work-oriented white-collar individuals. Blue-collar workers tend to be more dissatisfied with their jobs than white-collar workers. There are many explanations for this. Blue-collar workers often have fewer opportunities for interesting and varied tasks, and their working conditions are often poor (Herzberg, 1966). The group of people who feel the most dissatisfied with work are those who perceive they are being discriminated against (Blood & Hulin, 1967). For instance, many women report feelings of dissatisfaction related to discrimination. Related to this is the issue of whether or not opportunities have been provided at all. In fact, a lack of opportunity is related more to overall rate of unemployment than either personal dissatisfaction or employer dissatisfaction. Moreover, this lack of work opportunity interacts with a history of a lack of opportunities, for certain groups in particular, producing a situation of extreme concern. For instance, among high school dropouts, unemployment rates ranged from less than 5 percent for whites to greater than 20 percent for blacks during the late 1960s. Unemployment is highest for black men and women and lowest for white men. In contrast, unemployment rates during the same period for college graduates were between 1 and 4 percent. Despite fluctuation in rates, unemployment is higher for minority groups than for whites, for women than for men, and for younger age groups than for older (U.S. Department of Health, Education and Welfare, 1976).

Life Satisfaction. Historically, measures of life satisfaction have been viewed as general indicators of "successful aging." In turn, this issue was introduced by Cumming and Henry (1961) in relationship to their disengagement theory of aging. *Disengagement theory* postulates that both the individual and society prepare in advance for the disability and death of the individual through a mutual process of psychological and social disengagement. This process is seen as inevitable, gradual, and mutually satisfying. Thus disengagement theory does not simply propose that people fill fewer roles and become less active as they get older. Rather, it proposes that the decreased engagement of the older adult in society is mutually sought by the individual and society and is functionally advantageous to both.

Disengagement theory has been criticized by many researchers (Maddox, 1964; Rose, 1964). The counterposition has been labeled *activity theory.* This view does not deny the observation that later life is characterized by decreased social interaction. What it does deny is the assertion that this reduction in social interaction is voluntary or that it is functionally advantageous for the individual. According to activity theory, disengagement is imposed on the individual by the structural requirements of society. Further, according to activity theory, successful aging is reflected in the extent to which the individual is able to avoid this external pressure toward disengagement and maintain his or her involvement in the social context.

There is considerable evidence which suggests psychological and social disengagement increases gradually with age beginning in midlife. Havighurst, Neugarten, and Tobin (1968), for example, reported decreases in ego energy, role activity, and role investment for adults aged 50 to 70. In general, psychological disengagement appears in the fifties foreshadowing the social disengagement that appears in the sixties and seventies. These investigators also found that psychological and social engagement was associated with higher levels of life satisfaction. However, multiple patterns of activity may be associated with high life satisfaction. For example, as shown in Exhibit 13.7, Neugarten, Havighurst, and Tobin (1968) found that both disengagement and activity were associated with high life satisfaction for individuals with integrated personalities. More recent studies have found a number of variables to be associated with high life satisfaction, including health, income, and social activity (Flanagan, 1980; Markides & Martin, 1979; Palmore & Kivett, 1977).

Flanagan (1980), for instance, surveyed 2600 individuals in an effort to identify those factors which contribute to a person's quality of life. Nationally representative samples of 30-, 50-, and 70-year-old adults participated in the study. These individuals were asked to indicate how important fifteen dimensions were to them at the present time. Exhibit 13.8 shows the percentage reporting that the component is important to their quality of life (important or very important). According to these adults, health and personal safety, work, close relationship with a spouse, having and rais-

EXHIBIT 13.7   **Personality Type in Relation to Activity and Life Satisfaction**

Personality Type	Role Activity Type	Life Satisfaction
Integrated	Reorganized: Competent, engaged, and involved; substitutes new activities for old	High
	Focused: Integrated personality, medium levels of activity, centered in one or two role areas	High
	Disengaged: Low levels of activity and role involvement, voluntarily reduces role commitment, high self-esteem	High
Armored-defended	Holding on: Holds on to midlife roles and activities; when successful, maintains adequate levels of life satisfaction	High
	Constricted: Low to medium involvement in few areas; constricted role activity; preoccupied with losses and deficits	High to medium
Passive-dependent	Succor seeking: Medium activity level; maintains adequate levels of life satisfaction when strong dependency needs fulfilled	Medium
	Apathetic: Low role activity; passive and apathetic	Medium to low
Unintegrated	Disorganized: Deteriorated cognitive processes; poor emotional control	Low

Source: Based on Neugarten, Havighurst, and Tobin, 1968, Table 1, p. 174.

ing children and understanding yourself were perceived as the most important components contributing to the quality of life. In contrast, socializing, expressing yourself creatively, active recreation, passive recreation, and participating in local and national government were perceived as least important. In general, the importance of material comforts and close friends tended to increase with age, while the importance of work, learning, having and raising children, and close relationships with a spouse tended to decrease with age.

How well do these adults report that their needs have been met? Exhibit 13.9 shows the percentage reporting that their needs have been poorly met (moderately, slightly, or not at all). In spite of the fact that a substantial minority of individuals are dissatisfied, most people believe their needs are being met to a significant degree. The most problematic areas are learning, expressing yourself creatively, active recreation, and participating in local and national government—components which were perceived as among the least important for quality of life. Further, there does not appear to be a dramatic increase with age in the degree to which needs are not met.

These observations are confirmed by estimates of overall quality of life made by the 50- and 70-year-old groups. About 21 percent said it was excellent, 35 percent very good, 29 percent good, 13 percent fair, and only 3 percent poor. There were few differences in this rating according to age or sex.

Finally, Flanagan (1980) examined the extent to which the individuals' reports of how well their needs and wants were met predicted their overall life satisfaction. The most important contribution to overall life satisfaction was the extent to which the individual's needs and wants for material comforts were met. The second most

EXHIBIT 13.8   Percentages of a Sample of 1000 30-Year-Olds, 800 50-Year-Olds, and 800 70-Year-Olds Reporting Each of the 15 Quality-of-Life Components as Important or Very Important to Their Quality of Life°

	Male			Female		
Component	30 Years	50 Years	70 Years	30 Years	50 Years	70 Years
Physical and material well-being						
Material comforts	80	86	88	75	84	86
Health and personal safety	98	96	95	98	97	96
Relations with other people						
Relationships with relatives	68	64	62	83	76	79
Having and raising children	84	85	82	93	92	86
Close relationship with a spouse	90	90	85	94	82	43
Close friends	71	76	74	79	80	88
Social, community, and civic activities						
Helping and encouraging others	60	73	65	71	75	80
Participating in local and national government	47	64	65	42	60	57
Personal development and fulfillment						
Learning	87	69	52	81	68	59
Understanding yourself	84	85	81	92	91	87
Work	91	90	58	89	85	60
Expressing yourself creatively	48	40	37	53	54	60
Recreation						
Socializing	48	46	51	53	49	62
Passive recreation	56	45	53	53	56	64
Active recreation	59	52	47	50	52	51

* For all age groups the question read, "At this time in your life, how important to you is _____?"

Source: Flanagan, 1980.

important prediction was health, and work and active recreation were also important. The four poorest predictors were having and rearing children, participation in government and public affairs, helping and encouraging others, and relationships with relatives. It is interesting to note that those components which actually predicted life satisfaction were not necessarily those seen by the respondents as most important to them (e.g., material comforts, active recreation).

In general, other studies tend to be consistent with Flanagan's findings. For example, Palmore and Kivett (1977) examined changes in life satisfaction over a four-year period in 378 community residents aged 46 to 70. These investigators found few age- or sex-related changes in life satisfaction over the period. Further significant predictors of life satisfaction included self-rated health, sexual enjoyment, and social activity. Moreover, Palmore (1975) has shown that health and activity may be important determinants of life satisfaction even in Japan, a culture where older adults are more valued than in the United States.

The results we have reviewed in this section lend more support to activity theory than to disengagement theory. Nevertheless, neither of these "theories" is particularly useful for understanding social processes in adulthood. What is required is an examination of how adults adapt to various life events and role transitions which ultimately determine the life course and the individual's sense of satisfaction.

EXHIBIT 13.9 **Percentages of a Sample of 1000 30-Year-Olds, 800 50-Year-Olds, and 800 70-Year-Olds Reporting Their Needs as Moderately, Only Slightly, or Not at All Well Met for Each of the 15 Quality-of-Life Components°**

	Male			Female		
Component	30 Years	50 Years	70 Years	30 Years	50 Years	70 Years
Physical and material well-being						
Material comforts	26	27	24	24	32	26
Health and personal safety	14	17	15	14	18	20
Relations with other people						
Relationships with relatives	19	29	27	19	30	31
Having and raising children	20	14	18	17	14	16
Close relationship with a spouse	16	19	13	19	28	31
Close friends	19	19	20	18	21	22
Social, community, and civic activities						
Helping and encouraging others	39	28	28	38	26	26
Participating in local and national government	46	38	36	46	38	39
Personal development and fulfillment						
Learning	42	36	27	50	44	35
Understanding yourself	26	26	24	29	25	20
Work	21	26	24	21	32	24
Expressing yourself creatively	40	32	26	43	32	28
Recreation						
Socializing	27	28	27	26	31	27
Passive recreation	29	27	18	30	26	21
Active recreation	36	41	36	37	40	35

* For the 50- and 70-year-olds, the question read, "How well are your needs and wants being met in this regard?" For the 30-year-olds, the question read, "How satisfied are you with your status in this respect?"
Source: Flanagan, 1980.

PERSONALITY DEVELOPMENT IN ADULTHOOD AND AGING

How does the changing social context of adulthood influence the characteristics and behavioral tendencies, typically labeled personality, of the individual? The most widely held position is that personality patterns are established during childhood and adolescence and then remain stable during adulthood. Is this the case? Are our personalities in adulthood, shaped by events early in life, essentially constant? Or do our personalities undergo major changes as we encounter new events and situations in the process of growing older?

Types of Personality Change

In examining the issue of personality in adulthood it is important to distinguish between at least two change-related issues. On the one hand, the issue of *continuity* versus *discontinuity* refers to the degree of within-person change in personality dimensions over time. For example, if Jack is trusting at age 20 and trusting at age 40, we have an instance of continuity. If he is trusting at age 20 and suspicious at age 40, we have an instance of discontinuity. On the other hand, the issue of *stability* versus *instability* refers to the degree to which the individual retains the same relative position on a personality dimension over time. For example, if Rita is more self-assured than Pat at age 20 and also at age 40, we have an instance of stability. If Rita is more self-assured than Pat at age 20 but less self-assured than Pat at age 40, we have an instance of instability. Notice that it is possible to have continuity and stability (Jack is more trusting than José at age 20; they both retain the same level of trust over time; so at age 40, Jack is still more trusting than José), discontinuity and stability (Jack is more trusting than José at age 20; they both become less trusting over time, but at age 40, Jack is still more trusting than José), and discontinuity and instability (Jack is more trusting than José at age 20; Jack becomes less trusting over time and José becomes more trusting, so at age 40, José is more trusting than Jack).

The emphasis on these issues varies from one study to another. Thus whether one concludes that adult personality is characterized by change or lack of change depends, in part, on what type of change has been examined. Within this general context, some studies have supported the conclusion that discontinuity in adult personality is an intrinsic part of adult development. Other studies have supported the conclusion that change is limited and the adult personality is characterized by continuity and stability. Let us examine these two perspectives.

The Case for Discontinuity

The concept of discontinuity in personality is intrinsic to a number of theories of personality such as those of Bühler (1959), Erikson (1963), Havighurst (1953), Levinson (1978), and Gould (1978). These theorists support several general conclusions.

1 The interface between the individual and the larger social-cultural context is crucial to understanding personality. The demands of this context change during adulthood, thus imposing a changing set of developmental tasks on the individual which influence personality development.

2 Personality change may be understood as a sequence of stages, transitions, or transformations which reflect the emergence of qualitatively different characteristics at different points in the life span.

3 While individuals may differ in terms of specific behaviors and characteristics, the sequence of personality change defined by the various stages, transitions, or transformations is universal.

Let us examine some of these theories and related research stressing discontinuity in adult personality.

Erikson's Theory. According to Erikson (1959, 1963), personality is determined both by an inner-maturational "ground plan" and by the external demands of society. For Erikson, ego development involves a sequence of eight psychosocial stages. These stages are biologically based and constitute a fixed, universal sequence. Within each stage, however, a particular capability of the ego must be developed if individuals are to adapt to the demands placed on them by society at that point in the life span. If the capability is not developed within the allotted time, that aspect of the ego will be impaired. Each stage, then, constitutes a crisis—between attaining and sensing the attainment of the appropriate capability and not attaining and not sensing the development of the appropriate capability. As we have seen in earlier chapters, Erikson's first five stages concern childhood and adolescence. His last three stages, however, concern young adulthood, adulthood, and maturity.

In young adulthood, the crisis is between developing a sense of intimacy versus a sense of isolation. Hopefully, the person has achieved a sense of identity and now knows who he or she is. The society now requires the person to enter into an institution that will allow the society to continue to exist. Accordingly, the new family unit must be formed—for example, through marriage. The young adult must form a relationship with another person which will allow such an institution to prosper. Erikson argues that to enter into and successfully maintain such a relationship requires persons to give of themselves totally. Such openness is not limited to sexual relations. Rather, Erikson argues that all facets of one person (e.g., feelings, ideas, goals, attitudes, values) must be unconditionally available to the other person. Moreover, the person must be unconditionally receptive to these same things from the partner. To the extent that one can attain such an interchange, one will feel a sense of *intimacy*. If individuals cannot share and be shared, then they will feel a sense of *isolation*.

In adulthood the crisis is between developing a sense of generativity versus a sense of stagnation. In this stage, society requires the person to play the role of a productive, contributing member of society. *Generativity* requires that the individual contribute to the maintenance and perpetuation of society. One can be generative by creating products associated with the maintenance of society (e.g., goods and services), or by producing, rearing, and/or socializing children in order to perpetuate society. If the individual is unable to create products for the maintenance and perpetuation of society, then a sense of *stagnation* results.

In maturity, the crisis is between developing a sense of ego integrity versus a sense of despair. In this stage, individuals realize that they are reaching the end of life. If they have successfully progressed through the previous stages of development, then they will face old age with enthusiasm; they will feel that a full and complete life has been led. In Erikson's terms, the individual will achieve a sense of *integrity*. If not, he or she will feel a sense of *despair*. The person will feel that his or her life has been wasted.

Erikson proposes that the degree to which a sense of intimacy, generativity, and integrity is successfully achieved depends on whether a person has developed a strong sense of identity. With a strong sense of self, an individual is more likely to enter into adult roles in the family and community that reflect a mixture of adult freedom and responsibility. The adult life personality pattern therefore represents an identity that stabilizes during adolescence and young adulthood and interacts with how ego crises are experienced in later life.

There are some data to support these notions in young adulthood and, to a lesser extent, in adulthood. Constantinople (1969), for instance, developed a questionnaire designed to assess the individual's resolution of the first six crises in Erikson's theory. This measure was given to more than 900 first-year through senior college students in 1965 and to subgroups of this larger group in 1966 and 1967. Constantinople found that successful stage resolution was greater among seniors than among first-year students and among males than among females.

Following up on this work, Whitbourne and Waterman (1979) retested 147 of Constantinople's 1966 subjects in 1976 when they were an average of 30 years of age. They also gave the questionnaire to 224 men and women who were undergraduates in 1976. This design permitted a longitudinal comparison (college students in 1966 with these same individuals as adults in 1976), a cross-sectional comparison (college students in 1976 with alumni in 1976), and a time-lag comparison (college students in 1966 and college students in 1976). Taken together, the results of these analyses appear to support the hypothesis of personality development in the adult years. The longitudinal comparison showed increasingly successful resolution of crises over the ten-year period. In particular, increases were seen in industry versus inferiority, identity versus identity diffusion, and intimacy versus isolation. Resolution of the latter two crises would be predicted during late adolescence and early adulthood according to Erikson.

However, the time-lag analysis suggested that the sex differences reported by Constantinople (1969) may be partly due to cohort effects. That is, female undergraduates in 1976 scored substantially higher than female undergraduates in 1966. This difference may reflect changes in sex-role expectations for women or changes in recruiting practices at the university over the time period.

Other evidence relevant to Erikson's theory comes from data collected as part of the Grant Study (Vaillant, 1977). In this longitudinal study, 268 men were selected from the 1939 to 1941 and 1942 to 1944 classes of Harvard University on the basis of their academic performance and overall physical and psychological health. They were intensively studied as undergraduates and followed regularly by mail after graduation. Vaillant interviewed a random sample of ninety-four men from the classes of 1942 to 1944 when they were in their forties and fifties.

Vaillant interprets his data as both supporting and expanding Erikson's theory. He notes that from age 20 to 30 the Grant Study men focused on the crisis of intimacy versus isolation. "As soon as the men could win real autonomy from their parents and a sense of their own independent identities, they sought once more to entrust themselves to others" (Vaillant, 1977, p. 215). Often this change was represented by the event of marriage. Failure to achieve intimacy was important for later development. For example, twenty-eight of the thirty best-adjusted men at age 47 had achieved stable marriages before age 30 and had remained married until age 50. In contrast, twenty-three of the thirty worst-adjusted men had either married after age 30 or separated from their wives before age 50. The majority of the best marriages were contracted between the ages of 23 and 29.

From age 25 to 35 the Grant Study men focused on career consolidation and the development of their nuclear family. The focus is on work rather than play, outer life rather than inner life. "Once serious apprenticeship is begun, the refreshing openness . . . is lost; adolescent idealism is sacrificed to 'making the grade'—be it tenure, partnership, or a vice-presidency" (Vaillant, 1977, p. 216). Vaillant suggests this stage of development is not reflected in Erikson's theory.

Finally, Vaillant argues that during the late thirties and the forties, the Grant Study men underwent a "second adolescence" in which they reassessed and reordered what had occurred in adolescence and young adulthood. This is often a time of change and turmoil. However, part of the outcome of this process may be a sense of generativity—a concern for future generations. For example, by age 50, nineteen of the forty-four men who had entered business had become their own bosses. In doing so, their career patterns had broadened, and they assumed responsibilities for others. In contrast, others failed to achieve generativity, continuing to worry about "making it" and not about the welfare of others.

Vaillant concludes that the lives of the Grant Study men support both the basic stages outlined by Erikson as well as his assertion that a given stage of development can rarely be achieved until the previous one is mastered.

Levinson's Theory.   Recently Levinson (1978) and his colleagues have proposed a sequence of five eras and periods which span the male adult life cycle. The focus of Levinson's work is not on personality per se but on life course development. Personality is one facet of this development.

Levinson's (1978) theory is based on a biographical study of forty men. The men ranged in age from 35 to 45 years at the start of the study and were drawn from four occupational groups (hourly workers, executives, academic biologists, and novelists). Each man was seen from five to ten times over a period of ten to twenty months for a total of ten to twenty hours. A follow-up interview was conducted two years later. In addition to this primary sample, Levinson examined the lives of approximately 100 other men as depicted in autobiographies.

Levinson (1978) has focused on constructing the universal sequences which underlie the unique and diverse individual biographies of the subjects. Based on this research, Levinson has identified five eras within the life span, each of roughly twenty-five years' duration. These are not stages of biological, psychological, or social development, but represent a life-cycle macrostructure. The eras are (1) preadulthood, age 0 to 22; (2) early adulthood, age 17 to 45; (3) middle adulthood, age 40 to 65; (4) late adulthood, age 60 to 85; and (5) late late adulthood, age 80 plus. The evolution of these eras is structured by a series of developmental periods and transitions. The primary task of the stable periods is to build a life structure. The life structure involves three components: (1) the individual's sociocultural world including social structures (family, social class, occupational, political) and historic events (war, economic depression or prosperity); (2) participation in this world, including roles (husband, friend, worker, parent) and life events related to these roles (marriage, birth of a child, promotion, retirement); and (3) aspects of the self, including personality and abilities (talents, skills, fantasies, moods, values). Building a life structure involves making certain crucial choices and striving to attain particular goals. Stable periods ordinarily last six to eight years. The primary task of the transition periods is to terminate the existing life structure and initiate a new one. This involves a reappraisal of the current structure, exploration of new possibilities for change, and a movement toward crucial choices which will provide the basis for a new life structure. Transition periods ordinarily last four to five years. To date, Levinson's (1978) research has focused on the periods within early adulthood and middle adulthood. Within these eras, Levinson (1978) has identified eight periods as shown in Exhibit 13.10.

Early Adult Transition. This transition ordinarily begins at age 17 to 18 and extends until age 22 to 23. The Early Adult Transition represents a developmental link between preadulthood and early adulthood. It involves two major tasks. The first task is to terminate the adolescent life structure. This involves modifying relationships with the family and other persons, groups, and institutions significant to the preadult world. The second task is to make a preliminary step into the adult world. This involves making initial explorations and choices for adult living. Major life events within this transition may include graduation from high school, moving out of the family home, entering college, and graduation from college. "In this period a young man is on the boundary between adolescence and adulthood. The transition ends when he gets beyond the boundary and begins to create a life within the adult world" (Levinson, 1978, p. 57).

Entering the Adult World. This period begins in the early twenties and extends until the late twenties. The focus of the period is on exploration and provisional commitment to adult roles and responsibilities. The young adult faces two antithetical tasks according to Levinson (1978). On the one hand, he must explore alternate possibilities for adult living—keeping his options open and avoiding strong commitments. On the other hand, he must create a stable life structure—becoming responsible and "making something" of himself. Levinson (1978, p. 58) notes: "Find-

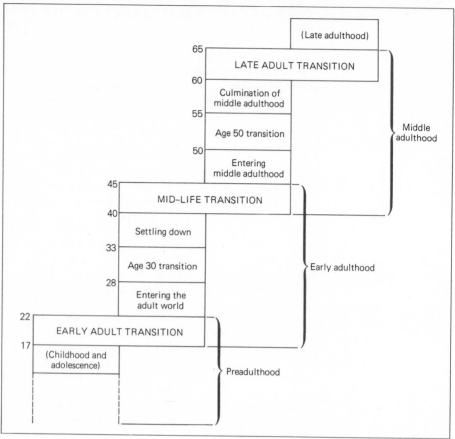

EXHIBIT 13.10   Developmental periods in early and middle adulthood.
(Source: Levinson, 1978.)

ing a balance between these tasks is not an easy matter. If the first predominates, life
has an extremely transient, rootless quality. If the second predominates, there is the
danger of committing oneself prematurely to a structure without sufficient exploration of alternatives." Examples of life events which are often crucial during this period include occupational choice, first job, marriage, and the birth of children.

Age Thirty Transition. The age range 28 to 33 years represents a transition period
between entering the adult world and the next period. This transition period provides an opportunity to modify the provisional adult life structure created earlier. As
Levinson notes, adult life is becoming more serious. "A voice within the self says:
'If I am to change my life—if there are things in it I want to modify or exclude, or
things missing I want to add—I must now make a start, for soon it will be too late'"
(Levinson, 1978, p. 58). Levinson argues that the life structure is always different at
the end of this transition than it was at the beginning, although the changes made
vary from man to man. Some men have a relatively smooth transition, building
directly on the past. The focus is on adjustment and enrichment. However, Levinson notes that for most men a moderate to severe crisis is common. Life events such

as divorce and occupational change are frequent during this period. Levinson concludes:

> *The shift from the end of the Age Thirty Transition to the start of the next period is one of the crucial steps in adult development. At this time a man may make important new choices, or he may reaffirm old choices. If these choices are congruent with his dreams, talents and external possibilities, they provide the basis for a relatively satisfactory life structure. If the choices are poorly made and the new structure is seriously flawed, he will pay a heavy price in the next period. (p. 59)*

Settling Down. This period begins in the early thirties and extends until about age 40. As implied in the name, this period emphasizes stability and security. The individual makes deeper commitments to his occupation, family, or whatever enterprises are significant to him. In addition, there is an emphasis on what Levinson calls "making it." This involves long-range planning toward specific goals within the context of a timetable for their achievement—an effort which Levinson labels a personal enterprise. Until the early thirties, the young man has been a novice adult. During Settling Down, his task is to become a full-fledged adult. Levinson notes that the imagery of a ladder is central to this period.

> *At the start of this period, a man is on the bottom rung of his ladder and is entering a world in which he is a junior member. His aims are to advance in the enterprise, to climb the ladder and become a senior member in that world. His sense of well-being during this period depends strongly on his own and others' evaluation of his progress toward these goals. (Levinson, 1978, p. 60)*

Levinson notes that most of his subjects fix on a key life event such as a promotion or a new job as representative of ultimate affirmation or devaluation by society.

During the last years of the Settling Down period, there is a distinctive phase which Levinson has designated as "Becoming One's Own Man." This phase ordinarily occurs at age 36 to 40. The major task of this phase is to achieve a greater measure of independence and authority in connection with the goals of the various enterprises. Levinson notes that a man often becomes sensitive about anything that interferes with these aims.

Midlife Transition. This transition spans a period of from four to six years, reaching a peak in the early forties. The Midlife Transition forms a developmental link between early adulthood and middle adulthood and is part of both eras. It represents a beginning and ending, a meeting of past and future. The transition may be relatively smooth but is more likely to involve considerable turmoil. However, this outcome is not entirely dependent on a man's previous success or failure in achieving goals. "It becomes important to ask: 'What have I done with my life? What do I really get from and give to my wife, children, friends, work, community—and self? What is it I truly want for myself and others?'" (Levinson, 1978, p. 60). The creation of a life structure in early adulthood involves a commitment to some goals and a rejection of others. No one life structure can permit the expression of all aspects of the self. A task of the Midlife Transition is to work on and partially resolve this discrepancy between what is and what might be. According to Levinson, the Midlife Transition is not prompted by any one life event or series of events. Rather, multiple processes and events are involved, including the reality and experience of bodily decline, changing relations among the various generations, and the evolution of career and other enterprises.

Entering Middle Adulthood. As the Midlife Transition ends, there is a new period of stability. A new life structure emerges which provides the basis for moving into Middle Adulthood. This period begins at about age 45 and extends until about age 50. Sometimes the start of this new life structure is marked by a significant life event —a change in job or occupation, divorce or a love affair, or a move to a new community. In other instances, the changes are subtler. As was the case following the Age Thirty Transition, the life structure that emerges following the Midlife Transition is crucial for the individual's adjustment.

> *Some men have suffered such irreparable defeats in childhood or early adulthood, and have been so little able to work on the tasks of their Mid-Life Transition, that they lack the inner and outer resources for creating a minimally adequate structure. They face a middle adulthood of constriction and decline. Other men form a life structure that is reasonably viable in the world but poorly connected to the self. Although they do their bit for themselves and others, their lives are lacking in inner excitement and meaning. Still other men have started a middle adulthood that will have its own special satisfactions and fulfillments. For these men, middle adulthood is often the fullest and most creative season in the life cycle. (Levinson, 1978, pp. 61–62)*

Subsequent Periods. Entering Middle Adulthood is the last specific period for which Levinson has data because of the current age of the men in his sample. However, he has projected a tentative view of subsequent periods during Middle Adulthood. An Age Fifty Transition, analogous to the Age Thirty Transition, is postulated to occur from age 50 to 55. In this period, a man can modify the life structure formed in the mid-forties. A stable period, analogous to Settling Down, is postulated to occur from age 55 to 60. This period is the culmination of Middle Adulthood. Finally, from age 60 to 65, a Late Adult Transition is postulated to terminate Middle Adulthood and provide a basis for living in late adulthood.

Like Erikson, then, Levinson suggests that adult development consists of a universal sequence of periods. Levinson's sample of men is small, and he acknowledges that an alternative sequence may be found for women. As a result, his work must be interpreted cautiously. However, as illustrated in Box 13.4, Gould (1978) has proposed a series of phases based on a separate investigation of men and women which are highly similar to Levinson's.

BOX 13.4

GOULD'S TRANSFORMATIONS

Gould (1972, 1978) has identified a sequence of four phases of adult life. Gould's work is based on two sets of data: psychiatrists' ratings of the concerns of different age groups of patients in therapy (Gould, 1972) and the responses of 524 nonpatients, aged 16 to 50, to a questionnaire.

Gould argues that the four phases are defined by the generic task of eliminating the distortions of childhood consciousness and its related protective devices that restrict the individual's life. Four major false assumptions, each associated with a phase, must be dismantled. These include:

1. "I'll always belong to my parents and believe in their world." **2.** "Doing things my parents' way, with willpower and perseverance, will bring results. But if I become too frustrated, confused, or tired or am simply unable to cope, they will step in and show me the right way." **3.** "Life is simple and controllable. There are no significant coexisting contradictory forces within me." **4.** "There is no evil or death in the world. The sinister has been destroyed."

According to Gould, these false assumptions become salient at different points in adult life.

Between the years of 16 and 22 the first false assumption must be challenged. During this phase, the adolescent or young adult is sensitive to authority figures, particularly parents. The dominant theme becomes "We have to get away from our parents." This involves an assessment of one's talents and goals, a division of loyalty between family and others, and acceptance of one's own sexuality. Despite the fact that concrete steps are taken toward these objectives (e.g., dating, work, college), most young adults do not feel completely independent of their parents.

Between the ages of 23 and 28 the second false assumption must be challenged. Major choices related to career, marriage, and family must be made. During this phase, young adults must accept the reality and responsibility of adult life. They become absorbed in the work of being adults, and an increased sense of self-confidence is experienced. By the late twenties or early thirties most adults have established the basic ingredients of an independent life.

Between the ages of 29 and 34 the young adult begins to have inner doubts. The third false assumption must be challenged. The contracts made at the beginning of adulthood are questioned, and careers and marriages come under particular scrutiny. As a result there is often a new sense of direction and a more realistic sense of one's own power.

During the early thirties, young adults make some changes and ignore others. Between the ages of 35 and 45, they see more clearly and realize they must act on their new vision of the world. Time is running out. During this period, the last false assumption must be challenged. Adults must cope with the repugnance of death, the hoax of life, and the evil within and around them. This process seems to lead to a mellowing of the personality. Gould says:

We live with a sense of having completed something, a sense that we are whoever we are going to be—and we accept that, not with resignation to the negative feeling that we could have been more and have failed, but with a more positive acceptance: "That's the way it is, world. Here I am! This is me!" And this mysterious, indelible "me" becomes our acknowledged core, around which we center the rest of our lives. (Gould, 1978, p. 311)

The Kansas City Studies. Other researchers have stressed discontinuities in adult personality which are universal but not reflective of a sequence of personality stages or life transitions. This approach is illustrated by the work of Neugarten and her colleagues at the University of Chicago (Neugarten & associates, 1964). The Chicago group conducted a sequence of interrelated cross-sectional studies over a ten-year period. Relatively large samples of healthy adults between the ages of 40 and 80 residing in the community of Kansas City during the 1950s participated in these studies. Generally, these studies found evidence for both continuity and change of adult personality. On the one hand, personality structure was stable. Four personality types—integrated, defended, passive-dependent, and unintegrated—emerged among respondents regardless of age. Similarly, characteristics dealing with the socioadaptional aspects of personality (e.g., goal-directed behavior, coping styles, life satisfaction) were not age-related. For example, Neugarten and her colleagues found no relationship between age and a measure of life satisfaction considered to reflect adaptive adjustment to aging. It seems, therefore, that the ways a healthy adult interacts with the environment are rather stable even though roles and statuses may alter. Neugarten (1964) points out that:

> *In a sense, the self becomes institutionalized with the passage of time. Not only do certain personality processes become stabilized and provide continuity, but the individual builds around him a network of social relationships which he comes to depend on for emotional support and responsiveness and which maintain him in many subtle ways . . . as individuals age, they become increasingly like themselves . . . the personality structure stands more clearly revealed in an old than in a younger person. (p. 198)*

On the other hand, Neugarten and her colleagues found marked age differences in the intrapsychic dimensions of personality—the individual's style of coping with the inner world of experience.

For example, there were age differences in the perception of the self in relation to environment. Whereas 40-year-olds felt in charge of their environment, viewed the self as a source of energy, and felt positive about risk taking, 60-year-olds saw the environment as threatening and even dangerous and the self as passive and accommodating (Neugarten, 1964; Neugarten & Datan, 1973). This change was described as a movement from *active mastery* to *passive mastery*. Gutmann (1964, 1974) has also carried out studies of aging men in preliterate societies to see whether the change from active to passive mastery observed in the United States was also found in other cultures. He found the same changes in men in four other societies: the Navajo of Arizona, the Lowland and Highland Mayans of Mexico, and the Druze of Israel.

Similarly, there appears to be an increased preoccupation with inner life among older adults. They engaged in greater introspection and self-reflection than younger adults and showed a general movement from an outer-world toward an inner-world orientation. Older persons tend to withdraw emotional investments, become less assertive, and avoid challenges (Rosen & Neugarten, 1964). This change was described as an increased *interiority* of the personality.

Along with the increase in passive mastery and interiority, Neugarten and her colleagues noted a sex difference among the older portion of their sample. On the one hand, older men seemed more receptive to their affiliative and nurturant impulses than younger men. On the other hand, older women seemed more receptive to their aggressive and egocentric impulses than younger women (Neugarten & Gutmann, 1958). This shift in sex-role perceptions only occurred for the older people in the sample. Perhaps this perception reflects the actual decreased authority older men experience with retirement and the increased authority older women experience with widowhood.

In summary, these studies point to significant age-related differences in the intrapsychic aspects of personality. Neugarten (1973) argues that these findings suggest basic developmental processes because the age differences occur well before decreases in social interaction which might affect such intrapsychic processes.

The Case for Continuity and Stability

In contrast to the views summarized above, evidence for the basic continuity and stability of adult personality has been accumulated by a number of major longitudinal studies as well as several recent sequential studies. More specifically, this cluster of investigations supports several conclusions:

1 There appear to be multiple patterns of personality development in adulthood. Like intelligence, personality is a multidimensional construct.

2 There is evidence for continuity and stability as well as discontinuity and instability in personality over the adult life span. However, the evidence for continuity and stability is more impressive than that for discontinuity and instability. Consistency is most obvious for those personality characteristics which are valued by the culture.

3 Sex-related and cohort-related differences are perhaps more important than age-related changes in understanding personality developments.

Let us examine some of the studies contributing to this case for continuity and stability of adult personality.

The Fels Study. Personality development from birth through early adulthood was the focus of a longitudinal study conducted at the Fels Research Institute (Kagan & Moss, 1962). Eighty-nine children were included in the study. Data based on behavioral observations, behavioral ratings, and interviews were available from four periods during childhood (birth to 3 years, 3 to 6 years, 6 to 10 years, and 10 to 14 years). In addition, seventy-two of these children participated in an intensive series of interviews and personality tests when they were young adults (19 to 29 years).

The most dramatic finding of this study was the emergence of a general pattern of stability in personal and social behavior by middle childhood. Passive withdrawal from stressful situations, dependency on family, ease of anger arousal, pattern of sexual behavior, involvement in intellectual mastery, sex-role identification, and social interaction anxiety in adulthood were each predicted by analogous behavioral dispositions during the early school years. The stability coefficients are shown in Exhibit 13.11.

In general, these results support the position that personality begins to take form during early childhood. However, the degree of stability in these behaviors exhibited from childhood to adulthood was highly related to cultural expectations for appropriate sex-role behavior held during the historical period studied (1929 to 1939 and 1957 to 1959). If a pattern of childhood behavior was consistent with sex-role expectations, it tended to remain stable over time. Passivity, for instance, was acceptable for females but not for males. Consistent with this, passivity was stable from childhood to adulthood for females but not for males. In contrast, aggression was acceptable for males but not for females. Consistent with this, aggression was stable from childhood to adulthood for males but not for females. Achievement, an acceptable behavior for both males and females, was stable for both sexes from childhood to adulthood.

When childhood behaviors conflicted with sex-role expectations, they appeared to find expression in consistent substitute behaviors that were socially more acceptable than the original behaviors. For example, passivity in boys predicted noncompetitiveness, sexual anxiety, and social apprehension in adult men, but not direct dependence on parents or others. Similarly, rage reactions in girls predicted intellectual competitiveness, masculine interests, and dependency conflict in adult women, but not direct aggression.

The Berkeley Studies. During the late 1920s and early 1930s, three longitudinal studies were begun at the University of California at Berkeley: the Berkeley Guidance Study, Berkeley Growth Study, and Oakland Growth Study. Each of these studies involved regular assessments during childhood and several follow-up as-

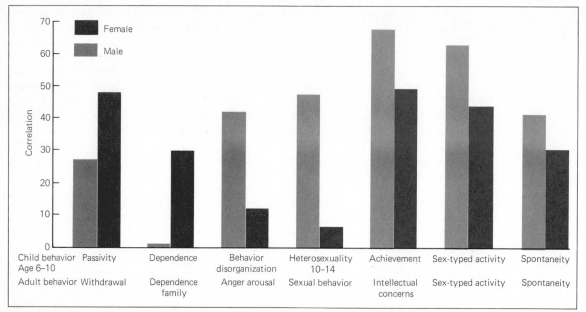

EXHIBIT 13.11   Relation between child behaviors (6 to 10 years of age) and similar adult behaviors. (Source: Kagan and Moss, 1962.)

sessments during adulthood. In addition, the parents of these children were studied —once when they were about 30 years of age and again when they were about 70 years of age.

Block (1971) combined data from the Berkeley Guidance Study and Oakland Growth Study to assess patterns of personality development from early adolescence to early adulthood. Data were available from the junior high school and high school periods, and Block collected follow-up data when the subjects were in their mid-thirties. Based on these data, Block identified five personality types for males and six personality types for females. He found evidence for substantial stability in these types over time. For example, the average correlation between personality types at senior high school and early adulthood was about 0.5. However, as new data on the same individuals at middle age were analyzed focusing on personality traits rather than types, evidence of change as well as stability in personality began to emerge (Haan, 1976; Haan & Day, 1974). Two major types of personality change were studied—ordered transpositions and experiential change. Ordered transpositions were defined as changes in degree of characteristics, but not in relative position within the group (e.g., Sue is more dependable than Joan in young adulthood; both become more dependable over time so that Sue is still more dependable than Joan in middle age). Experiental changes were defined as changes in both degree of a characteristic and relative position within the group (e.g., Sue is more insightful than Joan in young adulthood; both become more insightful over time so that Joan is now more insightful than Sue in middle age). These two types of change were contrasted with lack of change, which was labeled unaltered sameness (Haan, 1976).

A summary of the results of a variety of personality assessments of the Berkeley subjects during middle age is presented in Exhibit 13.12. There is evidence for both continuity and change. As shown in Exhibit 13.12, characteristics reflecting information processing and interpersonal relations tend to change, while those reflecting socialization and self-presentation tend to remain stable. Haan (1976) summarizes these changes by saying, "The life span seems typified by movement toward greater comfort, candor, and an objective sense of self that now characterizes these now forty- and fifty-year-old people" (p. 64).

In addition, 142 parents of the Berkeley subjects were assessed twice—initially when they were in their thirties and again when they were in their seventies (Maas & Kuypers, 1974). The major tool was an interview schedule divided into twelve sections, or "arenas" of living. The questions dealt with home, work, retirement and leisure activities, parenting, grandparenting, brothers and sisters (family of origin),

EXHIBIT 13.12   **Items Classified according to Various Developmental Patterns**

Unaltered Sameness	Ordered Transposition	Experiential Change
Information Processing		
Verbally fluent	− Brittle ego defenses + Philosophically concerned	− Self-defensive − Extrapunitive + Evaluates motivations − Fantasizing + Insightful − Projective
Interpersonal Relations		
Socially poised Assertive Aloof	− Gregarious	− Bothered by demands + Straightforward (mixed) interest in the opposite sex
Socialization		
Fastidious Undercontrolled Overcontrolled Sex-type behavior Rebellious Rushes limits	+ Dependable	− Feels guilty
Self-Presentation		
Rapid tempo Arouses liking Basic hostility Satisfied with self Physically attractive Cheerful Self-dramatizing Talkative Self-defeating		

Note: The plus (+) indicates that means over time generally increased; the minus (−), that they decreased.
Source: Haan, 1976.

marriage, friendships, formal organizational memberships which included church and various political roles, health, death, and perspectives on past life, or "life review." Several other instruments including self-report diaries were used as well. Based on this information, Maas and Kuypers examined different patterns of lifestyle and personality among these adults. Let us examine each of these patterns in turn.

In the case of lifestyles, Maas and Kuypers distinguished ten clusters—six for the mothers and four for the fathers. Several of these groups showed changes from young adulthood to old age. Work-centered mothers at the age of 30 expressed dissatisfaction with their economic and marital situations, but as they aged, relief from these distresses occurred. The relief was often associated with the loss of their spouses in their middle years. This event permitted the development of a new and gratifying lifestyle focused on their employment, independence, and new friends. Group-centered mothers, also, show a more psychologically compatible and satisfying life in later than earlier years. They stated that in old age they were more fully involved in extramarital activities and formal organizations than during their thirties because they were "freed from parental and, in some cases, a wife's responsibilities" (Mass & Kuypers, 1974, p. 123). The uncentered mothers were happy and healthy at age 30 but had lost their health and physical stamina by age 70. In fact, these mothers held the lowest-status economic positions of all the aging mothers. These women were strongly focused on family and home during their thirties, but with age came losses—of husband and home. At 70 they showed little interest in activities or roles such as that of grandparent. The disabled-disengaged mothers showed both continuities and discontinuities over their adult lives. A preoccupation with illness persisted, as did their parenting style. They also showed increased withdrawal from society. Finally, the husband-centered and visiting mothers showed few changes in lifestyle over the period.

In general, then, the women in this study evidenced quite a lot of change in lifestyles, with the most dramatic changes occurring for the work-centered mothers. For the fathers in the Berkeley Study, however, continuity was more characteristic. For example, unwell-disengaged fathers were both unwell and disengaged in early adulthood. Family-centered fathers had a lifelong focus on their families beginning in early adulthood.

In the case of personality, Maas and Kuypers distinguished seven groupings—four for the mothers and three for the fathers. In this case, the fathers, but not the mothers, tended to show changes in personality patterns from young adulthood to old age. The conservative-ordering fathers changed the most. As young adults, these fathers were shy, distant, withdrawn, and conflicted, reporting numerous marital problems. At 70, marital problems were not considered important, and they became conventional and controlling. Some changes also occurred for active-competent fathers. When they were in their thirties, they were exploitive, irritable, tense, and nervous. As older adults, they were direct, capable, charming, and conforming; spent leisure time alone; and were highly involved in the role of friend.

Thus in the case of lifestyle patterns, continuity between young adulthood and old age was more apparent among the fathers than among the mothers. In the case of personality patterns, continuity between young adulthood and old age was more apparent among the mothers than among the fathers. Maas and Kuyper's (1974) results, then, emphasize both continuity and change, as well as sex differences, in lifestyle and personality patterns. They conclude:

Most clearly, however, the majority of lives in this study run contrary to the popular and literary myth of inescapable decline in old age. Whether one con-

siders the women or the men, and whether one examines their psychological capacities and orientations or their styles of living, most of these aging parents give no evidence of traveling a downhill course. By far most of these parents in their old age are involved in rewarding and diversely patterned lives. And for those small proportions in our study whose personalities and life styles seem problematic, it is not merely old age that has ushered in the dissatisfactions and the suffering. In early adulthood these men and women were in various ways at odds with others and themselves or too constricted in their involvements. Old age merely continues for them what earlier years have launched. Finally, even when young adulthood is too narrowly lived or painfully overburdened, the later years may offer new opportunities. Different ways of living may be developed as our social environments change with time—and as we change them. In this study we have found repeated evidence that old age can provide a second and better chance at life. (p. 215)

The Normative Aging Study. This interdisciplinary longitudinal study, being conducted at the Veterans Administration Outpatient Clinic in Boston, involves a sample of approximately 2000 men ranging from the twenties through the eighties. Psychological measures have included global assessments of intelligence, personality, values, and attitudes and laboratory-based assessments of perception, decision making, and memory. Much of the information on personality from this study has been reported by Costa and his colleagues (e.g., Costa & McCrae, 1976, 1977, 1978, 1980).

Costa and McCrae (1980) propose that personality may be described by three broad dimensions: Neuroticism, Extraversion, and Openness to Experience. Each of these dimensions represents a cluster of *traits*—that is, generalized dispositions to think, feel, and behave in certain ways. Costa and McCrae suggest that each dimension reflects six traits (Neuroticism includes Anxiety, Hostility, Depression, Self-consciousness, Impulsiveness, and Vulnerability; Extraversion includes Attachment, Gregariousness, Assertiveness, Activity, Excitement Seeking, and Positive Emotions; Openness to Experience includes Openness to Fantasy, Aesthetics, Feelings, Actions, Ideas, and Values).

These characteristics predict a wide variety of attitudes, feelings, and behaviors (Costa & McCrae, 1977, 1978, 1980; Nuttal & Costa, 1975; McCrae, Bartone, & Costa, 1976). For example, men high on Neuroticism were more likely to drink and smoke, have more health problems and complaints, have sexual difficulties, be separated or divorced, and feel unhappy and dissatisfied with life. Extraversion appears to be particularly related to occupational choice. Men high on Extraversion were more likely to select jobs where dealing with people was more central than the task. Thus they preferred occupations such as social work, advertising, and law. Introverts selected professions that were task-oriented (e.g., architect). Occupational choices are also influenced by the third dimension. Men high on Openness to Experience were more likely to prefer the occupations of psychiatrist and psychologist and less likely to prefer those of banker and veterinarian. Open men were more likely to have higher theoretical and aesthetic values and lower religious and economic values. The lives of open men were also more eventful—for example, they were more likely to quit a job, be demoted, or change careers.

Longitudinal analyses of the traits associated with Neuroticism, Extraversion, and Openness to Experience suggest that adult personality is characterized by continuity and stability rather than by change (Costa & McCrae, 1980). On the one hand, Costa and McCrae (1978) report few significant mean changes in these personality dimensions over a ten-year period, thus suggesting quantitative continuity. Similarly, there is little evidence that these dimensions take on new relations to

Some researchers feel that personality characteristics such as extraversion and openness to experience remain stable over the life span. (Brody/Editorial Photocolor Archives.)

other variables, thus suggesting qualitative continuity as well (Costa & McCrae, 1977). On the other hand, the adult personality also appears to be stable. Ten-year retest correlations, for example, ranged from 0.58 to 0.69 for traits reflecting Neuroticism, from 0.70 to 0.84 for traits reflecting Extraversion, and from 0.44 to 0.63 for traits reflecting Openness to Experience (Costa & McCrae, 1977, 1978).

Similar evidence for the continuity and stability of adult personality based on an analysis of scores from the Guilford-Zimmerman Temperament Survey administered three times to several hundred men over a twelve-year period has recently been reported by Costa and his colleagues (Costa, McCrae, & Arenberg, 1980; McCrae, Costa, & Arenberg, 1980).

Terman's Study of the Gifted. As we noted in Chapter 2, Terman began a longitudinal study of 1528 grade school children with IQs of 135 or higher in 1921. The children averaged 11 years of age at the beginning of the study and were retested at regular intervals over a fifty-six year period—the most recent follow-up occurring in 1977 when they were in their mid-sixties. Terman's study was originally intended to follow the intellectual development of these gifted children. However, personality and social data were also obtained through various tests, rating scales, question-naires, and interviews. Of the many reports based on data from this study, we will focus on the Sears's investigation of sources of life satisfaction in these individuals at age 62 (Sears, 1977; Sears & Barbee, 1977).

In the case of men, Sears (1977) found occupational satisfaction was predictable as far back as early adulthood from individuals' expressions of feelings about their lives. Ambition, liking one's work and satisfaction with it in early adulthood, the feeling of choosing one's career rather than drifting into it, and a feeling of having lived up to one's potential were predictors of occupational satisfaction in later life. Satisfaction with family life had roots as far back as childhood. Social adjustment in grammar school, sociability in high school, and good mental health and marital satis-faction predicted a satisfying family life.

Sears and Barbee (1977) found measures of satisfaction to be more complex for women because of multiple family and career patterns. Again, however, themes of continuity were present. For example, general life satisfaction in 1972 was related to having positive relationships with parents in early childhood as well as to a favor-able self-concept and feelings of self-confidence.

These data suggest a significant degree of stability over a lengthy time span. Well-adjusted children appear to grow up to be well-adjusted adults.

The Britton and Britton Study. Focusing on personality change in very late life, Britton and Britton (1972) studied a group of elderly (65 to 85 years old) rural com-munity residents over a nine-year period. Interviews and objective and projective personality test data were obtained in 1956, 1962, and 1965. In general, the results emphasized continuity and stability. Interyear correlations were moderate and were generally higher for men than for women. In terms of change, Britton and Britton found no subjects who showed marked positive changes on measures of personality or adjustment. One-fifth of the men and one-third of the women showed marked negative change; the majority of the participants, however, were stable over time.

Sequential Studies. Sequential data collection strategies, as you will recall, consist of sequences of cross-sectional and/or longitudinal studies. Sequential strategies permit the separation of age- and cohort-related effects—a separation which is es-sential if the question of continuity versus change in adult personality is to be an-swered. Only a few sequential studies focused on personality have been completed to date. Generally, these studies have suggested that adult personality is character-ized more by continuity than by change (Douglas & Arenberg, 1978; Siegler, George, & Okun, 1979; Schaie & Parham, 1976; Woodruff & Birren, 1972).

Siegler, George, and Okun, for instance, administered the Cattell 16 Personality Factor test to 331 men and women. Twelve 2-year birth cohorts ranging in age from 46 to 69 years of age at the first time of measurement participated in the study. They were assessed four times over an eight-year period between 1968 and 1976. Siegler and her colleagues found that the strongest effects were related to sex. Females were less reserved and more submissive, tender-minded, naive, and tense than males. With the exception of factor B, which is a measure of intelligence, few reli-able age changes or cohort differences were evident.

In another recent sequential study, Douglas and Arenberg (1978) found evi-dence for age changes and cohort differences on the Guilford-Zimmerman Tem-perament Survey. The participants in this study were 915 men aged 17 to 98 years of age. Seven-year longitudinal data were available for 336 of the participants.

The Guilford-Zimmerman Temperament Survey provides an assessment of ten traits: General Activity (pace of activity), Restraint (emotional maturity), Ascen-

EXHIBIT 13.13   **Summary of Age-Related Changes, Culture-Related Changes, and Cohort-Related Differences in Personality**

	Source		
Personality Trait	**Age Related**	**Culture Related**	**Cohort Related**
General activity	Declined		
Masculinity	Declined		
Thoughtfulness		Decreased	
Personal relations		Decreased	
Friendliness		Decreased	
Restraint			Decreased
Ascendance			Decreased

Source: Based on Douglas and Arenberg, 1978.

dance (social assertiveness), Sociability (gregariousness), Emotional Stability (optimism), Objectivity (realistic outlook), Friendliness (agreeableness), Thoughtfulness (reflectiveness, introspection), Personal Relations (tolerance and cooperativeness), and Masculinity ("masculine" interests).

The data were analyzed using conventional cross-sectional and longitudinal as well as sequential arrangements designed to distinguish between age changes, cultural changes, and cohort differences (Schaie, 1965). Longitudinal changes were found on five scales, but only two of these were interpreted as age-related changes. Masculine interests (Masculinity) declined over the entire age range. Preference for rapidly paced activity (General Activity) declined with increasing age after the forties. Differences on three other scales were interpreted as cultural change effects. The tendency toward reflective and introspective thinking (Thoughtfulness), tolerance and cooperativeness with other people (Personal Relations), and agreeableness and thoughtfulness toward other people (Friendliness) declined during the period. Finally, differences on two other scales were interpreted as cohort effects. Earlier-born men were more serious-minded, persevering, and responsible (Restraint) and less socially assertive (Ascendance) than later-born men. These effects are summarized in Exhibit 13.13.

Douglas and Arenberg's (1978) findings contradict many of those from previous studies. For example, cross-sectional studies have suggested a movement from active to passive mastery and from an outer-world to an inner-world orientation with increasing age. In the present study, however, two traits that may be related to these characteristics (Thoughtfulness and Ascendance) showed cohort differences rather than age changes. Although the various measures involved may not be equivalent, Douglas and Arenberg's results suggest the need for caution in interpreting earlier studies.

Life-Course Analysis

Research examining the issue of discontinuity versus continuity and instability versus stability in adult personality has obviously produced mixed results. Evidence for both discontinuity and continuity and instability and stability exists. However, much of the research is plagued by difficult methodological problems. What does such a conclusion imply? Costa and McCrae (1980) suggest:

*The personologist interested in continuity or change should, we believe, refor-
mulate goals. Instead of looking for the mechanisms by which personality
changes with age, we should look for the means by which stability is maintained. Are traits genetically determined and therefore as stable as genetic influences? Do individuals choose or create environments which sustain the behavior that characterizes them? Are we locked into our nature by the network
of social expectations around us? Do early childhood influences continue to operate beyond the reach of corrective experiences? —and within any one of these
perspectives —genetic, behavioral, social role, or psychoanalytic —by what
mechanisms are we enabled to assimilate the changing experiences of a lifetime to our own nature? How do we cope, or adapt, or defend so as to preserve
our essential characteristics unchanged in the face of all the vicissitudes of
adulthood and old age? These are questions for the student of personality and
aging. (p. 81)*

In this context, Elder (1975, 1977) has suggested the need for a life course analysis of personality. This approach focuses on the examination of age- and sex-differentiated life patterns in relation to the historical context in which they are embedded. In particular, attention is directed toward an examination of the occurrence,
duration, and timing of events within the life course and the role of personality in
adapting to these events. In the next chapter we will examine the role of life events
and transitions in shaping the life course.

CHAPTER SUMMARY

Development is more than an individual process. It involves a dynamic interaction
between individuals and society. As individuals change over the life span from birth
to death, new cohorts of individuals are born replacing those who are dying off. The
succession of cohorts results in the formation of a series of age strata composed of
persons of similar age. The size and composition of these age strata vary over time,
creating a changing demographic profile within the society.

The age structure of a society is linked to social processes because society uses
age as a criterion for allocating roles and for evaluating performance of these roles.
All societies grade their members according to age. In general, the more modern the
society, the more age grades are distinguished. For example, in our culture we distinguish between adolescence, youth, early adulthood, middle age, the young-old,
and the old-old during the latter portion of the life span.

Within the general context of age grades, age is used as a criterion for entry into
and exit from the roles of society. Age may operate as a direct or as an indirect criterion. Age is also used as a criterion for evaluating behavior. People appear to be
aware of these age norms that constitute a normative social clock for the timing of
major events and that constitute a set of criteria for judging the appropriateness of
social behavior. In recent years, these age norms appear to have been weakened.

As a result of the age-related allocation of roles, the individual's participation in
different types of roles changes over the life cycle. Institutional roles increase
through midlife and then decrease during old age. Tenuous roles decrease during
adulthood and then increase during old age. Old age, then, becomes increasingly
nonnormative.

Society also differentially rewards various roles. Thus there are age-related inequalities in power, prestige, and resources within society. These differences may be understood within the framework of exchange theory. According to this theory, social interactions reflect exchanges between individuals based on power resources. Exchange theory postulates that power resources tend to be limited in young adulthood, reach their maximum in midlife, and decline in old age. This relative deprivation of the young and old is more a function of social and historical conditions than developmental processes.

The age-related allocation and evaluation of roles also interact with social change. In particular, disordered cohort flow—a discrepancy between the size and role preferences of the members of a cohort and the number and nature of the roles open to them in the society—will result in changes in the individual's development and in the social structure.

Research shows multiple patterns of role satisfaction and multiple factors which affect role satisfaction. Marital satisfaction appears to follow a curvilinear pattern: high during the newlywed period, low during the parental period, and high again during the middle-aged and aging periods. Occupational satisfaction appears to result when there is a good fit between the individual's abilities and interests and the characteristics and demands of the job. Life satisfaction appears to be related mostly to health, income, and social activity.

Within the area of personality, the major issue has been the question of whether significant personality change occurs during adulthood. This question is complicated because there is more than one type of change. The issue of continuity versus discontinuity refers to the degree of within-person change in a personality dimension over time. The issue of stability versus instability refers to the degree to which the individual retains the same relative position on a personality dimension over time.

Some studies have supported the conclusion that discontinuity in adult personality is an intrinsic part of adult development. These researchers (e.g., Erikson, Levinson, Gould) suggest personality change may be understood as a sequence of stages, transitions, or transformations which reflect the emergence of qualitatively different characteristics at different points in the life span.

In contrast, other studies suggest that the adult personality is characterized by continuity and stability. The results of many of the major longitudinal studies of personality (e.g., Fels study, Berkeley studies, normative aging study) tend to support this position.

The discrepancy in results within these two traditions suggests the need for an examination of the age- and sex-differentiated life patterns in relation to the historical context in which they are embedded.

chapter 14

Adulthood and Aging: life events and life transitions

CHAPTER OVERVIEW

Our lives are punctuated by transitions defined by life events. In this chapter, we will examine the role of life events in adulthood: their nature and influence on development. Adaptation to different types and sequences of life events will be considered.

CHAPTER OUTLINE

A LIFE-EVENT FRAMEWORK

Types of Life Events
The Timing, Sequencing, and Clustering of Life Events
Exposure to Life Events

ADAPTING TO LIFE EVENTS

Mediating Variables
The Adaptation Process

CHAPTER SUMMARY

ISSUES TO CONSIDER

How may a life-event framework be used to understand development?
What general types of life events may be identified?
How are the timing and sequencing of life events important for development?
What age and sex differences occur in exposure to and content of life events?
What cohort differences occur in exposure to and content of life events?
What factors mediate adaptation to life events?
What general steps characterize the adaptation process?

*C*onsider how you would answer the following question: *What was the last major event in your life that, for better or worse, interrupted or changed your usual activities?* How did adults of varying ages respond to this question? Consider the following excerpts of responses to it.

[Bob, age 30] *That would have to be the trouble with the business. I sell and service air conditioners. I just opened the business about four years ago, and there were a lot of start-up costs. But just as I was beginning to get on my feet, this energy thing hit for real. Sales of air conditioners have really fallen off because of the cost of electricity. Even people who already have it aren't using it. I'll tell a customer that he needs a couple of hundred bucks worth of repairs on his central system, and he'll tell me that he'll do without it this summer. Between the drop in business and this crazy inflation, we're really getting squeezed. Besides the business loan, we owe a couple of thousand on our car. Then there's the usual credit card bills. On top of everything else our roof has started to leak, and it looks like the whole thing will have to be replaced. I'm not sure where we go from here. We've talked about trying to get another loan from the bank, but I don't know what we'll be able to get. I do know that if business doesn't pick up soon we're going to be in real trouble.*

[Joan, age 42] *My son getting married, and then having my granddaughter. This happened a little over two years ago. Paul was 18 then and Suzie was only 17. They were still in high school and Suzie got pregnant. That part was very difficult. My husband passed away quite a few years ago, so Paul and I were quite close. At first, I was so afraid this was going to ruin his life, I guess I handled it badly. I tried to convince him he was making a mistake, and there were several weeks of tears and arguments. The final straw came when I told Suzie's father about their plans. You see, they had asked me not to tell him—but I thought he had a right to know. That was a mistake. He beat her up and threw her out of the house. I felt so badly that I stopped arguing with the kids, and they moved in with me. Since then, Paul has finished high school and is working in a TV repair shop. He's taking courses part-time at the University too. And, of course, there's Amy. She just had her second birthday last week, and she's as cute as she can be. I guess it was hard on all of us, but when I look at that little girl I know it worked out for the best.*

[Max, age 62] *Well, my heart attack, I guess. That certainly changed my activities. I used to work sixteen hours and smoke three packs of cigarettes a day. Now, I only work half days at the office, and no smoking. I guess it has been a change for the best. It was a pretty severe attack. I was in intensive care for three weeks, and it took almost six months before I could get around much. That was really frustrating. But I am beginning to get used to taking it slower. The doctor says if I go back to my old style I'll just do myself in. I still worry about the business though. My partner has been doing most of the work, and I know I am going to have to make some hard decisions soon. I guess I'm just not ready to retire yet.*

A LIFE-EVENT FRAMEWORK

We all experience various events during our lives. Indeed, adulthood and aging may be viewed as a series of transitions defined by such events. Riegel (1975), for instance, found that individuals recalling both their personal and cultural pasts focused on periods of transition defined by critical events rather than on periods of stability. A summary of the transitions and events noted by Riegel (1975) is shown in Exhibit 14.1. Such a listing of life events and transitions by itself is simply descriptive. However, when life events are viewed as important antecedents of behavior change during adulthood, a potentially powerful explanatory framework is generated (Hultsch & Plemons, 1979).

The framework includes four main elements which are illustrated in Exhibit 14.2: a set of antecedent life-event stressors, a set of mediating factors, a social-psychological adaptation process, and consequent adaptive or maladaptive responses. Within this framework, all life events are viewed as potential stressors to the extent that they require a change in the individual's customary patterns of behavior. This means that events which are typically thought of as positive (e.g., marriage, being promoted at work), as well as events which are typically thought of as negative (e.g., death of a spouse, being fired from work), are potentially stressful. Mediating factors include both internal resources (e.g., physical health, intellectual abilities) and external resources (e.g., income, social support from others). Social-psychological adaptation involves the application of coping strategies and resultant changes in behavior. This process may lead to either functional or dysfunctional outcomes.

The stress or crisis of life events does not reside within the event or within the

EXHIBIT 14.1

Levels and Events in Adult Life

Level (Years)	Gradual Changes					Sudden Changes
	Males		Females			
	Psychosocial	Biophysical	Psychosocial	Biophysical		
I (20–25)	College and first job		First job and college			
	First child		Marriage	First child		
II (25–30)	Second job		Loss of job			
	Other children			Other children		
	Children in preschool		Children in preschool			
III (30–35)	Move		Move			
	Promotion		Without job			
	Children in school		Children in school			
IV (35–50)	Second home		Second home			
	Promotion		Second career			
	Departure of children		Departure of children			
V (50–65)	Unemployment		Unemployment			Loss of job
	Isolation			Menopause		Loss of parents
	Grandfather		Grandmother			Loss of friends
	Head of kin		Head of kin			Illness
		Incapacitation				
VI (65+)	Deprivation			Widowhood		Retirement
		Sensory-motor deficiencies		Incapacitation		Loss of partner
						Death

Source: Riegel, 1975.

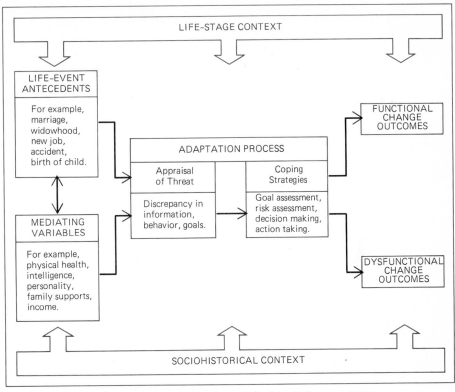

EXHIBIT 14.2   A life-event framework. (Source: Adapted from Hultsch and Plemons, 1979.)

individual. Rather, the crisis arises from an interaction between the individual and the situation—an asynchrony between change within the individual and change within the environment (Elder, 1974; Riegel, 1975). As illustrated in Exhibit 14.2, the life stage and sociohistorical context within which the event occurs are crucial; that is, two distinct time lines are involved in the life course—individual time and historical time. The same event occurring in a different context becomes, in many ways, a different event (e.g., becoming a widow at age 32 or at age 73; bearing six children in the 1940s or in the 1980s). The occurrence of a life event at an "inappropriate" time in the life cycle or history is likely to create more asynchrony.

Thus development may be characterized as adaptation to a series of crises, asynchronies, or transitions defined by life events. Such a perspective is useful since it forces us to focus on the interface between the individual and his or her world.

Types of Life Events

In the broadest sense, an event is a noteworthy occurrence. But what is a "significant," "stressful," or "critical" life event? Some investigators have restricted their definition of life events to "personal catastrophies" such as life-threatening illnesses (Hudgens, 1974). However, a more common approach has been to include a wide array of events "whose advent is either indicative of, or requires a significant change in, the ongoing life pattern of the individual" (Holmes & Masuda, 1974, p. 46). Such events encompass a variety of domains including school (started school, changed schools, graduated from school, failed school), work (changed jobs, was promoted, was laid off, retired), love and marriage (engaged, married, separated or divorced, spouse died), children (became pregnant, birth of a first child, had an abortion, child died), residence (remodeled home, moved to a better residence, moved to a worse residence, lost home through a disaster), finances (took out a mortgage, substantial increase in income, substantial loss in income, went on welfare), social activities (took a vacation, acquired a pet, broke up with a friend, close friend died), and health (became physically ill, was injured, health improved). Note that some of these events are "negative" in the sense that they are typically socially undesired, and some of them are socially desired and hence "positive." In either case, they require adaptation and change on the part of the individual experiencing them.

From a life-span developmental perspective, life events may be grouped into one of three major classes: normative, age-graded events; normative, history-graded events; and nonnormative events.

Normative, Age-Graded Events. Normative, age-graded events consist of biological and environmental events that are correlated with chronological age. Their occurrence depends, in part, on biological capacity and/or social norms. As a result, their timing, duration, and clustering tend to be similar for many individuals. Examples of normative, age-graded events include marriage, birth of a child, menopause, and retirement. The effect of these events is primarily evident with respect to the individual experiencing them, and secondarily, the individual's significant others. Normative, age-graded events constitute a social clock for the adult life cycle.

Normative, History-Graded Events. Normative, history-graded events consist of cultural events that are correlated with historical time. They are normative to the extent that they are experienced by most members of a cohort. Examples of normative, history-graded events include wars, political upheavals, mass emigrations or immigrations, and economic prosperity or depression. These events play a primary role in defining the developmental context of a particular birth cohort. Further, to the extent that they affect cultural change, they continue to exert their influence on later cohorts.

Nonnormative Events. When life events are related weakly to both individual and historical time indexes, they are called nonnormative events. Such events are relatively idiosyncratic in terms of occurrence or timing during the life cycle or are limited to a relatively small proportion of the total population. Examples of nonnormative events include accidents, illnesses, divorce, floods, droughts, and limited occupational layoffs.

The Timing, Sequencing, and Clustering of Life Events

The life-span developmental perspective emphasizes multidirectional changes as a result of the patterned interaction of many antecedents. This implies a sensitivity to the timing and sequencing of events over the life span. Thus, from this perspective, events do not have uniform meaning. When an event occurs is perhaps as important as whether it occurs at all. Elder's (1974) work on the impact of the Great Depression on men who experienced this event at different times in the life cycle is an example of this point. In the case of middle-class men, Elder reports that younger men were more negatively affected by the depression than older men. Individuals

(a)   *(b)*

a. Normative, age-graded events constitute a social clock for the adult life cycle. (Charles Gatewood.) *b.* The loss of one's limbs constitutes a nonnormative life event. (Charles Gatewood.)

who experienced the depression as young men were just beginning their work careers, while the older men had already established them. At a later date, men who had experienced the depression at an early age showed a much higher rate of career instability and disadvantage than men who experienced the depression at a later age. Elder notes that this pattern was reversed in the case of lower-class men. In this group, the depression had a more negative impact on men who were older at its onset than on men who were younger, reflecting a historical pattern of age discrimination in unskilled occupations.

The timing of history-graded and nonnormative events is largely idiosyncratic. However, as we discussed in Chapter 13, the timing of age-graded events is defined, in part, by age norms. According to theory, age norms specify appropriate times for certain life events such as leaving the family home, achieving economic independence, marriage, bearing children, retirement, and so forth (Neugarten & Hagestad, 1976). As individuals move through the life cycle, they are made aware of whether they are "early," "on time," or "late" with respect to these norms through a system of positive and negative sanctions. Again, from a life-span developmental perspective it is not only the presence or absence of an event—timing is also crucial. For example, Lowenthal, Thurnher, and Chiriboga (1975) report that the failure of an event to occur at the expected time was particularly stressful in the later stages of life. Middle-aged men, for instance, indicated that being "off time" with respect to promotions or salary increases was a major reason for a reduction in their life satisfaction. Similarly, Bourque and Back (1977) note that events such as the departure of children and retirement are perceived by respondents as most disruptive if they

BOX 14.1

THE MEASUREMENT OF LIFE EVENTS

How do life events differ from one another? Intuitively, we would expect that events such as the death of a spouse, changing schools, getting a promotion, and acquiring a pet are different in many ways. One major approach to this problem has been to measure the amount of stress or behavioral change associated with different events (Dohrenwend, Krasnoff, Askenasy, & Dohrenwend, 1978; Holmes & Rahe, 1967). Generally, researchers have done this by asking people to rate lists of life events compiled by the investigator. The widely used Social Readjustment Rating Scale developed by Holmes and Rahe (1967) is illustrative of this approach. Holmes and Rahe developed an approach to measuring the stress of life events based on the psychophysical procedure

of magnitude estimation. In this procedure, a designated target stimulus is given an assigned value, and judges are asked to rate the stimuli in relation to this target. Holmes and Rahe designated the event of marriage as the target and assigned it a value of 500. Individuals were then asked to provide judgments about the amount of readjustment required for each of the other events on the list in relation to marriage. The average rating (divided by 10) was then used to index the stressfulness of events reported by other individuals. The events of Holmes and Rahe's list and their average Life Change Unit scores are shown below. How many of these events have you, or others you know, experienced recently? Would you rank the events the same way?

Social Readjustment Rating Scale

Rank	Life Event	Mean Value	Rank	Life Event	Mean Value
1	Death of spouse	100	23	Son or daughter leaving home	29
2	Divorce	73	24	Trouble with in-laws	29
3	Marital separation	65	25	Outstanding personal achievement	28
4	Jail term	63	26	Wife begins or stops work	26
5	Death of close family member	63	27	Begin or end school	26
6	Personal injury or illness	53	28	Change in living condition	25
7	Marriage	50	29	Revision of personal habits	24
8	Fired at work	47	30	Trouble with boss	23
9	Marital reconciliation	45	31	Change in workhours or conditions	20
10	Retirement	45	32	Change in residence	20
11	Change in health of family member	44	33	Change in schools	20
12	Pregnancy	40	34	Change in recreation	19
13	Sex difficulties	39	35	Change in church activities	19
14	Gain of new family member	39	36	Change in social activities	18
15	Business readjustment	39	37	Mortgage or loan for lesser purchase (car, TV, etc.)	17
16	Change in financial state	38	38	Change in sleeping habits	16
17	Death of close friend	37	39	Change in number of family get-togethers	15
18	Change to different line of work	36	40	Change in eating habits	15
19	Change in number of arguments with spouse	35	41	Vacation	13
20	Mortgage or loan for major purchase (house, etc.)	31	42	Christmas	12
21	Foreclosure of mortgage or loan	30	43	Minor violations of the law	11
22	Change in responsibilities at work	29			

Source: From Holmes and Rahe, 1967.

BOX 14.2

THE GREAT DEPRESSION—A HISTORY-GRADED EVENT

The Great Depression was a major historical event of this century. Some historians, for example, see it as a watershed in the evolution of American society. How did people adapt to this event, and what impact did it have on their lives? Elder (1974) has examined this question through a reanalysis of longitudinal data collected as part of the Oakland Growth Study. This project, originally designed to investigate the physical, intellectual, and social development of boys and girls, was begun in 1931. The children were initially selected from the fifth and sixth grades of elementary schools in Oakland, California. The children and their families were studied intensively from 1932 to 1939 and were retested as adults in 1953 to 1954, 1957 to 1958, and 1964. Thus the children were preadolescents and adolescents during the depression decade and graduated from high school just before World War II.

Elder's (1974) reanalysis of these data illustrates the tremendous impact of a major cultural event. In part, the depression experience was determined by the degree of economic loss suffered by the family. Contrary to popular assumption, economic hardship in the depression was not a pervasive experience. While unemployment approached one-third of the work force in 1933, evidence suggests that the depression decade was not a time of great economic deprivation for at least half the population. Severe physical hardships were concentrated among the rural and urban poor, although status losses were common among the middle classes. Thus Elder (1974) divided his Oakland families into deprived and relatively nondeprived groups on the basis of relative income loss between 1929 and 1934. The average loss of the former group was 3 to 4 times that of the latter group.

In adapting to the economic losses of the depression, families often moved from a state of crisis to disorganization, to partial recovery through new modes of action, to eventual stabilization. Working mothers, boarders, money from relatives, and public assistance are examples of strategies used, particularly by deprived families. Public as-

Major historical events like the Great Depression often result in sweeping changes in the political, economic, and social institutions of a culture. (Wide World Photos.)

sistance, however, represented a strategy of last resort, often used following prolonged unemployment. Employment of the mother, while not socially acceptable, was more common. This strategy, as well as taking in boarders, resulted in striking changes in the division of labor within the family. In particular, they increased the power of the mother and decreased the power, prestige, and emotional significance of the father as perceived by the children. In addition, they shifted increasing responsibilities in the household economy to the children. Girls were most often involved in household tasks (food preparation, laundry, making clothing), while boys were most often involved in part-time jobs (newspaper carrier, store clerk, delivery boy). This downward extension of adultlike experience had considerable effect on the children's maturity. For example, deprivation and its associated familial adaptations were related to interest in becoming an adult, industriousness, financial responsibility, and an interest in persons outside of the family. These values affected later life events. For example, boys from deprived families made firmer vocational commitments in late adolescence than boys from nondeprived families. In adulthood, they were more likely to have followed the occupation which they preferred in adolescence, entered their career at an earlier age, and developed a more orderly career. In middle age, men from deprived families did not differ from men from nondeprived families in commitment to work. However, work had a different meaning to the deprived men. They were more likely to prefer job security and a modest income to the risk of obtaining a higher income. They also expressed more dissatisfaction with their incomes and working conditions.

Elder reports many other influences of this cultural event. From this sample it is clear, however, that the depression produced significant changes in family structure, behavior, and values, some of which were significant many years later. Interestingly, some of the adaptations produced by the Great Depression (e.g., working women, increased adultlike experience for children, etc.) may be increasing within the present historical context as a response to the severe inflation the economy has been experiencing.

---

occur at a nonnormative age. The departure of children from the home is least traumatic during the fifties; retirement is worst if it happens during the early sixties just prior to the normative age. However, being off time is not always a negative condition. Nydegger (1973), for instance, found that late fathers were more comfortable and effective in this role than either early or on-time fathers. By delaying this event, these men avoided a conflict between the demands of parenting and early career establishment.

Even when events occur within the limits of their normative age range, their temporal order (sequence) can be critical. Hogan (1978), for example, examined the effects of different orders of three life events experienced by men in young adulthood—completion of education, first job, and first marriage—on later marital stability. In the normative ordering of these events, a man first finishes his formal schooling, next becomes financially independent through employment at a full-time job, and finally marries (Hogan, 1978). Other events, however, particularly military service and the achievement of advanced levels of education, may disrupt the sequence. Hogan's analyses indicate that the nonnormative ordering of events substantially increased the likelihood that the first marriage would end in a separation or divorce.

Another consideration with respect to the timing of events concerns the clustering of events. Events may be particularly difficult to deal with when they occur close together, and there is considerable support for this notion. Holmes and Masuda (1974), for instance, asked physicians to provide an account of their life-event experiences and major health changes for the previous ten years. Reported life events were assigned their values from the Social Readjustment Rating Scale, and the Life Change Unit (LCU) scores of the life events were then plotted year by year along with the major health changes. Their results indicated that the vast majority of health changes were associated with a clustering of life events whose values summed to at least 150 Life Change Units per year. Further, the more severe the life

EXHIBIT 14.3 **Relationship of Life-Crisis Magnitude to Percentage of Life Crisis Associated with Health Changes**

Severity of Life Crises	Number of Life Crises		Total Number of Life Crises	Life Crises Associated with Health Changes (%)
	Associated with Health Changes	Not Associated with Health Changes		
Mild life crises (150–199 LCUs)	13	22	35	37
Moderate life crises (200–299 LCUs)	29	28	57	51
Major life crises (300+ LCUs)	30	8	38	79
Total	72*	58	130	55

* Some life crises were associated with more than one health change.
Source: Holmes and Masuda, 1974.

crisis, the greater the risk of an associated health change. Exhibit 14.3 shows that as the Life Change Unit score increased, so did the percentage of illness associated with the crisis. Of the mild life crises, 37 percent had an associated health change. This figure increased to 51 percent for moderate life crises and to 79 percent for major life crises.

Similarly, in a developmental study, Palmore and his colleagues (Palmore, Cleveland, Nowlin, Ramm, & Siegler, 1979) note that many adults adapted quite well to the occurrence of given events such as retirement, widowhood, departure of the last child from home, and illness. However, the simultaneous or consecutive occurrence of two or more life events did produce more serious problems of adaptation.

Finally, it is important to note that the issue of the pattern of events extends beyond a single individual. That is, it is not just the pattern of events within a single life cycle that is important, but how these events interact with events in the life cycles of significant others. Elder (1977) stresses this point in relation to the family. In this context, for example, scheduling problems often arise. For instance, the events of early career establishment may conflict with the events of childbearing and childrearing.

Exposure to Life Events

How frequently do individuals experience life events? What kinds of life events do they experience? We would expect to find both age and sex differences in exposure to and content of life events because of different normative expectations. Similarly, we would expect to find cohort differences in exposure to and content of life events because of cultural change. Let us examine these differences.

Age and Sex Differences in Exposure to Life Events. Information on the issue of age and sex differences in exposure to life events is available from Lowenthal, Thurnher, and Chiriboga's (1975) study of individuals at four stages of life (high school seniors, young newlyweds, middle-aged parents, and older adults about to retire). These investigators found that, generally, young persons (high school seniors and young newlyweds) reported more exposure to life events than older persons (middle-aged parents, adults about to retire). However, the two younger groups tended to report more positive stresses, while the two older groups tended to report more negative stresses. Age and sex differences were apparent in the nature of the stressful events as well (see Exhibit 14.4). The major cause of stress in the younger stages was education. This was followed by stresses related to dating and marriage. Statistical analyses showed that stresses related to dating and marriage as well as changes in residence were more significant for women, while stresses related to leisure activities and the military were more significant for men. Differences between the sexes were more dramatic for the older respondents. Indeed, there was no overall "major" source of stress for these groups. The most salient source of stress for men was work, while the most salient sources of stress for women were health and the family. In part, these differences are a function of the fact that older women appear to be stressed by events occurring to others, particularly their children. Thus events connected with their children's schooling (child dropped out of school), marriage (child separated or divorced), and occupation (child changed jobs) were significant stressors. These appeared to be particularly important for the middle-aged women.

Longitudinal data based on five- and seven-year follow-up interviews extend these findings by focusing on changes in stress (Chiriboga, 1978). Overall, middle-aged and older adults increased in negative stresses more than the two younger groups. More specifically, high school seniors and newlyweds showed little change in family stress, while the two older groups—especially the middle aged—showed increases in negative stresses in this area. In the case of positive stresses, the middle-aged group showed the greatest increase, while the older-age group showed the greatest decrease. Most of these changes were in the area of marriage. Middle-aged men and women showed the largest increases, while older and newlywed men and women showed decline on this dimension. These data suggest that middle age is a time of dramatic change incorporating both positive and negative components of stress. Later life, however, tends to be associated with an increasing amount of negative stress.

EXHIBIT 14.4   **Sources of Stress over Last Ten Years***

	Younger			Older		
Source of Stress	**Men**	**Women**	**Total**	**Men**	**Women**	**Total**
Education	71*	80	76	2	9	6
Residential	33	40	36	9	4	6
Dating and marriage	33	50	40	2	12	8
Friends	27	44	35	13	5	9
Family	24	36	30	13	26	20
Marriage	18	28	23	9	12	11
Health	10	22	16	13	38	27
Work	22	14	18	45	21	32
Leisure activities	35	12	23	6	2	4
Military	27	2	14	0	4	2
Death	14	12	13	18	19	19
Finances	6	4	5	18	11	14

* Percentage of respondents indicating area as a source of stress.
Source: Adapted from Lowenthal, Thurnher, and Chiriboga, 1975.

EXHIBIT 14.5   **Stress Typology**

	Perception of Stress	
Exposure to Stress	**High**	**Low**
High	Overwhelmed	Challenged
Low	Self-defeated	Lucky

Source: Adapted from Lowenthal, Thurnher, and Chiriboga, 1975.

However, the mere occurrence of an event is not the entire story. How a person interprets the event is critical. Thus what one person may experience as a catastrophe, another person may experience as a challenge. Lowenthal and her colleagues identified four types of persons based on this distinction between exposure to stress (self-reported incidence of life events) and perceived stress (preoccupation with themes of loss, stress, and deprivation in a life history interview). The four types are shown in Exhibit 14.5.

Within this framework, "overwhelmed" people were exposed to frequent or severe stress and perceived their lives as highly stressful. On the other hand, "challenged" people, similarly exposed to frequent or severe stress, did not perceive their lives as stressful. The "lucky" were exposed to infrequent or mild stress and perceived their lives as relatively unstressful. On the other hand, "self-defeating" people, similarly exposed to infrequent or mild stresses, perceived their lives as highly stressful.

The distribution of these types by age and sex is shown in Exhibit 14.6. Examination of the extreme right-hand column of this table indicates that more than 50 percent of the total sample were preoccupied with stress (overwhelmed or self-defeating), suggesting difficulty adapting to life events. The distribution of the four types was quite similar across the four age groups. Within each age group, however, there were dramatic differences between men and women (with respect to these differences the two younger and the two older groups resembled one another). For example, among the highly stressed young, men were more likely to be overwhelmed than women, but among the highly stressed old, women were more likely to be

EXHIBIT 14.6   **Stress Type by Stage and Sex (Percentages)**

	Age Group						
	Younger		Older		Younger and Older Combined		
Stress Type	**Men**	**Women**	**Men**	**Women**	**Men**	**Women**	**Total**
Considerable presumed stress							
Challenged	16	21	30	7	23	14	19
Overwhelmed	36	27	21	40	28	34	31
Light presumed stress							
Lucky	32	21	31	32	32	26	29
Self-defeating	16	31	18	21	17	26	21
Total	100	100	100	100	100	100	100

Source: Adapted from Lowenthal, Thurnher, and Chiriboga, 1975.

overwhelmed than men. These data suggest older women may have fewer resources than men to deal with high levels of stress. Whether this represents a developmental effect or a cohort effect is unknown at this point. Thus the data reviewed in this section suggests that there are both age and sex differences in life-event experiences. The sex differences, however, may be more significant than the age differences.

Cohort Differences in Exposure to Life Events. Of course, different cohorts are exposed to different history-graded events. Individuals born in 1920, for example, experienced World War II, while those born in 1960 did not. However, there are also cohort differences in exposure to and timing of age-graded events. These differences reflect cultural changes (e.g., industrialization, urbanization) and their accompanying demographic, technological, political, and economic developments.

For example, Uhlenberg (1979) has profiled the characteristics of three birth cohorts: 1870 to 1874, 1900 to 1904, and 1930 to 1934. The 1870 cohort reached age 65 in the late 1930s; the 1900 cohort reached this age in the late 1960s; and the 1930 cohort will not reach it until the late 1990s. Uniform changes in the characteristics and experiences of these three cohorts are evident as summarized in Exhibit 14.7. This sample of data suggests that:

1 With changing policies on immigration, successive cohorts are composed of fewer foreign-born individuals.

2 With industrialization, a decreasing proportion of individuals are living in a rural environment and an increasing proportion are living in an urban environment.

3 With changing economic patterns, a decreasing proportion of the male work force are in farming occupations and an increased proportion are in white-collar occupations.

4 With changing marital patterns, a decreasing proportion of women are remaining married and an increasing proportion are experiencing divorce.

5 With the decline in childlessness between the second and third cohorts, a decreasing proportion of women are entering old age childless.

6 With increasing sex differences in survival (and the tendency in our culture for women to marry men older than themselves), an increasing proportion of women are experiencing widowhood.

7 With changing economic patterns, an increasing proportion of men are experiencing retirement from the work force.

Thus Uhlenberg's (1979) data suggest that demographic and cultural changes since 1870 have significantly altered the life-event experiences of adults born just a generation apart.

Differences between cohorts are also seen in the timing of specific life events. For example, until recently, successive cohorts have exhibited a decrease in age at first marriage, birth of first child, and birth of last child. In the last few years, however, there has been an increase in age at first marriage along with a decline in the number of children born (Neugarten & Hagestad, 1976). Similarly, entry into the labor force has been increasingly delayed, and exit from the labor force has been increasingly accelerated for successive cohorts of men. The participation of women in the work force has also changed markedly. During the period 1950 and 1970, the proportion of middle-aged women in the work force doubled. During recent years, an increasing proportion of young mothers have remained in the work force (Neugarten & Hagestad, 1976).

EXHIBIT 14.7   Characteristics of Three Cohorts during Adulthood and Old Age

During Adulthood

Cohort Characteristics	Cohort of:		
	1870	1900	1930
Size when aged 25–29 (in thousands)	6529	9834	10804
Foreign born when aged 25–29*	17	11	3
Rural when aged 25–29*	47	37	28
Distribution of males by occupation when aged 35–39*			
White collar	NA	31	44
Blue collar (nonfarm)	NA	52	53
Farm	NA	17	3
Marital status of females:*			
Never married by age 50	10	8	5
Divorced by age 40–44 (of those ever married)	NA	11	21
Distribution of females by children ever born:*			
0	23	28	13
1–3	36	51	52
4+	41	21	35

During Old Age

Cohort Characteristics	Cohort of:		
	1870	1900	1930
Size when aged 65–69 (in thousands)	3807	6992	9023
Initial cohort surviving to age 65:*			
Males	37	50	63
Females	42	62	77
Foreign born when aged 65–69*	21	13	7
Sex ratio when aged 65–69	99	81	80
Average number of years of life remaining at age 65:			
Males	11.7	13.7	?
Females	12.8	17.1	?
Males in labor force when aged 65–69*	69	42	25
Ratio of age groups 70+/65–69, when aged 65–69	1.4	1.8	2.4

* In percentages.
Source: Adapted from Uhlenberg, 1979.

There are also differences between cohorts in sequences of life events. For example, in spite of the fact that entrance into the labor force has been increasingly delayed for successive cohorts, the entire transition from schoolboy to married adult has occurred more rapidly for recent cohorts (Winsborough, 1979). In this study, four events were examined: (1) completing high school; (2) taking a first full-time job; (3) entrance to and exit from the armed forces; and (4) entrance into first marriage. Exhibit 14.8 shows the time taken to complete these events for selected male cohorts. The top line for each birth year reflects duration of the school exit process. The line begins when 25 percent of the cohort had completed school, moves through the median age (indicated by the point on the line), and terminates when 75 percent of the cohort had completed school. In similar fashion, the second line reflects the duration of the first-job transition; the third line, duration of military ser-

Increasing sex differences in survival with historical time have resulted in an increasing proportion of women who experience widowhood. (Hella Hammid/Rapho/Photo Researchers, Inc.)

vice; and the fourth line, duration of the first-marriage transition. The entire process of moving over the four life events took approximately eighteen years for early cohorts and less than ten years for the recent cohorts.

ADAPTING TO LIFE EVENTS

In addition to the nature of the events themselves, there are two general factors that must be considered in order to understand how individuals adapt to life events. The first general factor consists of a variety of characteristics of the individual and the environment that mediate the impact of life events. These characteristics define the resources that will facilitate and the deficits that will interfere with the individual's adaptation to the demands of an event. These resources and deficits set the

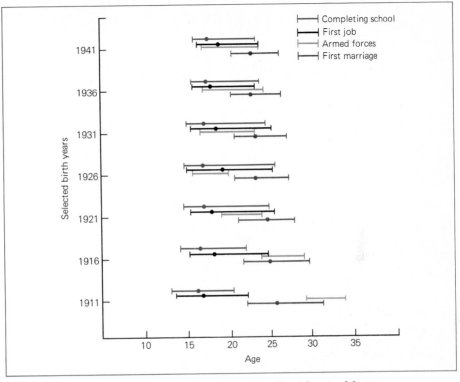

EXHIBIT 14.8 Quantities of the age distribution at school completion, first job, armed forces service, and first marriage for selected birth years. (Source: Winsborough, 1979.)

limits of the adaptational process. They are the tools the individual has to work with. The second general factor consists of the adaptation process itself. This process involves behaviors such as analyzing the problem, searching for solutions, evaluating the costs and benefits of alternative solutions, and acting to accomplish certain ends. In effect, these behaviors represent what people actually do when confronted with a life event. Let us briefly examine each of these factors.

Mediating Variables

There are many variables which may function as mediators of life events including biological, psychological, social, and physical characteristics of the individual and environment. We will not attempt to specify all these potentially relevant mediators but, instead, will present a sample of some of them. Within this context, three general points are important: (1) there are large between-person differences with respect to these variables, both within and between cohorts; (2) generally, the better the individual's resources, the better he or she is able to adapt to life events (Palmore et al., 1979)—however, it is also important to examine the relative distribution of resources and deficits (Lowenthal & Chiriboga, 1973)—and (3) the impact of resources and deficits varies over the life span (Lieberman, 1975; Lowenthal & Chiriboga, 1973). That is, what is a resource at one point in the life span may be irrelevant or a deficit at another point in the life span. The reverse is also true. Further, the balance of resources to deficits may be differentially important in different phases of the life cycle. Within this framework, let us briefly examine three sets of mediating variables.

Biological and Intellectual. It may be argued that biological and intellectual resources set a floor, or lower limit, on adaptation (Lieberman, 1975). That is, they are necessary but not sufficient for adaptation. For example, adaptation requires a certain expenditure of biological energy dependent on physical health. It is reasonable, then, to suppose that the physical capacities of the individual impose some limits on the ability to adapt.

Similarly, the intellectual capacities of the individual impose some limits on adaptational capacity. Various cognitive abilities (the ability to attend to, encode, store, and retrieve information; crystallized abilities reflective of knowledge of the culture; fluid abilities reflective of the perception of relationships in novel situations) are all probably required in order to appraise threat accurately and to select and implement coping strategies effectively. To the extent that the individual cannot perform these intellectual tasks, adaptation will be impaired.

There is evidence to support the lower-limit-setting role of biological and intellectual capacities. Lieberman (1975), for instance, examined the adaptation of elderly adults to radical changes in their environment (e.g., institutionalization). The results of these studies suggested that those who were physically and cognitively impaired could not adapt. In one study, for example, physical and cognitive characteristics accounted for 73 percent of the variance predicting breakdown. However, while inadequate physical and intellectual resources appear to predict breakdown, the reverse is not true. Adequate resources do not necessarily predict successful adaptation. Thus biological and intellectual resources set lower limits. Below these limits the individual cannot mobilize the basic resources necessary for successful adaptation. Above these limits other psychological and social variables appear to be critical.

Personal and Social. The individual's personality is an important mediator of life events. For example, as we noted in Chapter 13, Costa and McCrae (1980) have identified three broad domains of adult personality: Neuroticism, Extraversion, and Openness to Experience. In examining the role of these domains in relation to life events, Costa and McCrae found that virtually all events over which individuals had any control were related to personality.

Other findings stress that the role personality plays in mediating the individual's response to life events varies over the life cycle. For example, Lieberman (1975) reports that such traditionally positive characteristics as ego strength and impulse control were not predictive of adaptation in older adults experiencing radical changes in their life space. Indeed, Lieberman found that those elderly who were aggressive, narcissistic, demanding, and irritating were those individuals who were most likely to survive the crisis. Thus traditional views of psychological health were not applicable to older adults in this setting. Lieberman concludes: "A certain amount of magical thinking and perceiving oneself as the center of the universe, with a pugnacious stance toward the world—even a highly suspicious one— seemed more likely to insure homeostasis in the face of a severe crisis" (pp. 155– 156).

The impact of life events is also mediated by social factors such as socioeconomic status, income, and interpersonal support systems including family and friends. For example, supportive interpersonal relationships may serve as resources for the individual to the degree to which they provide physical, psychological, or financial support (Adams & Lindemann, 1974). The availability of these resources is likely to vary over the life cycle. For example, supportive frameworks and interpersonal relationships frequently decrease with increasing age. Older individuals tend to belong to fewer groups, have fewer friends, and see them less (Rosow, 1973). However, there is also evidence to suggest that interpersonal relationships among older adults are more likely to be complex and subtle (Lowenthal et al., 1975). In this sense, older adults may get more out of the relationships they do have. Among men in particular, interpersonal resources increasingly seem to serve as buffers against stressful events (Lowenthal et al., 1975). While the presence of a social network is often a resource for older adults, its absence may not be a deficit, at least for some individuals. For example, lifelong isolation is not necessarily associated with maladaptation in later life (Lowenthal et al., 1975). The lifelong isolate does not suffer from age-linked losses of social networks because he or she never had social ties. However, a pattern of marginal social relationships appears to be associated with poor adaptation at all points in the life cycle. Thus again we see that whether a characteristic is a resource or a deficit depends on one's stage in the life cycle.

Past Experience and Anticipatory Socialization. An individual's past behavior tends to be the best predictor of his or her future behavior. Responses to life events, then, are likely to depend, in part, on the individual's past experience with similar events (Lawton & Nahemow, 1973; Lieberman, 1975). Generally, it may be argued that successful adaptation in the past enhances the ability of the individual to cope with future events. Past experience may modify both affect (e.g., reducing anxiety) and coping strategies (e.g., goal setting, decision making). Persons *know* they can deal with this type of event. Of course, for past experience to be effective, the individual must be able to recognize the similarity between past and current events. In this regard, events which on the surface appear quite different probably require the application of similar analytic, problem-solving, and decision-making processes. This suggests the possibility of the development of generic life development and life-crisis skills which may be taught to individuals (Danish & D'Augelli, 1980).

One might expect that older adults may be able to use their past experience resource to a greater extent than younger adults. Indeed, there is some evidence to suggest that this resource is used by older adults and may underlie, at least partially, the finding that older adults reported fewer stresses than younger adults (Lowenthal et al., 1975). However, past experience may also be a deficit to the extent that the individual has not been able to cope with previous events. A history of past failure may lead the individual to perceive events as overwhelming and, therefore, to "avoid" life events (Levinson, 1978; Lowenthal et al., 1975).

Many life events can be anticipated. Thus many of the behaviors necessary for successful adaptation to such events may be learned through *anticipatory socialization*—a process of learning attitudes, values, beliefs, activities, and so forth of a situation one will encounter in the future. Such learning may occur through a variety of mechanisms including formal instruction, observation, modeling, and imitation. Anticipatory socialization for an event may serve as a resource to the extent that it decreases the ambiguity of the situation and increases the responses available to the individual (Albrecht & Gift, 1975). There appears to be less anticipatory socialization for events which are seen as involving defeats or losses. In addition, with respect to different points in the life span, there is evidence of less anticipatory socialization for events associated with aging than for events at other points in the life span (Rosow, 1973). Such anticipatory socialization requires a future orientation on the part of the individual. In this regard, a positive assessment of the future and an ability to introspect appear to be significant resources for older adults (Lieberman, 1975; Lowenthal et al., 1975).

The Resource-Deficit Balance. In addition to resources, individuals also have deficits. Further, deficits are not necessarily the absence of or the opposite of resources. The importance of the relative balance between resources and deficits is illustrated by Lowenthal, Thurnher, and Chiriboga's study of four life stages. These investigators measured both resources (e.g., trust, empathy, hopefulness, openness to change, physical health) and deficits (e.g., anxiety, depression, hostility, feelings of inferiority, self-criticism). Respondents were then ranked in the top, middle, or bottom third on their summed resource and deficit scores, thus producing a nine-cell typology (i.e., high resources–high deficits to low resources–low deficits). Lowenthal and her colleagues found that resources and deficits were not simply opposite ends of the same continuum. While 36 percent of the individuals fell into convergent categories (i.e., high resources–low deficits; low resources–high deficits), 46 percent fell into clearly divergent categories (i.e., high resources–high deficits; intermediate resources–high deficits).

Comparison of the nine groups of the typology on measures of happiness also suggests the interaction of resources and deficits as well as the age-specfic nature of adaptation. For the entire sample, the happiest people were at the extremes (high resources–high deficits and low resources–low deficits) and in one intermediate group (intermediate resources–low deficits). The least happy groups were low in resources and high in deficits; intermediate in resources and high in deficits; and intermediate in both. Developmentally, however, the pattern varied. Among the high school seniors, the happiest individuals were those with high resources and high deficits. However, the advantage of such complexity decreases over the life span. That is, the happiest newlyweds were those who had high resources and low deficits. Among middle-aged respondents, the happiest individuals were those with intermediate resources and low deficits. Finally, among the older adults, the happiest individuals were those with low resources and low deficits.

Thus the pattern appears to be one of movement from psychological complexity to psychological simplicity. Examination of differences among the groups on other variables supports this description. On the one hand, those high on both resources and deficits reported more life events and turning points in their lives, experienced and perceived more stress in their lives, and engaged in a broader range of activities than other groups. On the other hand, those low on both resources and deficits reported fewer life events and turning points in their lives, had few recent positive *or* negative emotional experiences, projected minimally into the future, and displayed little complexity in their goals for the next few years. The characteristics of the complex, therefore, resemble the cultural stereotype of the young, while the characteristics of the simple resemble the cultural stereotype of the old. Lowenthal and her colleagues suggest that convergence of optimum balance style and cultural stereotypes may not be coincidental.

The Adaptation Process

What do people actually do when confronted with a life event? In actuality they may engage in many specific behaviors. However, these may be classified into several general steps which characterize an *adaptation process* (see Exhibit 14.2). Box 14.3 provides an example of behaviors involved in adapting to divorce.

BOX 14.3

DIVORCE—A NONNORMATIVE EVENT

Divorce was relatively uncommon in the United States during the first half of the twentieth century. However, in recent years, particularly since the early sixties, divorce rates have escalated dramatically. They more than doubled from 1960 to 1977, and it has now been estimated that about half of all marriages involving young Americans will end in divorce (Glick & Norton, 1976). In spite of the fact that there is considerable evidence to suggest that divorce is a disruptive event, there have been few attempts to examine the adjustment process involved (Chiriboga, 1977; Hetherington et al., 1979; Spanier & Casto, 1979). A recent study by Spanier and Casto (1979) is illustrative of an attempt to examine this issue. Their analyses were based on interviews of adults aged 21 to 65 who had been separated or divorced during the two years preceding the interview. Spanier and Casto found that two related adjustments are involved in the process of separation or divorce. The first involves the adjustment to the dissolution of the marriage (e.g., working out property settlements and custody arrangements, informing and dealing with family and friends, and coping with feelings about the self and spouse, such as love, hate, guilt, self-esteem, and self-confidence). The second involves the adjustment to the process of creating a new lifestyle (e.g., finding a new place to live, making economic adjustments, adjusting to single parenthood, finding new friends, and resuming heterosexual dating). Several examples of these adjustments are given below.

Dependent children were often a source of significant adjustment problems. Many parents attempted to minimize the negative effects of the separation on the children. One father said:

We wanted there to be as little disruption as possible for the kids. Because both of the kids were loved very much by both of us and because of our agreement to share custody, I feel they suffered only minimally . . . now they don't have to compete with the other parent for attention and they don't have to live with constant quarreling. (p. 216)

In a minority of cases, however, the children were used to punish the other spouse or to obtain a better settlement. Once a separation or divorce had occurred, the parent receiving principle custody is faced with the task of performing multiple roles alone. Parents often felt trapped.

I hate feeling totally responsible for the kids. They're mine completely. At least when I was married, I could mentally not feel responsible at times. . . . It gets lonely with the kids in bed by 8:00. It's an ambiguous role. [I want to go out but] I don't want the kids to be stuck with a babysitter three or four nights a week. (p. 223)

Nevertheless, most parents were glad to have custody of the children. Those who did not have custody viewed their loss as a serious deprivation.

Spanier and Casto also note that all their respondents reported emotional problems as a result of the separation or divorce. The severity of these problems appeared to be significantly related to the degree to which the event was anticipated and the degree of attachment to the former spouse that remained. A sudden and unexpected separation often caused extreme reactions. A husband commented:

I was at first extremely distraught. I think that I was in a suicidal state for a while. I contemplated suicide at one time. After this period I just kind of went into a state of shock that lasted [a couple of months]. After that I just felt hurt that all of it had happened. (p. 218)

Similarly, a continued feeling of attachment was often a major cause of emotional problems. A young woman put it this way:

I still think about him. Knowing that he won't come home for dinner is sad. Dinner is the hardest time. We still talk and I wonder how he's doing . . . [there are other] memories like him cooking breakfast on Sunday mornings. And the family at home at Christmas. It's hard seeing his car downtown and thinking about he and I apart. (p. 219)

Related to this, Spanier and Casto found that individuals who developed new social relationships, including new heterosexual social relationships, had fewer adjustment problems than individuals who did not. For example, only 7 percent of those who were dating regularly, living with someone of the opposite sex, or remarried were having serious adjustment problems. In contrast, 45 percent of those with little or no heterosexual activity were having serious adjustment problems. The following comment was typical:

Getting involved with a woman has helped me to realize I can now communicate intimately and establish a relationship with someone. This has been very helpful. It's made the adjustment much easier. (p. 225)

Other factors predictive of adjustment included economic status and health status.

Spanier and Casto conclude that the adjustment to separation and divorce is more related to the process of creating a new lifestyle than to the process of dissolving the old one. This finding stresses the importance of mediating factors such as personality traits, social competence, and economic resources in coping with divorce.

Once an event occurs (or is anticipated), the individual assesses it with respect to its significance or degree of threat. The ability of the individual to make this assessment depends upon a certain level of adequacy in cognitive functioning. As noted previously, cognitive abilities appear to create a floor effect in this regard (Lieberman, 1975). Cognitive functioning below a certain level would incapacitate the individual's ability to conduct an appraisal of the event sufficient to determine the degree of threat posed. Above this particular level of adequacy, however, increasing cognitive ability would not be correlated with increased ability to carry out the appraisal process.

The degree to which the situation or event is assessed as threatening or stressful depends upon both event and mediating factors. As discussed previously, these will vary depending upon the type of life event, its historical and sociocultural context, the previous life history of the individual, and the point in the individual's life cycle at which the life event occurs. Regardless of these differences, the appraisal process is seen as being similar for all individuals. This process revolves around the degree of dissonance (Festinger, 1957; Lazarus, 1966) or incongruity (Moss, 1973) experienced by the individual as a result of the occurrence of the life event. One way of thinking about this process is to envision the life event as presenting the individual with certain information or requirements that either were not previously encountered or are in conflict with the individual's current information, values, goals, or behavior.

The outcomes of this initial appraisal of the event range from no threat or stress perceived to intense threat or stress perceived. If no stress is perceived, the individual is essentially unaffected by the life event. If the event is appraised as threatening or stressful, the individual begins to seek some means of resolving the conflict.

The first phase of the individual's attempts to recreate congruity with the environment involves the selection of *coping strategies* from among those already in his or her current repertoire. As we have mentioned, previous event sequences and the individual's mastery of these will have resulted in an established repertoire of coping strategies and tendencies. Thus one can expect between-person differences in the strategies available and in the selection of strategies.

Selection of coping strategies will also be influenced by the constraints perceived in the immediate context. For example, direct action or manipulation of the environment may be a feasible alternative under some circumstances and not others. Circumstances under which action tendencies are less likely would be either those in which their expression exposes the individual to threat from a different source or conditions in which direct action has little or no value. Conditions which exemplify the latter type of situation are severe injury, terminal illness, and death of a loved one. Here the individual is relatively helpless in the sense that there is little opportunity for direct action on the environment. Although these types of events may occur at any point in the individual's life span, as discussed previously, there is a prevalence of such events later in the life cycle. Older adults also tend to function within a more constricted environmental context in terms of decreased financial resources, decreased status, and decreased options for alternative support networks and employment opportunities (Rosow, 1973). This suggests that the use of intrapsychic coping strategies will tend to increase, while the use of direct action coping strategies will tend to decrease over the adult life span. Guttmann (1978), in a study of adult decision making in relation to life events, did find that action-takers tended to be younger than nonaction-takers. In particular, individuals 80 years of age and over tended to take only limited action or no action. Those with higher incomes and more education were also more likely to be action-takers.

Another important determinant of strategy selection is the degree of perceived threat. There is some evidence that a high degree of threat leads to a decrease in variability and an increase in stereotyped behavior (Lazarus, 1966). Thus when the degree of threat is intense, one might expect the selection of defensive maneuvers, including avoidance and denial. Such strategies may be partly successful in that they prevent the individual from involving himself or herself in problems too complex to handle. However, long-term use of such strategies tends to be maladaptive in that the behavioral and psychological flexibility of the individual becomes impaired. Some events simply cannot be avoided. In his studies of institutionalization, for instance, Lieberman (1975) reports that those who did not deny the threat and were able to engage in appraisal and resolution processes consistent with the threat fared better than those who did deny both the effect and acknowledgment of the threat.

Following the initial selection of coping strategies, attempts to cope with the situation using these strategies are undertaken. A variety of outcomes of this process are possible. The individual may find the initial selection of strategies to be adequate to cope with the events. In this case, the current behavior of the individual tends to remain stable. A second possibility is that the individual may find his or her usual strategies of coping inadequate or, at the most, marginally adequate. In this case, a phase of exploration may ensue. This is a time of searching for new solutions, seeking out and evaluating new information, and reappraising current assumptions and life goals. A highly unstable period, it is often accompanied by confusion, frustration, disruption of usual behavioral and psychological processes, and a general increase in the susceptibility to pathology (Moss, 1973).

The outcomes of the adaptation process depend, to a great extent, on the individual's balance of resources and deficits. With a high ratio of resources to deficits, the individual may be openly responsive to his or her current context and be able to mobilize the energy necessary to undertake learning of new behaviors and reorganization of current psychological structures. It is still possible, however, that this outcome will be associated with physical disease, depending upon the genetic and physiological predispositions of the individual. The outcome for the individual who possesses a low ratio of resources to deficits is more likely to be in the direction of physical or psychological dysfunction. This type of individual becomes so preoccupied with defining the self against the perceived threat that he or she is less able to be openly responsive to the environment and, thus, less able to find new solutions. The resulting state of perpetual defensiveness, nonresolution, and instability renders the person particularly susceptible to the development of physical or psychological dysfunction.

CHAPTER SUMMARY

Individuals experience many transitions during the life course defined by various life events. These events may be viewed as antecedents of developmental change. Within this framework, life events are seen as potential sources of stress. Through a social-psychological adaptation process, these events may lead to functional or dysfunctional outcomes. The impact of events is mediated by both internal and external resources. Thus development may be characterized as adaptation to a series of crises precipitated by life events.

Life events may be defined as occurrences that result in a change in the ongoing life pattern of the individual. Events may be age-graded (marriage, retirement), history-graded (war, economic depression), or nonnormative (divorce, illness).

From a life-span developmental perspective, the timing, sequencing, and clustering of life events are important. The timing of age-graded events is defined in part by age norms. As individuals move through the life cycle, they may be "on time" or "off time" with respect to various events. Similarly, individuals may experience different sequences or clusters of life events. All these factors appear to affect the way in which life events are experienced.

There are both age and sex differences in exposure to and content of life events. Generally, younger adults report more exposure to life events than older persons. Further, younger adults experience more positive stresses, while older adults experience more negative stresses. Sex differences appear to be even more significant than age differences. In particular, older women appear to have fewer resources than men for dealing with high levels of stress.

There are also cohort differences in exposure to and content of life events produced by cultural changes. These include differences in the frequency of events (e.g., divorce) and differences in the timing of events (e.g., more rapid exit from the labor force).

In addition to the nature of life events themselves, there are two factors which determine how individuals adapt to life events. The first consists of characteristics of the individual and environment that mediate the impact of events. These include biological (e.g., health), psychological (e.g., personality), social (e.g., family support), and experimental (e.g., anticipatory socialization) variables. The second consists of the social-psychological adaptation process itself. This includes a sequence of behaviors such as appraisals of threat and selection, application, and modification of coping strategies.

chapter 15

Adulthood and Aging: the life cycle

CHAPTER OVERVIEW

In this chapter we will examine the transitions and life events associated with three major periods of the adult life cycle: young adulthood, middle age, and old age. Events such as parenting, menopause, the empty nest, retirement, and widowhood will be considered. In addition, we will examine death—the final event of life.

CHAPTER OUTLINE

THE TRANSITION TO YOUNG ADULTHOOD

Commitment
Dating, Courtship, and Marriage
Parenting
Divorce
Occupation

THE MIDLIFE TRANSITION

The Case for Crisis
The Case for No Crisis
Conclusion: Crisis or No Crisis?
Physical Decline and Awareness of Death
Menopause
The Empty Nest
Careers

THE TRANSITION TO OLD AGE

Toward Integrity
Retirement
Grandparenthood
Widowhood
Relocation

DEATH AND DYING

The Meaning of Death
Age Differences in Death Perceptions and Attitudes
The Dying Process
Contexts for Dying
Ethical Issues
Grief and Bereavement

CHAPTER SUMMARY

ISSUES TO CONSIDER

What major commitments are made during young adulthood?

What factors influence mate selection?

What factors contribute to a successful marriage?

To what extent is parenting a stressful life event?

How are people affected by divorce?

What factors influence occupational choice?

To what extent may midlife be described as a crisis?

How does physical decline affect psychological functioning in midlife?

Is menopause a major turning point in life for most women?

How does the empty nest affect families?

How do evolving careers affect men and women at midlife?

What is involved in the development of a sense of integrity in old age?

How do people plan and adjust for retirement?

What does the grandparent role mean to older adults?

What factors influence the individual's adaptation to widowhood?

How stressful is relocation for older adults?

How do people's perceptions and attitudes about death vary according to age?

What are the stages of the dying process according to Kübler-Ross?

What are the characteristics of the phase theory of dying?

What is euthanasia and what ethical issues are related to it?

What factors characterize the grief process?

THE TRANSITION TO YOUNG ADULTHOOD

Commitment

*Y*oung adulthood is a time of making commitments. Individuals typically make their first major choices related to marriage, children, occupation, and style of life that will define their place in the adult world. This aspect of young adulthood is emphasized by various theories which view young adulthood as a distinctive phase of the life cycle.

Erikson (1959, 1963), for instance, emphasizes the need to make commitments which will allow the institutions of society to continue. In particular, a new family unit must be formed. This is usually accomplished by establishing an interpersonal relationship, socially sanctioning it with marriage, and allowing reproduction to occur. The psychosocial pressure for individuals during this stage of life is to form a close, stable interpersonal relationship. If stability is lacking, there is little assurance that children will be socialized appropriately; hence the perpetuation goal will not be served properly.

Erikson argues that to successfully maintain a close and stable relationship, each person must give of himself or herself totally. All aspects of a person such as ideas, goals, dreams, feelings, and values must be shared unconditionally with the other person who, in turn, must reciprocate. Each individual should no longer focus on self but should be more interested in meeting the needs of the other and deriving satisfaction from meeting those needs. The relationship will be maintained to the extent that such mutual interchange is attained. When a person cannot share or be shared, a sense of isolation will be felt which will erode the relationship.

Levinson (1978) proposes that there are four major tasks to be accomplished during young adulthood in the case of men: forming a dream; an occupation; mentor relationships; and love relationships, marriage, and family. According to Levinson, the dream is a vague sense of self in an adult world. It may take relatively dramatic form (e.g., the great scientist, artist, or athlete) or relatively mundane form (e.g., the skilled craftsman, effective father, respected community member). The task of the young man is to define the dream and find ways of living it out. Various aspects of the context, however, may interfere with the dream (e.g., parents, personality characteristics, lack of money or opportunity).

Levinson also argues that the formation of an occupation spans a large part of early adulthood. This involves sorting out interests, acquiring skills and credentials, and establishing oneself in the occupational role.

Finally, Levinson suggests that the young man must form significant relationships with other adults who will facilitate the dream. One of the most important others is the mentor. The mentor is ordinarily several years older than the young adult. He (or she) is often in the work setting (e.g., senior colleague) although neighbors, friends, or relatives may also play the role. The mentor essentially serves as teacher, sponsor, guide, or counselor. The most crucial task is to support and facilitate the young adult's dream. In addition, the young man must form an intimate relationship with a woman, and he must also accept the responsibilities of parenthood. Like Erikson, Levinson sees this commitment as crucial to the continuation of society.

Young adulthood, then, is characterized by the formation of multiple commitments which will establish the basic configuration of adult life. Let us examine some of these events.

Dating, Courtship, and Marriage

Although older adults date, it is typically viewed as an activity characteristic of adolescents and young adults (Winch, 1971). What are the characteristics, functions, and problems of dating?

Dating can function as a recreational or socializing activity, it can serve as a means to achieve status, and it can be a way to select a mate. Most people feel that the distinctions between casual dating and a serious relationship are not clear. De-Lora (1963) proposed, for instance, that dating is best viewed as a continuum from "casual" to engaged. Although many people contemplate marriage at some point during dating, not all people progress along the dating continuum. Also, not all dates serve a particular function. In a retrospective study, for instance, Hicks (1970) reported that among married couples at the Pennsylvania State University different forms of dating were perceived as serving different functions. Only casual or steady dating was considered fun and recreational. On the other hand, "going steady" served a socializing function: those informally engaged were planning for the future and couples formally engaged were busy making wedding plans. Dating, indeed, serves as a means of socializing and, therefore, is a way for individuals to discover whether they are fun to be with, overly emotional, too serious, and so on. As Winch (1970) states, "The opportunity to associate with those of the opposite gender gives a person the chance to try his own personality and to discover things about the personalities of others" (p. 351).

Mate selection is a primary function of dating. Although our system of mate selection is flexible, it is also competitive. Many complex rules, strategies, and goals are followed. For example, it is often considered a social norm for men to move a relationship toward sexual intimacy before making a final decision on selecting a mate. For women, on the other hand, the goal is to move the relationship toward commitment. Premarital sexual behavior, however, has increased during the last decade. In a cross-sequential study conducted from 1958 to 1968, Christensen and Gregg (1970) compared data among three groups: (1) Danish college students, (2) a conservative group of Mormon college students, (3) a midwestern college population. During the ten-year period, attitudes and behavior about premarital sex became more liberal, especially for women, even though males continued to be more liberal than females. Among the Danish population, increasing similarity between male and female premarital sexual behavior and attitudes was evidenced. Moreover, this intersexual convergence was apparent in that more than 95 percent of men and women reported having had a premarital sexual experience. Both American samples reported approving of premarital intercourse. Females in the American samples reported a threefold increase in premarital sexual intercourse over the ten-year period.

However, Americans seem to be maintaining a traditional view about sexual behavior prior to marriage. That is, premarital sexual behavior is said to occur only in the context of intimacy and love. Therefore, many young adults experience sex within a committed relationship. Nevertheless, attitudes still reflect a double standard. It is still more acceptable for males to have casual sexual experiences than it is for females (Kaats & Davis, 1970). However, casual sex for females met with greater approval in the 1970s than it did in the 1960s (Kaats & Davis, 1970). Despite some changing attitudes and behavior, it appears that we are not witnessing a revolution in premarital sexual behavior, but rather an evolution.

Of the findings dealing with mate selection in our culture, the clearest is that people marry people who are similar to themselves (Melville, 1977). This has been labeled *homogamy*. On the other hand, there is little evidence that opposites attract. That is, *heterogamy* does not appear to operate strongly in mate selection (Bowerman & Day, 1956). Within this general framework, there are a number of barriers to the beginning, continuation, and breakup of a specific relationship. Adams (1977) summarizes these as follows.

Propinquity, or residential closeness, facilitates the beginning of a relationship. Catton and Smircich (1964), for example, found that the probability of marriage decreases as the distance between premarital residences increases. Further, since people from the same racial, ethnic, religious, and social class categories tend to cluster together residentially, one effect of propinquity is to increase such categorical homogamy.

Once propinquity has limited the field of eligibles for marriage, there are a number of additional factors which form the basis for early attraction. The most significant of these is physical attractiveness (Elder, 1969). Other factors include surface behaviors and similar interests. This early attraction may be either perpetuated or reduced by additional factors. First the reactions of family and friends are important (Lewis, 1973). Favorable reactions increase the likelihood and unfavorable reactions decrease the likelihood that the relationship will be perpetuated. Second, disclosures made by the couple as they get to know one another better will either strengthen or weaken the relationship (Lewis, 1973). A positive result of self-disclosure is what Lewis calls "pair rapport." This involves feeling comfortable in each other's presence and is, in part, a function of perceiving similarities in each other.

Deeper attraction occurs as the relationship is perpetuated by favorable labeling by others and the development of rapport between the couple. The factor found to be most closely related to deeper attraction is value consensus, i.e., the orientation of the couple toward the same ideas, values, and goals (Coombs, 1966). Other factors leading to deeper attraction include a similar level of physical attractiveness, similar energy levels, and similar personalities (Murstein, 1972; Napier, 1971).

Mate selection is an important task of young adulthood. (Charles Gatewood.)

At this point the relationship may move toward marriage as the individual develops a conscious or unconscious feeling that the partner is "right for me." This is often reinforced by others who begin to regard the individuals as a "couple." The pair begins to work actively for the relationship's continuance (Bolton, 1961). One's identity becomes tied up in the relationship. Similarly, there are costs involved in terminating the relationship (hurting, disappointing, or embarrassing the partner and others, having to form alternative relationships, etc.). Such barriers to breakup, then, may be considered to be the final phase of the mate selection process. The next step is marriage itself.

Although homogamy predominates in terms of characteristics such as race, religion, age, education, and location of residence, people who marry do differ. Such differences lead to different types of marriages. After intensive interview sessions with 437 upper-middle-class married couples, Cuber and Harroff (1965) identified five types of marriages:

1 *Conflict habituated.* Arguing, bickering, nagging, or fighting prevail but are viewed by the couple as acceptable behaviors, which may even provide stability.

2 *Devitalized.* Although the couple is bored or disenchanted with the marriage, they believe that they still have love for each other.

3 *Passive-congenial.* Conflict is always minimized; marriage is considered a convenient arrangement, with partners not being involved with each other.

4 *Vital.* Partners are highly and actively involved with each other in all aspects of family life.

5 *Total.* There is a profound, intense involvement with each other. Involvement is so complete that the couple may be perceived as isolated and neurotic.

Parenting

In our culture individuals are socialized to believe that after marriage the role of parent should be assumed. Parenting is seen as normal and natural. Further, Americans also believe that proper parenting is normal and natural; it is not possible for "good" parents to have "bad" children. Parents are expected to provide appropriate experiences for their children and to have confidence in their judgments, feelings, and behavior toward their children.

These attitudes can create problems for adults who are experiencing the life event of parenthood, particularly for the first time. For instance, 50 to 80 percent of new parents reported that the birth of their first child was a moderate to severe crisis (Dyer, 1963; LeMasters, 1957). But recent evidence (e.g., Hobbs & Cole, 1976) suggests that the birth of the first child was viewed as "moderately stressful" by parents who also reported that many rewards were experienced with the advent of their new role.

There also appears to be a sex difference in how the individuals experience parenthood. Women who had borne their first child reported the postpartum period was characterized by emotional stress and negative mood changes, whereas men did not report such changes (Leifer, 1977).

There are many reasons why these data are contradictory. First, in earlier studies interviews, rather than questionnaires, were employed to obtain data. These interviews reflect higher crisis scores compared to the questionnaires. Hobbs and Cole (1976) expressed concern about the low return rates (less than 50 percent) for the questionnaires in their study. Second, reasons for having children could influ-

BOX 15.1

SEXUAL ACTIVITY IN MARRIAGE

One of the most important areas of interaction in a marriage is the sexual relationship. Research has shown that sexual activity among married couples has changed significantly over the years. In the most extensive survey of sexual activity in the United States since Kinsey's pioneering work, Hunt (1974) found that sex within marriage has become much more vital.

For example, the frequency of sexual intercourse has increased. As shown in the exhibit below, the median frequency of marital coitus for the youngest age group (18 to 24) increased from 2.45 times per week in the 1930s and 1940s to

Marital Coitus: Frequency per Week, Male and Female Estimates Combined, 1938 to 1946–1949 and 1972

1938 to 1946–1949 (Kinsey)		1972 (Hunt)	
Age	Median	Age	Median
16–25	2.45	18–24	3.25
26–35	1.95	25–34	2.55
36–45	1.40	35–44	2.00
46–55	0.85	45–54	1.00
56–60	0.50	55 and over	1.00

Source: Hunt, 1974.

3.25 times per week in the 1970s. Similar increases were noted for the older age groups as well.

Other indexes confirm the general increase in sexual activity. For example, Hunt found an increase in the proportion of married women experiencing orgasm on a regular basis. Similarly, there has been an increase in the length and types of foreplay as well as in the variety of positions used in coitus. To illustrate, a generation ago 75 percent of all males reported reaching orgasm within two minutes after initiating coitus (Kinsey, Pomeroy, & Martin, 1948). In contrast Hunt (1974) found that the median duration of coitus for the married couples in his sample was about ten minutes, with younger couples prolonging the act the longest. Similarly, a generation ago only 33 percent of married couples had engaged in intercourse with the female on top of the male. In contrast, Hunt found that 75 percent of the couples in his sample had used this position.

Perhaps most important, Hunt reports that the vast majority of married couples found their sexual activity provided a great deal of pleasure in their lives. "Sexual liberation has had its greatest effect . . . within the safe confines of the ancient and established institution of monogamous marriage" (Hunt, 1974, p. 194).

ence the way in which parenthood is regarded. Was the child conceived because the couple loves children? Did the couple want to remove the pressure of parents desirous of being grandparents? Did they desire a replacement for themselves, want to make a statement expressing their love for each other, or want to have someone take care of them in old age? Moreover, was the child planned or not? Third, differences in socioeconomic status could explain the equivocal findings. Russell (1974) found that middle-class couples stated that fewer rewards are accrued from parenting than did working-class couples. But middle-class parents read more books on childrearing than lower-class parents, and these books, though helpful, are often used as a standard by which parents judge themselves. As a result, middle-class couples who are parents for the first time may be more anxious than lower-class couples. This anxiety could contribute to their crisis perspective about parenthood.

In addition, since the husband-wife relationship is valued more by middle-class couples than lower-class couples, the birth of the first child could be more disruptive. For example, the amount of time spouses interact by talking with one another is reduced by as much as 50 percent after their first child arrives. Other sources of stress for middle-class American parents have been delineated by Rossi (1968):

1 *Cultural pressure.* Society sanctions commitment to parenthood as an event that is normal in an adult's life course. For couples not wanting children, feelings of "Are we abnormal?" must be dealt with.

2 *Inception.* Becoming a parent may not be planned or may not be planned well.

3 *Irrevocability.* The parental role is irrevocable. Almost any other decision relating to life events can be changed or reversed, but parenthood cannot.

4 *Preparation.* Individuals are not formally trained to be parents. It is assumed that because people have been children and interacted with caregivers they will know what to do with their children. The assumption is that parenthood is natural, normal, and right.

5 *Guidance.* There is no recipe for determining what are the best reasons for having a child or which childrearing practices will produce competent, mentally healthy adults.

Thus we can conclude that, for many reasons, parenthood is a crisis for some young adults, while for others it represents a positive transition to a newly acquired role. However, regardless of whether parenthood is perceived as a crisis or transition, as stressful or rewarding, men and women experience parenthood differently.

The decision about whether to be or not to be a mother is a difficult one for many young women in our culture. Many social sanctions exist pressuring women to make a commitment to motherhood. Many women have been socialized to believe that they will naturally love their child and will be able to nurture him or her properly. This view of the supposed norm derived from the notion that something exists known as the "maternal instinct"; that is, there is a natural, normal fit between women and motherhood.

The concept of maternal instinct was widely expressed in the psychoanalytic and ethological writings of the 1950s and 1960s. Bowlby (1969), for instance, portrayed the mother-child relationship as intimate, loving, dedicated, and natural: "What is believed to be essential for mental health is that an infant and young child should experience a warm, intimate, and continuous relationship with his mother (or permanent mother-substitute—one person who steadily mothers him) in which both find satisfaction and enjoyment" (Bowlby, 1972, p. 13). Although the mother's constant presence was thought to be critical for satisfactory child development, parenting goals were not explicitly advanced. Rather, Bowlby (1972) developed a deprivation thesis, which suggested that an impoverished child becomes a socially incapable adult and depriving parent. However, many developmentalists (e.g., Hoffman, 1974) have turned away from hypotheses supporting the notion of a maternal instinct and from investigations of maternal deprivation.

Just as we have moved from supporting the idea that only mothers can parent, so, too, have we recognized that some women may not want to have children or may not want to be the sole caregiver. College women who stated positive attitudes toward feminism, for instance, were more likely not to want to bear or rear children than women who held negative opinions about the women's liberation movement. Moreover, investigations of children reared in day-care centers demonstrate no significant difference on many developmental indicators (cognition, language, attachment, etc.) between children who received day care and those who were reared entirely at home (Caldwell, Wright, Honig, & Tannenbaum 1970; Kearsley, Zelazo,

Kagan, & Hartmann, 1975). Other researchers (e.g., Hoffman, 1974) have shown that maternal employment has no negative effects on children's development.

Also, it has traditionally been believed that fathers are important in the child-rearing context. However, studies have concentrated on the effects of their absence, and the effects—positive or negative—of their presence have received little research attention. On addressing the undesirable consequences of father absence, particularly to boys, many researchers (e.g., Lynn, 1974) accept the correlational evidence between father absence and juvenile delinquency, despite the controversy about the reasons for the relationship. Delinquency in boys could result from a masculine protest against feminine domination, from inadequate childrearing supervision, from the lack of a male model, and from the resulting loss of family cohesion. The assumption underlying this type of research has been that if a man is a good provider, he is a good father. Thus the research has been narrowly confined to the extreme situation of complete father absence either because of occupational necessities or because of emergencies such as wars.

According to Farrell (1974), men need to redefine their role in the family so they can become involved in childrearing and channel their energies in roles leading to increased fulfillment. Fathers of newborns were observed to be "engrossed" in their infants and stated that they were bonded, absorbed, and preoccupied with their child in a way that did not differ from mothers (Greenberg & Morris, 1974). In fact, the only finding (Parke & O'Leary, 1975) that differentiates mothers' and fathers' behavior toward the first newborn was the fact that fathers smiled less frequently than mothers. Men, in general, have the capacity for parenting just as women do, but their family roles often are limited because of the nature of their work roles (Pleck, 1975). Evidence also suggests that men who find themselves in a

Both mothers and fathers are important caregivers for their children.
(Charles Gatewood.)

single-father situation through desertion or divorce, or who are catapulted into a strong parenting role by unemployment, particularly when wives are employed, feel unaccepted, left out, and inadequate (Pleck, 1975).

Divorce

With a couple's desires to find an intimate, committed relationship can come the realization that the marriage is not working and that it needs to be dissolved. However, the law has been traditionally unwilling to make divorce easy. For years, court proceedings supported the idea that guilt must be assigned to one partner or another, with punitive actions being taken against the marital partner most responsible for the "breakup" of the marriage and social entitlements accorded to the "injured" partner and withheld from the guilty one. After the Matrimonial Causes Act of 1937, the number of divorces granted increased dramatically, and after the Divorce Reform Act of 1969 and the Matrimonial Proceedings Act of 1970, the bases on which divorce could be obtained were broadened. The rights of children were given no major role until the provisions of the Children's Act in 1975 required child representation and expeditious court proceedings.

Partly as a result of these legal changes, divorce reached a record high of 5 per 1000 population in 1976. The proportion of U.S. marriages ending in divorce, on a lifetime basis, is about 40 percent. These trends do not mean that we are abandoning the concept of marriage; in fact, of divorced males under 35 years of age, 50 percent remarry within a year, and of divorced women under 35 years of age, 50 percent remarry within a fourteen-month period. During a three-year period, 75 percent of divorced people remarry (U.S. Department of Health, Education and Welfare, 1973).

Divorce is generally a stressful event. The role of husband or wife is lost, as is an object of love and attachment. Divorced people demonstrate lower self-esteem, suffer from feelings of loneliness, are less productive at work, and are anxious over social situations. New experiences—financial arrangements, living arrangements, child-care problems, finding or maintaining a position, establishing a household, etc.—are also faced. The difficulty of the first year subsequent to a divorce (or separation) should not be underestimated. This period is the time when sex differences in adjustment are most dramatic. Women generally report greater difficulty adjusting to the process of divorce, while men report greater difficulties following the event. Family members and children of divorced or separated couples also express distress as a reaction to divorce. Finally, although the degree of trauma experienced is related to numerous factors (e.g., to length of marriage; age of the people involved; number and ages of children, if any; who suggested the divorce; and how the course of action was carried out), few divorced people feel they made a mistake. In fact, feelings of having exercised the right course of action increase with the period of time following the divorce (e.g., Bernard, 1973; Stinnett & Walters, 1977).

In a major study of divorce, Hetherington et al. (1979) studied family interaction among 144 divorced and married parents and their children. The mean ages of the four groups of parents ranged from 27 to 30. The investigators were primarily interested in the family relations and lifestyles of the divorced mothers and fathers as well as the coping strategies they used.

A key finding emerging from the interviews concerned practical problems. Divorced women felt overwhelmed by the quantity of tasks that faced them. They reported that there was not enough time or physical energy to deal with household maintenance, financial tasks, child care, and occupational and social demands. Compared to married women, divorced women spent considerably more time on all these tasks. Women reporting the most extreme "chaotic lifestyle" were those whose sex roles had been rigid in their previous marriage. Both divorced men and women, however, reported family disorganization, involving such things as eating at irregular times and performing household tasks erratically. This feeling of disorganization peaked for women during the first year of their divorce; it lessened during the second year. For divorced fathers, family disorganization was at its highest point during the first two months after the divorce. Compared with mothers and fathers from intact homes, divorced parents' ratings of family disorganization were consistently higher two months, one year, and two years following divorce. When female friends helped divorced mothers with household tasks and when divorced fathers employed a cleaning person, some relief from the stress of disorganization was reported.

Many differences between parents in divorced and intact families were found in terms of self-concept and emotional adjustment. These differences are considerably reduced within one to two years following divorce. Parents feel the most anxious, depressed, angry, rejected, and incompetent one year after being divorced.

Divorced fathers experience greater initial changes in self-concept than mothers, but the effects last longer in women. Feelings of unattractiveness and helplessness, and loss of identity associated with marital status, were reported by women. Men complained about not having structure in their lives. Parents who were older or married longer experienced the greatest changes in self-concept. Men seemed to cope with the problem of damaged self-concept by involving themselves in numerous social activities, altering their style of dress, and buying sports cars or motorcycles. Women showed greater weight gain, underwent cosmetic surgery, and altered their hairstyles. These activities appear to be temporary adaptive strategies, however, because after a satisfying, intimate heterosexual relationship was developed, they no longer occurred. Moreover, the establishment of a new relationship was associated strongly with an improved self-concept.

Divorced parents had negative attitudes toward social activities. They felt that the social world is a world of couples, not singles. Women felt this even more so than men, stating that they were "locked into a child's world" (Hetherington et al., 1979, p. 107). Although both parents reported a need for intimacy and were not satisfied with casual relationships, men were pleased with opportunities for sexual experiences immediately following divorce; women were not.

Divorced women reported feeling more depressed and incompetent than married women with respect to economic matters. They attributed these feelings to lack of experience, to discrimination encountered in financial dealings, and to sex-role ideology that encouraged incompetence in financial matters. Married women also stated that they were uninformed about economic matters, that their husbands "handled" them, and that they were not anxious about being incompetent. Surprisingly, feelings of incompetence were not strongly related to maternal employment.

The parent-child relations of divorced parents and their children were dramatically different from those of intact families. The differences were greatest during the first year. Father visitation was the biggest source of conflict. As time passed, fathers were increasingly late for visits, canceled visits, or did not appear when expected. Children reported distress, and mothers were infuriated. They coped by trying to console their children with presents they could not afford. Divorced parents in general demand less of their children, communicate less well, are less affectionate, and show inconsistent discipline and lack of control as compared with married parents.

The misuse of reinforcement appeared to be the major contributor to poor parent-child relations among divorced families. When children complied with parental requests, they received positive reinforcement less than half the time. Boys received significantly less than girls. Typical responses directed at boys were "You didn't do that very fast" or "You'd better shape up if you know what's good for you" (p. 117). Over the two-year period, mothers' responses to compliance tended to become more appropriate, while fathers' responses became less reinforcing. By the second year after divorce, therefore, disruptive behavior in girls had almost disappeared and was significantly reduced in boys. Children, especially sons, still exhibited more negative behavior in the presence of mothers than in the presence of fathers.

The study concluded by saying divorce is not victimless. Among the families studied, at least one member showed distress and disrupted behavior. In some situations, however, divorce is a positive solution. Thus divorce, although stressful, may represent a transition period and can lead toward improved individual and family functioning.

Occupation

What types of occupational commitments do young adults make? Are these commitments similar for most young adults, or are they different? Four distinctly different types of occupational commitment emerged in a study of college students (Marcia, 1966): (1) those not committed to an occupation, with no concern, (2) those not committed to an occupation, with expressed concern, (3) those attempting to find occupational commitment, and (4) those who did not address the issue of commitment. Among female young adults, those who select more nonstereotyped majors, such as biology or prelaw, are more resolute about their occupational commitment. Male young adults, regardless of major, were distributed equally across the four types of groups. There appears to be evidence suggesting that there is a differentiation in occupational commitment for young adults and that most young adults move toward establishing a commitment to an occupation.

Holland (1973) assessed the vocational preferences of high school and college students in a variety of ways and devised a personality-environment model for explaining career selection. Exhibit 15.1 summarizes the six types of personality environments he proposes and the specific characteristics of each. Within this framework, selection of a job is related to the degree of congruence between, or compatibility of, person and environment. Also, a person's vocational choice is likely to change when in a work environment that creates incongruence. Achievement and feelings of competency are thought to develop through a congruent match of person and work situation.

In general, occupational images are stable among adults and are perceived selectively depending upon one's social status, intelligence, and degree of involvement with the occupation in question (Holland, 1973). For example, several studies show that fifteen major occupations were perceived similarly by male and female high school students, college students, and faculty. However, bright high school students had more accurate perceptions of higher-level job stereotypes and less bright high school students had more accurate perceptions of lower-level job stereotypes.

It appears that entry into a vocation and sense of commitment are related to one's perceptions of the world, which are influenced by various familial experiences. Mothers' and fathers' occupations and childrearing experiences influence career selection and feelings of competency. Conventional types (both male and female) have mothers who were the most authoritarian. However, young adult males

EXHIBIT 15.1

Personality-Environment Model for Career Selection

Personality-Environment Types	Characteristics	
	Environment	Personality
Realistic	Stimulates people to perform tasks Fosters technical competencies and achievement Encourages a world view in simple, tangible, and traditional ways Rewards people for showing conventional values	Conforming, frank, persistent, practical, stable, uninsightful
Investigative	Stimulates people's curiosity Encourages scientific competencies and achievements Encourages a world view in complex, abstract, independent, and original ways Rewards people for displaying scientific values	Analytic, critical, curious, independent, intellectual, introspective, introverted, methodical, passive, pessimistic, precise, rational, reserved,
Artistic	Stimulates artistic activities Fosters artistic competencies and achievements Encourages a world view in complex, independent, unconventional, and flexible ways Rewards people for displaying artistic values	Unassuming, unpopular, complicated, disorderly, emotional, imaginative, impractical, impulsive, independent, introspective, intuitive, nonconforming, original
Social	Stimulates social activities Fosters social competencies Encourages a world view in flexibly cooperative social ways Rewards people for showing social values	Cooperative, friendly, generous, helpful, idealistic, insightful, kind, persuasive, responsible, sociable, tactful, understanding
Entertaining	Stimulates enterprising activities Fosters enterprising competencies and achievement Encourages a world view in terms of power, status, and stereotypical ways Rewards people for showing power, status, and money	Acquisitive, ambitious, argumentative, dependent, energetic, flirtatious, impulsive, pleasure seeking, self-confident, social
Conventional	Stimulates conventional activities Fosters conventional competencies and achievements Encourages world view in simple, constricted ways Rewards people for showing conventional values of money, dependability, and conformity	Conforming, conscientious, defensive, efficient, inhibited, obedient, orderly, persistent, practical, self-controlled, unimaginative

Source: Based on Holland, 1973.

classified as realistic types were more likely to have both parents with authoritarian attitudes. When both parents expressed democratic childrearing attitudes, young adult males were classified as investigative. Other associations between childrearing types and personality types were not significant (Holland, 1973). This suggests that familial experiences, e.g., parents as models, contribute to development of personality types. These, in turn, are thought to partially determine the way in which individuals select a work environment.

Holland's model suggests that a wide variety of factors will produce divergence in career choices as individuals move from adolescence to young adulthood. Abeles, Steel, and Wise (1980) point out that as such choices are made, the individuals are increasingly locked into and out of particular career lines for the rest of their lives. For example, investigating math-related careers, Steel and Wise (1979) found a linkage between course selection, math achievement, and career outcome. As one might expect, math achievement was a significant determinant of the realization of math-related career plans. However, differences in math achievement among students of equal initial abilities could be explained by differences in the number of math courses taken in high school. Seemingly routine decisions regarding whether or not to take additional math courses were critical. Students who decided against such additional courses in high school were effectively closing out their option to pursue a math-related career.

Thus an educational and occupational choice in adolescence and young adulthood has far-reaching implications. Often without knowing it, individuals make choices that are very difficult to reverse later on.

THE MIDLIFE TRANSITION

Until relatively recently, there has been little work which has examined development during the middle years of adulthood. The bulk of developmental research has focused on children and adolescents and, more recently, the elderly. Of course, research on middle age is not totally lacking. In the 1930s, for example, Bühler (1935) incorporated ideas about changes in midlife into her personality theory. Further, many developmental studies have included middle-aged individuals as a comparison group against which to evaluate the performance of younger or older adults (Borland, 1978). Nevertheless, little attention was devoted to middle age as a developmental phenomenon in its own right until about fifteen years ago.

As one might expect, given the short duration of this scientific interest, our understanding of midlife is limited. Conflicting images abound. On the one hand, some views of middle age emphasize the changes which occur during this period. Midlife is seen as a crisis which evokes feelings of frustration and dissatisfaction and precipitates major life events such as divorce and career changes. On the other hand, some views of middle age emphasize the stability of the period. The middle aged are seen as consistent, conservative, and responsible—the leaders of society and the parents of the next generation. Which of these two views is correct?

The Case for Crisis

One view of midlife is derived primarily from the literature of psychiatry. The major assumption of this view is that crisis is a critical mechanism in the intrapsychic development of the individual (Perun & Bielby, 1979). Crises are presumed to occur episodically throughout development, thereby defining the major transitions of the life cycle. Within this general framework, several theorists have characterized the middle years as a time of crisis (Erikson, 1959; Gould, 1978; Levinson, 1978; Vaillant, 1977).

Levinson (1978), for instance, feels the Midlife Transition spans a period of from four to six years, reaching its culmination in the early forties. Two fundamental tasks must be accomplished during this transition: (1) the individual must reappraise his past life structure and (2) must begin to make choices which will modify this early structure and provide a basis for living in middle adulthood.

We noted in Chapter 13 that Levinson's work is based on a study of forty men aged 35 to 45 who were interviewed over a two-year period. This study is open to criticism on a variety of points. The sample is small and, more importantly, is restricted to males. Moreover, Levinson's proposal of a universal sequence of periods, among which the Midlife Transition is only one, has been questioned (Brim, 1976). Nevertheless, this work is a good example of the transition and crisis view of midlife.

On the one hand, Levinson (1978) says it becomes important to question what one has done with one's life—to examine one's talents, accomplishments, and relationships. This process does not reflect a simple evaluation of success versus failure, although this is an important element. Rather, it is more a matter of judging the "goodness of fit" between the life structure and the self (Levinson, 1978). During young adulthood, certain choices were made as the life structure was developed. An individual may have made commitments to certain values and goals and to a particular spouse and occupation. The choices reflected certain aspects of the self. Other aspects, however, were ignored, rejected, or remained undiscovered. Levinson suggests that in midlife this discrepancy must be resolved regardless of the degree of success the individual has had in attaining the goals of early adulthood.

Similarly, Levinson indicates that during midlife, an individual discovers how much of this young adult life structure is based on illusions ("If I become vice president of Ajax and have a home in Forest Acres, I will be truly successful and happy"). An important part of the reappraisal process, then, is what Levinson labels "deillusionment." This involves reducing or losing some of the illusions of young adulthood. Levinson notes that illusions can be both helpful and harmful and that it is probably neither desirable nor possible to discard all of one's illusions. For example, in adolescence and young adulthood our illusions often help us to sustain commitments in the face of difficulty. In midlife, however, effective reappraisal and change requires an examination of these illusions.

According to Levinson (1978), a natural outcome of the reappraisal process is commitment to new choices or recommitment to old choices within a new framework. As the individual begins to make these commitments or recommitments, he leaves the transition phase and enters middle adulthood. For some men the modifications are extensive; for others they are limited. Levinson (1978) identifies several sequences during Early Adulthood and the Midlife Transition that reflect varying degrees of stability and success.

Advancement within a Stable Life Structure. The men in this group had established a stable life structure by the end of young adulthood. In addition, they had received moderate to notable success in their careers and other enterprises. The question now became, "Where do I go from here?" Fifty-five percent of the men were classified in this group.

Serious Failure or Decline within a Stable Life Structure. These men also established a stable life structure in young adulthood. However, before the end of this period, it became clear that they were doing badly in certain critical areas. They were unable to advance in their occupations; their marriages were failing; and they became physically ill. These failures had to be faced during the Midlife Transition. About 20 percent of the sample fit this pattern.

Breaking Out. Men in this group had established a stable life structure early in young adulthood. By the end of this period, however, the choices become unbearable and the individual "breaks out." This usually involves a major life event such as a divorce or job change. New commitments are made and then may quickly come under review as the individual enters the Midlife Transition. This sequence, often quite stressful, was shown by 13 percent of the subjects.

Advancement Which Produces a Change in Life Structure. Some men experienced such a major advancement during young adulthood that it resulted in a significant change in their life structure. This usually involved a promotion or a drastic increase in income. Often this proved to be a mixed blessing, and the changes had to be dealt with during the Midlife Transition. Three men experienced this sequence.

Unstable Life Structure. Finally, three men were unable to form a stable life structure during young adulthood. None of them sought this condition, and none of them found it a fulfilling state of affairs. They experienced frequent changes (e.g., jobs, residences, lovers, spouses) during their thirties when most men establish a relatively stable life. They were uniformly unable to cope with the tasks of the Midlife Transition.

Levinson (1978) estimates that for 80 percent of his subjects, the Midlife Transition was a time of moderate or severe crisis. He comments:

> *Every aspect of their lives comes into question, and they are horrified by much that is revealed. They are full of recriminations against themselves and others. They cannot go on as before, but need time to choose a new path or modify the old one. . . . A profound reappraisal of this kind cannot be a cool, intellectual process. It must involve emotional turmoil, despair, the sense of not knowing where to turn or of being stagnant and unable to move at all. . . . Every genuine reappraisal must be agonizing, because it challenges the illusions and vested interests on which the existing structure is based. (p. 199)*

The Case for No Crisis

A second view of midlife, held by a diverse group of psychologists and sociologists, suggests that the role of crisis in adult development is limited. According to this view, crises occur when developmental processes are interrupted. Crisis is not seen as an inherent part of development but as the result of "random" events. For these writers, development is organized by the age norms which regulate the individual's passage through the events of the life cycle (Neugarten & Hagestad, 1976). Crises occur when there are unanticipated interruptions in the rhythm of the life cycle. Crisis, then, plays only a limited role in this model. It suggests that major upheavals, such as those described by the crisis theorists, should be the exception rather than the rule.

Some research tends to support this view. For example, there appears to be little evidence that primary (e.g., anxiety, depression, physical complaints) or secondary (e.g., alcoholism, psychosis, suicide) crises peak in the middle years (Brim, 1976). Similarly, in one study Costa and McCrae (1980) developed a Mid-Life Crisis Scale reflective of the stresses of this portion of the life cycle as described by others. However, no age differences were found on this scale in a large sample of men aged 33 to 79. The scale did correlate significantly with Costa and McCrae's personality variables, particularly Neuroticism. Further, there were positive correlations between scores on the crisis scale and measures of Neuroticism obtained *ten years previously.* These findings led Costa and McCrae (1980) to suggest that crisis-prone men may have exhibited problems of adjustment for a long time. Some men do experience a crisis at midlife. Costa and McCrae suggest that these problems may be the result of unadjusted adolescents and young adults growing up to be unadjusted middle-aged adults rather than the result of a universal crisis confined to midlife.

Conclusion: Crisis or No Crisis?

Perhaps the most valid conclusion that can be drawn about midlife at present is that this period of the life cycle is characterized by change. For some individuals this period appears to be a crisis; for others it is not. Perhaps the question "Is midlife a crisis or not?" is the wrong question. Perhaps attention should be focused on how the various events which tend to occur in the middle years affect the individual. Let us examine some of these events.

Physical Decline and Awareness of Death

The peak of human biological functioning tends to be reached between the late teens and the late twenties, depending on the system involved. By the forties most individuals have experienced a noticeable decline in muscle strength, lung capacity, cardiac output, and other physiological capacities. These changes are accompanied by other alterations which affect our appearance, baldness, wrinkles, and weight gain being the most prominent.

These biological reminders of middle age occur gradually. Further, they do not interfere with the everyday lives of most individuals. The exceptions are those persons whose activities require exceptional strength, endurance, or skill, such as professional athletes. In such cases, a career adjustment is often required. For most adults, however, the physical changes associated with middle age do not in themselves significantly alter their life structure. Nevertheless, they do have psychological significance. For example, physical changes appear to serve as a marker of middle age itself, particularly in the case of men. In an interview study of 100 middle-aged men and women, Neugarten (1968) found that men commented more frequently than women on their decreased physical efficiency and other health-related concerns. Women were concerned with the physical health of their husbands —part of a behavior pattern Neugarten (1968) labels "rehearsal for widowhood."

The significance of physical decline in midlife and other cues which draw attention to the aging process appear to have an impact on the individual's time perspective. A number of studies have found that middle-aged individuals begin to restructure time so that life is thought of in terms of time left to live rather than time since birth (Jaques, 1965; Levinson, 1978; Neugarten, 1968). They realize that the time remaining to them is finite. They will die. Things will remain undone. Of course, we all realize that we will die; however, at midlife this realization appears to be qualitatively different. For the first time the inevitability of death becomes a psychological reality. One of Jaques's (1965) interviewees put it this way: "Up till now," he said, "life has seemed an endless upward slope, with nothing but the distant horizon in view. Now suddenly I seem to have reached the crest of the hill, and there stretching ahead is the downward slope with the end of the road in sight—far enough away, it's true—but there is death observably present at the end" (p. 506).

Jaques (1965) argues that it is the realization of the inevitability of one's own personal death which precipitates the midlife crisis. Others (Vaillant, 1977) reject this view and feel that this is only one factor involved in this transition. Nevertheless, there seems to be substantial evidence which points to a psychological shift of time perspective in midlife.

Menopause

During the midlife, there is a gradual decline in ovarian functions and their associated products—sex hormones and eggs. This process is labeled the *climacteric*. The cessation of menstruation refers to a particular event within this process—*menopause*. The age of menopause varies widely from the thirties to the sixties, with the median age being about 50 (McKinley, Jefferys, & Thompson, 1972). It is accompanied by a variety of physical symptoms. For example, vasomotor instability manifested as hot flashes, flushes, and perspiration is a common symptom experienced by a majority of women (McKinley & Jefferys, 1974). However, other symptoms often associated with menopause (sleeplessness, depression, weight gain) do not appear to be directly related to it.

What is the impact of this event on middle-aged women? Psychoanalytic therapists typically view it as a stressful event. From this perspective, feminine identity is strongly related to childbearing. When this possibility ends, a major source of the woman's self-esteem is removed and significant stress results. Deutsch (1949), for example, suggested menopause results in "narcissistic mortification." She also suggested that postmenopausal activities such as interest in work represent an attempt to compensate for the loss of femininity and sexuality. Similarly, Benedeck (1950) believed that menopause would be a more stressful event for women who never had children.

However, while specific clinical cases may be consistent with this view, more representative data sets suggest that menopause is not a particularly stressful event for most women. For example, Neugarten, Wood, Kraines, and Loomis (1963) asked women aged 21 to 65 to respond to a checklist designed to measure attitudes toward menopause. The findings indicated that few of these women saw menopause as a major turning point in their lives. Further, few of these women were concerned about the loss of reproductive capacity. Of the 100 women in the middle-age group (45 to 55 years), only four said, "Not being able to have children" was the worst thing about menopause. Finally, anxiety related to the anticipation of menopause was greater than anxiety related to the actual experience. In general, younger women anticipating menopause took a more negative attitude toward the process than women who were actually experiencing or had already experienced menopause. In a related study, Neugarten and Kraines (1965) found no relationship between menopausal status and a variety of personality characteristics. Similarly, Bart and Grossman (1978) found no relationship between menopausal status and self-evaluations in middle-aged women. The experience of menopause does appear to be influenced by social class and cultural context. For example, menopause is seen as more of a turning point in cultures which emphasize the centrality of the childbearing role for women (Datan, Maoz, Antonovsky, & Wijsenbeek, 1970). However, the bulk of the evidence suggests that menopause is not a particularly stressful event for most women in the United States.

In recent years, references to a "male menopause" have appeared in the popular press. Since menopause refers to the cessation of the menses, the use of the term in this way is incorrect. Although males do not experience menopause, they do experience age-related changes in the reproductive system. For example, there is a gradual decline in sperm production, although viable sperm are produced by the oldest men. Similarly, androgen levels decrease; erection occurs more slowly; the amount of seminal fluid decreases; the force of the ejaculation decreases; and it takes longer to achieve another erection following ejaculation. Any or all of these changes may cause the individual to believe he is losing his sexual abilities; however, there is no time in a man's life when he completely loses his sexual abilities as a result of aging per se.

The Empty Nest

In midlife the nuclear family is typically reorganized as the departure of the children results in an *empty nest*. Early writers suggested this stage of a woman's life was highly stressful and frequently resulted in severe depression, often accompanied by sleeplessness, loss of appetite, sexual desire, and ability to concentrate, and severely low levels of morale, self-esteem, and self-confidence. This behavior pattern was labeled the *empty nest syndrome*.

While cultural stereotype and early clinical evidence suggested that this syndrome was widespread, more recent and more representative data indicate it is relatively rare (Glenn, 1975; Lowenthal & Chiriboga, 1975; Neugarten, 1968; Maas & Kuypers, 1975).

For example, Lowenthal and Chiriboga (1975) interviewed twenty-seven men (average age 51) and twenty-seven women (average age 48) who were approaching the empty nest stage (i.e., their youngest child was about to graduate from high school). As part of the interview, the subjects were asked to complete a life evaluation chart, rating each year of their lives from 1 (absolute tops) to 9 (rock bottom). Among men, all life stages were characterized by a similar proportion of high points. Among women, the period of early marriage and parenthood was characterized by the greatest number of high points. But the empty nest period included more highs than early middle age (39 to 44 years) and the anticipated retirement period. Only five individuals indicated that a year in the empty nest period was the lowest point in their lives. Further, the reasons for the low ratings were not related to the empty nest in any of these cases. Finally, there was little evidence that the respondents' frustrations, marital difficulties, or perceptions of the major turning points in their lives were related to the departure of their children from the home.

Studies questioning women directly about the empty nest have typically revealed that they look forward to the departure of their children. Rubin (1979), for example, interviewed 160 women (average age 46.5 years) whose children were in various stages of leaving the nest. Only one of these women exhibited the classical symptoms of the empty nest. The great majority viewed the impending or actual departure of their children with a sense of relief, and this was particularly true among those whose children had already departed.

> *I can't tell you what a relief it was to find myself with an empty nest. Oh, sure, when the last child went away to school, for the first day or so there was a kind of a throb, but believe me, it was only a day or two. (Rubin, 1979, p. 15)*

> *When the youngest one was ready to move out of the house, I was right there helping him pack. We love having the children live in the area, and we love seeing them and the grandchildren, but I don't need for any of them to live in this house ever again. I've had as much as I need or want of being tied down with children. (Rubin, 1979, p. 16)*

Even those who had not yet had a child leave home were frequently ready to let them go.

> *From the day the kids are born, if it's not one thing, it's another. After all these years of being responsible for them, you finally get to the point where you*

*want to scream, "Fall out of the nest already, you guys, will you? It's time."
(Rubin, 1979, p. 13)*

*You know, when the kids get a little older, you can actually go away for a week-
end by yourselves, and that's great. But somehow, your head is still at home
worrying about what's going on there. I'm ready to be able to go away and
have all of me away. I'm ready. [Groping for words to express the depth of her
feeling] Oh hell, it's not just going away that I care about. I want all of me. It's
as if I want to take myself back after all these years—to give me back to me, if
you know what I mean. Of course, providing there's any "me" left. (Rubin,
1979, p. 18)*

Some women, typically those closer to the transition, did express ambivalence.

*It's complicated; it doesn't just feel one way or the other. I guess it's rather a
bittersweet thing. It's not that it's either good or bad, it's just that it's an era
that's coming to an end and, in many ways, it was a nice era. So there's some
sadness in it, and I guess I feel a little lost sometimes. But it's no big thing; it
comes and goes. [Suddenly straightening in her chair and laughing] Mostly, it
goes. (Rubin, 1979, p. 16)*

However, very often there had been anticipatory socialization for the event so that
by the time it occurred the ambivalence had been resolved.

*By the time my daughter left for college, I had already dealt with the issues.
From time to time in her senior year in high school, I'd get a pang thinking
about what was coming. I must admit, though, that by the time it actually happened, even I was surprised by how easy it was. I guess I had just grown accustomed to the idea by then. (Rubin, 1979, p. 20)*

In sum, the prevailing evidence suggests that while the empty nest may be a
negative event for a few women, it is a positive event for the great majority of
women. Little data exist in the case of men. Indeed, there is some evidence to suggest that the empty nest may be a more significant event for fathers than for mothers
(Rubin, 1979).

Hagestad (1980) points out that one of the major problems in this area has been
the lack of a conceptual framework for examining the positive effects of children
leaving home. She suggests that we should no longer view the event as a role loss
but as a role transition. That is, the mother-child dyad continues to exist, but the role
relationship is renegotiated as conditions change. As a result of these changes, Hagestad suggests that children increasingly become resource persons for their
mothers.

Careers

The conventional view of career has been limited by what Sarason (1977) calls the
"one life–one career imperative." According to this perspective, occupational
choice occurs in adolescence and young adulthood and represents a lifelong commitment. More recently, initial occupational choice has been seen as the first step in
a process of career development that spans adulthood (Sarason, 1977). Within this
context, increasing attention has been directed toward careers in middle age.

In the case of men, the emphasis has been on job and career changes during middle age. For example, in an effort to document career changes among professional men, Sarason (1977) tabulated the frequency of career shifts among entries in *Who's Who* for the years 1934–1935, 1950–1951, and 1974–1975. Of the roughly 2300 individual entries for each of these years, approximately 40 percent had experienced a career shift. Further, almost 10 percent of these shifts were radical changes from one professional field to another (e.g., business to medicine).

What factors influence career shifts among middle-aged men? Murray, Powers, and Havighurst (1971) developed a model which suggests that an individual is subject to two sources of pressure to change jobs—internal and external sources of pressure. Based on this notion, they developed a fourfold typology of potential work histories, shown in Exhibit 15.2. A study by Clopton (1973) provides some support for this model. Clopton compared two groups of men in their thirties and forties. One group made a midlife career shift, while the other did not. The two groups were matched on various background variables (e.g., birth order, college experiences). They were also similar in terms of their present family responsibilities and degree of success in their original careers. Clopton found that both personality factors and adult life experiences accounted for the differences between the "shifters" and the "persisters." Specifically, the shifters exhibited higher levels of self-acceptance and self-esteem than the persisters. In addition, the shifters had experienced more significant life events (e.g., divorce, religious conversion, sudden loss of employment) than the persisters. The shifters gave three major reasons for changing careers: (1) a gradual disenchantment with their first career; (2) discovery of a new career that promised more satisfaction (often through pursuit of an avocation); and (3) the reformulation of goals as a consequence of various life-event experiences.

In the case of women the emphasis has been on major shifts in participation in the labor force. For example, in the 1920s the typical working woman was single, under 30 years of age, and from the lower socioeconomic class. Today the typical working woman is married, has childrearing responsibilities, and represents the entire socioeconomic spectrum. Between 1950 and 1973, labor force participation rose 13 percent for never-married women and 80 percent for women living with their husbands. Many factors have influenced the increase in labor force participation by married women: (1) a desire for increased family income; (2) the need for independence, social contact, and self-actualization; (3) an interest in status, recognition, and achievement; (4) a more tolerant societal attitude toward working wives; (5) an

EXHIBIT 15.2   **Occupational Patterns as a Function of Personality and Situational Factors**

Personality Factors	Situational Pressures	Stability or Change in Occupation
1. Low self-direction	Little pressure to change	Routine career; advancement follows seniority
2. High self-direction	Little pressure to change	Flexible career; initiative for change and type of work assumed by person
3. Low self-direction	Much pressure to change	Disjointed career; technologically oriented reasons for unemployment (e.g., blue-collar workers), no skill or experience (e.g., widow)
4. High self-direction	Much pressure to change	Orderly, sequential career; well-planned effort in making changes assumed by person

Source: Murray, Powers, and Havighurst, 1971.

increase in the age of first marriage; (6) a decline in the birthrate; (7) increased availability of child-care services; and (8) reduced sex discrimination (Hoffman & Nye, 1974; Kreps, 1976).

In examining the careers of women, we must distinguish among several different groups:

1 Those who occupy an occupational role but not the role of spouse or mother
2 Those who occupy an occupational role and the role of spouse but not the role of mother
3 Those who occupy all three roles—worker, spouse, and mother
4 Those who occupy all three roles but whose children have left the home

Earlier studies reporting the negative consequences of the mother's working on children did not consider social class differences, maternal satisfaction or dissatisfaction, and provisions for child care. When these factors are considered, maternal

BOX 15.2

ARE WORKING WOMEN MORE SATISFIED?

Interestingly enough, there appears to be a sharp discrepancy in the literature examining the meaning of work for men and women (Wright, 1978). When the focus is on working men, work is generally pictured as degrading, alienating, and unsatisfying. In contrast, when the focus is on working women, work is generally pictured as enriching, liberating, and satisfying in spite of the fact that women are more likely to be found in less desirable occupations. The positive aspects of work for women are often contrasted with the role of the full-time housewife; the latter role is generally viewed as lonely, boring, and demeaning.

Who does lead the more satisfying life—women with jobs outside the home or full-time housewives? Many studies find little difference in overall satisfaction between the two roles (Blood & Wolfe, 1960; Wright, 1978). For example, in an analysis of national survey data collected between 1971 and 1976, Wright (1978) found no significant differences between working women and housewives in satisfaction with their lives in general or components thereof—e.g., their work, marriages, or families. Similarly, in a study of 142 middle-class women, Baruch and Barnett (1980) found no differences between working women and women at home in their role satisfaction or self-es-

teem. Thus in spite of the fact that some studies have found working women to be happier than nonworking women (Ferree, 1976), the bulk of the studies suggest that involvement in the multiple roles of wife, mother, and worker neither enhances nor diminishes general satisfaction with life.

Yet perhaps this is not the most important issue. For example, Baruch and Barnett (1980) found that the sources of satisfaction and self-esteem varied for employed and nonemployed women. Specifically, the well-being of nonemployed women was highly dependent on their husbands' approval of their activities. In contrast, the well-being of the employed women was significantly affected by their satisfaction with their jobs and careers as well as their interactions with their husbands. As Baruch and Barnett (1980) point out, these work-related variables are likely to constitute a more stable base for well-being since they are more directly under the woman's own control. Given the high rate of divorce in our contemporary society, building one's life on the approval of one's spouse can be a risky strategy. Baruch and Barnett (1980) conclude that the development of women is likely to be facilitated by the acquisition of occupational competence and the capacity for economic independence.

employment does not affect children adversely (e.g., Hoffman & Nye, 1974). For instance, role satisfaction is related to children's positive adjustment and success in school. "Better mothers," whether working or nonworking, were those who were satisfied with their roles. Also, adolescent daughters of working mothers were more likely to view women as competent and more likely to highly value women's accomplishments than daughters of nonworking mothers (Kreps, 1976).

Although many married women simply add the work role to their other ones, others have difficulty with employers who do not understand their need for "flex time," with family members who do not want to alter tasks, and with adequate child-care arrangements (Wilensky, 1968). In a study of middle-aged and older women, Kreps (1976) found that attention to their role as spouse was not reduced because of work. But attention to the roles of mother and homemaker was reduced when they were devoting energies to high work performance. In particular, employed women reported spending half the amount of time doing housework as unemployed women (Kreps, 1976). Consistently, women evidence career success and satisfaction when husbands are supportive of their careers (e.g., Kreps, 1976; Rapoport & Rapoport, 1971). Box 15.2 illustrates the meaning of work for women having multiple roles.

THE TRANSITION TO OLD AGE

Toward Integrity

Erikson (1959, 1963) proposes that the transition to old age is characterized by the crisis of integrity versus despair. As commitment to integrity develops, the ego strength of wisdom emerges. The individual is able to accept that life is coming to a conclusion. Erikson (1968) states, "It is the acceptance of one's one and only life cycle as something that had to be and that, by necessity, permitted no substitutions" (p. 87). Integrity provides the individual with the wisdom to understand his or her own life. According to Erikson, this understanding both balances the decreases of potency and performance in aging and allows the individual to serve as an example to the upcoming generation. Despair, on the other hand, represents a rejection rather than an acceptance of the past life and a fear of death resulting from the realization that there is insufficient time to make up for past mistakes. Finally, the individual may deny or fail to deal with the crisis at all. Indeed, there does seem to be evidence to suggest that the attempt to achieve a sense of completion is important during old age. In particular, reminiscence is thought to be an activity that facilitates feeling that life is complete.

Nevertheless, there are a number of questions that may be raised concerning the adequacy of integrity as an organizing concept for development in later adulthood. Clayton (1975), for example, suggests that relatively few individuals ever achieve a commitment to integrity, and, therefore, achieve wisdom in old age. In part, this failure reflects the tendency of individuals to become fixated at earlier stages of development—particularly the adolescent crisis of identity. It also has been noted that Erikson's last stage is the most intrapersonal of his stages (Glenwick & Whitbourne, 1978). The previous stages are rooted in the context of ongoing interpersonal interactions. The final stage is rooted in the person's past activities and future death. The research on reminiscence (cf. Butler, 1968) suggests this linkage of past and future is important. However, as a result of this focus, Erikson's theory tends to ignore the ongoing events in the individual's life and how these are embedded within the changing social context.

Another major theoretical framework has been developed which uses the concept of disengagement to explain how older adults exit from society. In its early formulation (Cumming & Henry, 1961), this view emphasized the adaptive consequences of the mutual disengagement occurring between the individual and society. As we noted in Chapter 13, this assertion produced considerable controversy, particularly with regard to the presumed mutuality of the process and its adaptive consequences (Havighurst, Neugarten, & Tobin, 1968; Maddox, 1964). Generally, it now has been concluded that while some individuals do disengage, the linkage of this behavior to successful aging is not useful. Neugarten (1973, p. 33) says, "In short, disengagement proceeds at different rates and different patterns in different people in different places and has different outcomes with regard to psychological well-being." Accordingly, while the concept of disengagement emphasizes the linkage of the individual with society, it focuses on a very narrow definition of social influence.

Past views of the transition to old age, then, are partially useful but fail on a number of crucial points. More recently, several writers have agreed that what is required is a systematic conceptualization which would emphasize the points of articulation between the individual and society (Lowenthal, 1977; Glenwick & Whitbourne, 1978). This brings us back to our emphasis on life events. In the following sections of the chapter, a perspective on the transition to old age is provided by an examination of several life events common to this part of the life cycle.

Retirement

Retirement is a phenomenon of modern industrial society. Only industrial societies have experienced an increase in the proportion of individuals designated as old, and only industrial societies have experienced a decrease in the proportion of the old who are active members of the labor force (Sheppard, 1976). How do people prepare for and adjust to retirement? How satisfied are they with it?

An accumulation of lifetime experiences influences attitudes toward retirement and retirement plans. Attitudes are influenced by a variety of factors such as income, educational level, and the type of work a person does. For example, people who find their jobs boring and unrewarding tend to hold very favorable attitudes toward retirement. Barfield and Morgan (1978) suggest this positive attitude reflects a desire to escape dissatisfying circumstances. Income is also a predictor of attitudes toward retirement (Shanas, 1972). Glamser (1976), for example, found that positive expectations of retirement were held by working men who felt they would have sufficient income to maintain their standard of living. Although high income appears to be a factor associated with favorable opinions toward retirement, it is difficult to separate the influences of this variable from others. Sheppard (1976), for instance, reports that highly educated men tend to be negative about retirement, but highly educated women are positive about retirement.

Retirement plans are also affected by a variety of factors. McPherson and Guppy (1979) suggest that people with certain lifestyles are most likely to plan for retirement. Most of the men in their study (aged 55 to 64 and representing a range of occupations and incomes) reported looking forward to retirement. However, men with higher socioeconomic status gave more thought to and made more specific retirement plans. In addition, higher levels of health, income, leisure interests, and life satisfaction were associated with retirement planning.

Lowenthal (1972) suggests that the way people have behaved throughout their work lives and the attitudes they hold will influence their adjustments to retirement.

A person who has been work-oriented throughout life will consider retirement traumatic. Substitutions need to be planned by such a person; otherwise, depression and fear of dependence may result. One such substitution is another job. Of those who said work was a major source of satisfaction, 73 percent returned to work after retirement from their other job (Streib & Schneider, 1971). These people were healthier, more educated, and held positions of higher status compared with those not returning to work. Others, who feel work is not an end in itself, but a means to an end (e.g., social acceptance, approval), appear able to adjust to retirement more easily than the work-oriented type.

Lowenthal (1972) describes three additional types of people—self-protective, autonomous, and receptive-nurturant. For the self-protective person retirement is a time when detachment and no responsibility for others are maintained. This person disengages, and retirement presents no particular problem unless it is experienced in the context of other events. The autonomous individual frequently has selected a career that allows him or her to decide when to retire. If the person can exercise this option, then the experience is usually positive. On the other hand, if retirement is mandatory, depression is a likely consequence and reorientation is required. Finally, the receptive-nurturant individual usually is female and has devoted her life to affect and intimacy. Her positive adjustment to retirement is contingent on her perceived quality of the marital relationship. To the extent that the marriage is "good," she adjusts easily to retirement and may even feel relieved that she can devote more time to her marital role. A theme cutting across Lowenthal's proposed retirement types is that a redistribution of energy is needed in order to accommodate to retirement and that this adaptation reflects the process used by the person throughout life.

How satisfied are people with retirement? A major research question deals with the relationship between level of activity or use of leisure time and life satisfaction. In a classic longitudinal study by Streib and Schneider (1971) of more than 4000 people who held a variety of jobs ranging from unskilled labor to professional positions, assessments of health status, economic status, and different psychosocial areas (e.g., satisfaction with life, feelings of usefulness, and adjustment to retirement) were gathered. The data indicated that most people were adjusted to retirement and satisfied with life. These individuals were healthy and stated their incomes were adequate. In fact, one-third of the sample said retirement was better than they thought it would be and only 4 to 5 percent felt that it was worse than they expected. However, 10 percent did return to work—again adding support to Lowenthal's conceptions.

Thus as long as income and health are maintained during retirement, life satisfaction and adjustment to retirement are positive. The only expressed concern differentiating men from women is that retired women were concerned with loss of social contacts (Fox, 1977).

Grandparenthood

The status of grandparent is a prevalent one in our society. During the last century, however, we have observed a quickening of the family life cycle (Neugarten & Moore, 1968). Marriage, the birth and departure of children, and grandparenthood all tend to occur earlier today than at the turn of the century. But what is the meaning of grandparenthood for various generations? How does the grandparent role vary for different individuals? How do grandparents influence their grandchildren?

There is relatively general agreement that the grandparent role is not clearly defined in today's society (Clavan, 1978; Kahana & Kahana, 1971; Robertson, 1977). Clavan (1978) suggests that, particularly for the middle class, it is a "roleless role." That is, there is a kinship status of grandparent, but few normative rights and responsibilities associated with it. As a result, grandparents, parents, and grandchildren must construct the role.

Given the diffuse nature of the expectations associated with grandparenthood, we would expect multiple styles of playing out the grandparent role. Several studies have examined these different patterns. In one study, Neugarten and Weinstein (1964) interviewed seventy pairs of grandparents about their relationship with their grandchildren. Three dimensions were examined: the degree of comfort with the role as expressed by the grandparent; the significance of the role to the grandparent; and the style with which the role is enacted by the grandparent. Neugarten and Weinstein's analysis indicated that the vast majority of grandparents expressed satisfaction with the role. However, a significant minority (approximately one-third) were experiencing sufficient difficulty in the role and said they felt discomfort or disappointment. These difficulties often were related to feeling uncomfortable in thinking of themselves as grandparents or conflict with the parents over rearing or interacting with the grandchild.

Within this context, Neugarten and Weinstein (1964) found grandparenthood had multiple meanings for people. Five themes were predominant. For some, grandparenthood was a source of *biological renewal and/or continuity.* It evoked feelings of youth (renewal) or extensions of the self and family into the future (continuity). For others, grandparenthood was a source of *emotional self-fulfillment.* It evoked feelings of companionship and satisfaction from the development of a relationship between adult and child often missing in earlier parent-child interactions. As shown in Exhibit 15.3, these themes were prominent for relatively large percentages of the respondents. Two other themes were suggested by relatively few grandparents. These were *resource person* and *vicarious achievement.* In the former case, satisfaction was derived from contributing to the development of the grandchild—

EXHIBIT 15.3   **Meaning and Style of Grandparenting**

	Grandmothers*	Grandfathers*
Meaning of Grandparent Role		
1. Biological renewal and continuity	42	23
2. Emotional self-fulfillment	19	27
3. Resource person	4	11
4. Vicarious achievement	4	4
5. Remote	27	29
Style of Grandparenting		
1. Formal	31	33
2. Fun seeking	29	24
3. Parent surrogate	14	0
4. Reservoir of family wisdom	1	6
5. Distant figure	19	29

* Responses in percentages. Difference between total for each column and 100 percent reflects insufficient data.
Source: Based on Neugarten and Weinstein, 1964.

for example, through financial aid, life experiences, and so on. In the latter case, grandparenthood was seen as providing an extension of the self in that the grandchild will accomplish what neither the grandparents nor the parents were able to. Finally, 27 percent of the grandmothers and 29 percent of the grandfathers felt *remote* from their grandchildren. These respondents saw relatively little effect of this role on their lives.

Neugarten and Weinstein (1964) also identified five styles of interaction which were somewhat independent of the meaning dimensions. As indicated in Exhibit 15.3, a *formal* style was relatively frequent for both grandmothers and grandfathers. This style involved performing what was regarded as a proper and prescribed role. While they maintained continued interest in the grandchildren and sometimes provided them with special treats, formal grandparents left parenting to the parents. They were careful not to offer advice on childrearing or otherwise interfere with the parental role. The *fun-seeker* style was characterized by informality and playfulness. Grandchildren were viewed as a source of leisure activity, and the emphasis was on mutual satisfaction. Authority was seen as irrelevant. The *surrogate parent* style was found only for grandmothers. In this case, the grandmother assumed actual caretaking responsibility for the grandchild while the mother worked. The *reservoir of family wisdom* occurred rarely and usually involved the grandfather. In this style, the grandparent was a dispenser of special skills or resources. Both parents and grandchildren were usually subordinate to an authoritarian grandparent. Finally, a substantial minority of both grandmothers and grandfathers were relatively *distant figures.* This style is benevolent, but contact is fleeting and infrequent. Of these styles, Neugarten and Weinstein found that the traditional was more prevalent among the older grandparents (over 65) while the fun seeker was more prevalent among the younger (under 65) grandparents. This difference may reflect either differences between cohorts of grandparents or changes with increasing age in the grandparent role.

More recently, Robertson (1977) examined the significance of the grandparent role in a sample of women. Overall, 80 percent of the respondents indicated they were excited, proud, and happy when they learned they were to be grandmothers. Further, the majority indicated their feelings had not changed over the years. However, similar to Neugarten and Weinstein (1964), it was found that this role has different meanings to different individuals. Robertson drew a distinction between a social and a personal orientation. Those emphasizing a social orientation focused on the normative expectations of grandparents (e.g., they should set a good example, encourage grandchildren to work hard and be honest, etc.). Those emphasizing a personal orientation focused on the joys and pleasures of grandparenthood (e.g., grandchildren make me feel young, less lonely, etc.). About 26 percent of Robertson's (1977) sample emphasized the social orientation, 17 percent emphasized the personal orientation, and 29 percent emphasized both aspects of the role. An additional 28 percent of the grandmothers were remote from their grandchildren, having little expectation or involvement in the role. These types tended to be predicted by the lifestyles of the family in general and the grandmothers in particular. For example, younger grandmothers tended to emphasize the social orientation, while older grandmothers tended to emphasize the personal orientation. This appears to be related to the fact that most of the younger grandmothers were married and working. As a result, they were involved with their own lives and placed less emphasis on grandparenting. In contrast, most of the older grandmothers were widowed and unemployed. As a result, grandparenting was a significant role. Indeed, these women had the greatest amount of interaction with their grandchildren.

What impact do grandparents have on their grandchildren? Very few studies have been directed toward this question. The few data that are available, however, suggest that grandparents are significant to their grandchildren, although the meaning of the relationship changes with the age of the children.

Kahana and Kahana (1970), for instance, examined the meaning of grandparents for middle-class children aged 4 to 5, 8 to 9, and 11 to 12 years. The youngest children (4 to 5) valued grandparents primarily for egocentric reasons, that is, for what the grandparents gave to the child in terms of love, food, and gifts. In contrast, the older children (8 to 9) mentioned characteristics of the grandparents (e.g., "He is a good man") and mutual activities (e.g., "We go to the ball game together"), as well as some of the egocentric responses as reasons for valuing grandparents. These changes probably reflect the increasing cognitive and emotional development of the child.

However, a study by Robertson (1976) suggests that grandparents maintain their significance in the lives of their young adult grandchildren. Robertson used a questionnaire to assess attitudes toward grandparents, perceptions of appropriate and/or expected grandparent behavior, grandchildren's responsibility toward grandparents, and conceptions of the ideal grandparent. Robertson (1976) found that young adult grandchildren held very favorable attitudes toward grandparents. For example, 92 percent agreed that children would miss much if there were no grandparents when they were growing up; 90 percent agreed that grandparents were not too old-fashioned or out of touch to be able to help their grandchildren; and 70 percent indicated that teenagers do not feel their grandparents are a bore.

Robertson's data suggest that adult grandchildren expect relatively little from grandparents other than gift giving (59 percent) and informing them of family history (56 percent). The majority did not expect them to provide financial aid (8 percent) or serve as a role model (23 percent), advisor (31 percent), or liaison between child and parent (29 percent). Young adult grandchildren, in turn, feel definite responsibilities toward their grandparents. About two-thirds felt grandchildren should help their grandparents, and 62 percent felt they should not expect money for such assistance. Over 50 percent said they visited with grandparents because they loved them or enjoyed being with them. In contrast, 20 percent visited with their grandparents because their parents did, and only 11 percent did so because it was expected of them. Robertson's (1976) data also show the importance of parents, with almost two-thirds of the respondents agreeing that parents set the pace of the grandparent-grandchild relationship. Finally, the ideal grandparent was seen as someone who loves and enjoys grandchildren, visits with them, and shows an interest in them. Characteristics receiving the highest ranking were loving, gentle, helpful, understanding, industrious, smart, a friend, talkative, and funny. Characteristics receiving the lowest ranking were lazy, childish, dependent, mediator, companion, and teacher.

The studies reviewed above suggest that the grandparent-grandchild relationship can be significant for both parties. There are multiple patterns of interaction, however, and the significance of the relationship appears to depend to a great extent on how much the various generations work at it. Further, there is evidence to suggest that the grandparent role may be undergoing change as individuals become grandparents at earlier ages or as changes in family functioning place new demands on it. Finally, it is possible that this role may become particularly significant for older adults. Grandparenthood is one of the few social roles potentially available to most older adults. Participation in it either as a biological or foster grandparent could prove highly significant for the social functioning of older adults.

Widowhood

In 1970, there were 2.11 million widowers and 9.64 million widows in the United States; that is, there was a 1:5 ratio of widowers to widows (U.S. Bureau of the Census, 1971). Of all women, 12.5 percent are widows, although the probability of remarriage is high for individuals (especially those aged 35 or younger). Men remarry faster than women. On the average, remarriage occurs 1.7 years after the death of a spouse for men and 3.5 years after the death of a spouse for women.

It appears that women who lose their spouse experience a different life event than men. For example, in a study of 403 community residents aged 62 and over, six major areas of life functioning were assessed: psychosocial needs, household roles, nutrition, health care, transportation, and education (Barrett, 1978). Widowers were found to experience lower morale, to feel lonelier and more dissatisfied with life, to

BOX 15.3

RETYING THE KNOT IN OLD AGE

To what extent does remarriage constitute a viable alternative for older widows and widowers? In terms of sheer numbers, remarriage among older adults has more than doubled in the last ten years. We know very little about these remarriages compared to marriages and remarriages between younger couples. Recently, however, Vinick (1978) interviewed remarried couples aged 60 to 84 in order to determine how these individuals experienced the event of remarriage.

She found that most people had originally chosen to live alone after the death of their spouses. Men remarried more quickly than women. Over half the men remarried within a year or less after becoming widowed, while only three women were remarried within two years after losing their previous spouses. Most of the couples had been introduced by a mutual friend or relative or had known one another when one or both of them had been married to their previous spouses. Usually, it was the men who took the initiative after the initial meeting. Often the relationship just seemed to grow gradually.

The most significant factor in the decision to remarry was the desire for companionship. Men also mentioned a desire for care, while women also mentioned the personal qualities of the prospective spouse.

Vinick's (1978) data suggest that these remarriages were very successful: 87 percent of the men and 80 percent of the women described themselves as "satisfied" or "very satisfied" with the relationship. The exceptions were those who married for "external" reasons. For example, one woman had remarried in order to attain financial security and then found her husband was autocratic and miserly. While the majority of the respondents indicated that adjustments had been required, the marriages were characterized by a serenity usually not found among young marrieds. Vinick (1978) comments:

Most people had a "live and let live" attitude toward the spouse. Time and again I heard that "it doesn't pay to get angry," "it takes two to make an argument" and that one should contain his or her feelings. Old age marriage is free from the strains of early marriage—child rearing, ambition for higher status, conflict with in-laws. (p. 362)

Vinick (1978) concludes that remarriage is a viable alternative for older adults. Many external conditions, of course, influence the practicality of this alternative. For example, the shorter average life span of men effectively limits the number of potential male partners.

consider community services more inadequate, to need more help with household chores, to have greater difficulty getting medical appointments, to eat more poorly, and to possess stronger negative attitudes about continued learning than widows. Moreover, widowers were more reluctant to talk about widowhood or death than widows were and stated that they did not want a confidant.

In our society, women outlive men. It is reported that half of all married women have been widowed by their early seventies and 80 percent have been widowed by their early eighties. However, only half of all married men are widowed in their early eighties (Atchley, 1977). Also, since there are more unmarried older women, widowers may be more likely to remarry. Further, one could argue that the loss of a husband would be more devastating than the loss of a wife because the male role is more prestigious in our society. Finally, most widows appear to be more strongly affected by the death of a spouse who is likely to be the sole breadwinner.

How does the event of widowhood affect people? Berardo (1968) reported that aged widowers were the most isolated, and he cited poor health as a contributing factor. Even when widowed individuals were compared to married persons of the same age and sex, widows and widowers were found to exhibit higher rates of mortality, mental disorder, and suicide than married people (Berardo, 1968).

Using a large national sample, Harvey and Bahr (1974) reported that economic status accounts for the negative impact of widowhood on life satisfaction. Data also indicate that widows with low incomes had lower social participation and greater loneliness (Atchley, 1975). There also was a greater likelihood for widows to have inadequate incomes compared to widowers. Atchley (1975) suggests that negative attitudes and social isolation of widows are influenced greatly by economic factors.

Morgan (1978) looked at the question of economic change at widowhood using data from the National Longitudinal Surveys Cohort of Mature Women (1967 to 1974). The 8000 women nationally sampled in 1967 were 30 to 44 years of age. The first question asked was whether there were cross-sectional differences in family income for particular years as a result of marital status. Various sources of family income across all family members were used to index family income. The results showed family income was lower (between 48 and 53 percent) for widows compared to married women for all years, although both groups reflected an increase in mean income over time due to inflation. For black widows, the mean income is lower in all years, but the ratios between widowed and married women are similar for blacks and nonblacks. Thus it appears that both groups are relatively equal over this period of time suggesting that economic cycles are similar with widows having substantially lower mean incomes than their married counterparts.

Researchers have also been concerned with events that occur immediately after the loss of a spouse and during the subsequent year. The bereavement process is referred to by some as "grief work" (Parkes & Brown, 1972). Grief work has been identified as a process leading to a new identity. The stages of the grief process are (1) numbness, (2) pining for the lost person, (3) depression, and (4) recovery. This process is thought to be essential after the loss of a spouse regardless of sex, age at widowhood, economic situation, or expectations of one's own death (Rux, 1976).

Studying a large urban sample of 301 widows of all ages Lopata (1973a, 1973b, 1973c) reported that reactions to widowhood reflected women's social-role backgrounds. Depending upon the type of the wife role held and other social roles assumed prior to the death of the husband, the type and intensity of the life-span disruption experienced after the death and following vary. For example, more highly educated middle-class women who were strongly involved in the role of wife reported "strong life disruption after the death of the husband" (Lopata, 1975, p. 229).

Other women living in more of a sex-segregated world where they were relatively independent of their husbands, such as engaging in neighboring or in jobs, stated they were lonely after the death but that their lifestyles did not change much.

Lopata (1975) proposed five types of widows: the "liberated widow" who is able to lead a multidimensional life; the "merry widow" who is socially active, dating and interacting with friends; the "working widow" who focuses on her job in a committed and individualistic way (a subtype exists: the returning-to-work widow who takes any job with minimal economic return); the "widow's widow" who remains in that role unwilling to relinquish it through remarriage, devotion to grandparenting, or other roles; and the "traditional widow" who devotes time to children, grandchildren, and siblings.

Adapting to new social roles also seems important for widows and widowers. In a study of 107 widowers and widows between the ages of 40 and 70, Rux (1976) found life satisfaction ratings were correlated significantly with current instrumental skills. Thus adjustment to widowhood seems to entail more than emotionally responding to the loss of a spouse; rather, it also involves being able to adapt to new roles and responsibilities.

Relocation

In recent years, considerable attention has been drawn to the impact of relocation in later life. Some of these studies have suggested that relocation has negative effects such as increased mortality and decreased morale (Aldrich & Mendkoff, 1963; Killian, 1970; Lieberman, 1975). Relocation has been found to be a major life event that potentially affects individuals of all ages. For example, Lowenthal and Chiriboga (1975) found that it was the second most frequent stress (after school) mentioned by female adolescents. Is relocation particularly stressful for older adults? Let us examine this question with respect to two types of relocation: relocation within the community and relocation from the community to an institution or from one institution to another.

The bulk of the research on relocation within the community has focused on planned housing for the elderly. One of the most extensive projects was conducted by Carp (1966) who studied 352 individuals who applied for public housing. Of these, 204 were ultimately relocated, while 148 were not. Self-report measures were obtained at the time of application and twelve to fifteen months following relocation. The overwhelming majority of the relocatees were highly satisfied with their new housing. This finding is perhaps not surprising considering that the tenants' previous accommodations were very inadequate—typically consisting of a small room on the second or third floor of an old house with communal bathroom facilities. The positive impact of relocation was seen in other measures as well, including increases in social relationships and activities, self-reported better health and morale, and decreases in the number of services needed.

Other studies generally have supported the conclusion that voluntary relocation within the community is not associated with significant negative effects (Lawton & Cohen, 1974; Storandt & Wittles, 1975; Wittles & Botwinick, 1974). For example, Lawton and Cohen (1974) found that several indexes of well-being (housing, satisfaction with status quo) were higher among older adults who had moved into housing projects than among those who had remained in the community. Wittles and Botwinick (1974) did not observe a relocation mortality effect in their study of voluntary relocation of elderly to new apartment complexes. Similarly, in a follow-up study, Storandt and Wittles (1975) found no decrements on measures of cognition (e.g., psychomotor speed, memory, intelligence), personality (e.g., neuroticism, depression, life satisfaction), self-rated health, and social activities.

There are exceptions to these largely positive findings. For example, the anticipation and experience of actually moving often result in a temporary decline in morale and other indicators of well-being (Schooler, 1975). Further, relocations which are involuntary or involve moving to a less desirable environment appear to have a negative impact (Brand & Smith, 1974; Kasteler, Grey, & Carruth, 1968).

Contrary to popular belief, the number of older adults living in institutions at any one time is relatively small—about 5 percent of the population over age 65. However, the total chance of being institutionalized at some time prior to death is greater than this. Palmore (1976) attempted to estimate this chance by analyzing the experience of individuals from the Duke First Longitudinal Study who were followed until their deaths. All 207 participants were community residents at the beginning of the study in 1955. Of these, fifty-four (26 percent) were institutionalized one or more times prior to death—suggesting a total chance of institutionalization of about 1 in 4, at least for this sample. Exhibit 15.4 shows a number of variables which

Among women, those who live alone, those who never married or are separated, and those who never had children have a greater chance of being institutionalized than others. (Miriam Reinhart/Photo Researchers, Inc.)

EXHIBIT 15.4   **Percentage Institutionalized in Selected Categories**

Variable	Category	(N)	Percentage Institutionalized	Percentage Not Institutionalized
. . .	All persons	(207)	26	74
Living arrangement	Alone	(42)	33	67
	With spouse	(119)	24	76
	With others	(46)	24	76
Marital status	Never married	(13)	54	46
	Separated	(10)	40	60
	Spouse present	(119)	24	76
	Widowed	(66)	20	80
Children living	None	(38)	34	66
	1 or 2	(74)	27	73
	3+	(95)	22	78
Sex	Women	(97)	33	67
	Men	(110)	20	80
Finances	Enough or better	(180)	28	72
	Cannot make ends meet	(23)	13	87
Education	7 years plus	(123)	30	70
	0–6 years	(84)	20	80
Race	White	(139)	32	68
	Black	(68)	15	85

Source: Palmore, 1976.

influence the chance of institutionalization. It can be seen that among women, those who live alone, those who never married or are separated, and those who never had children had higher rates of institutionalization. These factors appear to be related to the availability of care outside the institutional setting. Other factors appear to be related to the accessibility of institutional care. These include finances, education, and race.

What is the impact of institutionalization on older adults? Institutional relocation studies have focused both on the move from home to institution and from one institution to another. A major emphasis in this literarture has been the impact of relocation on mortality. The results of this research have been mixed. Some studies have shown institutionalization or institutional relocation to be associated with increased mortality rates (Aldrich & Mendkoff, 1963; Killian, 1970; Lieberman, 1975; Marlowe, 1973), while others have not (Gutman & Herbert, 1976; Borup, Gallego, & Heffernan, 1979). Despite such conflicting evidence, Lawton (1977) concludes that the relocation mortality effect has generally been upheld.

In spite of the significance of the relocation mortality issue, it may be argued that too much attention has been devoted to it. Of more significance for both theory and intervention are efforts to specifically isolate the variables with adaptive and maladaptive outcomes to institutional relocations.

Several variables appear to be important factors in adaptation to institutionalization or institutional relocation. One of the most important of these is the degree of environmental change. For example, Lieberman (1975) reports the results of four major studies involving four different types of relocation: (1) healthy older women forced to move from a small institution to a large institution; (2) community-dwelling older adults voluntarily relocated to homes for the aged; (3) highly selected older patients discharged from a mental hospital to community-based institutional and semi-institutional settings; and (4) elderly patients relocated from a state hospital to other institutional settings. In all instances, individuals were examined up to a year prior to the move and again a year after the move. In two studies, relocatees were compared with matched controls. Lieberman (1975) found significant declines in physical (including death) and behavioral functioning in roughly half (48 to 56 percent) of those experiencing relocation. The rate of change was considerably higher for those who moved than for those who did not. For example, in the study examining the transfer of mental patients, the death rate for relocatees was 3 times that of the controls.

Lieberman (1975) concludes that the most important predictor of adjustment to relocation was the degree of change involved. The frequency of breakdown roughly paralleled the degree of environmental change involved in the relocation. Further, within any one study the most important predictor of adjustment was the degree of similarity between the individual's two environments. Other research has supported this finding. For example, Schulz and Aderman (1973) found that the length of survival of terminal cancer patients depended significantly on the degree of disparity between their previous living arrangements and the institution. Patients who came from a similar institutional environment survived an average of almost one month longer than patients who came from a dissimilar home environment. Within this context, there is evidence to suggest that preparatory programs including visits to the institution and group or personal counseling can reduce the stress of relocation (Pastalan, 1976; Zweig & Csank, 1975).

The characteristics of the individual have also been found to be important. For example, Lieberman (1975) found that physical and cognitive resources set a lower limit on adaptation. That is, those who were physically or cognitively incompetent were unable to adapt. On the other hand, competence in these areas did not guarantee adaptation. Rather, other factors appeared to mediate the adaptation process. Lieberman (1975) found certain personality characteristics to be significant resources for coping with the institutional environment. In particular, those individuals who were aggressive, irritating, narcissistic, and demanding were most able to cope with the demand characteristics of the institution. Hope and the ability to introspect were also predictors of adaptation. Thus the research on institutionalization and institutional relocation suggests that this event may be a stressful experience, particularly when it produces a mismatch between the individual's competence and the demands of the environment.

DEATH AND DYING

The Meaning of Death

What does death mean? Kalish (1976) observes, "Death means different things to the same person at different times; and it means different things to the same person at the same time" (p. 483). This apparent double-talk is not meant to be elusive but rather to point out that the death of an individual is deeply personal. What do you think of when you think about death? Do you visualize a hospital? a corpse? a funeral home? a religious service? a cemetery? Do you sense absence rather than presence, coldness rather than warmth, or immobility rather than response? Do you believe that once you are dead, you will experience eternal life, a return to this world, a period of restful waiting, or nothing? As Kastenbaum (1975) notes, stimuli which represent death are found in a multitude of physical representations, verbal and nonverbal representations, and personal encounters.

Death "is the transition from the state of being alive to the state of being dead" (Kaas, 1971, p. 699). It is the final event of life. A certificate is issued marking the time and cause of death. Depending upon the situation surrounding death, death takes on different social symbolism. Death may be natural or not. Death may occur violently or not. Death may happen to an important person or to someone who does not have a major influence on society. Within a family, the death of one of its members has various implications for the remaining members. "The death event can fall like an iron gate, trapping some of the survivors in the past and liberating others to go forth to a new life style" (Kastenbaum, 1975, p. 25). As an event, death can challenge our value system, especially when it occurs off phase, under the wrong set of circumstances. Parents do not expect to bury children, and people do not anticipate that someone who is physically healthy and at the peak of life will die.

Although death is an inevitable universal event that occurs daily, our own death occurs only once. Death is a mysterious paradox, therefore, because it affects our developmental course even though we only experience it at the end of life. Becker (1973), for example, contends that the major concern of all ages is death and what it represents. If we did not have to face death, our lives probably would be transformed. We would have different views about the events we experience and the relevance of their timing.

Age Differences in Death Perceptions and Attitudes

Kalish (1976) points out that impending death alters the way in which we use time. To the elderly, death is an event that directly limits their personal future. In this regard, there seems to be an interesting contrast between the older adult's perception of retrospective time and ongoing time (Kastenbaum, 1966). When contemplated retrospectively as past years, months, and weeks, time seems to be speeding toward death with increasing rapidity. But when ongoing experienced time is considered, the minutes and hours move slowly because there is nothing meaningful to do—whatever is not done will remain unfinished.

In this regard, Back (1965) asked residents of rural communities what they would do if they knew they were going to die in one month. The older respondents' comments reflected that they were less likely than the younger respondents to alter their activities. This finding gains support from the results of a more recent study of 434 Greater Los Angeles area residents who represented three age groups (20 to 39, 40 to 59, and 60 plus). They were asked, "If you were told that you had a terminal disease and six months to live, how would you want to spend your time?" The results shown in Exhibit 15.5 indicate that more people in the oldest age group said they would not change their lifestyles. Moreover, 3 times as many older people compared with younger people said they would spend time in prayer, reading, contemplation, or other activities that reflect inner life, spiritual needs, or withdrawal (Kalish & Reynolds, 1976).

Persons facing death will also experience loss (Kalish, 1976). They will lose self (body, sensory awareness, opportunity to experience events) and others. In their study, Kalish and Reynolds (1976) asked their respondents what major losses would occur as a result of death. Again, age differences were found. Older respondents were less concerned with caring for dependents and causing grief to relatives and friends than younger respondents. Perhaps older people are more aware that they are likely to have less impact on others if only for the reason that they have fewer dependents. Older persons were also less concerned than younger persons about the loss of opportunity for experiences.

EXHIBIT 15.5   **Responses on Time Spent before Death, in Percentages**

	Age		
	20–39	**40–59**	**60+**
Marked change in lifestyle, self-related (travel, sex, experiences, etc.)	24	15	9
Inner life centered (read, contemplate, pray)	14	14	37
Focus concern on others, be with loved ones	29	25	12
Attempt to complete projects, tie up loose ends	11	10	3
No change in lifestyle	17	29	31
Other	5	6	8

Source: Kalish and Reynolds, 1976.

What antecedents account for these age differences in perceptions of death? To date, few data have been generated in attempts to answer this question. In light of what we know about adult development and aging, we could speculate that pressures for disengagement, a preference for an inward-directed personality, and physical changes, health-related changes, or both are possible explanations.

In a number of studies comparing death attitudes of the elderly and other age groups, data revealed that older people thought more and talked more about death, regardless of whether the index used was if they had thought about death in the last five minutes or how many times they had previously contemplated death (Kalish & Reynolds, 1976; Riley, 1970). Although the elderly seem preoccupied with death, consistent findings suggest that they also are less frightened by it than younger people (e.g., Feifel & Branscomb, 1973; Kalish & Johnson, 1972; Kalish & Reynolds, 1976; Kogan & Wallach, 1961; Martin & Wrightsman, 1965).

It might appear to be a contradiction that the elderly are less afraid of death and at the same time find it more salient than younger people. Kalish (1976) proposes three reasons for this attitude. First, the elderly recognize the limitations affecting their futures, e.g., health problems, role loss, and economic restrictions. Moreover, our social value system stresses the importance of "producers of tomorrow rather than those of yesterday" (p. 490). Second, the fact that people in industrial nations are provided with data that estimate life expectancy for various groups may affect their attitude about death. For instance, those who live longer than they might be expected to live may feel that they "have received their entitlement" (p. 490), and those who face death ahead of schedule may feel that they have been "cheated." Third, older people, more than younger people, have been socialized in such a way that they have become more accustomed to death. They have experienced death in various ways, e.g., through the loss of peers and family members. Thus death may bring relief from being in a constant state of bereavement. Although experiencing death does not mean one will develop a positive attitude about it, it does offer an opportunity to rehearse feelings about one's own death.

Many variables other than age mediate how death is perceived and what attitudes are held. For example, people who are more religious have less anxiety about death than those who are less religious (Martin & Wrightsman, 1965; Templar, 1972). More specifically, there appears to be a curvilinear relationship between fear of death and being religious. That is, the very religious have the least fear of death, the most nonreligious have a moderate fear of death, and irregular religious worshippers have the highest fear of death (Kalish, 1963; Nelson & Nelson, 1973).

The Dying Process

Undoubtedly the best-known description of the dying process is that proposed by Kübler-Ross (1969). Based on a set of clinical experiences with dying patients she proposed a series of five stages of dying: (1) denial, (2) anger, (3) bargaining, (4) depression, and (5) acceptance.

Denial is the first stage. The person resists the reality of impending death. In essence, the person says "No!" to death. Whether medical authorities have told the person of his or her terminal status or the person senses it, denial is displayed in various ways. For instance, some people sought other professional opinions and continued their search until they found a more favorable diagnosis. Others sought forms of religious assurance, while still other patients tried "miracle cures." Once they accepted that they were going to die, then anger and distress were expressed.

Anger is the second stage. During this time the person is saying, "Why me?" or "Why am I to die?" Their feelings are manifested in hostility, resentment, and even envy. They hate the fact that they are to die, resent their situation and eventuality, and envy all those who are not in such a predicament. These feelings may be directed at one or more targets, such as family, friends, the medical staff, aspects of the environment (e.g., a pen that won't work), or even God. It seems that the anger flourishes because the person feels frustrated by all that will remain unfinished. When it is realized that the question—"Why me?"—cannot be answered satisfactorily, the person begins to make deals with fate.

Bargaining is the third and middle stage. Here the person decides to change his or her strategy. That is, rather than saying "No!" and "Why me?" favors are asked to extend life or postpone death. Although such bargaining often is conducted between the individual and God in a covert fashion, sometimes it is evidenced overtly in interactions with others. For example, an individual might say, "If I rewrite my will and leave money to more people and charities rather than for a lavish funeral, then please let me live longer. I am a good person." Or "I will be a better person if I can have just a little more time." Typically, these bargains are not kept, and when the person outlives the bargain, another one is usually made.

Depression is the fourth stage. This occurs when the manifestations of the patient's terminal illness become too significant to ignore. Increasingly severe symptoms, hospitalization, and more surgery eventually lead to a realization that death cannot be avoided and to a sense of great loss.

Acceptance is the final stage. Actually the person is resolute about death, although not happy. Tired and weak physically, the person "is almost void of feelings. It is as if the pain had gone, the struggle is over, and there comes a time for 'the final rest before the long journey' as one patient phrased it" (Kübler-Ross, 1969, p. 100).

Kübler-Ross (1974) cautions that not all people go through this stage sequence and that we could potentially harm dying patients by viewing this series of feelings as invariant and universal. Unfortunately, her stages have become a prescription for dealing with dying patients—a kind of "pop death." In fact, Kübler-Ross has recently emphasized individual differences in the dying process and the importance of identifying and accepting a patient's response pattern. She stresses that the defenses of dying people should not be challenged or broken down by family, friends, or medical staff. For example, some individuals may exhibit denial throughout the dying process, and it is not productive to attempt to push them through the stages.

Other researchers prefer to represent the dying process as a series of phases. Pattison (1977), for instance, proposes three phases—an acute phase, a chronic living-dying phase, and a terminal phase. The acute phase is a crisis event. The individual becomes aware of impending death, manifests high levels of anxiety, denies, is very angry, and may even bargain. The chronic living-dying phase begins with a reduction in anxiety. During this phase, the person experiences various feelings, e.g., fear of the unknown, fear of loneliness, and anticipatory grief over the loss of friends, of body, of self-control, and of identity (Pattison, 1977). Levels of anxiety and sorrow may occur alternating or simultaneously with levels of hope, determination, and acceptance (Shneidman, 1973). The terminal phase is marked by the person withdrawing from the world. This phase is the shortest and is ended by death.

Contexts for Dying

Most people say that they want to die at home (Kalish & Reynolds, 1976), although the majority actually die in health-care institutions, primarily general hospitals. When individuals do die at home, family members typically report being glad that the person died at home rather than in an institution. These caregivers, however, also said that their burden of care often was excessive, and one-third of them wondered retrospectively whether the hospital would have provided a better environment (Cartwright, Hockey, & Anderson, 1973). At the present time, the leading causes of death are chronic diseases such as cardiovascular diseases and cancer. As a result, patients, family members, and health-care practitioners must often cope with a prolonged terminal phase of an individual's life.

For physicians and health-care providers, death often represents failure and defeat. In fact, considerable evidence indicates that the majority of physicians avoid patients once they begin to die (e.g., Kastenbaum & Aisenberg, 1972; Livingston & Zimet, 1965). This avoidance may be the result of several factors. For example, several investigators (e.g., Feifel, 1965; Feifel, Hanson, Jones, & Edwards, 1967) have suggested that the basic personality structure of physicians is responsible for their response to dying. That is, becoming a physician represents an attempt to overcome death. Although such a personality profile may exist, the attitude is probably reinforced by training. Moreover, physicians are not excluded from the negative associations with death. They are challenged to heal but are made aware of the temporal limits on their own lives. Lief and Fox (1963) propose that the physician-patient relationship is one of "detached concern." Imagine the difficulty of facing death personally with each patient. Also, there is the problem of allocating time (e.g., saving the 80-year-old versus the 40-year-old heart patient). Therefore, physicians, like others in society, are influenced by the fear and anxieties surrounding death and decisions regarding use of time.

The three major needs of terminally ill people are probably the need to control pain, to maintain dignity or feelings of self-worth, and to receive love and affection. Although the use of addicting drugs such as morphine loses its social consequence for people who are dying, medical practitioners still must decide who determines dosage. It has been argued that the patient should have input in this decision because only he or she is aware of fluctuations of internal states. The best solution is for the patient and medical staff to arrive at a situation where pain is prevented rather than simply alleviated (Neale, 1971).

Helping a patient feel a level of dignity or self-worth is often accomplished by letting the patient participate in decisions that affect treatment or outcomes. Control over one's life is thought to be as important during the terminal phase as during any other phase. Seligman (1975) reports that when individuals are not allowed to participate in such decisions, they experience feelings of helplessness, exhibit withdrawal, and evidence depression. Also, when institutionalized aged individuals were presented with an opportunity to control an aspect of their environment, they showed significant improvement in their physical and psychological status compared to others who were not afforded opportunities for self-control.

Love and affection for terminally ill patients often means holding, touching, or stroking. The medical staff usually has many opportunities to physically touch patients when caring for them. How the patient is handled can communicate a genuine concern. Also, by listening to the patient and supporting the patient's way of handling the situation, people, especially the family, can reassure the patient that he or she will not be abandoned (Schulz, 1978).

Saunders (1976), the director of St. Christopher's Hospice in London, a prominent figure in the hospice movement, maintains that the environment can be designed to facilitate freedom from pain, regaining a sense of self-worth or dignity, and experiencing a loving atmosphere. *Hospices* provide more than sympathetic attentive care for the dying. The primary goal of most hospices in the United States is to help people live as individuals during the weeks and months left to them and to help them die with as little discomfort and as much serenity as possible.

Specifically, the staff seeks to alleviate pain and allay fear. In hospitals, terminal patients are often heavily drugged and have little control over their medication. In contrast, hospices often administer a mixture called "Brompton's Cocktail" that consists of diamorphine (heroin) or morphine (this is used in the United States), cocaine, gin, sugar syrup, and chlorpromazine syrup in small doses around the clock before severe pain begins. It is felt that this approach allows patients to trust that they will not be in pain and consequently reduces their fears.

Another characteristic of the hospice is to treat the patient and family together. Thus visiting hours are not restricted, and family members are encouraged to perform practical services such as preparing special meals for their in-resident family member. Allowing family members to participate in the care of the patient is thought to minimize guilt during bereavement. Finally, the staff's involvement with the family does not end upon the patient's death. Family members are encouraged to contact hospice personnel if they experience problems during bereavement.

Ethical Issues

Sustained by extraordinary means of mechanical life support, Karen Ann Quinlan lay comatose while her case was considered by the New Jersey courts. This case helped bring the American public face to face with issues surrounding euthanasia— the act or practice of killing for reasons of mercy. In Karen Ann's case, her parents took legal action to permit her to live or die without the use of a respirator, and the courts upheld their second request. The court instructed that the respirator could be turned off if "there is no reasonable possibility" that she will return to a "cognitive sapient state" (Seligmann, 1976). The court also added that these standards be applied to all whose condition is progressive and incurable. But the decision to turn off the respirator does not end the story because at the time of this writing Karen Ann Quinlan still lives in a coma, breathing unaided.

There are two types of euthanasia—active and passive. *Active euthanasia* refers to deliberate actions (e.g., injecting air bubbles) to shorten an individual's life. *Passive euthanasia* refers to actions which allow a person to die because available preventive measures are not used. This means that treatment might be withdrawn or withheld, thus permitting a person to die earlier than if treatment were received.

Attitudes held by the public toward euthanasia have changed in the past two

To My Family, My Physician, My Lawyer and All Others Whom It May Concern

Death is as much a reality as birth, growth, maturity and old age—it is the one certainty of life. If the time comes when I can no longer take part in decisions for my own future, let this statement stand as an expression of my wishes and directions, while I am still of sound mind.

If at such a time the situation should arise in which there is no reasonable expectation of my recovery from extreme physical or mental disability, I direct that I be allowed to die and not be kept alive by medications, artificial means or "heroic measures". I do, however, ask that medication be mercifully administered to me to alleviate suffering even though this may shorten my remaining life.

This statement is made after careful consideration and is in accordance with my strong convictions and beliefs. I want the wishes and directions here expressed carried out to the extent permitted by law. Insofar as they are not legally enforceable, I hope that those to whom this Will is addressed will regard themselves as morally bound by these provisions.

Signed_____

Date _____

Witness_____

Witness_____

Copies of this request have been given to _____

_____

_____

_____

EXHIBIT 15.6   A living will. (Source: Concern for Dying, 1980. Reprinted with permission from Concern for Dying, 250 West 57th Street, New York, N.Y. 10107.)

decades. For example, in 1950, responding to a Gallup poll, 36 percent of the population said "yes" to the following question: "When a person has a disease that cannot be cured, do you think doctors should be allowed by law to end the patient's life by some painless means if the patient and his family request it?" By 1973, 53 percent of the population said "yes" to this question. At the American Medical Association convention in 1973 physicians also supported a "death with dignity proposition" and the use of "living wills" that allow people to choose their fate in conditions where death is imminent (see Exhibit 15.6).

Grief and Bereavement

Grief is an emotional response to loss. Many researchers feel individuals experience phases of grief (Glick, Weiss, & Parkes, 1974; Parkes, 1972). The initial phase begins when the death occurs and continues for a few weeks after the funeral. At

Grief is an emotional response to loss. (Arthur Tress/Photo Researchers, Inc.)

first, people react with shock and disbelief. People often state that they feel dazed, numb, empty, and confused. This reaction lasts several days and then gives way to an all-encompassing feeling of sorrow which often manifests itself in crying and weeping. With the passage of time, controlling emotions becomes associated with "doing well." Thus the bereaved person is encouraged to inhibit emotional responses. Some may turn to tranquilizers, sleeping pills, or alcohol, or they may report a variety of psychophysiological symptoms (e.g., shortness of breath, tightness in the throat, loss of appetite, irritability, muscular aches and pains, headaches, etc.). Typically, these symptoms are reduced after the first month of bereavement.

The intermediate phase of normal grief usually occupies the balance of the first year. Behavior in this phase consists of obsessional review, searching for an understanding of death, and searching for the presence of the deceased. The obsessional review is highlighted by dwelling on specific events associated with the death and berating oneself for not doing enough (e.g., "If I had only called the ambulance sooner" or "If I had only been with him in the hospital"). Searching for answers to understand death often ends with: "It's God's will." Finally, searching for the deceased occurs in various ways. Many activities remind the bereaved of the person and result in thoughts about or memories of the person. These behaviors decrease during the first year. Glick, Weiss, and Parkes (1974) reported, for example, that 60 percent of the widows in their study were beginning to feel more like their old selves after a few months and felt they had done quite well by the end of the year.

The second year of bereavement is referred to as the recovery phase, because the bereaved evidence a positive attitude toward life. They are alive and have a life to live. Most state that they have pride in having coped with and survived a traumatic, potentially devastating event. Although professional treatment is available for those who are bereaved, the majority who experience bereavement do so with the support of family and friends.

CHAPTER SUMMARY

Individuals experience many transitions during the life course defined by various life events. These events may be viewed as antecedents of developmental change.

The transition to young adulthood is characterized by the making of commitments. Major choices with respect to marriage, children, and occupation significantly define the individual's place in the adult world.

Although dating and courtship are processes leading to the event of marriage, they obviously serve as vehicles for establishing commitment. Again, cognitive, personality, and social variables influence who marries whom. Premarital and marital sexual behaviors are related to the development of interpersonal commitment.

Parenthood is a time of crisis for some young adults, and for others it represents a positive transition to a new role. In terms of parenting, investigators are turning from traditional ways of looking at mothering and fathering. Specifically, the idea of an innate "maternal instinct" is being questioned, as is the idea that fathering is limited to the role of provider.

Divorce appears to be a disruptive event in the lives of mothers, fathers, and children. Although individuals are affected differently, the first year is the most traumatic, and the mother-son relationship seems to involve the most problems. The greatest difficulty for divorced people is in reorganizing their lives. Practical tasks are especially difficult. Although divorce can be perceived positively, e.g., when it ends a conflict-ridden marriage, it takes a while (usually two years) for parent-child behavior and feelings to approach those of intact families.

Young adults vary greatly in their concern with and commitment to work. Career choice is influenced by a variety of factors including intelligence, personality, and family experiences.

The Midlife Transition has been characterized as a crisis. Levinson stresses that in midlife the individual must reappraise the past and make choices that will modify the future. This is a stressful experience for most people. However, other writers suggest that a crisis occurs only for those who experience unanticipated interruptions in the rhythm of their life course.

Middle-aged adults do appear to become increasingly aware of their physical aging. They also appear to become aware of death. Time is restructured to reflect time left to live rather than time since birth. Menopause, however, does not appear to be a major turning point for most women. Similarly, the empty nest appears to be a positive event rather than a negative event for most adults.

Career changes are often a significant aspect of midlife for men. Both personality and situational factors affect career change. Working women reflect various types of roles—from single-career role to the multiple roles of mother, wife, and worker. Groups of working women vary in terms of what mediating variables influence their career roles. For instance, unmarried career women were influenced greatly by supportive fathers who reinforced their non-gender-stereotyped behavior. Many women successfully handle multiple roles. The husband's support is critical in this regard. No negative effects of work on children were reported among women satisfied with their roles.

The transition to old age is characterized largely by role exits. Erikson says the task is to achieve a sense of integrity—an acceptance of one's life.

Attitudes toward and adjustment to retirement are related to various life experiences (e.g., being work-oriented or leisure-oriented, having a good marital relationship). In general, people feel positively about retirement as long as they have health, money, and ways of occupying their time meaningfully.

Grandparenthood is an event being experienced by people for a longer period of their life spans. It appears to be a "roleless role" in that members of the dyad must determine the role relationship. Older adults react differently to the role. Some see it as biological renewal or continuity, while others view it as a source of emotional satisfaction. They can have fun without childrearing responsibilities. Finally, grandchildren regard grandparents differently, depending upon the children's age. Most children, however, feel they would have missed much if they had not interacted with their grandparents. It therefore seems that the grandparent-grandchild relationship is significant for both parties.

Men experience the event of widowhood differently than women do. Also, different mediating variables (e.g., age, economic status, and education) have an effect on how the event is experienced. Ways in which people adapt to widowhood are important to consider because the death of a spouse is disruptive not only to the dyadic relationship but also to other relations organized around the marital role. Adjustment entails more than an emotional response and working through one's grief. It also involves adapting to new roles and responsibilities.

A relatively small portion of the population over age 65 live in institutions, and studies of the effects of institutionalization have produced mixed results. In general, mortality rates increase when people live in institutional settings; however, since people are often institutionalized because of physical problems to begin with, even this finding needs a cautious interpretation. Basically, to the extent that there is a mismatch between the person's competence and demands of the institutional environment, a stressful experience results.

The death of an individual is probably one of the most deeply personal events that occurs in one's life. But confronting how one feels about the death of others and self is very difficult for most people. Age differences in death perceptions and attitudes exist. On the one hand, older respondents said that if they knew they were going to die, they would not be likely to alter their activities or lifestyles, would increase inner-life activities, and would be less concerned with caring for dependents or causing grief to others than younger people. On the other hand, older people think more and talk more about death than younger people. Thus the elderly are less afraid of death while being more consumed by it than younger people.

The dying process can be thought of in terms of stages or phases. Kübler-Ross proposes that dying people go through five stages: (1) denial, (2) anger, (3) bargaining, (4) depression, and (5) acceptance. Not all individuals experience all these stages, however. As a result, the dying individual should not be pushed through the sequence. Proponents of a phase theory of dying suggest three phases: an acute phase, a chronic living-dying phase, and a terminal phase.

There are different contexts for dying. Although most people say they would rather die at home, the majority die of prolonged chronic diseases in general hospitals. Health professionals need to meet the needs of terminally ill patients by controlling pain, helping the patient retain dignity or feelings of self-worth, and providing affection and love. An alternative to dying in a hospital is to die at a hospice. The staff of a hospice does not follow a hospital model in its treatment but emphasizes the alleviation of pain and fear of death. The patient's family also plays an important role in the program of hospices. Family members can participate by living in.

Grief is an emotional response to loss. Phases of the normal process are (1) initial response of shock or disbelief, (2) intermediate phase of reviewing the factors associated with death, and (3) the recovery phase of realizing one has coped with and survived a devastating event.

acceptance According to the Kübler-Ross description of the dying process, the final stage in which a person is resigned to death.

accommodation In Piaget's theory, the alteration of existing cognitive structure to fit new stimulus objects.

accommodative power The ability to focus and maintain an image on the retina.

acculturation Adaptation to the cultural traits or social patterns of another group.

accumulation theory A nongenetic cellular theory which suggests that aging results from the accumulation of deleterious substances in the cells of the organism.

achievement test A test designed to measure what has been learned in school either for many subjects or for just one subject.

active euthanasia Deliberate, premeditated actions taken to shorten an individual's life, usually to prevent undue suffering.

active incidence The total amount of an activity which exists at one point in time.

active mastery Perception of the environment as under one's control and the self as a source of energy.

active organism paradigm A developmental model, rooted in the organismic world view, in which the individual is inherently dynamic and development is explained by the individual's action on the environment.

activity level In the New York Longitudinal Study, the category of temperament defined as the proportion of active periods to inactive ones.

activity theory Theory of aging which posits that disengagement is imposed on the individual by the structural requirements of society and that successful coping is dependent on the extent to which the individual is able to avoid this external pressure toward disengagement and maintain his or her involvement in the social context.

adaptability In the New York Longitudinal Study, the category of temperament defined as the ease with

which a child adapts to changes in his or her environment.

adaptation In Piaget's theory, the functional principle involving the processes of assimilation and accommodation.

adaptation process The process of adjusting to or coping with a life event.

adaptive Functional for survival; meeting demands of existence.

adolescence From the Latin, to grow into maturity; the period within the life span when most of the person's processes are in a state of transition from what typically is considered childhood to what typically is considered adulthood.

adolescent egocentrism Lack of ability to distinguish between one's own thinking and the thinking of others. *See also* imaginary audience; personal fable.

adrenalin Hormone produced by the adrenal glands in response to stress to induce fight-or-flight arousal.

affiliation Process of becoming a member of a group or belonging to a group, cause, or activity.

age-change function Specification of the relationship between chronological age and a behavioral change process.

age changes Within-person changes across chronological time.

age differences Age-related between-person differences.

age grades Age classes; for example, children, adults, the aged.

age norms Expected and acceptable behaviors for a person of a given age.

age-related phenomena Age-specific characteristics, behaviors, and abilities.

age-status system System of positions or locations in reference to others in a social network which is based on chronological age.

age strata The demographic structure of a population's age and birth cohorts.

ageism Inequality between age groups based on discrimination against the aging.

agency A set of behaviors involving

instrumental effectiveness, competence, and mastery.

ahistorical Pertaining to the study of behavior at one point in time with no regard as to how it came to be or what form it may take later.

alleles Variant forms of a gene at a given place on a chromosome.

alpha abundance Measure of the amount of time alpha rhythms are produced.

alpha frequency Measure of the frequency of alpha rhythms in cycles per second.

alpha rhythm Dominant brain-wave patterns with a frequency of 8 to 13 cycles per second, associated with a relaxed, awake state.

altruism Unselfish regard for or devotion to the welfare of others.

amenorrhea Abnormal absence of menstruation.

amino acids The chemical building blocks of proteins.

amniocentesis A medical procedure in which a needle is used to puncture the abdomen of a pregnant woman to remove fetal cells from the amniotic sac.

amniotic sac The protective membrane which is filled with fluid and which covers the developing fetus.

amorphous positions Role types that include de facto types in which objective circumstances prevent the individual from performing a role, such as being chronically unemployed, and role attrition types in which role responsibilities dwindle away or are lost.

anal stage In psychoanalytic theory, the second stage of psychosexual development during which bowel control is achieved and libidinal energy is focused on the functions of elimination.

androgen A male sex hormone secreted by the testes; testosterone is the principal androgen.

androgynous Pertaining to a male or female who identifies with both desirable masculine and desirable feminine characteristics.

anemia Shortage of oxygen-carrying material in the blood.

anesthesia Loss of sensation with or without a loss of consciousness.

anger According to the Kübler-Ross description of the dying process, the second stage in which the dying person manifests his or her feelings by hostility, resentment, and even envy.

angina pectoris Agonizing pain in the area of the heart, left shoulder, and arm; a symptom accompanying any interference with blood supply or oxygenation of the heart muscle.

anomalies Irregularities or deviations from general rules (e.g., of development).

anorexia nervosa Emotionally based eating disorder resulting in being seriously underweight.

anoxia Oxygen deprivation.

anterior Front part (e.g., of the pituitary).

anthropological Pertaining to the origins, physical and cultural development, social characteristics, social customs, and beliefs of humanity.

anticipation interval In a paired-associate learning task, the time span of the test phase in the anticipation method of presentation.

anticipation method Method of presenting a paired-associate task in which the stimulus is presented followed by the stimulus-response pair.

anticipatory socialization The process of learning attitudes, values, and activities that will be adaptive in a future situation.

anxiety Fear or worry as measured by self-reports.

apathetic Having or showing little interest, concern, or emotion.

Apgar score A scoring system developed by pediatrician Virginia Apgar to determine the physical and behavioral conditions of the newborn.

apnea Periods during which there is a pause in breathing during sleep.

approach-withdrawal In the New York Longitudinal Study, the category of temperament defined as the child's response to a new object or person.

aptitude test A test, such as an intelligence test, which has been designed to measure one's capacity to learn.

areola Elevated area surrounding the nipple of the breast.

arousal Autonomic nervous system activity in response to stress.

arteriosclerosis A group of processes involving thickening and loss of elasticity of the arterial walls.

assimilation In Piaget's theory, the alteration of an object, external to the person, to fit the person's existing cognitive structure.

associative learning Learning which involves the formation of stimulus-response (S-R) bonds.

associative shifting Same as classical conditioning.

associative stage Stage in the learning process of a paired-associate task during which the stimuli and responses are linked and recalled together.

atheoretical studies Studies designed to answer practical problems or to verify casual observations or satisfy curiosity.

atherosclerosis The development of arterial lesions accompanied by the accumulation of fat, cholesterol, and collagen at the lesion site.

attachment Proximity, contact-seeking, and/or maintaining behavior shown to one or a few specific others.

attention Concentration of the mind on a single object or thought.

attention span In the New York Longitudinal Study, the category of temperament defined as the amount of time devoted to an activity.

attitude A consistent cognitive disposition characterized by a readiness to react in a particular way toward specific objects, persons, or social issues.

attrition A decrease in the number of subjects in a study from the beginning of a study to the end through such causes as death, moving, loss of interest, etc.

atypical development Not typical; development which deviates from the norm for persons of the same sex and age.

auditory Pertaining to the sense of hearing.

authoritarian parent A parent who tries to shape, control, and evaluate the behavior of the child in accordance with a set, typically absolute standard of behavior.

authoritative parent A parent who tries to direct his or her child's activities through the use of a rational issue-oriented style.

authority and social-order-maintenance orientation The fourth stage in an early version of Kohlberg's theory of moral-reasoning development in which acts that are in accord with the maintenance of the rules of society and institutions of society are seen as moral.

autoimmunity The ability of the immune system to identify foreign material.

autonomic nervous system The division of the nervous system which controls visceral activities, smooth muscles, and endocrine glands.

autonomous morality The second phase in Piaget's theory of moral reasoning, involving the child's making subjective moral judgments.

axillary hair Underarm hair.

babbling An early stage of vocalization (from 3 to 4 months to 8 to 9 months) characterized by repetition of simple vowel and consonant sounds.

Babinski reflex A dorsal flexion of the big toe and an extension of the other toes in response to a gentle stroke of the side of the infant's foot from heel to toes.

Babkin reflex When lying on its back, the infant opens its mouth, closes its eyes, and returns its head to the midline in response to pressure applied on both of the baby's palms.

bargaining According to the Kübler-Ross description of the dying process, the third and middle stage in which the dying person negotiates covertly with God to extend life.

behavior-change processes A series of changes which form the bases of development.

behavior therapy A form of treatment, typically for individual-psychological disorders, that applies the principles of conditioning to clinical disorders.

behaviorism A theory derived from a mechanistic world view, which stresses that behavior may be understood by reference to external stimuli and responses.

beta rhythm Brain-wave patterns with a frequency of 18 to 30 cycles per second, associated with an attentive, alert state.

between-cohort differences in interindividual differences Differences in between-people differences from one cohort to another.

between-cohort differences in intraindividual change Differences in within-person change from one cohort to another.

biased sample A sample not representative of the population from which it is drawn.

biceps A muscle on the front of the arm responsible for bending the elbow.

biceps reflex The biceps muscle contracts, in response to a tap on the tendon of the muscle.

binaural Pertaining to both ears.

binocular Pertaining to the use of both eyes.

biofeedback A process of controlling one's bodily functions with the aid of a visual or auditory display of one's own brain waves, blood pressure, muscle tension, etc.

biological continuity The perception of grandparenting as a source of extending oneself and one's family into the future.

biological renewal The perception of grandparenting as a source of renewing one's youth.

biologically imperative Physically compelled (in a reproductive sense) to engage in those behaviors which are linked innately to one's biological status as a male or female.

birth canal Vagina.

birth cohort–related events Variables affecting subjects as a result of historical events intervening between the time of birth and the time of testing.

birth cohort Persons born in a given year.

birth order Birth position in relation to siblings (such as first, second, third child, etc.).

blink reflex Response (both eyelids closing) which occurs when a flash of light is presented to the eyes.

bloody show A sign of the onset of labor in which the mucus plug, blocking the opening of the cervix from the vaginal canal, is lost.

body-centered sex Sexual activity in which the emphasis is placed on the physical nature of sex.

body color One of five measurements included in the Apgar score: blue or pale body color and extremities = 0; pink body with blue extremities = 1; pink all over = 2.

Braxton-Hicks contractions Nonlabor contractions which may occur during the last trimester of pregnancy with little or no discomfort.

breast bud Small conelike protuberance of the breast caused by the raising areola.

breech delivery Birth process in which the feet or buttocks of the fetus appear first.

cardiovascular disease Diseases of the heart and blood vessels.

career An organized path undertaken by an individual that traverses time and space; consistent involvement in a particular occupational role over time.

caregiving practices Childrearing and disciplinary behaviors used by a person toward a child in his or her care.

castration anxiety Male child's fear that his father will punish him by removing his genitalia.

cataracts Opacities of the lens that obstruct light waves.

causal modeling techniques Statistical techniques which permit causal inference within the framework of nonexperimental designs.

causal relationship The situation in which one set of events produces or causes another set.

causality Reason or reasons why an event occurs.

central nervous system That part of the nervous system within the bony protection of the vertebral column and skull.

centration In Piaget's theory, a focus on, or embeddedness in, a particular point of view (usually one's own).

cerebral lateralization The specialization of the cerebral hemispheres according to the functional areas.

cervix The neck of the uterus which opens into the vagina.

cesarean delivery An operation in which a surgical incision is made through the abdominal wall and uterus in order to remove the infant.

chaining The linking together of simple learned behaviors to form a complex behavior.

chromosomes Thread-shaped bodies within the nucleus of a body cell, which carry the genes.

chronological age One's age in years.

classical conditioning A form of learning in which a neutral stimulus is paired with a meaningful (unconditioned) stimulus until eventually the neutral stimulus elicits the re-

sponse originally made only to the meaningful stimulus.

classical learning paradigm See classical conditioning.

climacteric The period of declining reproductive capability in men and women; includes menopause in women.

clitoris A small, sensitive sexual organ located above the urinary tract opening in the female.

cognition Knowledge; knowing; and the processes involved in its acquisition and utilization.

cognitive flexibility Ability to shift from one way of thinking to another.

cognitive social learning theory A theoretical approach to learning which emphasizes intrinsic, cognitive processes in the acquisition of behaviors.

cognitive structure A way of organizing knowledge that is constructed through interaction with the environment.

cohabitation A lifestyle in which a heterosexual couple live together without being married to one another.

cohort A group of persons experiencing some common event, e.g., being born in the same year.

cohort flow The progression of cohorts over historical time.

cohort life table A life table summarizing mortality rates which provides a longitudinal perspective, following the mortality experience of a single cohort.

coitus Sexual intercourse.

collagen An extracellular protein which surrounds blood vessels and cells and which slowly cross-links with age.

combinatorial thought Problem-solving ability in which one can imagine all possible solutions to a problem.

commitment An affirmation or pledge established toward an object, a person, or a course of action over time.

communion A set of behaviors characterized by interpersonal warmth, expressiveness, and sensitivity.

competence (1) Capacity to function in physiological, sensory, motor, and cognitive areas. (2) In Piaget's theory, the formal, logical representation of cognitive structures.

concept An abstraction; a term used

to represent or symbolize attributes of stimuli.

conception The moment of fertilization when the ovum and sperm unite.

concrete mortality *See* objective morality.

concrete operational egocentrism Lack of ability to differentiate between thoughts about reality and actual experience of reality.

concrete operations, stage of In Piaget's theory, the third stage of cognitive development; characterized by operations such as conservation, classification, and seriation.

conditioned response (CR) In classical conditioning, a response that is evoked by the conditioned stimulus after conditioning has occurred.

conditioned stimulus (CS) In classical conditioning, a previously neutral stimulus that, through pairing with an unconditioned stimulus, acquires the ability to produce a particular response.

conditioning A process by which stimuli and responses become associated, as in classical or operant conditioning; learning.

confounding A variable's influence on behavior that cannot be separated from another variable which could be influencing behavior at the same time.

congenital Existing at or present since birth.

congruence Compatibility, e.g., of person and environment.

conscience In Freud's theory, the component of the superego which represents the internalization of society's demands.

conscience (principle) orientation The sixth stage of the early version of Kohlberg's theory of moral-reasoning development in which acts that are consistent with one's own conscience are considered to be moral.

conservation In Piaget's theory, the ability to know that certain aspects of a stimulus remain unchanged even when other aspects have changed.

contamination Influence on data by variables extraneous to those which are central to the empirical study.

contextual Pertaining to the environment in which something exists or occurs.

contextual world view A world view which stresses that constant and completely interrelated changes characterize humans and their contexts.

contingent Conditional; dependent for existence upon something else.

continuity Behavior and/or its explanations stay the same over time.

continuous growth Smooth and nonabrupt changes in behavior.

contraception Prevention of conception; use of birth control devices to prevent pregnancy.

contraction The shortening of the muscles of the uterus to expel the fetus.

contractual legalistic orientation The fifth stage of the early version of Kohlberg's theory of moral-reasoning development in which acts which conform to those required by an implicit contract between self and society are considered moral.

contralateral Pertaining to the opposite side of the body.

control group A group of subjects who do not undergo experimental manipulation and who are used for comparisons with the experimental group to ascertain whether the subjects were affected by the experimental procedure.

control operations Means by which information is retrieved from one memory storage structure and entered into the next.

controlled observation The research situation is controlled by the researcher, but the behavior of the person is not directly manipulated.

conventional morality In Kohlberg's theory of moral-reasoning development, the second level of reasoning, involving judgments based on conformity to or maintenance of social rules and institutions.

cooperative play Type of play in which children engage in conversation and reciprocal, coordinated exchanges.

coping strategy The specific method used by an individual to adapt or adjust to a life event.

correlate *See* correlation.

correlation The degree of relationship between variables of measurements in a research study.

correlation coefficients A measure of the relationship between two variables.

correlative transformation Manner of cognitively manipulating a problem by relating it to other problems.

costs In exchange theory, unpleasant experiences or the necessity of abandoning other rewarding activities in order to pursue current activity.

counterfactual Not actually represented in the real world.

coupling Situation in which a labor contraction begins before a preceding one has actually terminated.

critical life events Occurrences that result in a change in the ongoing life pattern of the individual.

critical period A qualitatively discontinuous period of life which is fixed and universal and involves the need for a specific development in order for succeeding developments to proceed optimally.

cross-cultural Pertaining to the comparison of behaviors, development, and so forth in different cultural groups.

cross-linkages Bonds between components of the same molecule or between molecules which develop with the passage of time.

cross-modal Pertaining to the interrelating of information from two or more sense modalities.

cross-sectional design Differently aged individuals or groups are studied at one point in time.

cross-sectional sequences Successions of two or more cross-sectional studies completed at different times of measurement involving independent measures of different individuals.

crystallized intelligence General cognitive capacity reflecting the degree to which the individual has incorporated the knowledge and skills of the culture into thinking and actions; postulated to increase with age.

cultural transmission The transmission of knowledge, skills, or values throughout a given society over time.

culture A legacy or heritage comprising knowledge, beliefs, values, assumptions, and patterns of behavior.

cumulative incidence The total amount of an activity that has occurred over time.

current life table A life table summarizing mortality rates which provides a cross-sectional perspective, specifying the age-specific mortality rates for a population at a given time.

dark adaptation The ability to adjust vision when moving from high to low levels of illumination.

data sets A collection of factual information used as a basis for reasoning, discussion, or statistical analysis.

decay Loss of information from memory as a result of the erosion of the memory trace.

de facto types Role types in which the objective circumstances prevent the individual from performing a role, such as chronically unemployed.

delayed puberty Situation in which a person goes through the bodily changes associated with pubescence much later than usual.

delta rhythm Brain-wave patterns with a frequency of 0.5 to 4 cycles per second; associated with sleep.

denial According to the Kübler-Ross description of the dying process, the final stage of dying in which the person resists the reality of impending death.

deoxyribonucleic acid (DNA) Complex molecule, arranged in a double helix, bearing coded genetic information.

dependent variable An outcome variable.

depression According to the Kübler-Ross description of the dying process, the fourth stage in which the manifestations of the patient's illness become too significant to ignore and result in the realization that death cannot be avoided.

depth perception The ability to perceive and discriminate objects in the foreground and background and to determine relative distance between objects and oneself.

description A task of the life-span developmentalist which involves the specification of the identity and timing of behavioral change.

descriptive continuity Description of behavior at one point in the life span in the same way as behavior at another point.

descriptive discontinuity Description of behavior at one point in the life span differently than behavior at another point.

descriptive methods Methods for the identification and dimensionalization of within-person change and between-person differences in such change.

despair In Erikson's theory, the negative pole of the crisis of old age, characterized by a rejection of one's life as meaningless and wasted.

development The act or process of evolving, maturing, growing, and expanding.

developmental age The level at which a child can successfully perform age-specific tasks.

developmental approach See developmental view of development.

developmental disturbance Imbalance in development resulting from new phenomena that powerfully influence a person.

developmental psychology The study of how individuals and psychological processes develop and of factors influencing their development.

developmental quotient (DQ) A measure of a child's performance level on age-specific tasks derived by dividing developmental age by chronological age and multiplying by 100.

developmental tasks In Havighurst's conception of development, the tasks people must master during specific portions of their lives in order for life satisfaction and future task success to occur.

developmental view of development An approach of study which emphasizes individual history and goes beyond a primary focus on the task by considering the changes associated with the person over time.

dialectical Pertaining to a philosophical conception of reality (a world view) which stresses that constant, completely interrelated changes characterize humans and their contexts.

dialectical operations A proposed "fifth stage" of cognitive development which allows a person to recognize and deal with new problems and issues.

dialectical paradigm A developmental model, rooted in the contextual world view, in which the individual is continually changing and development is the result of reciprocal interactions between the changing individual and the changing context.

diethylstilbestrol (DES) A drug

taken by pregnant women during the 1950s and 1960s to prevent miscarriage, which has since been discovered to cause cervical cancer in the daughters of the women taking the drug.

differentiated crying Crying beginning the second month of life in which different patterns and pitches are discernible which signal hunger, pain, or distress.

differentiated phenomenon See differentiation.

differentiation The process of increasing specialization of functioning, complexity, and organization that is evident in both physical and psychological development.

dilate To enlarge, expand, and become wider.

dimension An imaginary continuum (or line) existing between two endpoints.

discipline A field of study.

discontinuity Changes in the descriptions and/or explanations involved in behavior over time.

discontinuity hypothesis The hypothesis that behavioral functions are affected only when biological functions reach a critical or limiting level.

discriminate To respond differentially to different stimuli.

discriminative stimulus(S^D) A stimulus that cues the occasion for a response.

disengagement The reduction of interaction between the individual and the environment; considered by some theorists to be one component of successful aging.

disengagement theory Theory which postulates that aged individuals and society prepare in advance for the disability and death of the individual through a mutual process of psychological and social disengagement.

disequilibrium The result of reproductive assimilation which disrupts equilibrium when the child continues to assimilate.

disjunctive reaction time The length of time before an individual makes a response to a task involving multiple signals and/or responses.

disordered cohort flow An imbalance between people and roles within an age grade which occurs when there is a discrepancy between the size or role preferences of a cohort's membership and the number and nature of the roles open in the role structure.

distant figure style A style of interaction between a grandparent and child in which contact is benevolent but fleeting and infrequent.

distractibility In the New York Longitudinal Study, the category of temperament defined as the degree to which extraneous stimuli alter behavior.

dizygotic twins Siblings born of the same pregnancy but from two separate ova, thus having different genotypes (also termed fraternal twins).

dorsal Pertaining to the back of a body part.

double-jeopardy hypothesis The idea that the minority aged suffer from the impact of both age and race discrimination.

Down's syndrome A form of mental retardation attributable to the presence of three number-21 chromosomes.

drive An energizer of behavior.

drive states Behaviors which are energizers of behavior such as hunger, sex, a drive to avoid pain.

dwarf Term applied to a person whose height deviates more than 40 percent below the norm for persons of the same sex and age.

dyad Two persons.

dynamic Pertaining to continuous, forceful, and changing action.

dynamic interaction An interrelation, among all levels of analysis involving humans and their contexts, which involves processes from each level being both a product and a producer of processes at all other levels.

dysfunction Impaired or abnormal functioning.

dysmenorrhea Difficult or painful menstruation.

early first-stage labor Stage during which regularly spaced contractions occur, the cervix becomes effaced and dilates from a closed state to about 4 centimeters, and the woman experiences mild cramping and lower backache and has a bloody show or rupturing of the amniotic sac.

early maturers People who go through their adolescent bodily changes faster than usual.

eating disorders Too great or too little of an intake of calories.

echolalia A stage of early vocalization (beginning 9 to 10 months) characterized by a conscious imitation of sounds.

ecological context See ecological milieu.

ecological milieu The naturally existing setting for an organism's development.

ecological validity A setting that corresponds to the naturally occurring one of an organism.

ectoderm The outermost of three primary germ layers of an embryo.

ectomorph Body type which is thin, has little muscle mass, and is frail-looking.

effaced To become thinned and stretched.

egg See ovum.

ego In Freud's theory, the personality structure whose sole function is to adapt to reality.

ego ideal In Freud's theory, an internal representation of the perfect man or woman; component of superego.

ego strength The degree of capability of the ego to perform its function (to meet the demands of reality).

egocentrism Embeddedness in one's own point of view; failure to differentiate between subject and object.

elaboration A control operation for retrieval from primary memory and entry into secondary memory which involves the processing of information at deeper, semantic levels.

elastin An extracellular protein which surrounds blood vessels and cells and which slowly cross-links with age.

electroencephalogram (EEG) A record of the electrical activity of the brain.

elicit In classical conditioning, the indication of the involvement of the autonomic nervous system and the body's involuntary musculature.

emancipated Free (e.g., from parental care) and independent enough to be responsible for one's own life.

embryo The fertilized egg from the time of implantation to the tenth or twelfth week of pregnancy.

embryological period The period from about the second week of life through the tenth to twelfth week of life in the human, when all organ systems of the organism emerge.

emotional self-fulfillment The perception of grandparenting as a source of companionship and satisfaction from the development of a relationship between adult and child.

empathic Pertaining to the identification with the feelings, thoughts, or attitudes of another person.

empathy The ability to take the perspective of another and feel what that person feels.

empirical Capable of observation; scientific; based on careful observation or research.

empirical ideas Concepts based upon experience or observation alone.

empiricism A view that knowledge is achieved through observation.

empty nest syndrome A sequence of significant life events associated with the process of launching children from the home.

encoding Formation of a code at the time of input into secondary memory.

endocrine glands Glands which have no external ducts.

endoderm The innermost of the germ layers of the embryo.

endomorph Body type which is characterized by a plump build, fatness, and a rounded appearance.

engaged Pertaining to the head of the fetus when it moves down toward the pregnant woman's cervix and faces toward her back.

enrichment An intervention strategy which attempts to optimize individuals' knowledge, skills, and development.

environment The context or system in which an individual develops; "nurture" in the nature-nurture controversy.

environmental press Demand and potential of a given environmental quality for activating behavior.

epidemiological Pertaining to the study of the incidence, distribution, and control of disorders and disease in a population.

epigenetic principle In Erikson's theory, the maturational ground plan underlying development which states that anything that grows has a ground plan from which parts arise, each at its own special time, until all parts form a functioning whole.

episiotomy Surgical cutting of the

skin and underlying tissues under the vaginal opening to ensure that the tissue surounding the vaginal opening will not be torn as the infant emerges from the birth canal.

epistemology The philosophy of knowledge.

equlibration In Piaget's theory, the balance between the activity of the person on the environment and the activity of the environment on the person.

erogenous zone Areas of the body sensitive to sexual stimulation.

error catastrophe An accumulation of RNA transcription errors resulting in cell death.

error of commission The act of providing an incorrect response (in paired-associate and serial learning tasks).

error of omission Failure to give a response (in paired-associate and serial learning tasks).

estrogen A female sex hormone produced by the ovaries.

ethnographic A descriptive anthropological account (of a culture).

etiological Causative.

euthanasia The act or practice of killing for reasons of mercy.

evolution Historical or biological changes characterizing a species.

exchange theory Theory of social interaction which postulates that individuals enter into and maintain social interactions because they find them rewarding. A basic assumption of the theory is that relationships entail both rewards and costs and the difference between rewards and costs equals profit.

exocrine glands Glands which have openings (ducts) to the world outside the body.

expansion A technique used by parents in parent-child language interactions in which parents repeat back to the child what the child has said with an expansion of the child's expression.

experiment Set of procedures designed to assess the consequences of an experimentor-controlled or naturally occurring event (treatment) which intervenes in the lives of the participants.

experimental Pertaining to or based upon a procedure for scientific investigation requiring the manipulation of some aspect of the environ-ment to determine what the effects of this manipulation might be.

experimental group A group of subjects who undergo experimental manipulation in order to observe its effects.

experimental mortality The loss of individuals from a sample during the course of research which may threaten internal or external validity; attrition.

experimental observation A type of scientific observation involving maximum control over the setting of observations and direct manipulation of variables.

experimental strategies Simulation strategies in which the independent causal variable is manipulated or controlled during the study.

explanation A task of the life-span developmentalist which involves the specification of the antecedents, or causes, and the conditions of behavioral change.

explanatory continuity The use of the same explanations to account for behavior across life.

explanatory discontinuity The use of different explanations to account for behavior across life.

explanatory methods Methods for the identification of the causes of within-person change and between-person differences in such change.

explanatory studies Studies that attempt to find the bases or causes of social and behavioral development.

extension The act of straightening a limb or other body appendage.

external validity The degree to which a relationship among variables observed in one data set can also be observed in other data sets.

extrinsic Not essential or inherent in the nature of a thing; operating or coming from without.

factor-analytic techniques Statistical techniques for determining the number and nature of underlying variables among a large number of measures.

familial Pertaining to the family.

family A unit of related individuals in which children are usually produced and reared.

fertilization The union of the ovum and sperm.

fetal alcohol syndrome A syndrome of malformations suffered by some infants whose mothers ingested alcohol during the prenatal period; often characterized by facial, limb, or organ defects; small physical size; and reduced intellectual ability.

fetal monitor Medical equipment used to detect cardiac and respiratory problems in the fetus during labor.

fetal period The period from about the third month through ninth month of gestation in humans, when organ systems continue to grow and functional characteristics appear.

fetus Term applied to the unborn child from about the third through ninth month of pregnancy.

fibroblast cells Connective tissue cells which normally divide.

fidelity In Erikson's theory, an emotional orientation in which one is committed to a role and an ideology.

first-stage labor Longest stage of labor, usually lasting from 10 to 14 hours and having three substages: early first stage, middle first stage, and transition.

fixated See fixation.

fixation In Freud's theory, an arrest of libidinal development.

fixity In development, the unchangeable character of a behavior.

flexion The act of bending a limb or other body appendages.

fluid intelligence General cognitive capacity reflecting the degree to which the individual has developed unique qualities of thinking independent of culturally based content; postulated to peak in early adulthood.

fluttering See quickening.

follicle-stimulating hormone (FSH) Hormone which encourages ovulation in females and spermatogenesis in males.

forceps Metal instruments used during difficult delivery to assist in guiding the infant's head down and out of the vagina.

Foreclosure An identity status of adolescents who have never experienced a crisis but who have adopted an identity which their parents want.

formal operations, stage of In Piaget's theory, the fourth and final stage of cognitive development; characterized by abstract, logical, and hypothetical reasoning and thought.

formal style A style of interaction between a grandparent and child which involves performing what is regarded as a proper and prescribed role and which leaves parenting to parents.

fraternal twins *See* dizygotic twins.

free-classification task Task in which individuals are required to group stimuli that are alike or that go together in some way.

free radicals Unstable chemical compounds within the body which react with nearby molecules, thus resulting in the alteration of the structure and function of cells.

free-recall task Task in which a series of items is presented to an individual who is then asked to recall as many of the items as possible in any order.

fun-seeker style A style of interaction between a grandparent and child which is characterized by informality and playfulness and in which grandchildren are viewed as a source of leisure activity.

functional Pertaining to contributing to maintenance and, thus, aiding survival; contributing to, maintaining, or furthering development optimally; aiding survival; adaptive.

functional analysis of behavior A view of behavior change which focuses on the relationship between externally manipulable stimuli and objectively verifiable responses.

functional assimiliation The process in which any cognitive structure brought about through assimilation continues to assimilate.

functional invariants Abilities which serve the same or similar purpose throughout development.

fundamental biological significance Having characteristics fitting the survival requirements for a particular environmental setting.

gender identity A person's identification as male or female.

general anesthesia Insensitivity to pain produced by inducing a loss of consciousness.

general developmental model System for incorporating assessment of markers of all sources of developmental change (age, cohort, and time) into one of several research designs.

general intelligence A broad ability domain postulated to permeate all cognitive tasks.

generalize In learning theory, to respond to a new stimulus in a manner similar to the way a previous stimulus had been responded to.

generation Group of people born during one period of history or span of time; varies from several years to three or four decades.

generation gap Significant disparities in attitudes, behaviors, or values between generations.

generativity In Erikson's theory, the positive pole of the crisis of middle age; a deep concern for and contribution to the maintenance and perpetuation of society.

generic Pertaining to characteristics of a whole group or class; general.

genes Chromosomal units of heredity.

genetic Produced by genes or pertaining to genetics.

genetic cellular theories Theories of aging which suggest that aging results from damage to the genetic information involved in the formation of cellular proteins.

genetic inheritance The process wherein each parent gives the offspring half the necessary numbers of chromosomes, thus ensuring that each parent will contribute a basis of the offspring's characteristics and future development.

genetics A branch of biology that deals with heredity and variations in organisms.

genital stage In psychoanalytic theory, the final stage of psychosexual development which is characterized by heterosexual interests.

genotypes The complement of genes transmitted to people at conception by the union of the sperm and ovum.

germ cells *See* ovum; sperm.

Gestalt "Totality."

gestate To carry children.

gestation period The time from fertilization until birth.

giant A person whose height deviates more than 40 percent above the norm for persons of the same sex and age.

gonadotropic hormone Hormone which acts on those glands in males and females that are associated with each sex, respectively.

gonads Sex glands in males (the testes) and in females (the ovaries).

good-person orientation The third stage in the early version of Kohlberg's theory of moral-reasoning development in which acts that help others, lend to the approval of others, or, given certain role expectations by society, should lend to social approval are judged as moral.

grammatical morphemes *See* morphemes.

graphemic Pertaining to a unit (letter) of a writing system.

gratification In Freud's theory, tension reduction resulting from appropriate stimulation.

grief An emotional response to loss characterized by somatic distress and feelings of sadness, guilt, anger, and depression.

grief work The process or task of shaping a new identity accompanying bereavement and mourning.

growth Change by tissue accretion.

growth spurt In adolescence, the occurrence of the peak velocity of growth in height and weight.

growth-stimulating hormone (GSH) Hormone which acts on all body tissues to stimulate their rate of growth and nourishment.

habits Behaviors which are repeated because they reduce drive states.

habituate *See* habituation.

habituation A decrease in the strength of a response to a stimulus, seen after repeated presentations of the stimulus.

heart rate One of five measurements included in the Apgar score: absent heart rate = 0; less than 100 beats per minute = 1; 100 to 140 beats per minute = 2.

hedonism The idea that pleasure or happiness is the chief good in life.

hemorrhage Loss of blood.

heredity That which is genetically transmitted in one's genotype; the complement of genes received at conception.

heterogamy Mate selection in which the partners are different on a wide array of social and psychological dimensions.

heteronomous morality In Piaget's theory of moral-reasoning development, the first phase, involving the child's being objective in moral judgments.

heterosexual Persons attracted sexually to persons of the other sex.

heterozygotic Having different alleles at a given place on a chromosome.

high associative strength One word frequently elicits another in a free-association task.

high forceps deliveries Deliveries in which the head of the fetus has not descended very far down the birth canal and forceps must be placed high into the birth canal to facilitate delivery.

hirsutism Excessive growth of hair.

historical studies Studies which focus on the time-related antecedents and/or consequences of behavior.

history effects A class of time-related threats to internal validity consisting of events external to the individual which are confounded with the presumed causal variable under scrutiny.

holistic Emphasizing the whole; an approach which considers the relationship between the parts and the whole.

homogamy Marriage or mate selection in which the partners are similar on one or several social and psychological dimensions.

homosexual Person attracted sexually to persons of the same sex.

homozygotic Having identical alleles at a given place on a chromosome.

homunculus In medieval Christian theology, a full-grown but miniature adult, believed to be present from birth in the newborn's head, which contains sin and basic depravity.

honorific positions Role types which constitute "social promotions" in which the honor is symbolic and no specific role activities beyond the most token are associated with the position (such as Nobel laureate).

hormone A substance secreted by endocrine glands that has the effect of stimulating activity in various organs of the body.

hospice A medical facility specifically designed for the dying patient; emotional as well as medical care is provided.

hyperactive Pertaining to a disorder marked by a chronic tendency to be excessively active, difficult to control, and unable to concentrate.

hypertension High blood pressure.

hypothalamus A structure of the brain which aids in initiation of bodily changes.

hypothesis A testable statement of the relation between two or more variables.

hypothetical thought Abstract and systematic thought about complex problems and propositions.

hypoxia Reduced level of oxygen.

id In psychoanalytic theory, an innate personality structure which is the center of the libido.

identical twins See monozygotic twins.

identity A sense of self-definition.

Identity achiever An adolescent who has had a crisis period but who is now committed to an ideology.

identity crisis In Erikson's theory, the nuclear (core) crisis of adolescence, involving the need to attain a sense of self-definition.

identity transformation Recognition of a problem in terms of its singular attributes.

ideology A set of rules, beliefs, attitudes, values, and behavioral prescriptions, usually for a particular role.

idiographic Of or about a single individual.

illegitimate birth Any birth outside of marriage.

imaginal function of language The use of language to pretend.

imaginary audience Component of adolescent egocentrism in which the individual believes others are as preoccupied with his or her appearance and behavior as he or she is.

imitation The performance of an act which was observed being performed by another.

immediate memory-span Task involving both primary- and secondary-memory components which assesses the longest string of items that can be immediately reproduced in the order of presentation.

imminent justice Belief that immediate punishment will be associated with any moral transgression.

immunological system System that protects the body against invading microorganisms and mutant cells by generating antibodies which react with the proteins of foreign organisms and forms cells which engulf and digest foreign cells.

implantation The attachment of the fertilized ovum in the lining of the uterus.

impregnate To make pregnant.

imprinting Supposedly innately based predisposition for early social and emotional attachments to be formed toward the first moving object seen.

incest Sexual intercourse of related individuals. Ordinarily forbidden by law or by social custom.

independent measures Different observational forms, indexes, or techniques used in observation of individuals.

independent variable A variable manipulated by a researcher to assess its influence on behavior.

individual life span One's development from birth to death; (see also ontogeny).

induce To bring about labor artificially by rupturing the amniotic membranes or by administering a drug which stimulates contractions.

induction Use of rationality and explanation in attempts to influence behavior.

infarction A localized area of tissue that is dying or dead due to a deprivation of blood because of an obstruction by an embolism or thrombosis.

informal role type Roles without status, such as heroes, villains, and playboys.

information function of language The use of language to communicate with others.

information processing theory A theory based on a computer model in which each cognitive development and advancement is believed to stem from a foundation of previously acquired prerequisite experiences.

innate Inborn; preexisting.

INRC group In Piaget's theory, the identity, negation, reciprocal, and correlative transformations (of cognitive problems) that characterize the structure of formal operational thought.

inspection interval In a paired-associate learning task, the time span of the study phase in the anticipation method of presentation.

instability The quality in which a person's rate of change relative to the others in the group changes over time.

instincts Preformed, innate potentials for behavior.

institutional role types Roles in which normative expectations are associated with definite positions or attributes.

instrumental function of speech The use of language to express wishes.

instrumental learning *See* operant conditioning.

integrity In Erikson's theory, the positive pole of the crisis of old age; an inner sense of peace and order which allows one to view one's personal life as meaningful and worthwhile.

intelligence quotient An index of an individual's measured intelligence derived by dividing mental age by chronological age.

intelligence test An instrument designed to measure a person's mental ability.

intensity of reaction In the New York Longitudinal Study, the category of temperament defined as the energy of response, regardless of its quality or direction.

interactionism A general term used in various theoretical contexts to express the ideas that two or more entities relate to each other in bidirectional or reciprocal ways.

interactive relation A relation between two or more elements wherein the function of any one element requires the presence of the other elements in the relation.

interactive view of development *See* dynamic interaction.

intercultural Between cultures.

interdependent A relation between two or more elements, wherein the function of one element (e.g., the behavior of one person) is both a product and producer of the function of the other elements in the relation.

interference Loss of information from memory as a result of information acquired prior to or following input.

intergenerational relationships Relationships between persons in different generations; parent-child and grandparent-child relations.

intergenerational solidarity The continuity of agreement between generations in behavioral orientations, attitudes, and values.

interindividual differences Differences between people.

interindividual differences in intraindividual change Between-person differences in within-person change.

interiority Tendency to respond to inner rather than outer stimuli.

internal validity The degree to which an observed relationship among variables is accurately identified and interpreted.

interpersonal function of speech The use of speech to interact with others.

intersensory integration The interrelating of information from two or more sense modalities.

intervening variable An inferred variable functionally connected with antecedents and consequences.

intervention Assistance to enhance an individual's or a group's personal, social, and physical developments; the attempt to help persons to change.

interventionists Persons who try to change human functioning in order to optimize it.

intimacy In Erikson's theory, the positive pole of the crisis of young adulthood; the quality of affection and rapport found in deep, personal relationships.

intragenerational relationships Relationships among members of the same generation; peer relations.

intraindividual Within a person.

intraindividual change Within-person change.

intramodal Within one sense modality.

intrapsychic process An inner-person (mental) process.

intrasensory integration Ability to relate two stimuli presented to the same modality.

intrinsic Pertaining to the nature or constitution of a thing.

in utero In the uterus.

invariant order A constant, unchanging pattern.

in vitro In an artificial environment.

in vivo Within living tissue.

ipsilateral Pertaining to the same side of the body.

isolation In Erikson's theory, the negative pole of the crisis of young adulthood; characterized by the inability to achieve a close personal relationship with another person.

judgmental sampling Choosing subjects who fit a specific definition for the study.

junior member Status of a person just entering an enterprise.

juvenile delinquency Violation of a law committed by a person prior to his or her eighteenth birthday, which would have been a crime if committed by an adult.

kibbutz A system of childrearing in Israel in which children are reared collectively by adults who are not necessarily their parents.

knee jerk reflex A quick extension or kick of the knee in response to a tap on the tendon below the kneecap.

knowledge actualization The use of memory for world knowledge acquired through real-life experience.

labia The folds of tissue on either side of the vaginal opening.

lactate To breastfeed.

lallation A stage of early vocalization (beginning at 6 to 8 months) characterized by infants' imperfect or accidental imitation of their own sounds and those of others.

Lamaze method A natural childbirth education program in which the woman is instructed in various breathing techniques that are used to allay pain during the various stages of labor.

late maturers People who go through phases of adolescent bodily change slower than usual.

latency stage In psychoanalytic theory, the period of psychosexual development which occurs between the phallic stage and puberty, during which sexual drives and feelings become lessened or nonexistent.

law A regular, predictable relationship among variables.

learning A relatively permanent change in behavior potentiality which occurs as a result of reinforced practice, excluding changes due to maturation, fatigue, and/or injury to the nervous system.

learning-performance distinction The fact that one cannot see learning per se but can see performance changes and infer that these changes represent learning.

learning theory principles Approach to learning which emphasizes the use of reward, punishment, and imitation in behavior acquisition or change.

legal blindness Corrected distance vision of 20/200 or worse in the better eye, or a visual field limited to 20° in its greatest diameter.

lexical Pertaining to the words or vocabulary of a language, separate from the grammatical and syntactical aspects.

lexical item A word or phrase.

libido In psychoanalytic theory, a psychic energy which governs life.

life-course analysis An approach to the study of personality which focuses on the examination of age- and sex-differentiated life patterns within particular historical contexts.

life event Any one of a wide array of events whose advent indicates or requires substantial change in the individual's life.

life expectancy The average length of life.

life review The process of reviewing one's life as the result of a realization of impending death.

life-span developmental approach An orientation to the study of human development, from conception to death, which is concerned with the description, explanation, and optimization of within-person changes in behavior and between-person differences in such changes in behavior.

life transition Change in pattern of behavior defined by major normative life events such as marriage, birth of children, and retirement.

lightening Situation which occurs about two to four weeks before the birth of first babies in which the fetus drops into its position in the uterus, typically moving its head closer to the cervix.

lipofuscin Dark-colored waste pigments in cells which increase at a constant rate with time.

locomotor ability The ability to move from one place to another.

locus of control The attribution made for the source of one's actions, that is, whether one's behavior is controlled by phenomena internal to, and thus dependent on, self versus events external to, and independent of, one's own control.

longitudinal design Research procedure in which the same persons are measured over time.

longitudinal sequences Successions of two or more longitudinal studies begun at different times of measurement involving repeated measures of the same individuals.

love withdrawal Temporary removal of love from the child because the child's actions have made the parent feel disappointed or ashamed.

low associative strength One word infrequently elicits another in a free-association task.

luteinizing hormone (LH) Hormone which encourages ovulation in females and testicular development in males.

majority, age of Adult legal status, usually 18 years of age or older.

manipulative studies Studies in which stimuli are varied to determine the effects on responses.

masturbation Erotic stimulation of one's own genitals.

maternal instinct The view that there is a natural, normal fit between women and motherhood.

matrix A complex plan, array, or interrelation of elements.

maturation Change by tissue growth and differentiation at any time in life; in adolescence, the physical and physiological changes, involving reproductive capability, associated with the prepubescent through postpubescent phases.

maturation effects A class of threats to internal validity consisting of events internal to the individual which are confounded with the presumed causal variable under scrutiny.

maturational readiness The belief that certain maturations must occur before experiential variables can exert any useful influence.

mature In a physiological sense, pertaining to when an organism is capable of reproductive function.

maximum life span The extreme upper limit of length of life.

mean Statistical average.

mean length of utterance (MLU) Index of language development which measures speech in terms of the number of morphemes.

measurement The act or process of determining the extent or quantity of an entity.

mechanistic Of or relating to a world view which stresses continuity and

reductionism; a machine model of human functioning.

mechanistic world view A world view which stresses continuity and reductionism; a machine model of human functioning.

mediation The formation of a covert response which forms a link between stimulus and response.

mediational strategies Covert responses which mediate between stimulus and response.

memory monitoring Prediction of the number of items that would be recalled following various memory tasks.

menarche The onset of menstruation in females.

menopause The cessation of menstruation, typically occurring during a two-year period at an average age of 47.

menstruation The monthly discharge of blood and tissue comprising the uterine lining in mature, premenopausal, nonpregnant women.

mental age A person's relative ability to think as compared to peers.

mesoderm The middle of three primary germ layers of an embryo.

mesomorph A body type which is muscular, has strong bones, and is athletic-looking,.

metamemory Knowledge of one's own memory processes.

middle first-stage labor Stage during which contractions become stronger and closer together and the cervix dilates from 4 to 8 centimeters. Pain become more intense in this stage.

milieu Context; environmental setting.

minor A person who has not yet reached adult legal status; usually a person under 18 years old.

miscarriage Loss of a fetus early in a pregnancy without outside interference, that is, a spontaneous abortion as opposed to an induced or therapeutic abortion.

mitosis The division of a single body cell into two identical cells.

modality One of the main avenues of sensation (such as vision, hearing, etc.).

model In certain learning theories, the person who displays the behavior a subject observes.

moderate interaction theory A theory which places equal emphasis on nature and nurture but sees the two sources as independent of each other.

molecular level A subordinate level of analysis; a lower or the lowest level of analysis.

mono Single; one.

monozygotic twins Twins who arise from the same zygote which splits after conception (also termed identical twins).

moral Refers to the ethical aspects of human behavior.

moral development Changes in knowing and understanding the rules of society.

moral dilemma A situation where there is no right or wrong answer.

moral rationality In Piaget's theory of moral-reasoning development, moral judgments based on the intentions underlying an act.

moral realism In Piaget's theory of moral-reasoning development, moral judgments based on the objective characteristics of an act.

moral relativism The idea that what is seen as moral behavior is defined in accordance with the cultural orientations of a particular society.

Moratorium Identity status of an adolescent who is in a state of search and actively trying to make a commitment.

morphemes The smallest units of speech that have meaning.

morphological Of or pertaining to the body.

mucus plug A thick, sticky secretion which blocks the opening of the cervix to deter foreign microorganisms from entering the uterus.

multidirectional Involving many directions or patterns of change.

multidirectional change Development which is characterized by multiple patterns of change differing in terms of onset, direction, duration, and termination.

multidirectional relations See dynamic interactions.

multidisciplinary Involving many disciplines.

multivariate Involving the simultaneous examination of two or more dependent variables.

muscle tone One of five measurements included in the Apgar score: limp and flaccid = 0; some flexion of the extremities = 1; good flexion, active motion = 2.

mutant cells Cells in which the genes and chromosomes have been altered by an external force (e.g., cancer cells).

mutation Any alteration of the genes or chromosomes of an organism.

myocardial infarction An area of dead tissue in the heart muscle resulting from excessive oxygen deprivation.

naive egoistic In Kohlberg's theory of moral-reasoning development, the second stage of this reasoning.

naively egotistic orientation Second stage in the early version of Kohlberg's theory of moral-reasoning development in which an act is judged right if it is involved with an external event that satisfies the needs of the person or someone close to the person.

nativistic Pertaining to behavior or development which is innately determined.

natural childbirth education Means by which parents prepare for the birth of their child by experiencing education (that is, in what is involved in labor and delivery) and special training (that is, in breathing and relaxation techniques designed to help the woman through the pains of the contractions).

natural selection Evolutionary process in which the characteristics of the natural setting determine which organism characteristics will lead to survival and which ones will not.

naturalistic observation Observation of behavior as it occurs in its natural setting.

nature Pertaining to heredity, maturation, or genes.

nature-nurture controversy Controversy over whether the primary determinant of human behavior and development is genetic or environmental.

nature-nurture issue See nature-nurture controversy.

nature theory Theory of human development which emphasizes an innate basis of development.

negation operation Manner of cognitively manipulating a problem by canceling the existence of the problem.

negative identity formation In Erikson's theory, adoption of a role which is self-destructive or socially disapproved.

negative reinforcer Stimulus which maintains behavior by its termination.

negatively reinforcing stimulus See negative reinforcer.

neonate A newborn infant.

neophyte A beginner.

nervous system In humans, the nervous system includes the brain, the spinal cord, all sensory pathways, and all sensory receptor cells such as those found in the eye and ear.

neurohormonal processes Mechanisms involved in the nervous system and endocrine system.

neurological Pertaining to the nervous system.

nocturnal emission Discharge of semen during sleep.

nomothetic Of or pertaining to a group; group analysis.

noncontingent Not dependent for existence upon something else.

nonexperimental strategies A simulation strategy in which the independent causal variable has occurred prior to the study.

nongenetic cellular theories Theories of aging which focus on changes that take place in the cellular proteins after they have been formed.

nonnormative, life-event influences Person-specific determinants of behavior such as illness, divorce, promotion, death of spouse.

nonnutritive nipple Nipple through which no nutrients are delivered.

nonobtrusive measure Any measure which naturally exists in the environment and which does not require direct interference with subjects.

nonreversal shift In a problem-solving task, a change in the stimulus dimension related to reward as compared to the previous task rather than a shift occurring within the dimension.

non-role type Refers to the absence of roles and statuses.

nonspecific transfer Transfer-of-training effects which are the result of general factors such as warm-up or learning to learn.

norm An average, a typical, or a modal characteristic of a group or population; social or biological expectations for behavior at certain ages.

norm of reaction Range of potential outcomes that could result from a given genotype's potentially infinite interactions with environments.

normative Pertaining to the average, typical, or modal characteristic.

normative, age-graded influences Biological and environmental determinants of behavior that are correlated with chronological age.

normative, history-graded influences Biological and environmental determinants of behavior that are correlated with historical time.

normative studies Studies which describe typical behavior of people of particular age levels and specific populations.

nubility Marriageable condition; ability to have sexual relations.

nurturance Affectionate care and attention.

nurture Pertaining to environment, experience, or learning.

nurture theory Theory of human development which emphasizes the role of the environment in shaping behavior.

obedience and punishment orientation First stage in the early version of Kohlberg's theory of moral-reasoning development in which an act is judged wrong or right if it is or is not associated with punishment.

obesity Eating disorder resulting in excess body fat; excessive overweight condition in which a person is 20 percent above the average weight associated with others of his or her height and sex.

object permanency In Piaget's theory, the child's realization that objects continue to exist even though they are no longer available to the senses.

objective morality In Piaget's theory of moral-reasoning development, judgment of an act in terms of its empirical consequences (associated with the first phase of such reasoning).

observational learning A type of learning stressing that behavior can be acquired through observation of a model's behavior. Reinforcement of the subject's behaviors is typically not seen as necessary from this viewpoint.

Oedipus complex In Freud's theory, the idea that a male experiences incestuous love for his mother and antagonism toward his father.

off time Age-inappropriate timing of life events.

on time Age-appropriate timing of life events.

ontogeny The life span of a single species from its conception to its death.

open growth systems The capacity of some species to continue to grow after reproductive maturity is reached.

operant conditioning A form of learning in which behavior is controlled by the consequences it produces.

operant learning paradigm See operant conditioning.

operation In Piaget's theory, an internalized action that is reversible.

operational definition A definition in terms of measurable statements.

operational structures See operation.

optimization A task of the life-span developmentalist which involves the modification and enhancement of behavioral change through intervention.

optimize To provide maximum enhancement; to make as effective as possible.

oral stage In psychoanalytic theory, the first stage of psychosexual development during which pleasure is focused on the mouth.

orchialgia Pain in the region of the testes due to unrelieved sexual tension.

organism A living creature; a biological entity, such as a person.

organismic Of or pertaining to a philosophy of science (a world view) which stresses emergence, qualitative discontinuity, nonreductionism, and holism.

organismic world view A world view which stresses emergence, qualitative discontinuity, nonreductionism, and holism.

organization (1) A control operation for retrieval from primary memory and entry into secondary memory involving the coding of items into higher-order units. (2) In Piaget's theory, the ability to arrange cognitive activity into a system of interrelated elements.

overgeneralization See overregularization.

overregularization Errors of grammar made by children when they develop simple rules of grammar and attempt to fit new words into the existing set of rules.

ovulation The process of releasing an ovum, or egg, from the ovary.

ovum Mature female reproductive cell (a female gamete or egg).

oxytocin A hormone used to induce labor.

paired-associate learning task Task in which the individual learns to associate pairs of items so that the second item of the pair is given as a response upon presentation of the first item of the pair.

palmar reflex The infant's grasping of the finger in response to pressure applied by the finger against the infant's palm.

palmarmental reflex See Babkin reflex.

panel design Longitudinal design; a design which follows the same individuals over time.

paradigm A model or example.

parallel play Type of play in which children play side by side, often engaging in similar behaviors but not really interacting with each other.

participant observation Observation in which a researcher becomes part of the natural setting he or she is systematically observing.

passive euthanasia Withdrawing or withholding treatment which permits a person to die earlier than if treatment were received.

passive mastery Perception of the environment as threatening and dangerous and the self as passive and accommodating.

passive role An accommodating behavior pattern.

patella Knee cap.

patellar tendon reflex See knee jerk reflex.

paternity The state of being a biological father.

patterned speech A stage of early vocalization beginning at about 1 year of age, at which time the child consciously produces adultlike intelligible sounds and uses them to communicate with those around him or her.

Pavlovian conditioning See classical conditioning.

peer One belonging to the same group in society; friend; person of similar age.

peer group People who are social equals or who are similar on characteristics such as age or grade level.

pendulum problem A task in which

children are asked to alter the speed of a pendulum and make an attribution for the result of their actions; used to illustrate quality of problem-solving abilities at different cognitive stages.

penis envy In Freud's theory, the female envies the male for his possession of a genital structure of which she has been deprived.

perception Sensation with meaning; the process of extracting, interpreting, and categorizing sensory stimulation.

perform To carry out an action or pattern of behavior.

performance The process by which available competence is assessed and applied in real situations.

perinatal At or about the time of birth.

period of the zygote Time from fertilization until the egg implants itself on the wall of the uterus (about ten to fourteen days after fertilization in humans).

peripheral nervous system That part of the nervous system which primarily lies outside the bony protective area of the vertebral column and skull.

peritoneum A membrane lining the abdominal cavity and enclosing the abdominal organs.

permissive parent A parent who attempts to behave toward a child's behaviors, desires, and impulses in a nonpunishing, accepting, and affirming manner.

perpetuation Reproduction of a species.

person-centered sex Sexual activity in which the emphasis is placed on the emotional relationship between the individuals who are engaged in the activity.

personal fable Component of adolescent egocentrism wherein a person believes that he or she is a unique person having singular feelings and thoughts.

personal function of speech The use of language to express one's individuality.

personal space The amount of distance one tends to maintain toward another person when interacting with him or her.

personal vocational role orientations Expectations held by children as to the vocation that they themselves will engage in as adults.

personological Pertaining to an orientation which focuses on an individual's development and behavior through sole reference to variables and processes at the individual-psychological level of analysis.

petting Engaging in kissing, embracing, and affectionate touching.

phallic stage In psychoanalytic theory, the third stage of psychosexual development during which sex-role behavior develops through the process of identification.

phenomena Observable objects or circumstances; experienced by the senses.

phenylketonuria (PKU) A disorder involving an inability to metabolize fatty substances because of the absence of a particular digestive enzyme.

philosophies General beliefs, concepts, and attitudes underlying a sphere of activity or thought.

phonemes The basic sound units of language.

phonemic Pertaining to a speech sound or utterance; the smallest units of speech.

phylogeny The evolutionary history of a species; phylogenetic; phyletic.

physiological theories Theories of aging which suggest that aging results from the failure of some physiological coordinating system, such as the immunological or endocrine system, to integrate bodily functions properly.

pituitary Major endocrine gland, stimulated by the hypothalamus.

placebo A substance having no known physiological effect but given to a research participant who supposes it to be a pharmacologic agent.

placenta The organ that nourishes the growing fetus in the uterus.

plantar Pertaining to the bottom of the foot.

plantar reflex A flexion of all toes in response to pressure applied by the finger against the balls of the infant's feet.

plasticity Ability or potential for change.

play Nonserious and self-contained activity engaged in for the sheer satisfaction it brings.

pleasure principle In Freud's theory, the id's striving for gratification of libidinal energy.

pluralism An orientation to science (e.g., regarding theory or method) that stresses the need to use multiple approaches to a topic (instead of only one).

population All possible members of a group of individuals or observations to be studied.

positive reinforcer A stimulus which maintains behavior by its production.

postconventional morality In Kohlberg's theory of moral-reasoning development, the third level of reasoning, involving the stage of contractual, legalistic moral reasoning and the stage of conscience and principle orientation.

postnatal After birth.

postpartum Pertaining to maternal occurrences after the birth of a baby.

postpubescent Pertaining to the period in adolescence after most bodily changes have occurred.

power In exchange theory, this term refers to the person who has more resources than the exchange partner.

power assertive discipline Controlling techniques used by parents such as physical punishments, threats of punishments, and other physical measures.

precocial birds Birds that can walk upon hatching.

precocious puberty Situation in which a person goes through the bodily changes associated with pubescence much earlier than usual.

preconventional morality In Kohlberg's theory of moral-reasoning development, the first level of such reasoning (involving the first two stages of moral reasoning).

preformationism A nativistic view of language acquisition which contends that language is innate.

prematurity The birth of an infant earlier than 37 weeks of gestation.

premenstrual syndrome (premenstrual tension) Cluster of symptoms which sometimes occur just prior to the onset of menstruation, including headache, anxiety, depression, emotional outbursts, insomnia, hypersensitivity, and crying spells.

prenatal Prior to birth.

preoperational egocentrism Lack of ability to differentiate between symbol and object.

preoperational stage In Piaget's theory, the second stage of cognitive development lasting from approximately ages 2 to 7 years. In this stage, a person can mentally represent absent objects and can use symbols to represent objects.

prepared childbirth Birth process utilizing techniques learned by the parent to reduce the need for drugs during labor and delivery.

prepubescent Pertaining to the period in adolescence when changes in one or more bodily characteristics have begun, but the majority of changes that will take place have not yet been initiated.

presbyopia A defect of vision characterized by recession of the near point of vision so that objects very near the eyes cannot be seen clearly.

preterm baby See prematurity.

prevention An intervention strategy which attempts to reduce the likelihood that behavioral dysfunctions or problems will occur.

primary graffian follicle A body which develops within the ovary and which later secretes hormones needed for reproduction.

primary memory A temporary maintenance system for conscious processing of information.

primary mental ability A dimension of intelligence that describes what is common to various intellectual tests and accounts for individual differences in performance on the tests.

primary process In Freud's theory, the fantasy or imaginary process associated with the id.

primary reinforcing stimulus A stimulus which is innately rewarding, such as food or sleep.

primary sexual characteristics Characteristics which are present at birth and involve the internal and external genitalia.

principled reasoning In Kohlberg's theory of moral-reasoning development, reasoning at the highest level of moral reasoning (involving stages 5 and 6).

profit In exchange theory, the difference between rewards and costs.

progesterone Primarily a female hormone which provides a basis for sexual maturation.

projective personality test Test which assumes that the person's responses reflect his or her underlying personality dynamics.

promiscuity (promiscuous) Condi-

tion of being sexually "free" or "loose"; having sexual relations with many partners or with little commitment or thought.

propinquity Residential closeness; nearness in place and time.

protoplasm The living material that makes up cells.

psychoanalytic Pertaining to psychoanalysis and the ideas of Sigmund Freud.

psychological approach See psychological view of development.

psychological discipline Nonphysical techniques (such as love withdrawal or induction) used by parents to control a child's behavior.

psychological view of development An approach which focuses on how people of a particular age group perform a specific task; little, if any, concern is shown to the history of changes in the person that contributed to task performance.

psychometric Measurement of mental or psychological functioning.

psychoprophylaxis See Lamaze method.

psychosexual Pertaining to the psychological aspects of sexual development.

psychosocial Pertaining to both psychological and social elements.

psychosocial crisis In Erikson's theory, a state of the ego wherein requisite societal demands within a stage are not yet met.

puberty The point at which the person is capable of reproduction; the event sometimes used to define the beginning of adolescence.

pubescent Pertaining to the period when most bodily changes that will eventually take place have been initiated.

qualitative approach Approach to the study of cognition which emphasizes the type of thinking a person uses to deal with information.

qualitative change Alterations in the kind or type of phenomenon that exists.

quality of mood In the New York Longitudinal Study, the category of temperament defined as the amount of friendly, pleasant, joyful behavior as contrasted with unpleasant, unfriendly behavior.

quantitative change Alterations in how much or how many of a phenomenon exist.

questioning function of speech The use of language to explore the world.

quickening The movement of the fetus during the fourth or fifth month of pregnancy which can be felt by the mother.

randomization The assignment of a group of individuals to subgroups in such a way that each individual has an equal probability of being assigned to each subgroup.

reaction time (RT) The length of time between the appearance of a signal and the beginning of a responding movement.

reactive model See reactive-organism paradigm.

reactive-organism paradigm A developmental model, rooted in the mechanistic world view, in which the individual is inherently passive and behavior is the result of external forces.

reactivity The tendency for a subject's responses or behavior to be influenced by the fact that he or she is participating in a research study.

reality principle In Freud's theory, the law governing the functioning of the ego; the ego requirement to adapt to reality.

recall Response mode in which the participant must produce the called-for response from memory.

recapitulation A repeating or mirroring.

recency effect Phenomenon in the free-recall procedure in which the last few items of a list are recalled first.

reciprocal transformation Manner of cognitively manipulating a problem by taking an opposite view.

recognition Response mode in which the participant must choose the correct response from among several alternatives.

reconstituted families Families formed following a remarriage.

reflex irritability One of five measurements included in the Apgar score.

reflexes Relatively invariant motor outputs (that is, muscular movements) in response to particular sensory input (that is, stimulation).

regional anesthesia Anesthesia produced in one area of the body only, reducing sensitivity to pain without causing loss of consciousness.

regulation function of speech The use of language to control others.

rehabilitation An intervention strategy which attempts to reduce the intensity or duration of a dysfunction which has already appeared.

reinforced practice Repeated acts which are reinforced.

reinforcement Any stimulus which produces or maintains behavior; a stimulus which increases the probability of behavior.

reinforcer See reinforcing stimulus.

reinforcing stimulus (S^R) A stimulus that increases the probability of a behavior.

reliability (reliable) The extent to which a measure gives consistent scores upon repeated administrations or the portions of a measure are internally consistent.

remediation An intervention strategy which attempts to correct or remove a behavioral dysfunction which has already appeared.

reminiscence Review and reconstruction of one's past life, particularly in old age.

repeated measures Observations of the same individuals repeatedly at different times.

representational ability Ability to represent internally an absent object; symbolic activity.

representational cognitive functioning Ability to use symbols to represent reality.

representative sample A sample of the population, the characteristics of which are typical of the entire population.

reproductive assimilation See functional assimilation.

research methods The set of specific procedures by which a scientist makes observations and collects data in order to examine relationships among variables.

reservoir of family wisdom style A style of interaction between a grandparent and child in which the grandparent is the dispenser of specific skills or resources.

resource person Term pertaining to the perception of grandparenting as a source of satisfaction derived from contributing to the development of the grandchild.

resource In exchange theory, anything which an exchange partner considers rewarding, such as money, knowledge, social position.

respiratory effort One of five measurements included in the Apgar score: no breathing for more than one minute = 0; slow and irregular breathing = 1; good breathing with normal crying = 2.

respondent conditioning See classical conditioning.

response Any physiological or psychological reaction resulting from stimulation.

response learning stage Stage in the learning process of a paired-associate task during which the responses are identified and made available for recall.

retina The inner, rear layer of the eye containing light-sensitive receptor cells.

retrieval Utilization of a secondary-memory input code at the time of output.

retrospection The process of reconstructing earlier, past events through remembering.

reversal shift Shift in a problem-solving task in which the same stimulus dimension is related to reward (as in a previous task), but the positive and negative stimuli within this same dimension is reversed.

reversibility of action Ability to understand that actions taken on an object, if reversed in sequence, will return the object to its original state.

rewards In exchange theory, pleasant experiences which maintain relationships providing that rewards exceed costs.

Rh factor Incompatibility of a particular chemical in the bloodstreams of the mother and her infant which can produce chemicals which interfere with the unborn infant's ability to carry oxygen in its own blood.

rhythmicity In the New York Longitudinal Study, the category of temperament defined as the regularity of hunger, excretion, sleep, and wakefulness.

ribonucleic acid (RNA) A complex molecule having the function of transcribing and transferring the information about protein synthesis contained in DNA.

rigidity Inflexibility in unlearning previous habits in order to gain new ones.

rites de passage Socially sanctioned ceremonies or events marking the transition of an individual from one status or age class to another.

role A socially expected behavior pattern; a position or status someone accepts.

role attrition Role types in which role responsibilities dwindle away or are lost.

rooting reflex The infant's turning of its head in the direction of the stimulation in response to light tactile stimulation of the cheek.

rubella A viral infection commonly known as German measles which can cause serious defects in an unborn child.

sample A group of individuals or observations investigated and assumed to be representative of a larger group of individuals or observations (populations).

savings effect Phenomena wherein some of the previous learning is retained thus making relearning of the material easier.

schemata A conceptual model for representing a phenomenon.

scheme An organized sensorimotor action sequence.

scientific method A set of empirically based techniques used by researchers to study the phenomena of the world.

scientific theory A system of statements about the ways in which variables are related to each other and from which hypothesis can be deduced.

scrotum The external sac below the penis containing the testes.

second-order abilities A set of higher-level dimensions of intelligence generated through factor analysis of first-order or primary abilities.

second-stage labor Stage in which the baby is born.

secondary memory A permanent maintenance system for information processing characterized by semantic content.

secondary process In Freud's theory, the set of abilities possessed by the ego to adapt to reality (e.g., cognition and perception).

secondary reinforcer See secondary reinforcing stimulus.

secondary reinforcing stimulus A stimulus which acquires its reinforcing power through association with primary reinforcers.

secondary sexual characteristics Physical and physiological characteristics which emerge during the prepubescent through postpubescent phases; the characteristics which are associated with reproductive capability and physical maturity.

secular trend The relation of adolescent bodily changes to historical time.

selection effects A class of threats to internal validity consisting of bias resulting from the differential selection of individuals in comparison groups.

selective survival effects Changes in the mortality curve from one cohort to another which may jeopardize internal or external validity.

self-concept The set of knowledge one maintains about oneself.

self-esteem Level of positive or negative evaluation one associates with the self-concept.

self-fulfilling prophecy A social climate created by attitudes about a target person's behaviors which channels the target's behavior in a way consistent with the attitudes.

semantic Pertaining to the meaning of words.

senior members Status of persons advanced in an enterprise.

sensation Reception of stimulation.

sensorimotor egocentrism Lack of ability to differentiate between self and external world.

sensorimotor stage In Piaget's theory, the first stage of cognitive development occurring from about birth to age 2 years. Thought during this stage involves developing the knowledge that objects continue to exist even when not sensed by the person.

sensory memory Modality-specific (e.g., visual, auditory) peripheral memory system involving a literal copy of information.

sequential method Method of repeatedly studying a cross-sectional sample of people from various cohorts over a fixed interval of time.

sequential research designs See sequential method.

sequential strategies See sequential method.

serial-learning task Task in which the individual is to learn a list of items so that on presentation of an item, the succeeding item is given as a response.

sex-appropriate Pertaining to behaviors which are socially acceptable for one's sex (e.g., girls playing with dolls, boys playing with trucks).

sex differences The degree to which biological, sociocultural, psychological, or social processes vary in relation to sex.

sex-linked Pertaining to a hereditary characteristic controlled by a gene carried on the sex-determining chromosome.

sex role A socially defined set of prescriptions for behavior for people of a particular sex group.

sex-role behavior Behavior functioning in accordance with sex-role behavioral prescriptions.

sex-role stereotypes Generalized beliefs that particular behaviors are characteristic of one sex group as opposed to the other.

sexual evolution Process in which change regarding sexual values follows from or is compatible with other societal changes.

sexual socialization The process of becoming sexual, taking on a gender (sex) identity, learning sex roles, and acquiring the knowledge, skills, and values which allow one to function sexually.

short-term memory The ability to recall material immediately or soon after it has been presented.

sibling Offspring of the same parents, i.e., brothers and sisters.

simple cross-sectional method Data collection method in which individuals of different ages are observed on a single occasion at the same point in time.

simple longitudinal method Data collection method in which the same individuals are observed at two or more points in time, thus yielding a comparison of the same individuals at different ages.

simple reaction time The length of time required by an individual to make a single response to a single signaling stimulus.

sleeper effect Term pertaining to a situation in which an important event may occur early in a person's life but the effect is not seen until later.

social Pertaining to human society; the interaction of the individual and the group.

social breakdown model Hypothesized negative spiral of breakdown created by interaction of the social environment with the diminished

self-concept and competence of the individual.

social clock Time norms in the life cycle by which people anticipate timing of major events.

social evolution Process in which the differences and similarities between generations result in gradual change.

social institutions Ongoing aspects of society around which many of life's most important activities revolve (the family, religion, the legal system, and the educational system being examples).

social learning See social learning theory.

social learning theory A type of mechanistic theory wherein the principles of conditioning and learning are used to explain the acquisition of personality and social behaviors.

social relationships Ongoing interactions between two or more people.

socialization Process by which members of one generation shape the behaviors and personalities of members of another generation; acquisition of knowledge, skills, and values which make us able members of our society.

socialized anxiety A learned anticipation of punishment for transgressing from society's rules, which involves an unpleasant state of feeling.

societal vocational role orientation The idea held by children about the roles in which the sexes could engage in society.

socioadaptation The adjustment of the individual to the demands and conditions of society.

sociocultural Involving or relating to a combination of social and cultural factors.

solitary play Play which involves no interaction between one child and another.

soma Body.

somatotropin Growth-stimulating hormone.

somatotypes Body types.

spatial perception Recognition of one's position and orientation in space as well as the distance relationship between oneself and other objects in one's context.

species-specific Pertaining to characteristics which are unique to a certain species.

specific transfer Transfer-of-training effects which are dependent on the similarity between tasks.

speeded task Simple assessment task in which the object is to complete the task as rapidly as possible.

sperm The male reproductive cell that can fertilize an ovum.

spermatogenesis Sperm production.

spinal anesthesia Anesthesia produced by the injection of an anesthetic into the spinal canal, reducing sensitivity to pain without causing loss of consciousness.

spontaneous abortion See miscarriage.

stability The quality in which a person's rate of change relative to others in the group stays the same.

stage A distinct period of development within a sequence of developmental levels.

stage progressions See age-related change.

stagnation In Erikson's theory, the negative pole of the crisis of middle age; characterized by self-indulgence and a sense of impoverishment.

statistical analysis The examination and interpretation of data using mathematical probability and organizational techniques.

statistical significance The likelihood that a finding could not have been due to chance more than a small portion of the time.

status One's position in a group, community, or society relative to others.

statutory rape Sexual intercourse with a minor, e.g., a person who has not reached the legal age of consent.

stepping reflex The infant's rhythmic stepping movements in response to being supported in an upright position and tilted forward and slightly to one side.

stereotype An overgeneralized belief or standardized, invariant depiction of an object.

still birth Birth in which the fetus is not born alive.

stimulation strategies The construction of a controlled, time-compressed, artificial situation designed to represent or model a real situation in order to permit the specification of causes.

stimuli Changes in energy in the environment affecting a person's functioning.

stimulus-response (S-R) connections The association or linkage of a response to a stimulus event.

stimulus substitution See classical conditioning.

storage Retention of a secondary-memory input code until the time of output.

storage structures Theoretical memory constructs modeled after computer processing units, including a sensory store, a short-term store, and a long-term store.

stratified random sampling A sampling in which the population is proportionately divided according to categories of interest and then randomly sampled from within those categories.

stridor A harsh vibrating sound heard in cases of obstruction of the lungs and air passages.

strong interaction position A stance on the nature-nurture controversy, rooted in the contextual world view, in which nature and nurture are viewed as completely interdependent.

structure The pattern of organization of the elements of a phenomenon; in physiology, the physical makeup and constitution of the body.

style of behavioral reactivity A person's style of responding or reacting to the world.

subjective morality In Piaget's theory of moral-reasoning development, judging moral acts in regard to the intentions of the actor (a characteristic of the second phase of this reasoning).

substitution A defense mechanism involving the replacement of an unobtainable object with an obtainable one.

sucking reflex The infant's rhythmical sucking in response to a finger inserted in the mouth.

Sudden Infant Death Syndrome (SIDS) The sudden death of an apparently healthy infant during the first year of life; largest single cause of death from unknown causes during infancy.

superego In Freud's theory, the structure of the personality composed of the conscience and the ego ideal.

surgent growth An abrupt spurt in growth.

surrogate parent style A style of interaction between a grandparent and child characterized by the grandparent assuming actual caretaking responsibility for the grandchild while the parent works.

survival of the fittest Idea that those organisms possessing characteristics fitting the adaptive requirements for a particular environmental setting will survive.

symbolic Pertaining to something which represents something else.

synthesis A meshing or blending of elements; in dialectics, a uniting of a thesis and an antithesis.

tabula rasa In Locke's empirical philosophy, the idea that the mind at birth is a blank slate and any knowledge attained is derived from experience.

tactile Pertaining to the sense of touch.

tautological True by virtue of its logical form alone; needlessly repetitive; obviously true.

telegraphic sentences Earliest sentences produced by the child which are abbreviated versions of adult speech and consist mainly of nouns and verbs.

temperament The stylistic component of behavior, the way one performs a behavior of any content; how one does whatever one does.

tendon Fibrous tissue which connect a muscle with a bone.

tenuous role type Refers to statuses without roles.

teratogens Substances capable of producing fetal abnormalities, such as certain drugs as well as alcohol and tobacco.

terminal drop Phenomenon in which an individual's intellectual performance shows a marked decline prior to death.

testalgia Pain in the region of the testes due to unrelieved sexual tension.

testes Two glands located in the scrotum of the male that produce sperm and testosterone.

testing effects A threat to internal validity consisting of the effects of taking a test more than once.

testosterone Male hormone that provides a basis for sexual maturation.

thalidomide A drug proscribed to some pregnant women in the 1950s which was later discovered to cause serious birth defects in their offspring.

theoretical model A type of model

which helps one to interpret, apply, and extend knowledge by suggesting appropriate and useful research questions.

theoretical studies Studies designed to test hypotheses.

theory An interrelated set of propositions which integrates existing facts and leads to the generation of additional facts.

theta rhythm Brain-wave pattern with a frequency of 5 to 7 cycles per second; associated with drowsiness.

third-stage labor Stage during which the placenta is delivered.

threshold Minimum amount of stimulation needed to evoke a response.

threshold of responsiveness In the New York Longitudinal Study, the category of temperament defined as the intensity of stimulation required to evoke a discernible response.

thrombus A blood clot.

time-lag design The study of one age level at different times in history.

time of measurement The time when behavior is measured.

time-related effects Variables in the milieu of subjects at the time they are tested which influence behavior.

tinnitus Ringing in the ears.

titular positions Role types which are honorary and nominal.

toe grasp reflex See plantar reflex.

toxemia Condition in which the blood contains poisonous products; often associated with pregnancy.

trait A relatively stable characteristic or quality.

transduce See intersensory integration.

transfer of training The effect of learning one task on the learning or retention of another task.

transform To alter or rearrange.

transition Last portion of the first stage of labor lasting about one to two hours. During this stage, the cervix is usually completely effaced and dilated from 8 to 10 centimeters; contractions are strong and long, making this the most painful stage of labor.

transmissiveness The amount of light reaching the eye.

transverse position Labor complication in which the fetus has settled crosswise in the uterus.

tropic hormone Hormone which stimulates other specific endocrine glands to produce their own specific hormones.

true experiment Design in which the participants are assigned to treatments in a random fashion.

tumultuous growth Period of development characterized by crisis, stress, and problems.

Type A behavior pattern In Rosenman and Friedman's typology, behavior characterized by excessive competiveness, hostility, an accelerated pace of life, and feelings of struggling against time and the environment.

Type B behavior pattern In Rosenman and Friedman's typology, behavior characterized by the relative absence of Type A behavioral tendencies, the ability to relax freely, and adaptative competiveness.

ultrasound Vibrations of the same physical nature as sound but with frequencies above the range of human hearing.

umbilical cord The organ which is the attachment between mother and offspring during pregnancy and childbirth.

unconditional response (UCR) See unconditioned response.

unconditional stimulus (UCS) See unconditioned stimulus.

unconditioned response (UCR) An unlearned response made to an unconditioned stimulus; often an inborn reflex, as in the case of salivation in response to food.

unconditioned stimulus (UCS) In classical conditioning, the stimulus that consistently elicits an unconditioned response.

unconscious In Freud's theory, the area of the mind which contains material most difficult to bring into awareness.

undifferentiated crying A stage of early vocalization (birth to 1 month) in which a child's cries of pain, hunger, or fear cannot be distinguished.

universal sequences A series of changes that apply equally to all people of all cultures at all times of measurement.

uterine Pertaining to the uterus.

uterus Pear-shaped organ in the female in which the unborn child develops.

validity The degree to which a measure assesses what it purports to assess.

variable A quantity or quality which changes.

vascular Pertaining to blood vessels.

vasocongestion Congestion (blood engorgement) in the pelvic region causing discomfort due to unrelieved sexual tension.

venereal disease An infection which is transmitted primarily by sexual intercourse.

verbal problem Problem which is presented verbally, the elements of which are not apparent to nor manipulated by the senses.

viable Capable of independent life.

vicarious Experienced through imaginative participation in the experience of another.

vicarious achievement Term pertaining to the perception of grandparenting as providing an extension of the self in that the grandchild will accomplish what the grandparent was not able to do.

visual acuity The accurate perception of small details in one's visual field.

visual-spatial perception The ability to visualize different perspectives of objects in space.

visualization Ability to organize and process visual information.

weak interaction position A stance on the nature-nurture controversy, rooted in the organismic world view, in which nature is viewed as the basic influence of development, with nurture determining only the rate of development or its end point.

wisdom The accumulation and integration of life knowledge.

withdrawal reflex Flexion of the leg in response to a pinprick applied to the sole of the infant's foot.

within-cohort interindividual differences in change Differences in change between individuals in a given cohort.

within-cohort intraindividual change Change within individuals in a given cohort.

womb See uterus.

world view The set of philosophical views a scientist holds that pertain to his or her ideas about human nature and/or about how the world is constructed.

zygote Fertilized egg.

Abeles, R. P., Steel, L., & Wise, L. L. Patterns and implications of life-course organization: Studies from Project TALENT. In P. B. Baltes & O. G. Brim, Jr. (Eds.), *Life-span development and behavior* (Vol. 3). New York: Academic Press, 1980.

Acredolo, L. P. Development of spatial orientation in infancy. *Developmental Psychology*, 1978, *14*, 224–234.

Adams, G. R. Physical attractiveness research: Toward a developmental social psychology of beauty. *Human Development*, 1977, *20*, 217–239.

Adams, G. R., & Crane, P. An assessment of parents' and teachers' expectations of preschool children's social preference for attractive or unattractive children and adults. *Child Development*, 1980, *51*, 224–231.

Adams, J. E., & Lidenmann, E. Coping with long-term disability. In G. V. Coelho, D. A. Hamburg, & J. E. Adams (Eds.), *Coping and adaptation*. New York: Basic Books, 1974.

Adelson, J. What generation gap? *New York Times Magazine*, Jan, 18, 1970, pp. 10–45.

Adelson, J. The political imagination of the young adolescent. *Daedalus*, 1971, *100*, 1013–1050.

Adelson, J., & O'Neil, R. P. Growth of political ideas in adolescence: The sense of community. *Journal of Personality and Social Psychology*, 1966, *4*, 295–306.

Ainsworth, M. D. S. Patterns of attachment behavior shown by the infant in interaction. In B. M. Foss (Ed.), *Determinants of infant behavior* (Vol. 2). New York: Wiley, 1963.

Ainsworth, M. D. S. Patterns of attachment behavior shown by the infant in interaction with his mother. *Merrill-Palmer Quarterly*, 1964, *10*, 51–58.

Ainsworth, M. D. S. The development of infant-mother attachment. In B. M. Caldwell & H. N. Ricciuti (Eds.), *Review of child development research* (Vol. 3). Chicago: University of Chicago Press, 1973.

Ainsworth, M. D. S., Blehar, M. C., Waters, E., & Wall, S. M. *The strange situation: Observing patterns of attachment.* Hillsdale, N.J.: Erlbaum, 1979.

Alan Guttmacher Institute. *11 million teenagers.* New York: The Alan Guttmacher Institute, Planned Parenthood Federation of America, 1976.

Albrecht, G. L., & Gift, H. C. Adult socialization: Ambiguity and adult life crises. In N. Datan & L. H. Ginsburg (Eds.), *Life-span developmental psychology: Normative life crises.* New York: Academic Press, 1975.

Aldrich, C., & Mendkoff, E. Relocation of the aged and disabled: A mortality study. *Journal of American Geriatrics Society*, 1963, *11*, 185–194.

Aleksandrowicz, M. K. The effect of pain relieving drugs administered during labor and delivery on behavior of newborn—A review. *Merrill-Palmer Quarterly*, 1974, *20*, 121–141.

Aleksandrowicz, M. K., & Aleksandrowicz, D. R. Obstetrical pain-relieving drugs as predictors of infant behavior variability. *Child Development*, 1974, *45*, 935–945.

Allen, D. I. Student performance, attitude and self-esteem in open-area and self-contained classrooms. *Alberta Journal of Educational Research*, 1974, *20*, 1–7.

Allport, G. W. *The nature of prejudice.* Reading, Mass.: Addison-Wesley, 1954.

Almquist, E. M. Sex stereotype in occupational choice: The case for college women. *Journal of Vocational Behavior*, 1974, *5*, 13–21.

Almquist, E. M., & Angrist, S. S. Role-model influences on college women's career aspirations. *Merrill-Palmer Quarterly*, 1971, *17*, 263–279.

American Humane Association *National Analysis of Official Child Neglect and Abuse Reporting.* Englewood, Colo.: American Humane Association, 1978.

American Psychological Association. Ethical principles for the conduct of research. *Manual of Ethical Principles*, Washington, D.C.: American Psychological Association, 1973.

American Social Health Association. *Today's VD control problem.* New York, 1975.

Ames, R. Physical maturing among boys as related to adult social behavior: A longitudinal study. *California Journal of Educational Research*, 1957, *8*, 69–75.

Anastasi, A. Heredity, environment, and the question "how"? *Psychological Review*, 1958, *65*, 197–208.

Anastasi, A. On the formation of psychological traits. *American Psychologist*, 1970, *25*, 899–910.

Andersen, A. E. Atypical anorexia nervosa. In R. A. Vigersky (Ed.), *Anorexia nervosa.* New York: Raven Press, 1977.

Anderson, B., Jr., & Palmore, E. Longitudinal evaluation of ocular function. In E. Palmore (Ed.), *Normal aging II: Reports from the Duke longitudinal studies, 1970–1973.* Durham, N.C.: Duke University Press, 1974.

Anderson, J. E. The limitations of infant and pre-school tests in the measurement of intelligence. *Journal of Psychology*, 1939, *8*, 351–379.

Anthony, E. J. The reaction of adults to adolescents and their behavior. In G. Caplan & S. Lebovici (Eds.), *Adolescence.* New York: Basic Books, 1969.

Anthony, E. J. The behavior disorders of children. In P. H. Mussen (Ed.), *Carmichael's manual of child psychology* (Vol. 2). New York: Wiley, 1970.

Anthony, E. J. The syndrome of the psychologically invulnerable child. In E. J. Anthony & C. Koupermik (Eds.), *The child in his family: Children at psychiatric risk* (Vol. 3). New York: Wiley, 1974.

Arenberg, D. Anticipation interval and age differences in verbal learning. *Journal of Abnormal Psychology*, 1965, *70*, 419–425.

Arenberg, D. *Memory and learning do decline late in life.* Paper presented at the Conference on Aging and Social Policy, Vichy, France, 1977.

Arenberg, D., & Robertson-Tchabo, E. A. Learning and aging. In J. E. Birren & K. W. Schaie (Eds.), *Handbook of the psychology of aging.* New York: Van Nostrand Reinhold, 1977.

Arguso, V. M., Jr. *Learning in the later years: Principles of educational gerontology.* New York: Academic Press, 1978.

Arlin, P. K. Cognitive development in adulthood: A fifth stage? *Developmental Psychology,* 1975, *11,* 602–606.

Atchley, R. C. Dimensions of widowhood in later life. *The Gerontologist,* 1975, *11,* 176–178.

Atchley, R. C. *The social forces in later life* (2d ed.). Belmont, Calif.: Wadsworth, 1977.

Ausubel, D. P., Montemayor, R., & Svajian, P. *Theory and problems of adolescent development* (2d ed.). New York: Grune & Stratton, 1977.

Bachman, J. G. *Youth in transitions: The impact of family background and intelligence on tenth-grade boys* (Vol. 2). Ann Arbor: University of Michigan Press, 1970.

Bachman, J. G., Green, S., & Wirtanen, I. D. *Youth in transition: Dropping out—problem or symptom?* (Vol. 3). Ann Arbor: University of Michigan Press, 1971.

Back, K. W. Meaning of time in later life. *Journal of Genetic Psychology,* 1965, *109,* 9–25.

Bacon, C., & Lerner, R. M. Effects of maternal employment status on the development of vocational-role perception in females. *Journal of Genetic Psychology,* 1975, *126,* 187–193.

Bakan, D. *The duality of human existence.* Chicago: Rand McNally, 1966.

Bakwin, H., & McLaughlin, S. M. Secular increments in height: Is the end in sight? *Lancet,* 1964, *2,* 1195–1196.

Balow, B., Rubin, R., & Rosen, M. J. Perinatal events as precursors of reading-disability. *Reading Research Quarterly,* 1976, *11,* 36–71.

Balter, P. B. Longitudinal and cross-sectional sequences in the study of age and generation effects. *Human Development,* 1968, *11,* 145–171.

Baltes, P. B. Prototypical paradigms and questions in life-span research on development and aging. *Gerontologist,* 1973, *13,* 458–467. (a)

Baltes, P. B. (Ed.) Strategies for psychological intervention in old age: A symposium. *The Gerontologist,* 1973, *13,* 4–38. (b)

Baltes, P. B. Life-span developmental psychology: Some converging observations on history and theory. In P. B. Baltes & O. G. Brim, Jr. (Eds.), *Life-span development and behavior* (Vol. 2). New York: Academic Press, 1979.

Baltes, P. B., Baltes, M. M., & Reinert, G. The relationship between time of measurement and age in cognitive development of children: An application of cross-sectional sequences. *Human Development,* 1970, *13,* 258–268.

Baltes, P. B., Cornelius, S. W., & Nesselroade, J. R. Cohort effects in behavioral development: Theoretical and methodological perspectives. In W. A. Collins (Ed.), *Minnesota symposium on child psychology* (Vol. 2). Hillsdale, N.J.: Erlbaum, 1978.

Baltes, P. B., & Danish, S. J. Intervention in life-span development and aging: Issues and concepts. In R. R. Turner & H. W. Reese (Eds.), *Life-span developmental psychology: Interventions.* New York: Academic Press, 1980.

Baltes, P. B., & Labouvie, G. V. Adult development of intellectual performance: Description, explanation, and modification. In C. Eisdorfer & M. P. Lawton (Eds.), *The psychology of adult development and aging.* Washington, D.C.: American Psychological Association, 1973.

Baltes, P. B., Reese, H. W., & Lipsitt, L. P. Life-span developmental psychology. *Annual Review of Psychology,* 1980, *31,* 65–110.

Baltes, P. B., Reese, H. W., & Nesselroade, J. R. *Life-span developmental psychology: Introduction to research methods.* Monterey, Calif.: Brooks/Cole, 1977.

Baltes, P. B., & Schaie, K. W. On life-span developmental research paradigms: Retrospects and prospects. In P. B. Baltes & K. W. Schaie (Eds.), *Life-span developmental psychology: Personality and social-*

ization. New York: Academic Press, 1973.

Baltes, P. B., & Schaie, K. W. Aging and IQ: The myth of the twilight years. *Psychology Today,* 1974, 7, 35–40.

Baltes, P. B., & Schaie, K. W. On the plasticity of intelligence in adulthood and old age: Where Horn and Donaldson fail. *American Psychologist,* 1976, *31,* 720–725.

Baltes, P. B., & Willis, S. L. Toward psychological theories of aging and development. In J. E. Birren & K. W. Schaie (Eds.), *Handbook of the psychology of aging.* Belmont, Calif.: Wadsworth, 1977.

Ban, P., & Lewis, M. Mothers and fathers, girls and boys: Attachment behavior in the one-year old. *Merrill-Palmer Quarterly,* 1974, *20,* 195–204.

Bandura, A. The stormy decade: Fact or fiction? *Psychology in the School,* 1964, *1,* 224–231.

Bandura, A. Influence of models' reinforcement contingencies on the acquisition of imitative responses. *Journal of Personality and Social Psychology,* 1965, *1,* 589–595.

Bandura, A. *Social learning theory.* Morristown, N.J.: General Learning Press, 1971.

Bandura, A. *Aggression: A social learning analysis.* Englewood Cliffs, N.J.: Prentice-Hall, 1973.

Bandura, A. *Social learning theory.* Englewood Cliffs, N.J.: Prentice-Hall, 1977.

Bandura, A. The self system in reciprocal determinism. *American Psychologist,* 1978, *33,* 344–358.

Bandura, A. Self-referent thought: A developmental analysis of self-efficacy. In J. H. Flavell & L. D. Ross (Eds.), *Cognitive social development: Frontiers and possible futures.* New York: Cambridge University Press, 1980. (a)

Bandura, A. The self and mechanisms of agency. In J. Suls (Ed.), *Social psychological perspective on the self.* Hillsdale, N.J.: Erlbaum, 1980. (b)

Bandura, A., & Huston, A. C. Identification as a process of incidental learning. *Journal of Abnormal and Social Psychology,* 1961, *63,* 311–318.

Bandura, A., & McDonald, F. The influence of social reinforcement and the behavior of models in shaping children's moral judgment. *Journal of Abnormal and Social Psychology*, 1963, *67*, 274–281.

Bandura, A., & Walters, R. H. *Adolescent aggression*. New York: Ronald Press, 1959.

Bandura, A., & Walters, R. H. *Social learning and personality development*. New York: Holt, Rinehart and Winston, 1963.

Bane, M. S. Marital disruption and lives of children. *Journal of Social Issues*, 1976, *32*, 103–117.

Barcus, F. E. Parental influence on children's television viewing. *Television Quarterly*, 1969, 8, 63–73.

Bardwick, J. M. *Psychology of women*. New York: Harper & Row, 1971.

Barfield, R. E., & Morgan, J. N. Trends in satisfaction with retirement. *The Gerontologist*, 1978, *18*, 19–23.

Barrett, C. J. Effectiveness of widows' groups in facilitating change. *Journal of Consulting and Clinical Psychology*, 1978, *46*, 20–31.

Barrett-Goldfarb, M. S., & Whitehurst, G. S. Infant vocalizations as a function of parental voice selection. *Developmental Psychology*, 1973, *8*, 273–276.

Barry, H., Bacon, M. K., & Child, I. L. A cross-cultural survey of some sex differences in socialization. *Journal of Abnormal and Social Psychology*, 1957, *55*, 527–534.

Bart, P., & Grossman, M. Menopause. In M. Notman & C. Nadelson (Eds.), *The woman patient*. New York: Plenum Press, 1978.

Bartlett, F. C. *Remembering*. Cambridge, Eng.: University Press, 1932.

Barton, E. M., Plemons, J. K., Willis, S. L., & Baltes, P. B. Recent findings on adult and gerontological intelligence: Changing a stereotype of decline. *American Behavioral Scientist*, 1975, *19*, 224–236.

Baruch, G. K., & Barnett, R. C. On the well-being of adult women. In L.A. Bond, & J. C. Rosen (Eds.), *Competence and coping during adulthood*. Hanover, N.H.: University Press of New England, 1980.

Baumrind, D. Child care practices anteceding three patterns of the preschool behavior. *Genetic Psychology Monographs*, 1967, 75, 43–88.

Baumrind, D. Authoritarian vs. authoritative parental control. *Adolescence*, 1968, *3*, 255–272.

Baumrind, D. Current patterns of parental authority. *Developmental Psychology Monographs*, 1971, *4*, 99–103.

Baumrind, D. Some thoughts about child rearing. In U. Bronfenbrenner (Ed.), *Influences on human development*. Hinsdale, Ill.: Dryden, 1972.

Bayley, N. Mental growth during the first three years: A developmental study of 61 children by repeated tests. *Genetic Psychology Monographs*, 1933, *14*, 7–92.

Bayley, N. The development of motor abilities during the first three years. *Monographs of the Society for Research in Child Development*, 1935, *1*.

Bayley, N. Consistency and variability in the growth of intelligence from birth to eighteen years. *Journal of Genetic Psychology*, 1949, *75*, 165–196.

Bayley, N. Individual patterns of development. *Child Development*, 1956, *27*, 45–74.

Bayley, N. Behavioral correlates of mental growth: Birth to thirty-six years. *American Psychologist*, 1968, *23*, 1–17.

Bayley, N. Development of mental abilities. In P. H. Mussen (Ed.), *Carmichael's manual of child psychology*. New York: Wiley, 1970.

Bayley, N., & Oden, M. H. The maintenance of intellectual ability in gifted adults. *Journal of Gerontology*, 1955, *10*, 91–107.

Beach, F. A. The Snark was a Boojum. *American Psychologist*, 1950, *5*, 115–124.

Bearison, D. J. The construct of regression: A Piagetian approach. *Merrill-Palmer Quarterly*, 1974, *20*, 21–30.

Becker, E. *The denial of death*. New York: The Free Press, 1973.

Becker, W. C., Peterson, D. R., Luria, Z., Shoemaker, D. J., & Hellman, K. Relations of factors derived from parent interview ratings to behavior problems of five-year olds. *Child Development*, 1962, *33*, 509–535.

Beckwith, J. B. Observations on the pathological anatomy of sudden infant death syndrome. In A. Bergman, J. Beckwith, & C. Ray (Eds.), *Sudden infant death syndrome: Proceedings of the Second International Conference on causes of sudden death in infants*. Seattle: University of Washington Press, 1970.

Beckwith, J. B. *The sudden infant death syndrome*. U. S. Department of Health, Education, and Welfare, 1977.

Bell, A. B. Role models of young adulthood: Their relationship to occupational behaviors. *Vocational Guidance Quarterly*, 1970, *18*, 280–284.

Bell, R. Q. Contributions of human infants to caregiving and social interaction. In M. Lewis & L. A. Rosenblum (Eds.), *The effect of the infant on its caregiver*. New York: Wiley, 1974.

Bell, S. W., & Ainsworth, M. D. Infant crying and maternal responsiveness. *Child development*, 1972, *43*, 1171–1190.

Belsky, J. Mother-father-infant interaction: A naturalistic observational study. *Developmental Psychology*, 1979, *15*, 601–607.

Belsky, J. Child maltreatment: An ecological integration. *American Psychologist*, 1980, *35*, 320–335.

Belsky, J. Early human experience: A family perspective. *Developmental Psychology*, 1981, *17*, 3–23.

Belsky, J., & Steinberg, L. D. The effects of day care: A critical review. *Child Development*, 1978, *49*, 920–949.

Belsky, J., Steinberg, L. D., & Walker, A. The ecology of day care. In M. E. Lamb (Ed.), *Childrearing in nontraditional families*. Hillsdale, N.J.: Erlbaum, 1982.

Belsky, J., & Tolan, W. J. Infants as producers of their own development: An ecological analysis. In R. M. Lerner & N. A. Busch-Rossnagel (Eds.), *Individuals as producers of their own development: A lifespan perspective*. New York: Academic Press, 1981.

Bem, S. L. The measurement of psychological androgyny. *Journal of Consulting and Clinical Psychology*, 1974, *47*, 155–162.

Bem, S. L. Sex-role adaptability: One consequence of psychological androgyny. *Journal of Personality and Social Psychology*, 1975, *31*, 634–643.

Bem, S. L. On the utility of alternative procedures for assessing psychological androgyny. *Journal of*

Consulting and Clinical Psychology, 1977, *45*, 196–205.

Bem, S. L. Theory and measurement of androgyny: A reply to the Pedhazur-Tentenbaum and Locksley-Colten critiques. *Journal of Personality and Social Psychology*, 1979, *37*, 1047–1054.

Benedek, T. Climacterium: A developmental phase. *Psychoanalytic Quarterly*, 1950, *19*, 1–27.

Bengtson, V. L. *The social psychology of aging*. New York: Bobbs-Merrill, 1973.

Bengtson, V. L., Dowd, J. J., Smith, D. H., & Inkeles, A. Modernization, modernity, and percpetions of aging: A cross-cultural study. *Journal of Gerontology*, 1975, *30*, 688–695.

Bengtson, V. L., & Kuypers, J. A. Generational differences and the developmental stake. *Aging and Human Development*, 1971, *2*, 249–260.

Bengtson, V. L., & Troll, L. Youth and their parents: Feedback and intergenerational influence in socialization. In R. M. Lerner & G. B. Spanier (Eds.), *Child influences on marital and family interaction: A life-span perspective*. New York: Academic Press, 1978.

Berardo, F. M. Widowhood status in the United States: Perspectives on a neglected aspect of the family cycle. *Family Coordinator*, 1968, *17*, 191–203.

Bergman, A. B., Ray, C. G., Pomeroy, M. A., Wahl, P. W., & Beckwith, J. B. Studies of sudden infant death syndrome in King County, Washington. 3. Epidemiology. *Pediatrics*, 1972, *49*, 860.

Berlyne, D. E. Curiosity and exploration. *Science*, 1966, *153*, 25–33.

Bernard, J. *Sex-role learning in children and adolescents*. Paper presented at the meeting of the American Association for the Advancement of Science, Washington, D.C., December, 1972.

Bernard, J. *The future of marriage*. New York: Bantam, 1973.

Berndt, T. J. The effect of reciprocity norms on moral judgment and causal attribution. *Child Development*, 1977, *48*, 1322–1330.

Berndt, T. J. Developmental changes in conformity to peers and parents. *Developmental Psychology*, 1979, *15*, 608–616.

Berscheid, E., & Walster, E. Physical attractiveness. In L. Berkowitz (Ed.), *Advances in experimental social psychology* (Vol. 7). New York: Academic Press, 1974.

Bertalanffy, von L. *Modern theories of development*. London: Oxford University Press, 1933.

Bertalanffy, von L. *Modern theories of development*. London: Oxford University Press, 1962.

Bijou, S. W. *Child development: The basic stage of early childhood*. Englewood Cliffs, N.J.: Prentice-Hall, 1976.

Bijou, S. W. *Some clarifications on the meaning of a behavior analysis of child development*. Paper presented at the Third Annual Midwestern Association of Behavior Analysis, Chicago, May 1977.

Bijou, S. W., & Baer, D. M. *Child development: A systematic and empirical theory* (Vol. 1). New York: Appleton-Century-Crofts, 1961.

Bijou, S. W., & Baer, D. M. *Child development: Universal stage of infancy* (Vol. 2). Englewood Cliffs, N.J.: Prentice-Hall, 1965.

Binet, A., & Simon, T. Sur la necéssité d'établir un diagnostic scientific des états inférieurs de l'intelligence. *L'Année Psychologique*, 1905, *11*, 162–190. (a)

Binet, A., & Simon, T. Methodes nouvelles pour le diagnostic du niveau intellectual des anormaus. *L'Année Psychologique*, 1905, *11*, 191–244. (b)

Birnbaum, J. A. Life patterns and self-esteem in gifted family oriented and career committed women. In M. T. S. Mednick, S. S. Tangri, & L. W. Hoffman (Eds.), *Women and achievement*. New York: Wiley, 1975.

Birren, J. E. Transitions in gerontology—From lab to life: Psychophysiology and speed of response. *American Psychologist*, 1974, *29*, 808–815.

Birren, J. E., & Botwinick, J. Age differences in finger, jaw, and foot reaction time to auditory stimuli. *Journal of Gerontology*, 1955, *10*, 429–432.

Birren, J. E., Butler, R. N., Greenhouse, S. W., Sokoloff, L., & Yarrow, M. R. (Eds.). *Human aging: A biological and behavioral study*. Washington, D.C.: U.S. Government Printing Office, 1963.

Birren, J. E., Riegel, K. F., & Morrison, D. F. Age differences in response speed as a function of controlled variations of stimulus conditions: Evidence of a general speed factor. *Gerontologia*, 1962, *6*, 1–18.

Birren, J. E., & Spieth, W. Age, response speed, and cardiovascular functions. *Journal of Gerontology*, 1962, *17*, 390–391.

Birren, J. E., & Woodruff, D. S. Human development over the life-span through education. In P. B. Baltes & K. W. Schaie (Eds.), *Life-span developmental psychology: Personality and socialization*. New York: Academic Press, 1973.

Bjorksten, J. The cross linkage theory of aging. *Journal of the American Geriatrics Society*, 1968, *16*, 408–427.

Black, L., Steinschneider, A., & Sheehe, P. R. Neonatal respiratory instability and infant development. *Child Development*, 1979, *50*, 561–564.

Blanchard, M., & Main, M. Avoidance of the attachment figure and social-emotional adjustment in day-care infants. *Developmental Psychology*, 1979, *15*, 445–446.

Blehar, M. C. Anxious attachment and defensive reactions associated with day care. *Child Development*, 1974, *45*, 683–692.

Block, J. *Lives through time*. Berkeley, Calif.: Bancroft, 1971.

Block, J. H. Conceptions of sex roles: Some cross-cultural and longitudinal perspectives. *American Psychologist*, 1973, *28*, 512–526.

Block, J. H. Issues, problems, and pitfalls in assessing sex differences: A critical review of *The psychology of sex differences*. *Merrill-Palmer Quarterly*, 1976, *22*, 283–308.

Blood, M. R., & Hulin, C. L. Alienation, environmental characteristics, and worker responses. *Journal of Applied Psychology*, 1967, *51*, 284–290.

Blood, R. O., & Wolfe, D. M. *Husbands and wives*. New York: Free Press, 1960.

Bloom, B. S. *Stability and change in human characteristics*. New York: Wiley, 1964.

Bloom, L., Hood, L., & Lightbow, P. Imitation in language development—If, when, and why. *Cognitive Psychology*, 1974, *6*, 380–420.

Bolles, R. C. Reinforcement, expectancy, and learning. *Psychological Review*, 1972, *79*, 394–409.

Bolton, C. D. Mate selection as the development of a relationship. *Marriage and Family Living*, 1961, *23*, 234–240.

Bond, E. K. Perception of form by the human infant. *Psychological Bulletin*, 1972, 77(4), 225.

Borland, D. C. Research on middle age: An assessment. *The Gerontologist*, 1978, *18*, 379–386.

Bornstein, M. H. Effects of habituation experiences on posthabituation behavior in young infants: Discrimination and generalization among colors. *Developmental Psychology*, 1979, *15*, 348–349.

Bornstein, M. H., Kessen, W., & Weiskopf, S. Color-vision and hue categorization in young infants. *Journal of Experimental Psychology*, 1976, *2*, 115–129.

Borup, J. H., Gallego, D. T., & Heffernan, P. G. Relocation and its effect on mortality. *The Gerontologist*, 1979, *19*, 135–140.

Botwinick, J. *Cognitive processes in maturity and old age.* New York: Springer, 1967.

Botwinick, J. Sensory-set factors in age differences in reaction time. *Journal of Genetic Psychology*, 1971, *119*, 241–249.

Botwinick, J. Intellectual abilities. In J. E. Birren & K. W. Schaie (Eds.), *Handbook of the psychology of aging.* New York: Van Nostrand Reinhold, 1977.

Botwinick, J., Robbin, J. S., & Brinley, J. F. Age differences in card-sorting performance in relation to task difficulty, task set, and practice. *Journal of Experimental Psychology*, 1960, *59*, 10–18.

Botwinick, J., & Storandt, M. *Memory, related functions and age.* Springfield, Ill.: Charles C. Thomas, 1974.

Botwinick, J., & Storandt, M. Recall and recognition of old information in relation to age and sex. *Journal of Gerontology*, 1980, *35*, 70–76.

Botwinick, J., & Thompson, L. W. Components of reaction time in relation to age and sex. *Journal of Genetic Psychology*, 1966, *108*, 175–183.

Bourque, L. B., & Back, K. W. Life graphs and life events. *Journal of Gerontology*, 1977, *32*, 669–674.

Bousfield, A. K., & Bousfield, W. A. Measurement of clustering and of sequential constancies in repeated free recall. *Psychological Reports*, 1966, *19*, 935–942.

Bower, T. G. R. Slant perception and shape constancy in infants. *Science*, 1966, *151*, 832–834.

Bower, T. G. R., & Paterson, J. G. Separation of place, movement, and object in the world of the infant. *Journal of Experimental Child Psychology*, 1973, *15*, 161–168.

Bowerman, C. E., & Day, B. R. A test of the theory of complimentary needs as applied to couples during courtship. *American Sociological Review*, 1956, *31*, 602–605.

Bowers, K. S. Situationalism in psychology. *Psychological Review*, 1973, *80*, 307–336.

Bowlby, J. *Attachment and loss* (Vol. 1). London: Hogarth Press and Institute of Psychoanalysis, 1969.

Bowlby, J. *Child care and the growth of love.* Hamondsworth, Eng.: Penguin, 1972.

Bowman, H. A., & Spanier, G. B. *Modern marriage.* New York: McGraw-Hill, 1978.

Boylin, W., Gordon, S. K., & Nehrke, M. F. Reminiscing and ego integrity in institutionalized elderly males. *The Gerontologist*, 1976, *16*, 118–124.

Brackbill, Y. Continuous stimulation and arousal level in infants: Additive effects. *Proceedings, 78th Annual Convention, American Psychological Association*, 1970, *5*, 271–272.

Brackbill, Y. Long-term effects of obstetrical anesthesia on infant autonomic function. *Developmental Psychology*, 1977, *10*, 529–536.

Brainerd, C. J. The stage question in cognitive-developmental theory. *The Behavioral and Brain Sciences*, 1978, *2*, 173–182.

Bransford, J. D., McCarrell, N. S., Franks, J. J., & Nitsch, K. E. Toward unexplaining memory. In R. Shaw & J. D. Bransford (Eds.), *Perceiving, acting, and knowing: Toward an ecological psychology.* Hillsdale, N.J.: Erlbaum, 1977.

Brazelton, T. B. Psychophysiologic reactions in the neonate: II. Effect of maternal medication on the neonate and his behavior. *Journal of Pediatrics*, 1961, *58*, 513–518.

Brazelton, T. B. Observation of the neonate. *Journal of the American Academy of Child Psychiatry*, 1961, *1*, 38–58.

Brazelton, T. B. Effect of prenatal drugs on the behavior of the neonate. *American Journal of Psychiatry*, 1970, *126*, 1261–1266.

Brazelton, T. B. *Neonatal behavioral assessment scale.* London: Heinemann, 1973.

Brazelton, T. B., Koslowski, B., & Main, M. The origins of reciprocity: The early mother-infant interaction. In M. Lewis & L. A. Rosenblum (Eds.), *The effect of the infant on its caregivers.* New York: Wiley, 1974.

Bremmer, J. G. Egocentric versus allocentric spatial coding in nine-month-old infants: Factors influencing the choice of code. *Developmental Psychology*, 1978, *14*, 346–355.

Brickman, P., & Bryan, J. H. Equity versus equality as factors in children's moral judgments of thefts, charity, and 3rd-party transfers. *Journal of Personality and Social Psychology*, 1976, *34*, 757–761.

Brim, O. G., Jr. Theories of the male mid-life crisis. *Counseling Psychologist*, 1976, *6*, 2–9.

Brim, O. G., Jr., & Kagan, J. Constancy and change: A view of the issues. In O. G. Brim, Jr., & J. Kagan (Eds.), *Constancy and change in human development.* Cambridge: Harvard University Press, 1980.

Brittain, C. V. Adolescent choices and parent-peer cross pressures. *American Sociological Review*, 1963, *28*, 385–391.

Brittain, C. V. A comparison of urban and rural adolescence with respect to peer versus parent compliance. *Adolescence*, 1969, *4*, 59–68.

Britton, J. H., & Britton, J. O. *Personality changes in aging.* New York: Springer, 1972.

Broadbent, D. E. *Perception and communication.* New York: Pergamon Press, 1958.

Brody, E. B., & Brody, N. *Intelligence: Nature, determinants, and consequences.* New York: Academic Press, 1976.

Brody, G. H., & Henderson, R. W. Effects of multiple model variations and rationale provision on moral judgments and explanations of young children. *Child development*, 1977, *48*(3), 1117–1120.

Bronfenbrenner, U. Freudian theories of identification and their derivatives. *Child Development*, 1960, *31*, 15–40.

Bronfenbrenner, U. Developmental theory in transition. In H. W. Stevenson (Ed.), *Child psychology.* Sixty-second yearbook of the National Society for the Study of Education, Part 1. Chicago: University of Chicago Press, 1963.

Bronfenbrenner, U. Toward an experimental ecology of human development. *American Psychologist,* 1977, *32,* 513–531.

Bronfenbrenner, U. *The ecology of human development.* Cambridge, Mass.: Harvard University Press, 1979.

Broverman, I. K., Vogel, S. R., Broverman, D. M., Clarkson, F. E., & Rosenkrantz, P. S. Sex-role stereotypes: A current appraisal. *Journal of Social Issues,* 1972, *28,* 59–78.

Brown, R. *A first language: The early stages.* Cambridge, Mass.: Harvard University Press, 1973.

Brown, R., & Bellugi, U. Three processes in the child's acquisition of syntax. In E. H. Lenneberg (Ed.), *New directions in the study of language.* Cambridge, Mass.: M.I.T. Press, 1964.

Bruch, H. *Eating disorders: Obesity, anorexia nervosa and the person within.* New York: Basic Books, 1973.

Bruch, H. Psychological antecedents of anorexia nervosa. In R. A. Vigersky (Ed.), *Anorexia nervosa.* New York: Raven Press, 1977.

Brückner, R. Longitudinal research on the eye. *Clinical Gerontology,* 1967, *9,* 87–95.

Brun-Gulbrandsen, S. *Kjonnsrolle og ungdomskriminalitet.* Oslo: Institute of Social Research, 1958 (mimeographed).

Bühler, C. *Der menschliche Lebenslauf als psycholgisches Problem.* Leipzig: Hirzel, 1933.

Bühler, C. The curve of life as studied in biographies. *Journal of Applied Psychology,* 1935, *19,* 405–409.

Bühler, C. Theoretical observations about life's basic tendencies. *American Journal of Psychotherapy,* 1959, *13,* 561–581.

Bundy, R. S. Discrimination of sound localization cues in young infants. *Child Development,* 1980, *51,* 292–294.

Bureau of Labor Statistics *Handbook of labor statistics, 1978, U.S. Department of Labor.* Washington, D.C.: U.S. Government Printing Office.

Burgess, R. L. Child abuse: A social interactional analysis. In B. B. Lakey & A. E. Kazdin (Ed.), *Advances in clinical child psychology.* New York: Plenum, 1979.

Burgess, R. L., & Conger, R. D. Family interaction in abusive, neglectful, and normal families. *Child Development,* 1978, *49,* 1163–1173.

Burgoon, C. F., Jr. Acne vulgaris. In V. C. Vaughn & R. J. McKay (Eds.), *Nelson textbook of pediatrics* (10th ed.). Philadelphia: Saunders, 1975.

Burns, S. M., & Brainerd, C. J. Effects of constructive and dramatic play on perspective taking in very young children. *Developmental Psychology,* 1979, *15,* 512–521.

Burt, C. The differentiation of intellectual abilities. *British Journal of Educational Psychology,* 1954, *24,* 76–90.

Burt, C. The evidence for the concept of intelligence. *British Journal of Educational Psychology,* 1955, *25,* 159–177.

Busse, E. W., & Obrist, W. D. Significance of focal electroencephalographic changes in the elderly. *Postgraduate Medicine,* 1963, *34,* 179–182.

Butler, R. N. The life review: An interpretation of reminiscence in the aged. *Psychiatry,* 1963, *26,* 67–76.

Butler, R. N. The life review: An interpretation of reminiscence in the aged. In B. L. Neugarten (Ed.), *Middle age and aging.* Chicago: University of Chicago Press, 1968.

Butler, R. N. The creative life and old age. In E. Pfeiffer (Ed.), *Successful aging.* Durham, N.C.: Duke University Center for the Study of Aging and Human Development, 1974.

Caldwell, B. M. The usefulness of the critical period hypothesis in the study of filiative behavior. *Merrill-Palmer Quarterly,* 1962, *8,* 229–242.

Caldwell, B. M., Wright, C. M., Honig, A., & Tannenbaum, J. Infant day care and attachment. *American Journal of Orthopsychiatry,* 1970, *40,* 397–412.

Campbell, A. The American way of mating: Marriage or children, only maybe. *Psychology Today,* May 1975, 39–42.

Campbell, D. T., & Stanley, J. C. Experimental and quasi-experimental designs for research on teaching. In N. L. Gage (Ed.), *Handbook of research on teaching.* Chicago: Rand McNally, 1963.

Campos, J. J., & Brackbill, Y. Infant state—Relationship to heart-rate, behavioral response, and response decrement. *Developmental Psychology,* 1973, *6,* 9–19.

Canestrari, R. E. Paced and self-paced learning in young and elderly adults. *Journal of Gerontology,* 1963, *18,* 165–168.

Canestrari, R. E. Age changes in acquisition. In G. A. Talland (Ed.), *Human aging and behavior.* New York: Academic Press, 1968.

Canning, H., & Mayer, J. Obesity—Its possible effect on college acceptance. *The New England Journal of Medicine,* 1966, *275,* 1172–1174.

Cantor, J. H., & Spiker, C. C. The effect of change in stimuli on the transfer of dimensional pretraining to the discrimination learning of kindergarten children. *Child Development,* 1978, *49,* 824–828.

Capon, N., & Kuhn, D. Logical reasoning in the supermarket: Adult females' use of a proportional reasoning strategy in an everyday context. *Developmental Psychology,* 1979, *15,* 450–452.

Caputo, D. V., & Mandel, W. Consequences of low birthweight. *Developmental Psychology,* 1970, *3,* 363–383.

Carns, D. E. Talking about sex: Notes on first coitus and the double sexual standard. *Journal of Marriage and the Family,* 1973, *35,* 677–688.

Caron, A. J., Caron, R. F., & Carlson, V. R. Infant perception of the invariant shape of objects varying in slant. *Child Development,* 1979, *50,* 716–720.

Carp, F. M. *A future for the aged: The residents of Victoria Plaza.* Austin: University of Texas Press, 1966.

Carpenter, D. G. Diffusion theory of aging. *Journal of Gerontology,* 1965, *20,* 191–195.

Cartwright, A., Hockey, L., & Anderson, J. L. *Life before death.* London: Routledge & Kegan Paul, 1973.

Casey, M. B. Color versus form discrimination learning in 1 year-old infants. *Developmental Psychology,* 1979, *15,* 341–343.

Cattell, P. *The measurement of in-*

telligence of infants and young children. New York: The Psychological Corporation, 1947.

Cattell, R. B. Theory of fluid and crystallized intelligence: A critical experiment. *Journal of Educational Psychology*, 1963, *54*, 1–22.

Cattell, R. B. *Abilities: Their structure, growth and action.* Boston: Houghton-Mifflin, 1971.

Catton, W. R., & Smircich, R. J. Comparison of mathematical models for the effect of residential propinquity on mate selection. *American Sociological Review*, 1964, *29*, 522–529.

Caudill, W., & Weinstein, H. Maternal care and infant behavior in Japan and America. *Psychiatry*, 1969, *32*, 12–34.

Cawood, C. D. Petting and prostatic engorgement. *Medical Aspects of Human Sexuality*, 1971, *5*, 204–218.

Chamove, A. S. Therapy of isolate rhesus: Different partners and social behavior. *Child Development*, 1978, *49*, 43–50.

Chand, I. P., Crider, D. M., & Willets, F. K. Parent-youth disagreement as perceived by youth: A longitudinal study. *Youth and Society*, 1975, *6*, 365–375.

Chandler, M. J., Greenspan, S., & Barenboim, C. Judgments of intentionality in response to videotaped and verbally presented moral dilemmas: Medium is message. *Child Development*, 1973, *44*, 315–320.

Charlesworth, R., & Hartup, W. W. Positive social reinforcement in the nursery school peer group. *Child Development*, 1967, *38*, 993–1002.

Chess, S., Thomas, A., & Birch, H. G. *Your child is a person.* New York: Viking, 1965.

Chiriboga, D. *Marital separation: A study of stress.* Paper presented at the Meeting of the Western Psychological Association, Seattle, 1977.

Chiriboga, D. *Life events and metamodels: A life span study.* Paper presented at the Meeting of the Gerontological Society, Dallas, November 1978.

Choate, R. *Testimony before the House Subcommittee on Communications.* United States House of Representatives, Washington, D.C.: Council on Children, Media and Merchandising, 1975.

Chomsky, N. *Language and mind.* New York: Harcourt, Brace & World, 1968.

Christensen, H. T., & Gregg, C. F. Changing sex norms in America and Scandinavia. *Journal of Marriage and the Family*, 1970, *32*, 616–627.

Chwast, J. Sociopathic behavior in children. In B. B. Wolman (Ed.), *Manual of child psychopathology.* New York: McGraw-Hill, 1972.

Ciaccio, N. V. A test of Erikson's theory of egio epigenesis. *Developmental Psychology*, 1971, *4*, 306–311.

Clarke, A. M., & Clarke, A. D. *Early experience: Myth and evidence.* New York: Free Press, 1976.

Clavan, S. The impact of social class and social trends on the role of grandparent. *Tha Family Coordinator*, 1978, *27*, 351–358.

Clayton, V. Erikson's theory of human development as it applies to the aged: Wisdom as contradictive cognition. *Human Development*, 1978, *18*, 119–128.

Clayton, V., & Overton, W. F. Concrete and formal operational thought process in young adulthood and old age. *International Journal of Aging and Human Development*, 1976, *7*, 237–245.

Clifford, M. M., & Walster, E. Effect of physical attractiveness on teacher expectations. *Sociology of Education*, 1973, *46*, 248–258.

Clopton, W. Personality and career change. *Industrial Gerontology*, 1973, *17*, 9–17.

Coates, B., Anderson, E. P., & Hartup, W. W. Interrelations in the attachment behavior of human infants. *Developmental Psychology*, 1972, *6*, 218–230.

Coates, B., Pusser, H. E., & Goodman, I. Influence of Sesame Street and Mister Rogers Neighborhood on children's social behavior in preschool. *Child Development*, 1976, *47*, 138–144.

Cochran, M., & Brassard, J. Child development and personal social networks. *Child Development*, 1979, *50*, 601–616.

Cohen, L. J. The operational definition of human attachment. *Psychological Bulletin*, 1974, *81*, 207–217.

Cohen, L. J., & Campos, J. J. Father, mother, and stranger as elicitors of attachment behaviors in infancy. *Developmental Psychology*, 1974, *10*, 146–154.

Cohen L. B., & Gelber, E. R. Infant visual memory. In L. Cohen & P.

Salapatek (Eds.), *Infant perception from sensation to cognition* (Vol. 1). New York: Academic Press, 1975.

Cohen, L. B., & Strauss, M. S. Concept acquisition in the human infant. *Child Development*, 1979, *50*, 419–424.

Colby, A. Evolution of a moral-developmental theory. *New Directions in Child Development*, 1978, *2*, 89–104.

Colby, A., Kohlberg, L., Gibbs, J., & Lieberman, M. A longitudinal study of moral judgment. Presentation at Center for Advanced Study in the Behavioral Sciences Summer Institute on Morality and Moral Development, 1979.

Coleman, J. S. *The adolescent society.* New York: Free Press, 1961.

Coleman, J. S. The adolescent culture. In I. J. Gordon (Ed.), *Human development.* Chicago: Scott, Foresman, 1965.

Colletta, N. D. *Divorced mothers at two income levels: Stress, support and child-rearing practices.* Unpublished thesis, Cornell University, 1978.

Collette-Pratt, C. Attitudinal predictors of devaluation of old age in a multi-generational sample. *Journal of Gerontology*, 1976, *31*, 193–197.

Collins, G. The good news about 1984. *Psychology Today*, January 1979, 34–38.

Collins, J. K., & Thomas, N. T. Age and susceptibility to same-sex peer pressure. *British Journal of Educational Psychology*, 1972, *42*, 83–85.

Collins, W. A. Effect of temporal separation between motivation, aggression, and consequenses: A developmental study. *Developmental Psychology*, 1973, *8*, 215–221.

Combs, J., & Cooley, W. W. Dropouts in high school and after school. *American Educational Research Journal*, 1968, *5*, 343–363.

Comfort, A. *Aging: The biology of senescence.* New York: Holt, Rinehart, and Winston, 1964.

Concern for Dying. *Revised living will.* New York: Concern for Dying, 1980.

Consortium for Longitudinal Studies. *Lasting effects after preschool.* ERIC/EECE, University of Illinois, College of Education, 1980.

Constantinople, A. An Eriksonian measure of personality development in college students. *Develop-*

mental Psychology, 1969, *1*, 357–372.

Conway, E., & Brackbill, Y. Delivery medication and infant outcome: An empirical study. *Monographs of the Society for Research in Child Development*, 1970, *35*, 24–34.

Cook, M., Field, J., & Griffiths, K. The perception of solid form in early infancy. *Child Development*, 1978, *49*, 866–869.

Cook, T. C., & Campbell, D. T. The design and conduct of quasi-experiments and true experiments in field settings. In M. D. Dunnette (Ed.), *Handbook of industrial and organizational research*. Chicago: Rand McNally, 1975.

Coombs, R. H. Value consensus and partner satisfaction among dating couples. *Journal of Marriage and the Family*, 1966, *28*, 166–173.

Coopersmith, S. *The antecedents of self-esteem*. San Francisco: Freeman, 1967.

Cornelius, S. W., & Denney, N. W. Dependency in day-care and home-care children. *Developmental Psychology*, 1975, *11*, 575–582.

Costa, P. T., & McCrae, R. R. Age differences in personality structure: A cluster analytic approach. *Journal of Gerontology*, 1976, *31*, 564–570.

Costa, P. T., & McCrae, R. R. Age differences in personality structure revisited: Studies in validity, stability, and change. *International Journal of Aging and Human Development*, 1978, *8*, 261–275.

Costa, P. T., & McCrae, R. R. Objective personality assessment. In M. Storandt, I. C. Siegler, & M. F. Elias (Eds.), *The clinical psychology of aging*. New York: Plenum Press, 1978.

Costa, P. T., & McCrae, R. R. Still stable after all these years: Personality as a key to some issues in aging. In P. B. Baltes and O. G. Brim, Jr. (Eds.), *Life-span development and behavior* (Vol. 3). New York: Academic Press, 1980.

Costa, P. T., McCrae, R. R., & Arenberg, D. Enduring dispositions in adult males. *Journal of Personality and Social Psychology*, 1980, *35*, 793–800.

Costanzo, P. R., & Shaw, M. E. Conformity as a function of age level. *Child Development*, 1966, *37*, 967–975.

Cowgill, D. O., & Holmes, L. D.

Aging and modernization. New York: Appleton-Century-Crofts, 1972.

Craik, F. I. M. Short term memory and the aging process. In G. A. Talland (Ed.), *Human aging and behavior*. New York: Academic Press, 1968. (a)

Craik, F. I. M. Two components in free recall. *Journal of Verbal Learning and Verbal Behavior*, 1968, *7*, 996–1004. (b)

Craik, F. I. M. Age differences in human memory. In J. E. Birren & K. W. Schaie (Eds.), *Handbook of the psychology of aging*. New York: Van Nostrand Reinhold, 1977.

Craik, F. I. M., & Lockhart, R. S. Levels of processing: A framework for memory research. *Journal of Verbal Learning and Verbal Behavior*, 1972, *11*, 671–684.

Craik, F. I. M., & Tulving, E. Depth of processing and the retention of words in episodic memory. *Journal of Experimental Psychology*, 1975, *104*, 268–294.

Crans, W. D., Kenny, J., & Campbell, D. T. Does intelligence cause achievement? A cross-lagged panel analysis. *Journal of Educational Psychology*, 1972, *63*, 258–275.

Cravioto, J., & DeLicardie, E. Longitudinal study of language development in severely malnourished children. In G. Serban (Ed.), *Nutrition and mental functions*. New York: Plenum Press, 1975.

Crawley, S. B., Rogers, P. P., Friedman, S., Iacobbo, M., Criticos, A., Richardson, L., & Thompson, M. A. Developmental changes in the structure of mother-infant play. *Developmental Psychology*, 1978, *14*, 30–36.

Crisp, A. H. Premorbid factors in adult disorders of weight, with particular references to primary anorexia nervosa (weight phobia). *Journal of Psychosomatic Medicine*, 1970, *14*, 1–22.

Crisp, A. H. Primary anorexia nervosa or adolescent weight phobia. *Practitioner*, 1974, *212*, 525–535.

Cronbach, L. J. *Essentials of psychological testing*. New York: Harper, 1960.

Cropper, D. A., Meck, D. S., & Ash, M. J. The relation between formal operations and a possible fifth stage of cognitive development. *Developmental Psychology*, 1977, *13*, 517–518.

Crowell, D. H., Blurton, L. B., Kobayashi, L. R., McFarland, J. L., & Yang, R. K. Studies in early infant learning—Classical conditioning of neonatal heart-rate. *Developmental Psychology*, 1976, *12*, 373–397.

Cuber, J. F., & Harroff, P. B. *Sex and the significant Americans*. Baltimore: Penguin, 1965.

Cumming, E., & Henry, W. E. *Growing old: The process of disengagement*. New York: Basic Books, 1961.

Cummings, E. M. Caregiving stability and day care. *Developmental Psychology*, 1980, *16*, 31–37.

Curtis, H. S. *Biological mechanisms of aging*. Springfield, Ill: Charles C. Thomas, 1966.

Dale, L. G. The growth of systematic thinking: Replication and analysis of Piaget's first chemical experiment. *Australian Journal of Psychology*, 1970, *22*, 277–286.

Danish, S. J., & D'Augelli, A. R. Promoting competence and enhancing development through life development intervention. In L. A. Bond & J. C. Rosen (Eds.), *Competence and coping during adulthood*. Hanover, N.H.: University Press of New England, 1980.

Darley, J. M., Klossen, E. C., & Zanna, M. P. Intentions and their contexts in moral judgments of children and adults. *Child Development*, 1978, *49*, 66–74.

Darwin, C. *The origin of species by means of natural selection or the preservation of favoured races in the struggle for life*. London: J. Murray, 1859.

Darwin, C. *The expression of emotions in man and animals*. London: J. Murray, 1872.

Dastur, D. K., Lane, M. H., Hansen, D. B., Kety, S. S., Butler, R. N., Perlin, S., & Sokoloff, L. Effects of aging on cerebral circulation and metabolism in man. In J. E. Birren, R. N. Butler, S. W. Greenhouse, L. Sokoloff, & M. R. Yarrow (Eds.), *Human aging: A biological and behavioral study*. Washington, D.C.: U.S. Government Printing Office, 1963.

Datan, N., & Ginsberg, L. H. (Eds.) *Life-span developmental psychology: Normative life crises*. New York: Academic Press, 1975.

Datan, N., Maoz, B., Antonovsky, A., & Wijsenbeek, H. Climacterium in three cultural contexts. *Tropical and Geographical Medicine*, 1970, 22, 77–86.

Datan, N., & Reese, H. W. (Eds.) *Life-span developmental psychology: Dialectical perspectives on experimental psychology.* New York: Academic Press, 1977.

Davis, A. *Socialization and the adolescent personality. Forty-third yearbook of the National Society for the Study of Education* (Vol. 43, Part 1). Chicago: University of Chicago Press, 1944.

Davison, M. L., Robbins, S., & Swanson, D. B. Stage structure in objective moral judgments. *Developmental Psychology*, 1978, 14, 137–146.

Dawe, H. C. The influence of size of kindergarten group upon performance. *Child Development*, 1934, 5, 295–303.

Dearden, R. F. The concept of play. In R. S. Peter (Ed.), *The concept of education.* London: Routledge, 1967.

DeCaspar, A. J., & Fifer, W. P. Of human bonding: Newborns prefer their mothers' voices. *Science*, 1980, 208, 1174–1176.

DeLora, J. R. Social systems of dating on a college campus. *Marriage and Family Living*, 1963, 25, 81–84.

Demos, J., & Demos, V. Adolescence in historical perspective. *Journal of Marriage and the Family*, 1969, 31, 623–639.

Denney, D. R., & Denney, N. W. The use of classification for problem solving: A comparison of middle and old age. *Developmental Psychology*, 1973, 9, 275–278.

Denney, N. W. Classification abilities in the elderly. *Journal of Gerontology*, 1974, 29, 309–314. (a)

Denney, N. W. Evidence for developmental changes in categorization criteria for children and adults. *Human Development*, 1974, 17, 41–53. (b)

Denney, N. W. Problem solving in later adulthood: Intervention research. In P. B. Baltes & O. G. Brim, Jr. (Eds.), *Life-span development and behavior* (Vol. 2). New York: Academic Press, 1979.

Denney, N. W., & Denney, D. R. Modeling effects on the questioning strategies of the elderly. *Developmental Psychology*, 1974, 10, 458.

Denney, N. W., & Lennon, M. L. Classification: A comparison of mid-

dle and old age. *Developmental Psychology*, 1972, 7, 210–213.

Despres, M. A. Favorable and unfavorable attitudes toward pregnancy in primaparae. *Journal of Genetic Psychology*, 1937, 51, 241–254.

Deutsch, H. *Motherhood: The psychology of women* (Vol. 2). New York: Grune and Stratton, 1949.

Deutscher, I. The quality of postparental life. *Journal of Marriage and the Family*, 1964, 26, 52–60.

deVilliers, J. G., & deVilliers, P. A. Competence and performance in child language: Are children really competent to judge? *Journal of Child Language*, 1974, 1, 11–22.

DeVries, R. Relationships among Piagetian, IQ, and achievement assessments. *Child Development*, 1974, 45, 746–756.

Dixon, R. A., Simon, E. W., Nowak, C. A., & Hultsch, D. F. Text recall in adulthood as a function of level of information, input modality, and delay interval. *Journal of Gerontology*, 1982, 37, 358–364.

Dobzhansky, T. Ethics and values in biology and cultural evaluation. *Zygon*, 1973, 8, 261–281.

Dohrenwend, B. S., Kransnoff, L., Askenasy, A. R., & Dohrenwend, B. P. Exemplification of a method for scaling life events: The PERI life events scale. *Journal of Health and Social Behavior*, 1978, 19, 205–229.

Dollard, J., Doob, L. W., Miller, N. E., Mowrer, O. H., & Sears, R. R. *Frustration and aggression.* New Haven, Conn.: Yale University Press, 1939.

Dollard, J., & Miller, N. E. *Personality and psychotherapy.* New York: McGraw-Hill, 1950.

Domey, R. G., McFarland, R. A., & Chadwick, E. Threshold and rate of dark adaptation as functions of age and time. *Human Factors*, 1960, 2, 109–119.

Dominick, J. R., & Greenberg, B. S. Attitudes toward violence: The interaction of television exposure, family attitudes, and social class. In G. A. Comstock & E. A. Rubinstein (Eds.), *Television and social behavior* (Vol. 3). *Television and adolescent aggressiveness.* Washington, D.C.: U.S. Government Printing Office, 1972.

Douglas, J. D., & Wong, A. C. Formal operations: Age and sex differences

in Chinese and American children. *Child Development*, 1977, 48, 689–692.

Douglas, K., & Arenberg, D. Age changes, cohort differences, and cultural change on the Guilford-Zimmerman Temperament Survey. *Journal of Gerontology*, 1978, 33, 737–747.

Douvan, E., & Adelson, J. *The adolescent experience.* New York: Wiley, 1966.

Dowd, J. J. Aging as exchange: A preface to theory. *Journal of Gerontology*, 1975, 30, 584–594.

Dowd, J. J. Exchange rates and old people. *Journal of Gerontology*, 1980, 35, 596–602.

Dowd, J. J., & Bengtson, V. L. Aging in minority populations: An examination of the double jeopardy hypothesis. *Journal of Gerontology*, 1978, 33, 427–436.

Doyle, A. B., Connolly, J., & Rivest, L. P. The effect of playmate familiarity on the social interactions of young children. *Child Development*, 1980, 51, 217–223.

Drachman, D. A., & Leavitt, J. Memory impairment in the aged: Storage versus retrieval deficit. *Journal of Experimental Psychology*, 1972, 93, 302–308.

Dubnoff, S. J., Veroff, J., & Kulka, R. A. *Adjustment to work: 1957–1976.* Paper presented at the meeting of the American Psychological Association, Toronto, August 1978.

Dubois, C. *The people of Alor.* Minneapolis: University of Minnesota Press, 1944.

Dussere, S. *The effects of television advertising on children's eating habits.* Unpublished master's thesis, Department of Health, University of Massachusetts, Amherst, 1974.

Duvall, E. M. *Family development.* Philadelphia: Lippincott, 1971.

Dwyer, J., & Mayer, J. Psychological effects of variations in physical appearance during adolescence. *Adolescence*, 1968–1969, 3, 353–380.

Dyer, E. D. Parenthood as crisis: A restudy. *Marriage and Family Living*, 1963, 25, 196–201.

Earhard, M. Retrieval failure in the presence of retrieval cues: A comparison of three age groups. *Canadian Journal of Psychology*, 1977, 31, 139–150.

Easterbrooks, M. A., & Lamb, M. E. The relationship between quality of infant-mother attachment and infant competence in initial encounters with peers. *Child Development,* 1979, *50,* 380–387.

Eato, L. E. *Perceptions of the physical and social environment in relation to self-esteem among male and female adolescents.* Unpublished master's thesis, The Pennsylvania State University, 1979.

Eckerman, C. O., Whatley, J. L., & Kutz, S. L. Growth of social play with peers during the second year of life. *Developmental Psychology,* 1975, *11,* 42–49.

Edwards, C. P. Society complexity and moral development: A Kenyan study. *Ethos,* 1975, *3,* 505–527.

Edwards, J. B. A developmental study of the acquisition of some moral concepts in children aged 7 to 15. *Educational Research,* 1974, *16,* 83–93.

Eisdorfer, C. Arousal and performance: Experiments in verbal learning and a tentative theory. In G. A. Talland (Ed.), *Human aging and behavior.* New York: Academic Press, 1968.

Eisdorfer, C., Nowlin, J., & Wilkie, F. Improvement in learning in the aged by modification of autonomic nervous system activity. *Science,* 1970, *170,* 1327–1329.

Eisdorfer, C. & Wilkie, F. Stress, disease, aging and behavior. In J. E. Birren, & K. W. Schaie (Eds.), *Handbook of the psychology of aging.* New York: Van Nostrand Reinhold, 1977.

Eisenberg-Berg, N. The relation of political attitudes to constraint-oriented and prosocial moral reasoning. *Developmental Psychology,* 1976, *12,* 552–553.

Eisenberg-Berg, N., & Geisheker, E. Content of preachings and power of the model-preacher: Effect on children's generosity. *Developmental Psychology,* 1979, *15,* 168–175.

Eisenberg-Berg, N., & Neal, C. Children's moral reasoning about their own spontaneous prosocial behavior. *Developmental Psychology,* 1979, *15,* 228–229.

Eisenberg, L. The "human" nature of human nature. *Science,* 1972, *176,* 123–128.

Eisenstadt, S. N. *From generations to generations.* New York: Free Press, 1956.

Ekstrom, R. B., French, J. W., Harman, H. H., & Dermen, D. *Manual for kit of factor-referenced cognitive tests.* Princeton, N.J.: Educational Testing Service, 1976.

Elder, G. H., Jr. Appearance and education in marriage mobility. *American Sociological Review,* 1969, *34,* 519–533.

Elder, G. H., Jr. *Children of the Great Depression.* Chicago: University of Chicago Press, 1974.

Elder, G. H., Jr. Age differentiation and the life courses. In A. Inkeles, J. Coleman, & N. Smelser (Eds.), *Annual review of sociology* (Vol. 1). Palo Alto, Calif.: Annual Reviews, 1975.

Elder, G. H., Jr. Family history and the life course. *Journal of Family History,* 1977, *2,* 279–304.

Elias, M. F., & Elias, P. K. Motivation and activity. In J. E. Birren & K. W. Schaie (Eds.), *Handbook of the psychology of aging.* New York: Van Nostrand Reinhold, 1977.

Elkind, D. Quantity conceptions in college students. *Journal of Social Psychology,* 1962, *57,* 459–465.

Elkind, D. Egocentrism in adolescence. *Child Development,* 1967, *38,* 1025–1034.

Elkind, D. Combinatorial thinking in adolescents from graded and ungraded classrooms. *Perceptual and Motor Skills,* 1968, *27,* 1015–1018.

Elkind, D., & Dabek, R. F. Personal injury and property damage in moral judgments of children. *Child Development,* 1977, *48,* 518–522.

Ellis, A. *Sex without guilt.* New York: Lyle Stuart, 1958.

Emmerich, W. Personality development and concepts of structure. *Child Development,* 1968, *39,* 671–690.

Endsley, R. C., Hutcherson, M. A., Garner, A. P., & Martin, M. J. Interrelationships among selected maternal behaviors, authoritarianism, and preschool children's verbal and nonverbal curiosity. *Child Development,* 1979, *50,* 331–339.

Engen, T., Lipsitt, L. P., & Peck, M. B. Ability of newborn-infants to discriminate sapid substances. *Developmental Psychology,* 1974, *10,* 741–744.

Entwisle, D. R., & Greenberger, E. Adolescents's views of women's work role. *American Journal of Orthopsychiatry,* 1972, *42,* 648–656.

Erber, J. T. Age differences in learning and memory on a digit-symbol substitution task. *Experimental Aging Research,* 1976, *2,* 45–53.

Erikson, E. H. Identity and the life cycle. *Psychological Issues,* 1959, *1,* 50–100.

Erikson, E. H. *Childhood and society* (2d ed.). New York: Norton, 1963.

Erikson, E. H. Inner and outer space: Reflections on womanhood. In R. J. Lifton (Ed.), *The woman in America.* Boston: Beacon, 1964.

Erikson, E. H. *Identity, youth and crisis.* New York: Norton, 1968.

Eron, L. D., Lefkowitz, M. M., Huesmann, L. R., & Walder, L. O. Does television violence cause aggression? *American Psychologist,* 1972, *27,* 253–263.

Escalona, S. *The roots of individuality.* Chicago: Aldine, 1968.

Esposito, N. J. Review of discrimination shift learning in young children. *Psychological Bulletin,* 1975, *82,* 432–455.

Eysenck, M. W. Age differences in incidental learning. *Developmental Psychology,* 1974, *10,* 936–941.

Fagan, F. J. Infant color-perception. *Science,* 1974, *183,* 973–975.

Fakouri, M. E. "Cognitive development in adulthood: A fifth stage?": A critique. *Developmental Psychology,* 1976, *12,* 472.

Falkner, F. Physical growth. In H. L. Bennett & A. H. Einhorn (Eds.), *Pediatrics.* New York: Appleton-Century-Crofts, 1972.

Fantz, R. L. The origin of form perception. *Scientific American,* 1961, *204,* 66–72.

Fantz, R. L., Ordy, J. M., & Udelf, M. S. Maturation of pattern vision in infants during the first six months. *Journal of Comparative and Physiological Psychology,* 1962, *55,* 907–917.

Farran, D. C., & Ramey, C. T. Social class differences in dyadic involvement during infancy. *Child Development,* 1980, *51,* 254–257.

Farrell, W. *Beyond masculinity.* New York: Random House, 1974.

Faust, M. S. Developmental maturity as a determinant in prestige of adolescent girls. *Child Development,* 1960, *31,* 173–184.

Feifel, H. The function of attitudes toward death. In Group for the Advancement of Psychiatry (Eds.),

Death and dying: Attitudes of patient and doctor. New York: Mental Health Materials Center, 1965.

Feifel, H., & Branscomb, A. B. Who's afraid of death? *Journal of Abnormal Psychology,* 1973, *81,* 282–288.

Feifel, H., Hanson, S., Jones, R., & Edwards, L. Physicians consider death. *Proceedings of the 75th Annual Convention of the American Psychological Association,* 1967, *2,* 201–202.

Feldman, N. S., Klosson, E. C., Parsons, J. E., Rholes, W. S., & Ruble, D. N. Order of information presentation and children's moral judgments. *Child Development,* 1976, *47,* 556–559.

Ferguson, C. P. *Preadolescent children's attitudes toward television commercials.* Austin, Tex.: Bureau of Business Research, University of Texas at Austin, 1975.

Ferree, M. Working class jobs: Housework and paid work as sources of satisfaction. *Social Problems,* 1976, *23,* 431–441.

Ferreira, A. *Prenatal environment.* Springfield, Ill.: Charles C. Thomas, 1969.

Feshbach, N. D., & Feshbach, S. Children's aggression. In W. W. Hartup (Ed.), *The young child: Reviews of research* (Vol. 2). Washington, D.C.: National Association for the Education of Young Children, 1972.

Feshbach, S., & Singer, R. *Television and aggression.* San Francisco: Jossey-Bass, 1971.

Festinger, L. A. *A theory of cognitive dissonance.* Stanford: Stanford University Press, 1957.

Field, J., Muir, D., Pilon, R., Sinclair, M., & Dodwell, P. Infants' orientation to lateral sounds from birth to three months. *Child Development,* 1980, *51,* 295–298.

Field, T. M., Dempsey, J. R., Hatch, J., Ting, G., & Clifton, R. K. Cardiac and behavioral responses to repeated tactile and auditory stimulation by preterm and term neonates. *Developmental Psychology,* 1979, *15,* 406–416.

Finch, C. E. The regulation of physiological changes during mammalian aging. *The Quarterly Review of Biology,* 1976, *51,* 49–83.

Finkelstein, N. W., Dent, C., Gallacher, K., & Ramey, C. T. Social behavior of infants and toddlers in a day-care environment. *Developmental Psychology,* 1978, *14,* 257–262.

Finkelstein, N. W., & Ramey, C. T. Learning to control the environment in infancy. *Child Development,* 1977, *48,* 806–819.

Fitzgerald, H. E., & Brackbill, Y. Classical conditioning in infancy: Development and constraints. *Psychological Bulletin,* 1976, *83,* 353–376.

Fitzgerald, J. M., Nesselroade, J. R., & Baltes, P. B. Emergence of adult intellectual structure. *Developmental Psychology,* 1973, *9,* 114–119.

Fiumara, N. J. Ineffectiveness of condoms in preventing veneral disease. *Medical Aspects of Human Sexuality,* 1972, *6,* 146–150.

Flanagan, J. C. Quality of life. In L. A. Bond & J. C. Rosen (Eds.), *Competence and coping during adulthood.* Hanover, N. H.: University of New England Press, 1980.

Flavell, J. H. *The developmental psychology of Jean Piaget.* New York: Van Nostrand, 1963.

Flavell, J. H. Cognitive changes in adulthood. In L. R. Goulet & P. B. Baltes (Eds.), *Life-span developmental psychology: Research and theory.* New York: Academic Press, 1970.

Flavell, J. H., & Wholwill, J. F. Formal and functional aspects of cognitive development. In D. Elkind & J. H. Flavell (Eds.), *Studies in cognitive development.* New York: Oxford University Press, 1969.

Flenner, D. E., & Cairns, R. B. Attachment behaviors in human infants: Discriminative vocalization on maternal separation. *Developmental Psychology,* 1970, *2,* 215–223.

Floyd, H. H., Jr., & South, D. R. Dilemma of youth: The choice of parents or peers as a frame of reference for behavior. *Journal of Marriage and the Family,* 1972, *34,* 627–634.

Fontana, V. J. Further reflections on maltreatment of children. *Pediatrics,* 1973, *51,* 780–782.

Fowles, B. *Testimony before the House Subcommittee on Communications.* Washington, D.C.: Governemnt Printing Office, 1975.

Fox, J. H. Effects of retirement and former work life on women's adaptation in old age. *Journal of Gerontology,* 1977, *32,* 196–202.

Fozard, J. L., & Popkin, S. J. Optimizing adult development: Ends and means of an applied psychology of aging. *American Psychologist,* 1978, *33,* 975–989.

Fozard, J. L., Wolf, E., Bell, B., McFarland, R. A., & Podolsky, S. Visual perception and communication. In J. E. Birren & K. W. Schaie (Eds.), *Handbook of the psychology of aging.* New York: Van Nostrand Reinhold, 1977.

Fraiberg, S. The development of human attachments in infants blind from birth. *Merrill-Palmer Quarterly,* 1975, *21,* 315–334.

Francis-Williams, J., & Davies, P. A. Very low-birth-weight and later intelligence. *Developmental Medicine and Child Neurology,* 1974, *16,* 709–728.

Frazier, T. M., Davis, G. H., Goldstein, H., & Goldberg, I. D. Cigarette smoking and prematurity: A prospective study. *American Journal of Obstetrics and Gynecology,* 1961, *81,* 988–996.

French, J. W., Ekstrom, R. B., & Price, L. A. *Kit of reference tests for cognitive factors.* Princeton, N.J.: Educational Testing Service, 1963.

Freud, A. Adolescence as a developmental disturbance. In G. Caplan & S. Lebovici (Eds.), *Adolescence.* New York: Basic Books, 1969.

Freud, S. *The ego and the id.* London: Hogarth, 1923.

Freud, S. *Outline of psychoanalysis.* New York: Norton, 1949.

Freud, S. Some psychological consequences of the anatomical distinction between the sexes. In *Collected papers* (Vol. 5). London: Hogarth, 1950.

Freud, S. *New introductory lectures on psychoanalysis.* New York: Norton, 1965.

Friedman, M., & Rosenman, R. H. *Type A behavior and your heart.* New York: Knopf, 1974.

Friedmann, E., & Havighurst, R. J. *The meaning of work and retirement.* Chicago: University of Chicago Press, 1954.

Friedrich-Cofer, L. K., Huston-Stein, A., Kipnis, D. M., Susman, E. J., & Clewett, A. S. Environmental enhancement of prosocial television content: Effects on interpersonal behavior, imaginative play, and self-regulation in a natural setting. *Developmental Psychology,* 1979, *15,* 637–646.

Friedrich, L. K., & Stein, A. H. Aggressive and prosocial television programs and the natural behavior of preschool children. *Monographs of the Society for Research in Child Development,* 1973, *38*(4, Serial No. 151), 1–64.

Friesen, D. Academic-athletic-popularity syndrome in the Canadian high school society (1967). *Adolescence,* 1968, *3,* 39–52.

Frodi, A. M., & Lamb, M. E. Child abusers' responses to infant smiles and cries. *Child Development,* 1980, *51,* 238–241.

Frodi, A. M., Lamb, M. E., Leavitt, L. A., Donovan, W. L., Neff, C., & Sherry, D. Fathers' and mothers' responses to the faces and cries of normal and premature infants. *Developmental Psychology,* 1978, *14,* 490–498.

Fuller, J. L., & Clark, L. D. Effects of rearing with specific stimuli upon post-isolation syndrome in dogs. *Journal of Comparative and Physiological Psychology,* 1966, *61,* 258–263. (a)

Fuller, J. L., & Clark, L. D. Genetic and treatment factors modifying the post-isolation syndrome in dogs. *Journal of Comparative and Physiological Psychology,* 1966, *61,* 251–257.

Furman, W., & Masters, J. C. Peer interactions, sociometric status, and resistance to deviation in young children. *Developmental Psychology,* 1980, *16,* 229–236.

Furman, W., Rahe, D. F., & Hartup, W. W. Rehabilitation of socially withdrawn preschool children through mixed-aged and same-age socialization. *Child Development,* 1979, *50,* 915–922.

Gaensbauer, T. J., & Emde, R. N. Wakefulness and feeding in human newborns. *Archives of General Psychiatry,* 1973, *28,* 894–897.

Gallatin, J. E. The development of political thinking in urban adolescents. (Final Report, Office of Education Grant 0-0554) Washington, D.C.: National Institutes of Education, 1972.

Gallatin, J. E. *Adolescence and individuality.* New York: Harper & Row, 1975.

Galst, J. P., & White, M. A. The unhealthy persuader: The reinforcing value of television and children's purchase-influencing attempts at the supermarket. *Child Development,* 1976, *47,* 1089–1096.

Garbarino, J. A preliminary study of some ecological correlates of child abuse: The impact of socioeconomic stress on mothers. *Child Development,* 1976, *47,* 178–185.

Garbarino, J. The human ecology of child maltreatment: A conceptual model for research. *Journal of Marriage and Family,* 1977, *39,* 721–736.

Garber, H., & Heber, R. The Milwaukee project: Early intervention as a technique to prevent mental retardation. University of Connecticut Technical Paper, 1973.

Gardner, H. *Developmental psychology.* Boston: Little-Brown, 1978.

Garrett, H. E. A developmental theory of intelligence. *American Psychologist,* 1946, *1,* 372–378.

Garry, R. Television's impact on the child. In S. Sunderlin (Ed.), *Children and T. V.: Television's impact on the child.* Washington, D.C.: Association for Childhood Association International, 1967.

Gengerelli, J. A. Graduate school reminiscence: Hull and Koffka. *American Psychologist,* 1976, *31,* 685–688.

Gerson, R. P., & Damon, W. Moral understanding and children's conduct. *New Directions for Child Development,* 1978, *2,* 41–59.

Gesell, A. L. Maturation and infant behavior pattern. *Psychological Review,* 1929, *36,* 307–319.

Gesell, A. L. The individual in infancy. In C. Murchison (Ed.), *Handbook of child psychology.* Worcestor, Mass.: Clark University Press, 1931.

Gesell, A. L. *An atlas of infant behavior.* New Haven, Conn.: Yale University Press, 1934.

Gesell, A. L. The ontogenesis of infant behavior. In L. Carmichael (Ed.), *Manual of child psychology.* New York: Wiley, 1946.

Gesell, A. L. The ontogenesis of infant behavior. In L. Carmichael (Ed.), *Manual of child psychology.* New York: Wiley, 1954.

Gesell, A. L., & Amatruda, C. S. *Developmental diagnosis: Normal and abnormal child development.* New York: Hoeber, 1941.

Gesell, A. L., & Amatruda, C. S. *Developmental diagnosis* (2d ed.). New York: Hoeber-Harper, 1947.

Gewirtz, J. L. Attachment, dependence, and a distinction in terms of stimulus control. In J. L. Gerwirtz (Ed.), *Attachment and dependency.* Washington, D.C.: Winston, 1972.

Gewirtz, J. L. The attachment acquisition process as evidenced in the maternal conditioning of cued infant responding (particularly crying). *Human Development,* 1976, *19,* 143–155.

Gewirtz, J. L., & Stingle, K. G. Learning of generalized imitation as the basis for identification. *Psychological Review,* 1968, *75,* 374–397.

Gibson, E. J., & Walk, R. R. The "visual cliff." *Scientific American,* 1960, *202,* 2–9.

Gibson, H. Early delinquency in relation to broken homes. *Journal of Abnormal Psychology,* 1969, *74,* 33–41.

Gil, D. G. *Violence against children: Physical abuse in the United States.* Cambridge, Mass.: Harvard University Press, 1970.

Gilbert, J. G. Age changes in color matching. *Journal of Gerontology,* 1957, *12,* 210–215.

Gladis, M. Age differences in repeated learning tasks in schizophrenic subjects. *Journal of Abnormal and Social Psychology,* 1964, *68,* 437–441.

Glamser, F. D. Determinants of a positive attitude toward retirement. *Journal of Gerontology,* 1976, *31,* 104–107.

Glenn, N. D. Psychological well-being in the post-parental stage: Some evidence from national surveys. *Journal of Marriage and the Family,* 1975, *37,* 105–110.

Glenwick, D. S., & Whitbourne, S. K. Beyond despair and disengagement: A transactional model of personality and development in late life. *International Journal of Aging and Human Development,* 1978, *6,* 261–267.

Glick, I. O., Weiss, R. S., & Parks, C. M. *The first year of bereavement.* New York: Wiley, 1974.

Glick, P. C. Updating the life cycle of the family. *Journal of Marriage and the Family,* 1977, *39,* 5–15.

Glick, P. C., & Norton, A. J. *Number, timing and duration of marriages and divorces in the United States: June 1975* (U.S. Bureau of the Census Population Reports, No. 297).

Washington, D.C.: U.S. Government Printing Office, 1976.

Glick, P. C., & Norton, A. J. Marrying, divorcing, and living together in the U.S. today. *Population Bulletin No. 5, Vol. 32*, Population Reference Bureau Inc., Washington, D.C., 1977.

Gold, D., & Andres, D. Developmental comparisons between ten-year-old children with employed and nonemployed mothers. *Child Development*, 1978, *49*, 75–84.

Gold, D., & Andres, D. Relations between maternal employment and development of nursery school children. *Canadian Journal of Behavioral Science*, 1978, *10*, 116–129. (b)

Goldberg, R. J. *Maternal time use and preschool performance.* Paper presented at the meeting of the Society for Research in Child Development, New Orleans, March 1977.

Goldberg, S. R., & Deutsch, F. *Life-span individual and family development.* Monterey, Calif.: Brooks/Cole, 1977.

Goldblatt, P. B., Moore, M. E., & Stunkard, A. J. Social factors in obesity. *Journal of the American Medical Association*, 1965, *192*, 1039–1044.

Goldfarb, W. Variations in adolescent adjustment in institutionally reared children. *Journal of Orthopsychiatry*, 1947, *17*, 449–457.

Goldfarb, W. Psychological privation in infancy and subsequent adjustment. *American Journal of Orthopsychiatry*, 1945, *15*, 244–257.

Goldstein, B. *Human sexuality.* New York: McGraw-Hill, 1976.

Golinkoff, R. M., & Ames, G. J. A comparison of fathers' and mothers' speech with their young children. *Child Development*, 1979, *50*, 28–32.

Gollin, E. S. A developmental approach to learning and cognition. In L. P. Lipsitt & C. C. Spiker (Eds.), *Advances in child development and behavior* (Vol. 2). New York: Academic Press, 1965.

Goodman, N., Dornbusch, S. M., Richardson, S. A., & Hastorf, A. H. Variant reactions to physical disabilities. *American Sociological Review*, 1963, *28*, 429–435.

Goodnow, J. J. A test of milieu differences with some of Piaget's tasks. *Psychological Monographs*, 1962, *76*, No. 36, Whole No. 555.

Goodnow, J. J., & Bethon, G. Piaget's tasks: The effects of schooling and intelligence. *Child Development*, 1966, *57*, 573–582.

Gordon, E. M., & Thomas, A. Children's behavioral style and the teacher's appraisal of their intelligence. *Journal of School Psychology*, 1967, *5*, 292–300.

Gordon, S. K., & Clark, W. C. Application of signal-detection theory to prose recall and recognition in elderly and young adults. *Journal of Gerontology*, 1974, *26*, 64–72.

Gorn, G. J., Goldberg, M. E., & Kanungo, R. N. The role of educational television in changing intergroup attitudes of children. *Child Development*, 1976, *47*, 277–280.

Gorsuch, R., & Barnes, M. Stages of ethical reasoning and moral norms of Carib youths. *Journal of Cross-Cultural Psychology*, 1973, *4*, 283–301.

Gottfried, A. W. Intellectual consequences of perinatal anoxia. *Psychological Bulletin*, 1973, *80*, 231–242.

Gottfried, A. W., & Rose, S. A. Tactile recognition memory in infants. *Child Development*, 1980, *51*, 69–74.

Gottfried, A. W., Rose, S. A., & Bridger, W. H. Effects of visual, haptic, and manipulatory experiences on infants' visual recognition memory of objects. *Developmental Psychology*, 1978, *14*, 305–312.

Gottfried, N. W., & Sealy, B. Early social behavior: Age and sex baseline data from a hidden population. *Journal of Genetic Psychology*, 1974, *125*, 61–69.

Gottman, J., Gonso, J., & Rasmussen, B. Social interaction, social competence, and friendship in children. *Child Development*, 1975, *46*, 709–718.

Gould, R. L. The phases of adult life: A study in developmental psychology. *American Journal of Psychiatry*, 1972, *129*, 521–531.

Gould, R. L. *Transformations: Growth and change in adult life.* New York: Simon and Schuster, 1978.

Goulet, L. R., & Baltes, P. B. (Eds.) *Life-span developmental psychology: Research and theory.* New York: Academic Press, 1970.

Graham, F. G., Matarazzo, R. G., & Caldwell, B. G. Behavioral differences between normal and traumatized newborns. *Psychology Monographs*, 1956, *70*, 427–438.

Graham, F. K., Ernhart, C. B., Craft, M., & Berman, P. W. Learning of relative and absolute size concepts in preschool children. *Journal of Experimental Psychology*, 1964, *1*, 26–36.

Gray, S. W., & Klaus, R. A. An experimental preschool program for culturally deprived children. *Child Development*, 1965, *36*, 887–898.

Gray, S. W., Ramsey, B. K., & Klaus, R. A. *The early training project in longitudinal perspective.* Paper presented at the Biennial Meeting of the Society for Research in Child Development, San Francisco, March 17, 1979.

Graziano, W., French, D., Brownell, C. A., & Hartup, W. W. Peer interaction in same-aged and mixed-aged triads in relation to chronological age and incentive conditions. *Child Development*, 1976, *47*, 707–714.

Graziano, W. G. Standards of fair play in same-age and mixed-age groups of children. *Developmental Psychology*, 1978, *14*, 524–530.

Green, J. A., Gustafson, G. E., & West, M. J. Effects on infant development on mother-infant interactions. *Child Development*, 1980, *51*, 199–207.

Greenberg, J. H. *Language universals.* The Hauge: Mouton, 1966.

Greenberg, M., & Morris, N. Engrossment: The newborn's impact upon the father. *American Journal of Orthopsychiatry*, 1974, *44*, 520–531.

Greulich, W. W. A comparison of the physical growth and development of American-born and native Japanese children. *American Journal of Physical Anthropology*, 1957, *15*, 489–515.

Griew, S. Uncertainty as a determinant of performance in relation to age. *Gerontologia*, 1958, *2*, 284–289.

Gross, L. The real world of television. *Today's Education*, 1974, *63*, 86–92.

Grove, W. R., & Tudor, J. F. Adult sex roles and mental illness. *American Journal of Sociology*, 1973, 78, 812–835.

Gubrium, J. F. Being single in old age. *International Journal of Aging and Human Development*, 1975, *6*, 29–41.

Guerney, B. G., Jr. *Relationship enhancement: Skill training programs for therapy, problem prevention, and enrichment.* San Francisco: Jossey-Bass, 1977.

Guilford, J. P. *The nature of human intelligence.* New York: McGraw-Hill, 1967.

Gump, P. V. Ecological psychology and children, In E. M. Hetherington (Ed.), *Review of child development research.* Chicago: University of Chicago Press, 1975.

Gump, P. V. The school as a social situation. *Annual Review of Psychology,* 1980, *31,* 553–582.

Gutman, G. M., & Herbert, C. P. Mortality rates among relocated extended-care patients. *Journal of Gerontology,* 1976, *31,* 352–357.

Gutmann, D. L. An exploration of ego configurations in middle and later life. In B. L. Neugarten & Associates (Eds.), *Personality in middle and later life.* New York: Atherton Press, 1964.

Gutmann, D. L. Alternatives to disengagement: The old men of the Highland Druze. In R. A. LeVine (Ed.), *Culture and personality: Contemporary readings.* Chicago: Aldine, 1974.

Guttmann, D. Life events and decision making by older adults. *The Gerontologist,* 1978, *18,* 462–467.

Haan, N. The adolescent antecedents of an ego model of coping and defense and comparisons with Q-sorted ideal personalities. *Genetic Psychology Monographs,* 1974, *89,* 273–306.

Haan, N. " . . . Change and sameness . . . " reconsidered. *International Journal of Aging and Human Development,* 1976, *7,* 59–65.

Haan, N. Two moralities in action contexts: Relationship to thought, ego regulation, and development. *Journal of Personality and Social Psychology,* 1978, *36,* 286–305.

Haan, N., & Day, D. A longitudinal study of change and sameness in personality development: Adolescence to later adulthood. *International Journal of Aging and Human Development,* 1974, *5,* 11–39.

Haan, N., Langer, J., & Kohlberg, L. Family patterns in moral reasoning. *Child Development,* 1976, *47,* 1204–1206.

Hadden, J. K. The private genera-

tion. *Psychology Today,* 1969, *3,* 32–35.

Hafez, E. S. E., & Evans, T. N. (Eds.) *Human reproduction.* New York: Harper & Row, 1973.

Hagestad, G. O. *Role change and socialization in adulthood: The transition to the empty nest.* Unpublished manuscript. The Pennsylvania State University, 1980.

Hahn, H. P. The regulation of protein synthesis in the aging cell. *Experimental Gerontology,* 1970, *5,* 323.

Hainline, L. Developmental changes in visual scanning of face and non-face patterns by infants. *Journal of Experimental Child Psychology,* 1978, *25,* 90–115.

Hakstian, A. R., & Cattell, R. B. An examination of adolescent sex differences in some ability and personality traits. *Canadian Journal of Behavioral Science,* 1975, *7,* 295–312.

Halbrecht, I., Sklorowski, E., & Tsafriv, J. Menarche and menstruation in various ethnic groups in Israel. *Acta Geneticae Medicae et Gemellologiae,* 1971, *20,* 384–391.

Hall, E. Acting one's age: New rules for old. An interview with Bernice Neugarten. *Psychology Today,* April 1980, 66–80.

Hall, G. S. *Adolescence.* New York: Appleton, 1904.

Hall, G. S. *Senescence: The last half of life.* New York: Appleton, 1922.

Halliday, M. A. K. *Learning how to mean: Exploration in the development of language.* London: Arnold, 1975.

Harker, J. O., Kent, C. R., Hartley, J. T., Finkle, T. J., & Walsh, D. A. Age-related differences in the patterns of reading and recalling discourse. In D. A. Walsh (Chair), *Age-related differences in the comprehension and recall of discourse.* Symposium presented at the 88th Annual Convention of the American Psychological Association, Montreal, Quebec, 1980.

Harlow, H. F., & Harlow, M. K. Social deprivation in monkeys. *Scientific American,* 1962, *207,* 137–146.

Harlow, H. F., & Harlow, M. K. Learning to love. *American Scientist,* 1966, *54,* 244–272.

Harlow, H. F., & Harlow, M. K. The young monkeys. In P. Cramer (Ed.),

Readings in developmental psychology today. Del Mar, Calif.: CRM Books, 1970.

Harlow, H. F., & Zimmerman, R. R. Affectional responses in the infant monkey. *Science,* 1959, *130,* 421–432.

Harman, D. Free radical theory of aging: Effect of free radical reaction inhibitors on the mortality rate of male LAF_1 mice. *Journal of Gerontology,* 1968, *23,* 476–482.

Harper, R. M., Hoppenbrowers, T., Bannett, D., Hodgman, J., Sterman, M. B., & McGinty, D. J. Effects of feeding on state and cardiac regulation in the infant. *Developmental Psychology,* 1977, *10,* 507–517.

Harris, D. B. Problems in formulating a scientific concept of development. In D. B. Harris (Ed.), *The concept of development.* Minneapolis: University of Minnesota Press, 1957.

Harris, L. The Life poll. *Life,* 1969, *66,* 22–23.

Harrison, G. A., Weiner, J. S., Tanner, J. M., & Barnicot, N. A. *Human biology* (2d ed.). London: Oxford University Press, 1977.

Hart, G. Role of preventive methods in the control of venereal disease. *Clinical Obstetrics and Gynecology,* 1975, *18,* 243–253.

Hartmann, D. P. Influence of symbolically modeled instrumental aggression and pain cues on aggressive behavior. *Journal of Personality and Social Psychology,* 1969, *11,* 280–288.

Hartup, W. W. Peer interaction and social organization. In P. H. Mussen (Ed.), *Carmichael's manual of child psychology* (Vol. 2). New York: Wiley, 1970.

Hartup, W. W. Peer interaction and behavioral development of the individual child. In E. Shopler & R. L. Reichler (Eds.), *Psychopathology and child development.* New York: Plenum, 1976.

Hartup, W. W. Perspectives on child and family interaction: Past, present, and future. In R. M. Lerner & G. B. Spanier (Eds.), *Child influences on marital and family interaction: A life-span perspective.* New York: Academic Press, 1978.

Hartup, W. W., & Coates, B. Imitation of a peer as a function of reinforcement from the peer group and rewardingness of the model. *Child Development,* 1967, *38,* 1003–1016.

Harvey, C. D., & Bahr, H. M. Widowhood, morale, and affiliation. *Journal of Marriage and the Family,* 1974, *36,* 97–106.

Hatcher, R. A., Stewart, G. K., Guest, F., Finkelstein, R., & Goodwin, C. *Contraceptive technology: 1976–1977.* New York: Irvington Publishers, 1976.

Havighurst, R. J. *Developmental tasks and education.* New York: Longmans, 1951.

Havighurst, R. J. *Human development and education.* London: Longmans, 1953.

Havighurst, R. J. Research on the developmental task concept. *School Review,* 1956, *64,* 214–223.

Havighurst, R. J. The leisure avtivities of the middle-aged. *American Journal of Sociology,* 1957, *63,* 152–162.

Havighurst, R. J. Social roles, work, leisure, and education. In C. Eisdorfer & M. P. Lawton (Eds.), *The psychology of adult development and aging.* Washington, D.C.: American Psychological Association, 1973.

Havighurst, R. J., Neugarten, B. L., & Tobin, S. S. Disengagement and patterns of aging. In B. L. Neugarten (Ed.), *Middle age and aging.* Chicago: University of Chicago Press, 1968.

Hay, D. F. Cooperative interactions and sharing between very young children and their parents. *Developmental Psychology,* 1979, *15,* 647–653.

Hayflick, L. The limited *in vitro* lifetime of human diploid cell strains. *Experimental Cell Research,* 1965, *37,* 614–636.

Healy, C. C. The relation of esteem and social class to self-occupational congruence. *Journal of Vocational Behavior,* 1973, *3,* 43–51.

Heber, F. R., Garber, H. L., Harrington, S., & Hoffman, C. *Rehabilitation of families at risk for mental retardation.* Unpublished progress reports, Research and Training Center, University of Wisconsin, Madison, Wis., 1971.

Heber, F. R., Garber, H. L., Harrington, S., & Hoffman, C. *Rehabilitation of families at risk for mental retardation.* Unpublished progress

reports, Research Training Center, University of Wisconsin, Madison, Wis., 1972.

Heglin, H. J. Problem solving set in different age groups. *Journal of Gerontology,* 1956, *11,* 310–317.

Heider, G. M. Vulnerability in infants and young children: A pilot study. *Genetic Psychology Monographs,* 1966, *73,* 1–216.

Hempel, C. G. *Fundamentals of concept formation in empirical science.* Chicago: University of Chicago Press, 1952.

Henry, W. E., Sims, J., & Spray, L. *The fifth profession: Becoming a psychotherapist.* San Francisco: Jossey-Bass, 1971.

Herrnstein, R. J. The evolution of behaviorism. *American Psychologist,* 1977, *32,* 593–603.

Herzberg, F. *Work and the nature of man.* Cleveland, Ohio: World, 1966.

Hess, R. D., & Shipman, V. C. *Cognitive elements in maternal behavior. Minnesota symposium on child psychology* (Vol. 1). Minneapolis: University of Minnesota Press, 1967.

Hess, R. D., & Torney, J. *The development of political attitudes in children.* New York: Aldine, 1967.

Hetherington, E. M. Divorce: A child's perspective. *American Psychologist,* 1979, *34,* 851–858.

Hetherington, E. M., Cox, M., & Cox, R. Stress and coping in divorce: A focus on women. In J. E. Gullahorn (Ed.), *Psychology and women in transition.* Washington, D.C.: V. H. Winston & Sons, 1979.

Hetherington, E. M., & Parke, R. D. *Child psychology.* New York: McGraw-Hill, 1975.

Hetherington, E. M., & Parke, R. D. *Child psychology: A contemporary viewpoint.* New York: McGraw-Hill, 1979.

Hewitt, L. S. The effects of provocation, intentions, and consequences on children's moral judgments. *Child Development,* 1975, *46,* 540–544.

Hiatt, S. W., Campos, J. J., & Emde, R. N. Facial patterning and infant emotional expression: Happiness, surprise, and fear. *Child Development,* 1979, *50,* 1020–1035.

Hickey, T., & Kalish, R. A. Young people's perceptions of adults. *Journal of Gerontology,* 1968, *23,* 215–219.

Hicks, D. L. Imitation and retention of film-mediated aggressive peer and adult models. *Journal of Personality and Social Psychology,* 1965, *2,* 97–100.

Hicks, M. W. An empirical evaluation about textbook assumptions about engagement. *Family Life Coordinator,* 1970, *19,* 57–63.

Hiernaux, J. Ethnic differences in growth and development. *Eugenics Quarterly,* 1968, *15,* 12–21.

Hill, K., & Enzle, M. Interactive effects of training domain and age on children's moral judgments. *Canadian Journal of Behavioral Science,* 1977, *9,* 371–381.

Hirsch, J. Behavior-genetic analysis and its biosocial consequences. *Seminars in Psychiatry,* 1970, *2,* 89–105.

Hite, S. *The Hite report.* New York: Macmillan, 1976.

Hobbs, D. F., & Cole, S. P. Transition to parenthood: A decade replication. *Journal of Marriage and the Family,* 1976, *38,* 723–732.

Hodgkins, J. Influence of age on the speed of reaction and movement in females. *Journal of Gerontology,* 1962, *17,* 385–389.

Hodos, W., & Campbell, C. B. G. Scala Naturae: Why there is no theory in comparative psychology. *Psychological Review,* 1969, *76,* 337–350.

Hoffman, L. W. The effects of maternal employment on the child—A review of research. *Developmental Psychology,* 1974, *10,* 204–228.

Hoffman, L. W. Maternal employment: 1979. *American Psychologist,* 1979, *34,* 859–865.

Hoffman, L. W., & Nye, F. I. *Working mothers.* San Francisco: Jossey-Bass, 1974.

Hoffman, M. L. Moral development. In P. H. Mussen (Ed.), *Carmichael's manual of child psychology* (Vol. 2). New York: Wiley, 1970.

Hoffman, M. L. Sex differences in moral internalization and values. *Journal of Personality and Social Psychology,* 1975, *32,* 720–729.

Hoffman, M. L. Moral development in adolescence. In J. Adelson (Ed.), *Handbook of adolescent psychology.* New York: Wiley, 1980.

Hoffman, R. F. Developmental changes in human infant visual-evoked potentials to patterned stimuli recorded at different scalp locations. *Child Development*, 1978, *49*, 110–118.

Hogan, D. P. The variable order of events in the life course. *American Sociological Review*, 1978, *43*, 573–586.

Hogan, R., Johnson, J. A., & Emler, N. P. A socioanalytic theory of moral development. *New Directions for Child Development*, 1978, *2*, 1–18.

Holland, J. I. *Making vocational choices: A theory of careers*. Englewood Cliffs, N.J.: Prentice-Hall, 1973.

Hollenbeck, A. R., & Slaby, R. G. Infant visual and vocal responses to television. *Child Development*, 1979, *50*, 41–45.

Hollingworth, H. L. *Mental growth and decline: A survey of developmental psychology*. New York: Appleton, 1927.

Holmes, T. H., & Masuda, M. Life change and illness susceptibility. In B. S. Dohrenwend & B. P. Dohrenwend (Eds.), *Stressful life events: Their nature and effects*. New York: Wiley, 1974.

Holmes, T. H., & Rahe, R. H. The social readjustment rating scale. *Journal of Psychosomatic Research*, 1967, *11*, 213–218.

Holstein, C. B. Irreversible, stepwise sequence in the development of moral judgment: A longitudinal study of males and females. *Child Development*, 1976, *47*, 57–61.

Holt, E. L., Jr. Energy requirements. In H. L. Barnett & A. H. Ernhorn (Eds.), *Pediatrics* (15th ed.). New York: Appleton-Century-Crofts, 1972.

Homans, G. C. *Social Behavior: Its elementary forms*. New York: Harcourt, Brace & World, 1961.

Hood, L., & Bloom, L. What, when, and how about why: A longitudinal study of early expressions of causality. With commentary by Charles J. Brainerd. *Monographs of the Society for Research in Child Development*, 1979, *44*(6, Serial No. 181).

Hooper, F. H., Fitzgerald, J., & Papalia, D. Piagetian theory and the aging process: Extensions and speculations. *Aging and Human Development*, 1971, *2*, 3–20.

Hooper, F. H., & Sheehan, N. Logical concept attainment during the aging years: Issues in the neo-Piagetian research literature. In W. F. Overton & J. M. Gallagher (Eds.), *Knowledge and development: Advances in the theory and research* (Vol. 1). New York: Plenum Press, 1977.

Horn, J. L. Organization of data on life-span development of human abilities. In L. R. Goulet & P. B. Baltes (Eds.), *Life-span developmental psychology: Research and theory*. New York: Academic Press, 1970.

Horn, J. L. Human ability systems. In P. B. Baltes (Ed.), *Life-span development and behavior* (Vol. 1). New York: Academic Press, 1978.

Horn, J. L., & Cattell, R. B. Age differences in primary mental ability factors. *Journal of Gerontology*, 1966, *21*, 210–220.

Horn, J. L., & Cattell, R. B. Age differences in fluid and crystallized intelligence. *Acta Psychologica*, 1967, *26*, 107–129.

Horn, J. L., & Donaldson, G. On the myth of intellectual decline in adulthood. *American Psychologist*, 1976, *31*, 701–719.

Horn, J. L., & Donaldson, G. Faith is not enough: A response to the Baltes-Schaie claim that intelligence does not wane. *American Psychologist*, 1977, *32*, 369–373.

Horn, J. L., & Donaldson, G. Cognitive development II: Adulthood development of human abilities. In O. G. Brim, Jr., & J. Kagan (Eds.), *Constancy and change in human development: A volume of review essays*. Cambridge: Harvard University Press, 1980.

Hornblum, J. N., & Overton, W. F. Area and volume conservation among the elderly: Assessment and training. *Developmental Psychology*, 1976, *12*, 68–74.

Horowitz, F. D., Ashton, J., Culp, R., Gaddis, E., Levin, S., & Reichman, B. The effects of obstetrical medication on the behavior of Israeli newborn infants and some comparisons with Uraguayan and American infants. *Child Development*, 1977, *48*, 1607–1623.

Hoyer, W. J., Labouvie, G. V., & Baltes, P. B. Modification of response speed deficits and intellectual performance in the elderly. *Human Development*, 1973, *16*, 233–242.

Hudgens, R. W. Personal catastrophe and depression: A consideration of the subject with respect to medically ill adolescents, and a requiem for retrospective life-event studies. In B. S. Dohrenwend & B. P. Dohrenwend (Eds.), *Stressful life events: Their nature and effects*. New York: Wiley, 1974.

Hulicka, I. M., & Grossman, J. L. Age-group comparisons for the use of mediators in paired-associate learning. *Journal of Gerontology*, 1967, *22*, 46–51.

Hulicka, I. M., Sterns, H., & Grossman, J. Age-group comparisons of paired-associate learning as a function of paced and self-paced association and response times. *Journal of Gerontology*, 1967, *22*, 274–280.

Hull, C. L. A functional interpretation of the conditioned reflex. *Psychological Review*, 1929, *36*, 498–511.

Hull, C. L. *A behavior system*. New Haven: Yale University Press, 1952.

Hultsch, D. F. Adult age differences in the organization of free recall. *Developmental Psychology*, 1969, *1*, 673–678.

Hultsch, D. F. Adult age differences in free classification and free recall. *Developmental Psychology*, 1971, *4*, 338–342.

Hultsch, D. F. Learning to learn in adulthood. *Journal of Gerontology*, 1974, *29*, 302–308.

Hultsch, D. F., & Craig, E. R. Adult age differences in the inhibition of recall as a function of retrieval cues. *Developmental Psychology*, 1976, *12*, 83–84.

Hultsch, D. F., & Hickey, T. External validity in the study of human development: Theoretical and methodological issues. *Human Development*, 1978, *21*, 76–91.

Hultsch, D. F., & Plemons, J. K. Life events and life span development. In P. B. Baltes & O. G. Brim, Jr. (Eds.), *Life-span development and behavior* (Vol. 2). New York: Academic Press, 1979.

Hunt, M. *Sexual behavior in the 1970's*. Chicago: Playboy Press, 1974.

Hurlock, E. B. *Adolescent development*. New York: McGraw-Hill, 1973.

Huston-Stein, A., & Baltes, P. B. Theory and method in life-span developmental psychology: Implications for child development. In H. W. Reese (Ed.), *Advances in child development and behavior* (Vol. 11). New York: Academic Press, 1976.

Huston-Stein, A., & Higgins-Trenk, A. Development of females from childhood through adulthood: Career and feminine orientations. In P. B. Baltes (Ed.), *Life-span development and behavior* (Vol. 1). New York: Academic Press, 1978.

Illingworth, R. S. *The development of the infant and young child: Normal and abnormal.* Edinburgh: Churchill Livingstone, 1975.

Inhelder, B., & Piaget, J. *The growth of logical thinking from childhood to adolescence.* New York: Basic Books, 1958.

Irwin, M., Engle, P. L., Yarbrough, C., Klein, R. E., & Townsend, J. The relationship of prior ability and family characteristics to school attendance and school achievement in rural Guatemala. *Child Development*, 1978, *49*, 415–427.

Izard, C. E., Huebner, R. R., Risser, D., McGinnes, G. C., & Dougherty, L. M. The young infant's ability to produce discrete emotion expressions. *Developmental Psychology*, 1980, *16*, 132–140.

Jackson, C. M. Some aspects of form and growth. In W. J. Robbins, S. Brody, A. F. Hogan, C. M. Jackson, & C. W. Green (Eds.), *Growth*. New Haven: Yale University Press, 1929.

Jackson, S. The growth of logical thinking in normal and subnormal children. *British Journal of Educational Psychology*, 1965, *35*, 255–258.

Jacobson, S. W. Matching behavior in the young infant. *Child Development*, 1979, *50*, 425–430.

Jaques, E. Death and the mid-life crisis. *International Journal of Psychoanalysis*, 1965, *46*, 502–514.

Jaffe, F. S., & Dryfoos J. G. Fertility control services for adolescents: Access and utilization. *Family Planning Perspectives*, 1976, *8*, 167–175.

Jarvik, L. F., & Cohen, D. A biobehavioral approach to intellectual changes with aging. In C. Eisdorfer & M. P. Lawton (Eds.), *The psychology of adult development and aging*. Washington, D.C.: American Psychological Association, 1973.

Jarvik, L. F., & Falek, A. Intellectual

stability and survival in the aged. *Journal of Gerontology*, 1963, *18*, 173–176.

Jenkins, J. J. Remember that old theory of memory: Well forget it. *American Psychologist*, 1974, *29*, 785–795.

Jennings, K. D., Harmon, R. J., Morgan, G. A., Gaiter, J. L., & Yarrow, L. J. Exploratory play as an index of mastery motivation: Relationships to persistence, cognitive functioning, and environmental measures. *Developmental Psychology*, 1979, *15*, 386–394.

Jennings, M. K., & Niemi, R. G. The transmission of political values from parent to child. *American Political Science Review*, 1968, *62*, 169–184.

Jensen, A. R. How much can we boost IQ and scholastic achievement? *Harvard Educational Review*, 1969, *39*, 1–123.

Joffe, J. M. *Prenatal determinants of behavior.* Oxford, Eng.: Pergamon, 1969.

Johns, E. B., Sutton, W. C., & Webster, L. E. *Health for effective living* (6th ed.). New York: McGraw-Hill, 1975.

Johnson, B., & Morse, H. A. Injured children and their parents. *Children*, 1968, *15*, 147–152.

Johnson, S., & Lobitx, G. The personal and marital adjustment of parents as related to observed child deviance and parenting behaviors. *Journal of Abnormal Child Psychology*, 1974, *2*, 193–207.

Johnston, L. D., & Bachman, J. G. Educational institutions. In J. F. Adams (Ed.), *Understanding adolescence* (3d ed.). Boston: Allyn and Bacon, 1976.

Jones, D. C., Richel, A. U., & Smith, R. L. Maternal child-rearing practices and social problem-solving strategies among preschoolers. *Developmental Psychology*, 1980, *16*, 241–242.

Jones, H. E. Problems of method in longitudinal research. *Vita Humana*, 1958, *1*, 93–99.

Jones, H. E. Adolescence in our society. In Anniversary Papers of the Community Service Society of New York. *The family in a democratic society*. New York: Columbia University Press, 1949, 70–82.

Jones, H. E. Intelligence and problem solving. In J. E. Birren (Ed.), *Handbook of aging and the individ-*

ual. Chicago: University of Chicago Press, 1959.

Jones. J. G., & Strowig, R. W. Adolescent identity and self-perception as predictors of scholastic achievement. *Journal of Educational Research*, 1968, *62*, 78–82.

Jones, K. L., & Smith, D. W. Recognition of the fetal alcohol syndrome in early infancy. *Lancet*, 1973, *2*, 999.

Jones, K. L., Smith, D. W., Ulleland, C. V., & Streissguth, A. P. Pattern of malformation in offspring of chronic alcoholic mothers. *Lancet*, 1973, *1*, 1267–1271.

Jones, M. C. The later careers of boys who were early- or late-maturing. *Child Development*, 1957, *28*, 133–138.

Jones, M. C. Psychological correlates of somatic development. *Child Development*, 1965, *36*, 899–911.

Jones, M. C., & Bayley, N. Physical maturing among boys as related to behavior. *Journal of Educational Psychology*, 1950, *41*, 129–148.

Jones, M. C., & Mussen, P. H. Self-conceptions, motivations, and interpersonal attitudes of early- and late-maturing girls. *Child Development*, 1958, *29*, 491–501.

Jones, W. H., Chernovetz, M. E. O'C., & Hansson, R. O. The enigma of androgyny: Differential implications for males and females. *Journal of Consulting and Clinical Psychology*, 1978, *46*, 298–313.

Kaats, G. R., & Davis, K. E. The dynamics of sexual behavior in college students. *Journal of Marriage and the Family*, 1970, *32*, 390–399.

Kacerguis, M. A., & Adams, G. R. Erikson stage resolution: The relationship between identity and intimacy. *Journal of Youth and Adolescence*, 1980, *9*, 117–126.

Kagan, J. Perspectives on continuity. In O. G. Brim, Jr., & J. Kagan (Eds.), *Constancy and change in human development*. Cambridge, Mass.: Harvard University Press, 1980.

Kagan, J. Inadequate evidence and illogical conclusions. *Harvard Educational Review*, 1969, *39*, 274–277.

Kagan, J. Resistance and continuity in psychological development. In A. M. Clarke & A. D. B. Clarke (Eds.), *Early experience: Myth and evidence*. New York: Free Press, 1976.

Kagan, J. Developmental categories

and the premise of connectivity. In R. M. Lerner (Ed.), *Developmental psychology: Historical and philosophical perspectives.* Hillsdale, N.J.: Lawrence Erlbaum Associates, in press.

Kagan, J., & Klein, R. E. Cross-cultural perspectives on early development. *American Psychologist*, 1973, *28*, 947–961.

Kagan, J., & Moss, H. A. *Birth to maturity: A study in psychological development.* New York: Wiley, 1962.

Kahana, B., & Kahana, E. Grandparenthood from the perspective of the developing grandchild. *Developmental Psychology*, 1970, *3*, 98–105.

Kahana, B., & Kahana, E. Theoretical and research perspectives on grandparenthood. *Aging and Human Development*, 1971, *2*, 261–268.

Kahn, R. L., Zarit, S. H., Hilbert, N. M., & Niederehe, G. Memory complaint and impairment in the aged. *Archives of General Psychiatry*, 1975, *32*, 1569–1573.

Kalish, R. A. An approach to the study of death attitudes. *American Behavioral Scientist*, 1963, *6*, 68–80.

Kalish, R. A. *Late adulthood: Perspectives on human development.* Monterey, Calif.: Brooks/Cole, 1975.

Kalish, R. A. Death and dying in a social context. In R. H. Binstock & E. Shanas (Eds.), *Handbook of aging and the social sciences.* New York: Van Nostrand Reinhold, 1976.

Kalish, R. A., & Johnson, A. Value similarities and differences in three generations of women. *Journal of Marriage and the Family*, 1972, *34*, 49–54.

Kalish, R. A., & Reynolds, D. K. *Death and ethnicity: A psychocultural study.* Farmingdale, N.Y.: Baywood, 1976.

Kandel, D. B., & Lesser, G. S. Paternal and peer influences on educational plans of adolescents. *American Sociological Review*, 1969, *34*, 213–223.

Kanin, E. J. Sex aggression by college men. *Medical Aspects of Human Sexuality*, 1970, *4*, 25–40.

Karniol, R. Children's use of intention cues in evaluating behavior. *Psychological Bulletin*, 1978, *85*, 76–85.

Kass, L. R. Death as an event: Commentary on Robert Morison. *Science*, 1971, *173*, 698

Kasteler, J. M., Gray, R. M., & Carruth, M. L. Involuntary relocation of the elderly. *The Gerontologist*, 1968, *8*, 276–279.

Kastenbaum, R. On the meaning of time in later life. *Journal of Genetic Psychology*, 1966, *109*, 9–25.

Kastenbaum, R. Is death a life crisis? On the confrontation with death in theory and practice. In N. Datan & L. H. Ginsberg (Eds.), *Life-span developmental psychology: Normative life crises.* New York: Academic Press, 1975.

Kastenbaum, R., & Aisenberg, R. *The psychology of death.* New York: Springer, 1972.

Katchadourian, H. *The biology of adolescence.* San Francisco: Freeman, 1977.

Kausler, D. H., & Lair, C. V. Associative strength and paired-associate learning in elderly subjects. *Journal of Gerontology*, 1966, *21*, 278–280.

Kay, H. The effects of position in a display upon problem solving. *Quarterly Journal of Experimental Psychology*, 1954, *6*, 155–169.

Kay, H. Infant sucking behavior and its modification. In L. P. Lipsitt & C. C. Spiker (Eds.), *Advances in child development and behavior* (Vol. 3). New York: Academic Press, 1967.

Kearsley, R. B., Zelazo, P. R., Kagan, J., & Hartmann, R. Separation protest in day care and home reared infants. In P. A. Mussen, J. J. Confer, & J. Kagan (Eds.), *Basic and contemporary issues in child developmental psychology.* New York: Harper & Row, 1975.

Keasey, C. B. The influence of opinion agreement and quality of supportive reasoning in the evaluation of moral judgments. *Journal of Personality and Social Psychology*, 1974, *30*, 477–482.

Keller, B. B., & Bell, R. Q. Child effects on adult's method of eliciting altruistic behavior. *Child Development*, 1979, *50*, 1004–1009.

Kelley, R. K. The premarital sexual revolution: Comments on research. *Family Coordinator*, 1972, *21*, 334–336.

Kemler, D. G. Patterns of hypothesis testing in children's discriminative learning: A study of the develop-

ment of problem-solving strategies. *Developmental Psychology*, 1978, *14*, 653–673.

Kendler, H. H., & Kendler, T. S. Vertical and horizontal processes in human concept learning. *Psychological Review*, 1962, *69*, 1–16.

Kerlinger, F. N. *Foundations of behavioral research* (2d ed.). New York: Holt, Rinehart, & Winston, 1973.

Killian, E. Effects of geriatric transfers on mortality rates. *Social Work*, 1970, *15*, 19–26.

Kimble, G. A. *Hilgard and Marquis' conditioning and learning.* New York: Appleton-Century-Crofts, 1961.

Kinney, D. K., & Kagan, J. Infant attention to auditory discrepancy. *Child Development*, 1976, *47*, 155–164.

Kinsey, A. C., Pomeroy, W. B., & Martin, C. *Sexual behavior in the human male.* Philadelphia: Saunders, 1948.

Kinsey, A. C., Pomeroy, W. B., & Martin, C. *Sexual behavior in the human female.* Philadelphia: Saunders, 1953.

Kintsch, W., & van Dijk, T. A. Toward a model of text comprehension and production. *Psychological Review*, 1978, *85*, 363–394.

Klaus, R. A., & Gray, S. W. The early training project for disadvantaged children: A report after five years. *Monographs of the Society for Research in Child Development*, 1968, *33*, (4, Serial No. 120).

Kleemeir, R. W. Intellectual change in the senium. *Proceedings of the Social Statistics of the American Statistical Association*, 1962, *1*, 290–295.

Klein, P. S., Forbes, G. B., & Nader, P. R. Effects of starvation in infancy (plyoric-stenosis) or subsequent learning abilities. *Journal of Pediatrics*, 1975, *87*, 8–15.

Kluckhohn, C., & Murray, H. Personality formation: The determinants. In C. Kluckhohn & H. Murray (Eds.), *Personality in nature, society, and culture.* New York: Knopf, 1948.

Kogan, N., & Wallach, M. Age changes in values and attitudes. *Journal of Gerontology*, 1961, *16*, 272–280.

Kohlberg, L. *The development of models of moral thinking and choice in the years ten to sixteen.*

Unpublished doctoral dissertation, University of Chicago, 1958.

Kohlberg, L. The development of children's orientations toward a moral order: I. Sequence in the development of moral thought. *Vita Humana*, 1963, 6, 11–33.

Kohlberg, L. A cognitive-developmental analysis of children's sex-role concepts and attitudes. In E. Maccoby (Ed.), *The development of sex differences*. Stanford, Calif.: Stanford University Press, 1966.

Kohlberg, L. Education for justice: A modern statement of the Platonic view. In T. Sizer (Ed.), *Moral education*. Cambridge, Mass.: Harvard University Press, 1970.

Kohlberg, L. From is to ought: How to commit the naturalistic fallacy and get away with it in the study of moral development. In W. Mischel (Ed.), *Cognitive development and epistemology*. New York: Academic Press, 1971.

Kohlberg, L. Continuities in childhood and adult moral development revisited. In P. B. Baltes & K. W. Schaie (Eds.), *Life-span developmental psychology: Personality and socialization*. New York: Academic Press, 1973.

Kohlberg, L. Moral stages and moralization: The cognitive-developmental approach. In T. Luckona (Ed.), *Moral development and behavior*. New York: Holt, Rinehart, & Winston, 1976.

Kohlberg, L. Revisions in the theory and practice of moral development. *New Directions for Child Development*, 1978, 2, 83–88.

Kohlberg, L., & Kramer, R. B. Continuities and discontinuities in childhood and adult moral development. *Human Development*, 1969, 12, 93–120.

Kohn, R. R. The heart and cardiovascular system. In C. E. Finch & L. Hayflick (Eds.), *Handbook of the biology of aging*. New York: Van Nostrand Reinhold, 1977.

Kolata, G. B. Behavioral teratology: Birth defects of the mind. *Science*, 1978, 207, 732–734.

Krantz, R. J., & Risley, T. R. Behavioral ecology in the classroom. In K. D. O'Leary & S. G. O'Leary (Eds.), *Classroom management*. New York: Pergamon, 1977.

Krause, H. D. Scientific evidence and the ascertainment of paternity.

Family Law Quarterly, 1971, 5, 252–281.

Kreps, J. M. (Ed.) *Women and the American economy: A look to the 1980's*. Englewood Cliffs, N.J.: Prentice-Hall, 1976.

Kübler-Ross, E. *On death and dying*. New York: Macmillan, 1969.

Kübler-Ross, E. *Questions and answers on death and dying*. New York: Macmillan, 1974.

Kuczaj, S. A. Children's judgments of grammatical and ungrammatical irregular past-tense verbs. *Child Development*, 1978, 49, 319–326.

Kuczaj, S. A. Evidence for a language learning strategy: On the relative ease of acquisition of prefixes and suffixes. *Child Development*, 1979, 50, 1–13.

Kuhn, D. Short-term longitudinal evidence for the sequentiality of Kohlberg's early stages of moral development. *Developmental Psychology*, 1976, 12, 162–166.

Kuhn, D., & Angelev, J. An experimental study of the development of formal operational thought. *Child Development*, 1976, 47, 697–706.

Kuhn, D., Ho, V., & Adams, C. Formal reasoning among pre- and late adolescents. *Child Development*, 1979, 50, 1128–1135.

Kuhn, T. S. *The structure of scientific revolutions*. Chicago: University of Chicago Press, 1962.

Kurtines, W., & Greif, E. B. The development of moral thought: Review and evaluation of Kohlberg's approach. *Psychological Bulletin*, 1974, 81, 453–469.

Kuypers, J. A., & Bengtson, V. L. Social breakdown and competence: A model of normal aging. *Human Development*, 1973, 16, 181–201.

Labouvie-Vief, G. Adult cognitive development: In search of alternative interpretations. *Merrill-Palmer Quarterly*, 1977, 23, 227–263.

Labouvie-Vief, G., & Gonda, J. N. Cognitive strategy training and intellectual performance in the elderly. *Journal of Gerontology*, 1976, 31, 327–332.

Lachman, J. L., & Lachman, R. Age and the actualization of world knowledge. In L. W. Poon, J. L. Fozard, L. S. Cermak, D. Arenberg, & L. W. Thompson (Eds.), *New directions in memory and aging: Proceedings of the George A. Talland

memorial conference*. Hillsdale, N.J.: Erlbaum, 1980.

Lachman, J. L., Lachman, R., & Thronesbery, C. Metamemory through the adult life span. *Developmental Psychology*, 1979, 15, 543–551.

Ladd, G. W., & Oden, S. The relationship between peer acceptance and children's ideas about helpfulness. *Child Development*, 1979, 50, 402–408.

Lair, C. V., & Moon, W. H. The effects of praise and reproof on the performance of middle-aged and older subjects. *Aging and Human Development*, 1972, 3, 279–284.

Lair, C. V., Moon, W. H., & Kausler, D. H. Associative interference in the paired-associate learning of middle-aged and old subjects. *Developmental Psychology*, 1969, 1, 548–552.

Lamb, M. E. The sociability of two-year-olds with their mothers and fathers. *Child Psychiatry and Human Development*, 1975, 5, 182–188.

Lamb, M. E. A re-examination of the infant social world. *Human Development*, 1977, 20, 65–85.

Lamb, M. E. (Ed.) *Social and personality development*. New York: Holt, Rinehart, and Winston, 1978.

Lamb, M. E., & Roopnarine, J. L. Peer influences on sex-role development in preschoolers. *Child Development*, 1979, 50, 1219–1222.

Landesman-Dwyer, S., Keller, S. L., & Streissguth, A. P. *Naturalistic observations of newborns: Effects of maternal alcohol intake*. Paper presented at the American Psychological Association Annual Meeting, San Francisco, 1977.

Landsbaum, J. B., & Willis, R. H. Conformity in early and late adolescence. *Developmental Psychology*, 1971, 4, 334–337.

Langlois, J. H., & Downs, A. C. Peer relations as a function of physical attractiveness: The eye of the beholder or behavior reality? *Child Development*, 1978, 50, 409–418.

Langlois, J. H., & Stephan, C. The effects of physical attractiveness and ethnicity on children's behavioral attribution and peer preferences. *Child Development*, 1977, 48, 1694–1698.

Langton, K. P., & Jennings, M. K. Political socialization and the high school civics curriculum in the

United States. *American Political Science Review*, 1968, 52, 852–867.

Larson, L. E. The influence of parents and peer during adolescence: The situation hypothesis revisited. *Journal of Marriage and the Family*, 1972, 34, 67–74.

Lasky, R. E. Serial habituation or regression to the mean? *Child Development*, 1979, 50, 568–570.

Latham, A. J. The relationship between puberal status and leadership in junior high school boys. *Journal of Genetic Psychology*, 1951, 78, 185–194.

Lathey, J. W. *Training effects and conservation of volume.* Child Study Center Bulletin. Buffalo, N.Y.: State University College, 1970.

Laurence, M. W., & Trotter, M. Effect of acoustic factors and list organization in multitrial free recall learning of college age and elderly adults. *Developmental Psychology*, 1971, 5, 202–210.

Lavin, D. E. *The prediction of academic performance: A theoretical analysis and review of research.* New York: Russell Sage Foundation, 1965.

LaVoie, J. C. Ego identity in middle adolescence. *Journal of Youth and Adolescence*, 1976, 5, 371–385.

LaVoie, J. C., & Adams, G. R. Physical and interpersonal attractiveness of the model and imitation in adults. *Journal of Social Psychology*, 1978, 106, 191–202.

Lawton, M. P. The impact of the environment on aging and behavior. In J. E. Birren & K. W. Schaie (Eds.), *Handbook of the psychology of aging.* New York: Van Nostrand Reinhold, 1977.

Lawton, M. P., & Cohen, J. The generality of housing impact on the well-being of older people. *Journal of Gerontology*, 1974, 29, 194–204.

Lawton, M. P., & Nahemow, L. Ecology and the aging process. In C. Eisdorfer & M. P. Lawton (Eds.), *The psychology of adult development and aging.* Washington, D.C.: American Psychological Association, 1973.

Lazarus, R. S. *Psychological stress and the coping process.* New York: McGraw-Hill, 1966.

Leahy, R. L. Development of preferences and processes of visual scanning in the human infant during the first three months of life. *Developmental Psychology*, 1976, 12, 250–254.

Lefkowitz, M. M., Eron, L. D., Walder, L. O., & Huesmann, L. R. Television violence and child aggression: A follow-up study. In G. A. Comstock & E. A. Rubinstein (Eds.), *Television and social behavior* (Vol. 3). *Television and adolescent aggressiveness.* Washington, D.C.: U.S. Government Printing Office, 1972.

Lehfeldt, H. Psychology of contraceptive failure. *Medical Aspects of Human Sexuality*, 1971, 5, 68–77.

Leifer, A. D., Leiderman, P. H., Barnett, C. R., & Williams, J. A. Effects of mother-infant separation on maternal attachment behavior. *Child Development*, 1972, 43, 1203–1218.

Leifer, A. D., & Roberts, D. F. Children's responses to television violence. In J. P. Murray, E. A. Rubinstein, & G. A. Comstock (Eds.), *Television and social behavior* (Vol. 2). *Television and social learning.* Washington, D.C.: U.S. Government Printing Office, 1972.

Leifer, M. Psychological changes accompanying pregnancy and motherhood. *Genetic Psychology Monographs*, 1977, 95, 55–96.

LeMasters, E. E. Parenthood as crisis. *Marriage and Family Living*, 1957, 19, 352–355.

Lenneberg, E. H. *Biological functions of language.* New York: Wiley, 1967.

Leonard, L. B., Schwartz, R. G., Folger, M. K., Newhoff, M., & Wilcox, M. J. Children's imitation of lexical items. *Child Development*, 1979, 50, 19–27.

Lerner, R. M. The development of stereotyped expectancies of body build–behavior relations. *Child Development*, 1969, 40, 137–141.

Lerner, R. M. The development of personal space schemata toward body build. *Journal of Psychology*, 1973, 84, 229–235.

Lerner, R. M. Showdown at generation gap: Attitudes of adolescents and their parents toward contemporary issues. In H. D. Thornburg (Ed.), *Contemporary adolescence* (2d ed.). Belmont, Calif.: Brooks/Cole, 1975.

Lerner, R. M. *Concepts and theories of human development.* Reading, Mass.: Addison-Wesley, 1976.

Lerner, R. M. Nature, nurture and dynamic interactionism. *Human Development*, 1978, 21, 1–20.

Lerner, R. M. A dynamic interaction concept of individual and social interaction relationship development. In R. L. Burgess & T. L. Huston (Eds.), *Social exchange in developing relationships.* New York: Academic Press, 1979.

Lerner, R. M., Benson, P., & Vincent, S. Development of societal and personal vocational role perception in males and females. *Journal of Genetic Psychology*, 1976, 129, 167–168.

Lerner, R. M., & Brackney, B. E. The importance of inner and outer body parts attitudes in the self concept of late adolescents. *Sex Roles*, 1978, 4, 225–238.

Lerner, R. M., & Busch-Rossnagel, N. A. Individuals as producers of their development: Conceptual and empirical bases. In R. M. Lerner & N. A. Busch-Rossnagel (Eds.), *Individuals as producers of their development: A life-span perspective.* New York: Academic Press, 1981.

Lerner, R. M., & Gellert, E. Body build identification, preference, and aversion in children. *Developmental Psychology*, 1969, 1, 456–462.

Lerner, R. M., & Iwawaki, S. Cross-cultural analyses of body-behavior relations: II. Factor structure of body build stereotypes of Japanese and American adolescents. *Psychologia*, 1975, 18, 83–91.

Lerner, R. M., & Iwawaki, S., & Chihara, T. Development of personal space schemata among Japanese children. *Developmental Psychology*, 1976, 12, 466–467.

Lerner, R. M., & Karabenick, S. A. Physical attractiveness, body attitudes and self-concept in late adolescents. *Journal of Youth and Adolescence*, 1974, 3, 307–316.

Lerner, R. M., Karabenick, S. A., & Meisels, M. One-year stability of children's personal space schemata towards body build. *Journal of Genetic Psychology*, 1975, 127, 151–152.

Lerner, R. M., Karabenick, S. A., & Stuart, J. L. Relations among physical attractiveness, body attitudes, and self-concept in male and female college students. *Journal of Psychology*, 1973, 85, 119–129.

Lerner, R. M., Karson, M., Meisels,

M., & Knapp, J. R. Actual and perceived attitudes of late adolescents and their parents: The phenomenon of the generation gaps. *Journal of Genetic Psychology*, 1975, *126*, 195–207.

Lerner, R. M., & Knapp, J. R. Actual and perceived intrafamilial attitudes of late adolescents and their parents. *Journal of Youth and Adolescence*, 1975, *4*, 17–36.

Lerner, R. M., & Korn, S. J. The development of body build stereotypes in males. *Child Development*, 1972, *43*, 912–920.

Lerner, R. M., & Lerner, J. V. Effects of age, sex, and physical attractiveness on child-peer relations, academic performance and elementary school adjustment. *Developmental Psychology*, 1977, *13*, 585–590.

Lerner, R. M., & Miller, R. D. Relation of students' behavioral style to estimated and measured intelligence. *Perceptual and Motor Skills*, 1971, *33*, 11–14.

Lerner, R. M., Orlos, J. B., & Knapp, J. R. Physical attractiveness, physical effectiveness, and self-concept in late adolescents. *Adolescence*, 1976, *11*, 313–326.

Lerner, R. M., & Pool, K. B. Body build stereotypes: A cross-cultural comparison. *Psychological Reports*, 1972, *31*, 527–532.

Lerner, R. M., & Ryff, C. D. Implementation of the life-span view of human development: The sample case of attachment. In P. B. Baltes (Ed.), *Life-span development and behavior* (Vol. 1). New York: Academic Press, 1978.

Lerner, R. M., & Schroeder, C. Kindergarten children's active vocabulary about body build. *Developmental Psychology*, 1971, *5*, 179.

Lerner, R. M., Schroeder, C., Rewitzer, M., & Weinstock, A. Attitudes of high school students and their parents toward contemporary issues. *Psychological Reports*, 1972, *31*, 255–258.

Lerner, R. M., & Spanier, G. B. (Eds.) *Child influences on marital and family interaction: A life-span perspective*. New York: Academic Press, 1978. (a)

Lerner, R. M., & Spanier, G. B. A dynamic interactional view of child and family development. In R. M. Lerner & G. B. Spanier (Eds.), *Child influences on marital and family interaction: A life-span perspective*. New York: Academic Press, 1978. (b)

Lerner, R. M., & Spanier, G. B. *Adolescent development: A life-span perspective*. New York: McGraw-Hill, 1980.

Lerner, R. M., Vincent, S., & Benson, P. One-year stability of societal and personal vocational role perceptions of females. *Journal of Genetic Psychology*, 1976, *129*, 173–174.

Lester, B. M. Cardiac habituation of the orienting response to an auditory signal in infants of varying nutritional status. *Developmental Psychology*, 1975, *11*, 432–442.

Leventhal, A. S., & Lipsitt, L. P. Adaptation, pitch discrimination, and sound localization in the neonate. *Child Development*, 1964, *35*, 759–767.

Levinson, D. J. Mid-life transition period in adult psychosocial development. *Psychiatry*, 1977, *40*, 99–112.

Levinson, D. J. *The season's of a man's life*. New York: Knopf, 1978.

Lewis, C. E., & Lewis, M. E. The impact of television commercials on health related beliefs and behaviors of children. *Pediatrics*, 1974, *35*, 431–435.

Lewis, C. N. The adaptive value of reminiscing in old age. *Journal of Geriatric Psychiatry*, 1973, *6*, 117–121.

Lewis, M. State as an infant-environment interaction: An analysis of mother-infant behavior as a function of sex. *Merrill-Palmer Quarterly*, 1972, *18*, 95–121.

Lewis, M., & Lee-Painter, S. An interactional approach to the mother-infant dyad. In M. Lewis & L. A. Rosenblum (Eds.), *The effect of the infant on its caregivers*. New York: Wiley, 1974.

Lewis, M., & McGurk, H. Evaluation of infant intelligence. *Science*, 1972, *178*, 1174–1177.

Lewis, M., & Rosenblum, L. A. (Eds.), *The effect of the infant on its caregiver*. New York: Wiley, 1974.

Lewis, M., Weinraub, M., & Ban, P. Mothers and fathers, girls and boys: Attachment behavior in the first two years of life (Res. Bulletin RB-72-60). Princeton, N.J.: Educational Testing Service, 1972.

Lewis, M., & Wilson, C. D. Infant development in lower class American families. *Human Development*, 1972, *15*, 112–127.

Lewontin, R. C. The fallacy of biological determinism. *The Sciences*, 1976, *16*, 6–10.

Liben, L. S., Patterson, A. H., & Newcombe, N. (Eds.) *Spatial representation and behavior across the life span*. New York: Academic Press, 1981.

Lieberman, M. A. Psychological correlates of impending death: Some preliminary observations. *Journal of Gerontology*, 1965, *20*, 181–190.

Lieberman, M. A. Adaptative processes in later life. In N. Datan & L. H. Ginsberg (Eds.), *Lifespan developmental psychology: Normative life crises*. New York: Academic Press, 1975.

Liebert, R. M., & Wicks-Nelson, R. *Developmental psychology*. Englewood Cliffs, N.J.: Prentice-Hall, 1981.

Lief, H. I., & Fox, R. C. Training for detached concern in medical students. In H. I. Lief & N. R. Lief (Eds.), *The psychological basis of medical practice*. New York: Harper & Row, 1963.

Light, R. Abuse and neglected children in America: A study of alternative policies. *Harvard Educational Review*, 1973, *43*, 556–598.

Livingston, P. B., & Zimet, C. N. Death anxiety, authoritarianism, and choice of specialty in medical students. *Journal of Nervous and Mental Disease*, 1965, *140*, 222–230.

Livson, F. B. Evolution of self: Personality development in middle-aged women. Unpublished doctoral dissertation, Wright Institute, 1974.

Livson, N., & Peskin, H. Perspectives on adolescence from longitudinal research. In J. Adelson (Ed.), *Handbook of adolescent psychology*. New York: Wiley, 1980.

Lloyd-Still, J. D., Hurwitz, I., Wolff, P. H., & Schwachmore, H. Intellectual development after severe malnutrition in infancy. *Pediatrics*, 1974, *54*, 306–311.

Long, B. H., Henderson, E. H., & Platt, L. Self-other orientations of Israeli adolescents reared in kibbutzim and moshavim. *Developmental Psychology*, 1973, *8*, 300–308.

Longstreet, B., & Orme, F. The unguarded house. In S. Sunderlin (Ed.), *Children and TV: Televi-*

sion's impact on the child. Washington, D.C.: Association for Children's Education International, 1967.

Looft, W. R. Egocentrism and social interaction in adolescence. *Adolescence,* 1971, *6,* 487–494.

Looft, W. R. The evolution of developmental psychology. *Human Development,* 1972, *15,* 187–201.

Lopata, H. Z. Self identity in marriage and widowhood. *Sociological Quarterly,* 1973, *14,* 407–418. (a)

Lopata, H. Z. Social relations of black and white widowed women in a northern metropolis. *American Journal of Sociology,* 1973, *74,* 1003–1010. (b)

Lopata, H. Z. *Widowhood in an American city.* Cambridge, Mass.: Schenkman, 1973. (c)

Lopata, H. Z. Widowhood: Societal factors in life-span disruptions and alternatives. In N. Datan & L. H. Ginsberg (Eds.), *Lifespan developmental psychology: Normative life crises.* New York: Academic Press, 1975.

Lorenz, K. *Evolution and modification of behavior.* Chicago: University of Chicago Press, 1965.

Lorenz, K. *On aggression.* New York: Harcourt, Brace & World, 1966.

Lougee, M., Grueneich, R., & Hartup, W. Social interaction in same- and mixed-age dyads of preschool children. *Child Development,* 1977, *48,* 1353–1361.

Lovell, K. A follow-up study of Inhelder and Piaget's "The growth of logical thinking." *British Journal of Psychology,* 1961, *52,* 143–153.

Lowenthal, M. F., & Chiriboga, D. Transition to the empty nest: Crisis, challenge or relief? *Psychiatry Digest,* 1975, *34,* 58.

Lowenthal, M. F. Some potentialities of a life-cycle approach to the study of retirement. In F. M. Carp (Ed.), *Retirement.* New York: Behavioral Publications, 1972.

Lowenthal, M. F. Toward a sociopsychological theory of change in adulthood and old age. In J. E. Birren & K. W. Schaie (Eds.), *Handbook of the psychology of aging.* New York: Van Nostrand Reinhold, 1977.

Lowenthal, M. F., & Chiriboga, D. Social stress and adaptation: Toward a life-course perspective. In C. Eisdorfer & M. P. Lawton (Eds.),

The psychology of adult development and aging. Washington, D.C.: American Psychological Association, 1973.

Lowenthal, M. F., Thurnher, M., & Chiriboga, D. *Four stages of life: A comparative study of women and men facing transitions.* San Francisco: Jossey-Bass, 1975.

Lubchenco, L. O. *The high risk infant.* Philadelphia: Saunders, 1976.

Luckey, E. B. Number of years married as related to personality perception and marital satisfaction. *Journal of Marriage and the Family,* 1966, *28,* 44–48.

Lyle, J. Television in daily life: Patterns of use (overview). In E. A. Rubinstein, G. A. Comstock, & J. P. Murray (Eds.), *Television and social behavior* (Vol. 4). *Television in day-to-day life: Patterns of use.* Washington, D.C.: U.S. Government Printing Office, 1972.

Lyle, J., & Hoffman, H. Children's use of television and other media. In E. A. Rubinstein, G. A. Comstock, & J. P. Murray (Eds.), *Television and social behavior* (Vol. 4). *Television in day-to-day life: Patterns of use.* Washington, D.C.: U.S. Government Printing Office, 1972.

Lynn, D. B. *The father: His role in child development.* Monterey, Calif.: Brooks/Cole, 1974.

Lytton, H. Socialization of 2-year-old boys: Ecological findings. *Journal of Child Psychology and Psychiatry and Allied Disciplines,* 1976, *17,* 287–304.

Maas, H. S., & Kuypers, J. A. *From thirty to seventy.* San Francisco: Jossey-Bass, 1974.

Maccoby, E. E. (Ed.) *The development of sex differences.* Stanford, Calif.: Stanford University Press, 1966.

Maccoby, E. E., & Feldman, S. S. Mother-attachment and stranger-reactions in the third year of life. *Monographs of the Society for Research in Child Development,* 1972, *37* (1, Serial No. 146).

Maccoby, E. E., & Jacklin, C. N. *The psychology of sex differences.* Stanford, Calif.: Stanford University Press, 1974.

Maccoby, E. E., & Masters, J. C. Attachment and dependency. In P. H. Mussen (Ed.), *Carmichael's*

manual of child psychology (Vol. 2). New York: Wiley, 1970.

MacDonald, J. M. False accusations of rape. *Medical Aspects of Human Sexuality,* 1973, *7,* 170–194.

MacDonald, N. E., & Silverman, I. W. Smiling and laughter in infants as a function of level of arousal and cognitive evaluation. *Developmental Psychology,* 1978, *14,* 235–241.

Maddox, G. L. Disengagement theory: A critical evaluation. *The Gerontologist,* 1964, *4,* 80–83.

Mandler, G. Organization and memory. In D. W. Spence & J. T. Spence (Eds.), *The psychology of learning and motivation* (Vol. 1). New York: Academic Press, 1967.

Marcia, J. E. Determination and construct validity of ego identity status. Unpublished doctoral dissertation, Ohio State University, 1964.

Marcia, J. E. Development and validations of ego-identity states. *Journal of Personality and Social Psychology,* 1966, *5,* 551–558.

Marcia, J. E. Ego identity status: Relationship to change in self-esteem, "general maladjustment," and authoritarianism. *Journal of Personality,* 1967, *1,* 118–133.

Marcia, J. E. Identity six years after: A follow-up study. *Journal of Youth and Adolescence,* 1976, *5,* 145–160.

Marcia, J. E., & Friedman, M. L. Ego identity status in college women. *Journal of Personality,* 1970, *38,* 249–263.

Maresh, M. M. A forty-five year investigation for secular changes in maturation. *American Journal of Physical Anthropology,* 1972, *36,* 103–110.

Markides, K. S., & Martin, H. W. Causal model of life satisfaction among the elderly. *Journal of Gerontology,* 1979, *34,* 86–93.

Marlowe, R. A. *Effects of environment on elderly state hospital relocates.* Paper presented at annual meeting of the Pacific Sociological Association, Scottsdale, Arizona, May 1973.

Marsh, G. R., & Thompson, L. W. Psychophysiology of aging. In J. E. Birren & K. W. Schaie (Eds.), *Handbook of the psychology of aging.* New York: Van Nostrand Reinhold, 1977.

Martin, D. S., & Wrightsman, L. The

relationship between religious behavior and concern about death. *Journal of Social Psychology,* 1965, 65, 317–323.

Martin, H. P., Beezley, P., Conway, E. F., & Kempe, C. H. The development of abused children. Part I. A review of the literature. Part II. Physical, neurologic, and intellectual outcome. *Advances in Pediatrics,* 1974, 21, 15–73.

Martin, J., & Redmore, C. A longitudinal study of ego development. *Developmental Psychology,* 1978, 14, 189–190.

Martin, R. M. Effects of familiar and complex stimuli on infant attention. *Developmental Psychology,* 1975, 11, 178–185.

Martorano, S. C. A developmental analysis of performance on Piaget's formal operations tasks. *Developmental Psychology,* 1977, 13, 666–672.

Mason, S. E. Effects of orienting tasks on the recall and recognition performance of subjects differing in age. *Developmental Psychology,* 1979, 15, 467–469.

Masters, R. D. Jean-Jacques is alive and well: Rousseau and contemporary sociobiology. *Daedalus,* 1978, 107, 93–105.

Masters, W. H., & Johnson, V. E. *Human sexual response.* Boston: Little Brown, 1966.

Matarazzo, J. D. *Wechsler's measurement and appraisal of adult intelligence* (5th ed.). Baltimore: Williams & Wilkins, 1972.

Matas, L., Arend, R. A., & Sroufe, L. A. Continuity of adaptation in the second year: The relationship between quality of attachment and later competence. *Child Development,* 1978, 49, 547–556.

Matteson, D. R. Exploration and commitment: Sex differences and methodological problems in the use of identity status categories. *Journal of Youth and Adolescence,* 1977, 6, 353–374.

Matteson, R. Adolescent self-esteem, family communication, and marital satisfaction. *Journal of Psychology,* 1974, 86, 35–47.

Maurer, D., & Salapatek, P. Developmental changes in scanning of faces by young infants. *Child Development,* 1976, 47, 523–527.

Mayr, E. Evolution. *Scientific American,* 1978, 239, 47–55.

McCall, R. B. Exploratory manipulation and play in the human life. *Monographs of the Society for Research in Child Development,* 1974, 39 (No. 155).

McCall, R. B. Challenges to a science of developmental psychology. *Child Development,* 1977, 48, 333–334.

McCall, R. B. Individual differences in the pattern of habituation at five and 10 months of age. *Developmental Psychology,* 1979, 15, 559–569.

McCall, R. B., Parke, R. D., & Kavanaugh, R. Imitation of live and televised models by children one to three years of age. *Monographs of the Society for Research in Child Development,* 1977, 42 (Serial No. 173).

McCandless, B. R. *Children.* New York: Holt, Rinehart and Winston, 1967.

McCandless, B. R. *Adolescents.* Hinsdale, Ill.: Dryden Press, 1970.

McCary, J. L. *Human sexuality* (3d ed.). New York: Van Nostrand, 1978.

McClelland, D. C. Testing for competence rather than for "intelligence." *American Psychologist,* 1973, 28, 1–14.

McCrae, R. R., Bartone, P. T., & Costa, P. T., Jr. Age, anxiety and self-reported health. *Aging and Human Development,* 1976, 7, 49–58.

McCrae, R. R., Costa, P. T., & Arenberg, D. Constancy of adult personality structure in males: Longitudinal, cross-sectional, and time-of-measurement analyses. *Journal of Gerontology,* 1980, 35(6), 877–883.

McElroy, W. D., Swanson, C. P., & Macey, R. I. *Biology and man.* Englewood Cliffs, N.J.: Prentice-Hall, 1975.

McGuire, I., & Turkewitz, G. Visually elicited finger movements in infants. *Child Development,* 1978, 49, 362–370.

McIntyre, J. J., & Teevan, J. J. Television violence and deviant behavior. In G. A. Comstock & E. A. Rubinstein (Eds.), *Television and adolescent aggressiveness* (Vol. 3). *Television and adolescent aggressiveness.* Washington, D.C.: U.S. Government Printing Office, 1972.

McKenzie, B. E., & Day, R. H. Infants' attention to stationary and moving objects at different distances. *Australian Journal of Psychology,* 1976, 28, 45–51.

McKenzie, B. E., Tootell, H. E., & Day, R. H. Development of visual size constancy during the 1st year of human infancy. *Developmental Psychology,* 1980, 16, 163–174.

McKinlay, S. M., & Jefferys, M. The menopausal syndrome. *British Journal of Preventive and Social Medicine,* 1974, 28, 108.

McKinlay, S. M., Jefferys, M., & Thompson, B. An investigation of the age at menopause. *Journal of Biosocial Science,* 1972, 4, 161–173.

McLeod, J. M., Atkin, C. K., & Chaffee, S. H. Adolescents, parents, and television use: Adolescent self-report measures from Maryland and Wisconsin sample. In G. A. Comstock & E. A. Rubinstein (Eds.), *Television and social behavior* (Vol. 3). *Television and adolescent aggressiveness.* Washington, D.C.: U.S. Government Printing Office, 1972. (a)

McLeod, J. M., Atkin, C. K., & Chaffee, S. H. Adolescents, parents, and television use: Self-report and other-report measures from the Wisconsin sample. In G. A. Comstock & E. A. Rubinstein (Eds.), *Television and social behavior* (Vol. 3). *Television and adolescent aggressiveness.* Washington, D.C.: U.S. Government Printing Office, 1972. (b)

McNeal, J. V. *Children as consumers.* Austin, Texas: Bureau of Business Research, The University of Texas at Austin, 1964.

McPherson, B., & Guppy, N. Pre-retirement life-style and the degree of planning for retirement. *Journal of Gerontology,* 1979, 34, 254–263.

Meacham, J. A. A transactional model of remembering. In N. Datan & H. W. Reese (Eds.), *Life-span developmental psychology: Dialectical perspectives of experimental research.* New York: Academic Press, 1977.

Medvedev, Z. A. Caucasus and Altay longevity: A biological or social problem? *The Gerontologist,* 1974, 14, 381–387.

Medvedev, Z. A. The nucleic acids in development and aging. In B. L. Strehler (Ed.), *Advances in gerontological research* (Vol. 1). New York: Academic Press, 1964.

Medvedev, Z. A. Aging and longevity: New approaches and new perspectives. *The Gerontologist,* 1975, *15,* 196–201.

Meichenbaum, D. Self-instructional strategy training: A cognitive prothesis for the aged. *Human Development,* 1974, *17,* 273–280.

Meisels, M., & Guardo, C. J. Development of personal space schemata. *Child Development,* 1969, *40,* 1167–1178.

Meltzer, H. Attitudes of workers before and after age 40. *Geriatrics,* 1965, *20,* 425–432.

Meltzoff, A. N., & Moore, M. K. Imitation of facial and manual gestures by human neonates. *Science,* 1977, *198,* 75–78.

Melville, L. *Marriage and family today.* New York: Random House, 1977.

Meredith, H. V. Somatic changes during prenatal life. *Child Development,* 1975, *46,* 603–610.

Meyer, B. J. F., & Rice, G. E. Information recalled from prose by young, middle, and old adult readers. *Experimental Aging Research,* 1981, *7,* 253–268.

Milewski, A. E. Visual discrimination and detection of configurational invariance in 3-month infants. *Developmental Psychology,* 1979, *15,* 357–363.

Millar, W. S., & Watson, J. S. The effect of delayed feedback on infant learning reexamined. *Child Development,* 1979, *50,* 747–751.

Miller, B. C. A multivariate developmental model of marital satisfaction. *Journal of Marriage and the Family,* 1976, *38,* 643–657.

Miller, D. R., & Swanson, G. E. *The changing American parent.* New York: Wiley, 1958.

Miller, G. A. The magical number seven, plus or minus two: Some limits on our capacity for processing information. *Psychological Review,* 1956, *63,* 81–97.

Miller, H. C., & Hassanein, K. Maternal factors in "fetally malnourished" black newborn infants. *American Journal of Obstetrics and Gynecology,* 1974, *118,* 62–67.

Miller, N., & Maruyama, G. Ordinal position and peer popularity. *Journal of Personality and Social Psychology,* 1976, *33,* 123–131.

Miller, N. E., & Dollard, J. *Social learning and imitation.* New Haven: Yale University Press, 1941.

Mischel, W. A social learning view of sex differences in behavior. In E. Maccoby (Ed.), *The development of sex differences.* Stanford, Calif.: Stanford University Press, 1966.

Mischel, W. Sex typing and socialization. In P. H. Mussen (Ed.), *Carmichael's manual of child psychology* (Vol. 2). New York: Wiley, 1970.

Mischel, W. On the future of personality measurement. *American Psychologist,* 1977, *32,* 246–254.

Mischel, W. Toward a cognitive social learning reconceptualization of personality. *Psychological Review,* 1973, *80,* 252–283.

Misiak, H., & Sexton, V. S. *History of psychology in overview.* New York: Grune & Stratton, 1966.

Moerk, E. L. Determiners and consequences of verbal behaviors of young children and their mothers. *Developmental Psychology,* 1978, *14,* 537–545.

Moltz, H., & Stettner, L. J. The influence of patterned-light deprivation on the critical period for imprinting. *Journal of Comparative and Physiological Psychology,* 1961, *54,* 279–283.

Monello, L. E., & Mayer, J. Obese adolescent girls: An unrecognized "minority group"? *American Journal of Clinical Nutrition,* 1963, *13,* 35–39.

Money, J., & Ehrhardt, A. E. *Man and woman, boy and girl.* Baltimore: John Hopkins University Press, 1972.

Monge, R. H., & Hultsch, D. F. Paired-associate learning as a function of adult age and the length of the anticipation and inspection intervals. *Journal of Gerontology,* 1971, *26,* 157–162.

Montagu, A. *Adolescent sterility.* Springfield, Ill.: Charles C. Thomas, 1946.

Montemayor, R., & Eisen, M. The development of self-conceptions from childhood to adolescence. *Developmental Psychology,* 1977, *13,* 314–319.

Moore, M. E., Stunkard, A. J., & Srole, L. Obesity, social class and mental illness. *Journal of the American Medical Association,* 1962, *181,* 962–966.

Moreau, T., & Birch, H. G. Relationship between obstetrical general anesthesia and rate of neonatal habituation to repeated stimulation. *Developmental Medicine and Child Neurology,* 1974, *16,* 612–619.

Morgan, L. A. Economic impact of widowhood in a panel of middle-aged women. Unpublished doctoral dissertation, University of Southern California, 1978.

Morrow, W. R., & Wilson, R. C. Family relations of bright high-achieving and under-achieving high school boys. *Child Development,* 1961, *32,* 501–510.

Moshman, D. Consolidation and stage formation in the emergence of formal operations. *Developmental Psychology,* 1977, *13,* 95–100.

Moss, G. E. *Illness, immunity, and social interaction.* New York: Wiley, 1973.

Muir, D., & Field, J. Newborn infants orient to sounds. *Child Development,* 1979, *50,* 431–436.

Murchison, C. (Ed.) *Handbook of child psychology.* Worcester, Mass.: Clark University Press, 1931.

Murray, F. B. Acquisition of conservation through social interaction. *Developmental Psychology,* 1972, *6,* 1–6.

Murray, J. R., Powers, E. A., & Havighurst, R. J. Personal and situational factors producing flexible careers. *The Gerontologist,* 1971, *11,* 4–12.

Murstein, B. I. Physical attractiveness and marital choice. *Journal of Personality and Social Psychology,* 1972, *22,* 8–12.

Mussen, P. H. (Ed.) *Carmichael's manual of child development.* New York: Wiley, 1970.

Mussen, P. H., Conger, J. J., & Kagan, J. *Child development and personality* (4th ed.). New York: Harper & Row, 1974.

Mussen, P. H., Conger, J. J., & Kagan, J. *Child development and personality* (5th ed). New York: Harper & Row, 1979.

Mussen, P. H., Conger, J. J., Kagan, J., & Geiwitz, J. *Psychological development: A life-span approach.* New York: Harper & Row, 1979.

Mussen, P. H., Harris, S., Rutherford, E., & Keasey, C. B. Honesty and altruism among pre-adolescents. *Developmental Psychology,* 1970, *3,* 169–194.

Mussen, P. H., & Jones, M. C. Self-conceptions, motivations, and interpersonal attitudes of late- and early-maturing boys. *Child Development,* 1957, *28,* 249–256.

Muuss, R. E. *Theories of adolescence.* New York: Random House, 1966.

Muuss, R. E. The philosophical and historical roots of theories of adolescence. In R. E. Muuss (Ed.), *Adolescent behavior and society: A book of readings* (2d ed.). New York: Random House, 1975. (a)

Muuss, R. E. (Ed.) *Adolescent behavior and society: A book of readings* (2d ed.). New York: Random House, 1975. (b)

Muuss, R. E. *Theories of adolescence* (3d ed.). New York: Random House, 1975. (c)

Namenwirth, J. Z. Failing in New Haven: An analysis of high school graduates and dropouts. *Social Forces,* 1969, *48,* 23–36.

Napier, A. Y. Marriage of families: Cross-generational complementarity. *Family Process,* 1971, *10,* 373–395.

National Center for Health Statistics. *Monthly Vital Statistics Report, Provisional Statistics.* U.S. Department of Health, Education, and Welfare, 1980.

National Society for the Prevention of Blindness. *Estimated statistics on blindness and vision problems.* New York: National Society for the Prevention of Blindness, 1966.

Neale, R. E. Between the nipple and the everlasting arms. *Archives of the Foundation of Thanatology,* 1971, *3,* 21–30.

Neimark, E. D. Intellectual development during adolescence. In F. D. Horowitz (Ed.), *Review of child development research* (Vol. 4). Chicago: University of Chicago Press, 1975.

Nelson, D. D. A study of personality adjustment among adolescent children with working and non-working mothers. *Journal of Educational Research,* 1971, *64,* 328–330.

Nelson, K. Structure and strategy in learning to talk. *Monographs of the Society for Research in Child Development,* 1973, *38,* No. 149.

Nelson, L. P., & Nelson, V. *Religion and death anxiety.* Paper presented

at Society for the Scientific Study of Religion and Religious Research Association, San Francisco, 1973.

Nesselroade, J. R., & Baltes, P. B. Adolescent personality development and historical changes: 1970–1972. *Monographs of the Society for Research in Child Development,* 1974, *39* (Whole No. 154).

Nesselroade, J. R., & Reese, H. W. (Eds.) *Life-span developmental psychology: Methodological issues.* New York: Academic Press, 1973.

Nesselroade, J. R., Schaie, K. W., & Baltes, P. B. Ontogenetic and generational components of structural and quantitative change in adult behavior. *Journal of Gerontology,* 1972, *27,* 222–228.

Neubeck, G. Getting older in my family: A personal reflection. *The Family Coordinator,* 1978, *27,* 445–447.

Neugarten, B. L. *Personality in middle and late life.* New York: Atherton Press, 1964.

Neugarten, B. L. The awareness of middle age. In B. L. Neugarten (Ed.), *Middle age and aging.* Chicago: University of Chicago Press, 1968.

Neugarten, B. L. Personality change in later life: A developmental perspective. In C. Eisdorfer & M. P. Lawton (Eds.), *The psychology of adult development and aging.* Washington, D.C.: American Psychological Association, 1973.

Neugarten, B. L. Personality and aging. In J. E. Birren & K. W. Schaie (Eds.), *Handbook of the psychology of aging.* New York: Van Nostrand Reinhold, 1977.

Neugarten, B. L., & Datan, N. Sociological perspectives on the life cycle. In P. B. Baltes & K. W. Schaie (Eds.), *Life-span developmental psychology: Personality and socialization.* New York: Academic Press, 1973.

Neugarten, B. L., & Gutmann, D. L. Age-sex roles and personality in middle age: A thematic apperception study. *Psychological Monographs,* 1958, *72* (Whole No. 470).

Neugarten, B. L., & Hagestad, G. O. Age and the life course. In R. H. Binstock & E. Shanas (Eds.), *Handbook of aging and the social sciences.* New York: Van Nostrand Reinhold, 1976.

Neugarten, B. L., Havighurst, R. J., & Tobin, S. S. Personality and patterns of aging. In B. L. Neugarten (Ed.), *Middle age and aging.* Chicago: University of Chicago Press, 1968.

Neugarten, B. L., & Kraines, R. Menopausal symptoms of women in various ages. *Psychosomatic Medicine,* 1965, *27,* 266–273.

Neugarten, B. L., & Moore, J. W. The changing age-status system. In B. L. Neugarten (Ed.), *Middle age and aging.* Chicago: University of Chicago Press, 1968.

Neugarten, B. L., Moore, J. W., & Lowe, J. C. Age norms, age constraints, and adult socialization. In B. L. Neugarten (Ed.), *Middle age and aging.* Chicago: University of Chicago Press, 1968.

Neugarten, B. L., & Peterson, W. A. A study of the American age-grade system. *Proceedings of the Fourth Congress of the International Association of Gerontology* (Vol. 3), 1957, pp. 497–502.

Neugarten, B. L., & Weinstein, K. K. The changing American grandparent. *Journal of Marriage and the Family,* 1964, *26,* 199–204.

Neugarten, B. L., Wood, V., Kraines, R., & Loomis, B. Women's attitudes toward the menopause. *Human Development,* 1963, *6,* 140–151.

Neumann, H. H., & Baecker, J. M. Treatment of gonorrhea: Penicillin or tetracyclines? *Journal of the American Medical Association,* 1972, *219,* 471–474.

New York State Education Department. Results of a four-year, longitudinal study of experimental prekindergarten programs operated in New York State, spring, 1980.

Nichols, I. A., & Schauffer, C. B. *Self-concept as a predictor of performance in college women.* Paper presented at the 83d Annual Convention of the American Psychological Association, Chicago, 1975.

Niemi, R. G. Political socialization. In J. N. Knutson (Ed.), *Handbook of political psychology.* San Francisco: Jossey-Bass, 1973.

Nisbet, J. D., Illsley, R., Sutherland, A. E., & Douse, M. J. Puberty and test performance: A further report. *British Journal of Educational Psychology,* 1964, *34,* 202–203.

Noller, P. Sex differences in the socialization of affectionate expression. *Developmental Psychology*, 1978, *14*, 317–319.

Novak, M. A. Social recovery of monkeys isolated for the first year of life: II. Long-term assessment. *Developmental Psychology*, 1979, *15*, 50–61.

Nowlis, G. H., & Kessen, W. Human newborns differentiate differing concentrations of sucrose and glucose. *Science*, 1976, *191*, 865–866.

Nuttall, R. L., & Costa, P. T., Jr. *Drinking patterns as affected by age and by personality type.* Paper presented at the scientific meeting of the Gerontological Society, Louisville, October 1975.

Nydegger, C. N. *Late and early fathers.* Paper presented at the meeting of the Gerontological Society, Miami Beach, November 1973.

Nye, F. I. Effects on mother. In L. W. Hoffman & F. I. Nye (Eds.), *Working mothers.* San Francisco: Jossey-Bass, 1974.

Obrist, W. D. The electroencephalogram of normal aged adults. *Electroencephalography and Clinical Neurophysiology*, 1954, *6*, 235–244.

Obrist, W. D. The electroencephalogram of healthy aged males. In J. E. Birren, R. N. Butler, S. W. Greenhouse, L. Sokoloff, & M. R. Yarrow (Eds.), *Human aging: A biological and behavioral study.* Washington, D.C.: U.S. Government Printing Office, 1963.

Obrist, W. D., & Bissell, L. F. The electroencephalogram of aged patients with cardiac and cerebral vascular disease. *Journal of Gerontology*, 1955, *10*, 315–330.

Obrist, W. D., Busse, E. W., Eisdorfer, C., & Kleemeier, R. W. Relation of the electroencephalogram to intellectual function in senescence. *Journal of Gerontology*, 1962, *17*, 197–206.

Obrist, W. D., Henry, C. E., & Justiss, W. A. Longitudinal study of EEG in old age. *Excerpta Medica International Congress Series*, 1961 (No. 37), 180–181.

O'Donnell, W. J. Adolescent self-esteem related to feelings toward parents and friends. *Journal of Youth and Adolescence*, 1976, *5*, 179–185.

Offer, D., & Howard, K. I. An empirical analysis of the Offer self-image questionnaire for adolescents. *Archives of General Psychiatry*, 1972, *27*, 529–533.

Offer, D., Ostrov, E., & Howard, K. I. The self-image of adolescents: A study of four cultures. *Journal of Youth and Adolescence*, 1977, *6*, 265–280.

O'Leary, V. E. Some attitudinal barriers to occupational aspirations in women. *Psychological Bulletin*, 1974, *81*, 809–826.

O'Leary, V. E. *Toward understanding women.* Belmont, Calif.: Brooks/Cole, 1977.

Orgel, L. E. The maintenance of the accuracy of protein synthesis and its relevance to aging. *Biochemistry*, 1963, *49*, 517–521.

Orians, G. H. *The study of life: An introduction to biology.* Boston: Allyn & Bacon, 1969.

Orlofsky, J. L., Marcia, J. E., & Lesser, I. M. Ego identity status and the intimacy versus isolation crisis of young adulthood. *Journal of Personality and Social Psychology*, 1973, *27*, 211–219.

Overton, W. F. On the assumptive base of the nature-nurture controversy: Additive versus interactive conceptions. *Human Development*, 1973, *16*, 74–89.

Overton, W. F., & Reese, H. W. Models of development: Methodological implications. In J. R. Nesselroade & H. W. Reese (Eds.), *Life-span developmental psychology: Methodological issues.* New York: Academic Press, 1973.

Oviatt, S. L. The emerging ability to comprehend language: An experimental approach. *Child Development*, 1980, *51*, 97–106.

Oyama, S. Concept of the sensitive period in developmental studies. *Merrill-Palmer Quarterly*, 1979, *25*, 83–103.

Pakizegi, B. The interaction of mothers and fathers with their sons. *Child Development*, 1978, *49*, 479–482.

Palmer, F. H. The effects of early childhood education intervention on school performance. Unpublished manuscript, July, 1977.

Palmore, E. *The honorable elders: A cross-cultural analysis of aging in Japan.* Durham, N.C.: Duke University Press, 1975.

Palmore, E. Total chance of institutionalization among the aged. *Gerontologist*, 1976, *16*, 504–507.

Palmore, E., Cleveland, W. P., Nowlin, J. G., Ramm, D., & Siegler, I. C. Stress and adaptation in late life. *Journal of Gerontology*, 1979, *34*, 841–851.

Palmore, E., & Kivett, V. Change in life satisfaction: Longitudinal study of persons aged 46–70. *Journal of Gerontology*, 1977, *32*, 311–316.

Palmore, E., & Whittington, F. Differential trends toward equality between whites and non-whites. *Social Forces*, 1970, *49*, 108–117.

Papalia, D. E. The status of several conservation abilities across the life-span. *Human Development*, 1972, *15*, 229–243.

Papalia, D. E., & Bielby, D. D. V. Cognitive functioning in middle and old age adults: A review of research based on Piaget's theory. *Human Development*, 1974, *17*, 424–443.

Papalia, D. E., Kennedy, E., & Sheehan, N. Conservation of space in noninstitutionalized old people. *Journal of Psychology*, 1973, *84*, 75–79.

Papalia, D. E., Salverson, S. M., & True, M. An evaluation of quantity conservation performance during old age. *Aging and Human Development*, 1973, *4*, 103–110.

Papousek, H. Experimental studies of appetitional behavior in human newborns and infants. In H. W. Stevenson, E. H. Hess, & H. L. Rheingold (Eds.), *Early behavior.* New York: Wiley, 1967.

Parke, R. D., Berkowitz, L., Leyens, J. P., West, S., & Sebastian, R. J. Film violence and aggression: A field experimental analysis. In L. Berkowitz (Ed.), *Advances in experimental social psychology* (Vol. 10). New York: Academic Press, 1977.

Parke, R. D., & Collmer, C. Child abuse: An interdisciplinary review. In E. M. Hetherington (Ed.), *Review of child development research* (Vol. 5). Chicago: University of Chicago Press, 1975.

Parke, R. D., & O'Leary, S. Father-mother-infant interaction in the newborn period: Some findings, some observations, some unresolved issues. In K. Reigel & J. Meacham (Eds.), *The developing in-*

dividual in a changing world: Social and environmental issues (Vol. 2). The Hague: Mouton, 1975.

Parke, F., & Sawin, D. *Infant characteristics and behavior as elicitors of maternal and paternal responsibility in the newborn period.* Paper presented at the biennial meeting of the Society for Research in Child Development, Denver, April 1975.

Parkes, C. M. *Bereavement: Studies of grief in adult life.* New York: International Universities Press, 1972.

Parkes, C. M., & Brown, R. Health after bereavement: A controlled study of young Boston widows and widowers. *Psychosomatic Medicine*, 1972, *34*, 449–461.

Parten, M., & Newhall, S. W. Social behavior of preschool children. In R. G. Barker, J. S. Kounin, & H. F. Wright (Eds.), *Child behavior and development.* New York: McGraw-Hill, 1943.

Pastalan, L. A. *Report on Pennsylvania nursing home relocation programs: Interim research findings.* Ann Arbor: Institute of Gerontology, University of Michigan, 1976.

Pastalan, L. A., Mautz, R. K., & Merrill, J. The simulation of age-related losses: A new approach to the study of environment barriers. In W. F. E. Preiser (Ed.), *Environmental design research* (Vol. 1). Stroudsberg, Pa.: Powden, Hutchinson, and Ross, 1973.

Patterson, G. R. Mothers: The unacknowledged victims. In T. H. Stevens & R. V. Matthews (Eds.), *Mother-child, father-child relations.* Washington, D.C.: National Association for the Education of Young Children, 1978.

Patterson, G. R. The aggressive child: Victim and architect of a coercive system. In E. Mash, L. Hamereynck, & L. Hangy (Eds.), *Behavior modification and families. I. Theory and research.* New York: Brunner/Mazel, 1977.

Patterson, G. R., Littman, R. A., & Bricker, W. Assertive behavior in children: A step toward a theory of aggression. *Monographs of the Society for Research in Child Development*, 1967, *32*(5, Serial No. 113).

Pattison, E. M. The dying experience—retrospective analysis. In E. M. Pattison (Ed.), *The experience of dying.* Englewood Cliffs, N.J.: Prentice-Hall, 1977.

Paul, E. W., Pilpel, H. F., & Wechsler, N. F. Pregnancy, teenagers, and the law, 1976. *Family Planning Perspectives*, 1976, *8*, 16–21.

Paulsen, E. P. Obesity in children and adolescents. In H. L. Barnett & A. H. Einhorn (Eds.), *Pediatrics.* New York: Appleton-Century-Crofts, 1972.

Peluffo, N. The notions of conservation and causality in children of different physical and sociocultural environments. *Archives de Psychologie*, 1962, *38*, 275–291.

Peluffo, N. Culture and cognitive problems. *International Journal of Psychology*, 1967, *2*, 187–198.

Pepper, S. C. *World hypotheses.* Berkeley: University of California Press, 1942.

Perlmutter, M. What is memory aging the aging of? *Developmental Psychology*, 1978, *14*, 330–345.

Perlmutter, M. An apparent paradox about memory aging. In L. W. Poon, J. L. Fozard, L. S. Cermak, D. Arenberg & L. W. Thompson (Eds.), *New directions in memory and aging: Proceedings of the George A. Talland memorial conference.* Hillsdale, N.J.: Erlbaum, 1980.

Perun, P. J., & Bielby, D. D. V. Mid life: A discussion of competing models. *Research on aging*, 1979, *1*, 275–300.

Peterson, L., Hartmann, D. P., & Gelfand, D. M. Developmental changes in the effects of dependency and reciprocity cues on children's moral judgments and donation rates. *Child Development*, 1977, *48*, 1331–1339.

Piaget, J. La pensée symbolique et la pensée l'enfant. *Archives of Psychology, Genève*, 1923, *18*, 273–304.

Piaget, J. *The psychology of intelligence.* New York: Harcourt Brace, 1950.

Piaget, J. *Judgment and reasoning in the child.* London: Routledge and Kegan Paul, 1951.

Piaget, J. *The origins of intelligence in children.* New York: International Universities Press, 1952.

Piaget, J. *The child's conceptions of numbers.* New York: Norton, 1965.

Piaget, J. *Six psychological studies.* New York: Random House, 1968.

Piaget, J. The intellectual development of the adolescent. In G. Caplan & S. Lebovici (Eds.), *Adoles-*cence: *Psychosocial perspective.* New York: Basic Books, 1969.

Piaget, J. Piaget's theory. In P. H. Mussen (Ed.), *Carmichael's manual of child psychology* (Vol. 1). New York: Wiley, 1970.

Piaget, J. Intellectual evolution from adolescence to adulthood. *Human Development*, 1972, *15*, 1–12.

Piaget, J., & Inhelder, B. *The psychology of the child.* New York: Basic Books, 1969.

Piersel, W. C., Brody, G. H., & Kratochwill, T. R. A further examination of motivational influences on disadvantaged minority group children's intelligence test performance. *Child Development*, 1977, *48*, 1142–1145.

Pineo, P. C. Disenchantment in the later years of marriage. *Marriage and Family Living*, 1961, *23*, 3–11.

Pleck, J. H. *Men's roles in the family: A new look.* Paper presented at Sex Roles in Sociology Conference, Merrill-Palmer Institute, Detroit, 1975.

Plemons, J. K., Willis, S. L., & Baltes, P. B. Modifiability of intelligence in aging: A short-term longitudinal training approach. *Journal of Gerontology*, 1978, *33*, 224–231.

Podd, M. H. Ego identity status and morality: The relationship between two developmental constructs. *Developmental Psychology*, 1972, *6*, 497–507.

Portney, F. C., & Simmons, C. H. Day care and attachment. *Child Development*, 1978, *49*, 239–242.

Powell, A. H., Eisdorfer, C., & Bogdonoff, M. D. Physiological response patterns observed in a learning task. *Archives of General Psychiatry*, 1964, *10*, 192–195.

Powell, L. F. The effect of extra stimulation and maternal involvement on the development of low-birthweight infants and on maternal behavior. *Child Development*, 1974, *45*, 106–113.

Prader, A. Growth and development. In A. Lobhart (Ed.), *Clinical endocrinology.* Springer-Verlag, 1974.

Pressey, S. L., Janney, J. E., & Kuhlen, R. G. *Life: A psychological survey.* New York: Harper, 1939.

Provence, S., & Lipton, R. C. *Infants in institutions.* New York: International Universities Press, 1962.

Provine, R. R., & Westerman, J. A. Crossing the midline: Limits of early eye-hand behavior. *Child Development*, 1979, *50*, 437–441.

Quinn, R., Staines, G., & McCullough, M. *Job satisfaction: Is there a trend?* (U.S. Department of Labor, Manpower Research Monograph No. 30). Washington, D.C.: U.S. Government Printing Office, 1974.

Rabbitt, P., & Birren, J. E. Age and responses to sequences of repetitive and interruptive signals. *Journal of Gerontology*, 1967, *22*, 143–150.

Rader, N., Bausano, M., & Richards, J. E. On the nature of the visual-cliff-avoidance response in human infants. *Child Development*, 1980, *51*, 61–68.

Ramey, C. T., Farran, D. C., & Campbell, F. A. Predicting IQ from mother-infant interactions. *Child Development*, 1979, *50*, 804–814.

Ramey, C. T., & Piper, V. Creativity in open and traditional classrooms. *Child Development*, 1974, *45*, 557–560.

Ramos, E. Imagen personal del adolescente nisei. *Revista Latinoamericana de Psicologia*, 1974, *6*, 229–234.

Rapoport, R., & Rapoport, R. N. *Dual career families*. Baltimore: Penguin, 1971.

Rauner, I. M. Occupational information and occupational choice. *Personnel and Guidance Journal*, 1962, *41*, 311–317.

Raymond, B. Free recall among the aged. *Psychological Reports*, 1971, *29*, 1179–1182.

Reese, H. W., & Lipsitt, L. P. (Eds.) *Experimental child psychology.* New York: Academic Press, 1970.

Reese, H. W., & Overton, W. F. Models of development and theories of development. In L. R. Goulet & P. B. Baltes (Eds.), *Lifespan developmental psychology: Research and theory.* New York: Academic Press, 1970.

Rehberg, R. A., & Westby, D. L. Parental encouragement, occupation, education, and family size: Artifactual or independent determinants of adolescent educational expectations? *Social Forces*, 1967, *45*, 362–374.

Reichle, J. E., Longhurst, T. M., & Stepanich, L. Verbal interaction in mother-child dyads. *Developmental Psychology*, 1976, *12*, 273–277.

Reinert, G. Comparative factor analytic studies of intelligence throughout the human life span. In L. R. Goulet & P. B. Baltes (Eds.), *Lifespan developmental psychology: Research and theory.* New York: Academic Press, 1970.

Reiss, I. L. *Premarital sexual standards in America.* New York: Free Press of Glencoe, 1960.

Rest, J. R., Cooper, D., Coder, R., Masanz, J., & Anderson, D. Judging the important issues in moral dilemmas—An objective measure of development. *Developmental Psychology*, 1974, *10*, 491–501.

Reuter, J., & Yunik, G. Social interaction in nursery schools. *Developmental Psychology*, 1973, *9*, 319–325.

Rheingold, H. L., Gewirtz, J. L., & Ross, H. W. Social conditioning of vocalizations in the infant. *Journal of Comparative and Physiological Psychology*, 1959, *52*, 68–73.

Rich, J. Effects of children's physical attractiveness on teacher's evaluations. *Journal of Educational Psychology*, 1975, *67*, 599–609.

Richards, O. W. Vision at levels of night road illumination: XII: Changes of acuity and contrast sensitivity with age. *American Journal of Optometry*, 1966, *43*, 313–319.

Richman, J. The foolishness and wisdom of age: Attitudes toward the elderly as reflected in jokes. *The Gerontologist*, 1977, *17*, 210–219.

Riegel, K. F. Dialectical operations: The final period of cognitive development. *Human Development*, 1973, *16*, 346–370.

Riegel, K. F. Toward a dialectical theory of development. *Human Development*, 1975, *18*, 50–64.

Riegel, K. F. The dialectics of human development. *American Psychologist*, 1976, *31*, 689–700. (a)

Riegel, K. F. From traits and equilibrium toward developmental dialectics. In W. J. Arnold & J. K. Cole (Eds.), *Nebraska symposium on motivation.* Lincoln: University of Nebraska Press, 1976. (b)

Riegel, K. F., & Riegel, R. M. Development, drop, and death. *Developmental Psychology*, 1972, *6*, 306–319.

Rieser, J. J. Spatial orientation of six-month-old infants. *Child Development*, 1979, *50*, 1078–1087.

Rigsby, L. C., & McDill, E. L. Adolescent peer influence processes: Conceptualization and measurement. *Social Science Research*, 1972, *37*, 189–207.

Riley, J. W., Jr. What people think about death. In O. G. Brim, Jr., H. E. Freeman, S. Levine, & N. A. Scotch (Eds.), *The dying patient.* New York: Russell Sage Foundation, 1970.

Riley, M. W. Age strata in social systems. In R. H. Binstock & E. Shanas (Eds.), *Handbook of aging and the social sciences.* New York: Van Nostrand Reinhold, 1976.

Riley, M. W. (Ed.) *Aging from birth to death: Interdisciplinary perspectives.* Boulder, Colo.: Westview Press, 1979.

Riley, M. W., Johnson, M. E., & Foner, A. *Aging and society: A sociology of age stratification* (Vol. 3). New York: Russell Sage Foundation, 1972.

Ringler, N., Trause, M. A., Klaus, M., & Kennell, J. The effects of extra postpartum contact and maternal speech patterns on children's IQs, speech, and language comprehension at five. *Child Development*, 1978, *49*, 862–865.

Roberge, J. J. Developmental analyses of two formal operational structures: Combinatorial thinking and conditional reasoning. *Developmental Psychology*, 1976, *12*, 563–564.

Roberge, J. J., & Flexer, B. K. Further examination of formal operational reasoning abilities. *Child Development*, 1979, *50*, 478–484.

Roberston, J. F. Significance of grandparents: Perceptions of young adult grandchildren. *The Gerontologist*, 1976, *16*, 137–140.

Robertson, J. F. Grandmotherhood: A study of role conceptions. *Journal of Marriage and the Family*, 1977, *39*, 165–174.

Robson, K. S., & Moss, H. A. Patterns and determinants of maternal attachment. *Journal of Pediatrics*, 1970, *77*, 976–985.

Roff, M. F., Sells, S. B., & Golden, M. M. *Social adjustment and personality development in children.* Minneapolis: University of Minnesota Press, 1972.

Rollins, B. C., & Cannon, F. Marital satisfaction over the family life cycle: A re-evaluation. *Journal of*

Marriage and the Family, 1974, *36,* 271–282.

Rollins, B. C., & Feldman, H. Marital satisfaction over the life cycle. *Journal of Marriage and the Family,* 1970, *32,* 20–28.

Rollins, B. C., & Galligan, R. The developing child and marital satisfaction of parents. In R. M. Lerner & G. B. Spanier (Eds.), *Child influences on marital and family interaction: A life-span perspective.* New York: Academic Press, 1978.

Rose, A. A current theoretical issue in social gerontology. *Gerontologist,* 1964, *4,* 46–50.

Rose, S. A., Gottfried, A. W., & Bridger, W. H. Cross-modal transfer in infants: Relationship to prematurity and socioeconomic background. *Developmental Psychology,* 1978, *14,* 643–652.

Rose, S. A., Schmidt, K., & Bridger, W. H. Changes in tactile responsivity during sleep in the human newborn infant. *Developmental Psychology,* 1978, *14,* 163–177.

Rosen, C. E. The effects of socio-dramatic play on problem solving behavior among culturally disadvantaged preschool children. *Child Development,* 1974, *45,* 920–927.

Rosen, J. L., & Neugarten, B. L. Ego functions in the middle and later years: A thematic apperception study. In B. L. Neugarten (Ed.), *Personality in middle and late life.* New York: Atherton Press, 1964.

Rosenberg, F. R., & Simmons, R. G. Sex differences in the self-concept in adolescence. *Sex Roles,* 1975, *1,* 147–159.

Rosenman, R. H. The role of behavior patterns and neurogenic factors in the pathogenesis of coronary heart disease. In R. S. Eliot (Ed.), *Stress and the heart.* New York: Futura, 1974.

Rosenman, R. H., & Friedman, M. Observations on the pathogeneis of coronary heart disease. *Nutrition News,* 1971, *34,* 9–14.

Rosenman, R. H., Friedman, M., Straus, R., Jenkins, C. D., Zyzanski, S., Jr., Wurm, M., & Kositchek, R. Coronary heart disease in the western collaborative group study: A follow-up experience of 4½ years. *Journal of Chronic Diseases,* 1970, *23,* 173–190.

Rosenman, R. H., Friedman, M., Straus, R., Wurm, M., Jenkins, C. D., Messinger, H. B., Kositchek, R., Hahn, W., & Werthessen, N. T. Coronary heart disease in the western collaborative group study: A follow-up experience of two years. *Journal of the American Medical Association,* 1966, *195,* 86–92.

Rosenman, R. H., Friedman, M., Straus, R., Wurm, M., Kositchek, R., Hahn, W., & Werthessen, N. T. A predictive study of coronary heart disease: The western collaborative group study. *Journal of the American Medical Association,* 1964, *189,* 15–22.

Rosenthal, R. *Experimenter effects in behavioral research.* New York: Appleton-Century-Crofts, 1966.

Rosenthal, T. L., & Zimmerman, B. J. *Social learning and cognition.* New York: Academic Press, 1978.

Rosow, I. The social context of the aging self. *The Gerontologist,* 1973, *13,* 82–87.

Rosow, I. *Socialization to old age.* Berkeley, Calif.: University of California Press, 1974.

Rosow, I. Status and role change through the life span. In R. H. Binstock & E. Shanas (Eds.), *Handbook of aging and the social sciences.* New York: Van Nostrand Reinhold, 1976.

Ross, M. B., & Salvia, J. Attractiveness as biasing factor in teacher judgment. *American Journal of Mental Deficiency,* 1975, *80,* 96–98.

Rossi, A. S. Transition to parenthood. *Journal of Marriage and the Family,* 1968, *30,* 26–39.

Rowe, E. J., & Schnore, M. M. Item concreteness and reported strategies in paired-associate learning as a function of age. *Journal of Gerontology,* 1971, *26,* 470–475.

Royce, J. R. The present situation in theoretical psychology. In J. R. Royce (Ed.), *Toward unification in psychology: The first Banff conference on theoretical psychology.* Toronto: University of Toronto Press, 1970.

Rubenstein, J. A. A concordance of visual and manipulative responsiveness to novel and familiar stimuli in six-month-old infants. *Child Development,* 1974, *45,* 194–195.

Rubenstein, J. L., & Howes, C. Caregiving and infant behavior in day care and in homes. *Developmental Psychology,* 1979, *15,* 1–24.

Rubin, L. B. *Women of a certain age: The midlife search for self.* New York: Harper & Row, 1979.

Rubin, K. H. Relationship between egocentric communication and popularity among peers. *Developmental Psychology,* 1972, *7,* 364.

Rubin, K. H., Attewell, P. W., Tierney, M. C., & Tumolo, P. Development of spatial egocentrism and conservation across the life span. *Developmental Psychology,* 1973, *9,* 432.

Rubin, K. H., & Maioni, T. L. Play preference and its relationship to egocentrism, popularity, and classification skills in preschoolers. *Merrill-Palmer Quarterly,* 1975, *21,* 171–179.

Rubin, K. H., Maioni, T. L., & Hornung, M. Free play behavior in middle-class and lower-class preschoolers. Parten and Piaget revisited. *Child Development,* 1976, *47,* 414–419.

Rubin, K. H., Watson, K. S., & Jambor, T. W. Free-play behaviors in preschool and kindergarten children. *Child Development,* 1978, *49,* 534–536.

Rubin, R. A., Rosenblatt, C., & Balow, B. Psychological and educational sequelae of prematurity. *Pediatrics,* 1973, *52,* 352–363.

Rudolph, A. H. Control of gonorrhea: Guidelines for antibiotic treatment. *Journal of the American Medical Association,* 1972, *220,* 1587–1589.

Ruff, H. A. Infant recognition of the invariant form of objects. *Child Development,* 1978, *49,* 293–306.

Ruff, H. A., & Halton, A. Is there directed reaching in the human neonate? *Developmental Psychology,* 1978, *14,* 425–426.

Rushton, J. P. Effects of prosocial television and film material on the behavior of viewers. In L. Berkowitz (Ed.), *Advances in experimental social psychology* (Vol. 12). New York: Academic Press, 1979.

Russell, C. S. Transition to parenthood: Problems and gratifications. *Journal of Marriage and the Family,* 1974, *36,* 294–302.

Russo, N. F. The motherhood mandate. *Journal of Social Issues,* 1976, *32,* 143–153.

Rust, L. D. The performance of arteriosclerotic and non-arteriosclerotic subjects on a mediation-interference task. Unpublished doctoral dissertation, State University of New York at Buffalo, 1965.

Rutter, M. Maternal deprivation, 1972–1978: New findings, new concepts, new approaches. *Child Development*, 1979, *50*, 283–305.

Rutter, M. Parent-child separation: Psychological effects on the children. *Journal of Child Psychology and Psychiatry*, 1971, *12*, 233–260.

Rux, J. M. Widows' and widowers' instrumental skills, socioeconomic status, and life satisfaction. Unpublished doctoral dissertation, The Pennsylvania State University, 1976.

Ryder, N. B. The cohort as a concept in the study of social change. *American Sociological Review*, 1965, *30*, 843–861.

Sachs, J., & Devin, J. Young children's use of age appropriate speech styles in social interaction and role playing. *Journal of Child Language*, 1976, *3*, 81–98.

Sahlins, M. D. The use and abuse of biology. In A. L. Caplan (Ed.), *The sociobiology debate*. New York: Harper & Row, 1978.

Salili, F., Maehr, M. L., & Gillmore, G. Achievement and morality: A cross-cultural analysis of causal attribution and evaluation. *Journal of Personality and Social Psychology*, 1976, *33*, 327–337.

Sameroff, A. The components of sucking in the human newborn. *Journal of Experimental Child Psychology*, 1968, *6*, 607–623.

Sameroff, A. Can conditioned responses be established in the newborn infant? *Developmental Psychology*, 1971, 5, 1–12.

Sameroff, A. Learning and adaptation in infancy: A comparison of models. In H. W. Reese (Ed.), *Advances in child development and behavior* (Vol. 7). New York: Academic Press, 1972.

Sameroff, A. Transactional models in early social relations. *Human Development*, 1975, *18*, 65–79.

Sameroff, A., & Cavanaugh, P. J. Learning in infancy: A developmental perspective. In J. Osofsky (Ed.), *Handbook of infant development*. New York: Wiley, 1978.

Sandberg, E. C., & Jacobs, R. I. Psychology of the misuse and rejection of contraception. *Medical Aspects of Human Sexuality*, 1972, *6*, 34–70.

Sanders, J. C., Sterns, H. L., Smith, M., & Sanders, R. E. Modification of concept identification performance in older adults. *Developmental Psychology*, 1975, *11*, 824–829.

Sanders, R. E., Sanders, J. C., Mayes, G. J., & Sielski, K. A. Enhancement of conjunctive concept attainment in older adults. *Developmental Psychology*, 1976, *12*, 485–486.

Sandgrund, A., Gaines, R. W., & Green, A. H. Child abuse and mental retardation: A problem of cause and effect. *American Journal of Mental Deficiency*, 1974, *79*, 327–330.

Sanford, N. Developmental status of the entering freshman. In N. Stanford (Ed.), *The American college*. New York: Wiley, 1962.

Santrock, J. W. Influence of onset and type of paternal absence on the first four Eriksonian developmental crises. *Developmental Psychology*, 1970, *3*, 273–274.

Santrock, J. W. Father absence, perceived maternal behavior, and moral development in boys. *Child Development*, 1975, *46*, 753–757.

Sarason, S. B. Jewishness, blackishness, and the nature-nurture controversy. *American Psychologist*, 1973, *28*, 962–971.

Sarason, S. B. Work, aging, and social change: Professionals and the one life-one career imperative. New York: The Free Press, 1977.

Saunders, C. St. Christopher's hospice. In E. Shneidman (Ed.), *Death: Contemporary perspectives*. Palo Alto, Calif.: Mayfield, 1976.

Scarr-Salapatek, S., & Williams, M. L. The effects of early stimulation on low birthweight infants. *Child Development*, 1973, *44*, 94–101.

Schachter, F. F., Shore, E., Hodapp, R., Chalfin, S., & Bundy, C. Do girls talk earlier?: Mean length of utterance in toddlers. *Developmental Psychology*, 1978, *14*, 388–392.

Schacter, B. Identity crisis and occupational processes: An intense exploratory study of emotionally disturbed male adolescents. *Child Welfare*, 1968, *47*, 26–37.

Schaffer, H. R., & Crook, C. K. Child compliance and maternal control techniques. *Developmental Psychology*, 1980, *16*, 54–61.

Schaffer, H. R., & Emerson, P. E. The development of social attachments in infancy. *Monographs of the Society for Research in Child Development*, 1964, *29* (3, Serial No. 94).

Schaie, K. W. Cross-sectional methods in the study of psychological aspects of aging. *Journal of Gerontology*, 1959, *14*, 208–215.

Schaie, K. W. A general model for the study of developmental problems. *Psychological Bulletin*, 1965, *64*, 92–107.

Schaie, K. W. A reinterpretation of age-related changes in cognitive structure and functioning. In L. R. Goulet & P. B. Baltes (Eds.), *Life-span developmental psychology: Research and theory*. New York: Academic Press, 1970.

Schaie, K. W. Methodological problems in descriptive developmental research on adulthood and aging. In J. R. Nesselroade & H. W. Reese (Eds.), *Life-span developmental psychology: Methodological issues*. New York: Academic Press, 1973.

Schaie, K. W. Translations in gerontology from lab to life: Intellectual functioning. *American Psychologist*, 1974, *29*, 802–807.

Schaie, K. W. The primary mental abilities in adulthood: An exploration in the development of psychometric intelligence. In P. B. Baltes & O. G. Brim, Jr. (Eds.), *Life-span development and behavior* (Vol. 2). New York: Academic Press, 1979.

Schaie, K. W., & Baltes, P. B. Some faith helps to see the forest: A final comment on the Horn and Donaldson myth of the Baltes-Schaie position on adult intelligence. *American Psychologist*, 1977, *32*, 1118–1120.

Schaie, K. W., Labouvie, G. V., & Buech, B. V. Generational and cohort-specific differences in adult cognitive functioning: A fourteen-year study of independent samples. *Developmental Psychology*, 1973, *9*, 151–166.

Schaie, K. W., & Labouvie-Vief, G. Generational and cohort-specific differences in adult cognitive behavior: A fourteen-year cross-sequential study. *Developmental Psychology*, 1974, *10*, 305–320.

Schaie, K. W., & Parham, I. A. Stability of adult personality: Fact or fable? *Journal of Personality and Social Psychology*, 1976, *34*, 146–158.

Schaie, K. W., & Parham, I. A. Co-hort-sequential analyses of adult intellectual development. *Developmental Psychology*, 1977, *13*, 649–653.

Schaie, K. W., & Strother, C. R. A cross-sequential study of age changes in cognitive behavior. *Psychological Bulletin*, 1968, *70*, 671–680.

Scheck, D. C., Emerick, R., & El-Assal, M. M. Adolescents' perceptions of parent-child external control orientation. *Journal of Marriage and the Family*, 1974, *35*, 643–654.

Schenkel, S. Relationship among ego identity status, field-independence, and traditional femininity. *Journal of Youth and Adolescence*, 1975, *4*, 73–82.

Schmidt, K., Rose, S. A., & Bridger, W. H. Effect of heartbeat sound on the cardiac and behavioral responsiveness to tactual stimulation in sleeping preterm infants. *Developmental Psychology*, 1980, *16*, 175–184.

Schneirla, T. C. The concept of development in comparative psychology. In D. B. Harris (Ed.), *The concept of development*. Minneapolis: University of Minnesota Press, 1957.

Schneirla, T. C. Instinct and aggression: Reviews of Konrad Lorenz, *Evolution and modification of behavior* (Chicago: The University of Chicago Press, 1965), and *On aggression* (New York: Harcourt, Brace & World, 1966). *Natural History*, 1966, *75*, 16.

Schneirla, T. C., & Rosenblatt, J. Behavioral organization and genesis of the social bond in insects and mammals. *American Journal of Orthopsychiatry*, 1961, *3*, 223–253.

Schneirla, T. C., & Rosenblatt, J. Critical periods in behavioral development. *Science*, 1963, *139*, 1110–1114.

Schonfeld, W. A. The body and the body image in adolescents. In G. Caplan & S. Lebovici (Eds.), *Adolescence: Psychosocial perspectives*. New York: Basic Books, 1969.

Schonfield, D., & Robertson, B. A. Memory storage and aging. *Canadian Journal of Psychology*, 1966, *20*, 228–236.

Schonfield, T. C., & Wenger, L. Age limitation of perceptual span. *Nature*, 1975, *253*, 377–378.

Schooler, K. K. Response of the elderly to environment: A stress-theoretic perspective. In P. G. Windley & G. Ernst (Eds.), *Theory development in environment and aging*. Washington, D.C.: Gerontological Society, 1975.

Schramm, W., Lyle, J., & Parker, B. *Television in the lives of our children*. Stanford, Calif.: Stanford University Press, 1961.

Schroeter, A. L., & Lucas, J. B. Gonorrhea diagnosis and treatment. *Obstetrics and Gynecology*, 1972, *39*, 274–285.

Schubert, J. B., Bradley-Johnson, S., & Nuttal, J. Mother-infant communication and maternal employment. *Child Development*, 1980, *51*, 246–249.

Schultz, N. R., Jr., & Hoyer, W. J. Feedback effects on spatial egocentrism in old age. *Journal of Gerontology*, 1976, *31*, 72–75.

Schulz, R. *The psychology of death, dying, and bereavement*. Reading, Mass.: Addison-Wesley, 1978.

Schulz, R., & Aderman, D. Effect of residential change on the temporal distance to death of terminal cancer patients. *Omega: Journal of Death and Dying*, 1973, *4*, 157–162.

Schwartz, A. N., Campos, J. J., & Baisel, E. J. The visual cliff: Cardiac and behavioral responses on the deep and shallow sides at five and nine months of age. *Journal of Experimental Child Psychology*, 1973, *15*, 86–99.

Schwartz, J. C., Strickland, R., & Krolick, G. Infant day care: Behavioral effects at preschool age. *Developmental Psychology*, 1974, *10*, 502–506.

Schwartz, M., & Day, R. H. Visual shape perception in early infancy. With commentary by Leslie B. Cohen. *Monographs of the Society for Research in Child Development*, 1979, *44* (7, Serial No. 182).

Schwebel, M. Logical thinking in college freshman. Final report, Project No. 0-B-105, Grant No. OEG-2-7-0039(509), April 1972.

Sears, R. R. Identification as a form of behavioral development. In D. B. Harris (Ed.), *The concept of development*. Minneapolis: University of Minnesota Press, 1957.

Sears, R. R. Relation of early socialization experience to self-concepts and gender role in middle child-hood. *Child Development*, 1970, *41*, 267–290.

Sears, R. R. Attachment, dependency, and frustration. In J. L. Gewirtz (Ed.), *Attachment and dependency*. Washington, D.C.: Winston, 1972.

Sears, R. R. Your ancients revisited. In E. M. Hetherington, J. W. Hagen, R. Kron, & A. H. Stein (Eds.), *Review of child development research* (Vol. 5). Chicago: University of Chicago Press, 1975.

Sears, R. R. Sources of life satisfaction of the Terman gifted men. *American Psychologist*, 1977, *32*, 119–128.

Sears, P. S., & Barbee, A. H. Career and life satisfaction among Terman's gifted women. In J. Stanley, W. George, & C. Solano (Eds.), *The gifted and the creative: Fifty-year perspective*. Baltimore, Md.: Johns Hopkins University Press, 1977.

Self, P. A. The further evolution of the paternal imperative. In N. Datan & L. H. Ginsberg (Eds.), *Life-span developmental psychology: Normative life crises*. New York: Academic Press, 1975.

Seligman, M. E. P. *Helplessness*. San Francisco: Freeman, 1975.

Seligmann, J. A right to die. *Newsweek*, 1976, *88*, 52.

Serunian, S., & Broman, S. Relationship of Apgar scores and Bayley mental and motor scores. *Child Development*, 1975, *46*, 696–700.

Sewell, W. H., & Shah, V. P. Parents' education and children's educational aspirations and achievements. *American Sociological Review*, 1968, *33*, 191–209. (a)

Sewell, W. H., & Shah, V. P. Social class, parental encouragement, and educational aspirations. *American Journal of Sociology*, 1968, *73*, 559–572. (b)

Shaffer, D. R. *Social and personality development*. Monterey, Calif.: Brooks-Cole, 1979.

Shah, F., Zelnik, M., & Kantner, J. F. Unprotected intercourse among unwed teenagers. *Family Planning Perspectives*, 1975, *7*, 39–43.

Shanas, E. Adjustment to retirement. In F. M. Carp (Ed.), *Retirement*. New York: Behavioral Publications, 1972.

Shappell, D. L., Hall, L. G., & Tarrier, R. B. Perceptions of the world of work: Inner-city versus suburbia.

Journal of Counseling Psychology, 1971, *18,* 55–59.

Sharaga, S. J. The effects of television advertising on children's nutrition attitude, nutrition knowledge and eating habits. *Dissertation Abstracts International,* 1974, *35B:7,* 3417–3418.

Shaw, M. E., & White, D. L. The relationship between child-parent identification and academic underachievement. *Journal of Clinical Psychology,* 1965, *21,* 10–13.

Sheldon, W. H. *The varieties of human physique.* New York: Harper & Row, 1940.

Sheldon, W. H. *The varieties of temperament.* New York: Harper & Row, 1942.

Sheppard, H. Work and retirement. In R. H. Binstock & E. Shanas (Eds.), *Handbook of aging and the social sciences.* New York: Van Nostrand Reinhold, 1976.

Shipman, W. G. Age of menarche and adult personality. *Archives of General Psychiatry,* 1964, *10,* 155–159.

Shirley, M. M. *The first two years.* Minneapolis: University of Minnesota Press, 1933.

Shneidman, E. S. *Deaths of man.* New York: New York Times Book Co., 1973.

Shnider, S. M., & Moya, F. Effects of meperidine on the newborn infant. *American Journal of Obstetrics and Gynecology,* 1964, *89,* 1009–1015.

Shock, N. W. Energy metabolism, caloric intake and physical activity of the aging. In L. A. Carlson (Ed.), *Nutrition in old age* (X symposium of the Swedish Nutrition Foundation). Uppsala, Sweden: Almqvist & Wiksell, 1972.

Shock, N. W. Biological theories of aging. In J. E. Birren & K. W. Schaie (Eds.), *Handbook of the psychology of aging.* New York: Van Nostrand Reinhold, 1977.

Shuttleworth, F. K. The adolescent period, a graphic atlas. *Monographs of the Society for Research in Child Development,* 1949, *14,* (1, Serial No. 49).

Siegel, L. S., & Brainerd, C. J. (Eds.), *Alternatives to Piaget: Critical essays on the theory.* New York: Academic Press, 1977.

Siegler, I. C., George, L. K., & Okun, M. A. Cross-sequential analysis of adult personality. *Developmental Psychology,* 1979, *15,* 350–351.

Sigman, M., & Parmelee, A. H. Visual preferences of four-month-old premature and full-term infants. *Child Development,* 1974, *45,* 959–965.

Simmons, R. G., & Rosenberg, F. Sex, sex roles, and self-image. *Journal of Youth and Adolescence,* 1975, *4,* 229–258.

Simmons, R. G. Rosenberg, F. R., & Rosenberg, M. Disturbances in the self-image in adolescence. *American Sociological Review,* 1973, *38,* 553–568.

Simpson, E. L. Moral development research: A case study of scientific cultural bias. *Human Development,* 1974, *17,* 81–106.

Sinex, F. M. The mutation theory of aging. In M. Rockstein (Ed.), *Theoretical aspects of aging.* New York: Academic Press, 1974.

Sinnott, J. D. Everyday thinking and Piagetian operativity in adults. *Human Development,* 1975, *18,* 430–443.

Sinnott, J. D., & Guttmann, D. Piagetian logical abilities and older adults' abilities to solve everyday problems. *Human Development,* 1978, *21,* 327–333.

Skeels, H. Adult status of children with contrasting early life experiences. *Monographs of the Society for Research in Child Development,* 1966, *31,* (No. 3).

Skinner, B. F. *The behavior of organisms.* New York: Appleton, 1938.

Skinner, B. F. Are theories of learning necessary? *Psychological Review,* 1950, *57,* 211–220.

Skinner, B. F. *Verbal behavior.* New York: Appleton-Century-Crofts, 1957.

Skolnick, A. *The intimate environment: Exploring marriage and the family.* Boston: Little Brown, 1973.

Slaby, R. G., & Hollenbeck, A. R. *Television influences on visual and vocal behavior of infants.* Paper presented at the biennial meeting of the Society for Research in Child Development, New Orleans, March 1977.

Slobin, D. I. Cognitive prerequisites for the development of grammar. In C. A. Ferguson & D. I. Slobin (Eds.), *Studies of child language development.* New York: Holt, Rinehart & Winston, 1973.

Sluckin, W. *Imprinting and early experience.* Chicago: Aldine, 1965.

Smith, A. D. Response interference with organized recall in the aged. *Developmental Psychology,* 1974, *10,* 867–870.

Smith, A. D. Aging and interference with memory. *Journal of Gerontology,* 1975, *30,* 319–325.

Smith, A. D. Age differences in encoding, storage, and retrieval. In L. W. Poon, J. L. Fozard, L. S. Cermak, D. Arenberg, & L. W. Thompson (Eds.), *New directions in memory and aging: Proceedings of the George A. Talland Memorial Conference.* Hillsdale, N.J.: Erlbaum, 1980.

Smith, P. K. A longitudinal study of social participation in preschool children: Solitary and parallel play. *Developmental Psychology,* 1978, *14,* 517–523.

Smith, T. E. Push versus pull: Intrafamily versus peer-group variables as possible determinants of adolescent orientations toward parents. *Youth and Society,* 1976, *8,* 5–26.

Snow, C. E., & Hoefnagel-Höhle, M. The critical period for language acquisition: Evidence from second language learning. *Child Development,* 1978, *49,* 1114–1128.

Snyder, E. E. High school student perceptions of prestige criteria. *Adolescence,* 1972, *6,* 129–136.

Sontag, L. W. The significance of fetal environmental differences. *American Journal of Obstetrics and Gynecology,* 1941, *42,* 996–1003.

Sontag, L. W. Differences in modifiability of fetal behavior and psychology. *Psychosomatic Medicine,* 1944, *6,* 151–154.

Sorell, G. T. *Adaptative implications of sex-related attitudes and behaviors.* Unpublished master's thesis. The Pennsylvania State University, 1979.

Soroka, S. M., Corter, C. M., & Abramovitch, R. Infants' tactual discrimination of novel and familiar tactual stimuli. *Child Development,* 1979, *50,* 1251–1253.

Sours, J. A. Anorexia nervosa: Nosology, diagnosis, developmental patterns, and power-control dynamics. In G. Caplan, & S. Lebovici (Eds.), *Adolescence: Psychosocial perceptives.* New York: Basic Books, 1969.

Spanier, G. B. *Sexual socialization and premarital sexual behavior: An empirical investigation of the im-*

pact of formal and informal sex education. Unpublished doctoral dissertation, Northwestern University, 1973.

Spanier, G. B. Perceived sex knowledge, exposure to eroticism and premarital sexual behavior: The impact of dating. *Sociological Quarterly*, 1976, *17*, 247–261.

Spanier, G. B. Sexual socialization: A conceptual view. *International Journal of Sociology of the Family*, 1977, *6*, 121–146.

Spanier, G. B., & Casto, R. F. Adjustment to separation and divorce: A qualitative analysis. In G. Levinger & O. Moles (Eds.), *Divorce and separation: Context, causes and consequences*. New York: Basic Books, 1979.

Spanier, G. B., & Glick, P. C. The life cycle of American families: An expanded analysis. *Journal of Family History*, 1980, *5*, 97–111.

Spanier, G. B., Lewis, R. A., & Cole, C. L. Marital adjustment over the family life cycle: The issue of curvilinearity. *Journal of Marriage and the Family*, 1975, *37*, 263–275.

Spearman, C. *The abilities of man*. New York: Macmillan, 1927.

Spelke, E. S. Perceiving bimodally specified events in infancy. *Developmental Psychology*, 1979, *15*, 626–636.

Spence, J. T., & Helmreich, R. The many faces of androgyny: A reply to Locksley and Colten. *Journal of Personality and Social Psychology*, 1979, *37*, 1032–1046.

Spence, J. T., & Helmreich, R. *Masculinity and femininity: Their psychological dimensions, correlates and antecedents*. Austin: University of Texas Press, 1978.

Speith, W. Slowness of task performance and cardiovascular diseases. In A. T. Welford & J. E. Birren (Eds.), *Behavior, aging, and the nervous system*. Springfield, Ill.: Charles C. Thomas, 1965.

Spitz, R. A. Hospitalism: An inquiry into the genesis of psychiatric conditions in early childhood. *Psychoanalytic Study of the Child*, 1945, *1*, 53–74, 113–117.

Spitz, R. A. Hospitalism: A follow-up investigation. *The Psychoanalytic Study of the Child*, 1946, *2*, 113–117.

Spitz, R. A. The role of ecological factors in emotional development in infancy. *Child Development*, 1949, *20*, 145–155.

Spitz, R. A., & Wolff, K. M. Anaclitic depression: An inquiry into the genesis of psychiatric conditions in early childhood, II. In A. Freud et al. (Eds.), *The psychoanalytic study of the child* (Vol. II). New York: International Universities Press, 1946.

Sprafkin, J. N, Liebert, R. M., & Paulous, R. W. Effects of a prosocial televised example on children's helping. *Journal of Experimental Child Psychology*, 1975, *20*, 119–126.

Staffieri, J. R. A study of social stereotype of body image in children. *Journal of Personality and Social Psychology*, 1967, *7*, 101–104.

Staffieri, J. R. Body build and behavioral expectancies in young females. *Developmental Psychology*, 1972, *6*, 125–127.

Stafford, R., Backman, E., & Dibona, P. The division of labor among cohabiting and married couples. *Journal of Marriage and the Family*, 1977, *39*, 43–58.

Stamps, L. E., & Porges, S. W. Heart-rate conditioning in newborn infants: Relationships among conditionability, heartrate variability and sex. *Developmental Psychology*, 1975, *11*, 424–431.

Starr, R. H., Jr. Child abuse. *American Psychologist*, 1979, *34*, 872–879.

Stayton, D. J., Ainsworth, M. D. S., & Main, M. B. Development of separation behavior in the first year of life: Protest, following, and greeting. *Developmental Psychology*, 1973, *9*, 213–225.

St. Clair, K. L. Neonatal assessment procedures: A historical review. *Child Development*, 1978, *49*, 280–292.

Stechler, G. Newborn attention as affected by medication during labor. *Science*, 1964, *144*, 315–317.

Steel, L., & Wise, L. L. *Origins of sex differences in high school mathematics achievement and participation*. Paper presented at the Annual Meeting of the American Education Research Association, San Francisco, California, April 1979.

Stein, A. H. The effects of maternal employment and educational attainment on the sex-typed attributes of college females. *Social Behavior and Personality*, 1973, *1*, 111–114.

Stein, A. H., & Bailey, M. M. The so-cialization of achievement orientation in females. *Psychological Bulletin*, 1973, *80*, 345–366.

Stein, A. H., & Friedrich, L. K. Television content and young children's behavior. In J. P. Murray, E. A. Rubinstein, & G. A. Comstock (Eds.), *Television and social behavior* (Vol. 2). *Television and social learning*. Washington, D.C.: U.S. Government Printing Office, 1972.

Stein, A. H., & Friedrich, L. K. Impact of television on children and youth. In E. M. Hetherington (Ed.), *Review of child development research* (Vol. 5). Chicago: University of Chicago Press, 1975.

Steinberg, L. D., & Hill, J. P. Patterns of family interaction as a function of age, the onset of puberty, and formal thinking. *Developmental Psychology*, 1978, *14*, 683–684.

Steinschneider, A. Implications of the sudden infant death syndrome for the study of sleep in infancy. In A. D. Pick (Ed.), *Minnesota symposium on child psychology* (Vol. 9). Minneapolis: University of Minnesota Press, 1975.

Stern, W. L. Uber die psychologischen methoden der Intelligenzprüfung. *Ber. V. Kongress Exp., Psychol.*, 1912, *16*, 10–160. American translation by G. M. Whipple, The psychological methods of testing intelligence. *Educational Psychology Monographs*, No. 13. Baltimore: Warwick & York, 1914.

Stevenson, H. W. Learning and reinforcement effects. In T. D. Spencer, & N. Koss (Eds.), *Perspectives in child psychology*. New York: McGraw-Hill, 1970.

Stevenson, H. W., & Weir, M. W. Developmental changes in the effects of reinforcement of a single response. *Child Development*, 1961, *32*, 1–5.

Stewart, A., & Reynolds, E. Improved prognosis for infants of very low birth-weight. *Pediatrics*, 1974, *54*, 724–735.

Stinnett, N., Carter, L. M., & Montgomery, J. E. Older persons' perceptions of their marriages. *Journal of Marriage and the Family*, 1972, *34*, 665–670.

Stinnett, N., & Walters, J. *Relationships in marriage and family*. New York: Macmillan, 1977.

Stolz, H. R., & Stolz, L. M. Adolescence related to somatic variation.

In *Adolescence: 43d yearbook of the National Committee for the Study of Education.* Chicago: University of Chicago Press, 1944.

Storandt, M., & Wittles, I. Maintenance of function in relocation of community-dwelling older adults. *Journal of Gerontology,* 1975, *30,* 608–612.

Strauss, M., Lessen-Firestone, J., Starr, R., & Ostrea, E. Behavior of narcotics-addicted newborns. *Child Development,* 1973, *46,* 887–893.

Strauss, S., Danziger, J., & Ramati, T. University students' understanding of nonconservation: Implication for structural reversion. *Developmental Psychology,* 1977, *13,* 359–363.

Strehler, B. L. *Time, cells, and aging.* New York: Academic Press, 1962.

Strehler, B. L. The mechanisms of aging. *The Body Forum,* 1978, *3,* 44–45.

Streib, G. F., & Schneider, C. J. *Retirement in American society: Impact and process.* Ithaca, N.Y.: Cornell University Press, 1971.

Streissguth, A. P. Maternal drinking and outcome of pregnancy: Implications for child mental health. *American Journal of Orthopsychiatry,* 1977, *47,* 422–431.

Stunkard, A. J., & Mendelson, M. Obesity and the body image: I. Characteristics of disturbances in the body image of some obese persons. *American Journal of Psychiatry,* 1967, *123,* 1296–1300.

Suomi, S. J., & Harlow, H. F. Social rehabilitation of isolate-reared monkeys. *Developmental Psychology,* 1972, *6,* 487–496.

Surber, C. F. Developmental processes in social inference: Averaging of intentions and consequences in moral judgment. *Developmental Psychology,* 1977, *13,* 654–665.

Surwillo, W. W. The relation of simple response time to brain wave frequency and the effects of age. *Electroencephalography and Clinical Neurophysiology,* 1963, *15,* 105–114.

Surwillo, W. W. Timing of behavior in senescence and the role of the central nervous system. In G. A. Talland (Ed.), *Human aging and behavior.* New York: Academic Press, 1968.

Sutton-Smith, B. *Child psychology.* New York: Appleton-Century-Crofts, 1973.

Swift, D. F. Family environment and 11+ success: Some basic predictions. *British Journal of Educational Psychology,* 1967, *37,* 10–21.

Talland, G. A. Age and the immediate memory span, In G. A. Talland (Ed.), *Human aging and behavior.* New York: Academic Press, 1968.

Tangri, S. S. Determinants of occupational role innovation in college women. *Journal of Social Issues,* 1972, *28,* 177–199.

Tanner, J. M. *Growth at adolescence* (2d ed.). Oxford: Blackwell, 1962.

Tanner, J. M. Physical growth. In P. H. Mussen (Ed.), *Carmichael's manual of child psychology* (Vol. 1, 3d ed.). New York: Wiley, 1970.

Tanner, J. M. *Foetus into man.* Cambridge, Mass.: Harvard University Press, 1978.

Tanner, J. M., Whitehouse, R. H., & Takaishi, M. Standards from birth to maturity for height, weight, height velocity and weight velocity: British children, 1965. *Archives of Diseases in Childhood,* 1966, *41,* 457–471, 613–635.

Taub, H. A. Mode of presentation, age, and short-term memory. *Journal of Gerontology,* 1975, *30,* 56–59.

Templar, D. I. Death anxiety in religiously very involved persons. *Psychological Reports,* 1972, *31,* 361–362.

Terkel, S. *Working: People talk about what they do all day and how they feel about what they do.* New York: Pantheon Books, 1972.

Terman, L. M. (Ed.), *Genetic studies of genius, I: Mental and physical traits of a thousand gifted children.* Stanford, Calif.: Stanford University Press, 1925. (a)

Terman, L. M. Mental and physical traits of a thousand gifted children. In L. M. Terman (Ed.), *Genetic studies of genius.* Stanford, Calif.: Stanford University Press, 1925. (b)

Terman, L. M., & Oden, M. H. *Genetic studies of genius, V: The gifted group at mid-life.* Palo Alto, Calif.: Stanford University Press, 1959.

Terman, L. M., & Tyler, L. E. Psychological sex differences. In L. Carmichael (Ed.), *Manual of child psychology.* New York: Wiley, 1954.

Tessman, L. H. *Children of parting parents.* New York: Aronson, 1978.

Thoman, E. B. Earliest behavioral developments of an infant who died of SIDS. (Abstract) *NIH SIDS Conference,* June 25–27, 1975.

Thomas, A., & Chess, S. Behavioral individuality in childhood. In L. R. Aronson, E. Tobach, D. S. Lehrman, & J. S. Rosenblatt (Eds.), *Development and evolution of behavior.* San Francisco: W. H. Freeman, 1970.

Thomas, A., & Chess, S. *Temperament and development.* New York: Brunner/Mazel, 1977.

Thomas, A., & Chess, S. *The dynamics of psychological development.* New York: Brunner/Mazel, 1980.

Thomas, A., Chess, S., & Birch, H. G. *Temperament and behavior disorders in children.* New York: New York University Press, 1968.

Thomas, A., Chess, S., & Birch, H. G. The origin of personality. *Scientific American,* 1970, *223,* 102–109.

Thomas, A., Chess, S., Birch, H. G., Hertzig, M. E., & Korn, S. *Behavioral individuality in early childhood.* New York: New York University Press, 1963.

Thompson, L., & Spanier, G. B. Influence of parents, peers, and partners on the contraceptive use of college men and women. *Journal of Marriage and the Family,* 1978, *40,* 481–492.

Thornburg, H. D. Student assessment of contemporary issues. *College Student Survey,* 1969, *3,* 1–5, 22.

Thornburg, H. D. Adolescence: A reinterpretation. *Adolescence,* 1970, *5,* 463–484.

Thornburg, H. D. Student assessment of contemporary issues. In H. D. Thornburg (Ed.), *Contemporary adolescence: Readings.* Monterey, Calif.: Brooks/Cole, 1971.

Thorndike, E. L. The newest psychology. *Educational Review,* 1904, *28,* 217–227.

Thurstone, L. L., & Thurstone, T. G. *Factorial studies of intelligence* (Psychometric Monograph No. 2). Chicago: University of Chicago Press, 1941.

Thurstone, L. L., & Thurstone, T. G. *SRA primary mental abilities.* Chicago: Science Research Associates, 1962.

Tobach, E. The methodology of sociobiology from the viewpoint of a comparative psychologist. In A. L.

Caplan (Ed.), *The sociobiology debate.* New York: Harper & Row, 1978.

Tobach, E., & Schneirla, T. C. The biopsychology of social behavior of animals. In R. E. Cooke & S. Levin (Eds.), *Biologic basis of pediatric practice.* New York: McGraw-Hill, 1968.

Toder, N. L., & Marcia, J. E. Ego identity status and response to conformity pressure in college women. *Journal of Personality and Social Psychology,* 1973, *26,* 287–294.

Tomlinson-Keasey, C. Formal operation in females from eleven to fifty-four years of age. *Developmental Psychology,* 1972, *6,* 364

Tomlinson-Keasey, C., Eisert, D. C., Kahle, L. R., Hardy-Brown, K., & Keasey, B. The structure of concrete operational thought. *Child Development,* 1979, *50,* 1153–1163.

Tompkins, W. T. The clinical significance of nutritional deficiences in pregnancy. *Bulletin of the New York Academy of Medicine,* 1948, *24,* 376–388.

Towler, J. O., & Wheatley, G. Conservation concepts in college students: A replication and critique. *Journal of Genetic Psychology,* 1971, *118,* 265–270.

Traeldal, A. The work role and life satisfaction over the life-span. Paper presented at the Biennial Meeting of the International Society for the Study of Behavioral Development, Ann Arbor, Mich., August 1973.

Treat, N. J., & Reese, H. W. Age, pacing, and imagery in paired-associate learning. *Developmental Psychology,* 1976, *12,* 119–124.

Trehub, S. A., & Curran, S. Habituation of infant's cardiac response to speech stimuli. *Child Development,* 1979, *50,* 1247–1250.

Troyer, W. G., Eisdorfer, C., Bogdonoff, M. D., & Wilkie, F. Experimental stress and learning in the aged. *Journal of Abnormal Psychology,* 1967, *1,* 65–70.

Tulkin, S. R. Social class differences in attachment of ten-month-old infants. *Child Development,* 1973, *44,* 171–174.

Tulving, E. Subjective organization in free recall of unrelated words. *Psychological Review,* 1962, 69, 344–354.

Turiel, E. Developmental processes in the child's moral thinking. In

P. H. Mussen, J. Langer, & M. Covington (Eds.), *Trends and issues in developmental psychology.* New York: Holt, Rinehart and Winston, 1969.

Uhlenberg, P. Demographic change and problems of the aged. In M. W. Riley (Ed.), *Aging from birth to death: Interdisciplinary perspectives.* Boulder, Colo.: Westview Press, 1979.

Uniform Crime Reports for the United States, 1975 Federal Bureau of Investigation. Washington, D.C.: U.S. Government Printing Office, 1976.

U.S. Bureau of the Census. *Number, timing, and duration of marriages and divorces in the United States: June 1975.* Current Population Reports, Series P-20, No. 297, October 1976. Washington, D.C.: U.S. Government Printing Office.

U.S. Bureau of the Census. *Statistical abstract of the United States, 1971* (92d ed.), 1971. Washington, D.C.: U.S. Government Printing Office.

U.S. Bureau of the Census. *Statistical abstract of the United States, 1977* (98th ed.), 1977. Washington, D.C.: U.S. Government Printing Office. (a)

U.S. Bureau of the Census. *Educational attainment in the United States: March 1977 and 1976.* Current Population Reports, Series P-20, No. 314, December 1977. Washington, D.C.: U.S. Government Printing Office. (b)

U.S. Bureau of the Census. *Characteristics of American children and youth: 1976.* Current Population Reports, Series P-23, No. 66, January 1978. Washington, D.C.: U.S. Government Printing Office. (a)

U.S. Bureau of the Census. *School enrollment—Social and economic characteristics of students: October 1976.* Current Population Reports, Series P-20, No. 319, February 1978. Washington, D.C.: U.S. Government Printing Office. (b)

U.S. Bureau of the Census. *Perspectives on American husbands and wives.* Current Population Reports, Series P-23, No. 77, December 1978. Washington, D.C.: U.S. Government Printing Office. (c)

U.S. Bureau of the Census. *Statistical abstract of the United States:*

1978 (99th edition). Washington, D.C.: U.S. Government Printing Office, 1978. (d)

U.S. Bureau of the Census. Unpublished data analyzed by Graham Spanier and Paul Glick, 1979.

U.S. Bureau of the Census. *Statistical abstract of the United States, 1980.* Washington, D.C.: U.S. Government Printing Office, 1980.

U.S. Department of Health, Education, and Welfare. *Marital status and living arrangements* (Current Population Reports, Series P-20, No. 255). Washington, D.C.: U.S. Government Printing Office, 1973.

U.S. Department of Health, Education, and Welfare. *The condition of education: A statistical report on the condition of education in the United States.* Washington, D.C.: U.S. Government Printing Office, 1976.

U.S. Department of Health, Education, and Welfare. *Vital statistics of the United States, 1976.* Hyattsville, Md.: National Center for Health Statistics, 1978.

U.S. Department of Health, Education, and Welfare. *Monthlty vital statistics report: Advance report, final mortality statistics, 1977.* Hyattsville, Md.: National Center for Health Statistics, 1979.

U.S. Department of Labor, Women's Bureau. *Working mothers and their children.* Washington, D.C.: U.S. Government Printing Office, 1977.

U.S. National Center for Health Statistics. *Vital Health Statistics* (Ser. 11, No. 147, 1975).

U.S. National Health Survey. *Monocular-binocular visual acuity of adults: U.S. 1969–1962.* Washington, D.C.: U.S. Department of Health, Education, and Welfare, 1968.

Vaillant, G. E. *Adaptation to life.* Boston: Little Brown, 1977.

Valdes-Dapena, M.A. *Sudden unexplained infant death 1970 through 1975. An evaluation in understanding,* 1977, DHEW Publication No (HSA) 78-5255.

Vandell, D. L. Effects of a playgroup experience on mother-son and father-son interaction. *Developmental Psychology,* 1979, *15,* 379–385.

Vander Maelen, A. L., Strauss, M. E., & Starr, R. H. Influence of obstetric medication on auditory habituation in the newborn. *Developmental Psychology*, 1975, *11*, 711–714.

Vaughn, B., Egeland, B., Sroufe, L. A., & Waters, E. Individual differences in infant-mother attachment at twelve and eighteen months: Stability and change in families under stress. *Child Development*, 1979, *50*, 971–975.

Veevers, J. E. Voluntarily childless wives: An exploratory study. *Sociology and Social Research*, 1973, *57*, 356–366.

Vernon, P. E. *The structure of human abilities*. London: Methuen, 1961.

Vigersky, R. A. (Ed.) *Anorexia nervosa*. New York: Raven Press, 1977.

Vinick, B. H. Remarriage in old age. *The Family Coordinator*, 1978, *27*, 359–364.

Vondracek, F. W., & Urban, H. B. Intervention within individual and family development: When? what kind? and how? In S. R. Goldberg & F. Deutsch (Eds.), *Life-span individual and family development*. Monterey, Calif.: Brooks/Cole, 1977.

Voss, H. L., Wendling, A., & Elliott, D. S. Some types of high-school dropouts. *Journal of Educational Research*, 1966, *59*, 363–368.

Waber, D. P. Sex differences in mental abilites, hemisphere lateralization, and rate of physical growth in adolescence. *Developmental Psychology*, 1977, *13*, 29–38.

Wadsworth, B. J. *Piaget's theory of cognitive development*. New York: McKay, 1971.

Wahler, R. G. Infant social development: Some experimental analyses of an infant-mother interaction during the first year of life. *Journal of Experimental Child Psychology*, 1969, *7*, 101–113.

Walford, R. L. *The immunologic theory of aging*. Baltimore: Williams & Wilkins, 1969.

Walker, L. J., & Richards, B. S. The effects of a narrative model on children's moral judgments. *Canadian Journal of Behavioral Science*, 1976, *8*, 169–177.

Walsh, D. A. Age differences in learning and memory. In D. S. Woodruff & J. E. Birren (Eds.), *Aging: Scientific perspectives and social issues*. New York: Van Nostrand, 1975.

Walters, J. Birth defects and adolescent pregnancies. *Journal of Home Economics*, 1975, *67*, 23–27.

Walters, R. H., & Thomas, E. L. Enhancement of punitiveness by visual and audio-visual displays. *Canadian Journal of Psychology*, 1963, *17*, 244–255.

Wang, H. S. Cerebral correlates of intellectual function in senescence. In L. F. Jarvik, C. Eisdorfer, & J. E. Blum (Eds.), *Intellectual functioning in adults: Psychological and biological influences*. New York: Springer, 1973.

Wang, H. S., & Busse, E. W. EEG of healthy old persons—A longitudinal study. I: Dominant background activity and occipital rhythm. *Journal of Gerontology*, 1969, *24*, 419–426.

Wang, H. S., Obrist, W. D., & Busse, E. W. Neurophysiological correlates of the intellectual function of elderly persons living in the community. *American Journal of Psychiatry*, 1970, *126*, 1205–1212.

Ward, S., & Wackman, D. B. Television advertising and intrafamily influence: Children's purchase influence attempts and parental yielding. In E. A. Rubinstein, G. A. Comstock, & J. P. Murray (Eds.), *Television and social behavior* (Vol. 4). Washington, D.C.: U.S. Government Printing Office, 1971.

Waring, J. M. Social replenishment and social change: The problem of disordered cohort flow. *American Behavioral Scientist*, 1975, *19*, 237–256.

Washburn, S. L. (Ed.), *Social life of early men*. New York: Wenner-Gren Foundation for Anthropological Research, 1961.

Waterman, A. S., & Goldman, J. A. A longitudinal study of ego identity development of a liberal arts college. *Journal of Youth and Adolescence*, 1976, *5*, 361–370.

Waterman, A. S., Kohutis, E., & Pulone, J. The role of expressive writing in ego identity formation. *Developmental Psychology*, 1977, *13*, 286–287.

Waterman, A. S., & Waterman, C. K. A longitudinal study of changes in ego identity status during the freshman year at college. *Developmental Psychology*, 1971, *5*, 167–173.

Waterman, C. K., Beubel, M. E., & Waterman, A. S. Relationship be-

tween resolution of the identity crisis and outcomes of the identity crisis and outcomes of previous psychosocial crises. *Proceedings of the 78th Annual Convention of the American Psychological Association*, 1970, *5*, 467–468.

Waterman, G., Geary, P., & Waterman, C. Longitudinal study of changes in ego identity status from the freshman to the senior year at college. *Developmental Psychology*, 1974, *10*, 387–392.

Waters, E. The reliability and stability of individual differences in infant-mother attachment. *Child Development*, 1978, *49*, 483–494.

Waters, E., Wippman, J., & Sroufe, A. L. Attachment, positive affect, and competence in peer group: Two studies in construct validation. *Child Development*, 1979, *50*, 821–829.

Watson, J. B. Psychology as the behaviorist views it. *Psychological Review*, 1913, *20*, 158–177.

Watson, J. B. *Psychology from the standpoint of a behavorist*. Philadelphia: Lippincott, 1918.

Watson, J. B. *Psychological care of infant and child*. New York: Norton, 1928.

Watson, J. B., & Raynor, R. Conditional emotional reactions. *Journal of Experimental Psychology*, 1920, *3*, 1–14.

Waugh, N. C., & Norman, D. A. Primary memory. *Psychological Review*, 1965, *72*, 89–104.

Weatherly, D. Self-perceived rate of physical maturation and personality in late adolescence. *Child Development*, 1964, *35*, 1197–1210.

Wechsler, D. *The measurement and appraisal of adult intelligence* (4th ed.). Baltimore: Williams & Wilkins, 1958.

Weiner, B., & Peter, N. A. Cognitive-developmental analysis of achievement and moral judgment. *Developmental Psychology*, 1973, *9*, 290–309.

Weinraub, M., Brooks, J., & Lewis, M. The social network: A reconsideration of the concept of attachment. *Human Development*, 1977, *20*, 31–47.

Weinstock, A., & Lerner, R. M. Attitudes of late adolescents and their parents toward contemporary issues. *Psychological Reports*, 1972, *30*, 239–244.

Weisler, A., & McCall, R. B. Exploration and play: Resume and redirection. *American Psychologist*, 1976, *31*, 492–508.

Welford, A. T. *Aging and human skill.* London: Oxford University Press, 1958.

Welford, A. T. Motor performance. In J. E. Birren & K. W. Schaie (Eds.), *Handbook of the psychology of aging.* New York: Van Nostrand Reinhold, 1977.

Wells, W. D. Television and aggression: Replication of an experimental field study. Unpublished manuscript, University of Chicago Graduate School of Business, 1973.

Wells, W. D., & LoSciuto, L. A. Direct observation of purchasing behavior. *Journal of Marketing Research*, 1966, *3*, 227–233.

Werner, H. *Comparative psychology of mental development.* New York: International Universities Press, 1948.

Wetherford, M. J., & Cohen, L. B. Developmental changes in infant visual preferences for novelty and familiarity. *Child Development*, 1973, *44*, 416–424.

Whitbourne, S. K. Test anxiety in elderly and young adults. *International Journal of Aging and Human Development*, 1976, 7, 201–210.

Whitbourne, S. K., & Waterman, A. S. Psychosocial development during the adult years: Age and cohort comparisons. *Developmental Psychology*, 1979, *15*, 373–378.

White, C. B. Moral development in Bahamian school children: A cross-cultural examination of Kohlberg's stages of moral reasoning. *Developmental Psychology*, 1975, *11*, 535–536.

White, C. B., Bushnell, N., & Regnemer, J. L. Moral development in Bahamian school children: A 3-year examination of Kohlberg's stages of moral development. *Developmental Psychology*, 1978, *14*, 58–65.

White, S. H. The learning-maturation controversy: Hall to Hull. *Merrill-Palmer Quarterly*, 1968, *14*, 187–196.

White, S. H. The learning theory tradition and child psychology. In P. H. Mussen (Ed.), *Carmichael's manual of child psychology* (3d ed.). New York: Wiley, 1970.

Wiggins, J. S. Personality structure. In P. R. Farnsworth, M. R. Rosenzweig, & J. T. Polepka (Eds.), *Annual review of psychology* (Vol. 19). Palo Alto, Calif.: Annual Reviews, 1968.

Wilensky, H. L. *Women's work: Economic growth, ideology and structure.* Berkeley: Institute of Indusrial Relations, University of California, 1968.

Wilkie, F., & Eisdorfer, C. Intelligence and blood pressure in the aged. *Science*, 1971, *172*, 959–962.

Wilkinson, A., Parker, T., & Stevenson, H. W. Influence of school and environment on selective memory, *Child Development*, 1979, *50*, 890–893.

Willems, E. P. Behavioral ecology and experimental analysis: Courtship is not enough. In J. R. Nesselroade & H. W. Reese (Eds.), *Life-span developmental psychology: Methodological issues.* New York: Academic Press, 1973.

Willemsen, E., Flaherty, D., Heaton, C., & Ritchey, G. Attachment behavior of one-year-olds as a function of mother versus father, sex of child, session, and toys. *Genetic Psychology Monographs*, 1974, *90*, 305–324.

Williams, L., & Golenski, J. Infant speech sound discrimination: The effects of contingent versus noncontingent stimulus presentation. *Child Development*, 1978, *49*, 213–217.

Williams, L., & Golenski, J. Infant behavioral state and speech sound discrimination. *Child Development*, 1979, *50*, 1243–1246.

Wilson, E. O. Sociobiology: The new synthesis. Cambridge, Mass.: Harvard University Press, 1975.

Winch, R. F. *The modern family* (3d ed.). New York: Holt, Rinehart, & Winston, 1971.

Winick, M., Karnig, K. M, & Harris, R. C. Malnutrition and environmental enrichment by early adaptation. *Science*, 1975, *190*, 1173–1175.

Winsborough, H. H. Changes in the transition to adulthood. In M. W. Riley (Ed.), *Aging from birth to death: Interdisciplinary perspectives.* Boulder, Colo.: Westview Press, 1979.

Witte, K. L. Paired-associate learning in young and elderly adults as related to presentation rate. *Psychological Bulletin*, 1975, *82*, 975–985.

Wittles, I., & Botwinick, J. Survival in relocation. *Journal of Gerontology*, 1974, *29*, 440–443.

Wolf, E. Glare and age. *Archives of Opthalmology*, 1960, *64*, 502–514.

Wolf, R. M. *The identification and measurement of environmental process variables related to intelligence.* Unpublished doctoral dissertation, University of Chicago, 1964.

Wolff, P. H. *The causes, controls, and organization of behavior in the neonate.* New York: International Universities Press, 1966.

Woodruff, D. S. Relationships among EEG alpha frequency, reaction time, and age: A biofeedback study. *Psychophysiology*, 1975, *12*, 673–681.

Woodruff, D. S. Brain electrical activity and behavior relationships over the life span. In P. B. Baltes (Ed.), *Life-span development and behavior* (Vol. 1). New York: Academic Press, 1979.

Woodruff, D. S., & Birren, J. E. Age changes and cohort differences in personality. *Developmental Psychology*, 1972, *6*, 252–259.

Worell, J. Sex roles and psychological well-being: Perspectives on methodology. *Journal of Consulting and Clinical Psychology*, 1978, *46*, 777–791.

Worell, J. Life-span sex roles: Development, continuity, and change. In R. M. Lerner & N. A. Busch-Rossnagel (Eds.), *Individuals as producers of their development: A life-span perspective.* New York: Academic Press, 1981.

Worell, J., & Stilwell, W. E. *Psychology for teachers and students.* New York: McGraw-Hill, 1981.

Wright, J. D. Are working women really more satisfied? Evidence from several national surveys. *Journal of Marriage and the Family*, 1978, *40*, 301–313.

Wylie, R. C. *The self-concept* (Vol. 1). Lincoln: University of Nebraska Press, 1974.

Yang, R. K., Zweig, A. R., Douthitt, T. C., & Federman, E. J. Successive relationships between maternal attitudes during pregnancy, analgesic medication during labor and delivery, and newborn behavior. *Developmental Psychology*, 1976, *12*, 6–14.

Yarrow, L. J. Research in dimensions

of early maternal care. *Merrill-Palmer Quarterly*, 1963, *9*, 101–114.

Yonas, A., Oberg, C., & Norcia, N. Development of sensitivity to binocular information for approach of an object. *Developmental Psychology*, 1978, *14*, 147–152.

Young, M. L. Problem solving performance in two age groups. *Journal of Gerontology*, 1966, *21*, 505–510.

Youniss, J., & Dean, A. Judgment and imagery aspects of operations: A Piagetian study with Korean and Costa Rican children. *Child Development*, 1974, *45*, 1020–1031.

Zablocki, B. *The joyful community: An account of the Burderhof, a communal movement now in its third generation.* Baltimore: Penguin, 1971.

Zajonc, R. B., & Marcus, G. B. Birth order and intellectual development. *Psychological Review*, 1975, *82*, 74–88.

Zaretsky, H., & Halberstam, J. Effects of aging, brain-damage, and associative strength on paired-associate learning and relearning. *Journal of Genetic Psychology*, 1968, *112*, 149–163.

Zeller, W. W. Adolescent attitudes and cutaneous health. *Journal of School Health*, 1970, *40*, 115–120.

Zelnik, M., & Kantner, J. F. Sexual and contraceptive experience of young unmarried women in the United States, 1976 and 1971. *Family Planning Perspective*, 1977, *9*, 55–71.

Zelnik, M., & Kantner, J. F. First pregnancies to women aged 15–19: 1976 and 1971. *Family Planning Perspectives*, 1978, *10*, 11–20.

Zuckerman, P., Ziegler, M., & Stevenson, H. W. Children's viewing of television and recognition memory of commercials. *Child Development*, 1978, *49*, 96–104.

Zussman, J. V. Relationship of demographic factors to parental discipline techniques. *Developmental Psychology*, 1978, *14*, 685–686.

Zweig, J., & Csank, J. Effects of relocation on chronically ill geriatric patients of a medical unit: Mortality rates. *Journal of American Geriatrics Society*, 1975, *23*, 133–136.

Name Index

Subject Index